Internal Corrosion of Water Distribution Systems

Internal Corrosion of Water Distribution Systems

Second Edition

American Water Works Association Research Foundation
DVGW-Technologiezentrum Wasser

Dedication

This book is dedicated to Heinrich Sontheimer and Fred Kuch for their contributions to corrosion research and for their active participation in the collaborative efforts that produced this book.

Long recognized and respected for his problem-driven approach to research, Dr. Sontheimer remained an inspirational teacher and mentor throughout his retirement years until his death. He was a driving force in producing the first edition of this book.

Dr. Kuch performed pioneering work to better understand the mechanisms of iron corrosion. His contribution to corrosion science is recognized posthumously as the "Kuch mechanism" by the authors of this book.

Colleagues and friends miss not only the valuable research of Drs. Sontheimer and Kuch but also their enthusiasm and advice.

Disclaimer

Library of Congress Cataloging-in-Publication Data

Internal corrosion of water distribution systems / American Water Works Association Research Foundation, DVGW-Technologiezentrum Wasser. -- 2nd ed.
xvi, 586 p. 17×25cm.
Includes bibliographical references and index.
ISBN 0-89867-759-9 (hardcover : alk. paper)
1. Water-pipes--Corrosion. I. AWWA Research Foundation. II. DVGW-Technologiezentrum Wasser.
TD491.I55 1996
628.1'44--dc20 95-53857
CIP

6666 West Quincy Ave.
Denver, CO 80235

Printed in USA

ISBN 0-89867-759-9

Printed on recycled paper

Contents

Preface to the First Edition

This report represents a cooperative effort of the AWWA Research Foundation (AWWARF) and the Engler-Bunte Institut (EBI) of Karlsruhe, West Germany, which was conducted over a two-year period. Its completion is a tribute to the authors' willingness to contribute their time and considerable expertise to the project and the collective belief that such a publication is an extremely useful document for the water supply industry throughout the world.

AWWARF became involved in this project after conducting several surveys of research needs within the water supply industry. In almost every survey, corrosion and its control were identified as the number one research priority. Many of the people surveyed indicated that no published report or guidance manual comprehensively addressed the subject of corrosion control in a water supply system. Many complained that the available material was either too basic, too complex, or too scattered throughout the literature. Considering the obvious need, AWWARF's Board of Trustees decided to assemble a group of independent experts to compile the existing material into one reference.

EBI joined in this effort because it has conducted extensive research on corrosion in cooperation with several water utilities throughout West Germany. EBI's efforts in this area have been internationally acclaimed as being among the most progressive and have resulted in several new approaches and understandings of the age-old problem of corrosion.

The experiences of the US and West German experts on corrosion of water supply systems have been integrated in this report, which is intended for engineers, utility managers, and experienced water professionals who need a practical guide on evaluation and resolution of corrosion problems within water supply systems. It has been written in such a manner and style so that the reader can easily obtain information and references on any particular problem or can become acquainted with the theory and practice for each type of corrosion problem and solution.

This entire report was made possible through the tireless efforts of the authors, who not only had to provide original material under extremely tight deadlines but also had to coordinate and integrate their manuscripts with those of other experts located on another continent. This comprehensive report was the result of international teamwork where each participant sacrificed individual recognition and worked toward the collective objective. The authors and their affiliations are shown in the respective chapters, the committee members are listed here collectively to recognize them as a group.

AWWARF Committee	EBI Committee
Vernon L. Snoeyink	Heinrich Sontheimer
John F. Ferguson	Ivo Wagner
Robert Ryder	Walter Kölle
Michael Schock	Heinrich Holtschulte
J. Edward Singley	Otto von Franqué
R. Rhodes Trussell	Alfred Kuch
Henry Cruse	Peter Wölfel
	Gerhard Werner

When the authors finished with the integration of ideas and approaches, the real work began. The entire report was edited and converted into a cohesive document by Sharon Crnkovich, the publications editor of AWWARF. Sharon's enthusiasm toward the completion of the report was matched only by Karen Potts' and Cheryl Anderson's ability to type and proof the copy.

The Foundation and the Institute owe a debt of gratitude to all the participants.

Kenneth J. Miller
Chairman, Board of Trustees
AWWA Research Foundation

Herr Prof. Dr. Heinrich Sontheimer
Engler-Bunte Institut

James F. Manwaring, P.E.
Executive Director
AWWA Research Foundation

Dr. Ivo Wagner
Engler-Bunte Institut

Preface to the Second Edition

Over the past few years concern about corrosion in water distribution systems has grown. The focus of this concern has been primarily on corrosion's effects on water quality and the implications for public health. With this growing concern, it became evident that the first edition of this report needed to be updated. So, true to its philosophy of assembling expert committees to address complex issues such as corrosion, the AWWA Research Foundation funded a cooperative project to revise the report. The Foundation again coordinated the work with the German Gas and Waterworks Association (Deutscher Verein des Gas- and Wasserfaches or DVGW) through its water technology center (Technologiezentrum Wasser or TZW).

The Foundation and TZW organized a volunteer committee of American and European corrosion experts, led by two excellent co-chairs—Professor Vern Snoeyink of the University of Illinois and Dr. Ivo Wagner of TZW. Working over the last several years, these experts expanded and improved the first edition's coverage of corrosion in water supply systems. These experts, who have so generously given of their time and expertise, are

American Committee	European Committee
Vernon L. Snoeyink (co-chair)	Ivo Wagner (co-chair)
Mark M. Benjamin	Otto von Franqué
John F. Ferguson	Pierre Leroy
Gregory J. Kirmeyer	Roger J. Oliphant
Steven H. Reiber	Heinrich Sontheimer (now deceased)
Robert A. Ryder	Eilen Vik
Michael R. Schock	
R. Rhodes Trussell	

The second edition consists of 10 chapters that cover corrosion principles, corrosion of various materials, including copper alloys and solder, mitigation of corrosion impacts, assessment technologies, and approaches to corrosion control studies. This new report differs in three significant ways from the original report: (1) There is now a separate chapter on copper alloys and solders; (2) the topics of inhibitors and low-mineral-content waters have been consolidated into a single chapter covering mitigation of corrosion impacts; and, (3) a new chapter describes a technical approach to studying corrosion and implementing a control program.

Throughout the book, the authors emphasize the importance of considering the secondary effects of approaches to corrosion control. Such effects can increase corrosion problems or adversely affect water quality. The authors have also greatly enhanced the usefulness of the report by developing numerous summary tables, which appear principally in chapters 8 through 10. These tables provide information on topics such as

- water quality conditions that affect corrosion of various types of materials
- chemicals used for corrosion control
- corrosion assessment options for metal plumbing materials
- coupon protocols for distribution system corrosion measures

- electrochemical corrosion assessment methodologies
- benefits and drawbacks of various levels of corrosion testing
- benefits and drawbacks of bench testing versus flow-through pipe loops

This report was edited most ably by Mary Kay Kozyra, the senior technical editor of the American Water Works Association; the Association publishes Foundation reports. Sharon Parks of the Foundation staff was the project manager.

The Foundation and TZW gratefully acknowledge the contributions of all the participants in this effort.

James F. Manwaring, P.E.
Executive Director
AWWA Research Foundation

Dr. Wolfgang Kühn
Executive Director
DVGW-Technologiezentrum Wasser

Acknowledgments

All the authors of this book sincerely thank the AWWA Research Foundation for its financial support and the Technologiezentrum Wasser of the German Gas and Waterworks Association for its organizational support. Without this support, this book would not have been written. Sharon Parks, the Foundation project manager, was responsible for organizing meetings and reminding overcommitted authors of their obligations and commitments to this project. Her efforts in bringing this project to completion are much appreciated, as are those of AWWA editors Mary Kay Kozyra and Deborah J. Lynes in making this a usable, first-class book.

Numerous people made major contributions to the development of the information in the first edition of chapter 3, which was the basis for the current chapter in this book. These include Gordon P. Treweek, Katherine P. Fox, Richard D. Pomeroy, John S. Lang, Carol H. Tate, Jerome F. Thomas, Robert L. Ray. All these researchers have been involved in one or more of several studies of galvanized pipe corrosion with R. Rhodes Trussell, especially addressing the issue of failure by pitting. Jerome F. Thomas and Robert L. Ray provided essential research on the significance of poor-quality weld seams and separation at the iron–zinc interface. Much of the interesting work discussed herein was developed as the result of a project funded by the US Environmental Protection Agency (USEPA) and the Metropolitan Water District of Southern California (MWD). Marvin Gardels served as project officer for the USEPA, Dale Newkirk served as project officer for MWD, and Gene Bowers offered useful advice. Additional information was derived from the independent work of W. Risner and L. Grasha at MWD.

In Germany, Dr.–Ing. Gerhard Werner at the Langenau Water Works of the Landeswasserversorgung has made many of the most important contributions to the information summarized in chapter 3. Of course, all the German work has been strongly influenced by the late Heinrich Sontheimer and his colleagues at the Engler–Bunte Institut in Karlsruhe.

The authors of chapter 4 would like to thank the individuals providing pipe specimens and some water quality data for the scale analyses reported in the chapter. They include John J. Collins (deceased) and Michael Covellone (Providence, Rhode Island); Peter Lassovzsky (USEPA, Office of Ground Water and Drinking Water), and Harry von Huben (USEPA Region V, retired) for Chicago, Illinois–area and Gary/Hobart, Indiana, samples; Thomas Bailey and Karen Himmelstein (Durham, North Carolina); John Eastman (LBJ Engineers/Architects) for Oakwood, Ohio, samples; Mort Anoushiravani, City of Portland, Oregon; Edward Beidler, City of St. Petersburg, Florida; Wayne A. Kopp, City of Piqua, Ohio; Kenneth S. Button, Columbus, Ohio; Don J. Heuer, City of Cleveland Division of Water, Cleveland, Ohio; and David Heumann, Los Angeles Department of Water and Power, California. The authors also thank Jarvis McFadzean of Strathclyde Water for information on lime orthophosphate treatment in Scotland and Ross Gregory of WRc for many helpful discussions on water treatment for control of plumbosolvency.

List of Authors

Mark M. Benjamin
Department of Civil Engineering
University of Washington
Seattle, WA 98195–2700

John F. Ferguson
Department of Civil Engineering
University of Washington
Seattle, WA 98195

Otto von Franqué
Labor für Korrosionsschutztechnik
Märkische Fachhochschule
Frauenstuhlweg 31
58644 Iserlohn, Germany

Gregory J. Kirmeyer
Economic & Engineering Services, Inc.
Bellevue, WA 98009

Pierre Leroy
CRECEP
F-75014 Paris, France

Roger J. Oliphant
Water Research Centre
Blagrove, Swindon
Wiltshire SN5 8YF, England, United Kingdom

Steven H. Reiber
HDR Engineering
Bellevue, WA 98004-6441

Robert A. Ryder
Environmental Engineering Consultant
Kennedy/Jenks/Chilton
San Francisco, CA 94107

Michael R. Schock
Research Chemist
USEPA
Water Supply and Water Resources Division
National Risk Management Research Laboratory
Cincinnati, OH 45268

Vernon L. Snoeyink
Department of Civil Engineering, MC-250
University of Illinois
Urbana, IL 61801

Heinrich Sontheimer (now deceased)

R. Rhodes Trussell
Montgomery Watson
Pasadena, CA 91101-7009

Eilen A. Vik
Aquateam-Norwegian Water
Technology Centre A/S
N-0604 Oslo 6, Norway

Ivo Wagner
DVGW-Technologiezentrum Wasser
76139 Karlsruhe, Germany

1

Principles of Corrosion of Water Distribution Systems

Vernon L. Snoeyink and Ivo Wagner

Corrosion is the "physicochemical interaction between a metal and its environment which results in changes in the properties of the metal" (ISO 1979). This definition agrees with that in the German standard DIN* 50900, part 1 (DIN 50900 1975).

Internal corrosion of water distribution systems leads to two major problems for water utilities. The first is the failure of the distribution system pipes. This failure results in water leakage and loss of hydraulic capacity caused by the buildup of corrosion products. The second problem is an unwanted change in water quality as the water is being transported through the distribution system. The change is caused by corrosion products entering the water. Examples of these problems include an increase in the concentration of lead and copper, which are released from pipes and solder and are very important because of their toxic properties; the presence of iron corrosion products, which can color the water and create nuisance conditions; and the occurrence of high pH, which results from leaching of lime from cement. Factors affecting both types of problems are discussed in this chapter.

This chapter presents the principles of corrosion of metals and cementitious materials and discusses the water quality factors that affect corrosion reactions in drinking water distribution systems. Succeeding chapters explain the application

---

NOTE: The first edition of this chapter was authored by Vernon L. Snoeyink and Alfred Kuch.

*Deutsches Institut für Normung (German Institute for Standardization), Burggrafenstr. 6, Berlin, Germany.

of these principles to corrosion of particular materials, including iron and steel, copper, lead, galvanized steel, cement, and other materials, as well as to corrosion inhibition. The factors that cause corrosion to occur are presented first, followed by discussions of the rate of corrosion reactions, the various forms of corrosion, water quality factors that affect corrosion, and corrosion indices.

ELECTROCHEMICAL CORROSION

The Corrosion Cell

Electrochemical corrosion is the destruction of a metal by electron transfer reactions. For this type of corrosion to occur, all components of an electrochemical cell must be present. The components include an anode and a cathode, which are sites on the metal that have different electrical potential; a connection between the anode and cathode for electron transport (internal circuit); and an electrolyte solution that will conduct ions between the anode and cathode (external circuit) (Figure 1-1). If any one of these components is absent, a corrosion cell does not exist and corrosion will not occur.

When corrosion takes place the metal is oxidized. This occurs at the anode. The electrons generated by the anodic reaction migrate through the internal circuit to the cathode, where they are discharged to a suitable electron acceptor, such as oxygen, by a chemical reduction reaction. The positive ions generated at the anode will tend to migrate through solution to the cathode, and the negative ions generated at the cathode will tend to migrate to the anode. This migration is in response to the concentration gradients and to maintain an electrically neutral solution.

The distribution of anodic and cathodic areas over the corroding material is the primary determinant of the form of corrosion. If the areas are microscopic, and very close to each other, and the potential differences between the areas are small, corrosion may be relatively uniform over the entire surface. However, if the areas are scattered and especially if the potential difference is large, pits may form, sometimes with the irregular deposits of corrosion products called tubercles (Larson 1975).

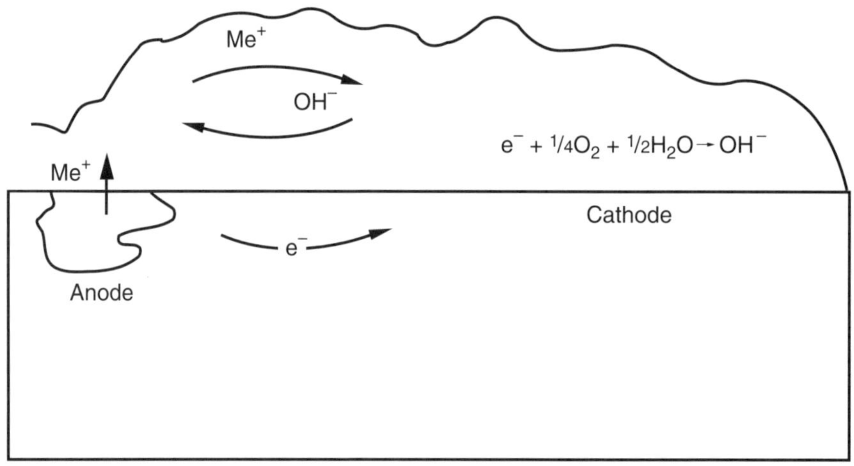

Figure 1-1 Schematic of a corrosion cell

The Electrochemical Potential

Under realistic conditions, the pure metals used in distribution systems are not in equilibrium with the solution that comes in contact with them. The driving force is always for the metal to be oxidized. At the phase boundary of a metal (electrode) in an electrolyte, a potential difference between the solution and the electrode surface exists. Its magnitude reflects the extent of disequilibrium of the reaction:

$$\mathrm{Me} \rightleftharpoons \mathrm{Me}^{z+} + z\,\mathrm{e}^- \tag{1-1}$$

At any instant, both the forward (oxidation) and reverse (reduction) reactions are occurring at various points on the electrode. The metal corrodes or dissolves in the forward reaction. As a result of this, electrons are produced. The current that results is called the anodic exchange current. In the reverse reaction, the metal ions are reduced by combination with electrons; the current that results in this case is called the cathodic exchange current. At equilibrium, the forward reaction proceeds at the same rate as the reverse reaction and the anodic current is equal to the cathodic current; thus, no net corrosion occurs at equilibrium. It should be noted that this equilibrium rarely occurs in natural systems, and that is why metals corrode.

Figure 1-2 shows schematically the dependency of the anodic, cathodic, and net electric current on the electrode potential. The net current (solid curve) represents the sum of the current produced by the oxidation and reduction half-reactions and is equal to

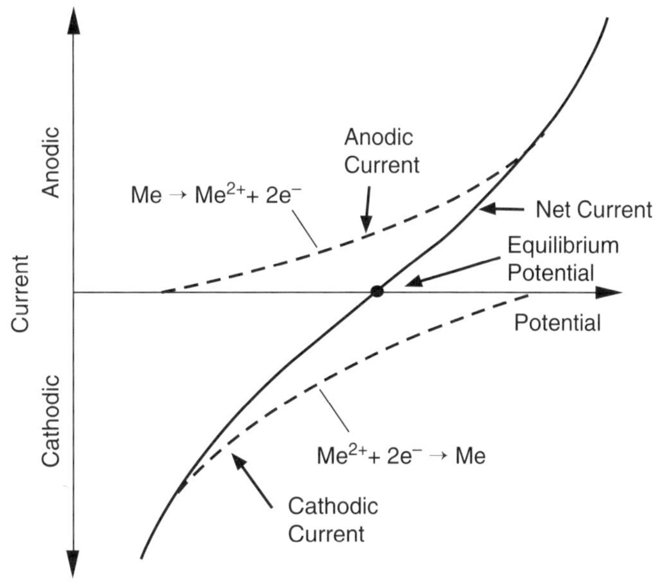

Figure 1-2 Potential versus current for an electrochemical reaction

the velocity of the reaction. The net current is zero when the rate of reduction is equal to the rate of oxidation and the system is at equilibrium. In natural waters, substances such as oxygen serve in place of the metal ion Me^{2+} to accept electrons (see the section, Corrosion Kinetics, later in this chapter). This condition raises the potential of the metal and prevents equilibrium from being achieved.

The Nernst equation. The equilibrium potential of a single electrode, as described previously, can be calculated by using the Nernst equation for the reaction:

$$E_{H,\mathrm{Me/Me}^{z+}} = E^0_{\mathrm{Me/Me}^{z+}} - \frac{RT}{zF} \ln a_{\mathrm{Me}^{z+}} \qquad (1\text{-}2)$$

Where:

E_H = the potential relative to a standard hydrogen electrode (SHE)
E^0 = the standard potential, a constant that can be obtained from tables of electrochemical data
R = the ideal gas constant
T = the absolute temperature, K
z = the number of electrons transferred in the reaction*
F = the Faraday constant
$a_{\mathrm{Me}^{z+}}$ = the activity of the ion Me^{z+}

From Figure 1-2 it can be seen that net dissolution of metal can only occur if the potential is above the equilibrium potential. In drinking water distribution systems, the reaction at the metal surface is coupled with another electrochemical reaction, usually involving H^+, dissolved oxygen (DO), or chlorine.

$$2H^+ + 2e^- \rightleftharpoons H_2 \qquad (1\text{-}3)$$

$$O_2 + 2H_2O + 4e^- \rightleftharpoons 4OH^- \qquad (1\text{-}4)$$

$$HOCl + 2e^- \rightleftharpoons Cl^- + OH^- \qquad (1\text{-}5)$$

---

*When the Nernst equation is written as in Eq 1-2 for a single electrode, it is assumed that the electrode is coupled with the normal hydrogen electrode at which the reaction

$$2H^+ + 2e^- \rightarrow H_{2\,(g)}\,;\, E^0 = 0.00$$

takes place, and reactants and products are at unit activity. Further the convention $\Delta G = -zFE$ is used, where ΔG = free energy change for the complete reaction. This practice is consistent with that of the International Union of Pure and Applied Chemistry (IUPAC) and is followed throughout this book. On occasion, in the discipline of corrosion, the convention $\Delta G = +zFE$ is used, leading to the Nernst equation of the form:

$$E_{H,\,\mathrm{Me/Me}^{z+}} = E^0_{\mathrm{Me/Me}^{z+}} + \frac{RT}{zF} \ln a_{\mathrm{Me}^{z+}}$$

The sign of the constant E^0 is also opposite for these two cases.

Application of the Nernst equation to these half-reactions shows that the potential associated with this reaction is a function of pH. For example, the Nernst equation for Eq 1-4 is:

$$E_{H,O_2/OH^-} = E^0_{O_2/OH^-} - \frac{RT}{4F} \ln \frac{(a_{OH^-})^4}{a_{O_2}} \quad (1\text{-}6)$$

Where:

$$a_{OH^-} = 10^{-(14 - pH)} \text{ at } 25°C$$

Pourbaix diagrams. Using the Nernst equation for appropriate electrochemical half-reactions, it is possible to construct potential–pH diagrams, which are also called Pourbaix diagrams. These diagrams give information about thermodynamically stable products under different potential and pH conditions. For instance, the Pourbaix diagram indicates the occurrence of insoluble corrosion products of a dissolved metal that limit the concentration of a free metal ion. The position of each regional boundary is also a function of the aqueous concentrations of ions that participate in the half-cell reactions.

Figure 1-3, which is a Pourbaix diagram of the iron–water system, shows that metallic iron and water are never thermodynamically stable simultaneously and that metallic iron is not stable in the presence of oxygen. Also, it shows that at low electrode potential, the iron will not corrode (i.e., it is immune). The iron potential can be reduced to the immune region, for example, by cathodic protection. To accomplish this, it is necessary to couple the iron with another, more easily corrodible material such as magnesium. The electrons produced by the corrosion of magnesium flow to the iron and serve to reduce the potential to the immune region. At low pH and intermediate-to-high potential, the diagram shows that the stable iron species is Fe^{2+} or Fe^{3+}. Corrosion will occur at a high rate under these conditions. In the high-potential and high-pH regions, solid products such as $Fe_2O_{3(s)}$ or $Fe_3O_{4(s)}$ may form and deposit as a layer on the surface of the iron. These solids decrease the corrosion rate by influencing anodic or cathodic reactions. The corrosion rate of a metal covered with a layer depends on the structure and chemical nature of the layer.

The diagram also shows the pH–potential regions for H_2–H_2O–O_2 stability. At very low potential, which can be achieved at the iron surface by using a battery and DC current to supply electrons to the surface, water is reduced to H_2. At high potential, water is oxidized to O_2. The stability of water thus limits the range over which the potential of a metal can be varied if it is in contact with water.

Development of the Pourbaix diagram depends on the availability of accurate thermodynamic data for each chemical reaction that will take place and for each solid that will form on the metal surface. Such data are not available for all of the solids that can form on iron when, for example, the solution contains a variety of inorganic ions (e.g., CO_3^{2-}, Cl^-, Ca^{2+}, etc.) that may become incorporated in the scale. Typical protective scales may contain a wide variety of crystal structures, and a diagram that would incorporate these would be hopelessly complicated even if the necessary thermodynamic data were available. Further, corrosion scales are not

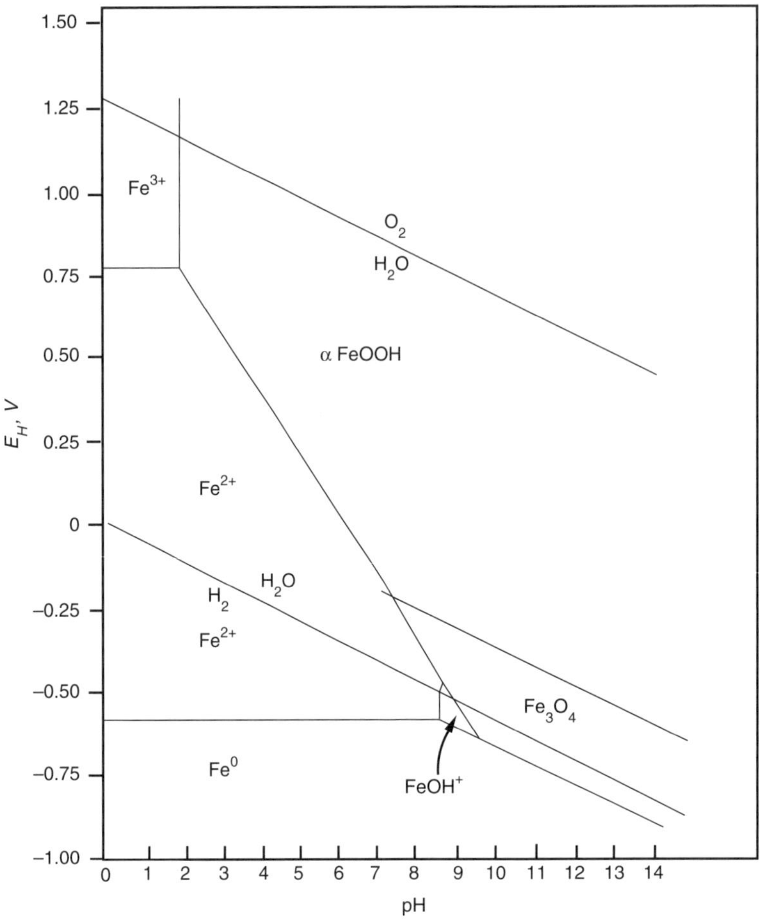

NOTE: Consider Fe, Fe_3O_4, and Fe_2O_3 as solid substances, and $[Fe^{2+}]$ or $[Fe^{3+}] = 10^{-6}M$.

Figure 1-3 Pourbaix diagram for the iron-water systems at 25°C

necessarily thermodynamically stable, so such solids would not appear on the diagram even though they might form in practice. Additionally, the Pourbaix diagram does not necessarily indicate the stable species near the pipe surface because the potential and the pH near the pipe surface are different from those of the bulk water phase and are usually not known. Finally, thermodynamic considerations alone never can give any information about the velocity of the corrosion process itself or the effectiveness of the solids deposited on the pipe at reducing the corrosion rate.

Corrosion Kinetics

Reaction mechanism. Corrosion can occur only if a system is not in the equilibrium state. During corrosion, the metal dissolves (Eq 1-1) and the electrons are accepted in cathodic reactions such as those involving the reaction of protons and oxygen (Eq 1-3 and 1-4). Other cathodic reduction reactions, for example, the reduction of H_2, Cl_2, HOCl, NH_2Cl, ClO_2, and Cu^{2+}, are also possible and can take place if the potential is in a suitable range. The total current produced by metal oxidation at the anode must equal the total current of the cathodic reactions. This is schematically illustrated in Figure 1-4 for the reduction of protons and oxygen. The corrosion potential E_{corr} lies between the equilibrium potentials of the anodic and cathodic reactions. The electrons produced by oxidation of Me at the anode are accepted by both the H^+ half-reaction and the O_2 half-reaction at the cathode.

Because the corrosion process is the result of an anodic and a cathodic process, it can be controlled either by anodic or cathodic reactions. For both anodic and cathodic reactions to take place, electrons and ions must be transferred. Further, the reactants must be transported from the bulk liquid and the products must be transported to the bulk liquid. Additional adsorption and desorption processes are also necessary for the corrosion process. The rate of corrosion is controlled by the slowest step in the series of processes that make up corrosion. In order to control corrosion, it is desirable to know the slowest step of the total process. The rate-controlling step, however, depends on conditions within the distribution system, and a change in these conditions can lead to a change in the rate-determining step. It is possible, for example, that a corrosion process under a high velocity of flow is controlled by an electron transfer process and that one under a low velocity of flow or stagnation conditions is controlled by a mass transport step.

The reaction mechanism must be known if all the factors that can affect the corrosion rate are to be understood. For example, the reaction for the oxidation of iron does not show that H^+ or OH^- is a reactant or product (see Eq 1-7 below), yet the reaction rate depends on pH (Bockris, Drazic, and Despic 1961). This dependency is explained by the series of steps by which iron corrosion occurs. The OH^- ion is part of the three steps of the reaction, thus accounting for the pH dependency.

1. $Fe + OH^- \rightarrow FeOH + e^-$
2. $FeOH \rightarrow FeOH^+ + e^-$ (rate-determining step)
3. $FeOH^+ \rightarrow Fe^{2+} + OH^-$

$$Fe \rightarrow Fe^{2+} + 2e^- \text{ (overall reaction)} \tag{1-7}$$

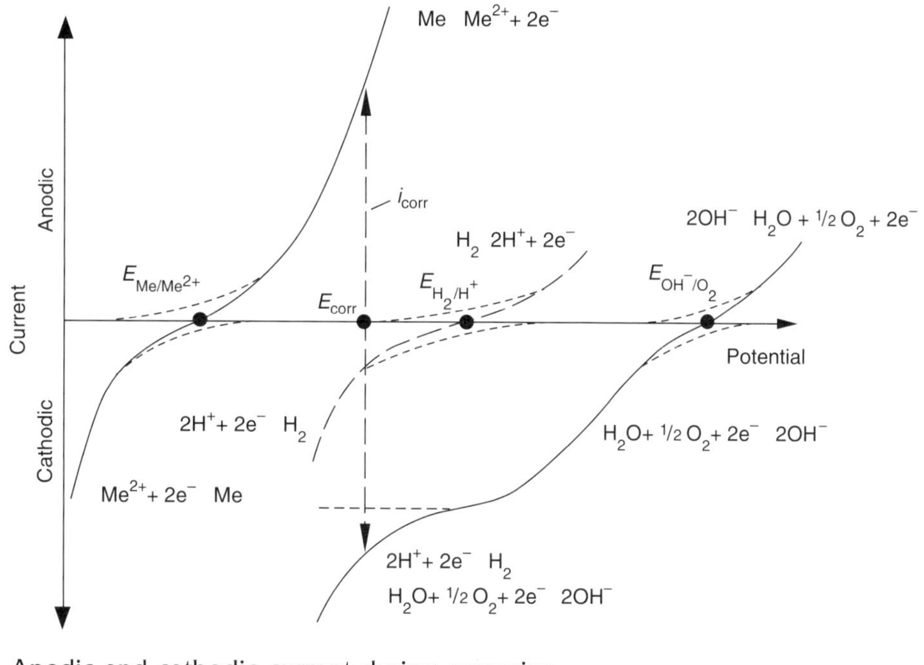

Figure 1-4 Anodic and cathodic current during corrosion

CHEMICAL REACTIONS AT THE METAL SURFACE AND CORROSION SCALES

Passivation and Protective Scales

A surface is passivated if it is covered with a dense oxide layer, or a layer of corrosion products, which serves to substantially reduce the rate of corrosion. Figure 1-5 shows schematically the anodic current potential dependency for iron during the change from the active to the passive state; this is for one type of passivation. This type of passivation occurs for iron alloys but not for many other metals of interest to the drinking water community, such as lead and copper. As potential increases to the critical potential E_{critical}, the rate of corrosion, shown by the anodic current, increases to a maximum value i_{max}. With a further increase in potential, the passivating oxide layer develops and provides resistance to the corrosion reactions; the corrosion current then decreases to i_{passive}. The magnitude of i_{passive} depends on the type of iron alloy. For spontaneous passivation, it is necessary that the cathodic reactions allow a higher anodic current than the maximum of the anodic curve. Passivating inhibitors, such as chromates, are used to create this condition in industrial applications.

The value of i_{max} is strongly influenced by alloy elements like chromium and nickel. These elements depress i_{max} so that spontaneous passivation in water can occur. Chloride ions, on the other hand, can cause localized disturbances of such films and thus accelerate pitting corrosion.

Scales such as $CaCO_{3(s)}$, iron carbonates, and iron oxides on corroding iron or steel are normally thicker and have higher porosity than the oxide layer described earlier. They do not decrease the corrosion rate as much as oxide films do, and the corrosion current–potential relationship shown in Figure 1-5 does not occur for such scales.

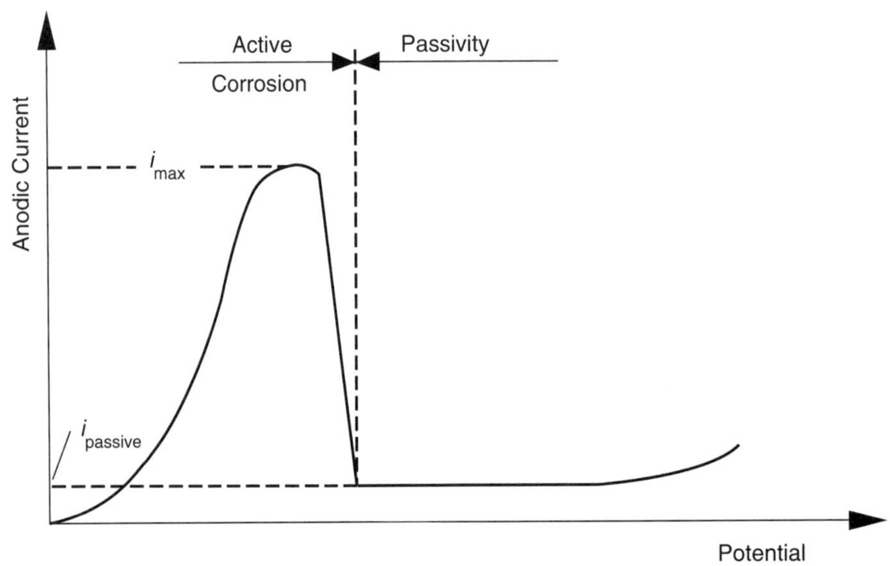

Figure 1-5 Corrosion current versus potential showing the passive state of iron

It is more common to see a gradual reduction in corrosion rate as the scale builds up, as shown in Figure 1-6, and has been observed for black steel (Sontheimer, Kölle, and Snoeyink 1981) and for copper (Reiber 1989). These scales often reduce the corrosion rate by decreasing the rate of oxygen transport to the metal surface, thus decreasing the rate of the cathodic reaction and the deposit on ferrous metals at potentials below $E_{critical}$. The behavior shown in Figure 1-5, however, results from an anodic-controlled corrosion reaction. Throughout this book, both types of surface coatings will be referred to as passivating layers.

Surface Reactions

Analysis of corrosion problems is complicated by the variety of chemical reactions that take place. For example, consider the reactions at an iron or steel surface. First, the primary reaction occurs.

$$Fe_{(s)} \rightarrow Fe^{2+} + 2e^- \quad (1\text{-}8)$$

The Fe^{2+} ions then diffuse into the water or they may undergo a number of secondary reactions.

$$Fe^{2+} + CO_3^{2-} \rightarrow FeCO_{3(s)(siderite)} \quad (1\text{-}9)$$

$$Fe^{2+} + 2OH^- \rightarrow Fe(OH)_{2(s)} \quad (1\text{-}10)$$

$$2Fe^{2+} + \tfrac{1}{2}O_2 + 4OH^- \rightarrow Fe_2O_3 \cdot H_2O + H_2O \quad (1\text{-}11)$$

The hydrated ferric oxides that form from reactions similar to that shown in Eq 1-11 are reddish and under some conditions may be transported to the consumer's tap, causing "red-water" complaints. Tertiary reactions may also occur at the surface. Possibilities are:

$$2FeCO_{3(s)} + \tfrac{1}{2}O_2 + H_2O \rightarrow 2FeOOH_{(s)} + 2CO_2 \quad (1\text{-}12)$$

$$3FeCO_{3(s)} + \tfrac{1}{2}O_2 \rightarrow Fe_3O_{4(s)(magnetite)} + 3CO_2 \quad (1\text{-}13)$$

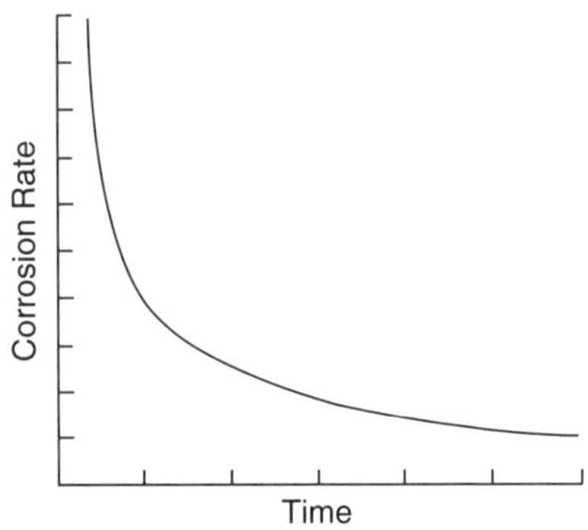

Figure 1-6 Corrosion rate versus time

Surface deposits that result from Eq 1-12 and 1-13 can reduce the flux of O_2 arriving at the anode, thus contributing to the cause of oxygen concentration cells (see the section, Concentration Cell Corrosion, later in this chapter). Other reactions that affect corrosion may take place, depending on the composition of the water and the type of metal, as discussed in the following chapters.

There are many possible cathodic reactions that may occur as well. Perhaps the most common in drinking water distribution systems is the acceptance of electrons by O_2.

$$e^- + \tfrac{1}{4}O_2 + \tfrac{1}{2}H_2O \rightarrow OH^- \quad (1\text{-}14)$$

This reaction causes an increase in pH near the cathode and triggers the following reactions:

$$OH^- + HCO_3^- \rightarrow CO_3^{2-} + H_2O \quad (1\text{-}15)$$

$$Ca^{2+} + CO_3^{2-} \rightarrow CaCO_{3(s)} \quad (1\text{-}16)$$

It is important to note that these reactions can cause $CaCO_{3(s)}$ to precipitate from solutions in which the bulk solutions are undersaturated with this solid. It is necessary only that the pH increase in the vicinity of the cathode be of sufficient magnitude to cause $CaCO_{3(s)}$ to precipitate.

The magnitude of the pH difference between the bulk solution and a corroding surface can be surprisingly large. Lewandowski et al. (1989) measured differences ranging from zero to more than four pH units at a cathodic steel surface. The magnitude of the difference depended on the rate of corrosion and the amount of buffering caused by the carbonate system. Model calculations by Dexter and Lin (1992) for cathodic surfaces in seawater showed pH differences between the surface and bulk solution ranging from zero to 1.5 units, depending on the rate of corrosion. The type of scale and its properties are expected to be a sensitive function of surface pH.

The nature of the scales or deposits that form on metals is very important because of the effect that these scales have on the corrosion rate. Figure 1-6, for example, shows the decrease in the rate of steel corrosion with time, presumably owing to the buildup of scale. A very long time, up to 18 months, may be required for the corrosion rate on iron and steel to stabilize because of the complex nature of the scales. A much shorter time may be sufficient for other metals. If a scale significantly reduces the rate of corrosion, it is said to be a "protecting scale," and if it does not, it is said to be "nonprotecting."

The scales that form may be a mixture of corrosion products that depend on the type of metal that is corroding as well as the composition of the water solution (e.g., $FeCO_{3(s)}$, $FeOOH_{(s)}$, $Pb_3(CO_3)_2(OH)_{2(s)}$, $Ca_2CO_3(OH)_{2(s)}$), precipitates that form because of pH changes that accompany corrosion (e.g., $CaCO_{3(s)}$), precipitates that form because the water entering the system is supersaturated (e.g., $CaCO_{3(s)}$, $Al_2O_{3(s)}$, $Mg(OH)_{2(s)}$, $MnO_{2(s)}$), and precipitates that form from inhibitors such as silicates or phosphates that are added to the water prior to transport (e.g., iron phosphates or silicates, lead hydroxyphosphates such as $Pb_5(PO_4)_3OH_{(s)}$). The specific minerals that form are thus a function of the type of metal and of the composition of the aqueous phase.

Scales that form on pipes may have deleterious effects in addition to the beneficial effect of protecting the metal from rapid corrosion. Water quality should be controlled so that the scale is protective but as thin as possible, because as thickness increases, the capacity of the pipe to carry water is reduced. Increased roughness of the pipe surface, owing to the formation of uneven deposits such as tubercles, also decreases the ability of the pipes to carry water. This effect is most often observed in iron or steel pipes.

The importance of scale is also demonstrated by the phenomenon of erosion corrosion. This type of corrosion is observed at points in the distribution system where there is a high velocity of flow or an abrupt change in direction of flow. The more intense corrosion that often is observed at such locations can be attributed to the abrasive action of the fluid (caused by turbulence, suspended solids, etc.), which scours away or damages the scale, and to the velocity of flow that carries away corrosion products before they precipitate and that supplies corrosion reactants to the surface more efficiently (Eliassen et al. 1956; Larson 1975; Fontana and Greene 1978; Schock and Neff 1982). As a result of this, the metal corrodes at a velocity that is not reduced by a corrosion-product scale, and localized damage at the point of corrosion occurs.

Some reactions at pipe surfaces may not affect corrosion but are important with respect to water quality. Leroy (1991) reports that NO_3^- is reduced to NO_2^- and NH_4^+ on iron surfaces when oxygen concentration decreases to zero. Reactions such as these make the determination of the concentrations of all forms of nitrogen during distribution system studies of great importance.

Factors Affecting Metal Release

Corrosion reactions can cause drinking water quality to deteriorate because of the metal ions that enter the water (i.e., metal ion release). The properties of corrosion scales, the composition of the water, the size of the distribution pipe, and the rate of flow through the pipe determine the solution concentration of the metals. The solubility of the deposits, or the maximum amount that can dissolve per unit volume of water, determines the maximum metal ion concentration that will be found dissolved in the water. Solubility is a very important controlling factor for heavy metals such as lead and cadmium, which have low drinking water quality limits.

Schock and Neff (1982) and EES and ISWS (1990) showed that equilibrium modeling of the solubility of scales that form during corrosion of some metals can be used to successfully predict the maximum concentration of corrosion product that will appear in the water and the conditions necessary for formation of a protective scale. It is necessary to know the types of solids that make up the scale, the composition of the solution, and the necessary thermodynamic data for the reactions that occur. This approach to lead corrosion problems (see chapter 4) does have promise. However, the use of this approach for other metals is limited by incomplete knowledge of the chemical composition and crystalline structure of the solids that make up the scale. The scales that form on iron pipe, for example, are too poorly understood to support this approach.

Scale solubility determines the maximum concentration of soluble metals, whereas the kinetics of dissolution and rate of mass transport into solution determine the concentration if the solubility limit has not been reached. The concentration of metals below the solubility limit depends on the surface-area-to-volume ratio of the transport pipe; whether the flow is laminar, turbulent, or stagnant; and the length of time the water is exposed to a particular type of scale.

The metal concentration in a sample of water taken from a pipe may be attributable to particles containing the metal as well as to metal ions in true solution. The solubility of a scale determines only the concentration of metal ions in true solution. Particles containing corrosion products may result from their precipitation after they are transported away from the pipe wall or from breakup of the scale itself. Loosely deposited scales may be scoured from the pipe surface during high flow, and differential thermal expansion properties of the scale may lead to its fragmentation as the pipe undergoes temperature changes. Control measures that can be used to limit the concentration of metals in true solution cannot be expected to work if the primary cause of high metal concentration is the formation of particulates.

Effect of pH and the Concentration of Carbonates

Other chapters in this book detail the importance of pH relative to corrosion and corrosion control for several important plumbing materials. For example, pH is often increased in order to reduce the concentration of lead and copper in tap water because of the effect of pH on the solubility of these metals. The concentration of calcium and carbonates in the water limits the pH change that can be made without causing excessive amounts of calcium carbonate to precipitate, however. For this reason, the Langelier index (LI), which is a measure of a water's pH relative to its pH of saturation with $CaCO_{3(s)}$, and the calcium carbonate precipitation potential (CCPP), which is a measure of the mass of $CaCO_3$ that will precipitate as the water comes to equilibrium with $CaCO_3$, are important.

The LI (Langelier 1939) is derived from thermodynamic data for the reaction:

$$Ca^{2+} + HCO_3^- \rightleftharpoons CaCO_{3(s)} + H^+ \tag{1-17}$$

and the equation:

$$LI = pH_a - pH_s \tag{1-18}$$

Where:

$$pH_s = pK - \log [Ca^{2+}] - \log [HCO_3^-] - \log \gamma_{Ca^{2+}} - \log \gamma_{HCO_3^-} \tag{1-19}$$

in which:

- pH_a = measured pH
- pK = –log K for Eq 1-16 at the temperature in question
- [ ] = the brackets indicate molar concentration of the free Ca^{2+} and HCO_3^- ions
- γ = the activity coefficient

The term pH_s represents the pH at which the solution is saturated with $CaCO_3$. The constants for calculating pH_s are those of Plummer and Busenberg (1982); these constants and procedures for calculating pH_s are given in *Standard Methods* (1992) and DIN 38404 (1995), as well as in many other sources (Schock 1984; Eberle 1989). A positive LI indicates that the water has a tendency to deposit $CaCO_3$, a negative LI indicates undersaturation with $CaCO_3$, and a value of zero indicates saturation with $CaCO_{3(s)}$.

The pH_s and LI calculated by this procedure are very useful indicators of the $CaCO_3$ saturation state in a water and of the limits of pH, alkalinity, and calcium concentration needed to avoid precipitation of $CaCO_3$. The calculation is based on the assumption that a particular solid form of $CaCO_3$ is present, either calcite, vaterite, or aragonite (the usual form is calcite). Both vaterite and aragonite are more soluble than calcite. Also, many of the analytical procedures used to determine calcium measure total calcium concentration rather than the free Ca^{2+} ion concentration. The total concentration term should be reduced by the amount of calcium that is complexed with inorganic ions, especially sulfate, and also with organic matter, such as humic substances. It is possible to calculate the concentration of inorganic complexes if a relatively complete inorganic analysis of the water is available. The concentration of natural organic matter complexes of calcium is more difficult to obtain, however. Free Ca^{2+} ion concentration can be measured using an electrode, but these measurements are quite rare and often less precise than measurements of total calcium. Fortunately, the amount of Ca^{2+} complexed by organic matter is expected to be small and the error introduced by neglecting this amount will probably be small. If no corrections are made for complex formation, the calculated pH_s will be lower than the true value and LI will be overstated. It is also possible that the crystallization kinetics are strongly affected by the presence of humic substances.

The CCPP indicates how much $CaCO_3$ will precipitate as the solution goes to equilibrium with $CaCO_3$. This quantity is in contrast to the LI, which indicates only whether precipitation will or will not occur. Larson (1975) showed, for example, that an LI of 0.9 is necessary to precipitate 10 mg $CaCO_3$/L at an alkalinity of 50 mg $CaCO_3$/L, but an LI of only 0.2 is necessary for the same amount of precipitation if the alkalinity is 200 mg/L. Precipitation of too much $CaCO_{3(s)}$ may result in reduction of the water-carrying capacity of the main, so it is important to be able to calculate and control the quantity of $CaCO_3$ that will precipitate. Calculation procedures are given by Caldwell and Lawrence (1953), Snoeyink and Jenkins (1980), and Sontheimer, Spindler, and Rohmann (1980), Schock (1990), Loewenthal and Marais (1976), Rossum and Merrill (1983), and *Standard Methods* (1989, Method 2330: Calcium Carbonate Saturation). The calculated values of CCPP are subject to the same limitations as pH_s and LI because the crystalline structure of $CaCO_3$ that forms must be known in order to make an exact calculation. Usually, the assumptions that the crystalline form of $CaCO_3$ is calcite and that the free Ca^{2+} concentration is equal to the total Ca^{2+} concentration are made.

CCPP next to the corroding pipe wall is also important but, to the author's knowledge, has not been quantified. The research of Dexter and Lin (1992) and Lewandowski et al. (1989) has shown that pH next to the pipe wall can be much higher than in bulk solution. As pH increases, HCO_3^- concentrations will decrease and CO_3^{2-} will increase. The amount of change will depend on the corrosion rate and scale characteristics, and as the corrosion rate decreases, the magnitude of the differences between surface and bulk solution will decrease. The total concentration of carbonate species will also play an important role. Because of the probable importance of $CaCO_3(s)$ precipitation on scale properties, research should be undertaken to quantify the factors that affect its precipitation at the pipe wall.

FORMS OF CORROSION

There are many different types of corrosion. The kind of attack depends on the material to become corroded, construction of the system, scale and oxide film formation, and hydraulic conditions. The corrosion forms range from uniform to intense localized attack.

Uniform Corrosion

When uniform corrosion of a single metal occurs, the surface may be referred to as a "polyelectrode" (Petrocelli 1960 [as discussed by Mancy 1966]). Any one site on the metal surface may be anodic one instant and cathodic the next. Because anodic sites shift or creep about the surface, the rate of metal loss is relatively uniform over the metal surface.

The reasons why corrosion cells develop on such metals are varied. The single metals themselves may be heterogeneous, with possible differences in potential existing between different areas because of differences in crystal structure or imperfections in the metal. Also, the concentrations of oxidants and reductants in solutions may be different, thus causing momentary differences in potential.

An alternative model for uniform corrosion is oxidation at a metal surface that is accompanied by electron transport through an adherent film (Figure 1-7). Oxygen reduction occurs at the film surface, and transport of ions to and away from the oxide film takes place. Electrons are probably not transported through an external portion of the metal. The overall rate of corrosion is controlled by the presence and properties of the film or by transport of reaction products, especially hydroxide ion, away from the film–metallic corrosion, notably by Ives and Rawson (1962) for copper.

Galvanic Corrosion

If the corroding system consists of different metals, the cathodic and anodic areas are fixed and so-called galvanic corrosion can occur. Galvanic corrosion results when two different types of metals or alloys contact each other and the elements of the corrosion cell are present. One of the metals serves as the anode, and thus deteriorates, while the other serves as the cathode. The arrangement of available metals and alloys in order of their tendency to be anodic is called the galvanic series. One such series modified from that given by Larson (1975) is shown in Table 1-1. It should be noted that the relative position of metals on this scale is sometimes influenced by water quality and temperature. For example, there are conditions under which iron is anodic to zinc in hot water.

If any two of the metals from the different groupings in this table are connected in an aqueous environment, the metal appearing first will tend to be the anode and the second will be the cathode. In general, the farther apart the metals are on the chart, the greater will be the potential for corrosion because the potential difference between them will be greater. However, the corrosion behavior of some metals is not dictated solely by differences in the electrode potential. The tendency of a metal to passivate in the environment in which it is used is a very important property that affects corrosion behavior. Some alloys of iron, such as stainless steel, are much more corrosion resistant in water than cast iron because of this property. Aluminum also tends to passivate rapidly when in contact with certain types of water.

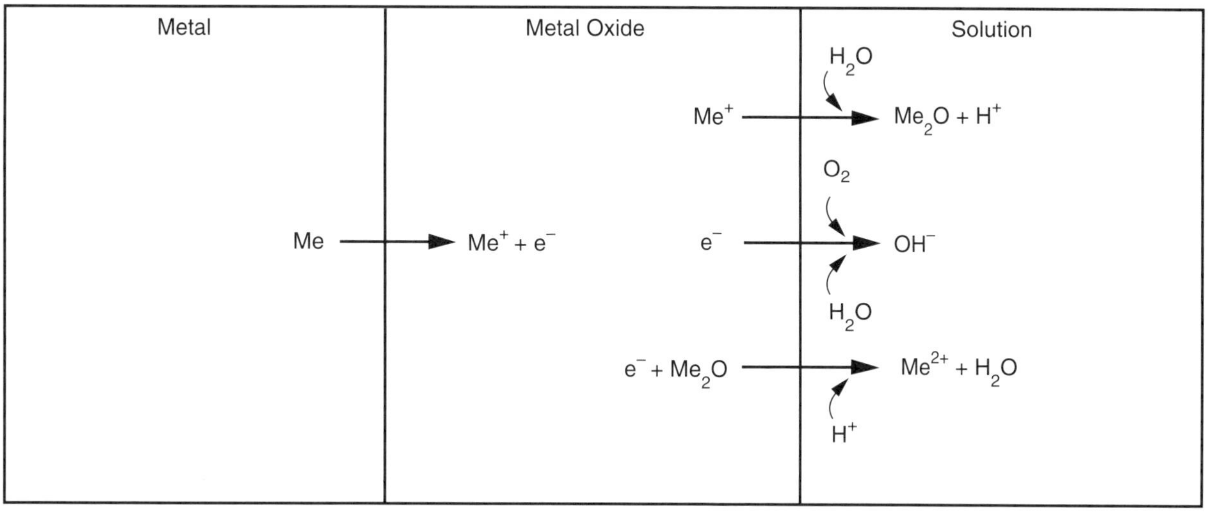

Figure 1-7 Uniform corrosion of a metal

The rate of galvanic corrosion is increased by greater differences in potential between the two metals. It is increased by large areas of cathode relative to the area of the anode, and it is generally increased by closeness of the two metals and by increased mineralization or conductivity of the water. Galvanic corrosion is often a great source of difficulty where brass, bronze, or copper is in direct contact with aluminum, galvanized iron, or iron. Copper-bearing metals are cathodic to aluminum, zinc, and iron, and their underwater contact very often results in corrosion of the latter metals. Galvanized (zinc-coated) steel is usually more serviceable than steel alone, because the iron exposed at joints and holidays (holes in the zinc) is protected at the expense of the zinc.

Lane and Neff (1969) noted that materials selection is the "first line of defense" in corrosion control in domestic cold- and hot-water distribution systems. To prevent galvanic corrosion, for example, they noted that only copper tubing should be used with copper-lined water heaters. Brass valves in contact with steel and galvanized plumbing in waters with high total dissolved solids cause corrosion of the steel and galvanized pipes. It is also possible for dissolved copper to plate out in spots on galvanized pipe, thereby causing copper–zinc galvanic cells (Kenworthy 1943).

Localized Corrosion

Localized attack resulting in pitting may occur, both with galvanic corrosion and with corrosion of a single-metal system, if there are imperfections in the metal or its oxide film or scale or if regions of high stress exist in the metal. Areas of high stress or of metal imperfection are usually anodic, and the potential difference between these regions and the rest of the metal is such that the anode remains at the same location. If the region of high potential is caused by continual stressing of a working metal part, the corrosion is called stress corrosion. Usually the anodic region is small relative to the cathodic area, and corrosion to the point of failure occurs rather rapidly compared to uniform corrosion. This type of corrosion also occurs if an anodic area is covered with a coating that is ruptured.

Table 1-1 Galvanic series of metals and alloys

Corroded end (anodic, or least noble)
Magnesium
Magnesium alloys
Zinc
Aluminum
Cadmium
Steel or iron
Cast iron
Iron alloys
Lead–tin solders
Lead
Tin
Nickel
Brasses
Copper
Bronzes
Titanium
Monel
Silver solder
Silver
Carbon (graphite)
Gold
Noncorroded end (cathodic, or most noble)

NOTE: The metals within each of the groupings have similar corrosion potentials.

Concentration Cell Corrosion

Corrosion potential is a function of the concentration of aqueous solution species that are involved in the reaction, as well as of the characteristics of the metal. A difference in the concentration of such species between two parts of the metal can drive the corrosion reaction. The corrosion process will always occur in a way that tends to equalize potential at the two locations. This type of localized corrosion is called concentration cell corrosion.

A common cause of corrosion is the presence of different concentrations of dissolved oxygen (DO) or hydrogen ion at different sites on a metal surface. When caused by dissolved oxygen, this is often referred to as "differential oxygenation corrosion." For example, the Nernst equation for corrosion of iron with oxygen as the electron acceptor is:

$$E = E^0 + \left(\frac{0.059}{4}\right)\log \frac{[Fe^{2+}]^2}{[O_{2(aq)}]\ [H^+]^4}$$

This equation can be used to show that if two sites on an iron pipe have the same pH and Fe^{2+} concentration, but the O_2 concentration is 5 mg/L at one site and 0.1 mg/L at the other, the potential difference between the two sites will be 25 millivolts. Some examples of this are illustrated in Figure 1-8. Common areas for differential oxygenation corrosion to occur are between two metal surfaces—under rivets, under washers, or in crevices.

The area beneath various types of surface adhesion, which may include a barnacle, a bacterial slime, or a piece of mill scale, is also a region where concentration cells develop. Since oxygen is not produced by either the anodic or cathodic reaction, corrosion will proceed in such a way that the higher oxygen concentration is reduced (Uhlig 1938) if pH is the same at both sites. Oxygen participates in the cathodic reaction; thus, the part of metal in contact with the higher oxygen concentration will be the cathode and the part of metal in contact with the lower oxygen concentration will be the anode and will corrode. This is consistent with the observation that corrosion takes place underneath adhesions on metal surfaces.

Oxygen may be depleted at a specific location because of the corrosion reaction, the reaction of oxygen with the corrosion product, or the microbiological reactions taking place within slime coatings. Resistance to the diffusion of oxygen to these locations by the slimes, chemical precipitates, deposits of debris, or simply distance of transport (into a crevice or to pipe threads, for example) and a sufficient supply of

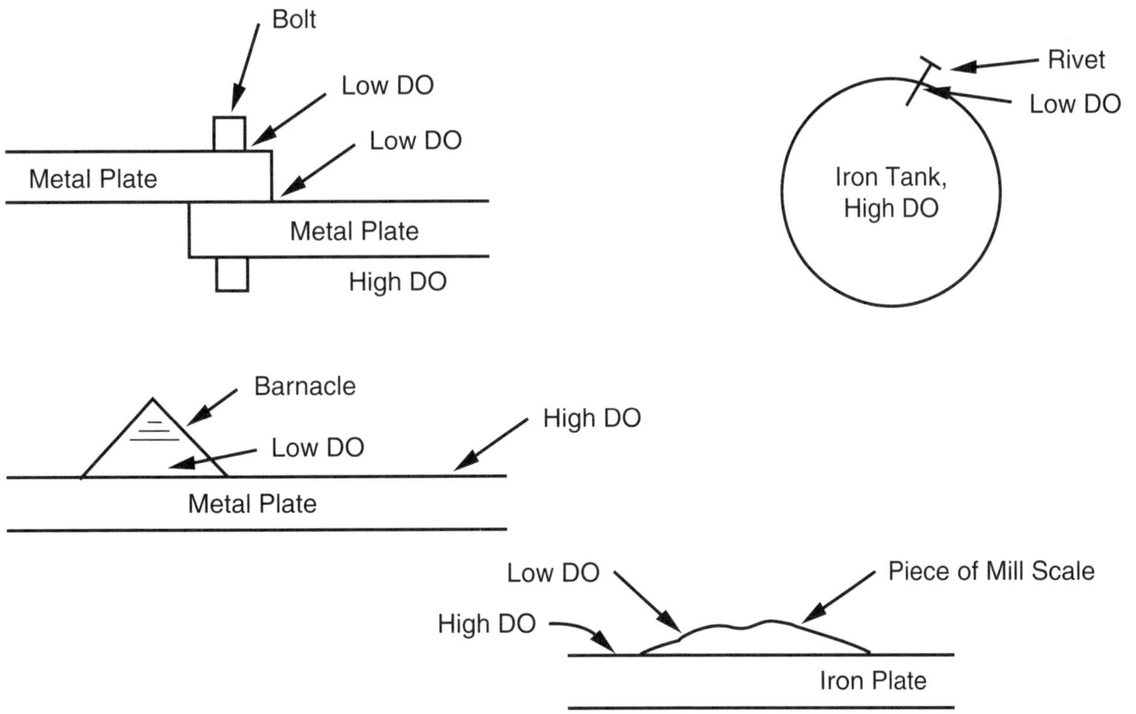

Source: Water Chemistry. *V.L. Snoeyink and D. Jenkins. Copyright 1980. Reprinted by permission of John Wiley & Sons, Inc.*

Figure 1-8 Examples of differential oxygenation corrosion

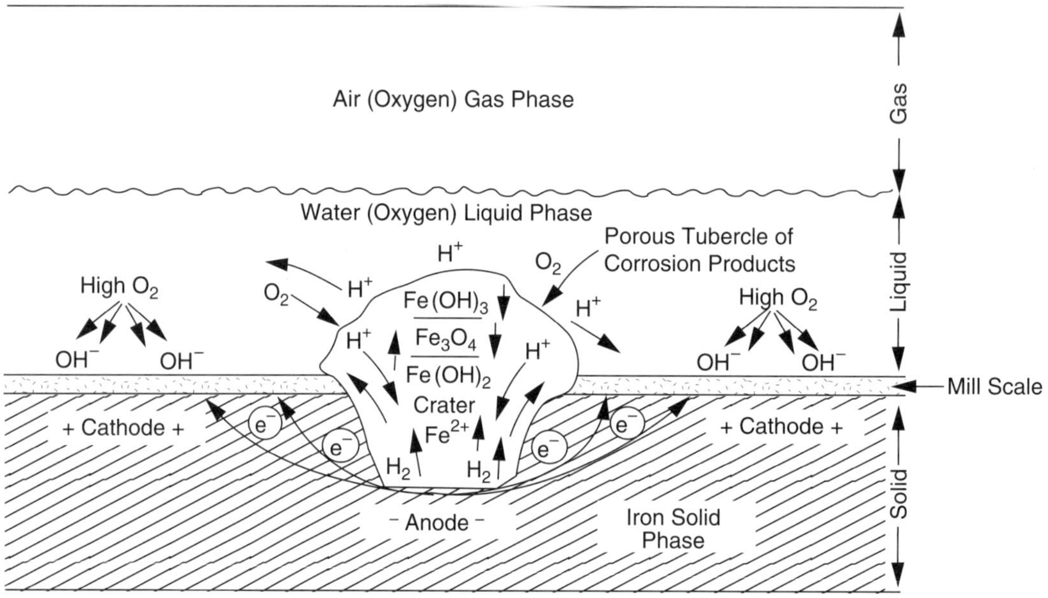

Source: After Obrecht and Pourbaix (1967).

Figure 1-9 Pitting of iron by tuberculation and oxygen-concentration cells

oxygen to adjacent cathodic sites maintain the oxygen concentration cell and the corrosion reaction. Continued attack at the low-oxygen site leads to pitting, the deepening of crevices, etc. Thus, oxygen concentration differences are frequently the cause of what is termed "crevice corrosion." A major cause of scale roughness is the formation of tubercles, which appear as nodules or knoblike prominences on the scale.

According to Larson (1975), uncontrolled tuberculation can double the cost of pumping. Obrecht and Pourbaix (1967) indicated that tuberculation on iron results from oxygen concentration cells and results in pitting and irregular deposits (Figure 1-9). According to this representation, the H^+ produced from the reaction of the Fe(II) corrosion products with oxygen can serve to neutralize the OH^- produced by the cathodic reaction and can accept electrons to form $H_{2(g)}$ within the pit. Reduction of O_2 by the Fe(II) corrosion products prevents O_2 from reaching the anodic sites, so the oxygen concentration cell is perpetuated.

$$O_2 + 4Fe^{2+} + 10H_2O \rightarrow 4Fe(OH)_{3(s)} + 8H^+ \quad (1\text{-}20)$$

BIOLOGICAL FACTORS AFFECTING CORROSION*

Growth of Microorganisms in Distribution Systems

Many different microorganisms have been found in distribution systems. Some may enter with the water that is introduced into the system and some may grow within the system. Certainly, not all microorganisms induce or aggravate corrosion or are associated with corrosion reactions. It is especially difficult to determine the exact

*Some material in this section was presented by W. Kölle in the first edition of this book.

role that microorganisms play in the corrosion process, but much can be inferred about possible effects based on what is known about growth requirements. In general, it appears as if some microorganisms have the ability to induce or promote corrosion and that some take advantage of corrosion deposits as growth habitats. Allen, Taylor, and Geldreich (1980) showed microscopic evidence of high populations of microorganisms at the water interface of corrosion-produced incrustations.

The typical distribution system has many locations, or niches, where microorganisms can grow. Niches, which consist of crevices, blisters, and incrustations such as tubercles and deposits of sediments, provide locations where growth can occur unhindered by disinfectants. Disinfectants can prevent the growth of microorganisms on clean pipe surfaces. However, a disinfectant may not always be present because it is consumed by corrosion and other reactions, especially during times of stagnation. Growth of microorganisms may occur on the pipe surface during these times; such growth is accompanied by extracellular bacterial secretions that provide protection from disinfectants and enable further growth.

Many factors affect the growth of microorganisms in distribution systems. Certainly the condition of the distribution system is important insofar as the number of niches present and the opportunity for growth is concerned. A heavily tuberculated system, or one with many sediment deposits, is more likely to have growth than a system that is clean and noncorroded. High concentrations of ammonia or biodegradable organic compounds and an absence of residual disinfectant promote growth and are responsible for what is called a "biologically unstable" water (Rittmann and Snoeyink 1984). A water that is inherently more corrosive will lose its residual disinfectant faster, due to corrosion, thus aggravating microbiological problems. Ainsworth, Oliphant, and Ridgeway (1980) noted that organic carbon appears to be a major factor influencing the numbers of microorganisms in a distribution system. They noted that the numbers of microorganisms increased as total organic carbon (TOC) and oxygen were used in the system. The microbial activity was found to be concentrated in the surface deposits and pipe sediments. In-plant treatment of the water to reduce or eliminate ammonia and biodegradable organic matter may be necessary to biologically stabilize the water and reduce disinfectant demand; this will in turn reduce the likelihood of extensive microbial growth in the distribution system. An important factor in reducing the disinfectant demand in the distribution system is the reduction in corrosivity of the water. Biodegradable compounds are usually measured using the assimilable organic carbon (AOC) test (van der Kooij, Visser, and Hijnen 1982) or the biodegradable organic carbon (BDOC) test (Joret et al. 1988).

Important factors other than biological stability, residual disinfectant concentration, and corrosivity of water that affect growth include temperature, flow rate, pH, oxygen concentration, and the presence of other substances, such as nitrate, nitrite, and sulfate, that can be reduced by microorganisms if oxygen and disinfectants are absent.

Microbially Influenced Corrosion

Microbes can influence corrosion in several ways. Growth of microorganisms in biofilms can create concentration cells that promote corrosion. They generally grow in patches along the surface in a way that can create adjacent regions with differential oxygen, hydrogen ion, and metal ion concentrations. This is due to the reactions they promote and the resistance the biofilm provides to transport of oxygen and other substances. These concentration differentials promote corrosion (see the section, Forms of Corrosion, earlier in this chapter). Lee, O'Connor, and Banerji (1980) found evidence that microbial growth promoted localized corrosion, consistent with concentration cell corrosion.

Microorganisms may catalyze reactions associated with the corrosion process. Bacteria that can oxidize or reduce different forms of iron and sulfur are especially important, as reviewed by Emde, Smith, and Facey (1992). Iron-precipitating bacteria such as *Gallionella* and *Sphaerotilis* convert Fe(II) to Fe(III) (also see Kölle and Rösch 1980) and influence the Fe(III) precipitate structure (see chapter 2). Organisms such as *Bacillus*, *Escherichia coli*, *Pseudomonas*, and *Citrobacter* have the ability to reduce Fe(III) to Fe(II) and have been found in tubercles, but the role these organisms play in the corrosion process has not been delineated. Organisms that catalyze all phases of the sulfur cycle in water have also been found, including *Desulfovibrio* and *Desulfotomaculum* that reduce sulfate to sulfite; *Clostridium*, *Bacillus*, and *Salmonella* that reduce sulfite to thiosulfite; *Enterobacteriaceae* such as *Citrobacter* that reduce thiosulfite to sulfide; and *Thiobacillus* that oxidizes sulfide to sulfate. Others (Larson 1966) have found that *Beggiatoa* oxidizes sulfide to sulfur. When these reactions occur, the changes in species concentrations that accompany them or the changes in precipitate mass or structure can all affect the corrosion process.

Bacteria that catalyze reactions of nitrogen are also important in water systems. Nitrosofication (conversion of ammonia to nitrite) and nitrification (conversion of nitrite to nitrate) are important if oxygen is present; denitrification (conversion of nitrate to nitrogen) is important in regions where oxygen is absent. These reactions affect oxygen concentration through utilization of oxygen, and the biofilms themselves attach to pipe walls, thereby creating oxygen concentration cells. The biofilms also affect disinfectant demand in the system. Larson (1939) reports a case history involving nitrate formation in the first part of a distribution system followed by denitrification in the outer reaches of the system.

The changes in solution concentration of corrosion reaction reagents by bacteria, and the biofilms themselves, probably severely aggravate the problem of distribution system corrosion.

CONTROL OF ELECTROCHEMICAL CORROSION

The information gained from corrosion studies is used to chose pipe materials and control water chemistry in order to minimize the occurrence of deleterious effects from corrosion. The extreme complexity of corrosion phenomena has prevented the development of general rules that can be used to predict when corrosion will or will not occur, except in a few limiting cases. Dissolved oxygen; pH; alkalinity; calcium; suspended solids; organic matter; buffer intensity; metal ions that can be reduced in cathodic reactions; total salt concentration; specific anions such as chloride, sulfate, phosphate, and silicate; biological factors; and temperature all have been shown to have an effect. Many of these factors have different effects on the corrosion of different metals. The indices used to predict the tendency for $CaCO_{3(s)}$ to precipitate have served useful purposes, but many have been misapplied, particularly when they are used without recognition of their limitations. The following discussion presents some of the more common indices and discusses their strengths and limitations.

Indices Related to $CaCO_3$

Langelier index. The LI is the most common of the "corrosion indices." Its determination was discussed earlier in this chapter.

A positive LI has widely been believed to indicate a "noncorrosive" water, ostensibly because of deposition of a protective "eggshell" layer of $CaCO_3$ at slightly positive LI values. While some water utilities have achieved relatively "noncorrosive" conditions for many materials as a result of LI adjustment, there are many reasons to believe that $CaCO_3$ saturation and deposition has little to do with protection. The limitations of the LI as a corrosion index are as follows:

- Determination of an LI value, using standard procedures, requires careful measurement of pH, alkalinity, calcium, temperature, and estimation of ionic strength.
- Complexation of Ca^{2+} and HCO_3^- with inorganic and organic substances is not accounted for, although it is possible to do so if needed analytical data are available.
- Even an accurate LI value provides little insight into the rate of calcite precipitation or dissolution.
- The crystalline form of $CaCO_{3(s)}$ is assumed to be calcite. However, the presence of another form, such as aragonite, for which there is a different equilibrium constant for Eq 1-16, has been observed. As reviewed by Schock and Neff (1982), the presence of other forms of $CaCO_{3(s)}$ may account for observations of substantial supersaturation with respect to calcite.
- Perhaps most important, a deposit of $CaCO_{3(s)}$ does not necessarily form an adherent protective film and may not aid in preventing corrosion, as discussed subsequently (Stumm 1960).

These are reasons why an LI value may not be an accurate indicator of whether $CaCO_{3(s)}$ will deposit, or how much will form, or how protective the deposit may be. Two reasons why the LI in many instances is not related to protective scale formation that does occur are

- $CaCO_{3(s)}$, if important, can also be deposited from waters with a negative LI because of the localized high pH, next to the pipe, generated by the cathodic reduction of O_2.
- Protective scales are usually formed with the metal cation forming a hydroxide, oxide, carbonate, phosphate, or perhaps, silicate precipitate. The mineralogy, scale structure, and chemistry of these deposition processes is quite complicated, as described in other chapters on the various metals.

Many types of solids other than $CaCO_{3(s)}$ provide resistance to corrosion, as reviewed by Sontheimer, Kölle, and Snoeyink (1981) for iron and steel and by many others for other types of pipe materials. Alkalinity and pH have an important role in precipitation chemistry for these deposits; calcium ion often has much less significance. The LI value characterizes only a small part of a very complex chemical system, and a positive LI gives no assurance that a protective scale is present or will form.

Finally, the preoccupation by many with maintaining a positive LI has led to excessive deposition of $CaCO_{3(s)}$ and to large decreases in the capacity of the system carry water. Calcium carbonate scaling problems may be encountered in hot-water systems, even if not found in the cold water. If calcium carbonate saturation is an objective, it should be used with the CCPP to show how much calcite can possibly precipitate per litre. If a thick layer is deposited, it should be protective but it will probably also reduce the ability of the pipe to carry water. Deposition of smaller amounts, in the form of an "eggshell" layer, is not often observed in water systems using LI adjustment.

Based on the preceding discussion, the following points seem relevant:

- Because $CaCO_{3(s)}$ is found in protective scales, notably on iron and steel, it is reasonable to evaluate control of the LI as part of a corrosion control program. In general, the higher the pH, alkalinity, and calcium the less corrosive a water will be, and higher values are associated with more positive LI values. However, a positive LI is not necessary, and an LI value of –0.1 or –0.2 may be sufficient and, in many cases, may lead to better protection against corrosion.
- Because of the difficulty in making an accurate calculation of LI and CCPP next to the corroding surface, these values should be calculated for the bulk solution, taking important inorganic species such as sulfate in addition to calcium, alkalinity, pH, and ionic strength effects into consideration. The changes in calcium and alkalinity concentrations through segments of the distribution system should be measured in order to monitor the quantity of $CaCO_{3(s)}$ that precipitates. These measurements, combined with visual and chemical observations of the scale on pipe segments taken from different parts of the system, should give important information about the kinetics of precipitation and permit adjustment of the water quality so that enough, but not too much, $CaCO_{3(s)}$ will precipitate.
- In high-hardness and high-alkalinity waters, the CCPP and LI values should be used to avoid excessive $CaCO_{3(s)}$ deposition.
- The LI should be used only as one of several pieces of data to indicate corrosivity or noncorrosivity. It should be supplemented with observations of pipe that has been in use, with analytical data on dissolved or particulate corrosion products, and with pipe loop or coupon studies, as described in chapter 9.

Other indices. Other indices related to the tendency of $CaCO_{3(s)}$ to precipitate, such as the Ryznar index and the aggressiveness index, were reviewed by Gardels and Schock (1981) and Singley (1981). They are based on principles similar to those used to develop the LI but are subject to many more limitations, and thus seem to have more limited application than the LI. Others such as Dye's momentary excess and McCauley's driving force index are related to the quantity of $CaCO_3$ that will precipitate at equilibrium and have little to add to the CCPP. They do not account for precipitation that takes place when the LI is less than zero because of high pH next to the corroding pipe wall.

Indices Based on Other Water Quality Parameters

Stumm (1960) and Sontheimer, Spindler, and Rohmann (1980) established the importance of the effect of buffer intensity on corrosion rate. Stumm (1960) noted that corrosion rate decreased as buffer intensity increased. He hypothesized that buffer intensity influenced the pH shift that occurs at the anode and cathode as a result of corrosion reactions. For example, a water with a high buffer intensity will have less of a pH increase at the cathode, and less of a pH decrease at the anode, than water with a low buffer intensity. Buffer intensity is a function of alkalinity and pH; it is defined as the resistance of a solution to pH change when a strong acid or strong base is added. It is given by:

$$\beta = \frac{dC_B}{d(\mathrm{pH})} = \frac{-dC_A}{d(\mathrm{pH})}$$

Where:

C_B = moles of strong base that are added per litre of solution
C_A = moles of strong acid that are added per litre of solution

When the carbonate system is the predominant weak acid/base system present:

$$\beta = 2.3\left([H^+] + [OH^-] + \frac{[H_2CO_3^*][HCO_3^-]}{C_{T,CO_3}} + \frac{[HCO_3^-][CO_3^{2-}]}{C_{T,CO_3}}\right)$$

or $$\beta = \beta_{H_2O} + \beta_{CO_3}$$

Where:

$\beta_{H_2O} = 2.3([H^+] + [OH^-])$ is the buffer intensity of water

$\beta_{CO_3} = 2.3\left(\frac{[H_2CO_3^*]\,[HCO_3^-]}{C_{T,\,CO_3}} + \frac{[HCO_3^-]\,[OH^-]}{C_{T,\,CO_3}}\right)$ is the buffer intensity of the carbonate species

In this equation, $[H_2CO_3^*] \simeq [CO_2]$, C_{T,CO_3} = the total concentration of carbonate species, and all concentrations are in moles per litre. Calculation procedures are given elsewhere (Ricci 1952; Stumm 1960; Snoeyink and Jenkins 1980; Sontheimer, Spindler, and Rohmann 1980). The β of a water with a C_{T,CO_3} of 0.5 m*M* is shown in Figure 1-10. The β decreases sharply as pH increases from 6.3 to about 8.3 and then increases again with increasing pH. Stumm (1960) showed that the rate of iron corrosion decreased in a uniform manner as β increased; exceptions were noted for those samples low in calcium and with an LI < –1.5. Buffer intensity thus seems to be an important indicator of corrosivity, although sufficient data are not available to indicate that it can reliably be used as a sole index. It is important to note that Stumm did not study waters with pH values above 8.5. Consequently, his results cannot be generalized to this region; he also did not study metals other than iron.

The effect of buffer intensity is closely linked with $CaCO_{3(s)}$ precipitation. Stumm demonstrated that hard water with a large buffer intensity and a slightly positive LI deposited less $CaCO_{3(s)}$ than soft water with a negative LI. Similar results were observed by Sontheimer, Kölle, and Rudek (1979). Scales with more than 5 to 10 percent (by weight) $CaCO_{3(s)}$ also have been found to be less protective than those with less $CaCO_{3(s)}$.

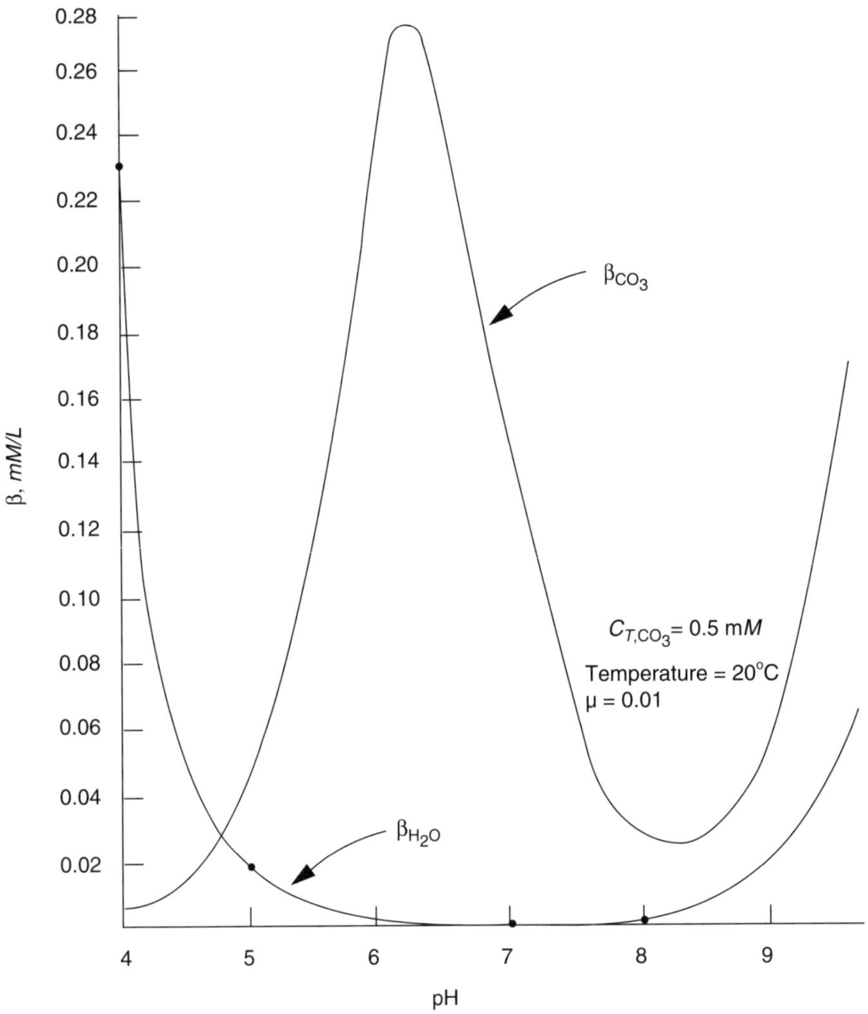

NOTE: β_{H_2O} = buffer intensity from water
β_{CO_3} = buffer intensity from carbonate species
$\beta_{TOTAL} = \beta_{H_2O} + \beta_{CO_3}$

Figure 1-10 Variation of buffer intensity with pH

Larson and Skold (1957) studied the effects of Cl^- and SO_4^{2-} on iron and mild steel corrosion and found that the effects are primarily related to their concentration relative to HCO_3^-. They proposed the relationships:

$$LI_1 \text{ (Larson index)} = \frac{(2[SO_4^{2-}] + [Cl^-])}{[HCO_3^-]} \tag{1-21}$$

and

$$LI_2 \text{ (Larson index)} = \frac{[Cl^-]}{[HCO_3^-]} \tag{1-22}$$

where LI represents the Larson index and the brackets indicate concentration in moles per litre. Observations of corrosion in water of different quality showed the rate of corrosion of mild steel increased significantly when LI_2 was greater than 0.3 (Larson 1975). A similar formula developed by Wagner, Gerber, and Kuch (1985) showed higher corrosion and iron release rates (see chapter 2) if the quotient was higher than 0.7. Further research is needed to fully establish the effect of Cl^- and SO_4^{2-} on corrosion of plumbing materials other than iron.

CORROSION OF CEMENTITIOUS MATERIALS

Many drinking water distribution systems contain inorganic materials, such as cement-based materials including concrete and asbestos–cement pipes and cement–mortar-lined steel and ductile iron pipes. The corrosion of these materials depends primarily on the quantity of calcium carbonate-aggressive CO_2 or CCPP. If the water is undersaturated with $CaCO_3$, the $CaCO_3$ layer will dissolve. This layer results primarily from the reaction of the alkaline part of the cement (mostly $Ca(OH)_2$) with the CO_2-containing species in the water. The result of a severe attack on the cement-based material by CO_2 is the decrease of the mechanical stability of the inner cement layer. As a consequence, there is the possibility that sand or asbestos fibers may wash off the inner surface of the pipe and that the cement layer may spall off, exposing the underlying steel.

If the distributed water is extremely low in minerals and does not contain sufficient hardness and carbonate species, the leaching of the alkalinity from the cement can lead to an increase in pH values; under unfavorable conditions, a pH as high as 12.5 can be reached with new mortar-lined pipe. Details on these effects are given in chapter 7.

SUMMARY

Most water quality parameters have been shown to have an effect on corrosion and often to affect the corrosion of the various piping systems in different ways. Specific aspects of the corrosion of each of the most important metals, as well as cement and asbestos cement, will be covered in the following chapters.

REFERENCES

Ainsworth, R.G., R. Oliphant, and J. Ridgeway. 1980. *The Introduction of New Water Supplies Into Old Distribution Systems*. Tech. Rept. TR146. Medmenham, England: Water Research Centre.

Allen, M.J., R.H. Taylor, and E.E. Geldreich. 1980. The Occurrence of Microorganisms in Water Main Incrustations. *Jour. AWWA*, 72(11):614.

Bockris, J. O'M., D. Drazic, and A.R. Despic. 1961. The Electrode Kinetics of the Deposition and Dissolution of Iron. *Electrochimica Acta*, 4:325.

Caldwell, D.H., and W.B. Lawrence. 1953. Water Softening and Conditioning Problems. *Ind. Eng. Chem.*, 45:535.

Dexter, S.C., and S.H. Lin. 1992. Calculation of Seawater pH at Polarized Metal Surfaces in the Presence of Surface Films. *Corrosion*, 48(1):51–60.

DIN 38404, Teil 10. 1995. Deutsche Einheitsverfahren zur Wasser-, Abwasser- und Schlammuntersuchung; Physikalische und physikalisch-chemische Stoffkenngrössen (Gruppe C); Calcitsättigung eines Wassers (C 10). Berlin: Beuth-Verlag GmbH.

DIN 50900, Teil 1. 1975. Korrosion der Metalle, Begriffe, Allgemeine Begriffe. Berlin: Beuth-Verlag GmbH.

Eberle, H.S. 1989. A Correctness Test of the Competition of Chemical Speciation for the Main Constituents of Natural Waters. *Wat. Res.,* 23:1373–1382.

EES and ISWS (Economic and Engineering Services, Inc., and Illinois State Water Survey). 1990. *Lead Control Strategies.* Denver, Colo.: AWWA Research Foundation.

Eliassen, R., et al. 1956. Effects of pH and Velocity on Corrosion of Steel Water Pipes. *Jour. AWWA,* 48:1005.

Emde, K.M.E., D.W. Smith, and R. Facey. 1992. Initial Investigation of Microbially Influenced Corrosion (MIC) in a Low Temperature Water Distribution System. *Wat. Res.,* 26(2):169–175.

Fontana, M.G., and N.D. Greene. 1978. *Corrosion Engineering.* 2nd ed. New York: McGraw-Hill.

Gardels, M.C., and M.R. Schock. 1981. Corrosion Indices: Invalid or Invaluable? In *Proc. of the Ninth Annual AWWA Water Quality Technology Conference.* Denver, Colo.: American Water Works Association.

ISO (International Standards Organisation) 8044-1986 (E/F/R). 1979. Manual of Symbols and Terminology for Physico-Chemical Quantities and Units. 2 Revision. *Pure and Applied Chemistry,* 51:141.

Ives, D.J.G., and A.E. Rawson. 1962. Copper Corrosion III: Electrochemical Theory of General Corrosion. *Jour. Electrochem. Soc.,* 109:452.

Joret, J.C., Y. Levi, T. Dupin, and M. Gilbert. 1988. Rapid Method for Estimating Bioeliminable Organic Carbon in Water. In *Proc. of the AWWA Annual Conference.* Denver, Colo.: American Water Works Association.

Kenworthy, L. 1943. The Problem of Copper and Galvanized Iron in the Same Water System. *Jour. Inst. Metals,* 69:67.

Kölle, W., and H. Rösch. 1980. Untersuchungen an Rohrnetzinkrustierungen unter mineralogischen Gesichtspunkten. *Vom Wasser,* 55:159.

Lane, R.W., and C.H. Neff. 1969. Materials Selection for Piping in Chemically Treated Water Systems. *Materials Protection,* 8:27.

Langelier, W.F. 1939. The Analytical Control of Anticorrosion Water Treatment. *Jour. AWWA,* 28(10):1500–1521.

Larson, T.E. 1939. Bacteria, Corrosion and Red Water. *Jour. AWWA,* 31:1186.

———. 1966. Deterioration of Water Quality in Distribution Systems. *Jour. AWWA,* 58(10):1307–1316.

———. 1975. *Corrosion by Domestic Waters.* Bulletin 59. Champaign, Ill.: Illinois State Water Survey.

Larson, T.E., and R.V. Skold. 1957. Corrosion and Tuberculation of Cast Iron. *Jour. AWWA,* 49:1294.

Lee, S.H., J.T. O'Connor, and S.K. Banerji. 1980. Biologically Mediated Corrosion and Its Effects on Water Quality in Distribution Systems. *Jour. AWWA,* 72:636.

Leroy, P. 1991. *Calcium et Corrosion.* Doctoral thesis, University of Paris, France.

Lewandowski, Z., W.C. Lee, W.G. Characklis, and B. Little. 1989. Dissolved Oxygen and pH Microelectrode Measurements at Water-Immersed Metal Surfaces. *Corrosion,* 45(2):92–98.

Loewenthal, R.E., and G.v.R. Marais. 1976. *Carbonate Chemistry of Aquatic Systems: Theory and Applications.* Ann Arbor, Mich.: Ann Arbor Science.

Mancy, K.H. 1966. Dissolved Oxygen—A Corrosion Inhibitor or Promoter in Water Distribution Systems. Paper presented at the National Association of Corrosion Engineers Meeting, Oct. 17–19, 1966, Boston.

Obrecht, M.F., and M. Pourbaix. 1967. Corrosion of Metals in Potable Water Systems. *Jour. AWWA,* 59:977.

Petrocelli, J.V. 1960. Electrochemistry of Dissolution Processes. In *The Surface of Metals and Semiconductors.* Edited by H.C. Gatos. New York: John Wiley and Sons.

Plummer, L.N., and E. Busenberg. 1982. The Solubilities of Calcite, Aragonite and Vaterite in CO_2-H_2O Solutions Between 0 and 90°C, and an Evaluation of the Aqueous Model for the System $CaCO_3$-CO_2-H_2O. *Geochim. Cosmochim. Acta,* 46:1011.

Reiber, S.H. 1989. Copper Plumbing Surfaces: An Electrochemical Study. *Jour. AWWA,* 81(7):114.

Ricci, J.E. 1952. *Hydrogen Ion Concentrations.* Princeton, N.J.: Princeton University Press.

Rittmann, B.E., and V.L. Snoeyink. 1984. Achieving Biologically Stable Drinking Water. *Jour. AWWA,* 76(10):106.

Rossum, J.R., and D.T. Merrill. 1983. An Evaluation of the Calcium Carbonate Saturation Indices. *Jour. AWWA,* 74(2):95.

Schock, M.R. 1984. Temperature and Ionic Strength Corrections to the Langelier Index—Revisited. *Jour. AWWA,* 76:72.

———. 1990. Internal Corrosion and Deposition Control. In *Water Quality and Treatment: A Handbook of Community Water Supplies.* New York: McGraw-Hill and American Water Works Association.

Schock, M.R., and C.H. Neff. 1982. Chemical Aspects of Internal Corrosion: Theory, Prediction and Monitoring. In *Proc. of the Eleventh Annual AWWA Water Quality Technology Conference.* Denver, Colo.: American Water Works Association.

Singley, J.E. 1981. The Search for a Corrosion Index. *Jour. AWWA,* 73:579.

Snoeyink, V.L., and D. Jenkins. 1980. *Water Chemistry.* New York: John Wiley and Sons.

Sontheimer, H., W. Kölle, and R. Rudek. 1979. Aufgaben und Methoden der Wasserchemie—Dargestellt an der Entwicklung der Erkenntnisse zur Bildung von Korrosionsschutzschichten auf Metallen. *Vom Wasser,* 52:1.

Sontheimer, H., W. Kölle, and V.L. Snoeyink. 1981. The Siderite Model of the Formation of Corrosion-Resistant Scales. *Jour. AWWA,* 73:572.

Sontheimer, H., P. Spindler, and U. Rohmann. 1980. *Wasserchemie für Ingenieur e.* Frankfurt: ZfGW-Verlag.

Standard Methods for the Examination of Water and Wastewater. 17th ed. 1989. Washington, D.C.: American Public Health Association, American Water Works Association, and Water Pollution Control Federation.

Stumm, W. 1960. Investigation on the Corrosive Behavior of Waters. *Jour. San. Engrg. Div.—ASCE,* 86:27.

Uhlig, H.H. 1938. *Corrosion and Corrosion Control.* New York: John Wiley and Sons.

van der Kooij, D., A. Visser, and W.A.M. Hijnen. 1982. Determining Concentration of Easily Assimilable Organic Carbon in Drinking Water. *Jour. AWWA,* 74:540–545.

Wagner, I., S. Gerber, and A. Kuch. 1985. Der Einfluss von Neutralsalzen auf das Korrosionsverhalten unlegierter Eisenwerkstoffe in gut gepufferten Trinkwässern. *Werkstoffe und Korrosion,* 36:64.

2

Corrosion of Iron and Steel

Mark M. Benjamin, Heinrich Sontheimer,
and Pierre Leroy

Iron, in one form or another, is the most common metal in water distribution systems, and the corrosion of iron in such systems is of utmost concern. Iron-based metallic surfaces that are likely to come into contact with potable water in its transit from the source to the consumer include the walls or working parts of well casings, pumps, mixing equipment, meters, pipes, valves, and fittings. Loss of pipe integrity, leading to leaks or, in extreme cases, collapse of the pipe wall, cannot be ignored as a possible consequence of iron corrosion. Fortunately such dramatic failures are rare in potable water systems. Rather, the principal iron corrosion problems in potable water systems are associated with the buildup and release of iron oxide corrosion products.

This chapter summarizes the current understanding of iron corrosion in potable water systems. In view of the vast literature on iron and steel corrosion, a comprehensive review is not possible. The authors have included those aspects that they feel are most important and have made reference to selected papers that address these issues. The chapter first describes the use of ferrous materials in potable water systems. The effects of water quality on the equilibrium and the kinetic aspects of iron corrosion are then discussed, emphasizing the different factors that might control corrosion rates as well as the type and extent of the scale layers that form on the metal surface. A new hypothesis explaining the formation of red water

NOTE: The first edition of this chapter was authored by J. Edward Singley, Heinrich Sontheimer, Alfred Kuch, Walter Kölle, and Abdul Ahmadi.

is presented, along with evidence supporting its validity. The chapter concludes with an attempt to integrate all the information presented in order to summarize the current state of the art and identify critical areas of research that can enhance the understanding of iron corrosion in distribution systems.

Although terms and concepts that are critical for an understanding of iron corrosion are introduced and defined in this chapter, the reader is referred to chapter 1 for a more detailed discussion of the fundamentals and vocabulary of corrosion. Additionally, a discussion of specific corrosion inhibitors is provided in chapter 8, and the reader is referred to that chapter for more information on inhibitor-based methods to control iron corrosion.

BACKGROUND

Ferrous Materials Used in Distribution Systems

The principal iron-based materials used in the drinking water community are cast iron, ductile iron, and steel. Although the properties of iron are certainly important in determining the behavior of all these materials, their mechanical properties are altered enormously by the relatively small (<10 percent by weight) portion of the bulk material that is *not* iron. In fact, the materials are distinguished based on differences in their nonferrous portions. Typical compositions and some mechanical properties of these materials are shown in Table 2-1.

The major chemical difference among the materials listed in Table 2-1 is that steel has a much lower carbon content and a much lower silicon content than either cast or ductile iron. Because the densities of carbon and silicon are both about one-third that of pure iron, the differences between steel and the other two materials on a volume basis are much greater than those shown, which are mass-based. For instance, the volume fraction of impurities in cast and ductile iron is about 18 percent, while that in steel is only a little over 1 percent. This fact is significant when considering corrosion reactions, since the fraction of the surface area occupied by impurities is expected to be about the same as the volume fraction of impurities. Thus, about 15 times as much surface area is occupied by impurities in cast or ductile iron as in steel.

Differences in mechanical properties usually control the choice of which material to use in a specific application. Unlined cast-iron mains have been in use for more than 300 years in Europe and since the early 19th century in the United States. However, virtually all the cast-iron mains installed in both the United States and Europe over the past few decades have been lined with bituminous or asphaltic material and/or with cement mortar. The linings are intended to prevent or at least minimize corrosion of the pipe surface. The most common linings in current use are asphaltic coatings over a cement layer.

General Observations on Iron and Steel Corrosion

The corrosion of steel and cast iron in distribution systems can be either general, i.e., occurring more or less uniformly over large portions of the surface, or localized. Uniform corrosion is most often attributed to the existence of numerous, short-lived anodic and cathodic sites. A small amount of corrosion takes place at the anode, after which the transient disturbance in the electrochemical potential at that point is eliminated and corrosion ceases, only to commence at another nearby site. Such a process might occur in a system with a protective film that is randomly breeched but then repairs itself rapidly. Localized corrosion, on the other hand, occurs in the presence of persistent local nonuniformities in the pipe or the water quality adjacent

Table 2-1 Typical compositions and properties of cast iron and steel

Sample	Type	Elemental Composition, %							Tensile Strength, $psi \times 10^3$ (*MPa*)	Elongation, %	Iron, %
		Carbon	Silicon	Manganese	Phosphorus	Sulfur	Nickel	Magnesium			
Ductile	1-in. (2.5-cm) block	3.70	2.50	0.45	0.03	0.01	0.60	0.05	80 (551)	12	92.66
	Pipe fitting	3.60	2.60	0.48	0.07	0.01	0.85	0.06	77 (530)	16	
Cast	1.2-in. (3.0-cm) bar, Class 30	3.35	2.25	0.55	0.20	0.12			30 (206)	0	93.18
Steel	1-in. (2.5-cm) round	0.15	0.17	0.50	0.03	0.03			62 (427)	32	99.12

Adapted from LaQue (1964).

to it. Such a condition may be caused by sulfide inclusions in steel pipes, the local breakdown of protective films that are not self-repairing, or a concentration cell that is maintained and exacerbated by the buildup of acidity or diffusion-limiting corrosion scales adjacent to the anode.

Localized corrosion of ferrous materials often leads to tuberculation, in which an actively corroding site is covered by growing mounds of corrosion products. Adjacent mounds may grow to the point that they are in contact, making individual sites of tuberculation indistinguishable and causing the pipe to appear to have a continuous layer of tuberculated corrosion products covering the surface. Tuberculation reduces the effective pipe size and increases the roughness of the pipe's inside surface, thereby restricting the flow of water. It may also provide attachment and growth sites for microorganisms.

If the products of iron corrosion do not remain attached to the pipe walls, they are released into the water supply. Iron release can lead to the staining of fixtures and the appearance of rusty or red water at the tap. The iron release rate is controlled by a complex combination of physical and water quality factors that may exert their influence during the time when the film is forming, as well as when red-water conditions are actually observed. As a result, the iron release rate often bears no simple relationship to the instantaneous overall corrosion rate, which is the rate at which metallic iron is being converted to an oxidized form. If most of the corroded iron forms an adherent film on the pipe surface, the corrosion rate can be much greater than the release rate. Conversely, previously immobilized iron can be released from the film at a time when the metallic iron at the pipe's surface is being oxidized slowly or not at all, in which case the short-term release rate would be greater than the corrosion rate. Since iron corrosion and iron release are independent processes, measures that limit iron release might not be effective at controlling the iron corrosion rate.

The failure to distinguish between the iron oxidation rate and the iron release rate has led to ambiguity in terminology (both are sometimes referred to as the "corrosion rate") and confusion about the effectiveness of various corrosion control techniques. The authors have attempted to make a clear distinction in this work and urge other researchers and practitioners to do so in the future.

Equilibrium, Pourbaix Diagrams, and Iron Corrosion

Some of the relevant chemical reactions of iron in a potable water system are listed in Table 2-2. Pourbaix diagrams for an idealized system containing only deionized water, iron, and other species that do not react with dissolved iron indicate a domain where the iron is immune (the unoxidized surface is thermodynamically stable) and other domains where it is subject to corrosion (these domains include all "nonimmune" domains) (Figure 2-1). When metals precipitate from solution, it is common for an amorphous, pseudo-stable solid to form first and then to convert to more stable, crystalline forms over time. Therefore, two Pourbaix diagrams are shown for this system: Figure 2-1A is based on the assumption that crystalline iron oxides can form in the system, whereas Figure 2-1B assumes that only amorphous hydrous oxides form. In each figure, the nonimmune region is further subdivided into areas where the corrosion product (oxidized iron) is substantially soluble and areas where the precipitation of a solid is expected to maintain total soluble iron at a concentration less than 0.056 mg/L (10–6 mol/L) under equilibrium conditions (higher soluble iron concentrations may exist if equilibrium is not attained). Such solids can partially protect the metal surface from further corrosion under

Table 2-2 Important chemical reactions associated with iron corrosion

Oxidation and Reduction Reactions*	
$Fe^0 \leftrightarrow Fe^{2+} + 2e^-$	Oxidation of metallic iron
$Fe^{2+} \leftrightarrow Fe^{3+} + e^-$	Oxidation of ferrous iron
$3Fe(OH)_{2(s)} \leftrightarrow Fe_3O_{4(s)} + 2H_2O + 2H^+ + 2e^-$	Partial oxidation of ferrous hydroxide to form magnetite
Precipitation and Dissolution Reactions*	
$Fe^{2+} + 2OH^- \leftrightarrow Fe(OH)_{2(s)}$	Precipitation of ferrous hydroxide
$Fe^{2+} + CO_3^{2-} \leftrightarrow FeCO_{3(s)}$	Precipitation of siderite
$3Fe^{2+} + 2PO_4^{3-} \leftrightarrow Fe_3(PO_4)_{2(s)}$	Precipitation of vivianite
$(1-x)\ Fe^{2+} + S^{2-} \leftrightarrow Fe_{1-x}S_{(s)}$	Precipitation of ferrous sulfide (variable stoichiometry)
$Fe^{3+} + 3OH^- \leftrightarrow Fe(OH)_{3(s)}$	Precipitation of amorphous ferric hydroxide
$2Fe(OH)_{3(s)} \leftrightarrow \alpha\text{-}Fe_2O_{3(s)} + 3H_2O$ $Fe(OH)_{3(s)} \leftrightarrow \alpha\text{-}FeOOH_{(s)} + H_2O$	Dehydration of amorphous ferric hydroxide to form hematite or goethite

*Reactions represent typical overall processes and are not intended to indicate detailed mechanisms.

certain conditions. However, the fact that an iron oxide solid is expected to form in a particular system indicates only that protection of the surface is a possibility; the precipitation of such a solid in no way ensures that the corrosion rate will be slowed significantly.

Although oxidants such as free or combined chlorine can influence the redox potential (E_H) of potable water significantly, most discussions in the literature consider the corrosion potential to be controlled jointly by the solution pH and the concentration of dissolved oxygen (DO). Thus, the E_H can be computed by:

$$E_H\,(\text{mV}) = 1.189 - 0.059\ \text{pH} + \frac{0.059}{4}\log(\text{DO}) \qquad (2\text{-}1)$$

Where:

DO = the dissolved oxygen concentration, mg/L

To locate a particular system on a Pourbaix diagram, the E_H and pH at the point of metal–water contact must be known. If one assumes, for simplicity, that the pH and DO concentration at the metal surface are about the same as in bulk solution, a Pourbaix diagram such as the one shown in Figure 2-1 can provide information about the tendency for the iron to corrode and about the iron species that would be dominant if the system reached equilibrium.

The E_H values of solutions that contain 2 mg/L DO or 0.2 mg/L total free chlorine residual (HOCl plus OCl^-) are indicated as a function of pH in Figure 2-1, for an assumed chloride ion concentration of 35 mg/L (1.0 meq/L). Significantly, there is no pH at which either metallic iron or divalent iron (Fe(II)) is stable in the presence of

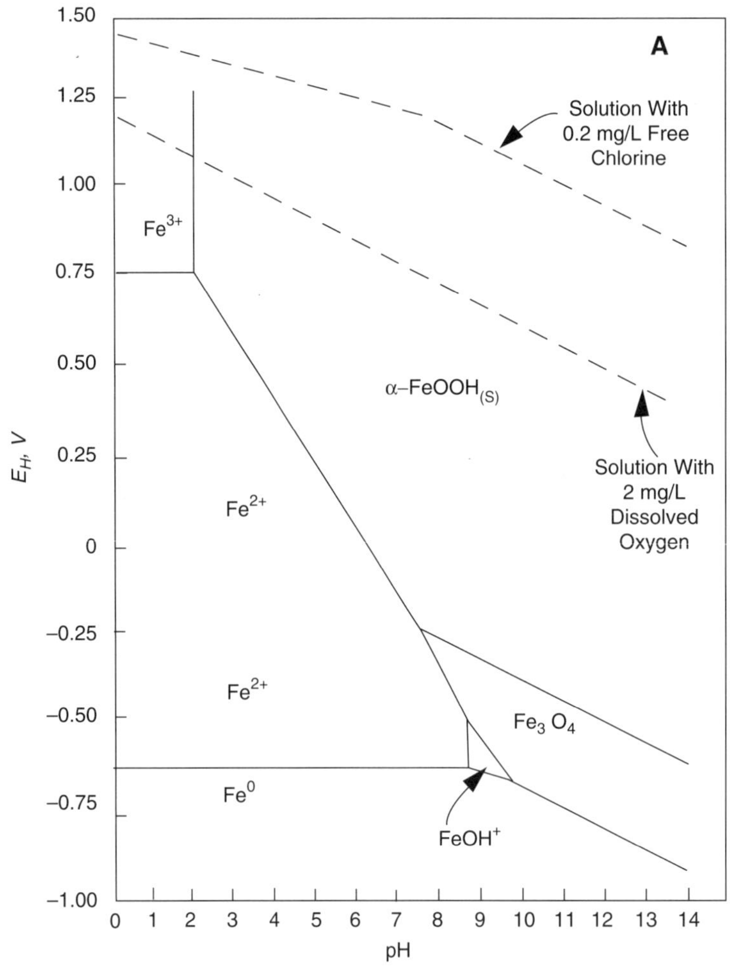

NOTE: Total soluble iron is $10^{-6}M$ (0.056 mg/L), and alkalinity is negligible. Figure 2-1A is based on the assumption that crystalline solids might form; Figure 2-1B assumes that only amorphous solids form. The dashed lines near the top of the diagrams indicate the E_H of solutions containing the indicated amounts of DO and free chlorine.

Figure 2-1 Pourbaix diagrams for iron

either one of these oxidants. The implication of this result is that if water containing the indicated concentration of oxygen or free chlorine contacted metallic iron, there would always be a driving force for the iron to be oxidized to the trivalent form (Fe(III)) and for the chlorine to be reduced to chloride and the oxygen to be reduced to H_2O or OH^-. Although it is not obvious from the diagram, it is easy to demonstrate that one would reach the same conclusions for any measurable values of DO or chlorine residual.

The fact that DO and free chlorine are not immediately consumed when they are in contact with iron pipes points out two limitations of Pourbaix diagrams. First, since the diagrams provide information about the chemical equilibrium conditions for the given water qualities, only factors that affect or are affected by chemical equilibrium can be displayed in or inferred from the diagrams. If, for instance, an increase in the alkalinity of a water changes the chemical stability of some dissolved and solid species in the system, these changes can be reflected in a Pourbaix diagram. However, alkalinity affects other aspects of corrosion reactions as well and, in particular,

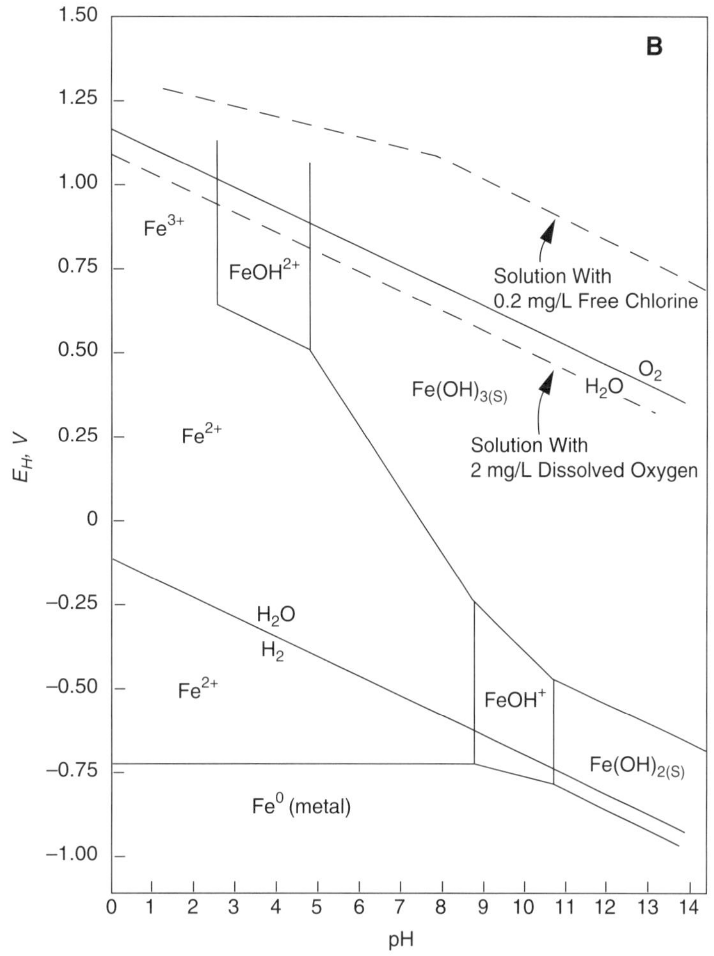

Figure 2-1 Pourbaix diagrams for iron (continued)

the kinetics of the reactions, e.g., by increasing the conductivity of the solution and by buffering the pH near actively corroding sites. These effects are not reflected in the diagrams. The nonequilibrium effects, which are often more important than the changes in the equilibrium conditions, are discussed in some detail later in this chapter (see the sections, Scale Formation and Corrosion of Iron Pipes and Iron Release, Red Water, and the Kuch Mechanism, later in this chapter). Biological reactions can also influence the corrosion of iron, but these effects are often very complex and less well understood. They are not discussed here in detail.

Second, the sharp distinctions on most Pourbaix diagrams between areas where solids are present and absent can be misleading: the location and even the existence of the boundaries depend on the species and equilibrium constants that have been chosen for the calculations, and this choice is often problematic. This difficulty can be illustrated by considering the variety of solids that can be identified generically as Fe(III) hydrous oxides. This group of solids, all of which can be represented by the chemical formula $Fe_2O_3 \cdot nH_2O$, includes goethite (α-FeOOH), lepidocrocite (γ-FeOOH), hematite (α-Fe_2O_3), and “amorphous” iron hydroxide (a generic label applied to ferric oxides with no identifiable crystallinity).

All of the iron oxides just listed can be found in nature, and even more solids with the same chemical formula have been prepared in laboratory environments. In addition to these products, other iron oxides exist that contain Fe(II) as well as Fe(III) (e.g., magnetite, Fe_3O_4). These different solids all have different solubilities and stabilities, with the amorphous hydroxide being least stable and hematite being most stable. However, many of these solids have been found in corrosion scales, often coexisting in close proximity. Therefore, the preparation of a Pourbaix diagram requires that choices be made about which of the solids to consider in the calculations. Such choices may be based on empirical evidence or informed speculation about the sequence of events and chemical conditions near the site of corrosion and precipitation. Different choices can lead to differences in the location of a boundary line on the diagram. For instance, Figure 2-1A indicates that at $E_H = 0$, an iron oxide solid will precipitate at pH 6.2; Figure 2-1B indicates that at the same E_H, the pH must be at least 8.0 for a solid to form. Neither diagram is inherently more correct than the other; they are simply based on different assumptions about which solid forms under a given set of conditions. In addition, the solubility constants for each individual solid are often known only to within a couple orders of magnitude, adding further uncertainty to the boundaries on the Pourbaix diagram.

The above discussion is not intended to diminish the importance of Pourbaix diagrams; such diagrams are appropriate for determining the thermodynamically stable, dominant iron species in a system where the water quality is known. However, it is important to remember that the diagrams are representations of available data, and the data available for iron-based systems are extremely incomplete. Additional investigations addressing the question of which solids are formed under various corrosion conditions are likely to lead to further advances in our understanding of the corrosion process.

MODELS FOR IRON CORROSION

It is a truism that all corrosion processes in water distribution systems are complex, owing to the range of possible oxidants and reductants, the dependence of the reactions on localized conditions, and the possibility that layers of corrosion products will gradually accumulate on the corroding surface. The situation is particularly complex in the case of iron corrosion, since the corroded iron can exist in at least two different oxidation states and in a dozen or more different soluble or precipitated forms, several of which might be present in a single system. As a result of this complexity, the same variable, e.g., the solution pH, might be of great significance in some systems and relatively unimportant in others. Under these circumstances, it is virtually impossible to describe, in concise generalizations, the effects of individual water quality factors on iron corrosion rates and iron release rates, since the details of the corroding system always have such a large influence on those effects.

In order to address this problem, the authors have divided the following discussion of iron corrosion into sections based on the dominant physical and chemical processes that are thought to be occurring in the corroding system. The first section describes the corrosion process at the pipe surface and addresses the oxidation of metallic iron. Although this process is frequently studied using new pipes, the results are thought to be applicable even in old pipes, at locations where the unoxidized metal is in contact with an aqueous solution, i.e., under the scale layers. The second section describes the fate of oxidized iron (Fe(II)) during the period immediately after its release into the solution adjacent to the pipe wall. During this time, the iron either precipitates and becomes part of the scale or it diffuses through the preexisting scale and enters the bulk solution. The third section describes changes that occur in scales as they age.

Processes occurring during each of these steps can affect both the corrosion rate and the iron release rate. In the discussion, an attempt is made to identify and explain the critical water quality variables that control the reactions in each step. An understanding of these interactions can help explain why any particular variable might be important in one system at one time but not in another system or in the same system under different flow conditions.

Oxidation of Metallic Iron

The discussion of the Pourbaix diagram for iron pointed out that metallic iron cannot coexist in an equilibrium system with water containing a measurable concentration of DO or free chlorine. However, as a practical matter, it is obvious that the corrosion reactions consuming the oxygen and chlorine often do not reach equilibrium, since finite concentrations of these oxidants can be maintained throughout distribution systems. The most likely explanation for these observations is that the corrosion rate is limited by the rate at which oxidants can be transported from bulk solution to the pipe surface.

The predominant anodic reaction on bare steel or cast iron surfaces in contact with potable water is oxidation of metallic iron and the predominant cathodic reaction on the same surfaces is reduction of oxygen (see reactions 1 and 2 listed on p. 41).

This observation supports the validity of using the oxygen reduction reaction to compute the E_H of the system. However, as a result of the reactions, the O_2 and OH^- concentrations and therefore the E_H at the pipe wall are different from those in the bulk water. This result applies even in the absence of a layer of corrosion scale, since in such a case there is still a barrier to diffusion imposed by the stagnant boundary layer next to the pipe wall. Thus, oxygen must be continuously supplied from the bulk solution for the reaction to proceed, and even on a scale-free surface, the corrosion rate could be limited by the rate at which oxygen diffuses to the surface and corrosion products diffuse from the surface.

One set of experiments in which the corrosion rate is controlled by diffusion of DO through the liquid boundary layer is presented in Figure 2-2. For a given temperature and water velocity, the rate of oxygen consumption is equivalent to the corrosion rate and is, to a first approximation, directly proportional to the oxygen concentration in the water, regardless of the concentrations of most other constituents dissolved in the water. Factors that affect the transport of oxygen from the bulk of water to the pipe surface, such as flow velocity and water temperature, also affect the corrosion rates of bare pipes. For instance, an increase in temperature leads to an increase of the diffusion coefficient and a decrease of the viscosity, both of which tend to increase the oxygen mass transfer coefficient and therefore the rate at which oxygen is supplied to the cathodes on the pipe surface.

Scales that develop on the pipe walls provide additional diffusion barriers to constituents approaching and leaving the surface and thereby decrease the corrosion rate. The effectiveness of these barriers is affected by the physical properties of the scale, such as porosity, tortuosity, and thickness. As the scales develop, their contribution to the overall diffusion barrier increases, and the influence of flow velocity on the mass transfer coefficient diminishes. This trend can be seen in Figure 2-3. Reactions within the scale layer might consume part of the DO as it diffuses toward the pipe wall, further reducing the rate at which oxygen reaches the cathodic sites. However, regardless of the number and types of ways in which oxygen transport to the pipe

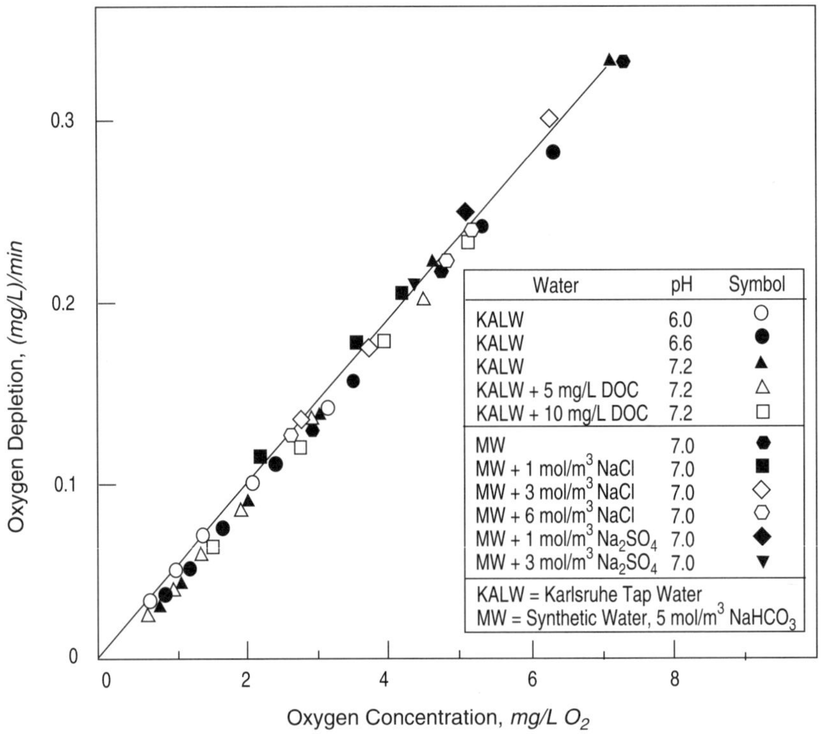

Figure 2-2 Oxygen consumption rate versus oxygen content for different water qualities in contact with a new active corroding steel surface (temperature = 12°C, flow velocity = 0.5 m/s)

surface is limited, it seems reasonable to assume that the fundamental process governing oxidation of the metal is the same in these situations as in the absence of the scale. That is, if one could measure the DO concentration at the metal surface, the linear relationship between DO concentration and iron oxidation rate would be confirmed. A similar statement might well apply to other oxidants, such as residual chlorine species, although fewer experimental data are available to test such a hypothesis at this time.

To summarize, when potable water containing dissolved oxidants is in contact with metallic iron, there is a driving force for active corrosion under any realistic water quality conditions. The dominant oxidant in most such systems is DO, and the rate of corrosion is probably limited by the rate at which oxygen is provided to the surface. This rate, in turn, is limited by molecular diffusion through the layers of stagnant water and scale adjacent to the metal. As discussed later in this chapter, oxygen reduction might also take place on the scale, rather than at the metal surface, if the scale is electrically conducting (so the electrons can travel from the pipe wall or from elsewhere in the scale to the site of reduction). Factors that affect the transport rate include temperature and water velocity. In the absence of scale, an increase in either of these variables increases the oxygen transport rate and the corrosion rate. Other water quality factors, including solution pH, alkalinity, and the concentration of other dissolved inorganic ions, have a negligible effect on corrosion rate under these conditions, but they may influence scale formation and, through this, the diffusion resistance and the corrosion rate at a later stage.

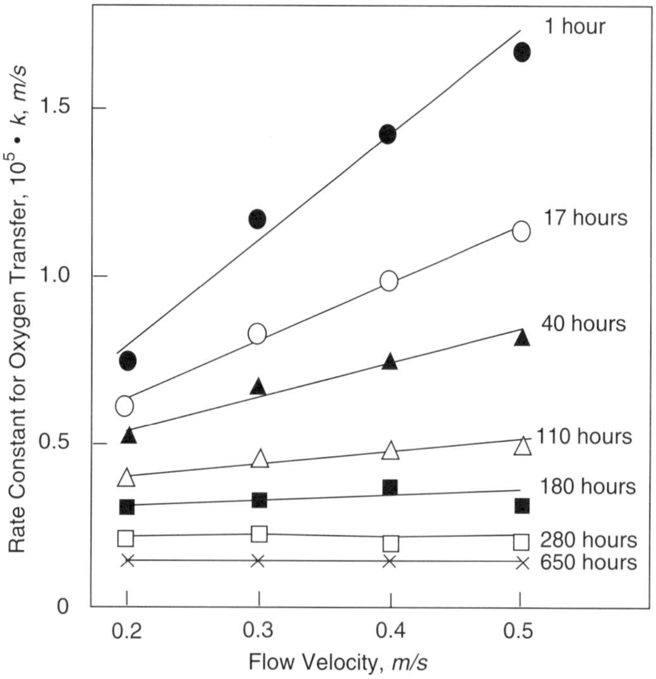

NOTE: Data obtained by oxygen consumption measurement on a steel pipe surface.

Figure 2-3 Dependency of the overall oxygen transfer coefficient on scale formation and flow velocity, where scale formation is expressed as exposure time.

Scale Formation and Corrosion of Iron Pipes

The formation, behavior, and effects of scales formed by corrosion products have been investigated extensively during the last century, both in the laboratory and in field environments. Despite a broad consensus regarding the qualitative importance of scales in corrosion protection, many critical issues about their behavior and fate are not understood in detail. These issues include identification of the processes controlling whether a scale forms, how and at what rate existing scales react with components of the solution or other scales, and whether a given scale is protective of the underlying pipe. To put the current understanding of such scale layers into perspective, a brief historical review of these studies follows.

Importance of corrosion scales: A historical review. About 100 years ago, Carl Heyer (1888), a chemist in Dessau, Germany, published a paper describing the corrosion of lead pipes and showing that high lead concentrations in the distribution water could be correlated with high concentrations of dissolved carbon dioxide (CO_2). He proposed reducing the "free CO_2" concentration by passing the water through a filter containing granular $CaCO_3$ and showed that such treatment could reduce lead solubility significantly. He also suggested that the ability of a water to react with $CaCO_3$ could be used as an indicator of the water's corrosivity and proposed a test to quantify that parameter. The test was widely adopted and became known as the Heyer test.

In an attempt to correlate the results of the Heyer test with experimental observations of the corrosion of cast-iron and steel pipes, Tillmans and his co-workers (Tillmans, Hirsch, and Weintraud 1927; Tillmans, Hirsch, and Schilling 1929) hypothesized that the mechanism of pipe protection is formation of a scale layer on the pipe surface. They postulated that only a mixture of $CaCO_3$ and rust could provide a good protective layer. Consequently, they suggested that a water that is to be distributed in steel or iron pipes should not have a CO_2 concentration higher than that in equilibrium with solid $CaCO_3$ (which they called "bound CO_2"), for the extant calcium concentration in the water. They reasoned that "aggressive carbon dioxide" (CO_2 in excess of the equilibrium concentration) might dissolve $CaCO_3$ and destroy the effectiveness of the protective layer. Tillmans also concluded, based on empirical observations, that the concentration of hydrogen carbonate (HCO_3^-) influences the formation of protective scales, but did not explain this dependence.

In 1936, Langelier defined the calcium carbonate saturation index, also called the Langelier index (LI), as the difference between the actual pH and the pH at which the solution would be in equilibrium with $CaCO_3$. He proposed using the index to describe the ability of a water to form protective layers on an iron surface. Since that time, the LI has been widely used for that purpose. Nomograms and various mathematical procedures have been developed to calculate the index, and instruments have been designed to measure it or similar parameters directly. The use of the LI for corrosion control has become common practice in most countries.

In spite of its popularity, the LI often fails to predict the corrosivity of water in contact with iron pipes. The first important papers demonstrating this failure were published over three decades ago by Larson and Skold (1957) and Stumm (1956, 1957, 1960). These authors showed that no reliable correlation exists between the corrosion rate and the bulk water LI. Their work and subsequent studies indicate that the same statement can be made concerning the calcium carbonate precipitation potential (CCPP). Stumm (1956), in particular, argued that one reason for the absence of a correlation is that the corrosion reactions change the pH near the metal surface and thus influence the LI at the locus of corrosion. The pH near the pipe wall can affect the tendency for $CaCO_3$ to form there, the distribution of anodic sites (Larson and Skold 1958), and the oxidation rate of Fe(II) to Fe(III) (Kooijmans 1938). With increasing pH, the anodic sites usually become larger in size and fewer in number, and the oxidation rate of Fe(II) increases. Both of these changes lead to more tuberculation and to formation of less protective scales. Stumm established that the buffer intensity is the most important factor for corrosion control, attributing its importance to the fact that a high buffer intensity maintains the pH at the pipe wall close to the pH in the bulk solution.

Direct analysis of scale constituents supports the conclusion that formation of a $CaCO_3$ layer does not ensure protection against corrosion. For instance, in a study of the influence of buffer intensity on corrosion, Kooijmans (1938) reported that $FeCO_3$ (siderite) often replaces calcite in protective scales. In fact, in an analysis of scales from operational distribution systems, he found that all of the protective scales contained at least 25 percent Fe(II) compounds (by weight) and not more than 10 percent calcite. Similarly, detailed studies for the Mannheim (Germany) waterworks conducted by the German Gas and Waterworks Association (DVGW) (EBI and RHE 1978) indicated that high levels of $CaCO_3$ in scales lead to formation of nonprotective, rough surface layers. Mineralogical investigations carried out by Kölle and Rösch (1978) at different waterworks showed that calcite crystals are often separated from other scale constituents and therefore are not directly involved in the formation of a dense protective layer, even when such a layer is present. In general, these studies supported the earlier findings of Stumm and others regarding the influence of buffer intensity on the properties of the scale. The results confirmed that the LI is not, in general, a useful indicator of the tendency for protective scales to form and prompted critical reconsideration of Tillmans' and related theories.

The siderite model for formation of protective scales. For many years, hydrous ferric oxide was practically the only oxidation product considered important in the context of corrosion of steel and cast-iron pipes. This conclusion was based on the fact that ferric oxide is the main component of corrosion products that have been removed from the pipes and dried. However, the drying process is almost always conducted in the presence of oxygen. Inspection of scales immediately after opening a pipe shows that the bulk of the material is not brown, like ordinary rust, but dark or even black. This observation suggests that, within the scales, many of the corrosion products are in reduced form, i.e., they contain ferrous (Fe(II)) iron. In light of this fact and the inability of Tillmans' theory (that protective scales are formed by a mixed layer of calcite and rust) to explain the beneficial effect that buffer intensity has on nearly all iron corrosion problems, Sontheimer, Kölle, and Rudek (1979) suggested that Tillmans had placed insufficient emphasis on the role of ferrous iron precipitates in the formation of protective scales, and they introduced the siderite model as an alternative.

The basic premise of the siderite model is that the formation of relatively insoluble solids containing reduced iron (Fe(II)), such as siderite (ferrous carbonate, $FeCO_3$) or the different types of green rust,* is the critical step in the formation of protective scales. Although the overall scales may build up over many years and may ultimately include a variety of precipitates containing both ferrous and ferric iron, the critical reactions are presumed to involve ferrous ions generated by the initial oxidation of the metal or the reductive dissolution of older scales and to depend strongly on the water quality adjacent to the pipe at the time. The model is based on the following reactions:

Primary Reactions			
Fe	$\leftrightarrow$	$Fe^{2+} + 2e^-$	Reaction 1
$\frac{1}{2}O_2 + H_2O + 2e^-$	$\leftrightarrow$	$2OH^-$	Reaction 2
$2HCO_3^- + 2OH^-$	$\leftrightarrow$	$2CO_3^{2-} + 2H_2O$	Reaction 3
Secondary Reactions			
$2Ca^{2+} + 2CO_3^{2-}$	$\leftrightarrow$	$2CaCO_{3(s)}$ (calcite)	Reaction 4
$2Fe^{2+} + 2CO_3^{2-}$	$\leftrightarrow$	$2FeCO_{3(s)}$ (siderite)	Reaction 5
$2Fe^{2+} + \frac{1}{2}O_2 + 4OH^-$	$\leftrightarrow$	$2FeOOH_{(s)}$ (goethite) + H_2O	Reaction 6
Tertiary Reactions			
$2FeCO_{3(s)} + \frac{1}{2}O_2 + H_2O$	$\leftrightarrow$	$2FeOOH$ (goethite, pseudomorph) + $3CO_3^{2-}$	Reaction 7
$3FeCO_{3(s)} + \frac{1}{2}O_2$	$\leftrightarrow$	Fe_3O_4 (magnetite) + $3CO_3^{2-}$	Reaction 8

As noted previously, in waters containing DO, the rate of the primary reactions depends strongly on the rate of oxygen transport to the metallic surface. In most practical situations, the dominant resistance to diffusion is provided by the

*Green rust is a commonly observed iron corrosion product that has been described as an intermediate product between ferric and ferrous hydroxides, often combined with additional anions such as chloride and sulfate (Tuovinen et al. 1980; McGill, McEnaney, and Smith 1976; Feitknecht and Keller 1950).

scale layer that forms from the products of the secondary and tertiary reactions. As discussed below, other solids may also be present in the scale layer, due to subsequent transformations of the products shown in these reactions.

The siderite model postulates that the density of the scale layer formed and its resistance to oxygen diffusion depend to a large extent on the competition between reactions 5 and 6. Direct and rapid oxidation (reaction 6) leads, in most cases, to formation of a very porous layer, whereas the formation of siderite (reaction 5) or of certain other Fe(II)-containing solids leads to a much denser, relatively nonporous scale that strongly interferes with oxygen transport. Furthermore, if the siderite or the green rust is eventually oxidized, as described by reactions 7 and 8, the oxidation product apparently retains the crystal structure of the siderite or green rust and therefore remains protective of the pipe.

The iron corrosion rate appears to depend on the buffer intensity when scale is present, but not in the absence of scale (as shown, for example, in Figure 2-2). According to the model, this dependence can be explained based on two facts: first, oxygen reduction at the cathode generates OH^- ions, increasing the pH near the pipe wall; and, second, the rate of oxidation of Fe(II) to Fe(III) increases dramatically with increasing pH. Thus, the cathodic reaction favors rapid oxidation of Fe(II), leading to formation of nonprotective scales. However, if the buffer intensity of the water is high, the OH^- ions generated at the cathode can be neutralized (by HCO_3^-), the pH near the wall does not increase as much, and the Fe(II) remains unoxidized long enough to form siderite, green rust, or other protective precipitates.

Some of the thermodynamic relationships relevant to the siderite model can be shown in Pourbaix diagrams. If one modifies either version of Figure 2-1 to account for the presence of 50 mg/L (1.0 meq/L) alkalinity as $CaCO_3$, a small region where siderite is the stable solid appears at near neutral pH and E_H less than 0 mV. Such a low E_H applies only in systems in which oxygen is completely depleted; if any DO were present, the stable form of iron would be Fe(III), which is also the case in a system with negligible alkalinity. However, if it is presumed that the oxidation of Fe(II) to Fe(III) is kinetically inhibited, so that all the oxidized iron remains in the ferrous form, the potential importance of siderite increases dramatically. For example, Figures 2-4A and 2-4B show Pourbaix diagrams for two such systems containing $10^{-4}M$ total soluble iron, one with negligible alkalinity and the other containing 50 mg/L alkalinity. In the system with alkalinity, siderite is predicted to precipitate at pH between 7.2 and 9.4, i.e., under conditions that are likely to be found near the metal surface in potable water systems.

According to this reasoning, any substance that interferes with the oxidation of Fe(II) would be expected to encourage the formation of siderite or other Fe(II)-containing solids, and thereby to encourage the formation of protective films. One such group of substances, which is present in all potable waters, includes humic and fulvic acids. These acids, which are constituents of natural organic matter, form complexes with both Fe(II) and Fe(III). It is well established that, in the presence of humic acids, Fe(II) ions are partially protected from oxidation by oxygen. The mechanism for this protection is probably formation of complexes between ferrous ions and the humics. However, the oxidation of Fe(II) to Fe(III) by oxygen is thought to involve the formation of perhydroxide free radicals (Weiss 1951), and the humics might also interfere with the progress of the oxidation by reacting with these free radicals.

Figure 2-5A shows the effects of both buffer intensity and humic substances on the corrosion rate of iron during the initial week of exposure of a pipe to various

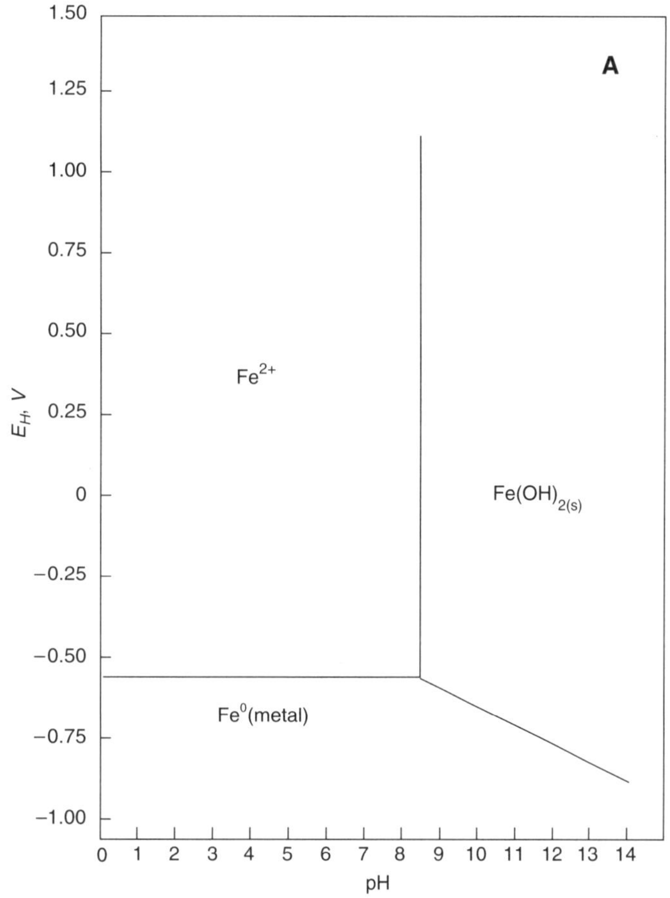

NOTE: Total soluble iron is 10^{-4} *M* (5.6 mg/L). Alkalinity is negligible in Figure 2-4A and equals 50 mg/L as $CaCO_3$ (1.0 meq/L) in Figure 2-4B (see next page).

Figure 2-4 Pourbaix diagrams for iron, assuming that the oxidation of Fe(II) to Fe(III) is prevented (continued next page)

waters. As would be expected, the corrosion rates decrease significantly as the test proceeds and a scale layer is formed. Furthermore, consistent with the model expectations, increasing either the buffer intensity or the concentration of humic substances reduces the corrosion rate. Figure 2-5B shows the composition of the scales in the same systems and confirms that changes that reduce the corrosion rate increase the percentage of Fe(II) in the scale.

Effect of water quality parameters unrelated to siderite formation. If the siderite model is valid, the presence of dissolved ions that do not affect siderite formation or Fe(II) oxidation should not affect the corrosion rate during the period when the initial protective films are forming. In particular, the direct effects of dissolved chloride, sulfate, and calcium on siderite formation are all expected to be negligible. However, there is evidence that some types of green rust play a role similar to that of siderite during scale formation, leading to formation of very dense and protective films. While sulfate is not expected to form precipitates directly with iron, bacterial reduction of sulfate to sulfide could lead to precipitation of ferrous sulfide in the scale. No such reaction would occur when a new pipe was put into service, but it could occur after a scale layer in which sulfate-reducing bacteria could form

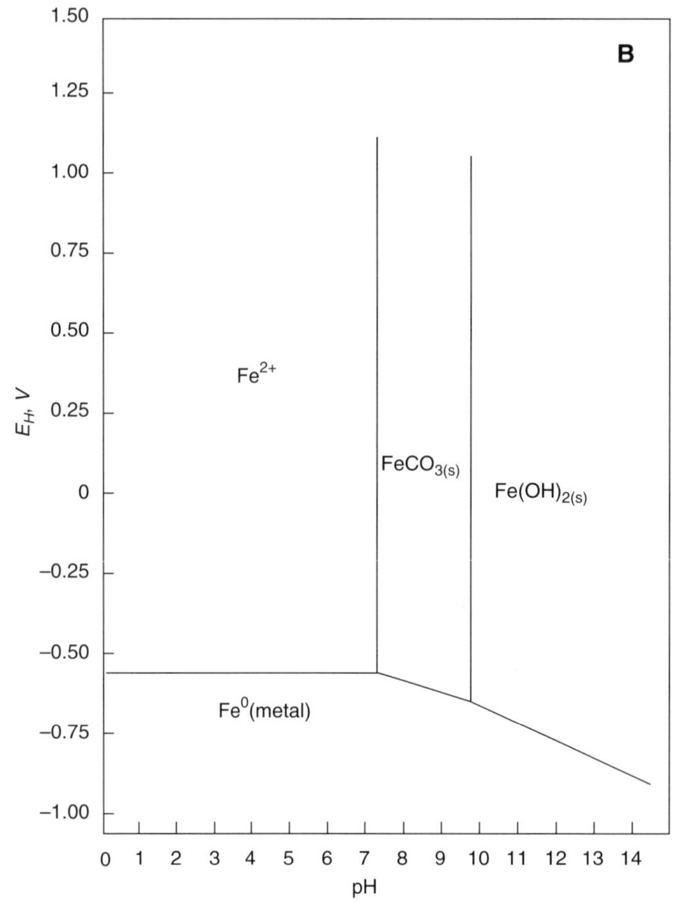

Figure 2-4 Pourbaix diagrams for iron, assuming that the oxidation of Fe(II) to Fe(III) is prevented (continued)

colonies had built up. Therefore, the chloride and sulfate concentrations in the water could both have some effect on the formation of protective scale in some situations.

Calcium might also influence film formation in several ways. For instance, calcite might form a protective layer, if it precipitates immediately adjacent to the pipe wall (Stumm 1956). There is also some evidence that calcium ions slow the oxidation rate of Fe(II), a process that could influence film formation and protectiveness.

Thus, the effect of water quality parameters other than solution pH and the inorganic carbon concentration on the formation of protective films has not been established conclusively. However, in most cases, their effect is not expected to be large. These neutral salt species do seem to have a very important influence on the formation of red water, as is discussed later in this chapter.

Summary. Interpreting the precipitation and subsequent conversion reactions that occur in a scale layer is very difficult, because scale formation is inherently a very complex process. Historically, the tendency to form or dissolve $CaCO_3$ has often been relied on to indicate a water's corrosivity to iron. This reliance reflects an assumption that was long taken for granted but is now considered incorrect, that $CaCO_3$ by itself could form a protective scale. There is extensive evidence that

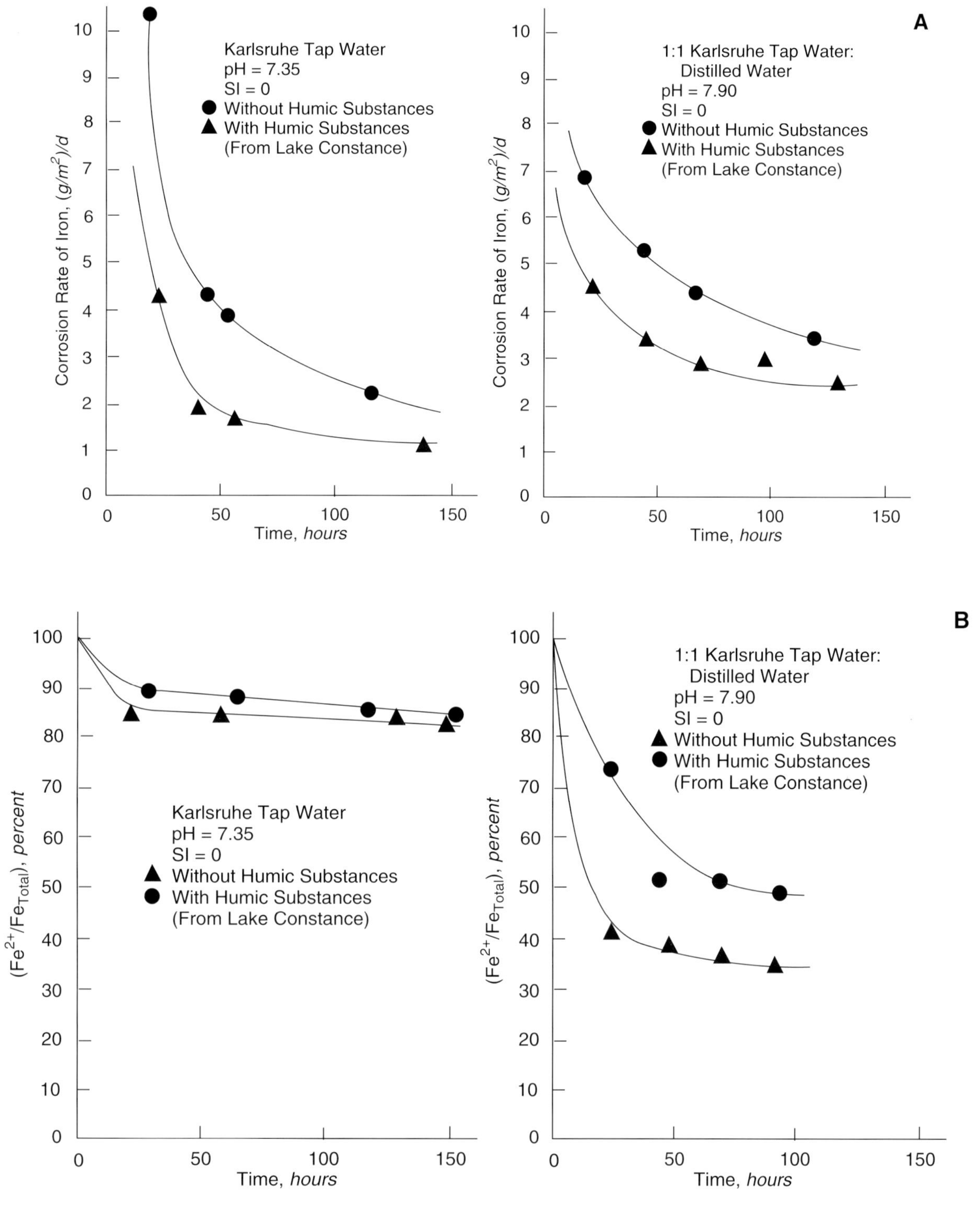

NOTE: The figure on the right in each case is for water prepared by diluting Karlsruhe tap water 1:1 with distilled water and then adding NaOH. These modifications decrease the buffer intensity by approximately a factor of seven.

Figure 2-5 Influence of buffer intensity and natural organic matter (i.e., humic substances) on (A) corrosion rate and (B) the percentage of the total iron in the scale that is ferrous iron both the LI and the CCPP are very unreliable indicators of a water's corrosivity to iron, and they should not be used for that purpose. They should only be used to indicate some limiting conditions under which $CaCO_3$ can precipitate from, or dissolve into, the bulk solution.

Despite the complex relationship between scale formation and corrosion rate, the siderite model, a relatively simple model that emphasizes the importance of the initial precipitates that form after Fe(II) is released from the pipe wall or generated by other processes within the scale layer, can account for many of the effects that water quality variables have on corrosion rates. According to this model, the corrosion rate in the presence of corrosion scales depends on the rate at which oxygen is transported to the pipe wall. This is also the case in the absence of scales. However, when scales are present, the transport rate is assumed to be limited more by the properties of the scale than by the DO concentration in the bulk water. Precipitates containing ferrous iron, in particular the mineral siderite ($FeCO_3$), are hypothesized to form tight scales that severely restrict oxygen transport, and any water quality factor that encourages formation of these compounds is expected to interfere with corrosion. Factors that encourage the rapid oxidation of Fe(II) to Fe(III), on the other hand, are expected to increase the formation of nonprotective layers and therefore allow more rapid corrosion. Increasing the buffer intensity (e.g., by increasing the alkalinity) or the concentration of humic substances (which stabilize Fe(II)) in the water encourages siderite formation. Chloride, sulfate, and calcium are not expected to have a direct effect on this process in most situations, although they might be important either indirectly or, under specialized conditions, directly. Their influence might be exerted, for instance, via a change in the conditions in the region where Fe(II) is first released, altering the identity of the initial Fe(II)-containing precipitate that forms or its tendency to oxidize. These processes, which are thought to influence iron release and red-water formation, are discussed later in this chapter.

Long-Term Reactions of Corrosion Scales: Structure and Composition of Old Scales

In general, the longer a pipe has been in service, the more heterogeneous and complex the scale layer becomes and the more difficult it is to characterize the reactions of the corrosion products. Thus, the relevant reactions and controlling factors for corrosion of bare metal are relatively straightforward, and those for the initial phases of scale formation are somewhat less certain but still reasonably well understood. By comparison, the understanding of scale transformations over exposure periods of months to years is much less thorough. Nevertheless, since the lifetime of a pipe in a water distribution system is usually more than 50 years and often more than 100 years, reactions of "old" scales are of interest. In this section, observations regarding the composition and structure of corrosion scales in older pipes are presented, along with some qualitative discussion of how such structures might affect the corrosion rate. The discussion serves both as a summary of existing observations and as the introduction for a new model of iron release, which is presented subsequently.

Figure 2-6 shows a simplified schematic for the mineralogical structure of the corrosion products in cast-iron pipes. The corrosion process leaves an iron-depleted zone in the the metal, called the topotactic layer or the graphitization zone, that contains original as well as oxidized iron components, possibly in addition to other components such as calcite. Magnetite (Fe_3O_4), goethite (α-FeOOH), siderite ($FeCO_3$),

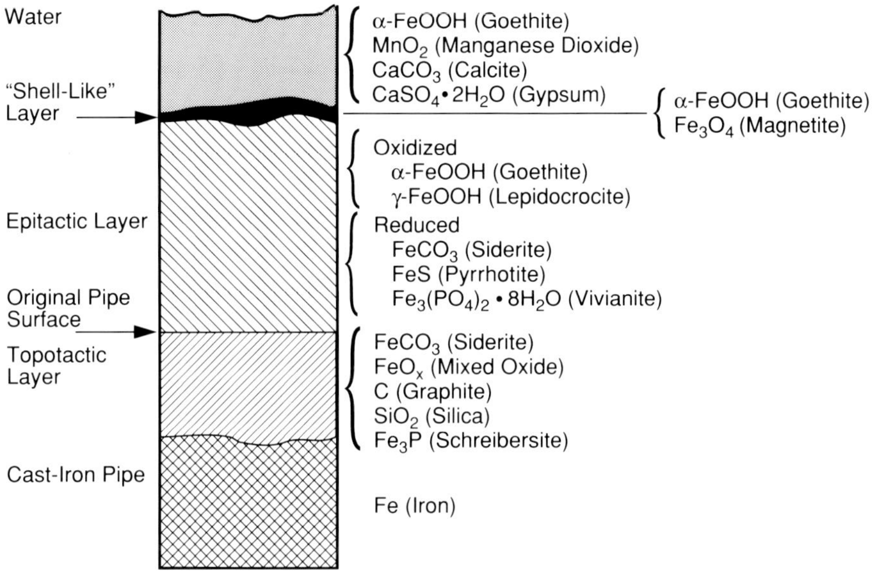

Figure 2-6 Schematic of scale on a cast-iron distribution pipe

lepidocrocite (γ-FeOOH), and calcite ($CaCO_3$) are all found in old scales, and their relative concentrations change with time as well as with water quality. Often, structures of corrosion products such as those shown schematically in Figure 2-7 are found in older pipes. Some examples can be seen in Figures 2-8, 2-9, and 2-10, which show that some of the corrosion products have formed enveloping, shell-like layers that enclose other "interior" corrosion products within the scale. The presence of "exterior" corrosion products, i.e., layers that cover the enveloping layer, is also common. Typical structures of this type have been described by Butler and Joon (1966), Rijkens (1967), and Weidmann (1970), and their importance has been reexamined by Kölle, Sontheimer, and their co-workers (Kölle 1977, 1978; Kölle and Sontheimer 1977; Sontheimer, Kölle, and Rudek 1979; Kölle and Rösch 1980; Sontheimer, Kölle, and Snoeyink 1981). Tuovinen, Mair, and Banovic (1984), as well as Chantereau (1977), considered similar structures in the context of sulfate reduction within tubercles.

The enveloping layer, a dark, relatively hard and brittle structure, usually consists of magnetite and some goethite. In the siderite model, this layer is envisioned to be formed by oxidation of Fe(II)-containing solids. The percentage of magnetite appears to depend on the distance from the metallic surface, the flow conditions in the pipe, and the water quality. The layer is a fairly good electrical conductor because of the magnetite content, and it may therefore act as a cathode in a sandwich-structured electrochemical cell, with the metallic (pipe) surface as the anode. The transformations shown in Figure 2-7 are based on this mechanism, which has been described by Butler and Joon (1966), Rijkens (1967), and Kölle et al. (1983).

An operating, sandwich-type cell introduces ferrous ions into the interior of the tubercle and releases hydroxide ions into the bulk of water. Maintenance of electroneutrality within the tubercle requires that anions pass through the enveloping layer and penetrate into the interior. Some such situations apparently favor the penetration of chlorides, in which case the solution adjacent to the pipe can acquire a relatively high concentration of ferrous chloride; concentrations up to 50 mM/L

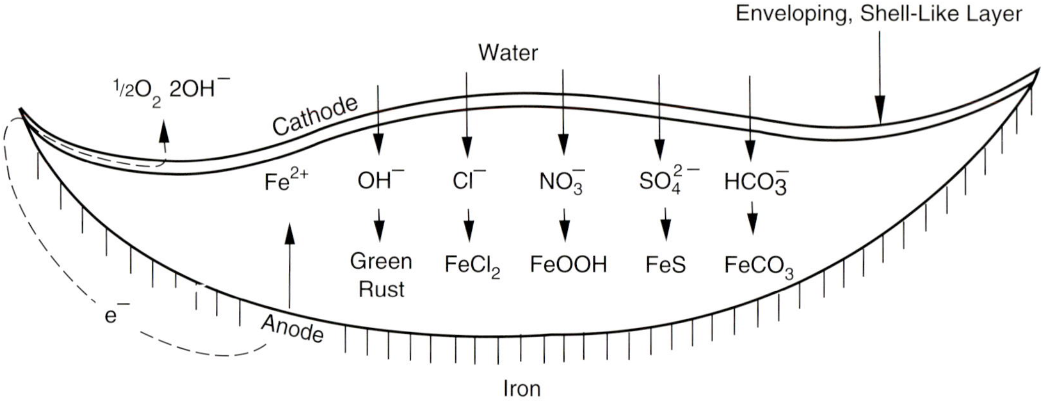

Figure 2-7 Scale-forming mechanisms (model concept)

Figure 2-8 Scale with enveloping, shell-like layer (favorable conditions)

have been reported (Kölle 1982). Anions other than chloride that might penetrate the enveloping layer to neutralize the charge of the Fe(II) include carbonate, nitrate, sulfate, and hydroxide, and, if it has been dosed into a water as a corrosion inhibitor, phosphate.

As the ferrous ions and the neutralizing anions accumulate, they may react either independently or with each other. For instance, ferrous carbonate (siderite) or ferrous phosphate (vivianite, $Fe_3(PO_4)_2 \cdot 8H_2O$) might precipitate, and these solids have indeed been found in interior scales. Siderite usually forms tight, microcrystalline layers of a characteristic structure (Figures 2-11 and 2-12). This compound has been observed as a scale component by Stumm (1960), Baylis (1926),

Figure 2-9 Scale with enveloping, shell-like layer (intermediate conditions)

Figure 2-10 Scale with enveloping, shell-like layer (unfavorable conditions) and filamentous structures of iron bacteria

Figure 2-11 Crystal crust of siderite

Figure 2-12 Crystal crust of siderite (red) with pyrrhotite intergrowth (brown)

Larson and King (1954), and McCauley and Abdullah (1958). As noted in the discussion of the siderite model, the siderite is expected to be gradually oxidized over time, but it can remain stable for relatively long periods. Vivianite occurs as radiated, crystalline needles, which are primarily white but become light blue and finally green after drying and aging (Figure 2-13).

There are also several pathways by which the ferrous ions or compounds in the scale could be oxidized, yielding a variety of end products. For instance, the direct oxidation of ferrous ions by oxygen that penetrates into the scale generally leads to the formation of lepidocrocite, crystals of which can be recognized in the scale as red, hexagonal plates of dimension on the order of 100 μm (Kölle and Rösch 1980). Oxidation of ferrous ions in the scale could also be microbially mediated. Either oxygen or nitrate might serve as the oxidant in these reactions. The reaction products in this case cling to the filamentous structures of the bacteria carrying out the reaction (Figure 2-14).

Microbiological oxidation of Fe(II) ions in scales can also be coupled to the reduction of sulfate. Since this reaction requires the redox potential to be considerably lower than that needed for reduction of nitrate, the products of these two reactions are spatially separated. Sulfate reduction leads to the formation of ferrous sulfide, of which pyrrhotite ($Fe_{1-x}S_{(s)}$) is the mineral form most commonly found in scales. However, troilite may also be observed. Detailed investigations of this reduction have been carried out by Chantereau (1977), Miller and King (1975), and Lee, O'Connor, and Banerji (1980). Pyrrhotite occurs as thin, hexagonal plates with a diameter of up to about 0.5 mm. Originally these plates are black, but they oxidize in the presence of oxygen to a mixture of goethite and lepidocrocite and to elemental sulfur. The oxidation occurs pseudomorphically, i.e., it does not alter the shape of the crystals (Figure 2-15). Green rust, which has physical characteristics

Figure 2-13 Vivianite crystals (dark)

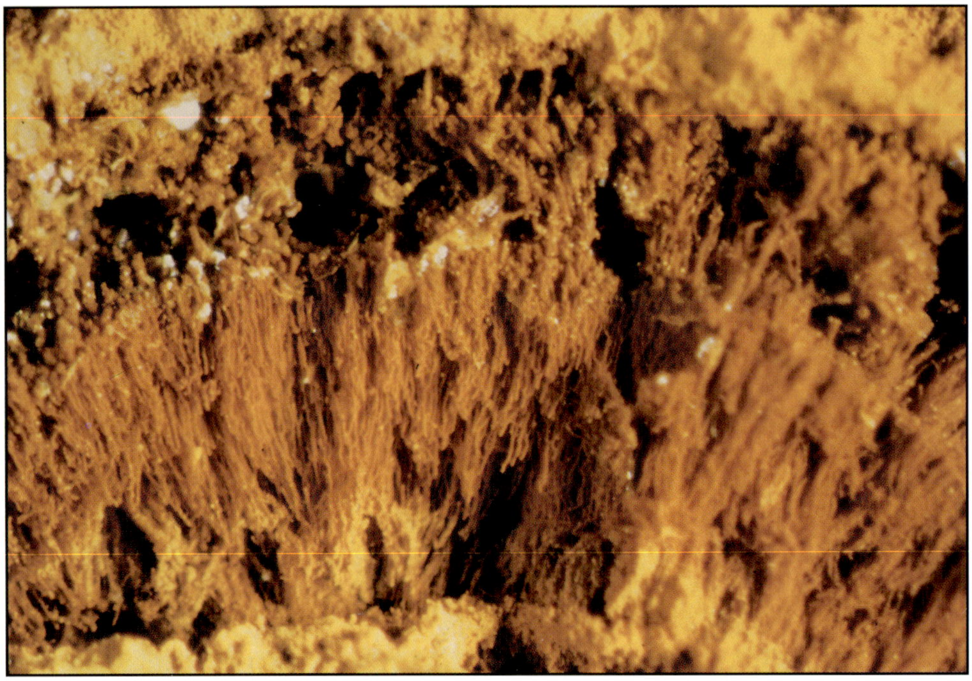

Figure 2-14 Filamentous structures of iron bacteria (nitrate content of drinking water, 20 mg/L)

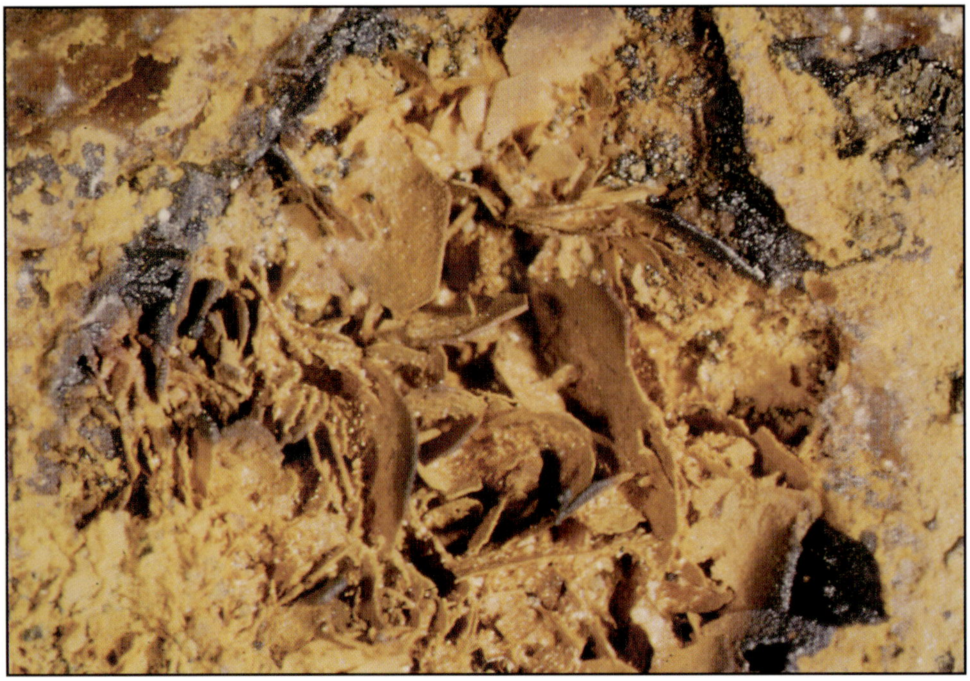

Figure 2-15 Hexagonal crystals of pyrrhotite, pseudomorphically oxidized

similar to those of pyrrhotite, also seems to be an important component of scale interiors. The crystals are also hexagonal, though smaller than those of pyrrhotite, and they are apparently easily oxidized in air, since they generally cannot be found in the scale once it has been dried in the presence of oxygen.

The dominant mineral found on the exterior scale, i.e., adjacent to the flowing water, is hydrous ferric oxide. This mineral is assumed to form via oxidation of ferrous ions or Fe(II)-containing solids that are exposed to the bulk water, with DO as the oxidant. In some cases, iron oxidizing bacteria may also mediate this oxidation. This ferric oxide might attach only weakly or not at all to the underlying scale, in which case it can be transported by the flowing water and lead to red-water problems, which are discussed in more detail in the following section.

In summary, numerous minerals, most of which contain iron, are present in old scales. They are often interspersed in ways that make their past history virtually impossible to decipher. Nevertheless, some common qualitative patterns can be observed in many pipes. Of these, the occurrence of a brittle layer containing magnetite that envelops a layer of other minerals between it and the pipe seems to be important. Corrosion of the pipe underlying this layer can lead to relatively high local concentrations of ferrous salts. The effect of water quality variables on the conversion of siderite or green rust to Fe(III) precipitates is not well studied, and it is possible that chloride, sulfate, or other ions affect this transformation. Additional study of this transformation and the factors that control it is a critical need in order to improve the overall understanding of iron corrosion.

IRON RELEASE, RED WATER, AND THE KUCH MECHANISM

Occurrence and Importance of Red Water

Many consumer complaints received by water utilities are initiated because of the appearance of red water at the tap. Failures of iron and steel pipes caused by internal corrosion are extremely rare, because the formation of protective scales in such pipes reduces the corrosion rates to very low values. For instance, Kölle and Sontheimer (1977) reported that the corrosion rates in systems that have been in service for more than one to three years typically cluster around 10 μm (0.4 mil) per year.

Portions of thick and irregular scales tend to be released from the pipe wall if sudden changes in flow velocity or water quality occur, and this may result in red-water problems. In most cases, however, ferric hydroxide flocs rather than crystalline scales are found in red waters. High concentrations of ferrous ions are also often observed in such waters, especially where no free chlorine residual is available. This suggests that many red-water problems arise through oxidation of ferrous ions in the bulk solution and the subsequent precipitation of ferric hydroxide, rather than by release of preformed particles from the existing scales. The ferrous ions could form at the metal surface or within the scales and then be transported into the bulk water phase by diffusion.

While studies of the effect of water quality on iron release are rare, a few detailed studies of the iron content in distributed drinking waters have been conducted. In some cases, these data can be correlated to water quality changes. For example, Madsen (1973) reported that changes in the chloride content of drinking water caused immediate changes in the total iron content of the water in Bremen,

Germany, for which the river Weser is the source. Similar results were found in Amsterdam for treated Rhine River water. Similarly, a study in Hannover, Germany, showed that an increase in the sulfate content from 70 mg/L in 1965 to 160 mg/L in 1983 correlated with an increase in complaints of red water in the distribution area (Heck 1974; Kölle, Strebel, and Böttcher 1985).

Table 2-3 shows the iron concentration that would be present in standing water after 24 hours of contact with old steel or cast-iron pipes, based on a typical corrosion rate of 10 µm/year for such pipes. These data indicate that fairly high iron concentrations can accumulate in the water in small pipes. This result is consistent with the findings of many water utilities that consumer complaints regarding rusty water are usually from locations with small pipes. In such pipes, long periods with no water flow are fairly common.

As noted previously, most old pipes have similar corrosion rates. It seems reasonable to assume that, like the corrosion rates in newer pipes, these rates are at least partially controlled by the transport of oxygen to the pipe surface, via diffusion of oxygen through a protective layer of oxide film. On the other hand, there is empirical evidence that the salt concentration influences the formation of red water, even though neutral salts like chlorides and sulfates do not influence oxygen diffusion rates significantly. Furthermore, in the few studies where corrosion rates based on weight loss have been compared directly with iron release rates, the two have not been strongly correlated. Based on these observations, it appears that the formation of protective scales and the formation of red water are independent processes, and that a change in one may not correlate at all with changes in the other.

Determination of Iron Release Rates

The rate at which ferrous ions are transported into flowing or standing water is difficult to determine. One approach that has been used is to evaluate the difference between the total amount of iron corroded and the amount of iron retained in the scale (Figure 2-16). The values determined using such an approach must be interpreted cautiously, since they represent small differences between relatively large values. The potential for error increases as the thickness of the scale increases.

To overcome this problem, Kuch (1984) designed an apparatus to directly measure iron release rates over a period of several hours (Figure 2-17). The apparatus is modeled after the closed-circuit system developed by Kuch, Sontheimer, and Wagner (1983) to measure iron release rates and oxygen depletion rates. In this apparatus, the water to be studied is pumped through the test pipe at a velocity that is,

Table 2-3 Calculated iron concentrations in old pipe systems*

	Percentage of Corroding Iron That Is Released to the Water		
	100%	10%	1%
Pipe Diameter, *mm*	Iron Concentration in the Water, *mg / L*		
20	40	4	0.4
100	8	0.8	0.08
800	1	0.1	0.01

*Based on a corrosion rate of 10 µm/year and a 24-hour period with no flow.

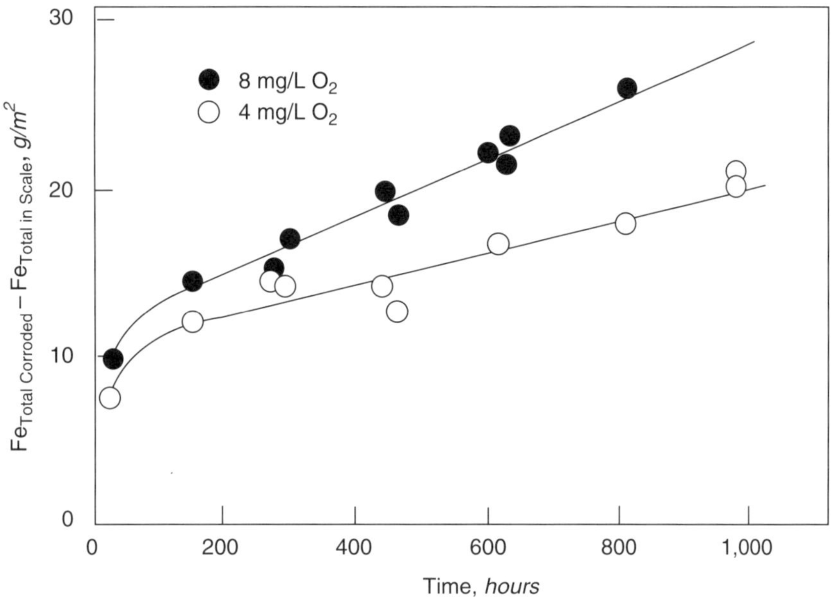

Figure 2-16 Influence of oxygen concentration on iron release (iron release calculated as the difference between weight loss and the iron content of the scale)

preferably, similar to those used in actual installations. As the corrosion reaction proceeds, the oxygen concentration of the water decreases and the iron content increases. The latter is monitored by analyzing samples automatically for ferrous and ferric iron concentrations, based on a photometric method using bathophenanthroline.

Results of Iron Release Measurements

The first studies with the apparatus described previously tested new pipes, using waters with different alkalinities and pH values or CO_2 concentrations. An unexpected result, and one that is of special interest for practical considerations, is that the iron release rate in oxygenated waters could be described solely as a function of the CO_2 concentration of the water (Figure 2-18), regardless of its alkalinity or pH. This result supports the early work of Heyer (1888) and Tillmans and his co-workers (Tillmans, Hirsch, and Weintraud 1927; Tillmans, Hirsch, and Schilling 1929). These researchers presented experimental evidence that showed that reducing the CO_2 content of water by a variety of methods always reduced problems associated with corrosion, i.e., red water in steel and cast-iron pipes in full-scale installations.

Other, more recent studies (Rothmann 1977; Schmitt and Rothmann 1977, 1978; Schwenk 1974; De Waard and Williams 1975; Fischer and Siedlarek 1977) have reported that corrosion rates in oxygen-free waters depend on the pH only. For a fixed total inorganic carbon concentration, this result is consistent with a dependence on the CO_2 concentration only. However, Kuch (1984) showed that the corrosion rate and iron release rate in pipes exposed to oxygen-free waters depend on the bicarbonate concentration in addition to the carbonic acid concentration.

Kuch (1984) suggested that the results for iron release from new pipes could be explained by assuming that the diffusion rate of iron species into the bulk water depends on the iron concentration at the metal surface. According to this hypothesis, if a solid containing Fe(II) (e.g., green rust, ferrous hydroxide, or siderite) is

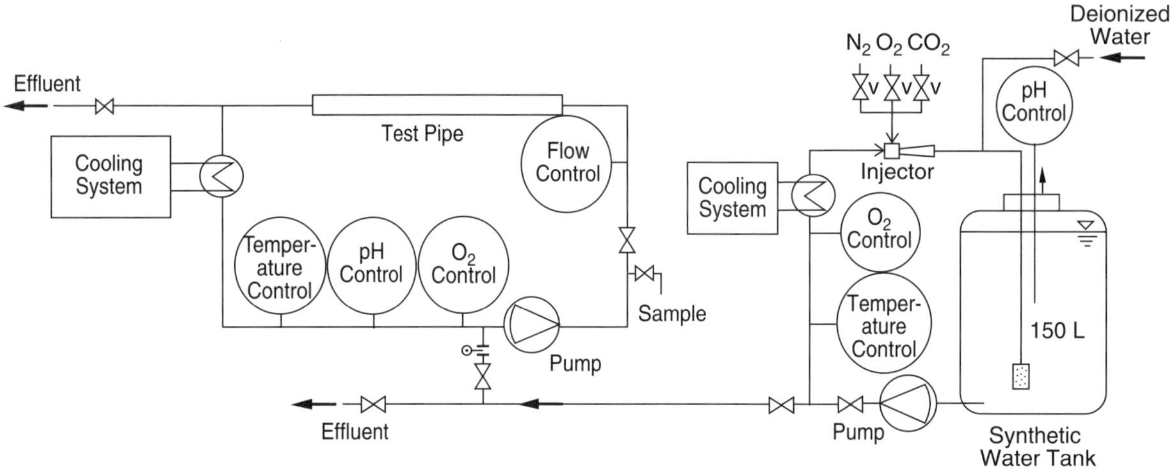

NOTE: The apparatus allows the testing of various pipe probes exposed to different water qualities.

Figure 2-17 Schematic diagram of the closed-circuit system for analyzing iron release rates in different waters

precipitated near the pipe wall, the maximum concentration of dissolved iron at the pipe surface and within the scale is limited, and the rate at which iron reaches the bulk water is less than if the precipitate were absent. It is for this reason that the iron release rate does not necessarily increase when the corrosion rate increases.

The likelihood that such a solid will precipitate, in turn, depends on water quality conditions near the pipe wall, in particular the pH and the DO concentration. Specifically, as noted previously in the discussion of the siderite model, oxidation of Fe(II) leads to precipitation of insoluble hydrous ferric oxide solids, and the rate of this reaction increases with increasing pH and increasing DO concentration. Thus, according to the proposed model, increasing the pH or DO concentration near the pipe wall should encourage precipitation of the oxide and reduce iron release. Similarly, increasing the HCO_3^- concentration in oxygen-free waters has essentially the same effect, by increasing the likelihood that ferrous carbonate will precipitate.

The increase in pH near the wall relative to that in the bulk water depends on the relative rates at which hydroxide ions are generated (by the cathodic reduction of oxygen) and consumed (primarily through the conversion of dissolved CO_2 to bicarbonate ion, $CO_2 + OH^- \rightarrow HCO_3^-$). Increasing the CO_2 content of the water increases the rate at which OH^- is consumed, limiting the pH increase and the likelihood of iron precipitation in oxygenated water.

Based on this discussion, the oxidation of Fe(II) to Fe(III) and the precipitation of Fe(III) oxides would be expected to increase if the solution pH or DO concentration were increased, other factors being held constant. According to Kuch's hypothesis, either of these changes would reduce iron release.

Because the CO_2, HCO_3^-, CO_3^{2-}, H^+, and OH^- concentrations in a solution are all interrelated, it is not possible to change one of the concentrations without changing at least one of the others. This interconnectedness makes it difficult to test the hypothesis that the rate of OH^- consumption by reaction with CO_2 is a critical factor controlling the iron release rate. To overcome this difficulty, Kuch performed experiments in which the enzyme carbonic anhydrase was added to the water. The enzyme increased the rate at which CO_2 and HCO_3^- equilibrated with

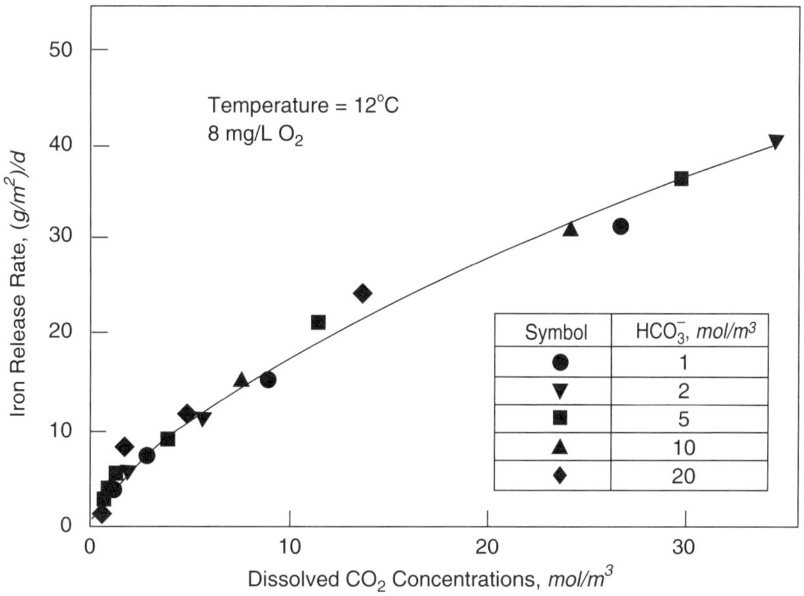

Figure 2-18 Influence of CO_2 concentration on iron release rates in new steel pipes, for waters of various alkalinities

one another* and therefore increased the rate at which OH^- ions were neutralized. After a small amount of the enzyme was injected into the closed-circuit loop, the rate of iron release increased (Figure 2-19). When the catalyst was destroyed by the addition of chlorine, the iron release rate returned to its previous value.

The experiments just described used slightly aged pipes. In water distribution systems, the pipe surfaces are virtually always covered with scales formed by corrosion products. Iron release rates in pipes with developing corrosion scales are shown in Figures 2-20 and 2-21. For these tests, the initial scales were formed in Karlsruhe tap water at a pH of 7.5. Comparison of the release rates in these figures with that in Figure 2-18 indicates clearly that the release rate decreases as the pipes age and scales develop. The figures also show that, consistent with Kuch's hypothesis, higher values of pH or of DO concentration decrease iron release.

The stoichiometry of corrosion is such that 3.5 mg of Fe(II) is produced per milligram of O_2 consumed. Thus, if iron release were directly and exclusively related to O_2 consumption, and if the oxygen reduction rate were directly proportional to the DO concentration, as described previously in this chapter, then the iron release in the experiment described by Figure 2-21 would be given by the dashed line in that figure. Clearly, factors other than O_2 consumption affect the iron release rate. This result means that there must be a process in addition to the "normal" corrosion reaction (oxidation of metallic iron coupled to reduction of oxygen) that is providing iron to the flowing water. Results of electrochemical measurements in similar

*The overall reaction $CO_2 + H_2O \leftrightarrow HCO_3^- + H^+$ proceeds in two steps: hydration of dissolved CO_2 to H_2CO_3, followed by release of H^+ from H_2CO_3 to form HCO_3^-. The first of these two steps is known to be much slower than the second, and the enzyme increases the rate of the first step.

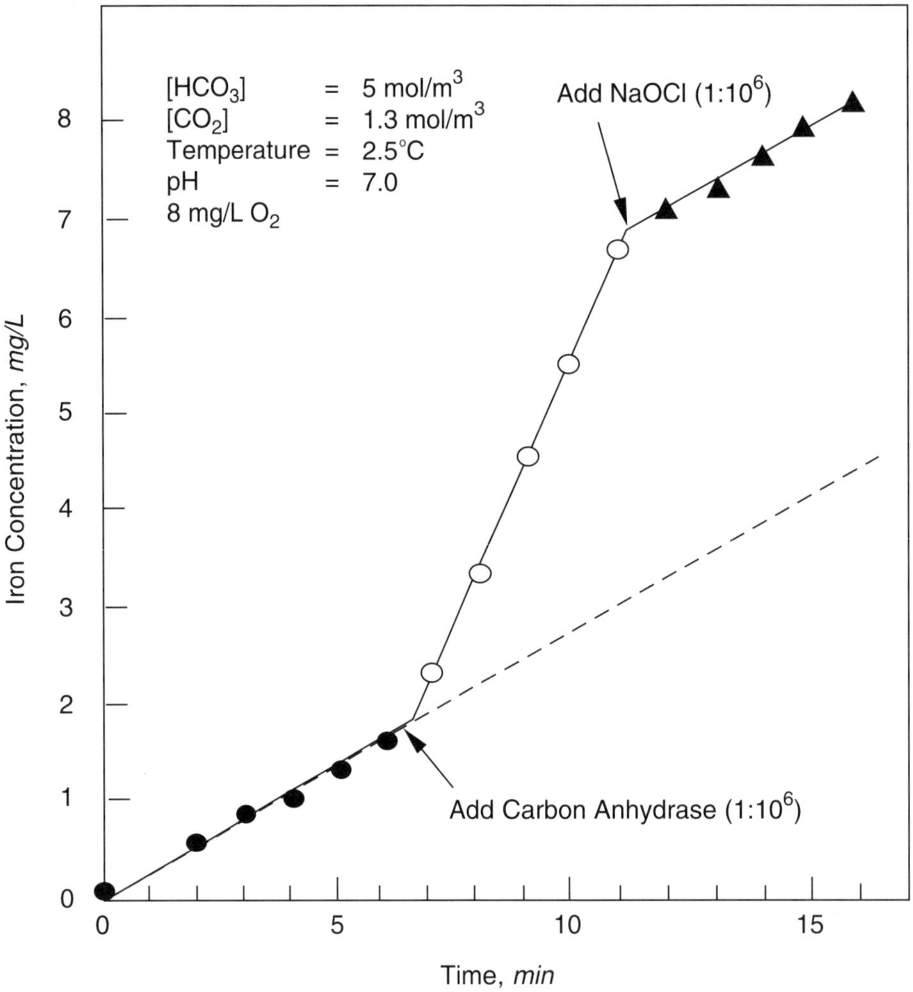

Figure 2-19 Changes in the iron release rate associated with the addition and subsequent destruction of carbon anhydrase (a catalyst for the formation of H_2CO_3 from CO_2 and water)

systems (Figure 2-22) provide supporting evidence that the corrosion mechanism changes when the oxygen becomes depleted.

Additional evidence that, under some conditions, the iron release rate can be greater than the corrosion rate is provided in Figure 2-23, which shows the loss of mass from iron pipes for different operational conditions in a steel pipe system. The different pipes used in parallel for this test were exposed to Karlsruhe tap water for alternating periods with and without flow. At selected times during the test, the weight loss in a small pipe section was determined. Weight loss was greater for the pipes in which water was flowing only 50 percent of the time than for those in which flow was continuous. Since it is likely that the DO was depleted near the metal surface very shortly after the water flow stopped, some reduction reaction other than the conversion of O_2 to OH^- must have been active to account for the higher weight loss in the pipes with intermittent flow. A similar argument can be made for the pipe with flow for only one hour per day, in which the weight loss was nearly the same as in the pipe with continuous flow.

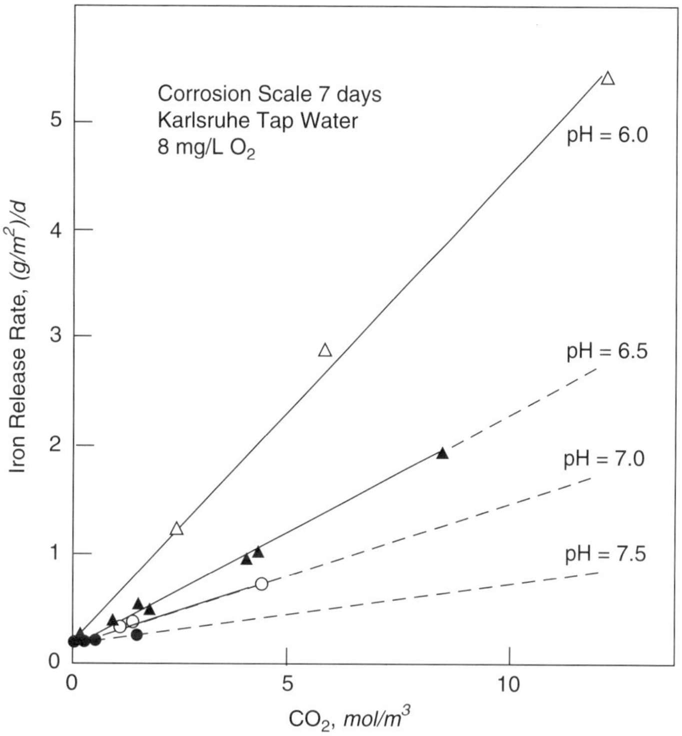

Figure 2-20 Influence of pH and CO_2 concentration of synthetic waters on iron release rate in steel pipes with scale (scale formed by exposure to Karlsruhe tap water at pH 7.5 for seven days)

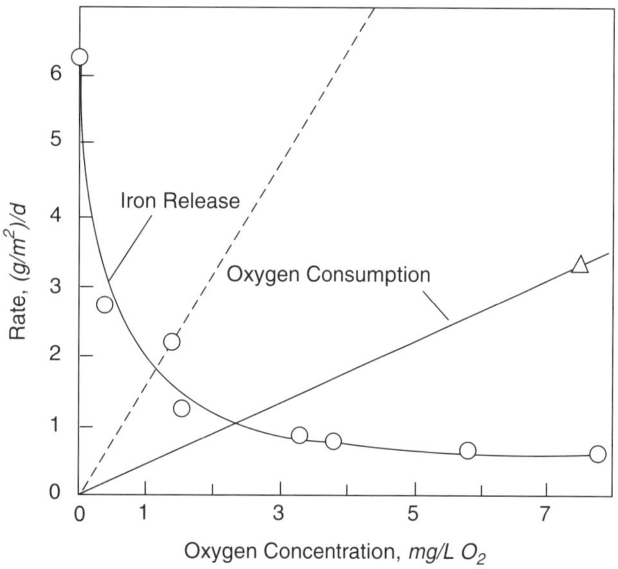

NOTE: The dashed line represents the predicted iron release rate if iron release depended only on the oxygen consumption rate.

Figure 2-21 Influence of oxygen concentration on the corrosion (oxygen depletion) rate and the iron release rate in steel pipes with scale (scale formed by exposure to Karlsruhe tap water at pH 7.5 for seven days)

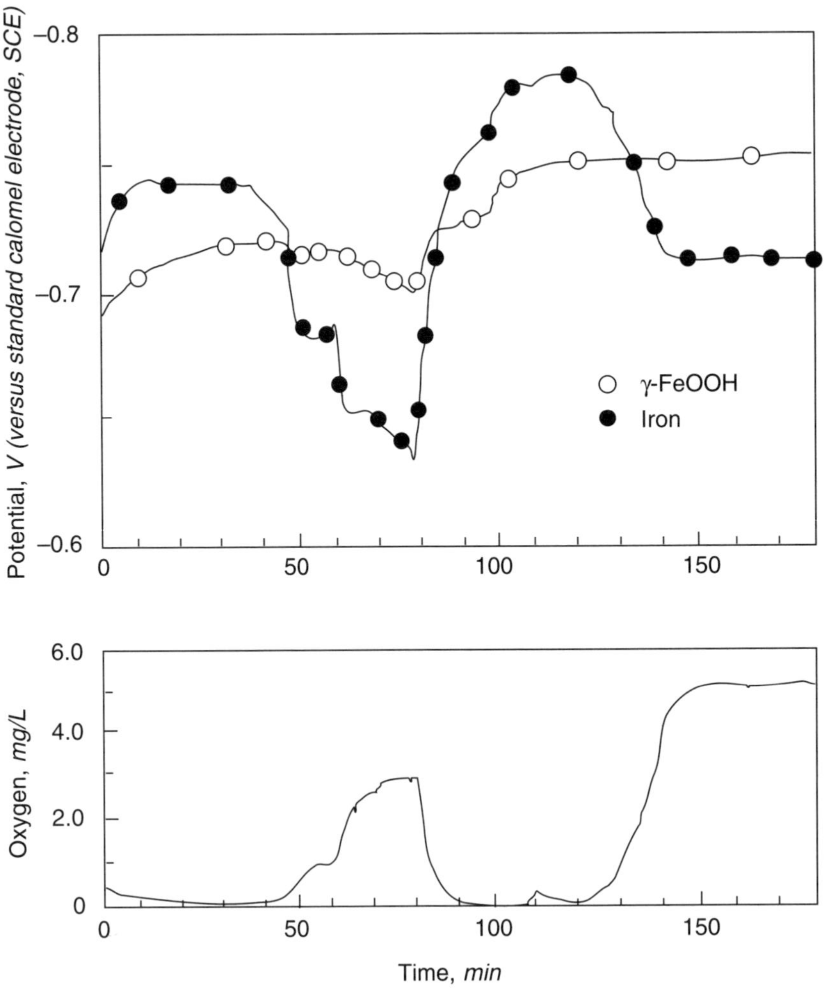

Figure 2-22 Electrical potential measurements in metallic iron and γ-FeOOH, both versus a standard calomel electrode, at different oxygen concentrations

The Kuch Mechanism* for Red-Water Formation

The observations and results presented thus far indicate that in flowing, oxygenated water corrosion rates are directly related to the oxygen content of the water and that corrosion can continue even in situations where oxygen is virtually totally depleted near the metal. Thus, a change in the corrosion mechanism must occur after some time in standing water, allowing metal dissolution and weight loss to continue in the absence of oxygen. Considering the limited number of available oxidants and the high ferrous concentrations found in many waters after periods of stagnation, ferric oxides in the scale

*Fred Kuch died very unexpectedly in 1988. In order to honor his important contribution to the understanding of corrosion in potable water systems, the authors of this volume have agreed to call the mechanism responsible for the formation of red or brown water resulting from the change between flowing and standing water the "Fred Kuch mechanism," or Kuch mechanism. Data presented in this chapter suggest that this mechanism is extremely important in controlling the release of corrosion products into drinking water distribution systems with iron pipes.

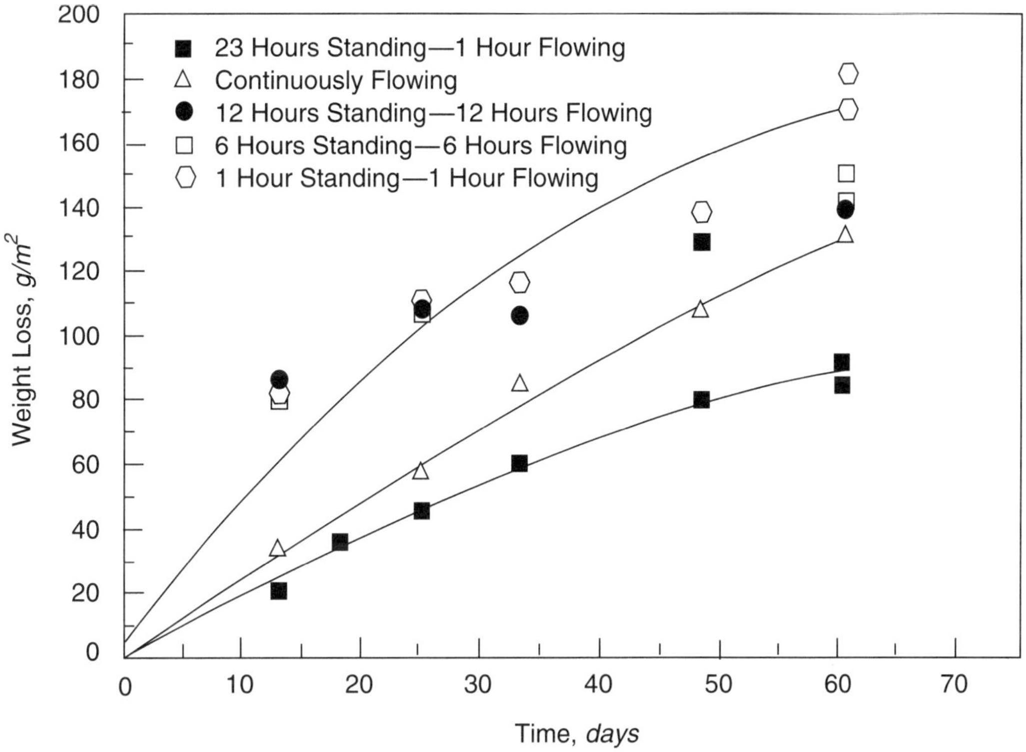

Source: Assad and Sontheimer (1991).

Figure 2-23 Results of weight loss measurements for pipes exposed to Karlsruhe tap water, using exposure cycles with different periods of standing and flowing water

layer are logical candidates to serve as the alternative oxidant once the oxygen is depleted. The cathodic reduction of these oxides leads to the formation of ferrous compounds.

According to the proposed mechanism, oxygen diffusion through the scale might be the rate-determining step for the corrosion reaction during periods of flow but not during periods of stagnation. During periods of stagnation, even though the bulk water may retain some DO, the oxygen diffusion rate is insufficient to maintain oxygenated conditions near the pipe wall. Under such conditions, the following corrosion reactions are proposed to occur on steel pipe walls and in the adjacent scales:

Anodic reaction: $Fe \leftrightarrow Fe^{2+} + 2e^-$

Cathodic reaction: $2FeOOH_{(s)} + 2H^+ + 2e^- \leftrightarrow 2Fe^{2+} + 4OH^-$

Both reactions produce Fe^{2+} ions, which may remain in solution or precipitate as green rust, siderite, etc. The ferrous ions that do not precipitate are transported into the standing water, where they are likely to be oxidized, leading to the formation of red water. The resumption of flow allows oxygen-rich water to permeate the scale layer once again, and the ferrous products within the scale can be oxidized. If the flow stops again, the reaction mechanism reverts to ferric reduction after some time, and so on. The overall cycle between oxidized and reduced iron products is summarized in the schematic in Figure 2-24.

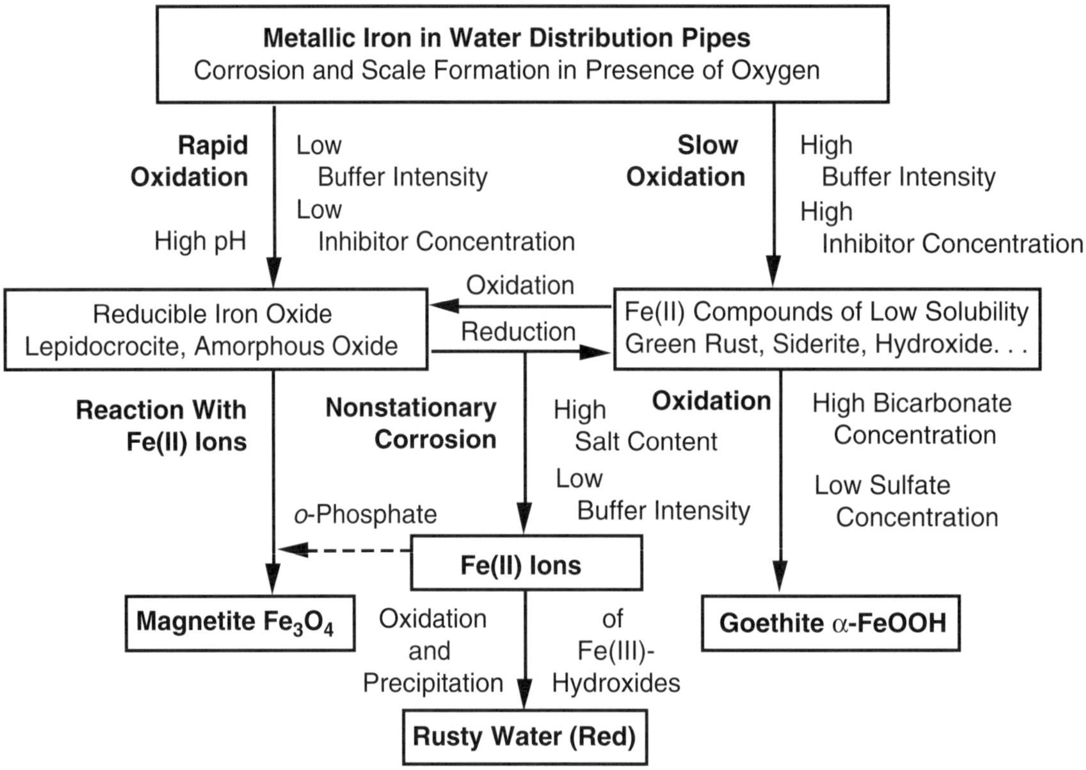

Figure 2-24 Schematic of the hypothesized oxidation-reduction cycle involving components of the scale layer

The proposed scheme for continued corrosion in the absence of DO requires the presence of reducible iron oxides in the scale, a path for the electrons generated by the anodic corrosion reaction to reach the cathodic sites, and transport of anions to complete the electrical circuit. Under these conditions, electrons released by the oxidation of pipe metal might be transported to the outer surfaces of the scale, where recently precipitated, easily reducible Fe(III) oxides are likely to be found. Electron transport through the scale might be through the enveloping scales discussed earlier.

Water quality could have both a direct and an indirect effect on the corrosion rate in standing water. The direct effect reflects the requirement for transport of ions through the water to complete the electrical circuit. This transport process is the most likely rate-determining step in the corrosion reaction in standing water, so the corrosion rate, and therefore the rate of formation of ferrous compounds, may depend significantly on the salt concentration. To test this hypothesis, iron release was evaluated in standing water in a pipe exposed to full-strength Karlsruhe tap water and to the same water diluted by a factor of 5 with distilled water (Figure 2-25). The dilution reduced both the salt concentration and the buffer intensity. The former change tends to decrease the release rate, while the latter tends to increase it. Since the overall effect was a decrease in release rate, the reduction of the salt concentration apparently dominated the overall effect. A different quantitative result might be obtained for different exposure conditions or different source waters. However, the result does support the hypothesis that the corrosion rate under stagnant conditions and the release rate when flow begins again depend on the salt content of the water.

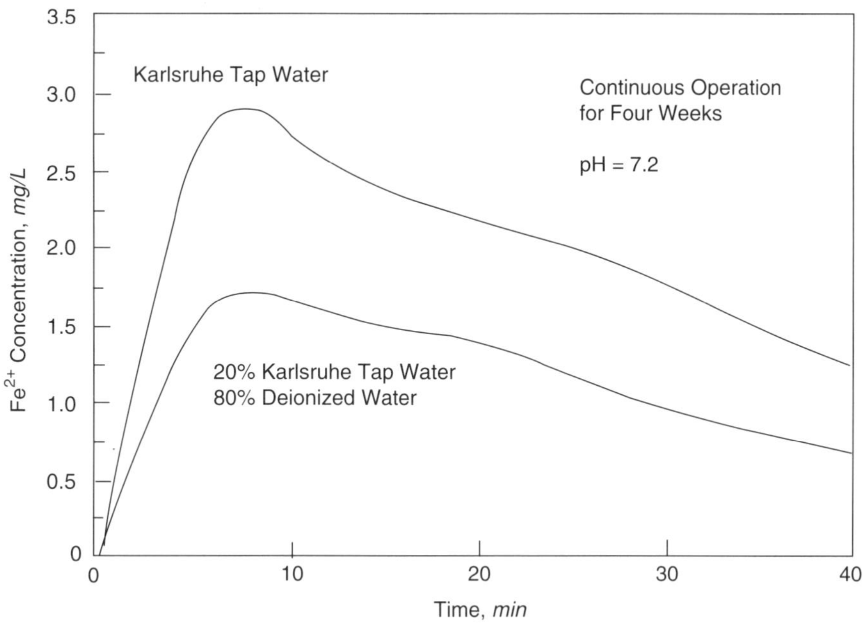

NOTE: For the measurements given, samples of the circulating water were taken and were replaced by fresh water containing no iron. As a result, the measured iron concentrations eventually decreased.

Figure 2-25 Iron release for Karlsruhe tap water after 16 hours of standing time and for the same water after a 5× dilution with deionized water

The hypothesized change in the corrosion mechanism when oxygen is depleted is also supported by electrochemical measurements, as shown in Figure 2-22. In this figure, the electrical potential relative to a standard calomel electrode (SCE) is shown for two test electrodes: one made of metallic iron and another made of lepidocrocite (γ-FeOOH). The measurements were conducted in solutions containing various oxygen concentrations, as shown in the lower part of the figure. It is evident from these results that there must be a reverse of the electron flow at low oxygen concentration, since the metallic iron is more noble than the oxide at high oxygen concentrations and less so at low oxygen concentrations.

Water quality factors other than the oxygen concentration can also have a long-term, indirect effect on the potential for iron release from standing water, since the water quality may determine which iron oxides are present in the scale. Goethite and magnetite are probably not reducible under the conditions expected in the scale (Sontheimer and Kuch 1989). Other iron oxides, such as lepidocrocite, are much easier to reduce. As shown in Figure 2-26, the sulfate concentration and pH value can have a large influence on the relative amounts of the different ferric oxides that form. For the water quality conditions tested and shown in the figure, lepidocrocite formed preferentially at higher sulfate concentrations and lower pH values. To the extent that these results are generally valid, more severe red-water problems are expected under such conditions. These results conform with practical experience. Valid, more severe red-water problems are expected under such conditions. These results conform with practical experience.

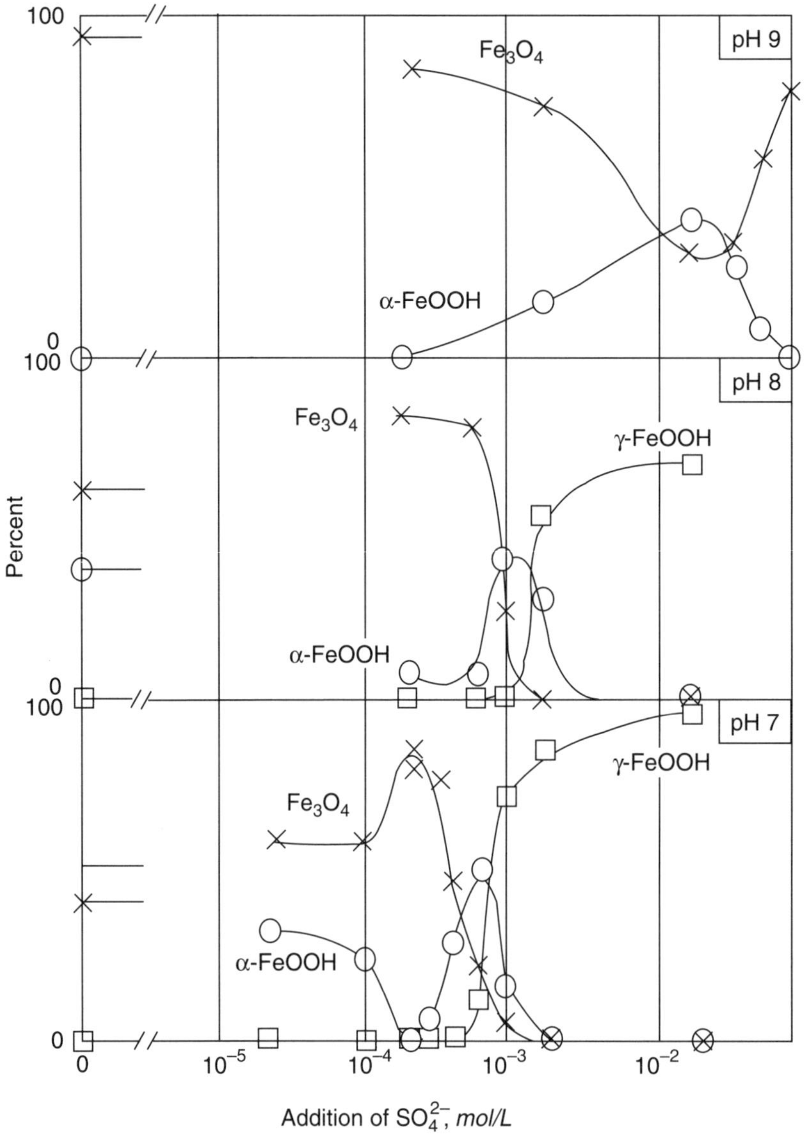

Figure 2-26 Influence of the sulfate concentration and the pH value on the formation of different modifications of ferric oxide

Other Causes of Red-Water Problems

The previous discussion presents a novel hypothesis for the genesis of red-water problems. The data supporting the potential importance of this mechanism are compelling, and it is likely to be a significant source of red water in parts of systems that are subject to frequent, alternating periods of flow and stagnation. As noted earlier, prior investigations that addressed iron release often led to ambiguous results, perhaps because the distinctions between the release rate and the corrosion rate were not fully appreciated or because the studies only evaluated long-term, average release rates that could not be associated with the water quality conditions at the time the release was actually occurring. It is possible and indeed likely that several other mechanisms for iron release might be significant in some systems. Some mechanisms that seem likely include the sloughing of crystalline iron oxides or of iron-oxidizing bacteria, to which amorphous iron oxide particles are often attached, and the precipitation of hydrous ferric oxides near the pipe wall shortly after the release of ferrous ions, if oxygen is present and the pH is high enough. Alternatively, if oxygen diffusion to the pipe wall is restricted by the presence of scale or by reactions with scale constituents or if the pH of the system is low enough to prevent rapid oxidation of ferrous ions, the ferrous ions released at the wall might diffuse into the solution and be oxidized in the flowing water. Such a process is similar to the mechanism described earlier, but it does not involve any assumptions about the water being stagnant for part of the process.

Physical examination of the iron particles that are causing a red-water problem might provide information about the source of the problem and thus provide insight on how to solve the problem. Additional investigations of mechanisms leading to iron release and the appearance of red water are encouraged, in order to expand the database for interpreting and overcoming these problems.

Conclusions

This section presented evidence supporting the concept that red-water problems can be a result of cyclic periods of flow and stagnation in a water pipe. It is the authors' opinion that such cycles are probably the most frequent cause of red-water problems occurring in practice. During periods of stagnation, oxygen is not available to act as the oxidant, and it has been proposed that reducible Fe(III) oxides in the scale serve as the alternative oxidants. In such cases, Fe(II) is generated by both the oxidation and reduction reactions. This Fe(II) might either precipitate or remain soluble. The fraction that remains soluble migrates toward the bulk solution, where it can be oxidized, precipitated, and carried to the consumer's tap when flow is reinitiated.

Water quality can affect this process in several ways. First, it might control the identity and properties of the minerals that are present in the scale. In general, lepidocrocite is more easily reduced than either goethite or magnetite (Sontheimer and Kuch 1989), and water quality factors that encourage formation of lepidocrocite increase the likelihood that red-water problems can occur. Second, the salt content of the water might control the reaction rate in standing water, thereby directly affecting the amount of Fe(II) generated during such periods. Finally, the DO concentration affects the amount of time that elapses before Fe(III) reduction becomes favorable. Thus, the DO concentration, along with pH and alkalinity, affects the solubility of the Fe(II) that is released from both the pipe wall and the preexisting scales.

SUMMARY

This chapter has described the critical processes affecting corrosion of iron and steel in potable water systems. Equipment composed primarily of iron and steel is abundant throughout water supply networks, and corrosion of this equipment can lead to leaks, loss of pipe carrying capacity, and deterioration of water quality. In iron and steel pipes, the major corrosion problems are associated with the buildup of corrosion products on the pipe walls and the release of corrosion products into the water. The characteristic pattern of iron corrosion leads to the formation of tubercles on the pipe surface. Although the tubercles are generated by localized corrosion, they may eventually overlap to give the pipe surface the appearance of having a fairly uniform layer of scale.

Corrosion of iron to Fe(II) and subsequent oxidation of Fe(II) to Fe(III) is thermodynamically favorable when the metal is in contact with any potable water. In most water supplies, the dominant oxidant is DO, at least when the water is flowing. Under such conditions, the rate of oxidation is proportional to the oxygen concentration adjacent to the metallic surface and is minimally dependent on other water quality variables. The concentration of DO at that location, however, might depend on several factors other than the DO concentration in the bulk water. In particular, the rate at which oxygen can diffuse from the bulk water to the pipe surface is likely to be affected by the velocity of flow; the temperature; and the thickness, porosity, and other physical properties of the scale that accumulates between the pipe and the flowing solution.

In the past, it has been assumed that $CaCO_3$ was a critical component of protective scales, and water utilities have relied on the LI as an indicator of a water's corrosivity toward iron. In light of much empirical evidence contradicting the presumed connection between the LI and corrosion, this practice should be abandoned.

Rather, in many cases, the precipitation of siderite ($FeCO_3$) or of green rust soon after the Fe(II) is generated seems to be critical for the formation of a protective scale layer, i.e., one that is a good barrier to oxygen diffusion and that therefore severely limits the corrosion rate. Siderite formation adjacent to the pipe wall leads to the formation of a tight scale, while the rapid oxidation of Fe(II) to Fe(III) and the formation of ferric oxides adjacent to the pipe generates a more porous, brittle, and less protective scale. If siderite or other reaction products containing ferrous iron form initially, the protectiveness of the layer does not seem to diminish later, even if the iron in the solid is subsequently oxidized to form a ferric oxide.

The major water quality factors that determine whether siderite forms are the solution pH and buffer intensity (or, equivalently, pH and alkalinity). High pH in the water and, in particular, near the wall favors rapid oxidation of Fe(II) and formation of nonprotective scales. Since OH^- ions are generated by the reduction of oxygen, there is a natural tendency for the pH of the water near the pipe wall to increase relative to that in bulk solution. A high buffer intensity counteracts this tendency, keeping the pH in a range where siderite formation might be kinetically favored over iron oxidation. Natural organic compounds, phosphates, and possibly silicates interfere with the oxidation of Fe(II) and thereby encourage the formation of protective scale layers. Other constituents dissolved in the water might affect this stage of corrosion, but there is no clear indication of such effects at this time.

Over a time frame of months to years, much of the corroded iron forms identifiable minerals in the scale layer. Both ferrous and ferric solids have been identified in the scale, including hydrous oxides, sulfides, carbonates, phosphates, and other solids (e.g., green rust). Nonferrous solids, such as calcite, are also often found. These solids might form via purely chemical processes or the reactions might be mediated by bacteria. One structure that is commonly observed in older scales and that might have special significance includes a layer of goethite and magnetite that envelops a group of interior corrosion products, separating them from the bulk water. Reactions occurring on and inside this layer can generate localized solutions that have dissolved ionic concentrations much higher than the flowing water. These solutions might exert a strong influence on the types of scale that ultimately dominate the system and on the tendency for red-water problems to develop.

Iron corrosion can proceed in a water supply system even under conditions where there is negligible DO adjacent to the pipe, conditions that are most likely to develop when the water is stagnant. Under such conditions, the apparent oxidant is ferric oxide scale that formed previously in the system, under oxic conditions (with oxygen present). Ferrous iron is generated at both the anode and the cathode in these cases, and, if not precipitated in situ, it is likely to migrate toward the bulk water. When flow of oxygenated water resumes in the pipe, this iron can be taken up by the water and cause red water to appear at the tap. The rate of ferrous iron production during the period of stagnation is related to the initial DO concentration, the salt content of the water, and the properties of the scale layer, including the availability of reducible iron minerals. The tendency for the ferrous iron that is generated during such a period to migrate into the solution depends on the alkalinity and pH of the water in addition to the factors listed above. As a result, the prediction of red-water events is quite difficult.

Figure 2-24 is an attempt to synthesize the key features of the siderite model for formation of protective scales with the Kuch mechanism for formation of red water. The schematic drawing makes it apparent that corrosion rates can be controlled through influences on the anodic as well as on the cathodic reactions. While the siderite model attributes the effectiveness of protective layers to their ability to limit oxygen diffusion, the Kuch mechanism explains red-water problems as the result, in most cases, of a change in the cathodic reaction when oxygen is unavailable.

Although an attempt has been made to show the combined influences of several variables on both processes in the diagram, the overall process is very complex and does not lend itself to easy generalizations. Nevertheless, the model does explain many empirical observations and provides a framework for addressing iron corrosion problems. Certainly, the model cannot serve as a substitute for corrosion testing in any particular system, and it is unlikely that any corrosion problem can be solved based purely on theoretical analyses of the situation. Nevertheless, the authors believe that an understanding of the mechanisms leading to iron corrosion problems can greatly aid in the search for an appropriate solution, and that belief has been the guiding principle in the preparation of this chapter.

Several mechanisms other than the one discussed in detail in this chapter might account for the formation of red water in parts of systems that are not subject to cyclic periods of stagnation and flow, including some mechanisms that involve biological activity. Although some such mechanisms have been proposed, their significance is not established. Attempts to elucidate these mechanisms should distinguish clearly between the release rate and the corrosion rate and should focus more strongly than has been done in the past on the short-term conditions in the pipe that might lead to iron release.

Solving Iron Corrosion Problems

When a utility is faced with an iron corrosion problem, the first step should be to identify the problem(s) clearly. The most common problem encountered is likely to be red water at the consumer's tap. In such a case, a relatively simple examination should indicate whether the iron oxide particles generating the color are ferric hydroxide flocs or crystallized iron oxides. Additional information that should be gathered includes the frequency of the problem—Is it fairly continuous or intermittent? Does it occur exclusively near dead-end pipes or also near large pipes with continuously flowing water? This type of information should allow one to draw some preliminary conclusions about the mechanism responsible for the problem.

Information about both the corrosion rate and the iron release rate is desirable when assessing an iron corrosion problem. In most situations, the corrosion rate can be equated with the rate at which oxygen is consumed in the system. In cases where oxygen is not available, other oxidants might support the corrosion reaction. To understand the major reactions occurring in a particular system, testing equipment that allows one to measure both oxygen depletion and iron release in pipe samples that have been exposed to the water of interest for various amounts of time should be used. Often, a few simple tests with such testing devices can provide a large amount of useful information.

Once the problem is identified, a strategy can be developed to overcome it. Often, chemical inhibitors can limit oxygen diffusion and also reduce red-water formation. Additional information on instruments for measuring corrosion rates and release rates and for evaluating corrosion inhibitors is presented in chapters 9 and 8, respectively.

REFERENCES

Baylis, J.R. 1926. Prevention of Corrosion and "Red Water." *Jour. AWWA,* 15:598.

Butler, G., and H.C.K. Joon. 1966. *Corrosion and Its Prevention in Waters.* London: Leonard-Hill.

Chantereau, J. 1977. *Corrosion Bactérienne, Bactéries de la Corrosion.* Paris: Technique et Documentation.

De Waard, C., and D.E. Williams. 1975. Carbonic Acid Corrosion of Steel. *Corrosion—NACE,* 31:177.

EBI and RHE. 1978. Untersuchungen zur zentralen Teilenthärtung des Trinkwassers in Mannheim. Mannheim, Germany: EBI and RHE.

Feitknecht, W., and G. Keller. 1950. Über die dunkelgrünen Hydroxyverbindungen des Eisens. *Z. Anorg. Chem.,* 262:61.

Fischer, W., and W. Siedlarek. 1977. Wasserstoffentwicklung aus CO_2-haltigen wässrigen Elektrolyten. *Werkstoffe und Korrosion,* 28:822.

Heck, R. 1974. Deterioration in the Physical and Chemical Quality of Water in the Distribution System. Paper presented at the International Water Supply Association Tenth Congress, at Brighton, England.

Heyer, C. 1888. *Ursache und Beseitigung des Bleiangriffs durch Leitungswasser.* Dessau: Verlagsbuchhandlung Paul Baumann.

Kölle, W. 1977. Corrosion in Drinking Water Systems. Paper presented at the Society of Chemical Industry Sixth European Congress on Metallic Corrosion, at London.

———. 1978. Corrosion in Drinking Water Systems. Paper presented at the International Water Supply Association Twelfth Congress, Kyoto.

———. 1982. Unpublished data.

Kölle, W., and H. Rösch. 1978. Korrosionsvorgänge in Gussrohren. *Vom Wasser,* 51:161.

———. 1980. Untersuchungen an Rohrnetz-Inkrustierungen unter mineralogischen Gesichtspunkten. *Vom Wasser,* 55:159–177.

Kölle, W., and H. Sontheimer. 1977. Untersuchungen zur Schutzschichtbildung in Gussrohren. *Vom Wasser,* 49:277–294.

Kölle, W., O. Strebel, and J. Böttcher. 1985. Formation of Sulfate by Microbial Denitrification in a Reducing Aquifer. Paper presented at the International Water Supply Association Specialized Conference, Berlin.

Kölle, W., P. Werner, O. Strebel, and J. Böttcher. 1983. Denitrifikation in einem reduzierenden Grundwasserleiter. *Vom Wasser,* 61:125–147.

Kooijmans, J. 1938. Korrosion und Schutzschichtbildung durch Leitungswasser. *gwf—Das Gas- und Wasserfach,* 81:611–615, 628–633.

Kuch, A. 1984. Untersuchungen zum Mechanismus der Aufeisenung in Trinkwasserverteilungssystemen. Ph.D. diss., Universität Karlsruhe, Germany.

Kuch, A., H. Sontheimer, and I. Wagner. 1983. Die Messung der Aufeisenung im Trinkwasser—Ein neues Verfahren zur Beurteilung von Deckschichten in schwarzen Stahlrohren. *Werkstoffe und Korrosion,* 34:107–111.

Langelier, W.F. 1936. The Analytical Control of Anticorrosion Water Treatment. *Jour. AWWA,* 28:1500–1521.

LaQue, F.L. 1964. Corrosion Characteristics of Ductile Iron. *Jour. AWWA,* 56:1433.

Larson, T.E., and R.M. King. 1954. Corrosion by Water at Low-Flow Velocity. *Jour. AWWA,* 46:1.

Larson, T.E., and R.V. Skold. 1957. Corrosion and Tuberculation of Cast Iron. *Jour. AWWA,* 49:1294.

———. 1958. Laboratory Studies Relating Mineral Quality of Water to Corrosion of Steel and Cast Iron. *Corrosion—NACE,* 14:43–46.

Lee, S.H., J.T. O'Connor, and S.K. Banerji. 1980. Biologically Mediated Corrosion and Its Effects on Water Quality in Distribution Systems. *Jour. AWWA,* 72:636–645.

Madsen, S. 1973. Mischwasserversorgung in Bremen aus der Sicht des Wasserchemikers. Hamburg: DVGW-Aussprachetagung Aumühle.

McCauley, R.F., and M.O. Abdullah. 1958. Carbonate Deposits for Pipe Protection. *Jour. AWWA,* 52:1419.

McGill, J.R., B. McEnaney, and D.C. Smith. 1976. Crystal Structure of Green Rust Formed by Corrosion of Cast Iron. *Nature,* 259:200.

Miller, J.D.A., and R.A. King. 1975. Biodeterioration of Metals. In *Microbial Aspects of the Deterioration of Materials.* Edited by D.W. Lovelock and R.J. Gilbert. Society for Applied Bacteriology, Technical Series 9. New York: Academic Press.

Rijkens, B.A. 1967. Verhalten der Polyphosphate bei der Korrosion der Metalle. *Wasser, Luft und Betrieb,* 11:728–733.

Rothmann, B. 1977. Zum Mechanismus der Korrosion von Stahl in kohlendioxidhaltigen, sauerstoffreien Wässern. Ph.D. diss., Rheinisch-Westfälische Technische Hochschule, Aachen.

Schmitt, G., and B. Rothmann. 1977. Untersuchungen zum Korrosionsmechanismus von unlegiertem Stahl in sauerstoffreien Kohlensäurelösungen. Teil I, Kinetik der Wasserstoffabscheidung. *Werkstoffe und Korrosion,* 28:816–922.

———. 1978. Untersuchungen zum Korrosionsmechanismus von unlegiertem Stahl in sauerstoffreien Kohlensäurelösungen. Teil II, Kinetik der Eisenauflösung. *Werkstoffe und Korrosion,* 29:98–100.

Schwenk, W. 1974. Korrosion von unlegiertem Stahl in sauerstoffreien Kohlensäurelösungen. *Werkstoffe und Korrosion,* 25:643–646.

Sontheimer, H., and A. Kuch. 1989. Einfluß der Sulfatkonzentration in Trinkwässern auf die Deckschichtbildung und die Aufeisenung (Rostwasserbildung) in Stahlrohren. Bereich Wasserchemie am Engler-Bunte Institut der Universität Karlsruhe, Heft 33.

Sontheimer, H., W. Kölle, and R. Rudek. 1979. Aufgaben und Methoden der Wasserchemie—Dargestellt an der Entwicklung der Erkenntnisse zur Bildung von Korrosionsschutzschichten auf Metallen. *Vom Wasser,* 52:1–18.

Sontheimer, H., W. Kölle, and V.L. Snoeyink. 1981. The Siderite Model of the Formation of Corrosion-Resistant Scales. *Jour. AWWA,* 73:572–579.

Stumm, W. 1956. Calcium Carbonate Deposition at Iron Surfaces. *Jour. AWWA,* 48:300–310.

———. 1957. Corrosion Studies. *Public Works,* 88:78–80.

———. 1960. Investigation on the Corrosive Behavior of Waters. *Jour. San. Engrg. Div.—ASCE,* 86:27.

Tillmans, J., P. Hirsch, and K. Schilling. 1929. Die selbsttätige Schutzschichtbildung in eisernen Rohren. *gwf—Das Gas- und Wasserfach,* 72:49–53, 78–82, 689–690.

Tillmans, J., P. Hirsch, and W. Weintraud. 1927. Die Korrosion von Eisen unter Wasserleitungswasser. *gwf—Das Gas- und Wasserfach,* 70:845–849, 877–884, 898–904, 919–925.

Tuovinen, O.H., D.M. Mair, and J. Banovic. 1984. Chlorine Demand and Trihalomethane Formation by Tubercles from Cast Iron Water Mains. *Envir. Tech. Letters,* 5:97–108.

Tuovinen, O.H., K.S. Button, A. Vuorinen, L. Carlson, D.M. Mair, and L.A. Yut. 1980. Bacterial, Chemical, and Mineralogical Characteristics of Tubercles in Distribution Pipelines. *Jour. AWWA,* 72:626–635.

Weidmann, G. 1970. Aufbau und chemische Zusammensetzung von Rosten in eisernen Rohrleitungen. *gwf—Wasser/Abwasser,* 111:713–715.

Weiss, J. 1951. Electron Transfer Reactions in the Mechanism of Oxidation-Reduction Process in Solution. *Jour. de Chimie-Physique,* 48:C6–C10.

3

Corrosion of Galvanized Pipe

R. Rhodes Trussell and Ivo Wagner

Galvanized pipe is one of the oldest and most common plumbing materials used to transport domestic drinking water. For most of this century, it has been the dominant material used for domestic plumbing in the United States. However, during the past three decades, copper tubing has gradually displaced galvanized pipe as the leading material for domestic plumbing in the most developed nations. This is in part due to a relaxation in the required wall thickness for copper tubing and also due to the rising cost of labor (installation of galvanized pipe is more labor intensive). Plastic plumbing materials, particularly polyvinyl chloride (PVC), are also becoming more common. Still, galvanized pipe is used for domestic plumbing in developed countries and it remains the material of choice in lesser developed and developing countries where the cost of labor is less important than the cost of materials.

In the United States and northern Europe, galvanized pipe is only occasionally used. This is particularly true in the United Kingdom, Germany, the Netherlands, and Denmark where copper is dominant. In central eastern Europe, galvanized pipe is more common, comprising between 30 and 40 percent of new installations. In eastern Europe and in former Soviet bloc countries, it is still the dominant plumbing material. An increasing concern about the impact of corrosion of consumer plumbing on the aesthetic and health aspects of domestic drinking water has prompted greater interest in the corrosion of galvanized pipe and the corrosion by-products that result. This chapter summarizes current knowledge of galvanized pipe and the problems it presents to the drinking water community today.

NOTE: The first edition of this chapter was authored by R. Rhodes Trussell and Ivo Wagner.

PIPE MANUFACTURE AND GALVANIZING

In order to understand some of the problems in pipe quality that result in corrosion problems, it is important to understand the methods used to manufacture the pipe. Manufacturing practice will vary from one plant to another and the differences can be significant. The following discussion gives a general overview of a typical process and highlights issues that can be important to corrosion.

Manufacturing Process

The first step in the manufacturing process is to make the steel pipe. This is accomplished by one of three methods: (1) continuous butt-welding (type F), (2) electrical resistance welding (type E), or (3) continuous extrusion (type S). The process begins in the steel mill with slabs of refined steel. For continuous butt-welding and electrical resistance welding, the slabs are sent to the hot strip mill where they are converted into long, thin, coiled sheets called "coiled skelp." The skelp is then formed to the proper width and thickness to roll directly into pipe. In a butt-welding pipe mill, the skelp is fed continuously through a heating furnace where the two edges are rolled up, heated, and fused together under conditions of high temperature and pressure using a forming and welding mill. In an electrical resistance-welding mill, the process is much the same, except the weld itself is produced not by the heat of the furnace alone but by the heat obtained from the resistance of the pipe to the flow of an electrical current. Both of these processes form a continuous butt weld using the native materials in the pipe. In continuous extrusion, which is much like the process used to produce copper pipe, the slab itself is heated in a press and forced through a mandrel to produce a seamless steel tube called "wrought steel seamless pipe."

After welding or extruding, the finished steel pipe is cooled and sent to the galvanizing shop. There, the pipe is first dipped, usually as a loose bundle, into an acid pickling bath and rotated to obtain good exposure. Pickling is followed by rinsing and then a preflux dip, usually of zinc ammonium chloride. Then individual pieces of pipe are fed into the zinc galvanizing bath and then sent to an annealing furnace. Immediately after the annealing, hot steam is blown around the outside and through the inside of the pipe to remove excess zinc. After steaming, the pipe is cooled and rolled down a ramp onto a platform where it is inspected at both ends. The zinc-coated pipe is then dipped in chromate to give it an attractive finish. Lacquer is sometimes used instead of chromate. If lacquer is used, it is usually applied only on the outside; chromate is applied as a dip that coats both the inside and the outside. As a result, the lacquer does not protect the inside of the pipe. Because of environmental constraints, chromate dipping is increasingly rare (it is no longer allowed in Germany). Sodium nitrite is sometimes used as a substitute, but usually the final dip is eliminated altogether. The final step is to stamp the product with the following information: manufacturer, type of pipe, ASTM* or DIN† specification number, etc. Pipe smaller than 1.5 in. (40 mm) in diameter is sometimes bundled and tagged rather than stamped.

---

*American Society for Testing and Materials, 100 Barr Harbor Dr., West Coshohocken, PA 19428-2959.

†Deutsches Institut für Normung (German Institute of Standardization), Burggrafenstr. 6, Berlin, Germany.

Manufacturing Problems

With a basic understanding of the pipe manufacturing process, some of the problems that occur during that process can be examined. Sometimes the skelp used to form the pipe is cut too wide and the excess width is forced into the weld, creating a noticeable bead. This practice can result in a distinct weld seam (Figure 3-1). Likewise, the skelp can be cut too narrow and the weld seam will be deficient in material, leaving an intermittent groove in the seam. Pitting corrosion problems are often associated with both of these characteristics. When the pipe is not rotated during pickling, the surface is not uniformly cleaned and old corrosion deposits can prevent the zinc layer from properly adhering.

If temperature conditions during galvanizing or annealing are too hot or annealing continues for too long, the diffusion process will progress too far, the zeta layer (iron–zinc diffusion zone) will become larger, and the zinc-rich eta layer can almost disappear. A coating with no eta layer does not have the same protective properties as a proper hot-dip zinc coating. (A more detailed description of the different layers can be found on p. 74 and in Figure 3-2.) Excessive amounts of carbon, phosphorus, manganese, or silicon cause this same effect by accelerating the growth of the compound layers in the alloy. In Europe, these phenomena are referred to as "galvannealing."

If the temperature is too cold, a heavy, uneven galvanizing will result. When the steam-blowing step is delayed too long and some of the zinc begins to harden or when steam at inappropriate temperatures or pressures is used, an uneven zinc

Figure 3-1 Pipe section samples indicating variations in weld seams

coating can occur. If air is substituted for steam, which is sometimes done, a porous and relatively unstable zinc coating can result.

If the chromate dip is too concentrated, it will give the pipe a yellowish hue. This does not reduce the corrosion resistance of the pipe but it may result in excessive leaching of chromium into the water. Pipe that uses a lacquer coating rather than a chromate dip is not as stable. If the galvanizing coat is too thin, a dull appearance may result, often a harbinger of a limited life for the coating. Many of these phenomena will be illustrated further in the following text.

STANDARDS USED FOR GALVANIZED PIPE

In the United States, pipe quality specifications are generally governed by the standards of ASTM. In Germany, specifications are generally governed by the DIN. The standards issued by these organizations that apply to galvanized pipe will be briefly discussed.

ASTM

There are two basic types of ASTM specifications of interest in this chapter: (1) those that specify the quality of the materials and the galvanizing process (ASTM A53-93A, ASTM A385-80, and ASTM B6-87 [92 E1]), and (2) those that outline the tests that may be used to evaluate the product (ASTM A90/A90M-93, ASTM A239-89, and ASTM A123-89A E1).

Three ASTM specifications are of principal importance in characterizing the attributes of properly manufactured seamless galvanized steel pipe: ASTM A53 characterizes the pipe itself, ASTM B6 characterizes the slab zinc that is used in the galvanizing process, and ASTM A385 outlines recommended practice for galvanizing. ASTM A53, which supersedes ASTM A120, allows that pipe be furnace buttwelded, electric-resistance-welded, or be wrought steel seamless pipe. The pipe is to be coated inside and out with the hot-dip process, using zinc that meets the quality requirements of ASTM B6, to an average thickness of not less than 550 g/m^2 (1.8 oz/ft^2). This figure is the sum of weight on both sides of the pipe and corresponds to a thickness of about 3 mil, or 75 μm. A hydrostatic test is also required. For the sizes commonly used in consumer plumbing, the test pressure is 4.83 MPa (700 psi). The specification requires that the weld seam of electrical resistance-welded pipe be heat treated after welding so that no martensite remains.

ASTM B6 establishes three grades of slab zinc—special high grade, high grade, and Prime Western. Each has a specified maximum level for lead, iron, and cadmium and a minimum level for zinc (by difference). Prime Western, which is normally used in the manufacture of galvanized pipe for domestic plumbing, has the following limits:

Lead	1.4%
Iron	0.05%
Cadmium	0.20%
Aluminum	0.01%
Copper	0.02%
Total impurities	2.0%

ASTM A385 describes the structure and metallurgy of a normal hot-dip zinc coating, showing the presence of five separate metallurgical layers: (1) the base steel; (2) the gamma layer at the steel-coating interface; (3) the delta layer, which is iron-rich zinc; (4) the zeta layer, which can be characterized as the iron–zinc diffusion zone; and (5) the eta layer, which is virtually pure zinc in composition. Impurities such as lead and cadmium are generally found at their highest concentration in the eta layer. A cross section identifying these layers is shown in Figure 3-2. ASTM A385 also notes the need for the use of steels meeting certain chemical requirements. Specifically, maximum limits are set for carbon at 0.25 percent, phosphorus at 0.05 percent, manganese at 1.3 percent, and silicon at 0.05 percent. Levels in the base steel above these limits may result in abnormal annealing, which was discussed earlier. Often such impurities also result in a more brittle and less adherent coating with a dull or mottled appearance. Microscopic examination of such coatings will show a dominance of the zeta layer and a greatly reduced or eliminated eta layer, the result of accelerated growth of the compound or alloyed zeta layer. This appearance is the "galvannealing" phenomena described earlier.

Where the zinc layer itself is concerned, ASTM A53 requires only that the zinc used meet the requirements of ASTM B6 and the weight requirements mentioned earlier. There are, however, three ASTM tests that are useful in evaluating galvanized pipe: (1) ASTM A90 (1993), which is used to characterize the weight of the zinc coating; (2) ASTM A239 (1992), the Preece test, which can be used to evaluate the uniformity of the coating; and (3) ASTM A123 (1992), which addresses questions of finish, appearance, and adherence of galvanized materials. In ASTM A90, the coating is evaluated by weighing a specimen, stripping the coating with an acid solution, and weighing the specimen again. The weight of the coating is determined by difference. ASTM A53 requires two specimens having an average coating weight of no less than 550 g/m^2 (1.8 oz/ft^2), with neither sample having a coating weight of less than 490 g/m^2 (1.6 oz/ft^2). The weight of the coating is determined by the combined weight of coating on the inside and outside of the pipe. ASTM A239, the Preece test, employs repeated exposure of the coated surface to a saturated copper

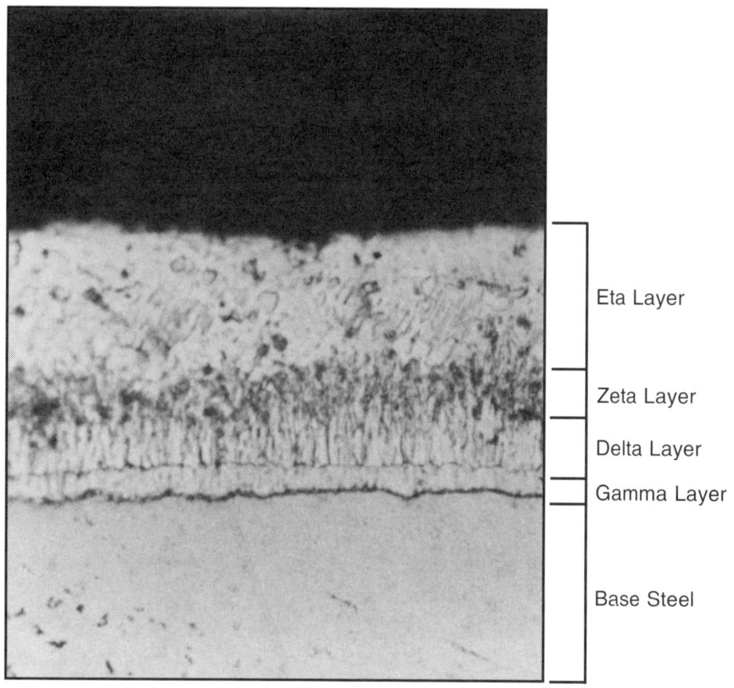

Figure 3-2 Metallurgical layers in a normal hot-dip zinc coating

sulfate solution until an area of the base metal of 7.7 mm^2 or larger is exposed. Normally the base metal is easily recognized because ferrous metals will form a bright, adherent copper deposit, whereas zinc will not. For a particular test piece, the number of dips required to expose the iron beneath is a measure of the minimum thickness of the zinc layer. The variation from one sample to the next gives an indication as to the uniformity of the coatings. A good-quality galvanized pipe will normally require two dips or more. Although not intended for galvanized pipe used in domestic plumbing, ASTM A123 does include descriptions of the required finish and appearance that are useful. ASTM A123 also includes the stout knife test, which is useful in evaluating poor adhesion associated with improper pickling prior to the galvanizing process.

DIN

Like US standards, those developed in Germany were designed to reduce failures in material, hence they specifically address certain corrosion issues that have developed in German experience. Four standards now apply: DIN 2440 (1978) and DIN 17100 (1966), which characterize the pipe before galvanizing; DIN 2444 (1984), which characterizes the quality of the galvanizing; and national German legislation (TrinkwV 1990), which limits the use of galvanized pipe to certain water qualities.

DIN 2440 requires that the steel pipe have a clean and complete weld seam (the weld can protrude no more than 0.3 mm plus 5 percent of the pipe wall thickness), a scale-free surface, and that it meet certain mechanical tests on bending stress. Requirements on the weld seam derive from experience with pitting corrosion associated with poor quality pipe. DIN 17100 requires that the base steel contain no more than 0.06 percent phosphorus and no more than 0.05 percent sulfur by weight.

DIN 2444 emphasizes the zinc layer itself. The zinc used for galvanizing must be 98.5 percent pure and can contain no more than 0.01 percent cadmium and no more than 0.8 percent lead. The coating must have a weight of at least 600 g/m^2 (2.0 oz/ft^2), it must cover the whole surface and be free of local accumulations, excess zinc must be removed by steam and not air, the production must be supervised by an independent testing institute, and the pipe must be durably marked with the manufacturer, the specification, the type of manufacture, and a seal of approval. The cadmium and lead requirements were in response to concern about the exposure of consumers to these metals in drinking water containing corrosion by-products from galvanized pipe.

As a result of the European Community (EC) Directive Relating to the Quality of Water (EC 1980), recent German legislation (TrinkwV 1990) now includes a requirement that the pipe must not release more than 5 mg/L of zinc upon 12 hours of exposure to the water under stagnant conditions. To obtain this, the use of galvanized pipe for domestic plumbing has been limited by DIN 50930, T.3 (1993), which limits the use of galvanized pipe to waters that have a level of CO_2 of less than 22 mg/L. If CO_2 data are not available, the pH shall be greater than 7.3 *and* the total alkalinity shall not exceed 250 mg/L as $CaCO_3$.

METHODS FOR CHARACTERIZING GALVANIZED PIPE

The methods for characterizing galvanized pipe fall into three broad categories: (1) qualitative visual examination, (2) measures of coating thickness and uniformity (the ASTM and DIN tests), and (3) microscopic and detailed chemical examination of the zinc galvanizing layer. The following discussion uses available US and German standards to illustrate the concepts.

The results of tests conducted by Fox, Tate, and Bowers (1983) and other tests conducted by the authors will be used to serve as examples in this section and those that follow. Fox, Tate, and Bowers (1983) tested galvanized pipe from three different manufacturers, for testing purposes identified as A, K, and U. Each manufacturer represented certain market niches in the United States at that time, with the pipe from manufacturer A being the most expensive, the pipe from manufacturer K the least expensive, and the pipe from manufacturer U being in between. Because of its low cost, the K pipe had been used in several large projects. The local dealers indicated that the A pipe was considered to be of the highest quality. The U pipe appeared to fall in between K and A, both in quality and cost.

In order to conduct the testing, several 6.4-m (21-ft) lengths of each brand of pipe were purchased at local pipe yards. Pipes were examined visually and were subjected to the two-sided thickness test (ASTM A90) and to the Preece test. Chemical and microscopic examinations were also performed.

The surface differences among the three pipes were quite distinct. While pipe A appeared relatively smooth and uniform, the K pipe contained distinctive, intermittent blemishes and showed a clear ridge along the weld (see Figure 3-1). A few spots on the K pipe showed rust. The U pipe showed the most surface variability. The zinc coating on this pipe showed a range of colors from shiny silver-gray, through yellowish silver-gray, on to dark blue and black. Several pieces contained a series of wavelike ridges, suggesting that the zinc coating may have been too cool when it was blown out with steam. These characteristics were also observed along the length of individual pieces of pipe; that is, each A pipe was smooth and uniform, each K pipe was generally smooth with isolated blemishes, and each U pipe showed color and texture variations along its entire length.

Quantitative ASTM Tests

Two separate coating tests were conducted on several samples of each type of pipe. ASTM A90 was used to determine the weight of the zinc coating on the steel by the difference in weight of the specimen before and after stripping with an acid solution. This test was used in combination with ASTM A120, which calls for specimens to have a minimum weight of 1.6 oz/ft^2 (490 g/m^2) and an average weight of 1.8 oz/ft^2 (550 g/m^2) or more. ASTM A239, the Preece test, was also used. The Preece test is a qualitative test of uniformity. Galvanized samples are dipped in saturated copper sulfate for 1 min and then examined. The process is repeated until a shiny copper spot appears, indicating that the base steel has been exposed. The number of "dips" required is a measure of the coating's thickness. Samples were taken from the middle and one end of each 6.4-m (21-ft) piece of pipe.

Figure 3-3 shows the results of the two-sided coating thickness test. The A pipe samples showed a high degree of uniformity, all 14 samples falling within a narrow range of coating thickness and all meeting ASTM coating requirements. All U pipe samples met specifications, but the range of coating thicknesses was highly variable. This appears to be a result of the deposits of excess zinc on several U samples. This

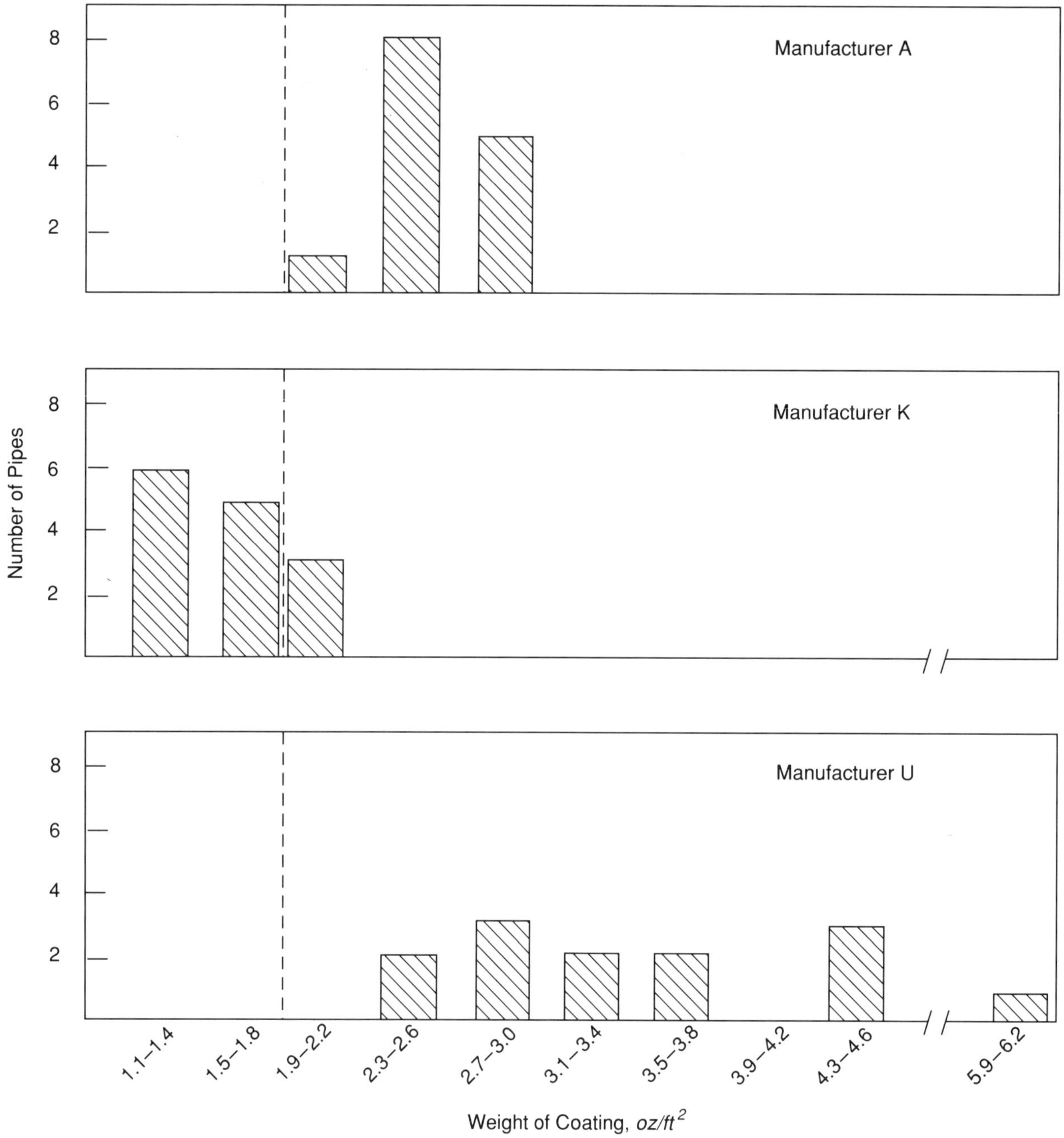

NOTE: All samples for manufacturer A are above the minimum requirement. All samples for manufacturer U are above the minimum requirement but lack uniformity and average nearly twice the required level. For manufacturer K, 13 of the 16 samples tested were below the required average.

Source: Fox, Tate, and Bowers (1983).

Figure 3-3 Results from two-sided coating thickness test

extra zinc resulted in high apparent weights. In one case, the calculated weight of coating was 5.9 oz/ft^2 (1,800 g/m^2), three times the standard. The K pipe was more uniform than the U pipe, but a significant number of the 14 samples did not meet the weight specification.

Single-sided coating tests were also conducted, and the results of these tests indicated that the inside coating of zinc may be slightly thicker than the outside coating for all three types of pipe tested. If this difference is present in pipes from other manufacturers, builders may want to consider specifying the inside and outside coating weights separately.

Figure 3-4 shows the results of the Preece test, which confirmed the pattern observed in the thickness test. As Figure 3-4 shows, the A and K pipes have relatively

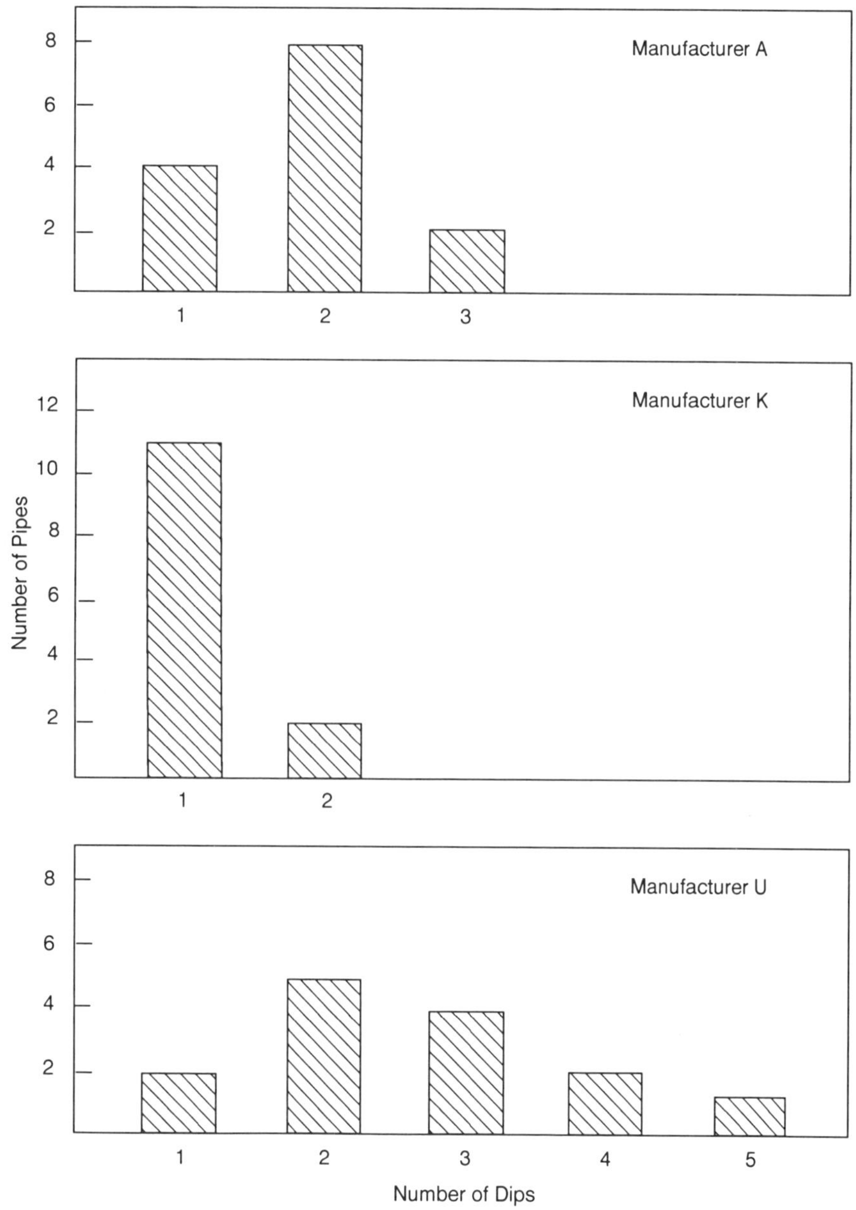

Source: Fox, Tate, and Bowers (1983).

Figure 3-4 Results from Preece test for locating the thinnest spot in galvanized coating uniform coating thicknesses, while the U pipe is decidedly nonuniform. However, it should be noted that this nonuniformity is in the direction of greater coating weight. Of the 14 samples of U tested, 12 required more than one dip to remove the thinnest portion of the zinc coating. This is in contrast with the K pipe for which only 2 out of the 14 samples required more than a single dip. The A pipe again performed well. The majority of these samples required two dips, four required only one dip, and two required three dips.

Chemical Examination

Chemical examinations were also conducted on the steel pipe itself and on the galvanized zinc coating. Table 3-1 shows the results of a chemical analysis of the surface of the base steel for each of four types of pipe. These results, which are in comparison to ASTM and DIN specifications, are for the three types of pipe tested (A, K, and U) plus another type produced by manufacturer G. The analyses for the A, K, and U pipes were conducted by using emission spectroscopy at a depth of about 0.02–0.025 in. (0.5–0.6 mm) below the surface of the inside diameter. The data for the G pipe were analyzed by using the entire pipe section. The results from the different methods are not strictly comparable.

Using samples of exposed pipe, it is possible to obtain a reasonably accurate assessment of the base steel. Where the zinc coating is concerned, however, this examination is more limited in its effectiveness, because many of the contaminants of greatest concern in the slab zinc are common in treated water. This problem is illustrated in Table 3-2, which compares analyses of the zinc on the surface of exposed A, K, and U pipes with that for new, unexposed G pipe. For the A, K, and U pipes, these analyses had been conducted on pipe exposed to treated water at the

Table 3-1 Chemical analysis of base steel compared with requirements of ASTM A385 and DIN 17100

Element	ASTM* Specification, *Max.* %	DIN† Specification, *Max.* %	Pipe K, %	Pipe U, %	Pipe A, %	Pipe G, %
Carbon	0.25	—	0.03	0.12	0.14	0.12
Manganese	1.3	—	0.35	0.46	0.42	0.44
Phosphorus	0.05	0.60	0.014	0.029	0.008	0.012
Silicon	0.05	—	0.006	0.02	<0.005	0.19
Sulfur	—	0.50	0.016	0.035	0.020	0.013
Copper	—	—	0.02	0.02	0.03	0.03
Chromium	—	—	0.01	0.01	0.05	0.02
Nickel	—	—	0.02	0.01	0.01	0.01
Molybdenum	—	—	<0.005	<0.005	0.005	0.01
Aluminum	—	—	0.03	0.007	0.005	0.04
Vanadium	—	—	<0.005	<0.005	<0.005	—
Boron	—	—	<0.005	<0.005	<0.005	—

*American Society for Testing and Materials.
†Deutsches Institut für Normung (German Institute of Standardization).

Table 3-2 Chemical analysis of new and exposed zinc coatings compared with the requirements of ASTM B6 and DIN 2444

Element	ASTM* Specification, *Max.* %	DIN† Specification, *Max.* %	Pipe K Exposed, %	Pipe U Exposed, %	Pipe A Exposed, %	Pipe G New, %
Lead	1.4	0.8	0.08	0.10	0.08	0.58
Iron	0.05	—	0.18	0.14	0.16	—
Cadmium	0.2	0.01	<0.005	<0.005	<0.005	0.003
Aluminum	0.01	—	1.0	0.80	0.30	0.042
Calcium	—	—	5.0	3.0	1.0	—
Silicon	—	—	1.5	1.5	1.5	—
Copper	0.02	—	0.40	0.15	0.25	0.022
Magnesium	—	—	0.30	0.30	0.15	—
Manganese	—	—	0.03	0.009	0.009	—
Strontium	—	—	0.02	0.01	0.002	—
Barium	—	—	0.03	0.02	0.02	—
Titanium	—	—	0.008	0.002	0.002	—
Tin	—	—	0.004	0.020	0.004	—
Chromium	—	—	0.002	<0.001	<0.001	0.054
Nickel	—	—	0.001	<0.001	<0.001	—
Bismuth	—	—	0.0008	0.008	0.008	—
Silver	—	—	0.0002	0.001	0.0002	—

*American Society for Testing and Materials.
†Deutsches Institut für Normung (German Institute of Standardization).

Weymouth plant of the Metropolitan Water District of Southern California (MWD) for a period of 14 months. The specimens for analysis were removed from the corroded pipe sample by deforming the pipe and spalling the coatings free. They were then analyzed by emission spectroscopy, again limited to a depth of 0.02–0.025 in. (0.5–0.6 mm). Analysis to this shallow depth shows the quality of the eta layer only. The data seem to indicate the presence of unacceptably high levels of iron, aluminum, and copper. All of these, however, are probably due to the exposure conditions. The presence of copper is due to the fact that copper was being added to the water in the particular corrosion experiments being conducted at MWD to simulate copper-induced pitting. Because the samples were being exposed to raw water, the presence of aluminum was probably due to aluminosilicate clays rather than filter alum used in water treatment. The iron may have resulted from the corrosion of exposed portions of the base metal and upstream conduits. The values shown for the new G pipe are more representative of a new galvanic coating.

Microscopic Examination

Figure 3-5 shows the general nature of the inside coatings on each of the three types of pipe examined. The uniformity of the coating on the K pipe is striking when compared to the others, particularly the U pipe, which has a significantly thicker coating. The somewhat reduced thickness of the K pipe's coating is also evident. The photo also shows the nonuniform nature of the U pipe extends to the microscale. Both of the other coatings appear to be relatively uniform on a microscopic scale.

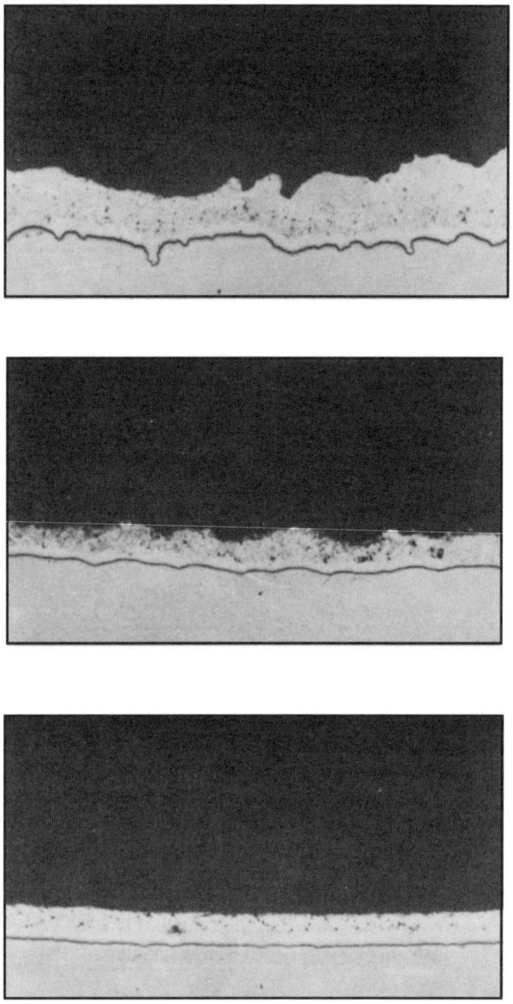

Manufacturer A

Manufacturer K

Manufacturer U

Figure 3-5 Typical view of coating on the inside of new pipe (magnification 40×)

Photomicrographs at higher magnifications were also taken. These are shown in Figures 3-6 and 3-7. Both sets of photomicrographs clearly show all the features outlined in Figure 3-2. There are some differences where the base-metal coating interface is concerned. The coatings shown for manufacturers A and U are generally normal, though the delta layer is dominant for manufacturer A, indicating a tendency toward galvannealing. Though four of the five layers for manufacturer K appear normal, the interface with the base metal does not appear normal, showing a distinct separation zone, particularly those in Figure 3-7. Although the specific cause of this poor interface is not known, it would appear to be a result of manufacturing practice. Such a condition may result in failure during service.

The three makes of pipe also showed significant differences where weld quality was concerned. Figure 3-8 shows a transverse section of a sample of each of the three types of pipe. Once again, the more heterogeneous nature of the coating from manufacturer U is evident. Also evident is the rather distinctive weld seam (see arrow) in pipe K. Both weld-seam protrusion and differences in microstructure are

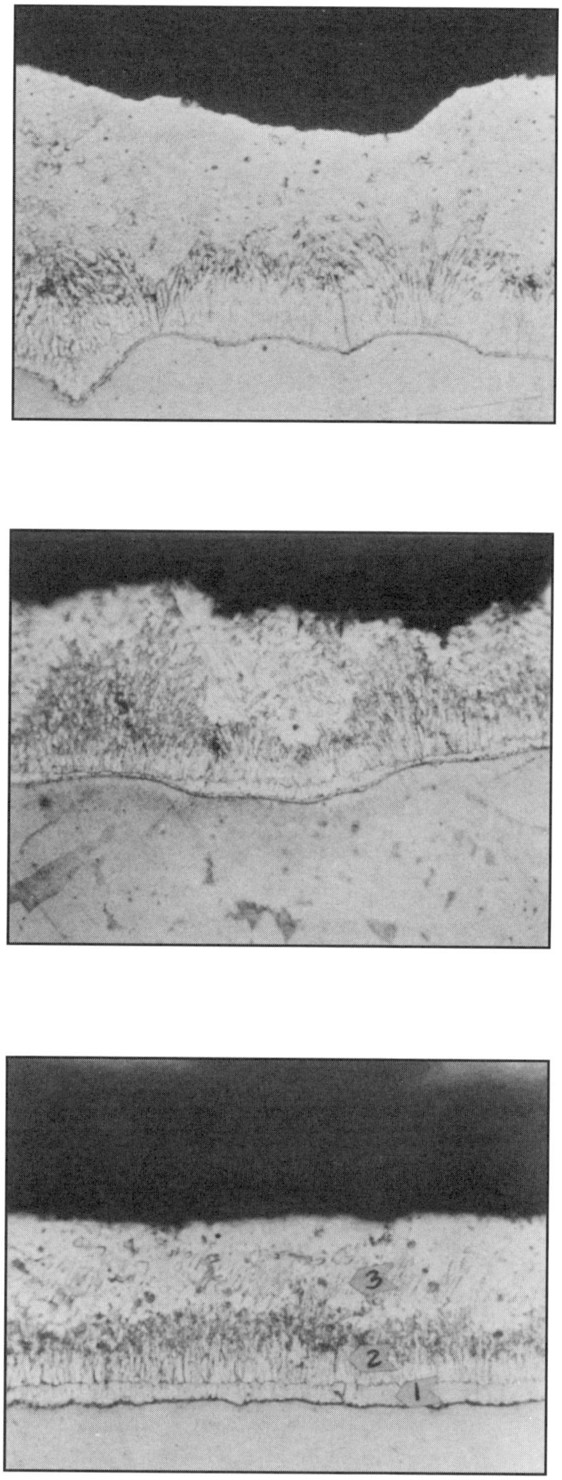

Manufacturer A

Manufacturer K

Manufacturer U

Source: Thomas and Ray (1984).

Figure 3-6 Steel–zinc interface of new pipe under high magnification (magnification 425×)

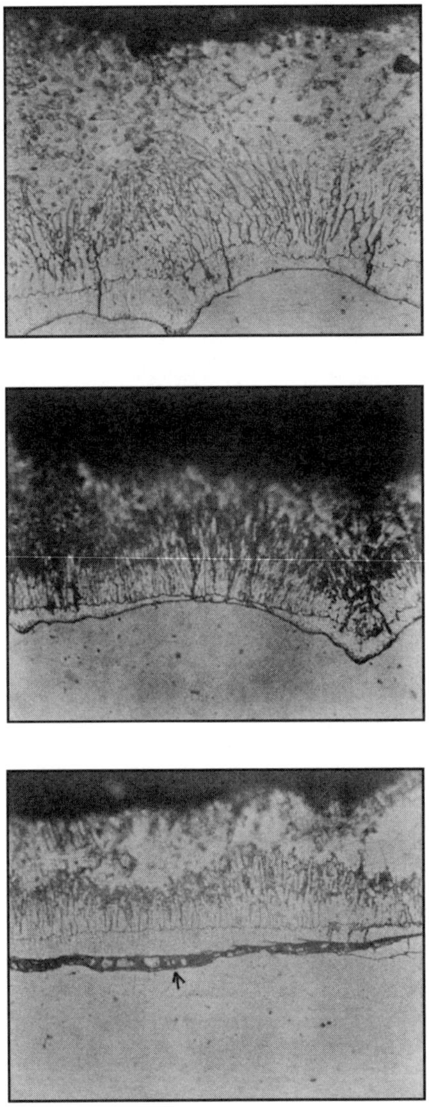

Manufacturer A

Manufacturer K

Manufacturer U

Source: Thomas and Ray (1984).

Figure 3-7 Steel-zinc interface of new pipe under high magnification (magnification 450×)

evident. The local microstructure differences may be an artifact of electrical resistance welding, since the pipe is only heated locally. A barely perceptible weld seam is visible in pipe U, and no seam is visible in pipe A. Figure 3-9 shows the weld zone from a section of new K-type galvanized pipe. The photomicrographs show the nital-etched sample from the internal wall of new, unexposed K pipe. Examination suggests that as it is used on this pipe, electrical resistance welding produces an undesirably large seam that exhibits discontinuities in the microstructure of the weld flash. Both of these characteristics would be expected to exacerbate corrosion problems in the vicinity of the weld seam. This pipe would probably not meet the "no martensite" requirement of ASTM A53.

Photomicrographs such as those just discussed can be used to characterize other manufacturing problems. Two examples of air-blown coatings are shown in

Figure 3-8 Transverse sections of new galvanized steel pipe

Figure 3-10. These examples illustrate the properties of the galvanized layer that is produced in the manufacturing process when air is substituted for steam, which is normally used to blow the excess molten zinc from the pipe surface. The entrained air bubbles, which result from this practice, greatly reduce the stability of the zinc coating, shortening the life of the pipe. It is not difficult to envision that corrosion problems may occur.

MECHANISM OF PROTECTION

In classical discussions of corrosion, the protection of the zinc layer is attributed to its more active potential. That is, zinc, being lower on the oxidation–reduction (redox) scale, will corrode before the more noble iron, providing the latter with cathodic

Magnification
Approximately
50×

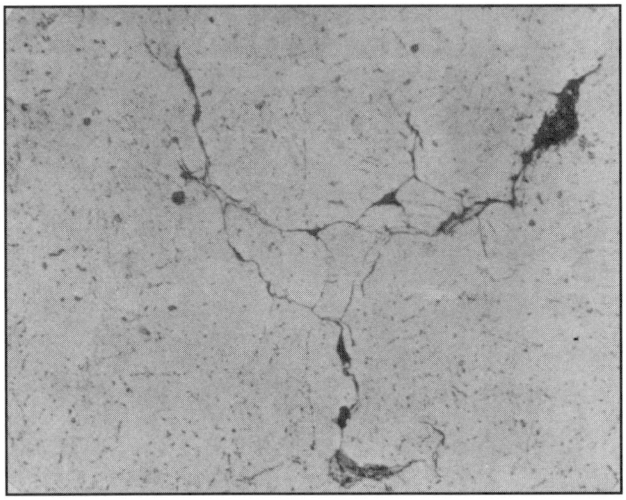

Magnification
Approximately
450×

Figure 3-9 Microstructure of the weld zone (pipe from manufacturer K)

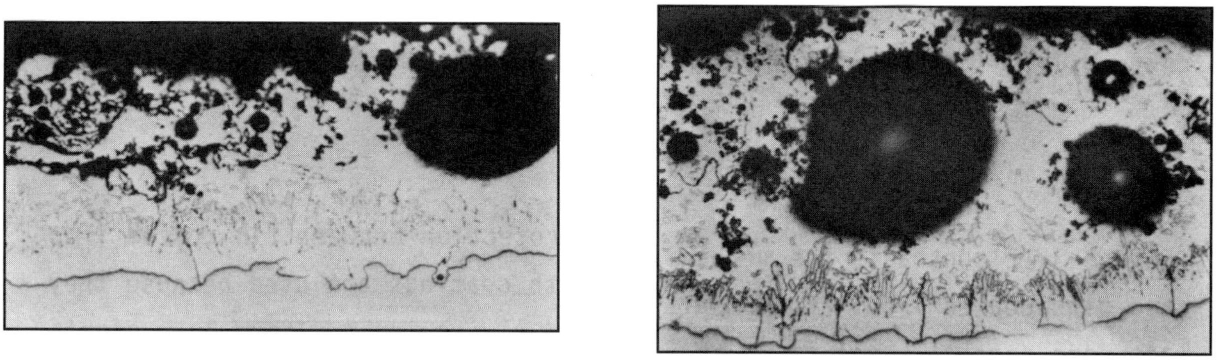

Figure 3-10 Photomicrographs of an air-blown galvanized steel pipe coating (magnification 200×)

protection (see Table 1-1). This is in contrast to tin and chrome coatings where a more noble metal is used to form a barrier between the iron and the bulk solution.

The two coating strategies operate quite differently if a holiday occurs in the protective coating. With tin and chrome, the coating operates as a cathode to the iron and a holiday in the coating results in pitting corrosion. With a galvanized coating, the coating acts as an anode to the iron and a holiday in the coating does not necessarily result in pitting corrosion. These principles are illustrated in Figure 3-11. This basic mechanism is oversimplified, but it does serve to illustrate the important difference between galvanized coatings and coatings made from more noble metals. It is critical that a tin or chrome coating be uniform and continuous, because holidays in the coating can result in serious short-term failures. By contrast, lack of perfection in a galvanized coating, though important, does not necessarily result in serious short-term failures.

Protection From Zinc Oxide

In the case of the noble metals, the passive potential of the metal itself can, in part, explain its corrosion resistance. Considering the relative positions of zinc and iron on the redox scale, it might appear that zinc's more active potential would also cause it to corrode faster than the iron, eliminating the benefit of cathodic protection after a short time. But there is another significant difference between zinc and iron where corrosion is concerned. Under many water quality conditions, the rate of corrosion is retarded more by the surface layer that forms on zinc than by the surface layer that forms on iron. In other words, zinc has a lower corrosion rate than bare iron or mild steel in many environments. As a result, the zinc coating can last a long time. Nevertheless, the probable life of the zinc coating is an important issue in considering the use of galvanized pipe.

Table 3-3 summarizes corrosion rates for black iron pipe and galvanized pipe reported for 12 different waters in the United States. Except for those of the Owens and Colorado rivers (Streicher 1956), these values were all determined by using minor variations of the machined nipple test in ASTM D2688 (1992). Most of these results represent six months of exposures. Values for the Tolt and Cedar rivers are from the Seattle Internal Corrosion Study (Kennedy Engineers 1976), and the remainder are from work by Trussell (1985). Most of these waters are rather soft and would not be ideally suited for the use of galvanized pipe.

Nevertheless, in every case the corrosion rate of the zinc is less than the corrosion rate of the iron. The life of a 3-mil-thick zinc coating projected from the

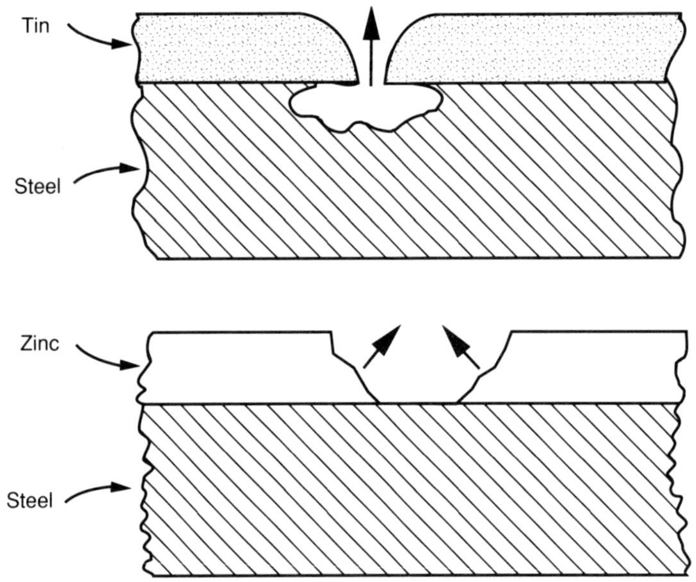

Source: M.G. Fontana and N.D. Greene. Corrosion Engineering. *Copyright 1978. McGraw-Hill, Inc. Reproduced with permission.*

Figure 3-11 Classic comparison of protection of iron and zinc when a holiday is present

corrosion rates shown in Table 3-3 would range from 60 years for the Potomac River to slightly less than eight months for the Tolt River. These data confirm that the zinc layer will protect the iron pipe in most cases. At the same time, it is worth noting that the life of the zinc coating would be less than five years for most of the water qualities shown in the table. This also corresponds to German experience.

Protection of Uncovered Iron

There is also a limit to the amount of exposed iron that can be cathodically protected by a given amount of zinc. Assuming that zinc maintains a more active potential, its ability to protect iron depends not only on its low corrosion rate but also on the relative areas of exposed zinc and iron and on their relative corrosion rates in the particular water. The amount of current required for cathodic protection depends on the relative corrosion rates of zinc and iron in the medium in question. Based on the results in Table 3-3, a zinc coverage of at least 25 percent would be necessary for most of the waters shown. That is, once the zinc layer is corroded enough so that it covers less than 25 percent of the pipe surface, the pipe would corrode much as if it were a typical iron pipe.

Work conducted in the early 1980s by Bächle et al. (1981) demonstrates this principle. The change in composition of the scale layer on a galvanized pipe over a two-year period of exposure to a water with a pH of 7.2 and a velocity of about 1.5 m/s is shown in Figure 3-12A. Early in the pipe's life, zinc compounds were the principal component, but gradually the iron component increased until, after one and one-half years or so, the zinc was depleted and the iron became dominant. The level of calcium in the scale did not vary a great deal. The peak level of zinc in the scale corresponded to the point at which the original zinc coating was completely corroded away (compare Figures 3-12A and 3-12B). Bächle and co-workers concluded that the zinc products then provided additional protection against red water while a protective iron-based coating was formed.

Figure 3-12B also demonstrates that, as would be expected, once the zinc layer is corroded away, galvanized steel pipe behaves in the same manner as black iron pipe with no galvanizing. In the example shown in the figure, two pipes were exposed.

Table 3-3 Relative corrosion rates of galvanized iron pipe and black iron pipe in several US water supplies: A comparison with water quality and common corrosion indices

	Water Quality							Indices			Corrosion			
Water Supply	pH	Calcium, *mg* Ca^{2+}/L	Alkalinity, *mg* $CaCO_3/L$	Chloride, *mg* Cl^-/L	Total Dissolved Solids, *mg/L*	Total Organic Carbon, *mg C/L*	Phosphate, *mg* PO_4^{3-}/L	$CaCO_3$ Saturation Index	Larson's Ratio, *meq/meq*	Buffer Intensity, *meq/pH unit*	Galvanized Iron, *mil/year*	Black Iron, *mil/year*	Relative Corrosion Rate*	Life of Zinc, *years*
Contra Costa Canal	8.4	17	70	52	215	3.5	0.07	0.1	1.1	0.07	1.6	5.7	4.6	1.9
Owens River	8.3	25	116	15	255	3	n.d.†	0.5	0.2	0.11	2.5	6.4	3.3	1.2
Colorado River	8.2	81	125	92	697	3	n.d.	0.9	1	0.13	2.2	7	4.1	1.4
Tolt River	6.1	3.4	3	2.9	30	3	<0.01	−4.2	1.4	0.09	4.6	13	3.6	0.7
Cedar River	7.3	6	13	4.2	40	1.5	0.01	−2.1	0.4	0.07	4.2	13	4.0	0.7
Bull Run River	7	1.8	13	2.2	23	1.2	<0.01	−2.9	0.2	0.12	0.6	3.7	8.0	5.0
Potomac River	7.8	52	61	47	246	1.2	<0.01	0	1.1	0.11	0.05	0.12	3.1	60.0
Mokulumne River	7.2	3.2	13	1	30	1.7	<0.002	−2.5	0.11	0.08	1.3	2.5	2.5	2.3
American River	6.9	5	19	2	39	1.6	<0.01	−2.4	0.15	0.2	1.4	4.8	4.4	2.1
Sacramento River	7.3	10	48	5	86	2.1	0.08	−1.3	0.15	0.24	2.7	8.5	4.1	1.1
Clifton Court	7.4	16	53	39	190	4.5	0.07	−1	1	0.22	2.5	7.3	3.8	1.2
Indian Slough	7.4	19	70	58	255	4.5	0.07	−0.8	1.2	0.29	3.3	6.2	2.4	0.9

*Ratios taken after conversion to $\mu A/cm^2$.

†n.d. = not determined.

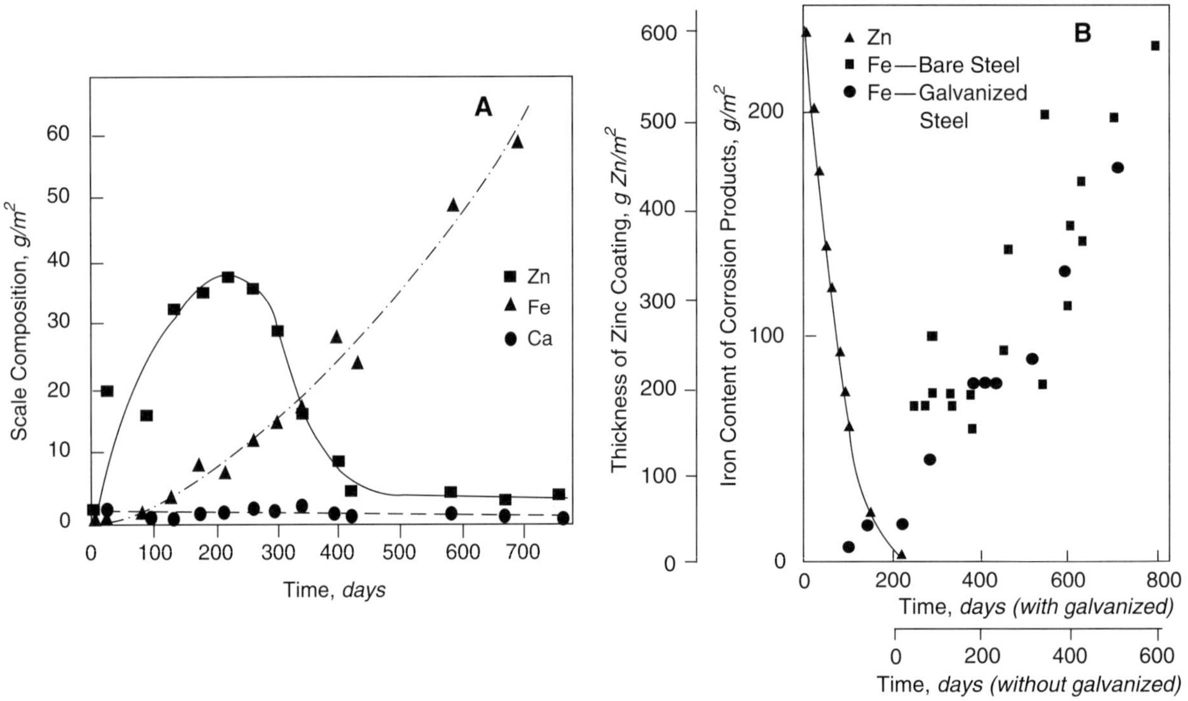

Source: Bächle et al. (1981).

Figure 3-12 Rapid loss of zinc in an aggressive water with offset time scales: (A) Rapid decline in thickness of zinc layer on galvanized specimen and (B) Comparison of iron content of corrosion products on galvanized steel and bare steel pipes

One pipe had a galvanized layer and the other did not. For the galvanized pipe, about 200 days were required to corrode away the zinc layer. Superimposing a plot of the scale's iron content on results for both pipes, so that the beginning of the exposure of the black iron pipe corresponds to the time when the zinc layer on the galvanized pipe was eliminated, demonstrates that the scales on both pipes have a similar character.

Long-Term Overlook

This discussion pertains to the rate of corrosion of the galvanized pipe itself, that is, the rate at which the pipe cross section is reduced by the corrosion process. Where the consumer's drinking water quality is concerned, however, the transfer of corrosion products into the solution is also of principal importance. When iron is used, stagnant periods long enough to eliminate oxygen generally result in red-water problems (see discussion of Kuch-type corrosion in chapter 2). In large mains, the frequent passage of fresh water reduces this problem and also allows for the formation of protective scales on the surface of the iron, making it suitable for service under these conditions. In consumer plumbing, however, long stagnant periods are unavoidable and a much longer time is required to establish a protective layer. Zinc is not prone to Kuch-type corrosion. As a result, it does not present as much of a consumer problem during periods of stagnation with new plumbing. Even though the zinc does not last for the full life of the pipe, by the time the zinc layer has corroded away, a protective layer is usually established on the remaining iron and service is satisfactory.

CHEMISTRY OF ZINC

To understand the influence water quality has on the behavior of galvanized pipe, it is important to understand the behavior of the oxidized zinc layer. Zinc, along with cadmium and mercury, is a group II metal. It has only two stable oxidation states, Zn and Zn(II). It is not very abundant in nature compared to metals like aluminum but has long been recognized because it is easily refined. Zinc is found in nature as zinc blend, ZnS; smithsonite, $ZnCO_3$; willemite, Zn_2SiO_4; zincite, ZnO; and other minerals. Small amounts of cadmium are often found in zinc ores. Zn(II) forms chloride complexes, but not as strongly as Fe(III) or Pb(II), and it forms ammonia complexes but not as strongly as Cu(II). Zn(II) also forms sulfate complexes (probably $ZnSO_4^0$ and $Zn(SO_4)^{2-}$) that are not very strong; orthophosphate complexes that are apparently not very strong; and carbonate complexes ($ZnHCO_3^-$, $ZnCO_3^0$, and $Zn(CO_3)^{2-}$), some of which are reasonably strong, though not as strong as those of copper.

Zn(II) Solids

Zinc precipitates from bulk solution to form an amorphous hydroxide that eventually undergoes further hydrolysis to form zinc oxide or zincite. Baes and Mesmer (1976) also reported six additional crystalline modifications of zinc hydroxide that can form under different circumstances. The solubility of these lies between that of the oxide and the amorphous hydroxide. In 1967, Schindler reported on the solubility of the oxide, and his estimate of the solubility product has been generally accepted since that time. In reviewing zinc solubility, Baes and Mesmer (1976) accepted Schindler's constant for zincite and provided an estimate of the solubility of the amorphous hydroxide, also based on Schindler's earlier work (Schindler, Althaus, and Feitknecht 1964). These solubility data, along with those reported in a recent updated review (Morel and Hering 1993), are shown in Table 3-4.

There are also two solid phases of zinc carbonate of interest, zinc carbonate itself (smithsonite) and basic zinc carbonate (hydrozincite). The solubility of both of these was also reported by Schindler (1967), but neither has been widely studied since that time. These values are also reported in Table 3-4.

It is important to understand which Zn(II) solids are likely to form on the surface of galvanized pipe. The analysis by Stumm and Morgan (1981) suggests that zinc carbonate is the least soluble species at pH values below 8, hydrozincite at pH values

Table 3-4 Zinc solubility products

Reaction	Solubility Product, $log_{10}(K_{so25°C})$	Reference
$ZnO_{(s)} + 2H^+ = Zn^{2+} + 2H_2O$	11.2	Schindler 1967
	11.2	Baes and Mesmer 1976
	11.2	Morel and Hering 1993
	11.2	This analysis*
$Zn(OH)_{2(c)} + 2H^+ = Zn^{2+} + 2H_2O$	12.5	Baes and Mesmer 1976
	12.5	Morel and Hering 1993
	12.5	This analysis*
$5Zn^{2+} + 6H_2O + 2CO_3^{2-} = Zn_5(CO_3)_2(OH)_{6(s)} + 6H^+$	2.54	Schindler 1967
$Zn^{2+} + CO_3^{2-} = ZnCO_{3(s)}$	10.2	Schindler 1967

*The value used in this book.

of 8 to 12, and the hydroxide at higher pH values. A recent study by Paulson, Benjamin, and Ferguson (1989) used X-ray diffraction to demonstrate that hydrozincite is the stable species in two-day-old precipitate from zinc carbonate solutions for pH values below 10. Poorly crystalline hydrozincite was observed at pH values above 8 and the oxide at pH values above 10.5. The work of Pisigan and Singley (1985) (discussed in more detail later in this section) also shows hydrozincite and zinc oxide on the surface of corroding galvanized pipe (see Figure 3-16 in that section). Although stability calculations suggest that zinc carbonate may appear at pH values of less than 7 or 8, no direct evidence of this phenomena is available, and Paulson's work suggests that hydrozincite is the metastable species under these conditions as well. Generally, it would appear that zinc oxide and hydrozincite are the long-term stable species on the surface of galvanized pipe and that, under some conditions, a metastable amorphous hydroxide may also be important.

Hydrolysis of Zn(II)

Schindler's 1967 discussion is still the fundamental basis for our understanding of zinc hydrolysis. Schindler's work summarized data on the stability of the first, third, and fourth hydrolysis products of zinc as well as for zinc oxide, or zincite. In 1976, Baes and Mesmer also conducted a critical review of this thermodynamic data, confirming Schindler's estimate for the first hydrolysis constant and making slight revisions to constants for the third and fourth hydrolysis steps and the solubility of the oxide. Baes and Mesmer also proposed constants for the second hydrolysis step and for the solubility of the freshly precipitated hydroxide and introduced somewhat more limited evidence for the polynuclear hydrolysis products: $Zn_2(OH)_3^+$ and $Zn_2(OH)_6^{2-}$. These latter complexes would not be of significance under drinking water conditions. More recently, Morel and Hering (1993) included zinc hydrolysis in their critical review and proposed essentially no changes. Table 3-5 summarizes these results, noting the values that were chosen for the purposes of calculations in this chapter.

Zinc and Carbonates

A number of zinc carbonate complexes have also been proposed in various studies. Table 3-6 summarizes the results of those efforts. There is a significant amount of literature on the simple ion pairs $ZnHCO_3^+$ and $ZnCO_3^0$ and there are three reports of a two-ligand carbonate complex $Zn(CO_3)_2^{2-}$. Ferri et al. (1985), working with high concentrations of carbonate, also report more complicated zinc carbonate complexes, but these are not important under conditions normally observed in drinking water. Five values for the stability constant for the bicarbonate ion pair appear in the literature. Most of them are reasonably close; the value from Ferri et al. being about half the others. The choice for the analysis in this discussion is the average of the values excluding Ferri's. There are also five values for the carbonate ion pair. This ion pair is significant and, with the exception of Mattigod and Sposito (1977), the reported values for the stability constant are in pretty good agreement. The choice for the analysis in this discussion is the median value of the values listed, excluding Mattigod and Sposito. There are three reported values for the two-ligand carbonate complex, and two of them are in fairly good agreement. The choice for the analysis in this discussion is the average of the two most recently reported values rounded to one significant figure.

Table 3-5 Zinc hydrolysis/oxide constants

Reaction	Hydrolysis Constant, $\log_{10}[K_{25\,°C}]$	Reference
$Zn^{2+} + H_2O = Zn(OH)^+ + H^+$	–8.96	Schindler 1967
	–8.96	Baes and Mesmer 1976
	–9.0	Morel and Hering 1993
	–9.0	This analysis*
$Zn^{2+} + 2H_2O = Zn(OH)_2^0 + 2H^+$	–16.9	Baes and Mesmer 1976
	–16.9	Morel and Hering 1993
	–16.9	This analysis*
$Zn^{2+} + 3H_2O = Zn(OH)_3^- + 3H^+$	–28.1	Schindler 1967
	–28.4	Baes and Mesmer 1976
	–28.4	Morel and Hering 1993
	–28.4	This analysis*
$Zn^{2+} + 4H_2O = Zn(OH)_4^{2-} + 4H^+$	–40.9	Schindler 1967
	–41.2	Baes and Mesmer 1976
	–41.2	Morel and Hering 1993
	–41.2	This analysis*

*The value used in this book.

Table 3-6 Zinc carbonate stability constants

Reaction	Stability Constant, $\log_{10}[K_{25\,°C}]$	Reference
$Zn^{2+} + HCO_3^- = ZnHCO_3^+$	2.1	Zirino and Yamamoto 1972
	1.4	Ryan and Bauman 1978
	0.8	Ferri et al. 1985
	2.2	Fouillac and Criaud 1984
	1.6	Stanley and Byrne 1990
	1.8	This analysis*
$Zn^{2+} + CO_3^{2-} = ZnCO_3^0$	5.3	Zirino and Yamamoto 1972
	4.7	Bilinski and Stumm 1976
	6.6	Mattigod and Sposito 1977
	4.8	Fouillac and Criaud 1984
	4.8	Stanley and Byrne 1990
	4.8	This analysis*
$Zn^{2+} + 2CO_3^{2-} = Zn(CO_3)_2^{2-}$	9.6	Mattigod and Sposito 1977
	7.3	Stanley and Byrne 1990
	6.8	Ferri et al. 1985
	7	This analysis*

*The value used in this book.

Dominant Species of Soluble Zn(II)

Figure 3-13 illustrates the distribution of zinc species as a function of pH and total carbonate concentration (C_t) values of 0.1, 1, and 10 m*M*/L. It would appear that zinc chemistry is adequately modeled considering only Zn^{2+}, $Zn(OH)_2^0$, $Zn(OH)^+$, and $ZnCO_3^0$. The first hydrolysis product, $Zn(OH)^+$, is of little significance, reaching a maximum of 10 percent of the soluble zinc at a pH of approximately 8.5. The bicarbonate ion pair, $ZnHCO_3^+$, and the two-ligand carbonate complex, $Zn(CO_3)_2^{2-}$, only seem significant at the highest C_t levels. Subsequent calculations will be based primarily on the chemistry of these eight species.

Zinc Solubility

Using the information just outlined, Zn(II) solubility diagrams have been constructed and are shown in Figure 3-14. Again, C_t values of 0.1, 1, and 10 m*M*/L were used. Only the oxide, the hydroxide, and the hydrozincite were considered to be the solid species in these calculations. The solid that corresponds to the lowest equilibrium level of zinc ion is the most stable under the conditions specified. This is illustrated on the left half of the diagram. Generally, hydrozincite is stable at low pH and the oxide at elevated pH values. In waters with C_t = 0.1 m*M*/L, hydrozincite is stable at pH 7.5 and below and the oxide above that. When C_t = 1 m*M*/L, the breakpoint moves a little above pH 8. For very high alkalinity waters, hydrozincite dominates over the entire range of normal pH values.

The diagrams in Figures 3-14A, B, and C do not illustrate the effects of complex formation. In Figures 3-14D, E, and F, the total solubility of Zn(II) is shown, along with the concentration of each of the six major species, Zn^{2+}, $Zn(OH)^+$, $Zn(OH)_2^0$, $ZnCO_3^0$, $ZnHCO_3^+$, and $Zn(CO_3)_2^{2-}$. In the pH range shown, $Zn(OH)^+$, $ZnHCO_3^+$, and $Zn(CO_3)_2^{2-}$ are of little significance, and Zn^{2+}, $Zn(OH)_2^0$, and $ZnCO_3^0$ seem to be the most important. At high C_t levels, the formation of $ZnCO_3^0$ significantly increases solubility.

Zinc Redox Behavior

It is useful to view the equilibrium chemistry of the zinc surface in terms of its redox behavior, as well as in terms of solubility alone. The Pourbaix diagram (Pourbaix 1949, 1966) is often used for this purpose. Figure 3-15A shows Pourbaix's version of the diagram, using a slightly different set of constants and giving consideration only to oxide species. The diagram suggests that zinc will be unstable at low pH values as a result of the increased solubility of the hydroxide. The boundary between zinc hydroxide and zinc ion is based on a solubility of 0.0001*M* for zinc. The $Zn(OH)_3^-$ ion is not important in the pH range of interest in domestic water supply. The diagram also shows the hydroxide as the stable solid phase for pH values slightly over 7. The boundaries for ZnO are not clearly shown.

A revision of Pourbaix's diagram (Figure 3-15B) after considering the solubility data described earlier (C_t = 0.001*M*) shows hydrozincite as the dominant phase for pH values of 7 to 8.5 and the hydroxide as the dominant phase for pH values above 8.5. It is assumed that the oxide does not form and that the hydroxide is the only stable phase, and the transition occurs at a slightly higher pH. Thus, thermodynamic data currently available suggest that hydrozincite is the most stable solid phase over the pH range of greatest interest. Even this form becomes thermodynamically unstable at pH values below 7. Further, if oxide formation does not take place, and there is some evidence of this, then the zinc surface is also unstable at

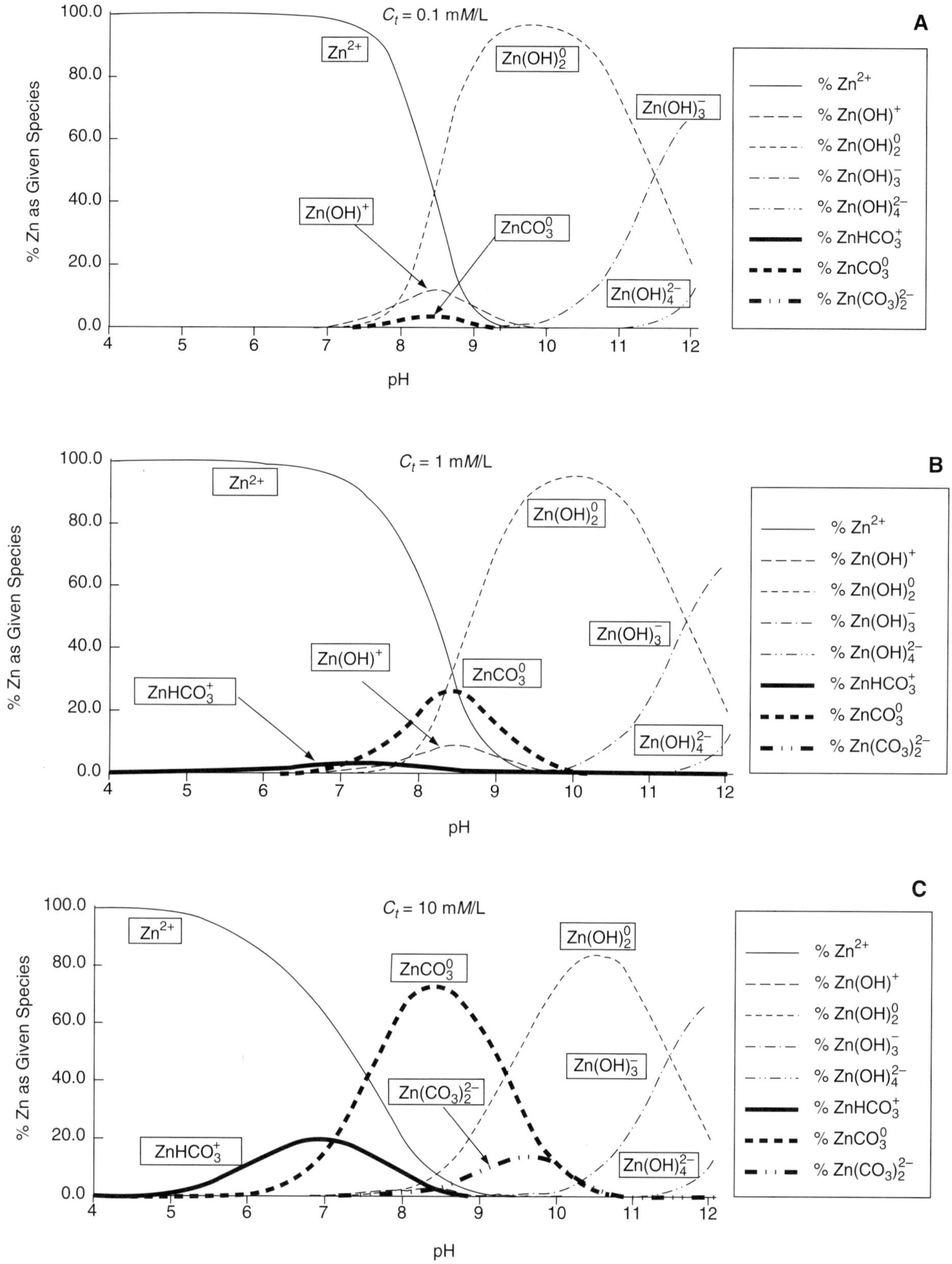

Figure 3-13 Distribution of soluble zinc species as a function of pH and total carbonate concentration: (A) C_t = 0.1 mM/L; (B) C_t = 1 mM/L; and (C) C_t = 10 mM/L

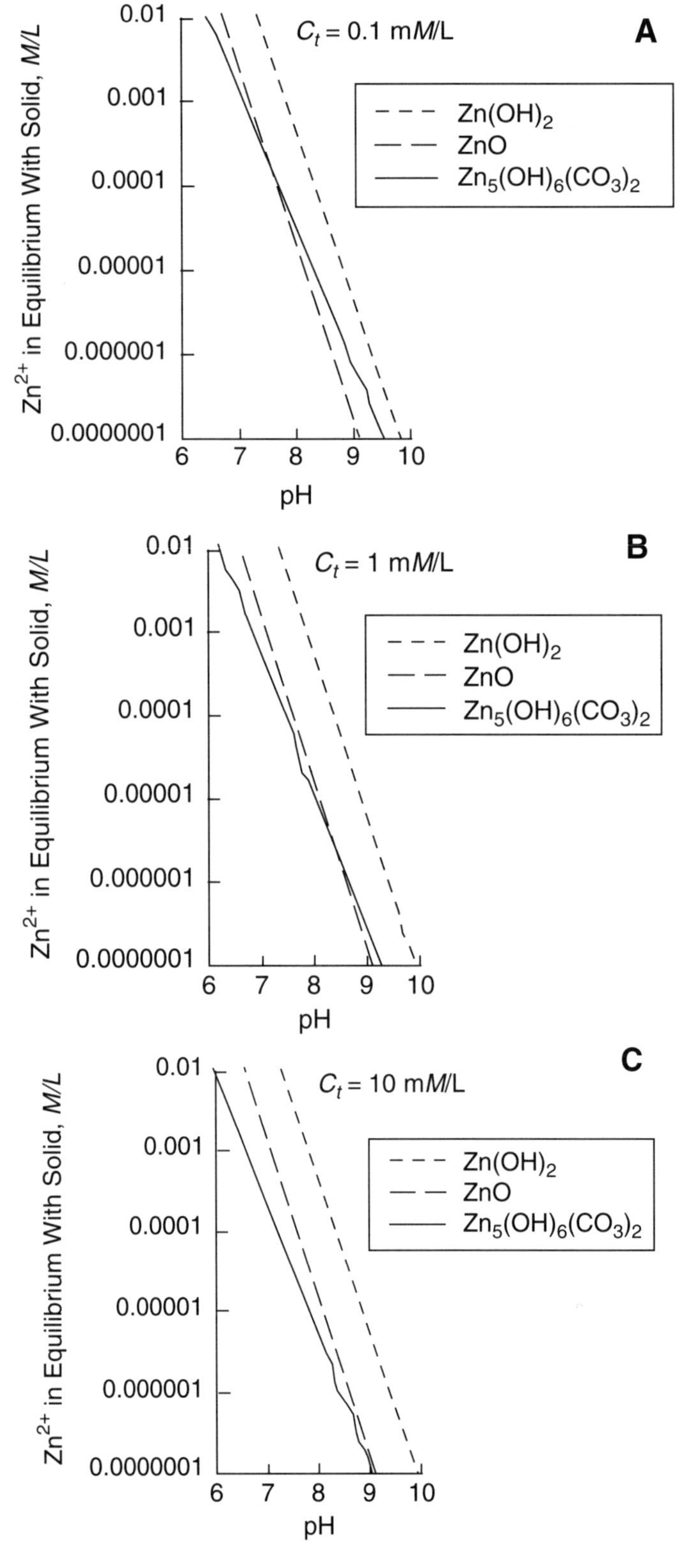

Figure 3-14 Zinc solubility as a function of pH and C_t of three solids: ZnO, $Zn(OH)_2$, and $Zn_5(OH)_6(CO_3)_2$. (A, B, and C show the solubility of Zn^{2+} ion in equilibrium with each solid for a range of pH values and C_t values of 0.1, 1.0 and 10 mM/L. D, E, and F show overall zinc solubility and major soluble zinc species under the same conditions). (continued next page)

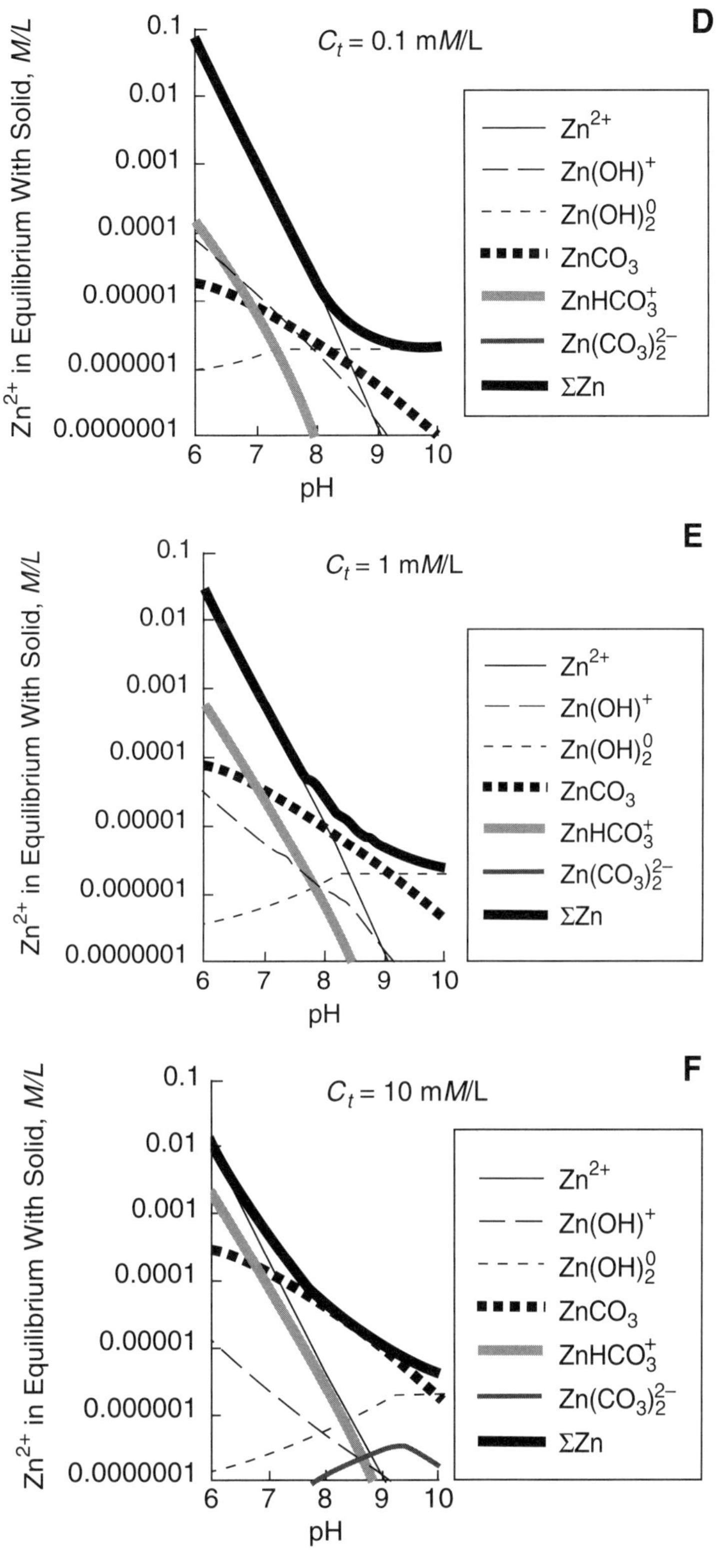

Figure 3-14 Zinc solubility as a function of pH and C_t of three solids: ZnO, $Zn(OH)_2$, and $Zn_5(OH)_6(CO_3)_2$ (continued). (A, B, and C show the solubility of Zn^{2+} ion in equilibrium with each solid for a range of pH values and C_t values of 0.1, 1.0 and 10 mM/L. D, E, and F show overall zinc solubility and major soluble zinc species under the same conditions).

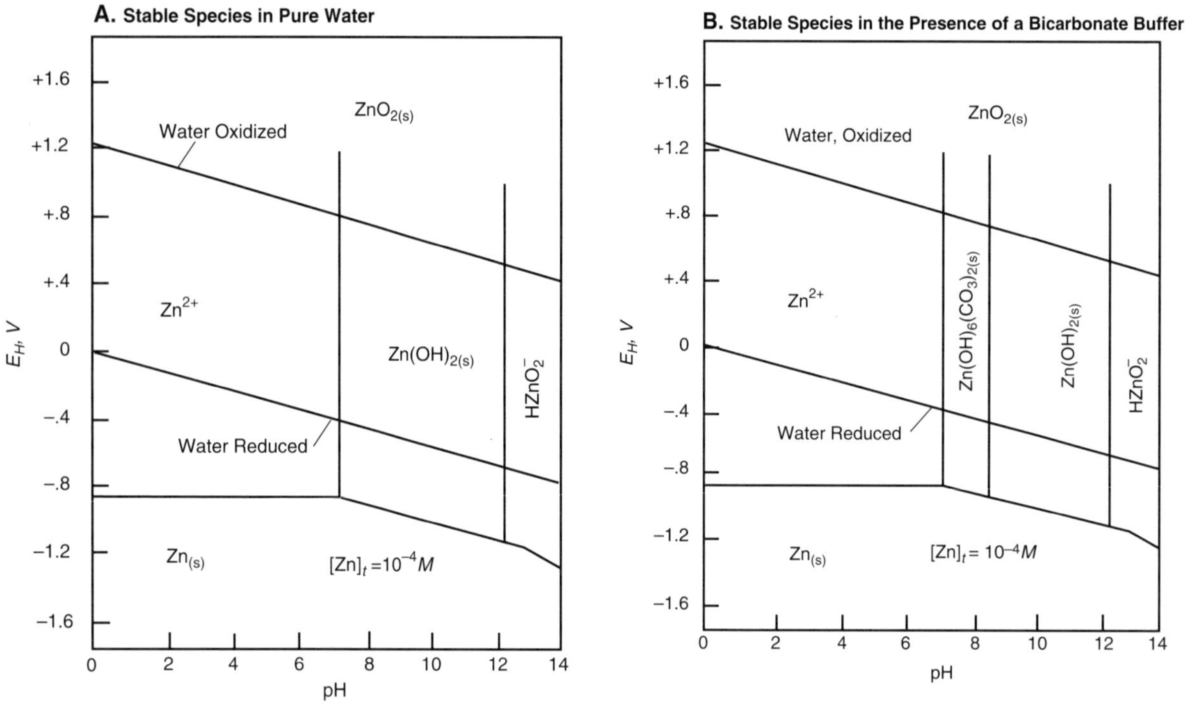

Source: Pourbaix (1966).

Figure 3-15 Two Pourbaix diagrams of zinc

pH values above 8.5 to 9. Field experience seems to show that the behavior of galvanized pipe is most satisfactory in near neutral to slightly alkaline waters between pH 7.0 and 8.5.

Have any direct observations been made of the scale materials on a galvanized pipe exposed to water that could resolve this issue? The International Lead Zinc Research Organization conducted studies in the late 1960s (ILZRO 1967) that examined the nature of the corroding zinc surface in distilled water. Using X-ray diffraction and other means, the researchers concluded that the scale on the surface of the corroding pipe consisted of ZnO and $Zn(OH)_2$. Some evidence suggested that the ZnO was the initial corrosion product, with the $Zn(OH)_2$ being formed later. The ratio of ZnO to $Zn(OH)_2$ in the scale was found to be a function of the temperature and the length of exposure. These results generally agree with the earlier equilibrium discussion.

Pisigan and Singley (1985) prepared X-ray diffractograms of the corrosion products on the surface of galvanized steel pipes after 40 days of exposure in a recirculating loop. Figure 3-16A gives results for pipe exposed to distilled water; Figure 3-16B gives results for a pipe exposed to a water at pH 7.7 with 150 mg/L of alkalinity and 100 mg/L $CaCO_3$. These diffractograms also suggest that most of the equilibria described above are appropriate. In the zinc-distilled water system, eta zinc hydroxide (η-$Zn(OH)_2$) and zincite (ZnO) are the dominant species. This is much as one would expect from redox data and it confirms the work of ILZRO. The diffractogram in Figure 3-16B shows the presence of hydrozincite ($Zn_5(CO_3)_2(OH)_6$), aragonite

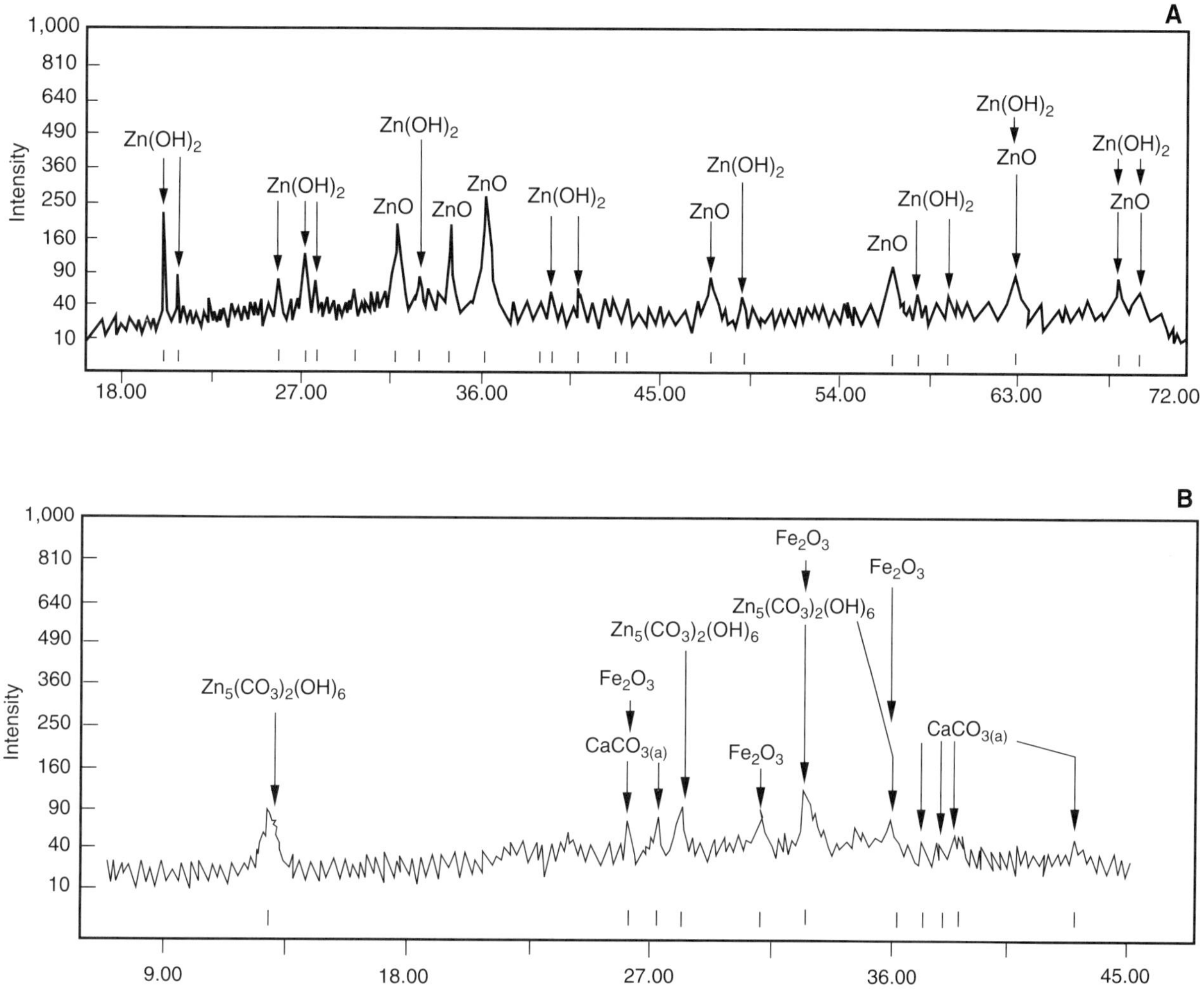

Source: Pisigan and Singley (1985).

Figure 3-16 X-ray diffraction of corrosion products: Diffractograms conducted on pipe exposed to (A) pure, deionized water and (B) water at pH = 7.7, alkalinity = 150 mg/L, and hardness = 100 mg $CaCO_3$/L

($CaCO_3$), and hematite (Fe_2O_3). The presence of hydrozincite also confirms the analysis described earlier. Under the conditions of exposure, calcite would be the most stable form of calcium carbonate, but aragonite was observed instead.

Others have also reported aragonite under similar circumstances (ILZRO 1967), suggesting that kinetic mechanisms on the zinc surface may favor the formation of this metastable compound. The fact that calcium carbonate deposited in the water regardless of the Langelier index (LI) value (i.e., negative or positive) suggests that chemical conditions on the microscale near the pipe surface may be equally or more important than those of the bulk solution. Trussell, Russell, and Thomas (1977) proposed a buffer intensity index for the supersaturation ratio as a way to assess the potential for this kind of activity.

MODES OF FAILURE

Four basic processes stimulate interest in the corrosion of galvanized pipe, and although they are related indirectly, each process requires the development of different information. The processes are: (1) the general, uniform corrosion of the pipe; (2) the release of metals into solution; (3) pitting; and (4) tuberculation. Galvanic couples can also play a role in the deterioration of galvanized plumbing systems; however, most of these problems can be controlled by proper design of the plumbing system.

General, uniform corrosion is the parameter that is measured by most traditional corrosion testing techniques, such as those specified in ASTM D2688 (1994). Although mechanical failure due to uniform corrosion is rare, it is important because under normal conditions the uniform corrosion rate is broadly indicative of the expected life of the conduit, the potential for corrosion by-products, the risk of tuberculation, and the likelihood of pitting. When the general or uniform corrosion rate of galvanized pipe is high, the pipe will often fail by rapid loss of the zinc layer and subsequent tuberculation.

The release of metals into solution is important because these metal concentrations may have a health impact. Concern about these corrosion by-products has generated much of the recent interest in corrosion of consumer plumbing. Zinc and iron are among the corrosion by-products most commonly found in consumer plumbing, and aesthetic issues usually are the principal concern, especially where red ferric corrosion products are involved. In addition, both lead and cadmium are found in the slab zinc used in the galvanizing process and have also been reported as corrosion by-products in low-alkalinity waters. Also chromate has been used in the finishing step of the galvanizing process.

Pitting is the rapid perforation of pipe in a period of time far less than its normal life (often 10-fold less). Pitting destroys the pipe's effectiveness as a water conduit long before the end of the pipe's normal economic life. Also of concern are the leaks that result when pitting occurs. Not only must the plumbing be replaced, but the leaks also result in the destruction of other property.

Tuberculation is the development of rounded bulges or protuberances on the pipe surface. When tuberculation occurs on the inside of the pipe, the resulting occlusion reduces the pipe's carrying capacity. Tuberculation is common in most ferrous plumbing materials and is a direct result of the chemistry of iron and the character of the corrosion by-products that it produces. Like pitting, tuberculation limits the economic life of the pipe, though it normally does not result in destruction as rapidly as is observed with pitting. Tubercles are usually pits, but the pits formed are shallow and broad. As a result, tuberculated pipe usually fails due to occlusion before perforation of the pipe wall occurs.

Galvanic couples can impact the deterioration of galvanized pipe. The typical potential between zinc and the various forms of copper and brass can be as great as 700 to 800 mV. Galvanic couples are particularly important when poor design results in a large area of a more noble material (such as brass) in contact with a much smaller area of a more active material (such as zinc in a galvanized coating). It is less important when a brass fixture is attached to a long galvanized pipe.

Each of these modes of corrosion will be discussed in the following paragraphs.

General Corrosion

In the past, relatively little work was done to study the impact of water quality parameters on the general rate of corrosion of galvanized pipe. More recently, however, research has been conducted at the Wasserwerk Rheinau in Mannheim, Germany (EBI and RHE 1978), at the Landeswasserversorgung in Stuttgart, Germany (Werner, Wurster, and Sontheimer 1973; Werner 1976, 1984), and at the University of Florida at Gainesville (Pisigan and Singley 1985).

Studies at Rheinau. At Rheinau, research concentrated on the character of the scale on the corroding surface and on the influence of pH and velocity on the corrosion rate. Tests were conducted with waters of three different qualities.

The waters were passed through test coupons constructed from ¾-in. (20-mm) galvanized pipe for a two-year period. The composition of the scale layers on the inside surface of the pipe as the exposure progressed was measured. Figure 3-12, discussed earlier, illustrates the general change in composition of the layer. Figures 3-17A and B demonstrate the impact of water quality and velocity on the corrosion of the zinc layer. Figure 3-17A shows that at a velocity of 0.5 m/s (1.6 ft/s), the zinc layer had a greater life in the waters with higher pH values in spite of their softer characteristics. Figure 3-17B shows that for the hard water, increasing the velocity from 0.5 to 1.0 m/s (1.6 to 3.2 ft/s) had little impact, but that a further increase to 1.5 m/s (4.8 ft/s) resulted in substantially more rapid corrosion. The report by Bächle et al. (1981) included a summary of data gathered on the corrosion rates for 26 different water qualities (Figure 3-18). These data were collected with pipe rigs similar to those used at Rheinau and at a permanent flow rate of 0.5 m/s but at pH values ranging from 7.0 to 8.1. In the pH range of 7.0 to 8.0, these data seem to indicate a strong relationship between the corrosion rate of zinc and the pH. Assuming that an average zinc coating is 490 g/m^2 (1.6 oz/ft^2), these data would seem to indicate that under permanent flow conditions, the coating would have a life of between 3 and 8 months at a pH of 7.0 and between 1.4 and 4.5 years at a pH of 8.0.

Studies at Landeswasserversorgung. Using the oxygen-based corrosion rate apparatus developed in the mid-1970s by the laboratory of the Landeswasserversorgung and described by Kuch, Sontheimer, and Wagner (1983), Werner (1984) studied the influence of several parameters on corrosion rate. As a result of this study, the authors developed a model of the corrosion process for galvanized pipe.

Figure 3-19A illustrates the dependence of the corrosion rate of the zinc layer on the flow velocity. The corrosion rate of a fresh zinc surface, which has had no time to develop a coating, is strongly dependent on velocity. Once sufficient time has passed for the formation of a normal coating, this dependence is significantly reduced. Tests with the zinc–iron alloy layer (zeta layer) showed corrosion depended very little on velocity.

Figure 3-19B gives a slightly different perspective to the same problem. Here, two phases to the corrosion process are evident: (1) an early phase of coating formation and decelerating corrosion, and (2) a later phase with a stable coating and constant corrosion rate. The damage that occurs during the early stages of coating formation is dependent on velocity; higher velocities require longer periods for coating formation. Once the coating is stabilized, the corrosion rate also stabilizes until the zinc layer is removed. Even after the coating is relatively stable, however, increases in velocity result in increased corrosion rates. This effect is attributed to the thickness of the diffusion boundary layer, including the coating itself.

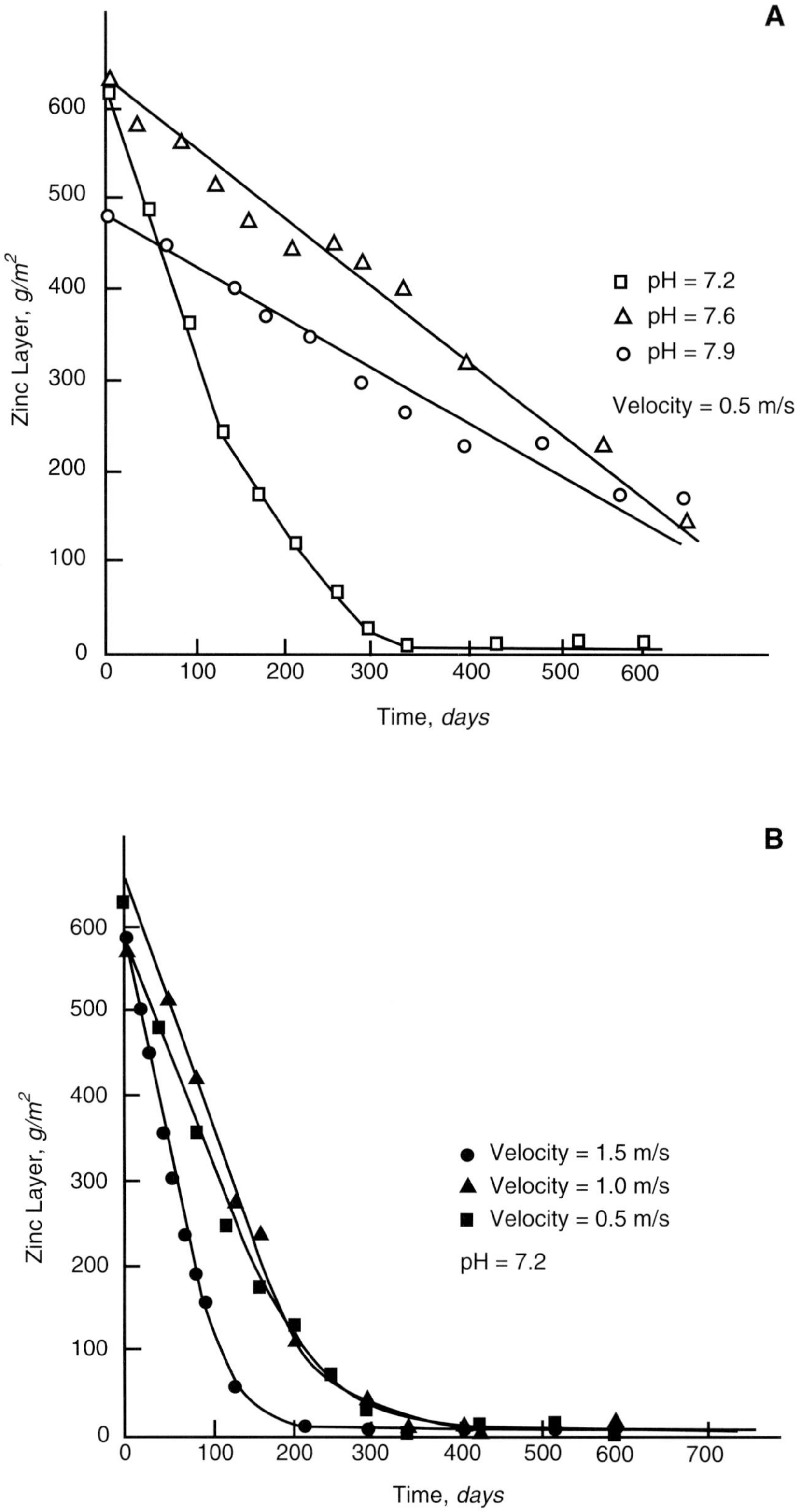

Figure 3-17 Rheinau studies: Influence of pH (A) and velocity (B) on corrosion rate of the zinc layer

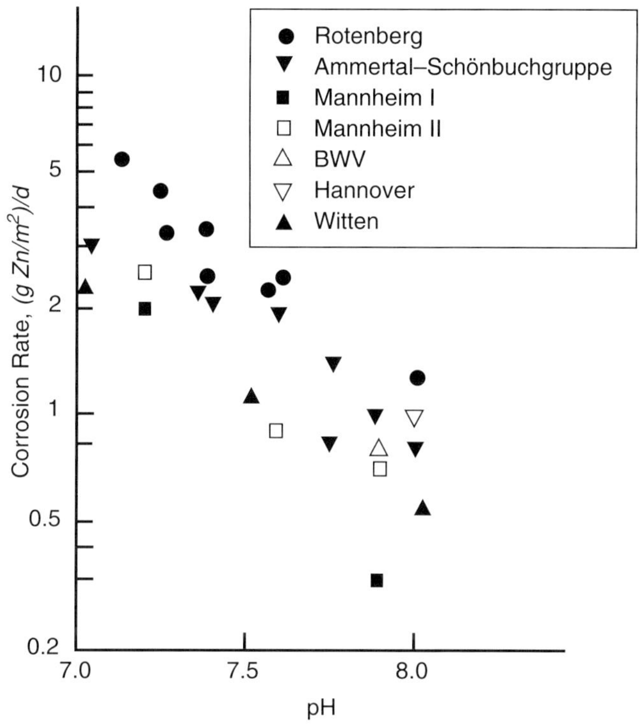

Source: Bachle et al. 1981.

Figure 3-18 Influence of pH on corrosion rate of galvanized steel pipe

The work at Landeswasserversorgung also yielded data that demonstrate the impact of pH on the corrosion rate of zinc (Werner, Wurster, and Sontheimer 1973). These data, shown in Figure 3-20, are supported by the work of Bächle et al. (1981), demonstrating increasing corrosion rates with decreasing pH. The pH corrosion rate relationship was correlated by Kruse (1983), with a model that relates the corrosion rate to the level of carbon dioxide in the water.

Apart from the hydrogen ion, there are only a few natural water constituents that increase the corrosion rate of the pure zinc layer (eta layer) of galvanized pipe as they increase in concentration. One of the most important of these is the carbonate hardness. Werner (1984) observed a nearly linear relationship between carbonate hardness and the corrosion rate of zinc.

Studies at the University of Florida. Pisigan and Singley (1985) also investigated the influence of water quality, particularly pH and the LI, on the corrosion rate of galvanized pipe. Tests were conducted in a five-day jar test. Figure 3-21 illustrates their results with the corrosion rate of zinc and the release of zinc products in the short-term test as a function of pH. The results regarding corrosion rate show a rapid increase in corrosion rate for both deionized water and Gainesville water (total dissolved solids, 190 mg/L; hardness, 112 mg/L) at pH values of 7 and below. Contrary to the longer-term German tests, these short-term tests exhibit little change in the corrosion rate at pH values above 7. Increasing the pH of the solution has a dramatic impact on the amount of dissolved zinc corrosion product in the solution. This is consistent with the equilibria presented earlier. In general, the constituents in the natural water from Gainesville resulted in reduced corrosion and reduced levels of corrosion products when compared to a demineralized water control.

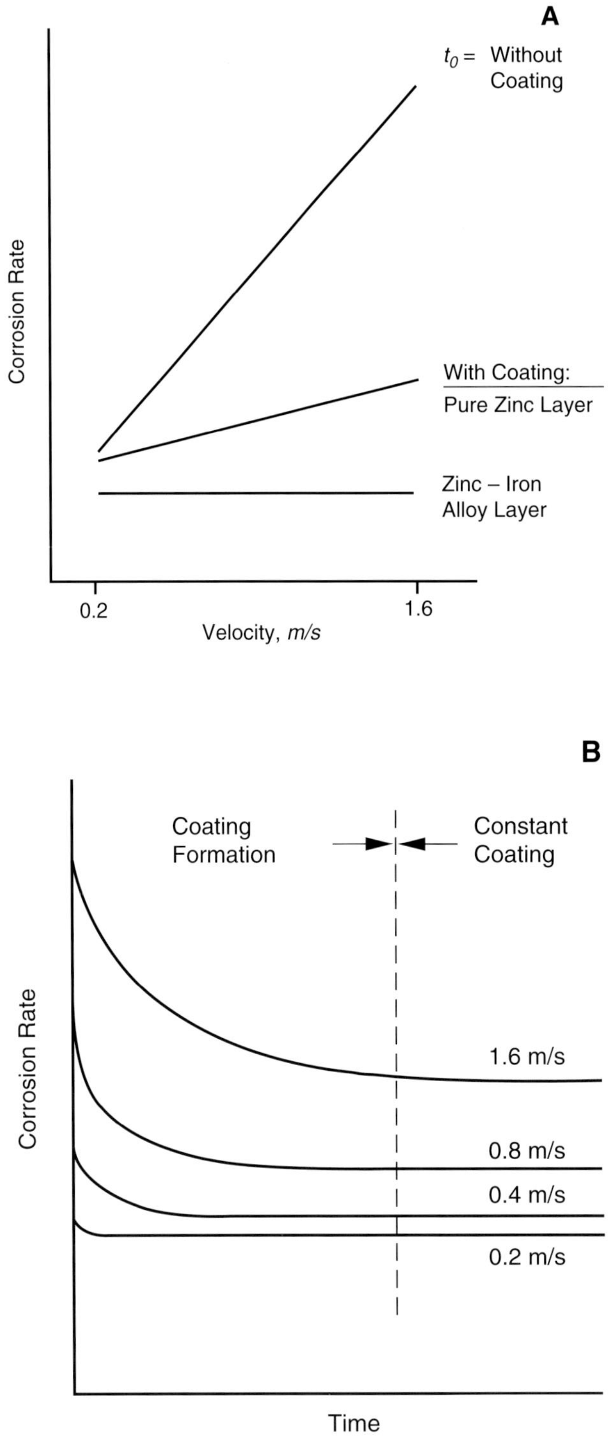

NOTE: Studies conducted by the Landeswasserversorgung in Langenau, Germany, have resulted in a model of the process of corrosion of galvanized iron. The model is illustrated in the two figures above. The basic argument is that a protective coating must be formed on the pipe surface before the pure zinc layer is corroded away. The zinc serves as temporary protection while the long-term protective scale is formed. Figure 3.19A illustrates that the corrosion of the unprotected zinc layer is more sensitive to velocity than it is once a protective scale is formed. Figure 3.19B illustrates the corrosion rate at several velocities, showing that the corrosion rate rapidly drops during the coating formation period but stabilizes once a constant or steady-state coating is achieved. Again velocity has less impact once the steady-state coating is formed.

Figure 3-19 Landeswasserversorgung studies: Corrosion and coating formation model

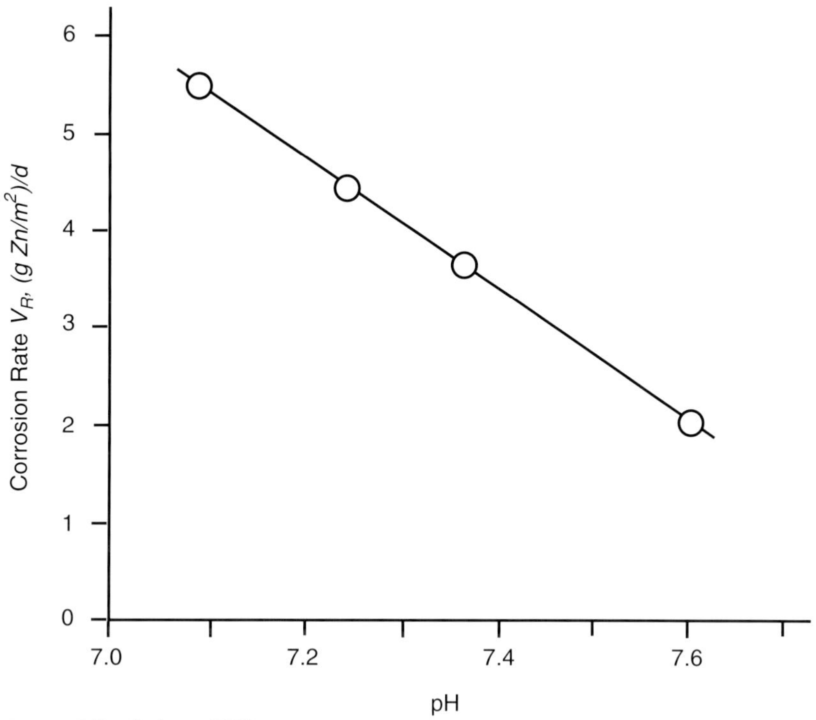

Source: Werner, Wurster, and Sontheimer 1973.

Figure 3-20 Landeswasserversorgung studies: Influence of pH on the corrosion rate of zinc

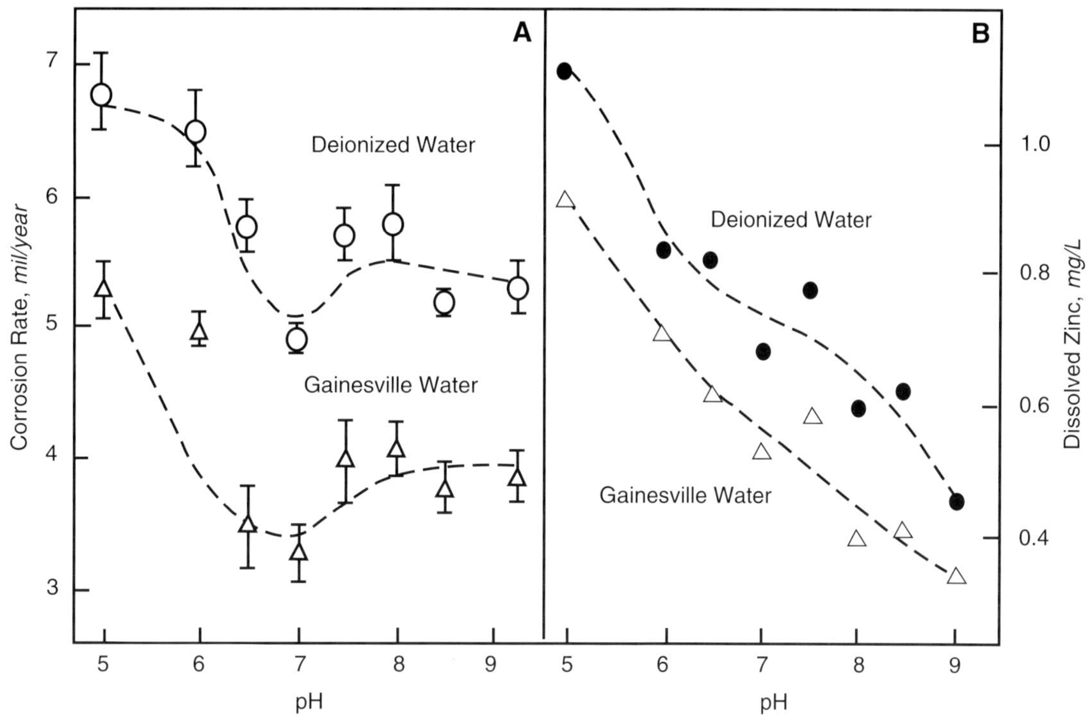

Source: Pisigan and Singley (1985)

Figure 3-21 Gainesville studies: Influence of pH on the corrosion rate of zinc

Metal Leaching

As a result of concerns about corrosion by-products in consumer plumbing in the early 1980s, the US Environmental Protection Agency's (USEPA) Office of Drinking Water issued *Water Supply Guidance Number 73*, a document that recommended that surveys of consumer plumbing be conducted whenever a water had corrosive characteristics (USEPA 1982). This document was a precursor to the current Lead and Copper Rule (USEPA 1991), which encourages surveys of both running and standing water in all types of plumbing materials in order to characterize the spectrum of qualities the consumer may experience. Two significant surveys conducted around that time were those in Seattle, Washington (Dangel 1975; Seattle Water Metals Committee 1974; Kennedy Engineers 1976), and Portland, Oregon (JMM 1982). Both studies were of low-alkalinity water supplies in the northwestern United States and both studies included a substantial number of homes plumbed with galvanized pipe. Approximate qualities of the three waters involved (the Cedar River and Tolt River in Seattle and the Bull Run River in Portland) are shown in Table 3-3. Both studies showed consistently elevated levels of iron and zinc in standing and running water, particularly standing water. Table 3-7 summarizes the data on standing water from the Portland study, showing that iron and zinc are the principle corrosion by-products in taps connected to galvanized pipe.

Meyer (1980) studied leaching of heavy metals from galvanized pipe under conditions of 144 min of intermittent flow each day at a velocity of 0.5 m/s (1.6 ft/s). The levels of metals leached from the surface were measured over a range of stagnation times to determine the time corresponding to the maximum rate. Studies showed that much like the results of continuous-flow experiments, the amount of material leached into the water increased with decreasing pH. As might be expected, the levels of lead and cadmium also followed this trend. Although lead and cadmium levels can exceed German drinking water standards when new pipe is exposed to waters of certain quality, these levels decrease with running time, as illustrated in Figure 3-22. Note the peak levels of cadmium are somewhat higher than those observed in Portland.

Another, more unusual form of metal leaching has occurred in Germany. The problem occurred in well-buffered groundwaters where galvanized pipe normally behaves reasonably well (Figure 3-23A). Pipes with an irregular zinc layer that resulted from using air to remove excess zinc in the galvanizing process behaved differently (see Figure 3-10 and Figures 3-23B and C). When water was drawn from these pipes, parts of the corrosion scale were dislodged and appeared in the drinking water as reddish-yellow sand particles. For this reason, the Germans refer to the problem as "Zinkgeriesel," or zinc sand (see Figure 3-23D).

Table 3-7 Levels of metals in taps connected to galvanized pipe, Portland, Oregon

Metal	Median Level, *mg / L*	90% Level, *mg / L*
Zinc	0.5	1.7
Iron	1.1	3.0
Lead	0.01	0.02
Cadmium	<0.001	0.002

Source: Data from JMM (1982).

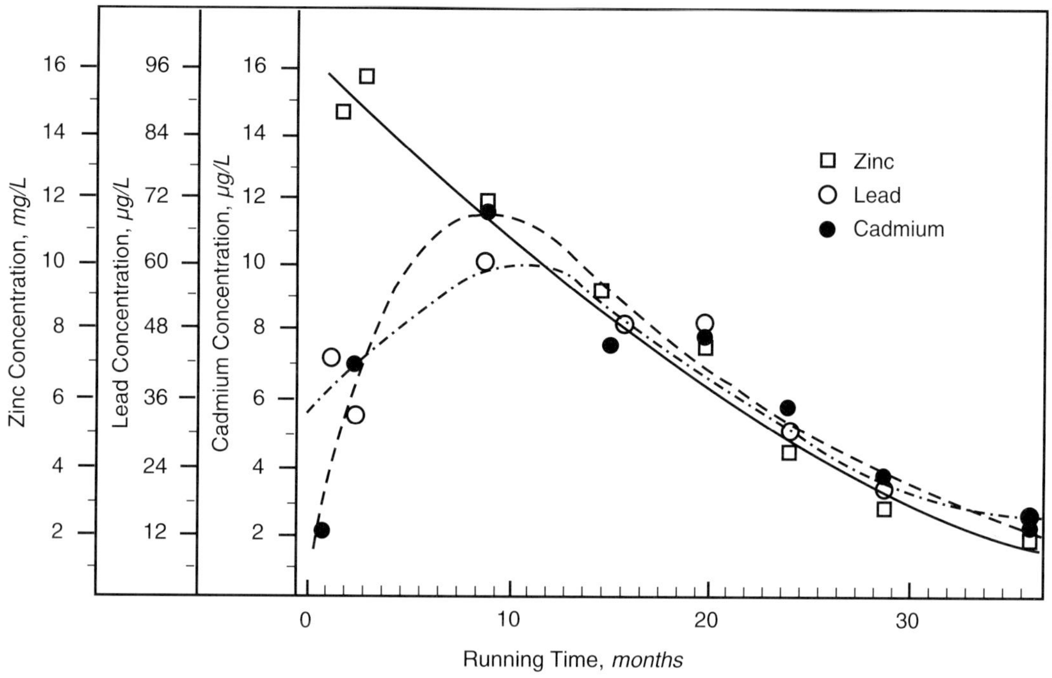

Source: Meyer (1980).

Figure 3-22 Dependence of zinc, lead, and cadmium levels on the running time of the test pipe rig in water of pH 7 after 8 hours of stagnation

Shortly after installation of the new pipe, whitish zinc corrosion products appear, as is commonly experienced with the installation of galvanized pipe in aggressive waters. In later stages, however, the Zinkgeriesel phenomenon is often accompanied by the appearance of red water. An investigation of nearly 1,000 households with this type of problem demonstrated that the red water was usually accompanied by Zinkgeriesel particles (Werner 1983). The zinc content in the corrosion product appearing at the tap was usually between 30 and 80 percent. The level of iron in the red water associated with this problem was significantly higher than that caused by other problems with galvanized pipe. Levels of up to 200 mg/L for both iron and zinc have been observed.

After some study, the following hypothesis was developed to explain the evolution of the problem. At sites with thick deposits of corrosion products, the mass transfer of oxygen to the pipe wall is retarded, resulting in an oxygen concentration cell. Portions of the surface where the corrosion products are thinner remain cathodic and the areas with a thicker scale become anodic. Under the thick deposits, first the alloy layer and finally the base metal itself are corroded to form broad, shallow pits. The Fe(II) ions formed there diffuse through the porous zinc corrosion products and are oxidized to Fe(III). As a result, the water often contains large amounts of fine, flocculent iron hydroxide as well as the sandy yellow particles derived from the corroding zinc itself.

Zinkgeriesel and its associated red water are primarily due to poor-quality galvanized pipe. Occurrence of the problem was also found to respond to water quality and to certain treatment strategies. For example, Zinkgeriesel was usually observed in pipes using groundwater and only infrequently in pipes using surface water. Studies also showed that the addition of orthophosphate can reduce the

A. Normal Zinc Layer

B. Air-Blown Galvanized Pipe

C. Zinkgeriesel Structure on Air-Blown Pipe

D. Zinkgeriesel Particles

E. Air-Blown Pipe With Inhibitors

Figure 3-23 The "Zinkgeriesel" phenomenon

problem. The Zinkgeriesel structures are finer following the addition of this material (Figure 3-23E). Adjusting the pH did not show much improvement. Research at Darmstadt, Germany (Wagner and Dannöhl 1985), also showed that high levels of nitrate may be associated with the Zinkgeriesel problem.

Zinkgeriesel has been limited to poor-quality imported pipe, which was on the market in Europe in large quantities between 1960 and 1972. High-quality galvanized pipe, which complies with DIN 2444, has a uniform galvanized layer (see Figures 3-23A). In this pipe, a smooth, white coating forms on the eta zinc layer and Zinkgeriesel particles are rarely found.

Failure by Pitting and Tuberculation

For as long as galvanized pipe has been in use, there has been concern about failure by pitting and tuberculation. Although pitting of galvanized pipe does occur in cold water, pitting in hot water is more common. A great deal of research was done on this subject between 1935 and 1950, and some research has been carried out since then (Schikorr 1939; Kenworthy 1943; Kenworthy and Smith 1944; Gilbert 1948; Hoxeng and Prutton 1949; Hoxeng 1950; Guest 1956; Weast and Shulman 1962; Gouda, Khedr, and Shams El Din 1967; Mihailov and Iovchev 1981; Haney 1982). Probably the majority of this work has discussed the problem in terms of potential reversal, which occurs between zinc and iron (that is, zinc becoming noble with respect to iron) under certain conditions of water quality and temperature. Significant research has also been done on copper-induced pitting, poor quality of manufacture, and galvanic couples. Studies on potential reversal at elevated temperature began because of an interest in pitting problems in water heaters. Also, pitting has, in many instances, been associated with poor-quality manufacturing processes. These four subjects—potential reversal, copper-induced pitting, pitting resulting from poor-quality manufacturing, and pitting or tuberculation by galvanic couples—will be discussed in the following paragraphs.

Pitting as a result of potential reversal. Schikorr (1939) was the first to demonstrate that the relative potentials of zinc and iron can be reversed at higher temperatures. He also demonstrated that the reversal is accomplished primarily by shifts in the potential of the zinc surface and that iron's potential remains relatively stable. Figure 3-24 summarizes some of Schikorr's results, showing the current flow between iron and zinc electrodes (briefly connected for each measurement) and the temperature and velocity during the test. The authors of this chapter have confirmed this phenomenon in their laboratories, as have most of the investigators cited earlier. It has been shown to be a function of water quality, oxygen tension, and temperature.

Perhaps the most thoughtful analysis of this hot-water, potential reversal pitting problem is that of Gilbert (1948), who integrated the work of others with his own to develop an explanation. Gilbert's analysis follows.

In cold water, zinc is normally anodic to steel. Depending on conditions, pitting does sometimes occur, but the compound or alloy layers beneath the eta layer (Figure 3-2) have a more noble potential than does the eta layer itself. As a result, the pits will normally stop once they reach the alloy. Often the pits begin to widen at this point. Because of its more noble potential, the alloy does not offer much protection to the base steel. As a result, the iron will eventually pit, but normally this does not occur until more than half of the surface is corroded to the alloy layer.

In hot water, the surface scale of zinc apparently changes, forming a scale that more effectively supports the reduction of oxygen at the cathode and that "heals" some of the anodic sites, reducing the anodic area. Gilbert suggested that zinc might have a hydroxide scale at low temperatures and a zinc oxide scale at higher temperatures. This process has little effect on the overall corrosion rate of zinc, but the increased polarization of the anodic sites due to their reduced area and the more effective depolarization of the cathodic sites due to the greater conductivity of the new scale substantially increase the potential of the zinc, making it more noble than the iron. As a result, any exposed iron becomes anodic and severe pitting begins. In soft water, no scale is deposited and the pitting attack will continue until the pipe wall is perforated. In hard, scale-forming water, if the conditions are

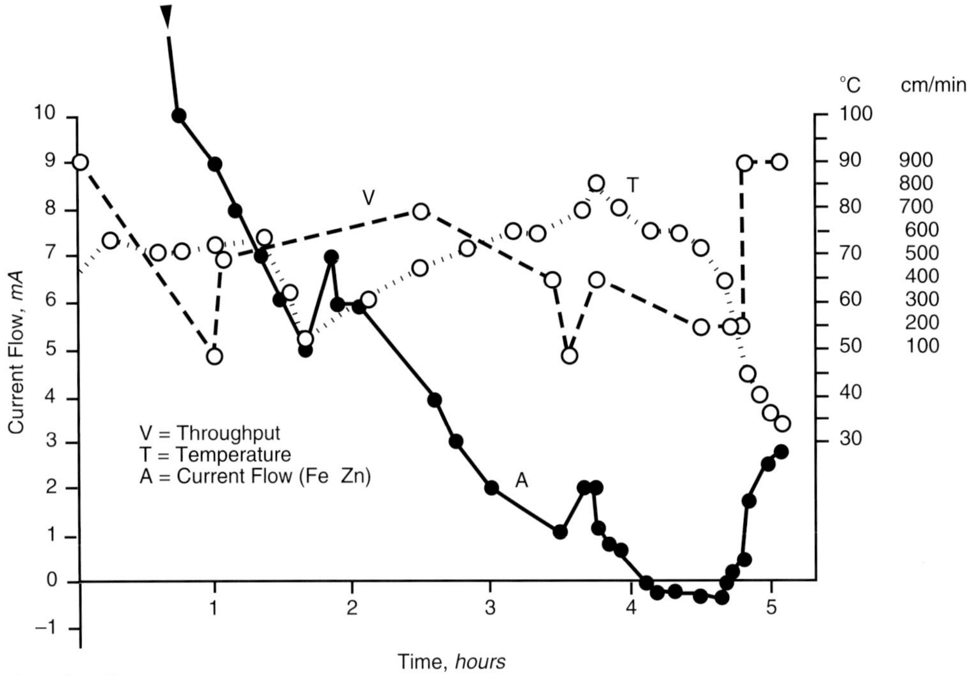

Source: Schikorr (1939).

Figure 3-24 Potential reversal of iron and zinc in a circulating hot-water loop

particularly severe, perforation may occur rapidly. However, if no failure occurs during these early stages, a scale will be deposited and pitting will be arrested.

In the late 1960s, the ILZRO examined some of the basic assumptions on which Gilbert's model is based (ILZRO 1967). Examining the scale formed on galvanized pipe corroding in distilled water, they found zincite (ZnO), beta zinc hydroxide, and eta zinc hydroxide on the pipe surface. On the surface of the relatively pure zincs, equal amounts of beta and eta $Zn(OH)_2$ were found in scale formed at temperatures below 5°C, but ZnO dominated at higher temperatures. Tests confirmed Gilbert's suggestion that the zincite formed will serve as a semiconductor.

Unfortunately, these studies were not extended to determine the nature of the scale formed in the presence of the carbonate buffer found in tap water. As discussed earlier, recent work at room temperature (Pisigan and Singley 1985) confirmed the work of ILZRO with distilled water but showed that hydrozincite is dominant in the presence of carbonate. The hydrozincite scale may break down at high temperatures in certain water qualities, leaving a zincite scale and facilitating the mechanisms described by Gilbert.

A number of factors influence the potential reversal phenomenon. First, oxygen must be present. Without oxygen, cathodic polarization is limited by the hydrogen overpotential and the potential of the zinc surface remains active with respect to the iron. Gilbert noted that the zinc potential is influenced by chloride, sulfate, and bicarbonate. Chloride and sulfate decreased the potential of the zinc, making polarity reversal less likely; bicarbonate ennobled the zinc, enhancing reversal. Studies by others (Hoxeng and Prutton 1949; Hoxeng 1950; Gouda, Khedr, and Shams El Din 1967; Haney 1982) confirmed this information and also established that potentials can be reduced by calcium and silicates and increased by the addition of monobasic sodium phosphate. Gouda, Khedr, and Shams El Din demonstrated that the influence of the aggressive ions (chloride and sulfate) on the potential of the zinc surface could be described by the following relationship:

$$E = a - b \log (C)$$

Where:

C = the concentration of the ion

Generally, the behavior of these ions is consistent with Gilbert's model of the process. Aggressive ions like chloride and sulfate would continue to attack the anodic sites, maintaining larger anodic areas at higher temperatures. Calcium, bicarbonate, carbonate, and silica would be expected to maintain an active potential on the zinc surface by laying a calcium or silicate scale on the cathodic sites, preventing the more rapid depolarization of oxygen even though the nature of the zinc scale itself may become more conductive.

Kruse (1983) provided a complete summary of the hot-water pitting problem. Results reviewed by Kruse indicated that ennobling can occur at temperatures as low as 35°C and that the problem of potential reversal is highly dependent on water quality. In response to these problems, the Germans limit the use of galvanized steel pipe to applications in which the temperature is less than 60°C (DIN 50930).

Copper-induced pitting. According to Kenworthy (1943), published references to the problem of copper-induced pitting of galvanized pipe can be found as far back as 60 years ago. At that time, McKee (1932) cautioned against the use of copper sulfate as an algicide in cooling systems using galvanized pipe. The bulk of the research and experience with the problem, however, deals with the failure of galvanized pipe and tankage in recirculating hot-water systems that are comprised of both copper and galvanized materials. Kenworthy argued that concentrations of copper in the recirculating water in excess of 0.1 mg/L are sufficient to cause greatly accelerated attack. Figure 3-25 shows some of Kenworthy's results, demonstrating the relationship between the corrosion of galvanized pipe in stagnant and recirculating systems and the concentration of copper in solution.

More recently, Cruse (1971) reported several case histories in which copper was implicated in the pitting of iron and galvanized pipe. Cruse demonstrated an empirical correlation between the level of copper deposited on the inside of the galvanized pipe and the maximum rate of pitting (Figure 3-26). Levels of copper exceeding 10 mg/dm^2 (1 g/m^2) were proposed to be significant. This same study also reported copper sulfate algicide as a source of copper associated with certain pitting incidents. Cruse (1971) and Treweek, Trussell, and Pomeroy (1978) reported incidents in which the copper applied as an algicide was implicated in the rapid pitting of pipe in hundreds of dwellings constructed with new galvanized pipe. More recently, MWD (Grasha et al. 1980; Grasha, Risner, and Davis 1981) conducted research to evaluate the conditions surrounding this phenomenon.

The principle presumed to be behind copper-induced pitting of galvanized pipe is relatively straightforward. At first, the copper ions from the water are precipitated on the less noble zinc surface.

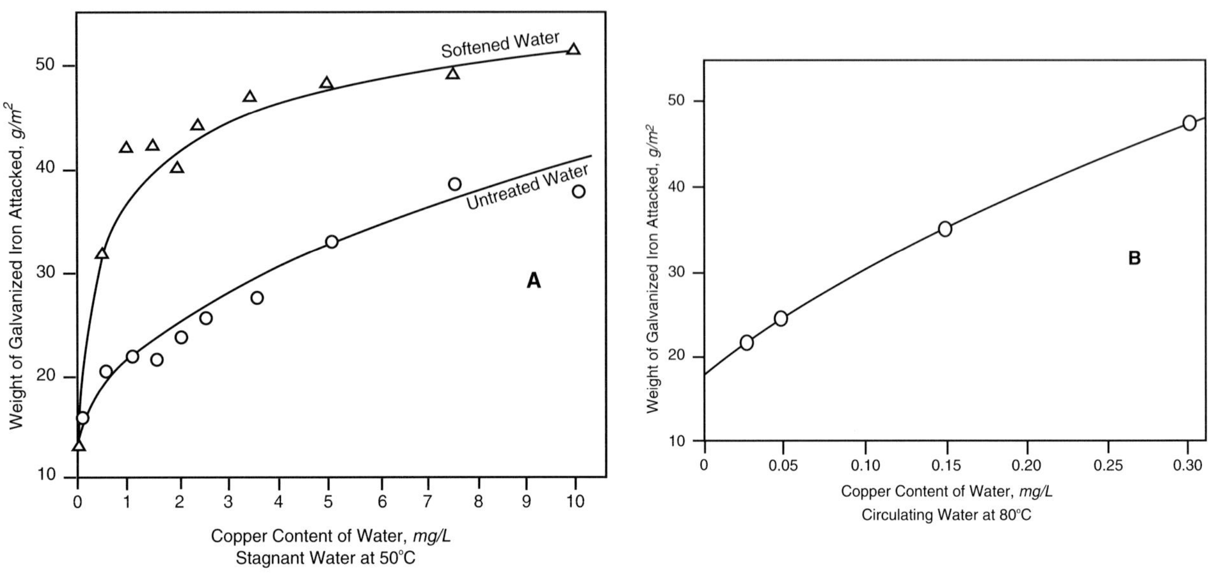

Source: Kenworthy (1943).

Figure 3-25 Attack of galvanized iron by copper in (A) stagnant water and (B) circulating water

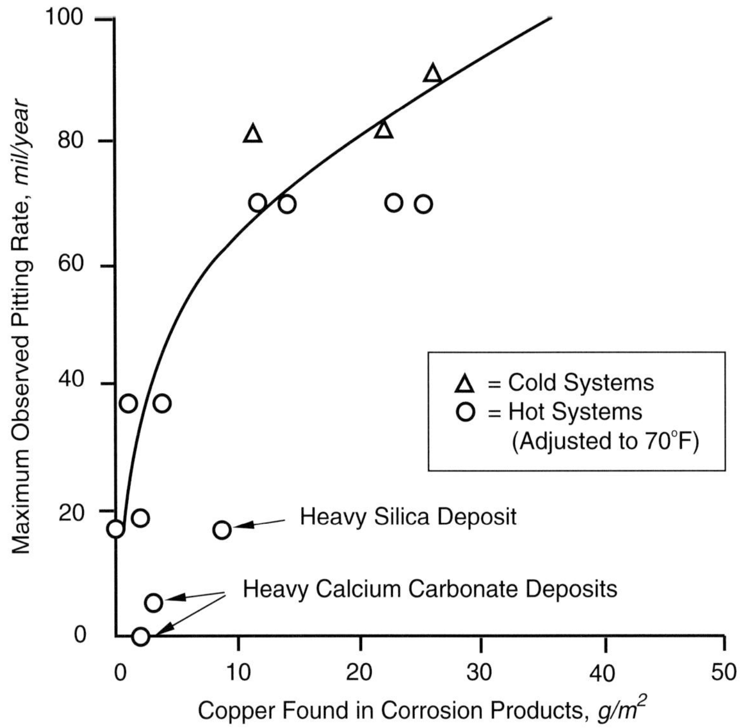

Source: Cruse (1971).

Figure 3-26 Copper-induced pitting of galvanized steel

Once the copper is deposited, however, any copper in contact with the zinc metal will act as a cathode, creating many small short-circuited galvanic cells. If the temperature is also elevated enough, the phenomenon that Gilbert described will reduce the anode area and rapid pitting will result. The more conductive ZnO scale may also increase the access of the deposited copper to the zinc surface beneath.

Copper can be present in water from a number of sources. These include the use of copper sulfate or citrate-complexed copper as an algicide and copper released from the corrosion of copper or copper alloys. Figure 3-27 illustrates the results of a German test conducted to evaluate the release of copper by brass fittings. Brass fittings were installed in a galvanized pipe system designed to run in two modes: (1) with once-through flow, and (2) with recirculating flow. The system was first put in the recirculating mode for 24 hours and the level of copper in the recirculating solution observed. Then the system was put in the once-through mode and run for one month. Following this exposure, it was put back in the recirculating mode for another 24-hour period, and so on. The results suggest that such fittings may be expected to release significant levels of copper to downstream plumbing. These results are particularly significant with respect to copper and brass fittings on galvanized systems, especially those associated with hot-water tanks and recirculating systems.

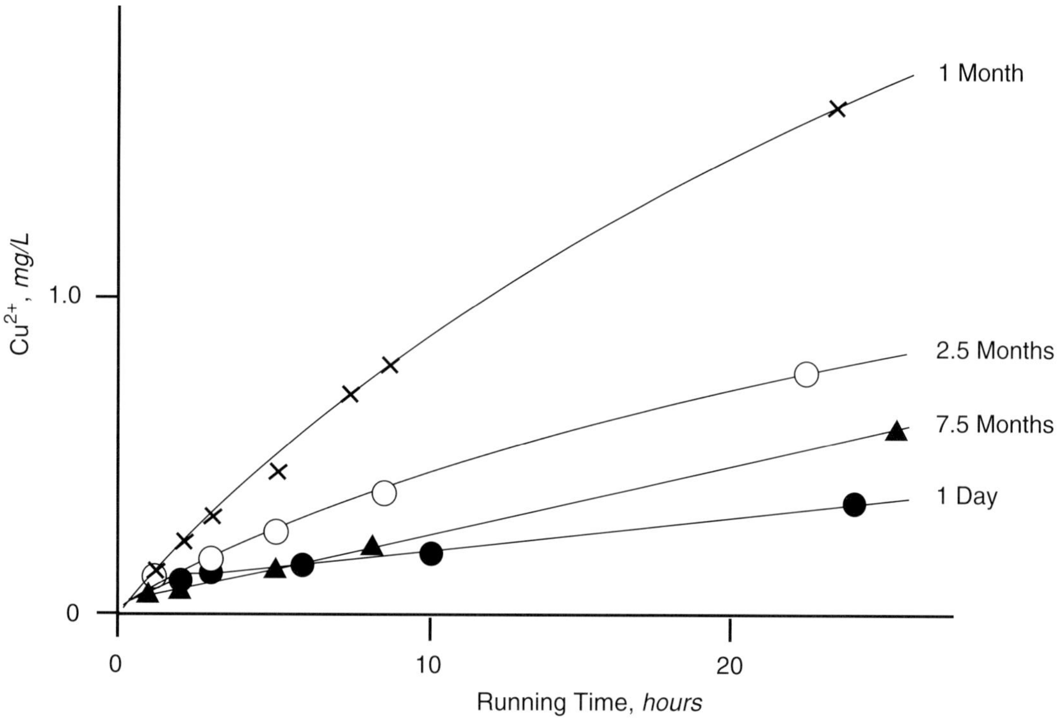

NOTE: This figure demonstrates the leaching of copper from brass surfaces in recirculating systems. Brass fittings were exposed to tap water for periods of 1 day, 1 month, 2.5 months, and 7.5 months. Copper levels were then monitored in a recirculating system containing fresh water of the same quality. Substantial levels of copper were achieved even after 7.5 months of exposure.

Figure 3-27 Leaching of copper from brass fittings

Quality of manufacture. Whenever rapid pitting of galvanized pipe occurs, the quality of manufacture of the pipe should be considered. Inadequate galvanizing, incomplete or poor-quality welds, an excessive weld seam, a poor zinc–steel interface, and other manufacturing defects can all contribute to the problem. Problems of this kind, when combined with the factors previously discussed, increase the risk of failure by pitting. Fortunately, most of these manufacturing problems can be overcome by inspecting and testing the pipe as it is installed to ensure that it meets commonly recognized standards like those of ASTM or DIN outlined earlier in this chapter. Several examples of pipe failure resulting from quality of manufacture follow.

One of the most common defects is an inadequate weld seam (see Figures 3-1 and 3-8). Figure 3-28 shows two photographs of corrosion failure on the weld seam, which occurred in galvanized pipe used in a recirculating hot-water system in the United States (Thomas and Ray 1984). A slip of paper has been inserted through the point of failure, for illustration.

Similar problems have occurred in Germany (Figure 3-29), where they have been attributed to weld seams that protrude inside the surface of the steel pipe after production (as shown in Figures 3-1 and 3-29). As a result, DIN 2440 and DIN 2444 require that weld seams provide a smooth, uniform surface.

A second manufacturing defect can result from problems in alloy formation during the galvanizing process. Undesirable expansion of the alloy layer, or "galvannealing," is the most common among the defects of this kind. The zeta, or "alloy" layer, has been demonstrated to have a more noble potential than the eta, or "pure zinc" layer (Gilbert 1948), in fact, more noble than the iron itself. As a result, manufacturing practices that encourage galvannealing also encourage potential reversal between the zinc (alloy) layer and the base iron, and pitting will occur once the pipe is put in service. Figures 3-30A and 3-30B illustrate the result obtained from long-term studies done at Biberach, Germany, and Darmstadt, Germany (Schneider et al. 1982; Wagner and Dannöhl 1985).

A third defect is a poor bond between the galvanizing layers and the base metal, as shown in Figure 3-7. Figure 3-31 is a photograph of the surface of an exposed sample of the K pipe described earlier. Close examination of photographs of samples exposed for about five months to Colorado River water shows pits of

Exterior View

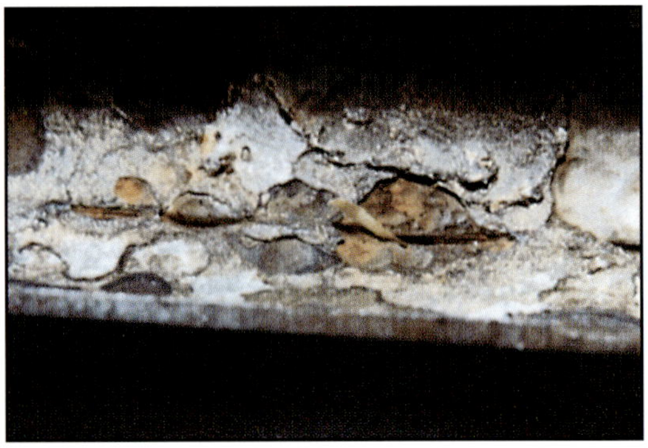

Interior View

Source: Thomas and Ray (1984).

Figure 3-28 Corrosion failure at weld seam

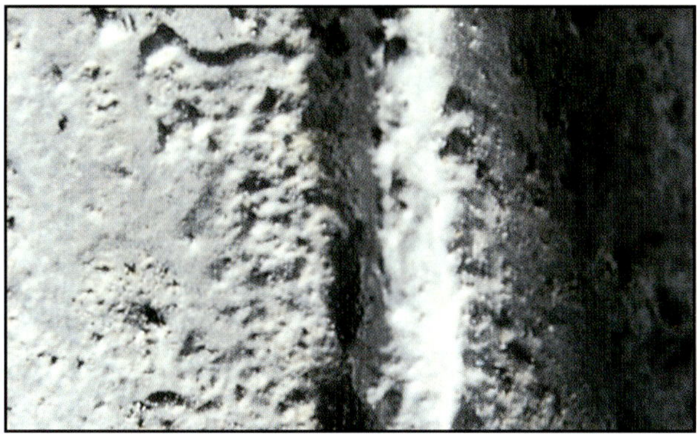

A. Weld Seam in New Pipe

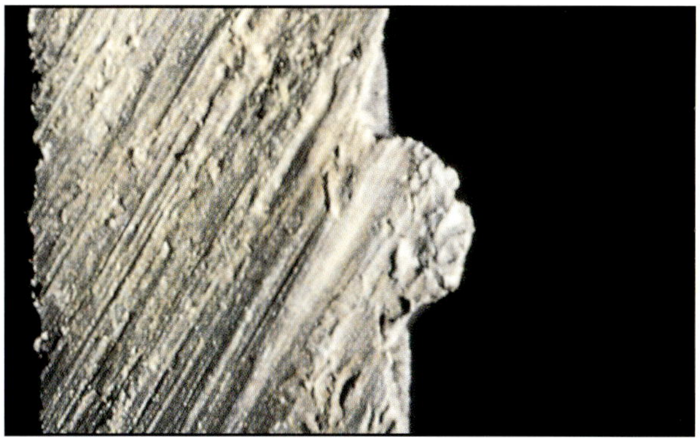

B. Cross Section of A in Magnification

C. Failed Pipe Section From House Installation in Service for Three Years

D. Cross Section of C, With Corrosion Attack Along Weld Seam Clearly Visible

Figure 3-29 Failure at weld seam

40–60 mil in diameter, which correspond to the areas with a loose hot-dip coating (Thomas and Ray 1984). The two large circular areas overlie pits. These pits probably originated from portions of the pipe surface that experienced separation between the iron and zinc layers.

Galvanic couples. Whenever galvanized pipe is used in consumer plumbing, the possibility of brass–zinc galvanic couples arises. Most US plumbing codes (UPC 1990) require that all materials used in a particular water system be of "like material"; however, an exception is normally included for valves and similar devices. In fact, the use of brass valves with galvanized pipe is common practice throughout the United States.

Stern (1958) and Mansfield (1971) developed the principles for characterizing the relationship between the relative areas of two metals in a galvanic couple and the corrosion current density on each. Jones (1984) extended this work to the brass–steel galvanic couple and developed data that can be used to analyze the behavior of the brass–zinc galvanic couple under very saline conditions. Figure 3-32 from Jones (1984) shows the corrosion potential of brass materials as a function of copper content. There is a sharp transition in the potential at a copper content of about 40 percent. Beneath this level, the potential of the material is nearly the same as the potential of zinc. Above 40 percent copper, the potential is essentially that of copper.

NOTE: The upper panel is an enlarged photo of the surface of a galvannealed pipe after three years of exposure. The lower panel shows the same surface after the scale and corrosion products were removed by etching.

Source: Schneider et al. (1982).

A. Surface of a Galvannealed Pipe After Three Years of Exposure. Scale and Corrosion Products Were Removed by Etching

Source: Wagner and Dannöhl (1985).

B. Pitting of Galvannealed Pipe Exposed to a High-Nitrate Groundwater in Darmstadt, Germany, for a Period of Two Years

Figure 3-30 Galvannealing and pitting-type corrosion

NOTE: This photomicrograph shows the surface of type K galvanized pipe that was exposed to Colorado River water. The two large circular areas overlie pits. These pits probably originated from portions of the pipe surface that experienced separation between the iron and zinc layers.

Source: Thomas and Ray (1984).

Figure 3-31 Surface of exposed galvanized pipe

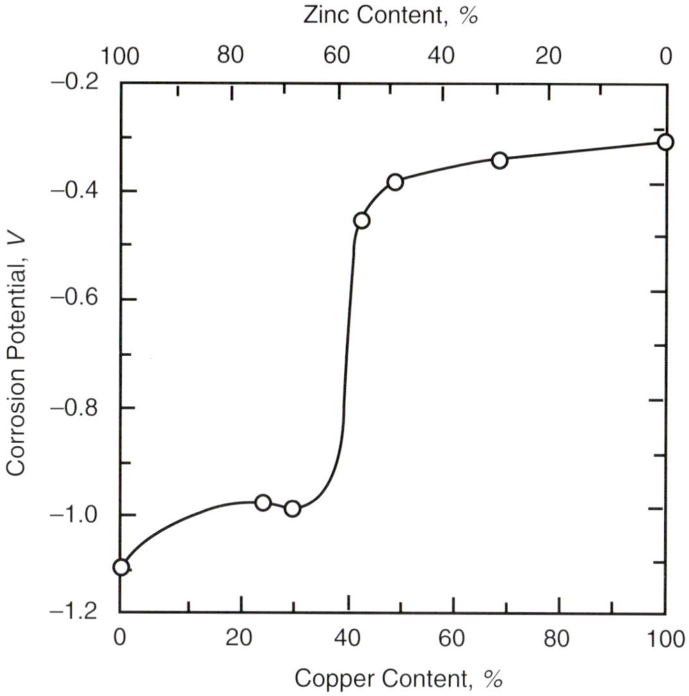

Source: Jones (1984); copyright NACE International.

Figure 3-32 Effect of brass composition on corrosion potential

Jones's measurements were taken in 20 percent seawater, but classical potential data suggest that the potential difference between the two materials in fresh water would be approximately 800 mV, similar to that in seawater. Thus, a significant potential will exist between most galvanized and brass surfaces. The influence of this potential decreases with increasing distance from the point of contact between the dissimilar metals. The extent of influence of this potential is a function of the conductivity of the water being used. As the conductivity of the water decreases, galvanic corrosion will be increasingly focused near the point of interconnection between the two metals. At the same time, the overall galvanic current (hence the total metal corroded) increases with increasing conductivity. This principle is illustrated by Figure 3-33 from Uhlig (1948), which contrasts the character of corrosion near a galvanic connection in seawater and that in distilled water.

Figure 3-34 illustrates the polarization curves that were obtained by Jones (1984) for various brass alloys, steel, and zinc in a similar solution. It should be noted that the cathodic polarization curves for both brass and zinc show mass-transfer limitations at a current density of approximately 100 μA/cm^2 (corresponding to a corrosion rate of about 0.075 mm/year). This is a common level of current density encountered for the mass transfer of oxygen to metal surfaces under normal laboratory conditions (Fontana and Greene 1978). It is also consistent with available information on both zinc and copper surfaces, indicating that their rate of corrosion is normally limited by mass transfer to the cathodic reaction.

Under normal conditions, the corrosion rate of copper (expressed as μA/cm^2) is several fold lower than the corresponding corrosion rate for zinc. As a result, a small cathodic surface such as a valve in a home plumbing system will normally have only a local impact on the corrosion of galvanized pipe in soft waters. On the other hand, when the ratio of the areas of brass and zinc become too unfavorable (such as when a small galvanized fitting is used in portions of a system dominated by copper or brass), sufficient cathodic activity is available on the copper surface to

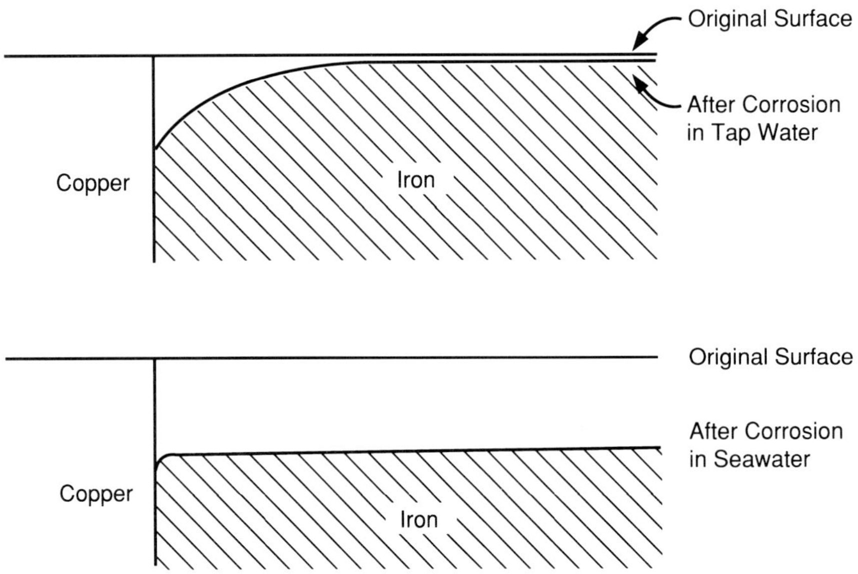

Source: Uhlig (1948).

Figure 3-33 Influence of conductivity on corrosion at a galvanic interface

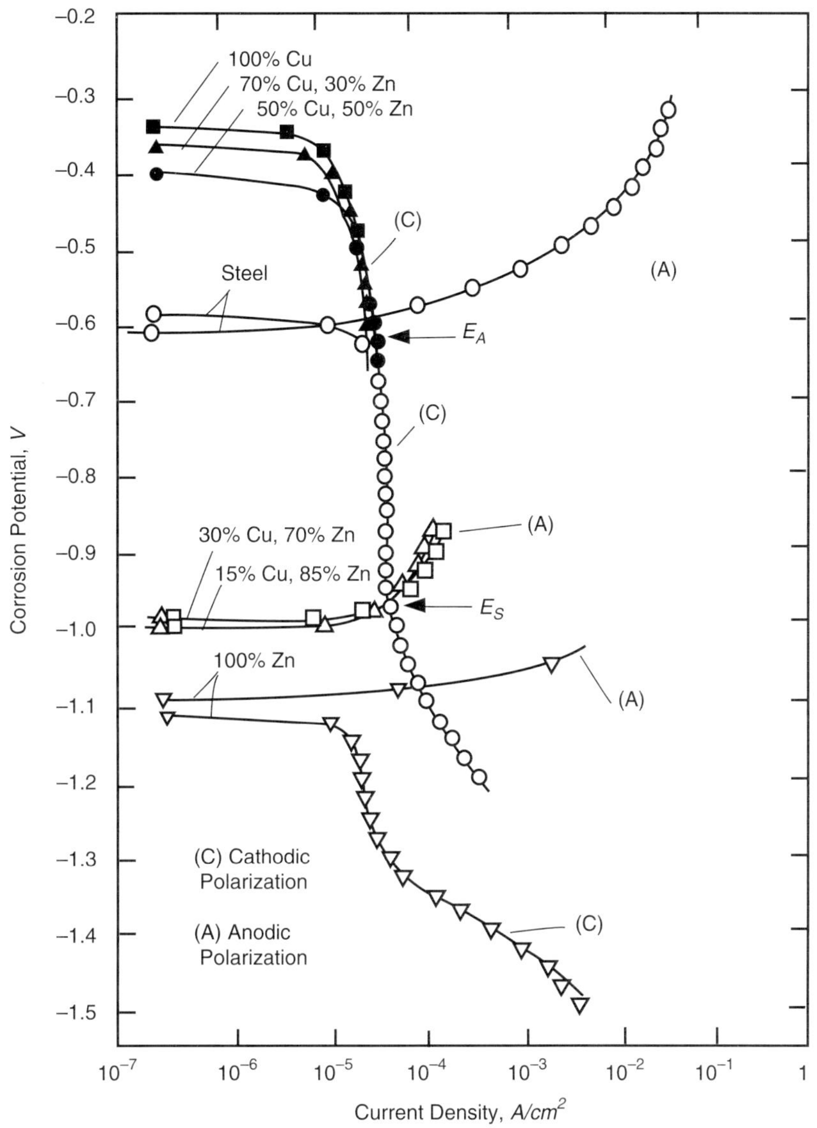

Source: Jones (1984); copyright NACE International.

Figure 3-34 Polarization curves for copper, zinc, steel, and brass

greatly accelerate corrosion of the galvanized surface. In other words, as the ratio of the area of brass to zinc increases, the corrosion potential of the couple increases and the zinc, being the anode, corrodes more rapidly. This principle is illustrated for the case of the brass–steel couple in Figure 3-35.

Corrosion Problems in Soft Waters

Rapid loss of zinc and subsequent tuberculation commonly occur in soft acid waters that have a pH below 7. The data in Table 3-3 illustrate that, under these conditions, the life of the zinc coating is only a few months. Once the zinc layer is corroded away, corrosion of the iron pipe proceeds at a more rapid rate and

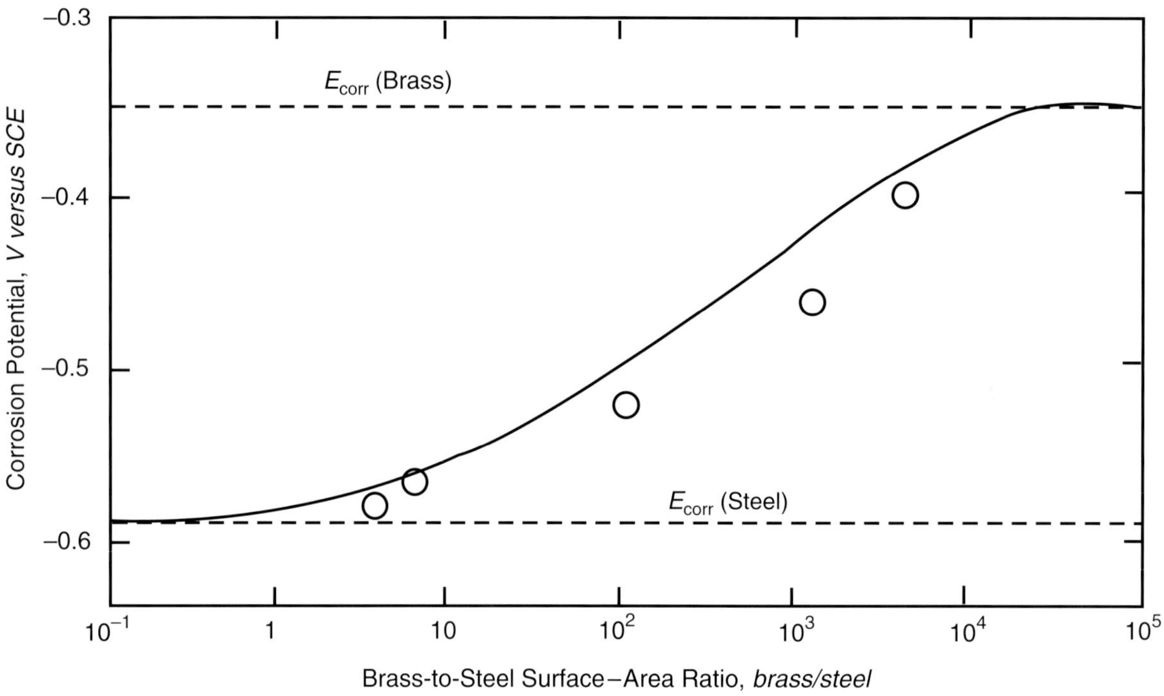

Figure 3-35 Brass-steel ratio and corrosion potential

tuberculation begins to accelerate as well. Red water does not normally occur unless the pH is particularly low. This phenomenon is largely due to the poor protection provided by iron corrosion products. Typically, the pipes are completely occluded in 10 to 20 years or the wall thickness is reduced to the point at which mechanical failure occurs. In Germany, these experiences have led to the CO_2 and pH requirements of DIN 50930 (1993) mentioned earlier.

TREATMENT INFORMATION

Generally, the most effective treatment alternatives for controlling the corrosion of galvanized pipe involve the formation of effective protective scales on the surface of the pipe. Because the chemistry of each water supply is different, no single treatment remedy can be successful for every case. Rather, tests should be conducted on the water supply in question to determine the effectiveness of the alternatives that have shown some success. Generally, the most effective alternatives found for controlling the corrosion of galvanized pipe are pH adjustment and the use of silicates or orthophosphates. Each of these will be briefly discussed.

Adjustment of pH

The chemistry and the long-term corrosion results shown earlier suggest that pH adjustment is often useful when dealing with cold-water corrosion. Evidence indicates that zinc hydroxide, zinc oxide, zinc hydroxycarbonate, and calcium carbonate (often aragonite) are the principal compounds of the protective scale formed on the surface of a galvanized pipe exposed to natural water. The solubility of all of these materials is influenced by the pH. As discussed earlier, the solubility of the zinc compounds is adversely affected by pH values below 7. It would appear galvanized pipe would normally be expected to give its best service at pH values between 7 and 8.5, depending on the alkalinity of the water. Fortunately, this rather narrow pH range is common in drinking water systems. These pH effects have already been discussed in the section on general corrosion in this chapter.

A related issue is calcium carbonate saturation. Work conducted in a recirculating loop using ½-in. (15-mm) sections of galvanized pipe as test sections (Pisigan and Singley 1985) showed no clear relationship between the level of calcium carbonate saturation at the beginning of the test period and the rate of corrosion of galvanized pipe (Figure 3-36). At the same time, no attempt was made to control the quality of the water in the loop during the course of the experiment. There is also a possibility that the buildup of zinc in the loop may result in a reduction in the rate of corrosion of the galvanized surface (Schwenk 1976).

Although calcium carbonate compounds do form on the surface of corroding galvanized pipe, it is not clear what their role is in protecting the pipe. Even if they play an important role, the work of Pisigan and Singley (1985) confirms the position of Stumm (1960) and Trussell, Russell, and Thomas (1977) that microscale effects may result in the formation of calcium carbonate deposits, even when the bulk solution is not saturated.

These same questions can be evaluated by plotting the data from Table 3-3. Figures 3-37 and 3-38 show the measured corrosion rates for black iron and galvanized steel versus the LI and the Ryznar index. Neither index shows a very strong correlation, particularly the Ryznar index. It is not clear that these indices should be used as often as they are for estimating the corrosivity of a water.

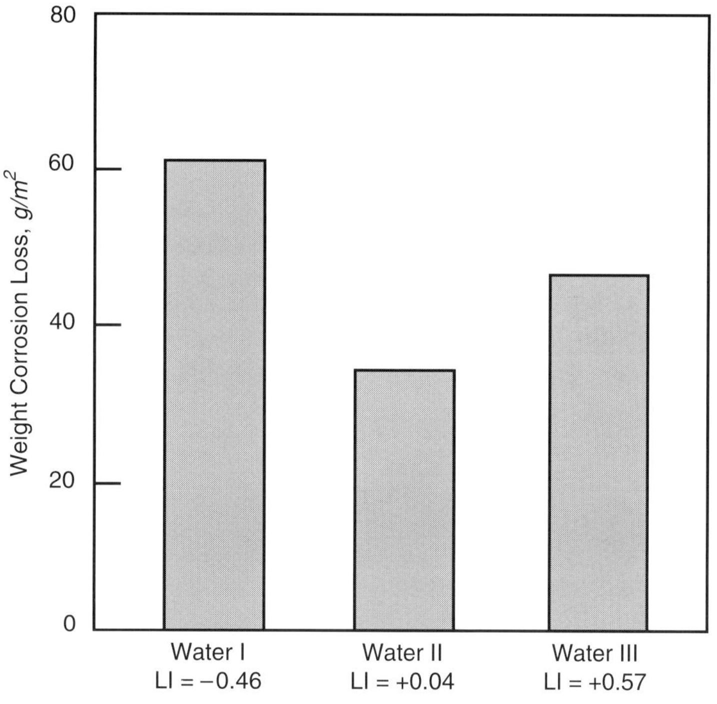

Source: Pisigan and Singley (1985).

Figure 3-36 Gainesville studies: Influence of $CaCO_3$ saturation on corrosion rate of galvanized pipe

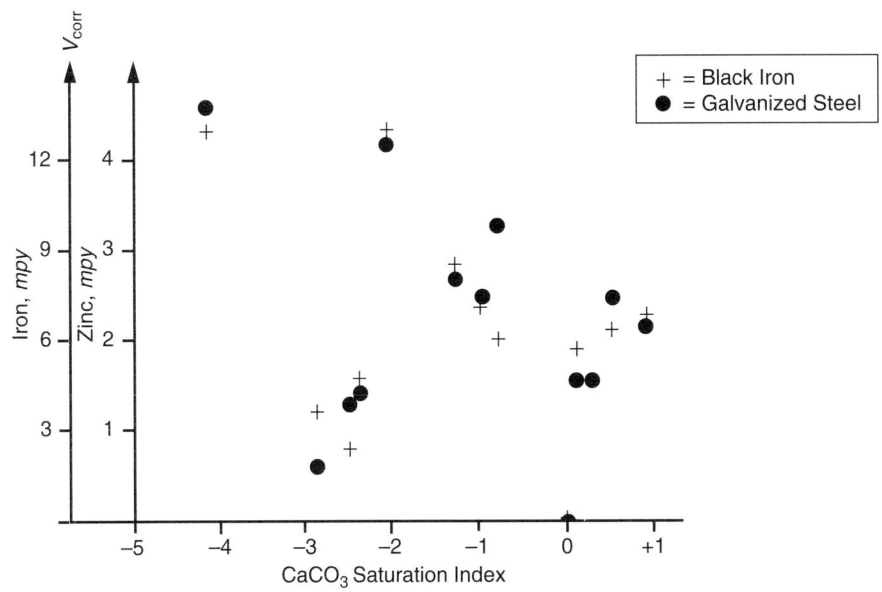

Figure 3-37 Measured corrosion rates (Table 3-3) for black iron and galvanized steel versus the $CaCO_3$ saturation index

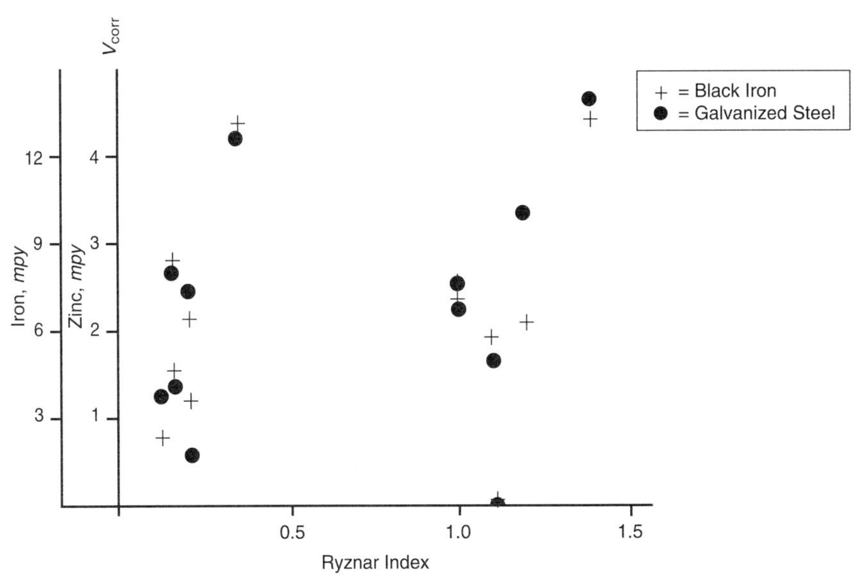

Figure 3-38 Measured corrosion rates (Table 3-3) for black iron and galvanized steel versus the Ryznar index

As suggested by Trussell (1985), the potential for the formation of a calcium carbonate deposit is a complex phenomenon. It is, perhaps, best estimated by a combination of three indices: (1) a bulk solution equilibrium index like Langelier's (1936) or the free energy driving force of the reaction itself; (2) a capacity index such as the calcium carbonate precipitation potential (CCPP) proposed by Merrill and Sanks (1977a, 1977b, 1978); and (3) some measure of the buffer intensity of the solution with respect to calcium carbonate saturation, such as that proposed by Trussell, Russell, and Thomas (1977).

Silicates

Since the early work of Thresh (1922), corrosion inhibiting properties have been attributed to the silicates. Much of this is due to the efforts of Speller (1926) and Cox (1934), who employed silicates to control corrosion problems in the soft waters of the state of New York. Speller, Cox, and their colleagues (Streicker 1938) also ascribed to silicates' protective properties for galvanized materials. Since that time, the work that has been done on the use of silicates to inhibit the corrosion of galvanized pipe has focused on its effectiveness in hot-water systems where the potential reversal problem often causes pitting (Lehrman and Schuldener 1951, 1952; Schuldener and Lehrman 1957; Lane et al. 1973; Lane, Larson, and Schilsky 1977). Generally, this work shows that although silicates appear to be an effective means for controlling hot-water corrosion of galvanized pipe, the levels of silicate required are rather high (8 to 25 mg/L), limiting their use to certain large industrial and institutional applications.

Orthophosphates

The first successful use of orthophosphate for controlling corrosion of iron pipe, with and without galvanizing, occurred in Long Beach, California. The city had experienced serious problems with corrosion of these materials following the introduction of Colorado River water to their supply. Murray (1971) proposed an inhibitor formulation composed of zinc sulfate, orthophosphate, and sulfamic acid and showed data demonstrating its effectiveness. Since that time, most of the published studies concerning the use of this inhibitor combination (Kelley, Kise, and Steketee 1973; Mullen 1974) have been concerned with its effectiveness in controlling the corrosion of iron pipe. Generally, the inhibitor has proven successful in hard waters and a little less successful in softer waters, though there are exceptions to both. The sulfamic acid has been found to be unnecessary.

Because orthophosphates generally form dense, rather insoluble scales with iron, zinc, and calcium, their use as corrosion inhibitors has been considered by practitioners for some time (Gilbert 1948). Nancollas (1983) demonstrated that precipitation kinetics and the phenomena that influence it can provide important insight into the conditions under which orthophosphate inhibitors can be expected to work successfully.

Recent work in Germany indicates that the use of orthophosphates can, indeed, be a useful means of controlling the corrosion of galvanized pipe used in domestic plumbing, particularly the Zinkgeriesel mentioned earlier (AWWA Research Foundation and DVGW, Engler-Bunte Institut 1985; Werner 1984). The studies conducted at Landeswasserversorgung are particularly interesting.

Werner (1984) investigated the influence of pH, humic substances, carbonate hardness, and silicates on corrosion of galvanized materials, as well as the effect of orthophosphates. Chapter 8 on mitigation summarizes some of these results, showing that orthophosphates were the most effective, but all of them had some effect on the corrosion rate.

After several years of work, the German investigators developed the following equation. It is used to estimate the relationship between the corrosion rate of galvanized pipe, the level of natural inhibitor, the level of orthophosphate, and the pH.

$$V_L = K \frac{1}{C_0 + C_{PO_4}} \cdot \frac{10^7}{10^{pH}}$$

Where:

V_L	=	corrosion rate of the zinc–iron alloy layer, (g Zn/m^2)/d
K	=	correction factor, which accounts for the flow velocity
C_0	=	basic inhibition (natural inhibition of water)
C_{PO_4}	=	concentration of orthophosphate

Figure 3-39 from Werner (1984) illustrates the relationship between orthophosphate concentration and corrosion rate at one particular pH. It also illustrates the use of the above equation for the waters Werner tested. The relationship shown is based primarily on work with groundwaters in southern Germany, and additional work is required to confirm its use in other water qualities. Though it may not be clear how effectively these specific results and equations may translate to experience elsewhere, it is clear that inhibition with orthophosphate should probably be considered for situations in which corrosion of galvanized materials is excessive.

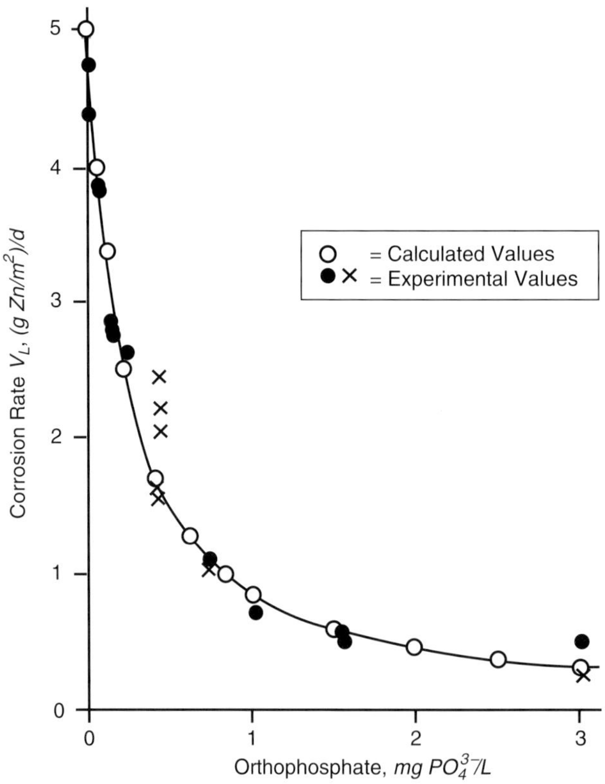

Source: Werner (1984).

Figure 3-39 Influence of phosphate on corrosion of galvanized pipe

CONCLUSIONS

1. Galvanized pipe can be expected to give good service in hard or soft waters in the pH range of 7 to 8.5. Generally, better service is achieved with hard water. In waters that are more acid than this, galvanized pipe should be used only as a last resort.

2. When installing galvanized pipe, the responsible party should perform some testing to ensure that the quality of the pipe meets the appropriate standards. Experience has shown that inferior quality pipe is often distributed in the market.

3. When investigating corrosion problems where pipe quality may be an issue, it is important to learn about the manufacturing process and the characteristics that are associated with common errors in the process.

4. Galvanized pipe should be used with caution in hot-water systems above approximately 50°C because potential reversal and pitting can result. The practitioner should seek to find the experiences of others using the same water and materials.

5. In recirculating hot-water systems containing brass or copper fittings or in other systems where copper is used as an algicide, there is an increased risk of rapid failure due to pitting.

6. Generally, galvanized pipe will not give good service in waters where the corrosion rate of both zinc and iron are high.

REFERENCES

ASTM A53-93A. *Standard Specification for Pipe, Steel, Black and Hot-Dipped, Zinc-Coated, Welded and Seamless.* Philadelphia, Pa.: American Society for Testing and Materials.

ASTM A90/A90M. 1993. *Standard Test Method for Weight (Mass) of Coating on Iron and Steel Articles With Zinc or Zinc-Alloy Coatings.* Philadelphia, Pa.: American Society for Testing and Materials.

ASTM A123-89A E1. *Standard Specification for Zinc (Hot-Dip Galvanized) Coatings on Iron and Steel Products.* Philadelphia, Pa.: American Society for Testing and Materials.

ASTM A239-89. *Standard Test Method for Locating the Thinnest Spot in a Zinc (Galvanized) Coating on Iron or Steel Articles by the Preece Test (Copper Sulfate Dip).* Philadelphia, Pa.: American Society for Testing and Materials.

ASTM A385-80 (91). *Standard Practice for Providing High-Quality Zinc Coatings (Hot-Dip).* Philadelphia, Pa.: American Society for Testing and Materials.

ASTM B6-87(92 E1). *Standard Specification for Zinc.* Philadelphia, Pa.: American Society for Testing and Materials.

ASTM D268894. *Standard Test Methods for Corrosivity of Water in the Absence of Heat Transfer (Weight Loss Methods).* Philadelphia, Pa.: American Society for Testing and Materials.

AWWA Research Foundation and DVGW, Engler-Bunte Institut. 1985. *Internal Corrosion of Water Distribution Systems.* 1st ed. Denver, Colo.: American Water Works Association Research Foundation (pp. 593–604).

Bächle, A., et al. 1981. Corrosion of Galvanized and Unalloyed Steel Tubes in Drinking Water of Different Hardness and Increased Neutral Salt Content. *Werkstoffe und Korrosion,* 32:435–442.

Baes, C., and R. Mesmer. 1976. *The Hydrolysis of Cations.* New York: Wiley-Interscience.

Bilinski, H., and W. Stumm. 1976. Determination of the Stability Constants of Some Hydroxo and Carbonato Complexes of Lead(II), Copper(II), Cadmium (II), and Zinc(II) in Dilute Solutions by Anodic Stripping Voltammetry and Differential Pulse Polarography. *Anal. Chem. Acta,* 84:157.

Cox, C.R. 1934. Equipment for the Chlorination of Small Water Supplies. *Jour. AWWA,* 26:1587–1601.

Cruse, H. 1971. Dissolved-Copper Effect on Iron Pipe. *Jour. AWWA,* 63:79–81.

Dangel, R.A. 1975. *Study of Corrosion Products in Seattle Water Department Tolt Distribution Systems.* EPA 670/2-75-036. National Environmental Research Center. Cincinnati, Ohio: US Environmental Protection Agency.

DIN 2440, Ausgabe Juni 1978. Stahlrohre; Mittelschwere Gewinderohre. Berlin, Germany: Beuth-Verlag GmbH.

DIN 2444, Ausgabe Januar 1984. Zinküber-Züge auf Stahlrohren. Qualitats- norm für die Feuerverzinkung von Stahlrohren für Installationszwecke. Berlin: Beuth-Verlag GmbH.

DIN 17100, Ausgabe September 1966. Allgemeine Baustähle; Gütevorschriften. Berlin, Germany: Beuth-Verlag GmbH.

DIN 50930, Teil 3. Ausgabe Februar 1993. Korrosion der Metalle; Korrosion metallischer Werkstoffe im Innern von Rohrleitungen, Behältern und Apparaten bei Korrosionsbelastung durch Wässer; Beurteilung der Korrosionswahrscheinlichkeit Feuerverzinkter Eisenwerkstoff. Berlin, Germany: Beuth-Verlag GmbH.

DVGW and RHE. 1978. *Untersuchungen zur Zentralen Teilenthärtung des Trinkwassers in Mannheim.* Mannheim, Germany: RHE.

EBI and RHE. 1978. *Untersuchungen zur zentralen Teilenthärtung des Trinkwassers in Mannheim.* Mannheim, Germany: EBI and RHE.

EC (European Community). 1980. Council Directive of 15th July 1980 Relating to the Quality of Water Intended for Human Consumption. (80/778/EEC) *Off. Jour. Eur. Commun.*, 23(L229/11).

Ferri, D., I. Grenthe, S. Hietanen, E. Néher-Neumann, and F. Salvatore. 1985. Studies on Metal Carbonate Equilibria. 12. Zinc(II) Carbonate Complexes in Acid Solutions. *Acta Chim. Scand.,* 39:347–353.

———. Fontana, M.G., and N.D. Greene. 1978. *Corrosion Engineering.* 2nd ed. New York: McGraw-Hill.

Fouillac, C., and A. Criaud. 1984. Carbonate and Bicarbonate Trace Metal Complexes: Critical Re-evaluation of Stability Constants. *Geochem. Jour.,* 18:297–303.

Fox, K., C. Tate, and E. Bowers. 1983. The Interior Surface of Galvanized Steel Pipe: A Potential Factor in Corrosion Resistance. *Jour. AWWA,* 75:84–86.

Gilbert, P. 1948. An Investigation Into the Corrosion of Zinc and Zinc-Coated Steel in Hot Waters. *Sheet Metal Industries,* pp. 2003–2012, October; pp. 2243–2254, November; pp. 2441–2460, December.

Gouda, V., M. Khedr, and A. Shams El Din. 1967. Role of Anions in the Corrosion and Corrosion-Inhibition of Zinc in Aqueous Solutions. *Corros. Sci.,* 7:221–230.

Grasha, L., W. Risner, and M. Davis. 1981. *Corrosive Effects of Dissolved Copper on Galvanized Steel, Phase II.* Internal Report. LaVerne: Metropolitan Water District of Southern California.

Grasha, L., et al. 1980. *Corrosive Effects of Dissolved Copper on Galvanized Steel.* Internal Report. LaVerne: Metropolitan Water District of Southern California.

Guest, R. 1956. The Potentials of Zinc and Steel in Tap Water. *Can. Jour. Tech.,* 34:245–247.

Haney, E. 1982. The Zinc-Steel Potential Reversal in Cathodic Protection. *Materials Performance,* 21:44–51.

Hoxeng, R. 1950. Electrochemical Behavior of Zinc and Steel in Aqueous Media—Part II. *Corrosion,* 6:308–312.

Hoxeng, R., and C. Prutton. 1949. Electrochemical Behavior of Zinc and Steel in Aqueous Media. *Corrosion,* 5:330–338.

ILZRO (International Lead Zinc Research Organization). 1967. *Surface Characteristics of Zinc in Aqueous Media.* Final Rept. International Lead Zinc Research Organization Project ZE-20.

JMM (J.M. Montgomery Consulting Engineers, Inc.). 1982. *Internal Corrosion Mitigation Study, Final Report.* Report prepared for the Portland, Oregon, Water Bureau.

Jones, D. 1984. Polarization Studies of Brass–Steel Galvanic Couples. *Corrosion,* 40:181–185.

Kelley, M., M. Kise, and F. Steketee. 1973. Zinc/Phosphate Combinations Control Corrosion in Potable Water Distribution Systems. *Materials Protection and Performance,* 12:28–30.

Kennedy Engineers. 1976. *Seattle Internal Corrosion Study*. Study conducted for the Seattle Water Department.

Kenworthy, L. 1943. The Problem of Copper and Galvanized Iron in the Same Water System. *Jour. Inst. Metals,* 69:67–90.

Kenworthy, L., and M. Smith. 1944. Corrosion of Galvanized Coatings and Zinc by Waters Containing Free Carbon Dioxide. *Jour. Inst. Metals,* 70:463–489.

Kruse, C.-L. 1983. Korrosionsverhalten von Zink und feuerverzinktem Stahl in kaltem und erwärmtem Trinkwasser unter besonderer Berücksichtigung des Transportes in Rohrleitungen. Ph.D. diss., Technische Universität, München.

Kuch, A., H. Sontheimer, and I. Wagner. 1983. Die Messung der Aufeisenung im Trinkwasser—Ein neues Verfahren zur Beurteilung von Deckschichten in schwarzen Stahlrohren. *Werkstoffe und Korrosion,* 34:107–111.

Lane, R., T. Larson, and S. Schilsky. 1977. The Effect of pH on the Silicate Treatment of Hot Water in Galvanized Piping. *Jour. AWWA,* 69:457–461.

Lane, R., et al. 1973. Silicate Treatment Inhibits Corrosion of Galvanized Steel and Copper Alloys. *Materials Protection and Performance,* 12:32–37.

Langelier, W.F. 1936. The Analytical Control of Anticorrosion Water Treatment. *Jour. AWWA,* 28:1500.

Lehrman, L., and H. Schuldener. 1951. The Role of Sodium Silicate in Inhibiting Corrosion by Film Formation on Water Piping. *Jour. AWWA,* 43:175–189.

———. 1952. Action of Sodium Silicate as a Corrosion Inhibitor in Water Piping. *Ind. Eng. Chem.,* 44:1765–1769.

Mansfield, F. 1971. Area Relationships in Galvanic Corrosion. *Corrosion,* 27:436–442.

Mattigod, S.V., and G. Sposito. 1977. Estimated Association Constants for Some Complexes of Trace Metals With Inorganic Ligands. *Jour. Amer. Soc. Soil Sci.,* 41:1092.

McKee, C. 1932. Effect of Copper Sulphate on Pipe Materials. *Power Plant Eng.,* 36(17):658–659.

Merrill, D.T., and R.L. Sanks. 1977a. Corrosion Control by Deposition of $CaCO_3$ Films: Part 1, A Practical Approach for Plant Operators. *Jour. AWWA,* 69 (11):592–599.

———. 1977b. Corrosion Control by Deposition of $CaCO_3$ Films: Part 2, A Practical Approach for Plant Operators. *Jour. AWWA,* 69(12):634–640.

———. 1978. Corrosion Control by Deposition of $CaCO_3$ Films: Part 3, A Practical Approach for Plant Operators. *Jour. AWWA,* 70(1):12–18.

Meyer, E. 1980. Beeinträchtigung der Trinkwassergüte durch Anlagenteile der Hausinstallation—Bestimmung des Schwermetalleintrags in das Trinkwasser durch Korrosionsvorgänge in metallischen Rohren. *DVGW-Schriftenreihe Wasser Nr.,* 23:113–131.

Mihailov, G., and M. Iovchev. 1981. Corrosion Studies of Galvanized Steel and Zinc in Soft Water at Various Temperatures. In *Proc. of the Eighth International Conference on Metallic Corrosion.* Mainz, Germany.

Morel, F., and J. Hering. 1993. *Principles and Applications of Aquatic Chemistry.* New York: John Wiley and Sons.

Mullen, E. 1974. Potable Water Corrosion Control. *Jour. AWWA,* 66:474–479.

Murray, B. 1971. A Corrosion Inhibitor Process for Domestic Water. *Jour. AWWA,* 63:659–662.

Nancollas, G.H. 1983. Phosphate Precipitation in Corrosion Protection: Reaction Mechanisms. *Corrosion,* 39:77–82.

Paulson, A., M. Benjamin, and J. Ferguson, J. 1989. Zn Solubility in Low Carbonate Solutions. *Wat. Res.,* 23:1563–1569.

Pisigan, R.A., Jr., and J.E. Singley. 1985. Effects of Water Quality Parameters on the Corrosion of Galvanized Steel. *Jour. AWWA,* 77(11):76–82.

Pourbaix, M. 1949. *Thermodynamics of Dilute Aqueous Solution, With Applications to Electrochemistry and Corrosion.* London: Edward Arnold and Co.

———. 1966. *Atlas of Electrochemical Equilibria in Aqueous Solutions.* Oxford, England: Pergamon Press.

Ryan, M.P., and J.E. Bauman, Jr. 1978. Thermodynamics of the Zinc Bicarbonate Ion Pair. *Inorg. Chem.*, 17(12):3329.

Schikorr, G. 1939. The Cathodic Behavior of Zinc Versus Iron in Hot Tap Water. *Trans. Electrochem. Soc.*, 76:247–258.

Schindler, P. 1967. Heterogeneous Equilibria Involving Oxides, Hydroxides, Carbonates, and Hydroxide Carbonates. In *Equilibrium Concepts in Natural Water Systems.* Edited by W. Stumm. American Chemical Society Advances in Chemistry Series, No. 67. Washington, D.C.: American Chemical Society.

Schindler, P., H. Althaus, and W. Feitknecht. 1964. Löslichkeitsprodukte und Freie Bildungsenthalpien von Zinkoxid, amorpherm Zinkhydroxid, β_1-, β_2-, γ-, δ- und ε-Zinkhydroxid. *Helv. Chim. Acta*, 47:982.

Schneider, Th., et al. 1982. Zinkgeriesel? Eine Nachbetrachtung mit Ergebnissen aus einer Trinkwasser-Versuchsinstallation der Stadtwerke Biberach an der Riss. *Neue Deliwa Zeitschrift ndz.* (H. 3):90–96.

Schuldener, H., and L. Lehrman. 1957. Influence of Bicarbonate Ion on Inhibition of Corrosion by Sodium Silicate in Zinc–Iron Systems. *Jour. AWWA*, 49:1432–1440.

Schwenk, W. 1976. Beitrag über die Kinetik der Zinkauflösung in fliessenden Leitungswässern. *Werkstoffe und Korrosion,* 27:157–159.

Seattle Water Metals Committee. 1974. *Seattle Water Metal Survey—Interim Report—Sampling Period, Nov. 1 to Dec. 1974.* Seattle Water Department, Seattle, Wash. Unpublished.

Speller, F. 1926. *Corrosion Causes and Prevention*. New York: McGraw-Hill.

Stanley, J., and R. Byrne. 1990. Inorganic Complexation of Zinc(II) in Seawater. *Geochim. Cosmochim. Acta,* 54:753–760.

Stern, M. 1958. Surface Area Relationships in Polarization and Corrosion. *Corrosion,* 14:P329t.

Streicher, L. 1956. Effects of Water Quality on Various Metals. *Jour. AWWA*, 48:219.

Streicker, W. 1938. Sodium Silicates in Water to Prevent Corrosion. *Ind. Eng. Tech.*, pp. 348–351.

Stumm, W. 1960. Investigation on the Corrosive Behavior of Waters. *Jour. San. Engrg. Div.*—ASCE, 86:27.

Stumm, W., and J.J. Morgan. 1981. *Aquatic Chemistry, An Introduction Emphasizing Chemical Equilibria in Natural Waters.* 2nd ed. New York: Wiley-Interscience.

Thomas, J., and L. Ray. 1984. Corrosion—Galvanized Pipe. A report to James M. Montgomery Consulting Engineers, Inc., Pasadena, Calif.

Thresh, J. 1922. The Action of Natural Waters on Lead. *Analyst* (Br.), p. 459.

Treweek, G.P., R.R. Trussell, and R.D. Pomeroy. 1978. Copper-Induced Corrosion of Galvanized Steel Pipe. In *Proc. of the Sixth Annual AWWA Water Quality Technology Conference.* Denver, Colo.: American Water Works Association.

TrinkwV. 1990. Verordnung über Trinkwasser und über Wasser für Lebensmittelbetriebe (Trinkwasserverordnung—TrinkwV). vom 12. Dezember 1990. BGBl. I, 2613–2629.

Trussell, R.R. 1985. *Corrosion. In Water Treatment: Principles and Design.* New York: Wiley-Interscience.

Trussell, R.R., L. Russell, and J. Thomas. 1977. The Langelier Index. In *Proc. of the Fifth Annual AWWA Water Quality Technology Conference.* Denver, Colo.: American Water Works Association.

Uhlig, H. 1948. *Corrosion Handbook.* New York: John Wiley and Sons.

UPC (Uniform Plumbing Code). 1990. Walnut, Calif.: International Association of Plumbing and Mechanical Officials.

USEPA (US Environmental Protection Agency). 1982. *Water Supply Guidance Number 73.* Washington, D.C.: Office of Water Supply.

———. 1991. Lead and Copper in Drinking Water, Final Rule. 56 FR 26460 7. Washington, D.C.: Government Printing Office.

Wagner, I., and R. Dannöhl. 1985. Der Einfluss unterschiedlicher Nitratgehalte auf das Korrosionsverhalten von verzinkten Stahlrohren. *Werkstoffe und Korrosion*, 36:17.

Weast, R., and S. Shulman. 1962. Corrosion of Zinc in Pressurized Dilute Aqueous Solution Systems. *Corrosion*, 18:417.

Werner, G. 1976. Untersuchungen zum Korrosionsverhalten feuerverzinkter Installationsrohre in kalten Trinkwässern. Ph.D. diss., Universität Karlsruhe, Germany.

———. 1983. LW-Bericht Nr. 1983-11, Zweckverband Landeswasserversorgung, Stuttgart.

———. 1984. LW-Bericht Nr. 1984-4. Zweckverband Landeswasserversorgung, Stuttgart.

———. Werner, G., E. Wurster, and H. Sontheimer. 1973. Korrosionsversuche des Zweckverbandes Landeswasserversorgung mit feuerverzinkten Stahlrohren. *gwf—Das Gas- und Wasserfach*, 114:105.

Zirino, A., and S. Yamamoto. 1972. pH-Dependent Model for the Chemical Speciation of Copper, Zinc, Cadmium, and Lead in Seawater. *Limnol. Oceanogr.*, 17(5):661.

4

Corrosion and Solubility of Lead in Drinking Water

Michael R. Schock, Ivo Wagner,
and Roger J. Oliphant

The first part of this chapter provides an overview of the occurrence of lead plumbing in drinking water distribution systems in the United States and parts of Europe, particularly the United Kingdom and Germany. Following this overview is a survey of historical studies that are important in developing a general understanding of the issues of lead corrosion and plumbosolvency (the actual dissolution of lead into drinking water as opposed to the electrochemical deterioration of the material). Many of the studies covered in this section are very old and often obscure enough that they are virtually invisible to most practicing engineers and scientists outside of the research community. Some of these historical studies provide a basis of practical experience in implementing various treatment programs to reduce the corrosivity of water supplies to lead piping through central water treatment.

The next section focuses on combining the results of mainly experimental studies, to develop or summarize solubility and mass-transfer models that can semi-quantitatively or quantitatively predict the response of plumbosolvency to various water chemistry factors. Evidence of lead control mechanisms is then shown by analyses of pipe corrosion deposits obtained from relevant field and laboratory studies. The final section summarizes the predictions of the lead solubility and mass-transfer

NOTE: The first edition of this chapter was authored by Michael R. Schock and Ivo Wagner.

models (as the state of the art permits) for developing various water treatment strategies that are targeted for plumbosolvency reduction and for estimating their limitations.

OCCURRENCE AND SIGNIFICANCE OF LEAD

Regulation of Lead Levels

Unlike other plumbing materials such as galvanized steel and copper, the corrosion of which is important primarily for economic reasons (high costs of frequent replacement), lead dissolution produces an extremely toxic corrosion product. Lead's influence on water quality is described as a "corrosion damage" by a German standard on corrosion (DIN* 50900 1975).

In the United States, drinking water lead levels are restricted by the Lead and Copper Rule, promulgated in 1991 (*Federal Register* 1991a, 1991b, 1992). This highly complex regulation requires a treatment technique and uses an "action level" of 0.015 mg/L at the 90th percentile to trigger various responses, depending on utility size. Guidance and detailed interpretations of the rule are found in various documents (AWWA 1991; USEPA 1991, 1992). There is still considerable interest in lowering the lead standard even further, and it has become a matter of major debate (Pontius and Warburton 1992).

The "directive relating to the quality of water intended for human consumption" of the European Community (EC 1980) has a maximum admissible concentration value for lead of 50 µg/L in a running water sample. There is an extra clause that states that samples taken either directly or after flushing shall not ". . . frequently or to an appreciable extent" exceed 100 µg/L. However, under this directive, national governments can enact different standards, provided that they are more stringent than those in the directive. For example, in the United Kingdom, the water quality standards are set by the 1989 and 1990 Water Acts. These acts set 50 µg/L as the standard for *any* water sample. Germany decided in its national regulation that 40 µg/L would be the maximum contaminant level (MCL) (TrinkwV 1990).

Recently, the World Health Organization (WHO) revised its guidance for an acceptable level of lead in drinking water to an *average* value of 10 µg/L (WHO 1993). This is intended to protect the most vulnerable members of society, i.e., bottle-fed infants, who are estimated to get 50 percent of their lead exposure from this source. It is difficult to relate this guidance to the standard approach in the United Kingdom, which is a *maximum* value rather than an average. However, should the WHO guidance be adopted as a maximum lead level standard, it would most likely be necessary to remove virtually all possible sources of lead from distribution and domestic plumbing systems in many water supply areas.

An important problem to focus on when determining lead levels is the protocol for sampling the water to determine the lead level. Stagnation time, flushing, and the specific conditions of the installation under consideration have an important influence on the result (Bailey et al. 1986; Britton and Richards 1981; Kuch and Wagner 1983; Schock 1990a). The measurement of the corrosion rate of lead is a relatively minor concern.

*Deutsches Institut für Normung (German Institute of Standardization), Burggrafenstr. 6, Berlin, Germany.

Sources of Lead

In addition to the corrosion of lead interior or service line piping, lead may also enter potable water through the use of lead in pipe jointing compounds and soldered joints; the dissolution of brass and bronze plumbing fixtures; and the use of lead for goosenecks, valve parts, or gaskets in water treatment plants or distribution mains. The extent of lead exposure from water faucet assemblies is an extremely important area to investigate. It is particularly important in controlling the exposure to lead from water for school children and in allowing utilities to meet the 0.015-mg/L 90th percentile action level in the United States. The corrosion of other lead-containing alloys and solders is covered in chapter 6.

An often overlooked area of lead contamination from drinking water is that resulting from lead in cistern materials, or from accumulation of rainwater in cisterns. Some aspects of this problem have been studied by a variety of researchers (Olem and Berthouex 1989; Yaziz et al. 1989; Young and Sharpe 1984). Little published research is available for direct lead contamination from cistern or water storage tanks in the United States, although it is generally acknowledged to be a problem in some areas.

An additional source of lead, according to McFarren et al. (1977), could be the impurities in the coating on galvanized pipe (see chapter 3). Because most historical sampling programs have included water passing through brass faucets, the relative contribution of various materials within a water line is usually ambiguous. Theoretically, several water treatment chemicals could also be a potential source for contamination of the distributed water. Ainsworth et al. (1977) estimated this contribution to be ≤ 1 μg Pb/L. In the United States, these levels are controlled to a large extent by the testing of chemicals and subsequent listing in Standard 60, developed by the National Sanitation Foundation for the Drinking Water Additives Program (McClelland, Gregorka, and Carlton 1989).

Uses of Lead

Historically, lead has been considered a convenient and suitable material for the conveyance of water (Akers and Fellow 1979). Lead is easily formed, cut, and jointed, and its flexibility provides resistance to subsidence and frost. The thickness of lead pipe and its resistance to pitting corrosion also make it a desirable and durable material. Although commercial lead manufacturing processes can produce lead with a purity in excess of 99.99 percent, lead in service normally contains small amounts of other metals that reduce the grain size and improve the strength (Walker and Oliphant 1982). Occasionally, the presence of more noble metals can accelerate the attack of water on the lead pipe (Wolffhügel 1887; Carles 1900).

The extent to which lead has been used in the United States for residential or institutional potable water lines has not been well documented, causing difficulty in determining the locations of target monitoring sites under the Lead and Copper Rule. The use of lead for service line installation was prohibited in the United States by the Lead Contamination Control Act of 1988 (Public Law 100-572). The act also restricted lead in solder to ≤ 0.2 percent and to ≤ 8 percent in other materials in contact with drinking water (e.g., brass faucets).

US surveys. An early survey of the extent of lead service line use in the United States was done by Donaldson (1924). At that time, lead was the second most popular material (behind galvanized pipe) for service lines in the cities surveyed, although it was neither a complete survey nor necessarily a statistically representative one. The greatest relative concentrations of lead service lines were found in New England, New York, the Midwest, Texas, Oklahoma, and Montana.

A later survey of the use of lead in drinking water systems was compiled by Chin and Karalekas (1985). Each of the 10 US Environmental Protection Agency (USEPA) regional water supply branches sent questionnaires or conducted telephone interviews in their respective regions. A total of 153 public water supplies in 41 states was surveyed, with 91 of the systems serving populations of more than 100,000 persons. Approximately 73 percent of the systems stated that lead service lines had been installed in the past. Almost 62 percent reported the use of lead goosenecks and almost half the use of lead sweat joints.

At the time, few water suppliers had active ongoing replacement programs for lead materials that had been previously installed in the distribution system. Nearly all of the responses to the question of whether there was a lead problem were negatives. The authors believed that the extremely limited monitoring program (once per year) accounted for many of the negative responses.

In 1990, a comprehensive survey of lead service lines was commissioned by the American Water Works Association (AWWA). The study was principally a benefit–cost analysis conducted by R.F. Weston, Inc., and Economic and Engineering Services, Inc. (Weston and EES 1990). In gathering information for the analysis, much important information was revealed regarding the occurrence and geographic distribution of lead service lines in the United States. The study used several sources for information, including a USEPA survey conducted by the water supply branches of the 10 USEPA regions and a database compiled from an AWWA lead information survey. Information reported covered a wide range of utility sizes—from ≤3,300 population through the largest cities, including New York, Boston, Chicago, and Los Angeles.

The nationwide occurrence of lead connections between drinking water mains and residences was extrapolated from the survey results. The results indicated that as of 1990 there were approximately 6.4 million lead connections (e.g., pigtails, goosenecks) and 3.3 million lead service lines. The states having the largest number (>100,000) of lead service lines were: Illinois, New York, Minnesota, Michigan, Wisconsin, Ohio, Indiana, Missouri, and Pennsylvania. The survey also indicated that 61,000 lead service lines are being removed on a yearly basis, through various voluntary programs operated by the water utilities.

European surveys. A survey on lead service lines and house installations was made in the United Kingdom in 1975–1976 (DoE 1977). The survey found that about 50 percent of all installations were either completely or partially of lead. Another survey of 167 drinking water supply areas in eastern England found 40 supplies that had a lead solubility problem (Greene 1984). Random daytime samples from more than 2.5 percent of the properties surveyed in these problem supplies had concentrations exceeding 0.10 mg Pb/L. More than 50 percent of the implicated properties were in "hard" surface water supply areas where the problem had not been anticipated.

In 1990, DoE updated the original survey for England and Wales (Chambers and Hitchmough 1992). The new survey found that on average 41 percent of houses still had lead service pipes. In the southeast of England, approximately 19 percent of the houses had lead service branches, while in the northwest part of England approximately 62 percent of the houses had lead service branches. For internal plumbing, the occurrence ranged from approximately 8 percent in the east to approximately 75 percent in the northwest part of England. Also in 1990, 97 percent of all water samples complied with the standard maximum allowable concentration for lead of 50 μg/L for random samples. If the standard had been a maximum of 10 μg/L, the compliance rate would have been 81 percent.

Similar reports do not exist for Germany, but there are rough estimates of the lead use in the larger German water utilities in the part of the country that was formerly West Germany (Wagner and Kuch 1981). Since 1980, the number of known service lines in the former West Germany dropped to less than 100,000 and a significant reduction (approximately 30 to 40 percent) in the house installations took place during the last decade. However, the number for the reunified Germany rose to about 500,000 lead service lines and increased to more than 600,000 house installations with lead.

HISTORICAL BACKGROUND

Investigations in Germany

At the end of the 18th century, the Duke of Württemberg in Germany warned against use of lead pipes in drinking water distribution systems (Württemberg 1793). In 1878, in the same German state, lead pipes were outlawed (Württemberg 1878). Several other German states also outlawed lead pipes between the end of the 19th century and the beginning of the 20th, while in other parts of Germany, mainly in Prussia and Saxony, lead pipes were in use until the 1940s.

In 1886, acute lead intoxications occurred in the German town of Dessau. Wolffhügel (1887) published the case and stated that human diseases could clearly be connected with the high lead concentrations found in the water. The reasons for these lead concentrations were the content of oxygen and free carbon dioxide, chemical composition of water parameters, age and quality of lead pipes (e.g., alloyed metals, copper–lead combinations), stagnation time, flow velocity (contact time), and water temperature.

The lead leaching problem at Dessau was solved by Heyer (1888), who added limestone, and later soda ash, to the water. Cases similar to that in Dessau occurred in Calau, Clausthal–Zellerfeld, Wilhelmshaven, Emden, Offenbach, Helgoland, and Leipzig (Proskauer 1893; Rapmund 1889; Liebrich 1898; Tergast 1899; Pullmann 1887; Höll 1934; Kruse and Fischer 1930; Stich 1930; Fuchss, Bruns, and Haupt 1938). The problems were finally solved by using water treatment that included reduction of free carbon dioxide, which also increased the pH of water, or by ruling out lead as a plumbing material.

In 1981, Wagner and Kuch developed a monograph about various aspects of lead solubility in drinking water and factors governing lead levels at the tap (Wagner and Kuch 1981). They presented a model for the increase in lead with standing time or flow through pipes of various diameters. These results are discussed later in this chapter.

Investigations of the effect of chloride and sulfate on lead corrosion in Berlin tap water and the nature of existing surface corrosion films in the pipe were published by Feller et al. (1984, 1988). They found $PbCO_3$, $Pb_3(CO_3)_2(OH)_2$, and PbO as important components in the somewhat porous surface film. They indicated the stabilization of lead by aqueous complexes such as $PbHCO_3^+$, $PbCO_3^0$, and $Pb(CO_3)_2^{2-}$ may contribute to the scale porosity. They also found that the buffer capacity of the water influenced the current density for the anodic reaction, lower current being produced at higher buffer intensities.

Arts et al. (1987) conducted an intensive study of lead levels produced in multistory apartment buildings. An important conclusion of their study was that flushing water before consumption was of little or no use in reducing lead concentrations to safe levels in buildings having lead interior plumbing. This phenomenon results from the prolonged contact of the water with the lead pipe as it flows to the tap.

In 1989, Wagner presented the results of a study comparing the effect of three different inhibitor formulations on lead pipe in several groundwaters or "bank-filtered" waters (Wagner 1989). The sodium orthophosphate chemical dosed at approximately 0.4 to 3 mg PO4/L substantially reduced the lead concentrations in long-standing samples from excavated lead service lines installed in pipe racks at two water supplies (Table 4-1). This chemical outperformed the orthophosphate–silicate–polyphosphate blend when applied at a third water system, where both chemicals had a similar total phosphate concentration when applied to the water (Table 4-2).

Table 4-1 Lead levels in two water supplies before and after orthophosphate treatment

	Groundwater		Bank Filtration	
Date	Max. Lead Level, *μg/L Pb*	Orthophosphate, $mg/L\ PO_4^{3-}$	Max. Lead Level, *μg/L Pb*	Orthophosphate, $mg/L\ PO_4^{3-}$
July 1983	240	<0.05	200	<0.05
Apr. 1984	210	<0.05	160	<0.05
Oct. 1984	220	<0.05	220	<0.05
July 1985	65	1.5	40	3.0
Jan. 1986	45	1.5	20	3.0
July 1986	65	1.5	40	3.0
Jan. 1987	75	0.7	25	1.5
Sept. 1987	85	0.7	35	1.5
Mar. 1988	100	0.4	20	1.3
Aug. 1988	60	0.4	20	1.3
Dec. 1988	20	0.7	20	1.3

Table 4-2 Lead levels in bank-filtered water depending on inhibitor dosage

	Lead Level, *mg/L*		
Date	No Dosage	Phosphate/Silicate	Orthophosphate
Oct. 1985	180	100	40
Dec. 1986	150	90	25

Investigations in the Netherlands

In 1980, Elzenga and Graveland published a summary of field and experimental studies of lead pipe dissolution in the Netherlands (Elzenga and Graveland 1980). They estimated the existence of more than 1.2 million lead service pipes and more that 1.5 million lead household installations in that country. Analyses were conducted of corrosion deposits in the existing pipes, and of tap water samples after 16 hours of stagnation from various water supplies. The highest pH values represented were approximately 8.7 in a 1950 study cited by the researchers and 8.2 in their newer study. A distinct correlation was found between pH and lead levels, with the lowest lead concentrations associated with the highest pH values. They found no significant correlation between lead levels and either hardness or calcium carbonate saturation state. Scales in lead pipes removed after 30 to 80 years of exposure were predominantly composed of the basic lead carbonates, $2PbCO_3{\cdot}Pb(OH)_2$ and $3PbCO_3{\cdot}Pb(OH)_2$; the normal lead carbonate, $PbCO_3$; and PbO. The scales varied with the water composition. In tin-lined pipes, SnO and SnO_2 were found, in addition to some lead solids. Deposit analyses were done by X-ray diffraction.

Complementing their field investigation were several pipe rack studies, including an attempt to use multiple-phase linear regression analysis to explain the sources and impacts of major water quality parameters on metal lead, copper, and zinc levels. Stagnation curves were also mathematically developed along the same lines as the work by Kuch and Wagner (1983).

In 1984, Haring published an extensive compilation of data from a variety of studies of lead in drinking water in the Netherlands. The compilation included studies of human exposure, chemical corrosion studies, and impacts of copper and lead originating from household corrosion on wastewater treatment and sludge metal concentrations (Haring 1984).

Proportional sampling devices were installed at 200 homes in a pilot study in Amsterdam. The lead concentrations were found to be skewed to the right, indicating a distribution tending toward lognormal. Good correlations were found for higher lead levels with longer lengths of lead pipes and longer stagnation times. The use of proportional sampling was extended then to a wider survey of 19 communities representing different water qualities throughout the country.

In this extended survey, using data from 50 sites in each community, Haring concluded that

- variability between communities was generally greater than variability within a given community
- pH had the greatest influence on overall plumbosolvency levels
- the variability within each community could not be significantly related to any single factor, e.g., length of piping, consumption pattern, etc.

Data on the water qualities in the communities covered by this survey and the relationship of average lead concentration to pH are shown in Table 4-3 and Figure 4-1. It was concluded that decreasing plumbosolvency by increasing pH sometimes in combination with partial chemical softening of the water was the best short-term measure for reducing water lead exposure.

Haring also conducted experiments in which he used a polarization resistance measurement system to determine corrosion rates of lead specimens in Amsterdam water supplies. He tested both untreated "hard" water and water that had been partially softened. Contrary to direct measurement of metal levels in the water,

Table 4-3 Geometric mean of lead concentration in tap water and general chemical characteristics of the waters

Community	Pipe Type*	Geometric Mean Lead, μg/L	% Samples <50 μg/L	pH	Calcium, mg/L	HCO_3, mg/L as $CaCO_3$°	Temperature, °C	LI†
Leeuwarden	L,LT	75	20	7.55	105	405	11	+0.16
The Hague	L	64	27	7.70	89	245	12	–0.16
Rotterdam	L	32	69	8.05	41	139	14	–0.14
Roosendaal	L,LT	42	54	7.60	75	305	11	+0.06
Arnhem	LT	8	98	7.33	20	89	12	–1.28
Tilburg	LT	3	100	7.95	66	274	11	+0.28
Gouda	L	19	90	7.70	85	232	11	–0.06
Haarlem	L	50	36	7.30	88	313	15	–0.17
Zwolle	LT	5	100	7.70	78	207	11	–0.13
Bolsward/Sneek	LT	7	98	7.55	112	480	11	+0.27
Katwijk/Leiden	L	40	63	7.41	112	285	16	–0.07
Utrecht	LT	4	100	7.60	40	146	17	–0.42
Groningen	L	34	82	7.54	62	227	17	–0.14
Maastricht	LT	11	94	7.30	117	408	10	+0.02
Amsterdam	L	13	90	8.00	87	244	10	+0.10
Zutphen	L	68	16	7.60	94	373	11	+0.10
Oudenbosch/Zevenbergen	L,LT	15	74	7.65	99	408	11	+0.26
Eindhoven	LT	5	98	8.15	37	152	11	–0.03
Wasiwijk	L	79	13	6.95	31	98	10	–1.51

Source: After Haring (1984).

*L = lead, LT = lead with tin lining.

† LI = Langelier index.

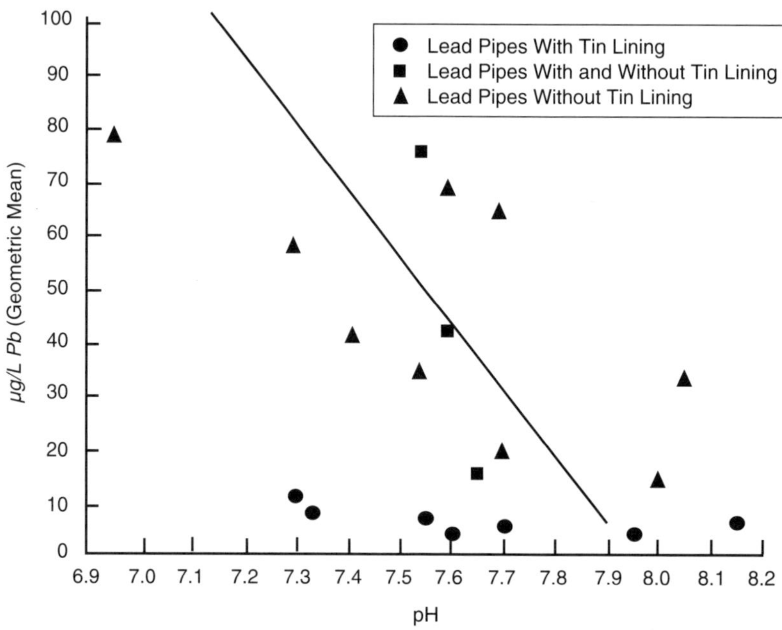

Source: Haring (1984).

Figure 4-1 Relationship of geometric mean lead concentration to pH in several Dutch water supplies which showed the higher-pH softened water to be less plumbosolvent, the electrochemical analysis indicated the untreated harder water was less corrosive. He noted that a source of the disagreement was probably that the polarization resistance technique gives information about the total anodic dissolution of lead, even though the liberated lead may precipitate on the pipe surface and thereby not be determined through concentration measurements. In addition to reducing lead and copper concentrations in drinking water, Haring suggested that considerable benefit from partial central water softening would be derived from reducing metal content in wastewater treatment sludges.

The problems encountered with investigations of corrosivity and corrosion control based on tap sampling led to a new experimental approach recounted by van den Hoven (van den Hoven 1985, 1987; van den Hoven and van Eekern 1988). A "lead-pipe-test" was developed using old lead pipes carefully removed from homes or the ground and installed in standardized rack systems. Initial erratic performance was encountered after disturbance of the pipes dislodged particles or damaged the passivating films in the pipes. With months of continuous flushing, restabilization took place, and the systems could be used for plumbosolvency evaluation. Even minor adjustments to the pipe racks or the addition of replicate loops or other equipment could disturb the surface films and necessitate restabilization.

After a viable systematic and reliable protocol for testing was established, tests were conducted to determine lead levels in different water qualities after various lengths of stagnation. From stagnation curves (discussed in more detail later in this chapter), two important parameters were inferred that would make the test systems relevant to conditions at the tap.

The parameter PbMAX was defined as the plateau "equilibrium" value after long stagnation. A second parameter T_{50} was defined as the time at which a concentration of 50 μg/L was achieved. Curves were fitted following the diffusion-based model of Kuch and Wagner (Kuch and Wagner 1983). For most of the pipe rigs, close agreement was attained for the different pipes. Correlations were made between PbMAX and T_{50} values for different water supplies and the results of studies using proportional sampling, as described by Haring (Haring 1984). The relationships for the water supplies tested were

$$\text{PbTAP} = 0.349 \ \text{PbMAX} - 22.3 \tag{4-1}$$

$$\text{PbTAP} = 902/T_{50} + 17.1 \tag{4-2}$$

Where:

PbTAP = the mean lead concentration at the tap in households with lead plumbing, μg/L

In all 30 areas examined with lead pipe tests, PbMAX exceeded 50 μg/L.

To gain insight into the causes of differences among PbMAX values, a multiple regression analysis was performed. The independent variables were taken to be: pH, total inorganic carbon (TIC), dissolved oxygen (DO), chloride, sulfate, and temperature. The relationship

$$\text{PbMAX}(\mu\text{g/L}) = -141 \ \text{pH} + 12 \ \text{T} + 1{,}135 \tag{4-3}$$

Where:

T = temperature, in degrees Celsius

provided the best fit but only explained about 37 percent of the variability. The trends were essentially as predicted by the Water Research Centre (WRc) in Great Britain and USEPA model calculations (Gregory and Jackson 1984; Jackson and Sheiham 1980; Schock 1980, 1981; Schock and Gardels 1983). Ultimately, a recommendation that pH be adjusted to at least 7.8 for the first phase of treatment of plumbosolvency was produced from these studies (van den Hoven and van Eekern 1988).

Investigations in the United States

US studies before 1975. In the United States, the health significance of lead corrosion in plumbing systems was recognized at least as early as 1845 in the northeastern United States (Karalekas et al. 1983; Bunker 1921). Numerous studies between 1900 and 1930 in the northeastern United States observed the attack on lead pipe by waters that were generally acidic, with low alkalinities (Donaldson 1924; Bunker 1921; Weston 1920; AWWA 1886; Mason 1904). In many cases, the plumbosolvency was not considered significant despite concern about health consequences (Horsford 1849). Some of these studies have been reviewed by Karalekas et al. (1983), who also noted that as early as 1920, Weston demonstrated that plumbosolvency at Milford, Massachusetts, could be decreased through lime addition. Presumably, the lime was successful because it increased the pH of the water.

In 1923, Howard reported data for the 1909–1922 period in which the New Hampshire Board of Health frequently analyzed private drinking water supplies for lead (Howard 1923). The lead typically originated from small aqueducts and transport systems from hillside springs. Howard showed that, in this period, the percentage of sites showing at least 0.5 mg Pb/L dropped from 49.5 to 18.7, still a sizable number. In the 1909–1910 period, approximately one fifth of the total number of private sources examined were condemned because of "dangerous" amounts of lead. An examination of numerous pipe sections indicated that the white coating usually present on the lead piping materials consisted of the "ordinary hydrated carbonate" of lead, not "lime salts" as was commonly assumed. Further, some of the pipe sections had a very thick and bulky brownish super-coating that was "vegetable" in nature and loosely adherent. Therefore, with any sudden vibration of the pipe or with sudden water velocity increases (e.g., as a result of turning a faucet on and off), the coating could easily be dislodged, possibly causing increased dissolution of freshly exposed pipe surface.

Donaldson (1924) reported several simple lead solubility experiments described by Thresh (1922a). Distilled water, free of dissolved gases, had "no action" on lead. Dissolved oxygen in distilled water produced the formation of the colloidal hydrate. If excess oxygen was present, a definitely crystalline deposit formed. When carbon dioxide was present, in addition to oxygen, the "oxycarbonate" was formed (which is probably what others call basic lead carbonate, i.e., $Pb_3(CO_3)_2(OH)_2$) and the solution had a white turbidity. Addition of excess carbon dioxide resulted in clarification of the water and redissolution of the white precipitate. Unfortunately, the pH values of the experiments were not reported. It was not analytically possible to measure pH at that time. While several authors at that time discussed the role of free carbonic acid content or some organic acids, Donaldson correctly pointed out the importance of pH in conjunction with other factors. He suggested that the use of limestone contactors or soda ash might be helpful.

In 1926, Baylis conducted several experiments by using lead wool and two water sources—a mineralized spring water and Baltimore, Maryland, tap water (Baylis 1926). He found that lead rapidly formed a coating of either oxide or basic carbonate ($Pb_3(CO_3)_2(OH)_2$) in waters with high levels of free carbon dioxide. The pH rose several units over three days of contact with the lead pipe, indicating corrosion, but lead levels were not published.

In 1928, Rees and Elder published a study on the attack on lead pipe in several Midwestern water supplies and also performed an extensive literature survey of the problem (Rees and Elder 1928). They found several occurrences of lead in water samples in excess of what is now the MCL, but the sensitivity of the analytical method was not up to today's standards, and interpretations must be made with caution. An apparent association of greater leaching with alkalinities in excess of 200 mg $CaCO_3$/L was observed. However, no pH data were available, and the variability of water hardness, chloride, and total dissolved solids (TDS) makes correlations difficult.

At least two studies, reported by Rees and Elder (1928), indicated that silicates could prevent or decrease lead corrosion. Watkins (1886) cited the inhibiting effect of (apparently adsorbed) alumino silicates in Mississippi River water on the solubility of lead. Clark (1927) postulated that either an increase in water hardness or the addition of sodium silicates could be beneficial in this regard.

In 1942, Schaut published an investigation that attempted to look at the action of a water supply on lead pipes under realistic conditions (Schaut 1942). The water had a pH of near neutrality, an alkalinity of 35–50 mg $CaCO_3$/L, and a total hardness of approximately 120 mg/L. He collected evidence that the rate and extent of lead dissolution increased with temperature. Further, he found that increasing the alkalinity increased the lead content at a pH of approximately 7 and that free residual chlorine also appeared to increase the lead solubility. He also examined some old pipe sections and found a white coating that he presumed to be basic lead carbonate.

While knowledge of lead contamination of drinking water supplies in New England dates back to the 1800s, comprehensive monitoring programs followed by remedial treatment changes were not developed until approximately 1975. There is an apparent gap in the scientific literature in the United States between the publication of Schaut's paper (1942) and this time, with the exception of a lead poisoning case study in Nebraska (Jolliff et al. 1959). Three serious cases of lead poisoning were traced to lead service plumbing in multiple-family dwellings. Water analyses showed approximately 0.03 mg/L in flushed samples and 0.4 mg/L upon standing.

The classic New England studies of the 1970s and 1980s. Two medical surveys of lead in potable water supplies found a link to the body burden (Worth et al. 1975; Craun and McCabe 1975), clearly demonstrating the need for immediate action. As part of one of these studies, which analyzed samples from 383 households in three Massachusetts cities (Worth et al. 1975), Karalekas et al. (1976) investigated the role of the plumbing systems in governing the levels of trace metals in the waters.

The three cities selected for the study were Boston, Cambridge, and Somerville. Many houses in those cities were known to have lead or lead-lined water service piping. The treated water lead concentrations entering the homes were relatively small, making an examination of the input of lead solely from household plumbing systems possible. Approximately half of the buildings in Somerville and Cambridge were thought to still have some lead service pipe (Karalekas et al. 1976).

The water supply for Cambridge was slightly harder than that for Boston and Somerville, had a higher alkalinity, had slightly different chemical characteristics (Table 4-4), and underwent different treatment processes. The sources of water supply for the area of the Cambridge studies were the Hobbs Brook and Stony Brook reservoirs. Treatment consisted of alum coagulation, sedimentation, rapid sand filtration, aeration, chlorination, fluoridation using hydrofluosilicic acid, and pH adjustment using sodium hydroxide (Karalekas et al. 1976). Boston and Somerville purchased water on a wholesale basis from the Metropolitan District Commission (MDC), which pipes water from Quabbin Reservoir to Wachusett Reservoir, and then on to the metropolitan area. Treatment consisted only of chlorination and ammoniation.

The procedure used to identify households for sampling and to elicit cooperation has been described by Karalekas et al. (1976). They also developed a useful sampling procedure to isolate water from the interior plumbing and service line that has been applied to many other subsequent studies.

Of the 936 samples taken for the study in which lead was analyzed, 15.4 percent were found to exceed the MCL for lead at that time of 0.05 mg/L. One or more of the samples from a household exceeded 0.05 mg/L in 30 percent of the Somerville residences, 25.5 percent in Boston, and 14.5 percent in Cambridge. Approximately 4 percent of the houses that had lead in excess of 0.05 mg/L had no record of lead pipe. This observation may be caused by several factors in addition to the most obvious one (unknown sections of lead pipe). Karalekas et al. (1976) suggested lead goosenecks, inadequate caulking of poured lead joints in cast-iron pipe, lead gaskets on flanges or for joining large valves and pipes, brass and bronze fixtures, and tin–lead solder as potential sources of lead in the drinking water.

Table 4-4 Chemical characteristics of finished water supplies before distribution

	City	
Characteristics*	Cambridge, Mass.	Boston, Mass., and Somerville, Mass.
pH units	6.9–8.0	6.0–7.0
Dissolved solids	170	30
Chloride	50	7
Sulfate	30	10
Specific conductance, μS	280	62
Hardness (as $CaCO_3$)	56	14
Cadmium	<0.002	<0.002
Copper	<0.010	<0.010
Iron	0.03	<0.006
Lead	<0.021†	<0.013†
Manganese	0.05	<0.010
Zinc	<0.006	<0.006

Source: Karalekas et al. (1976). Reprinted courtesy of New England Water Works Association Journal and Peter Karalekas, Jr.

*All concentrations are in milligrams per litre, unless otherwise specified.

†Detection limit levels at the time of sampling.

Following the initial phase of the Boston lead study, Karalekas et al. (1978) continued their field investigation and broadened it to cover 16 water supplies during the period September 1975 to August 1977. In their report, they stated that their main objectives were to determine what treatments were currently in use, determine how effective they were in controlling lead corrosion in each particular system, and provide some general guidance for water systems faced with similar problems. The sampling protocol was similar to that previously presented (Karalekas et al. 1976). Three systems, aside from the Boston metropolitan area, will be focused on here, to represent some of the important findings of the study. Summaries of the general chemical characteristics and treatment processes used in Chatham and New Bedford, Massachusetts, and Providence, Rhode Island, are given in Table 4-5.

The source water for Chatham was a well water, and the only known installations of lead pipe were goosenecks. Considering the short sections of lead pipe that were present, the finding that 20 of the 30 samples taken showed an increase in lead concentration, with 4 exceeding 0.05 mg/L, is very important.

New Bedford used a soft water supply, and the only forms of treatment were chlorination for disinfection and soda ash for pH adjustment. All 10 houses sampled had lead service lines, and high lead levels were found in the water. A total of 14 samples (of 30) exceeded the 0.05-mg/L MCL, while 20 showed an increase in lead relative to that in the distributed water. The highest value was 0.26 mg/L. As with Chatham, the treatment program was obviously inadequate to halt lead corrosion.

The situation in Providence was different. Although possessing a low alkalinity and a medium hardness, the finished water showed low plumbosolvency and copper release because of its high pH. Only 3 of 30 samples taken showed detectable lead, with the highest value being exactly 0.05 mg/L. The highest copper value reported was only 0.14 mg/L, which is an important factor in evaluating corrosion control effectiveness in systems with mixed plumbing materials.

Table 4-5 Chemical characteristics of several water supplies in New England

	City		
Characteristic*	Chatham, Mass.	New Bedford, Mass.	Providence, R.I.
pH (units)	6.3	7.3	10.1
Hardness (as $CaCO_3$)	20	12	40
Alkalinity (as $CaCO_3$)	3	24	20
Specific conductance, μS	107	109	114
Sodium	11.9	16.5	7.4
Zinc	<0.02	<0.02	<0.02
Copper	<0.02	<0.02	<0.02
Lead	<0.005	0.005	<0.005
Treatment†	A	B	C
Number of houses sampled	10	10	10

Source: Karalekas et al. (1978). Reprinted courtesy of New England Water Works Association Journal and Peter Karalekas, Jr.

*All concentrations are in milligrams per litre, unless otherwise specified.

†Treatment:

A = sodium hexametaphosphate, 0.4 mg/L.

B = chlorination, plus soda ash.

C = aeration, lime, coagulation, sedimentation, filtration, fluoridation, chlorination.

Subsequent to the studies of Worth et al. (1975) and Karalekas et al. (1976), O'Brien (1976) described an independent study of Boston water by city inspectors from June 1975 to December 1975. In O'Brien's study, 300 random samples were taken from neighborhoods throughout the city and analyzed for lead. Nearly 30 percent of all samples collected contained lead above the 0.05-mg/L MCL. A companion survey of nine households in Worcester, Massachusetts, found seven of the nine equaled or exceeded the MCL in samples of "standing water" (O'Brien 1976). Lead levels in running water samples from all nine sites were less than 0.005 mg/L, however, assuming 0.005 would be rounded to 0.01 in O'Brien's table. At that time, O'Brien suggested "increasing the carbon alkalinity" to 15–20 mg $CaCO_3$/L as a way to reduce the lead levels, in combination with pH adjustment to bring about a pH of approximately 8.6 at the consumer's tap.

Further sampling was performed in Boston and Somerville, to continue the monitoring program that was started in 1974. The previous studies (Karalekas et al. 1976; Worth et al. 1975) had established a baseline to which various subsequent treatment strategies could be compared (Karalekas et al. 1978). In June 1976, the Boston MDC began dosing a commercial zinc orthophosphate corrosion inhibitor (Karalekas et al. 1978). No special pH adjustment was made to accompany addition of the inhibitor. For several weeks, a passivation dosage of 13 mg/L (presumably as Zn, but no units were given) was applied, followed by reduction to a maintenance dosage of 3.2 to 4.5 mg/L until termination of the test in December 1976. During the period of February 1976 to August 1977, the water pH ranged from 5.9 to 6.8 and alkalinity ranged from 1 to 10 mg $CaCO_3$/L, although details of their variation in conjunction with the orthophosphate treatment were not published (Karalekas et al. 1978). This pH range has been subsequently shown to be too low for plumbosolvency control. The lead levels observed during the orthophosphate test and subsequent treatment programs are presented in Figure 4-2 (Karalekas et al. 1983). General water characteristics are summarized in Table 4-6. The average lead concentrations in the households studied did not drop to the MCL, although there is possible evidence of some benefit just before and shortly after the end of the testing period. The authors also noted that the lead increase during the summer of 1976 might have been the result of the temperature effect on lead solubility and might not have been the result of the orthophosphate exacerbating the corrosion. Problems with algal growth in storage reservoirs, which may have been related to the use of the orthophosphate, plus the inability to reduce the average lead levels to below the MCL prompted discontinuation of that treatment.

Six months after the end of the orthophosphate test, the application of sodium hydroxide for pH adjustment was instituted. A sizable drop in average lead concentrations took place shortly thereafter, and the downward trend continued through the most recent sampling session reported (August 1981). The MDC treatment involved adding 50 percent sodium hydroxide at a dose of 14 mg/L, which was sufficient to raise the pH high enough to maintain a pH of well over 8 into many of the consumers' taps. Figure 4-2 shows clearly that the lowest lead levels are associated with the highest pH readings. The lowering of the action level for lead by the 1991 regulation, however, has forced a reevaluation of the treatment program to try to further reduce lead levels in consumers' plumbing.

One of the most gratifying findings of the treatment program at the time of sampling was that the treatment program reduced both iron and copper levels in addition to those of lead (Karalekas et al. 1983). The results of the Boston area study are in excellent agreement with the Glasgow, Scotland, case history to be presented subsequently, where the water was very similar in chemical character.

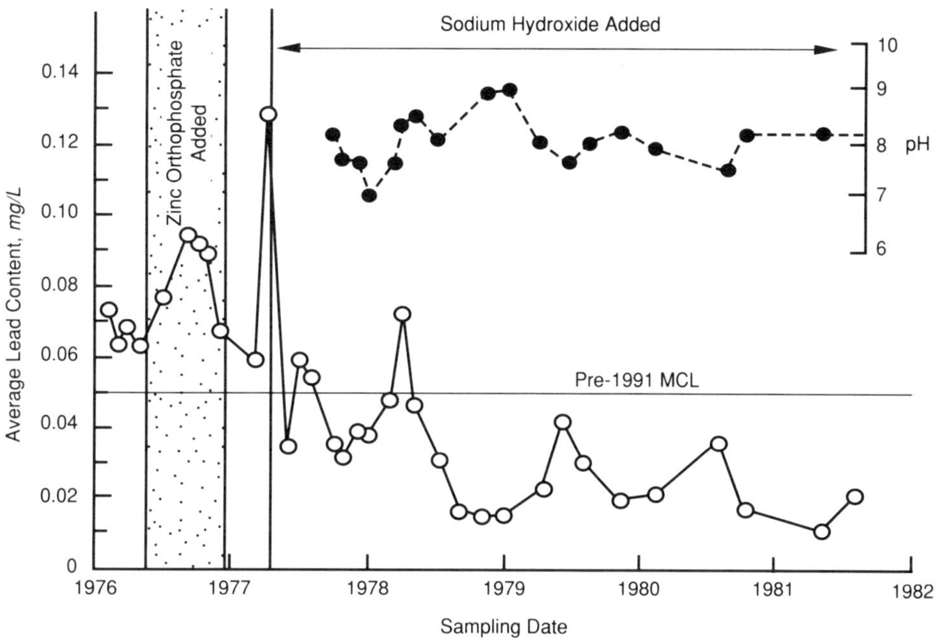

Source: Karalekas et al. (1983).

Figure 4-2 Mean levels of lead in samples taken from Boston, Massachusetts, and Somerville, Massachusetts, 1976-1981

Table 4-6 Chemical characteristics of raw and finished water for the studies of Karalekas et al. (1983)

Characteristic	Water Sample*	
	Raw Water†	Finished Water‡
Hardness (as $CaCO_3$)	12	12
Alkalinity (as $CaCO_3$)	8	12
Total dissolved solids	37	46
Calcium	3.2	3.4
Sodium	5.5	9.7
Sulfate	<15	<15
Chloride	<10	<10
Specific conductance, μS	59	78
pH (units)	6.7	8.5
Copper	<0.02	<0.02
Iron	<0.10	<0.10
Zinc	<0.02	<0.02
Lead	<0.005	<0.005

*All values are in milligrams per litre, unless otherwise specified.

†Water source—Shaft 4, Southborough, Massachusetts.

‡Water source—Norumbega Reservoir, Weston, Massachusetts.

Patterson and O'Brien (1979) conducted coupon test experiments in soft waters with several carbonate levels, using initial pH values of 7 and 8 and exposures of 9 and 30 days. Sodium hydroxide was used to adjust the pH, and sodium bicarbonate was used for carbonate addition. The tests were made under static conditions in sealed containers. Both the pH and alkalinity of the samples were found to be important and interrelated in controlling lead corrosion.

The highest lead levels were found in the absence of carbonate. The pH 8 experiments resulted in lower lead at all levels of alkalinity investigated, and the particular alkalinity (6–95 mg $CaCO_3$/L) did not make much difference in the final lead levels at this initial pH. The higher alkalinities reduced the lead dissolution when the initial pH was 7.0.

Additional experiments were carried out with sections of pipe removed from the distribution system in Boston. Three treatments were investigated: caustic addition for pH adjustment to initial pH 8.5, sodium bicarbonate addition for carbonate adjustment to initial 20 mg $CaCO_3$/L, and caustic plus sodium bicarbonate addition (pH 8.5 and carbonate alkalinity = 20 mg $CaCO_3$/L).

Patterson and O'Brien concluded that the best results were obtained through optimization of both pH and carbonate. However, they suggested that carbonate supplementation alone (which would also raise the pH of these naturally soft waters) provides significantly better results than the classic technique of pH adjustment alone. Whether caustic addition alone can work depends on the initial carbonate concentration of the water. Caustic addition has been shown to be effective in several of the field studies reported in this chapter.

USEPA studies of carbonate and pH relationships. In 1977, the USEPA initiated research into various treatments applicable to lead dissolution control. A lead pipe loop was set up and several field studies were developed. The initial pipe loop experiments were described by McFarren et al. (1977). A detailed schematic and description of the USEPA pipe loop experimental system was given by Schock and Gardels (1983) and Schock (1981).

Three general water qualities were used initially in the USEPA experimental studies. These were chosen based on the assumption that certain models were correct in predicting that the minimization of lead solubility could be achieved at pH 8.3–8.5 by continuing to increase carbonate dosage (Patterson and O'Brien 1979; Mason 1904). In this pH range, total alkalinities of 10–20 mg $CaCO_3$/L, 50 mg $CaCO_3$/L, and 100 mg $CaCO_3$/L were used in the experiments. Later, two additional experiments were conducted at pH levels of approximately 9.25 and 9.5, with alkalinities of approximately 19 and 17 mg $CaCO_3$/L, respectively (Schock and Gardels 1983). The alkalinities of the these two high-pH experiments were selected to maintain a consistent level of dissolved inorganic carbon (DIC) with the low-alkalinity experiment of pH 8.3–8.5, because DIC is independent of pH while alkalinity is not. This is illustrated in Figure 4-3. The chemical and mathematical relationships between DIC and total alkalinity are discussed in detail elsewhere (Loewenthal and Marais 1976; Schock 1980, 1989, 1990b; Snoeyink and Jenkins 1980; Stumm and Morgan 1981). Table 4-7 summarizes the approximate background water quality used in the pipe loop experiment.

Each experimental run lasted three to eight months. In contrast to previous predictions, the lead content of samples standing overnight or of samples standing longer than 24 hours increased with the carbonate content of the water. Although lead levels in very few of the samples remained below 0.05 mg/L, the lead level was found

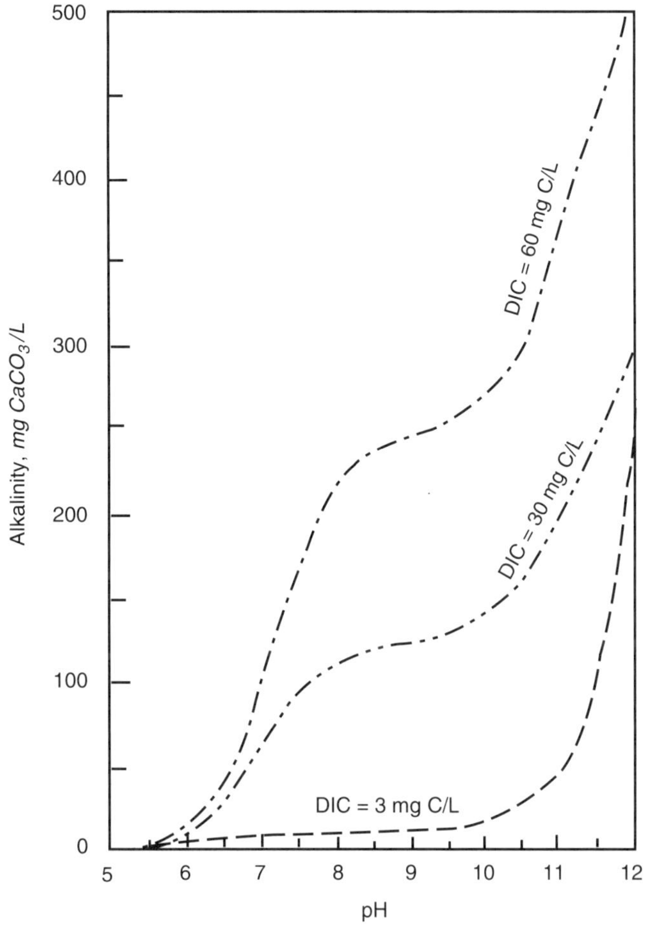

Figure 4-3 Relationship between DIC concentration and alkalinity at 15°C, I = 0.01 mol/L

Table 4-7 Approximate background water quality used in USEPA pipe loop experiment

Constituent*	Concentration, *mg/L*
Calcium	3–6
Magnesium	0.5–1
Sodium	5–50†
Potassium	<0.2
Sulfate	<8
Chloride	2–5
Orthophosphate	<0.2‡
Lead	<0.005
Dissolved inorganic carbon (as C)	3.5–24

*Temperature, 22–25°C.

†The wide ranges in sodium concentration reflect the control of alkalinity by sodium bicarbonate addition.

‡Detection limit for this study.

to be nearest that level in the runs that were over pH 9. The lead levels were also relatively more consistent than in the pH 8.3–8.5 experiments.

Near-equilibrium lead concentrations were apparently reached after "overnight" standing of 7 to 9 hours, which differed little from those standing 21 to 88 hours. X-ray diffraction analysis showed $Pb_3(CO_3)_2(OH)_2$ (hydrocerussite) to be the lead solid coating of the pipe interior in the pH >9 runs.

Data from the low-DIC experiments for standing times >7 hours are presented in Figure 4-4. The pH values shown represent those directly corresponding to the lead concentrations shown (i.e., after standing). The trend of the data closely corresponds to that for theoretical equilibrium with hydrocerussite ($Pb_3(CO_3)_2(OH)_2$).

When the data from the flowing samples were examined, the lowest carbonate level studied gave the lowest and most consistent lead values as well. Sporadic sloughing-off of corrosion product was apparently observed in the 50- and 100-mg/L alkalinity runs, which demonstrated that they not only provided no increased advantage but were actually detrimental. Figures 4-5 and 4-6 show data for running samples in the USEPA studies (Schock 1980; Schock and Gardels 1983). The samples from the pH 9.2–9.5 experiments show the most consistent and effective reduction in the lead levels of flowing samples.

Several samples were also taken after both short and very long standing times, to look at the rate of lead dissolution in the pipe. A rapid release was seen, with 0.05 mg/L exceeded within 1.5 hours. This rate was also seen to decrease at pH >9. These data were fit by the diffusion-based stagnation curve model of Kuch and Wagner (1983) and are discussed in a later section of this chapter.

The data from the USEPA studies were used to develop a working chemical model for lead solubility that agreed extremely well with the experimental evidence (Schock 1980, 1981; Schock and Gardels 1983). A computer program was developed to calculate the equilibrium solubility of divalent lead in the presence of carbonate, orthophosphate, sulfate, and chloride, at different pH values and ionic strengths. The essential feature of the computer model, LEADSOL, was the inclusion of previously neglected aqueous lead carbonate complexes and of the basic lead carbonate solid (hydrocerussite). The total lead solubility was defined as:

$$\begin{aligned} S_t = {} & [Pb^{2+}] + [PbOH^+] + [Pb(OH)_2^0] + [Pb(OH)_3^-] + 2[Pb_2OH^{3+}] + 3[Pb_3(OH)_4^{2+}] \\ & + 4[Pb_4(OH)_4^{4+}] + 6[Pb_6(OH)_8^{4+}] + [PbCO_3^0] + [Pb(CO_3)_2^{2-}] \end{aligned}$$

where brackets indicate molar concentration, which enabled generally accurate predictions to be made. For practical purposes, the polymeric lead species are present in extremely low concentrations and can be neglected over the normal range of pH and DIC in potable waters. The pipe loop data, interpreted in this new solubility framework, were shown to be consistent with earlier published precipitation and coupon experiments of other researchers (e.g., Patterson and O'Brien 1979; Patterson et al. 1977), as well as with most of the field experiences in the United States and Europe. This new picture of lead solubility was much different than the classical $PbCO_{3(s)}$ precipitation models, such as those hypothesized by O'Brien (1976), Patterson et al. (1977), and Pourbaix (1966). As a result of additional computer modeling, a suggestion was made that orthophosphate could indeed be a viable corrosion control alternative, but only within certain pH and carbonate concentration constraints (Schock 1980, 1981; Gardels and Schock 1981). Previously the pH constraints of orthophosphate addition had not been emphasized.

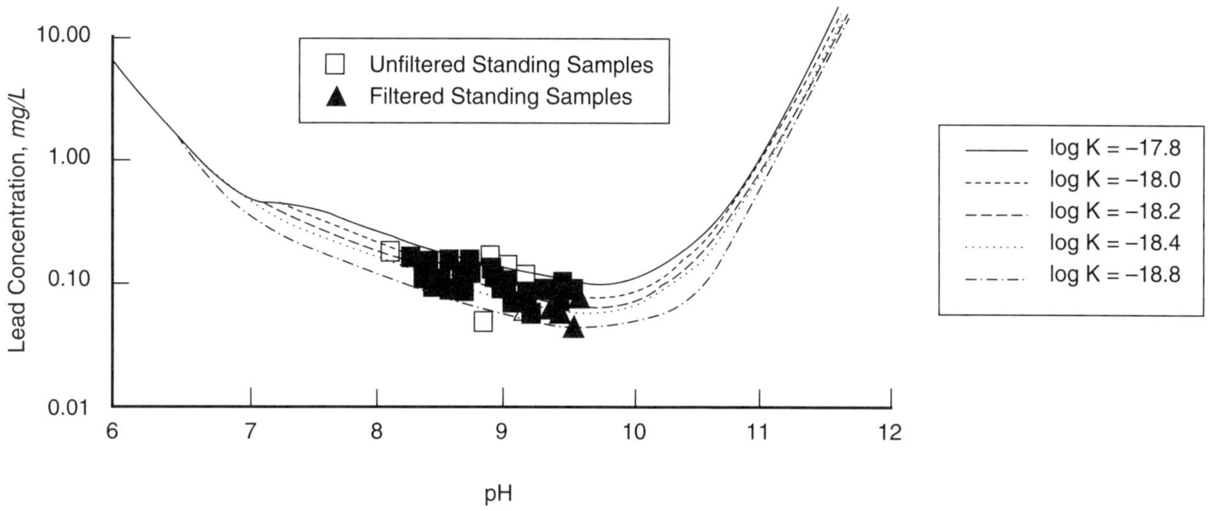

NOTE: Reference lines represent theoretical solubility curves for DIC = 3.6 mg C/L, I = 0.0075 mol/L at two different assumed K_{sp} values.

Source: Adapted from Shock and Gardels (1983).

Figure 4-4 Data for standing time >7 hours in USEPA lead pipe loop experiments

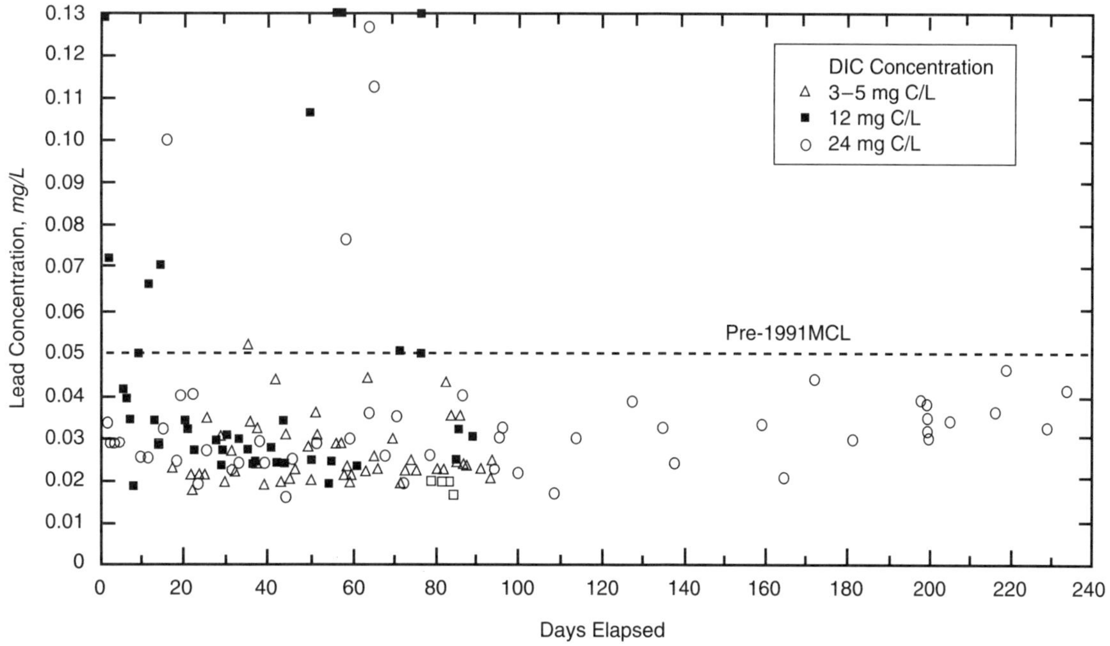

Source: Schock (1980).

Figure 4-5 Total lead concentration in flowing samples from the USEPA pipe loop experiments, taken during the first three experimental runs for DIC = 3-24 mg C/L

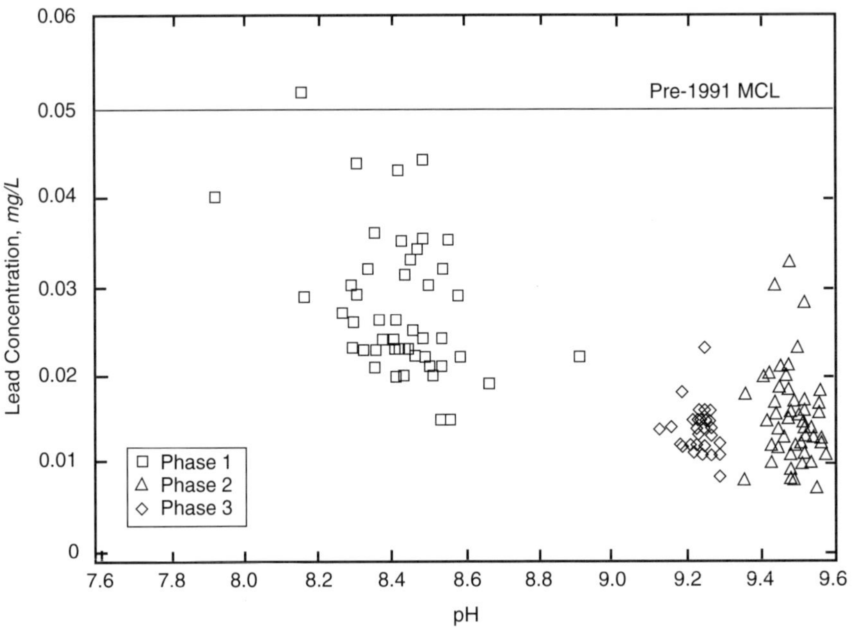

Source: Schock and Gardels (1983).

Figure 4-6 Total lead concentrations in flowing samples from the USEPA pipe loop experiments with DIC concentrations of 2.4–4.8 mg C/L

USEPA investigations of orthophosphate and silicate effects. From 1979 through 1981, the USEPA conducted several long-term lead solubility control experiments. These experiments included a brief look at the effect of silicate (added as reagent sodium silicate) at pH 8.2 in a low-hardness water with an alkalinity of 20 mg $CaCO_3$/L and orthophosphate dosage of approximately 0.5 mg/L as PO_4 at pH 7, 8, and 8.5 in the same water. All orthophosphate was dosed by using analytical reagent-grade dipotassium hydrogen phosphate. Periodically, orthophosphate salt was added to replenish that lost by deposition and/or adsorption. A recirculation system involving a 100-gal (380-L) stainless steel tank with polyvinyl chloride (PVC) piping and a floating lid was used, patterned after that used by Buelow et al. (1980) and Schock and Buelow (1981) for asbestos–cement (A–C) pipe research. The system used for metallic corrosion experiments has been previously described, along with some of the conclusions (Schock and Wagner 1985). All experiments used new 0.75-in.–ID (1.91-cm–ID) pipe. An attempt was made to maintain a reasonable free chlorine residual in all tanks, with most residuals between 0.2 and 0.5 mg Cl_2/L.

Silicate was found to possibly be beneficial to plumbosolvency control at a pH of approximately 8.2, with dosages of greater than 10 mg SiO_2/L, particularly after several months of contact with the inhibitor. No other pH levels were tested. Hydrocerussite, $Pb_3(CO_3)_2(OH)_2$, could be identified on the control loop interior surface, but no solids could be identified on the silicate-treated pipe. Data from these experiments will be discussed in more detail in a later section of this chapter.

In 1980, the theoretical solubility of divalent lead was computed for a system containing several levels of orthophosphate at a DIC concentration of approximately 4 mg C/L, similar to that in the USEPA pipe loop studies (Schock 1980, 1981). Assuming that the solids $PbHPO_4$, $Pb_3(PO_4)_2$, $Pb_4O(PO_4)_2$, and $Pb_5(PO_4)_3OH$

could precipitate and that the aqueous complexes $PbHPO_4^0$ and $PbH_2PO_4^+$ were possible, the calculations indicated that substantial reduction of lead solubility could take place from approximately pH 7 to 9 with an orthophosphate concentration of only 0.5 mg PO_4/L. The effect of 0.5 mg PO_4/L in the presence of carbonate species activities of 20 mg $CaCO_3$/L on the potential–pH diagram (Pourbaix diagram) for lead was also shown. A literature review indicated that lead should be controlled by the extremely insoluble solid chloropyromorphite ($Pb_5(PO_4)_3Cl$). However, although "thermodynamically stable" based on the published solubility constant and geochemical arguments (Nriagu 1973), scant evidence has been uncovered that chloropyromorphite is an important solubility-controlling precipitate in potable waters. Some of the other solids included in the solubility calculations may also be unrealistic, and revisions of the computations will be in order as more information is developed on scales formed by treated drinking waters.

A comparison of solubility calculations for a zinc–carbonate–orthophosphate water system with the equivalent systems for lead predicted that low levels of orthophosphate ion (≤1 mg PO_4/L) alone could control lead solubility (Schock 1980, 1981). The calculations indicated that the solubility of $Zn_3(PO_4)_2 \bullet 4H_2O$ is great enough that lead dissolution levels will be controlled by the orthophosphate ion activity alone, unless high concentrations of free zinc ion, free orthophosphate ion, or both are present. This cast doubt on the belief that was widely accepted at that time that orthophosphate "inhibitor" formulations must necessarily control metal dissolution through the formation of a solid zinc orthophosphate film that would cover up the lead pipe surface.

The orthophosphate recirculation system tests showed that at all three pH levels tested, the lead levels were consistently below the control lead level. The test runs at pH 8.5 showed the lowest dissolved lead levels, but the total lead concentration was comparable to that present with both of the other pH levels. Tests conducted will be reported here. Lead levels generally remained below 0.05 mg/L, even at pH 7.0. In the pH 7 experiments, the lead dissolved rapidly, but then dissolution slowly decreased until approximately stable lead concentrations were attained after approximately five months. Thus, slow formation of a passivating film was indicated. Dissolution was slower at pH 8 and 8.5. The amount of undissolved or colloidal lead was found to increase with pH from 7 to 8.5 when orthophosphate was added. Unfortunately, these experiments did not test any pH between 7 and 8, which is the optimum region shown by solubility modeling calculations (Schock 1989). More results from these experiments will be reported in a later section of this chapter.

X-ray diffraction analyses were performed on deposits scraped from lead pipe specimens from the orthophosphate experiments. The solid $Pb_3(PO_4)_2$ was found on all test specimens. Hydroxypyromorphite, $Pb_5(PO_4)_3OH$, was found on specimens from duplicate experiments at pH 7 and on one specimen from one of the runs at pH 8 and 8.5. Basic lead carbonate, $Pb_3(CO_3)_2(OH)_2$, was found in deposits from the duplicate pH 7 experiments and one of the pH 8 runs. Massicot, PbO, was found in deposits from the pH 7 and 8.5 runs. Even though galvanized pipe was also in the system, providing substantial amounts of dissolved zinc into the water, no zinc compounds were identified on any lead specimen.

Modern studies from throughout the United States. In 1987, Heumann described a domestic sampling program and lead corrosion investigation in Los Angeles, California (Heumann 1987). Most of the lead appeared to be contributed by soldered joints and faucets, more so than lead goosenecks, in the 36 residences and

7 schools sampled. He found considerable variation in lead levels with usage pattern and stagnation time, as would be expected. He found no discernible correlation of lead levels with Langelier index (LI) values, also as would be expected. He also reported many sampling protocol problems and confounding contributions of lead from faucet materials that would be useful in designing future studies.

A 1988 study in the western part of Fairbanks, Alaska, showed 43 percent of the standing tap water samples in excess of 0.05 mg Pb/L (Townsend and Pollen 1988). The water supply was characterized by a pH of close to 7, an alkalinity of approximately 190 mg $CaCO_3$/L, a calcium concentration of 40 mg/L, TDS of approximately 235 mg/L, and a temperature during the sampling period of about 5°C. A mixed ortho/polyphosphate chemical was then added to attempt to reduce plumbosolvency and copper corrosion. A dosage of 6.9 mg/L of chemical was used, resulting in a measured 1.5 mg/L of orthophosphate for a passivating dosage, with later treatment reduced to 2.3 mg/L (0.5 mg/L orthophosphate). While the program initially showed a reduction of first-draw lead levels to below 0.05 mg/L, with the average being 0.03 mg/L, long-term results were not available. Once again, the pH for optimal orthophosphate effectiveness is somewhat higher than that used in the system.

In 1988, Bennett and co-workers reported a summary of lead monitoring and treatment data for 47 US municipalities (Bennett et al. 1988). A particularly interesting observation was that the lead levels in many homes were substantially lower than would be predicted by lead solubility chemistry. This is somewhat in contrast to many reports from the United Kingdom, where there are many occasions of unusually high lead occurrences. Differences in water chemistry and the nature of passivating pipe deposits are likely causes of the differences observed.

An attempt was made to statistically analyze the influence of several chemical and physical factors on lead levels. No significant consistent trends with temperature could be found for the six utilities having multiseasonal monitoring data. Major problems in data interpretation were caused by inconsistency in sampling protocols and inadequate information on background chemistry, ages and types of plumbing materials, small number of samples, etc.

Trends in the field data indicated the most promising water chemistry targets for lead control were a pH between 8 and 10, with a carbonate–bicarbonate alkalinity of 30–150 mg $CaCO_3$/L. However, they suggested utilities must do their own studies to determine the most appropriate control strategies for their specific conditions.

In 1989, follow-up studies in Los Angeles to those previously reported focused on lead contamination from the 15,000 gooseneck connections then present (Heumann 1989). Static fill-and-draw tests in the laboratory indicated further field studies were warranted. Logistical problems finally precluded residential testing, so a laboratory experimental system was constructed. An interesting sampling protocol was developed to try to simulate and follow the dispersion of a slug of lead leached from a gooseneck during the slug's travel to the sample tap. By integrating the area under curves obtained by small-volume sequential sampling, the experimental protocol showed agreement with ±10 percent of the tap and direct gooseneck sampling. Figure 4-7 shows an example of the dispersion of lead originating from the pigtail by the time it reached the sampling tap.

The study concluded that in addition to water temperature and water chemistry, the nominal diameter and size of the gooseneck was important. Somewhat surprisingly, the flow rate of the water did not have much impact on the lead levels. Heumann further concluded that the LI was of little predictive value for

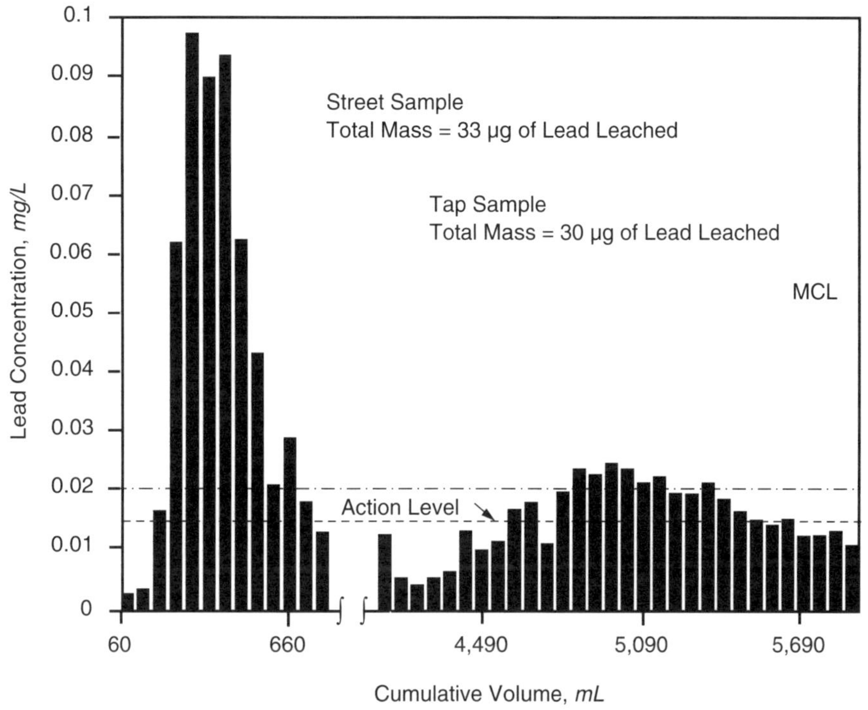

NOTE: Flow Rate = 1.62 L/Min.
Van Norman lead gooseneck, 1 in.
Despite the different profiles, the total lead content of each slug is similar (30μg versus 33 μg).
Source: Heumann (1989).

Figure 4-7 Lead concentration profiles for two different sampling points at the Van Norman site in the Los Angeles lead dispersion study

lead corrosion in the water supplies studied and that, unlike soldered joints, lead piping continues to provide a consistently strong source of lead after many years of water contact.

A comprehensive study of lead levels at the tap in 94 water companies and district members of the American Water Works Service Company was published by Lee, Becker, and Collins (1989). For nonlead service lines, the sampling program consisted of a 100-mL first-draw sample, followed immediately by a 1-L sample. These were taken after at least 6 hours of stagnation. Following the 1-L sample, water was flushed for 2–3 minutes and another 1-L sample was taken. When lead service lines were present, the initial sample was 1,000 mL, taken after flushing the volume of water from the closest tap to the service line, plus 500 mL. Then, the second sample was taken after 2–3 minutes' flushing.

For all sites sampled, information was gathered on the type and age of plumbing materials, electrical grounding, and water quality characteristics. From each district or company, water quality and water treatment data were gathered. Each of the companies or districts collected a sample from one site for every 1,000 customers, which resulted in a total of 1,484 sites. Of these, 1,451 did not have lead service lines, 27 did, and 6 were unknown.

The original article should be consulted for the details of the statistical analyses and data breakdowns. Conclusions on the impact of several variables on 1-L first-draw lead levels are summarized in Table 4-8.

Table 4-8 Conclusions on the impact of different variables on first-draw lead levels for American Water Works Service Company study

Variable	Correlation Found, *Yes / No*	Type of Correlation
pH	Y	Not direct; pH >8 resulted in lower lead levels
Alkalinity	N	
Calcium hardness	Y	
Total hardness	N	
Zinc orthophosphate	Y	Dosages 0.4–1 mg/L were lowest lead level group
Chlorine residual	N	
Sampling site age	Y	Newer copper plumbing had higher lead levels without corrosion control
Sampling site grounding	Y	When pH between 7 and 8, grounding increased lead levels
Length of stagnation	Y	>10 hours standing, lead levels were higher than 6–10 hours standing

Source: Data from Lee et al. (1989).

The interrelationships among pH, alkalinity, DIC, and orthophosphate concentrations were examined in some detail in a review on plumbosolvency control strategies by Schock (1989). The complexation of lead by carbonate was shown to limit the amount of reduction in lead solubility that was theoretically possible through orthophosphate dosing. The complicated nature of the solubility showed possible adverse impacts from zinc insolubility, depending on the formulation of orthophosphate inhibitor used. Many secondary impacts of lead control treatment on other aspects of water use, corrosion of other materials, and waste disposal were also noted. The optimum pH for orthophosphate dosing was shown to be in the range of approximately pH 7.2 to 7.8, depending on the DIC and hardness of the water and depending on the accuracy of existing thermodynamic data for solids and complexes.

Several studies were conducted by the University of North Carolina–Asheville Environmental Quality Institute (UNC–EQI) on North Carolina public schools and tap water samples (Maas and Patch 1990a, 1990b; Maas et al. 1991). Many samples were submitted voluntarily. By 1990, 50 percent of the studies' samples came from outside of North Carolina, but 72 percent of the sites included in the reports were from North Carolina. Samples returned to UNC–EQI were accompanied by answers to a detailed questionnaire to enable statistical categorization. Some of the conclusions from these studies are summarized as follows (Maas and Patch 1990a, 1990b):

- An estimated 26 percent of North Carolina residences had first-draw 500-mL sample lead levels in excess of 0.010 mg/L. Of these, 93 percent can reduce the levels below that level by a 1-min flush.
- Roughly 40 percent of residential tap water sample lead was attributed to brass fixtures. Because of this, tap water lead levels from PVC-plumbed household systems were about as likely to exceed 0.010 mg/L as those from copper-plumbed systems.
- For residences with high 8-hour-standing first-draw lead levels, approximately 34 percent of the lead level was attained after only 15 min of stagnation.

- Increased pH, use of orthophosphate inhibitors, and increasing hardness were found to be beneficial in reducing lead levels.
- Two North Carolina cities with very soft water (≤ 3 mg/L hardness) and low pH (6.8–7) using sodium hexametaphosphate chemicals did not seem to have elevated plumbosolvency. For hexametaphosphate-dosed systems, pH did not appear to have much effect.
- Systems using silicate/phosphate blends showed variation in effectiveness with pH (higher pH seemingly more beneficial).
- Plumbing systems between 3 and 10 years of age (as of 1990) had the highest likelihood of tap water lead contamination.
- A higher percentage (10 percent or more) of homes receiving water from private wells or springs had elevated lead levels, compared to those receiving municipal drinking water.

Schock published two papers on factors contributing to the high variability of lead levels observed in tap water monitoring programs (Schock 1990a; Schock, Levin, and Cox 1988). An examination of data from several field investigations, plus laboratory lead pipe loop experience, indicated the following general observations:

- Physical factors, time, and chemical mass transfer are significant in governing lead levels in nonequilibrium systems, such as those conditions represented by many sampling programs.
- The diameter and length of the pipe is important, as well as the age and chemical history of the solder joints and brass fixtures.
- Within-site and between-site variability far exceeds analytical uncertainty.
- Variability at most sites tends to be characterized by background values that were interrupted by occasional high values ("spikes").
- Variability for sites in studies where comprehensive repetitive sampling was done for "overnight standing" samples (e.g., Glasgow, Boston area) showed approximate relative standard deviations on the order of ± 50 to 75 percent in both treated and untreated water, almost regardless of concentration (Figure 4-8).
- Data distribution was essentially non-normal approaching lognormal for lead in the studies examined. It is especially magnified for small numbers of samples.

From these observations, it was concluded that monitoring studies must include controls for the causes of both between-site and within-site variability. Further, to ensure that results will not be unrepresentative and irreproducible, a statistically valid number of samples and frequency of sampling must be determined and the plan followed.

Vaccari and Mueller (1991) conducted a study that included laboratory coupon experiments, computer solubility calculation, and statistical analysis of monitoring data obtained by the New Jersey Department of Environmental Protection. In their coupon studies with several different water chemistries, they found that the LI and pH were relatively poor predictors of the weight loss of the coupons or lead concentrations in the solutions. They found there was a correlation with the theoretical

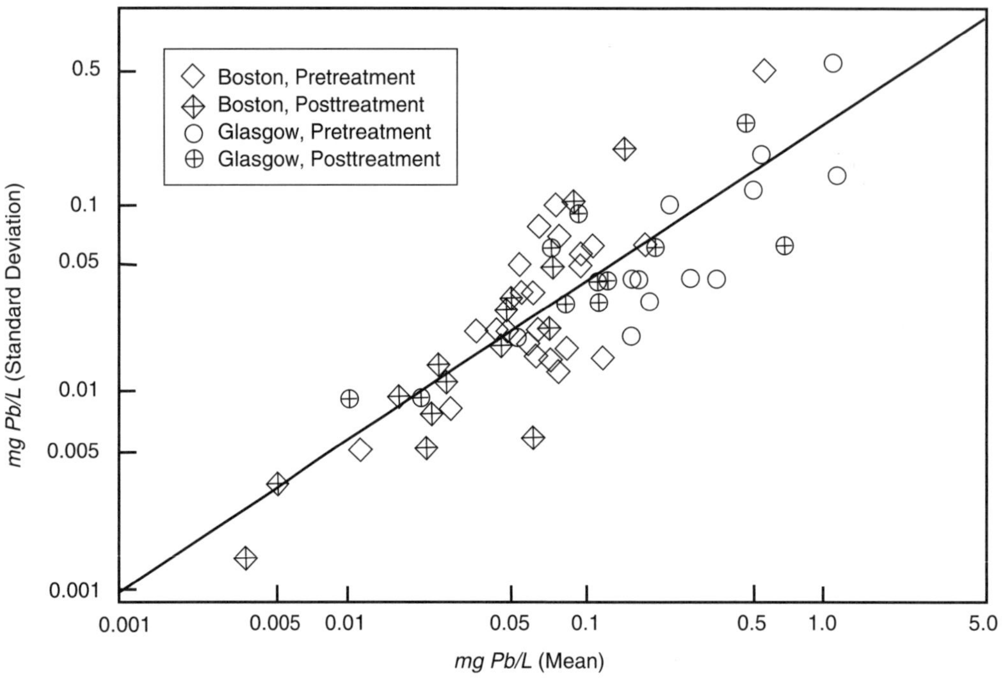

Source: Schock (1990a). Reprinted by permission of Kluwer Publishers.

Figure 4-8 Regression of log-within-site standard deviation on log-within-site mean of Boston and Glasgow data, for the same sites before and after treatment

lead solubility. Their solubility model included significant hydroxide and carbonate complexes of lead.

The field data included descriptive information about the specific sites, including types of plumbing materials, age of the materials, number of soldered joints in the plumbing line, along with water quality information. Both "total" and "soluble" lead were analyzed and incorporated into the database. A step-wise multiple linear regression technique was used to try to find trends and associations.

For first-draw total lead concentrations, none of the measured variables provided a significant association. For first-draw soluble lead, the most significant correlations were with LI (interchangeable with pH in the analysis because of the induced correlation), age of the plumbing, and pipe length. As expected, the trend was for lower lead levels with higher pH–LI, shorter pipe, and older plumbing material. Only about 26 percent of the variability was explained by this relationship, however. Only the pH–LI relationship showed even marginal significance for the soluble lead in the flushed samples. They concluded that particulate material is the frequent cause of excessively high lead levels. They suggest that in soft water systems, much of the lead problem may be virtually "mechanical" in origin, e.g., separation of particles of lead salts from the scale on the pipe surface through vibration, flexing of beams as people walk above the plumbing, thermal expansion and contraction, etc.

Howard (1992) described the general results for the pH–carbonate adjustment program initiated by Bennington, Vermont, in 1977 for their soft, low-pH water supply. The data presented show that lead levels in first-draw tap samples were still decreasing slowly after the initial, almost immediate 90 percent drop when the program was instituted in 1977. The final pH and alkalinity were not given, though presumably they are in the neighborhood of pH 8.5 and alkalinity of 20–25 mg/L as $CaCO_3$. The importance of consistently maintaining the chemical feed program was noted.

The need for constant and consistent treatment was shown in pipe rack studies at Philadelphia (Pennsylvania) Suburban Water Company (Gittelman et al. 1992). They addressed the problem of multiple water sources in their experiments, which tested pH adjustment from about 7.1 to 7.7, plus lime stabilization to a calcium carbonate precipitation potential (CCPP) of 4 mg/L. They provided good guidance on the structure of pipe rack and field sampling for systems with multiple water sources and also showed the distinct tendency toward lognormality of the lead data.

In 1993, the popular magazine *Consumer Reports* conducted a study of first-draw and fully flushed samples from 2,643 individuals across the country (Anon. 1993a). Approximately 1,000 samples were selected to provide a wide geographic distribution. The remaining samples were selected to concentrate on eight major cities expected to yield a range of results.

Nationally, 61 percent of the first-draw samples and 90 percent of the purged-line samples were below 0.002 mg Pb/L. Only 4 percent of the first-draw samples exceeded 0.015 mg/L. When viewing the database for tendency, both Chicago, Illinois, and Boston, Massachusetts, had large numbers of samples having high lead levels. This was logical because of the widespread occurrence of lead service lines and lead plumbing in those areas. In Boston, 12 percent had greater than 0.05 mg/L. A tendency toward higher-than-average numbers of samples with elevated lead levels was also found for Washington, D.C., San Francisco, California, and New York, New York.

Investigations in the United Kingdom

Early investigations. In the early 1850s, concern was raised over the use of Loch Katrine to provide water to the residents of Glasgow, Scotland (Anon. 1993b). Among others, Dr. Frederick Penny, professor of chemistry at the Andersonian University in Glasgow, objected that the soft, acidic water available from the loch would readily dissolve lead from the widely used lead piping.

In 1913, Heap published a series of papers on the effects of various water quality parameters on lead pipe solubility (Heap 1913a, 1913b, 1913c). He included an insightful critical review of many studies dating from 1834, many of which contained important observations that could be explained later in a more accurate mechanistic manner using more developed solution chemistry theory. Early studies that were cited even indicated that some carbon dioxide in the water tended to passivate lead, but that increased levels solubilized it, a phenomena now clearly explained by the formation of aqueous carbonate and bicarbonate complexes such as $PbHCO_3^+$, $PbCO_3^0$, and $Pb(CO_3)_2^{2-}$.

Heap used several types of experiments to elucidate the effects of different chemical parameters on plumbosolvency, including what was essentially an early "pipe loop" consisting of 10-yd (9-m) lengths of ½-in. (1.3-m) water pipe. He observed poor performance with U-shaped pipe, suggesting that inconsistent strain from pipe to pipe complicates the reproducibility of the experiments. He also was concerned with the variation in purity of the lead materials used, and he observed that lead pipes could not be reused in different water quality tests because of pipe surface changes. Many of his observations of the experimental procedure are equally relevant today.

During tests with lake water followed by distilled deaerated water, he observed redissolution of films established in earlier experiments, though not immediately. Thus, the sequence of exposure is important in influencing short-term test results. He found that the tested lake water took longer to reach an equilibrium lead concentration than distilled water, because of formation of a coating on the pipe interior.

Heap tested various additions to the lake water that formed the background for his tests. These substances included: "peaty matter," sodium carbonate, and various salts of calcium and magnesium. Irreproducibility problems (from pipe to pipe with the same water) caused him to pursue additional studies with metal foil strips. These experiments concluded:

- phosphate had the most effect, followed by carbonate
- nitrate salts tended to attack the lead vigorously
- some attack was seen from sulfate, chloride, and acetate
- certain ammonium compounds, particularly ammonium nitrate, severely attacked the lead
- calcium oxide also tended to attack the pipe

He went on to study the effects of various natural waters and made several important observations:

- the action of naturally hard waters was highly varied and seemed related to whether carbonate or bicarbonate was present (tending toward protection)
- the action of waters with high organic content was somewhat varied and depended on various other properties of the water, such as the acidity of the water and type of organic matter present
- aluminum sulfate could enhance the dissolution of the lead
- increasing temperature of lake water increased the attack on lead
- waters containing certain forms of organic matter eventually (on standing) produce a loosely attached, somewhat protective precipitate on the lead surface

In the published discussion immediately following the paper by Heap, an interesting observation was made that the addition of calcium phosphate to the water greatly reduced the attack of carbon dioxide on lead pipe in experiments by a scientist at a different water supply.

In 1922, Thresh published two papers that detailed numerous field observations and laboratory experiments involving the determination of important water constituents that could accelerate or decrease the solubility and corrosion of lead (Thresh 1922a, 1922b). The natural waters tested by Thresh were found in several areas of England, Scotland, and Wales. Thresh's conclusions included the following:

- natural levels of chlorides found in the moorlands neither enhance the solubility of lead, nor slow its oxidation
- the small natural quantities of sulfate may slightly slow the rate of oxidation of the lead but do not appreciably change the solubility or type of deposit found

- carbonate and bicarbonate do not retard the oxidation of lead but are involved in film formation
- phosphates have a marked action in decreasing lead solubility where present in appreciable amounts but not at the levels found in moorland waters
- nitrates, particularly ammonium nitrate, accelerate the early stages of lead oxidation when present in substantial concentration
- an acid present in moorland organic matter serves to prevent deposition of a film on lead and to retain lead in solution
- sodium silicate in sufficient quantity can retard lead oxidation and produce a very insoluble and adherent deposit on the pipe

Not all of Thresh's interpretations have been confirmed, particularly the roles of silicate, sulfate, and nitrate. Interestingly, many of his other observations were remarkably similar to those in the German research from 1886 through approximately 1905.

In 1948, Miles published a summary of his investigation into the effect of organic matter on lead dissolution by moorland and peat bog water (Miles 1948). Miles concluded that increasing the organic content of the water accelerated the degree of initial attack on the lead.

In 1973, Moore published a study wherein he attempted to evaluate the effects of pH, temperature, and water hardness on lead leaching during 1 hour of contact time with a used piece of lead pipe (Moore 1973). The experiments with distilled water showed that the rate of lead dissolution increased exponentially with temperature at pH 6.2 (up to the testing limit of 60°C) and that the plumbosolvency was at a minimum in the pH range of 6 to 8 for the 1-hour contact period.

Walker and Oliphant (1982) have reviewed numerous early studies on lead corrosion in the United Kingdom. They described a 1944 investigation by Liddiard and Bankes of the corrosion of lead in a hard public water supply and by distilled water. Liddiard and Bankes concluded that carbonates, bicarbonates, and silicates are protective; that some inhibition is provided by phosphates and sulfates; and that an increase in the carbon dioxide content could increase lead solubility.

Studies of pH control and phosphate dosing in the Strathclyde region. During the 1960s, extensive sampling in the Strathclyde region of Scotland sparked renewed awareness of plumbosolvency problems previously described by Miles (1948). By 1968, the use of lead materials in new work was prohibited in that area. Numerous steps to correct the existing lead problem were taken and have been described in detail by Richards et al. (1980) and Britton and Richards (1981).

Two approaches were initially adopted in the case of the Loch Katrine supply: pH adjustment, usually through lime addition, and elimination of existing lead facilities, whenever feasible. Since 1970, the Glasgow District Council has removed all lead piping to the taps used to supply water for human consumption in houses being modernized. Further, individual property owners are strongly encouraged to replace all lead lines whenever pipes are replaced or repaired.

In 1973, laboratory experiments indicated that raising the pH from 6.3 to 7.8 would considerably reduce the plumbosolvent character of the water. A plant was designed, following limited experimentation in one section of the supply area, to dose lime up to the level of 1 ton/day (2–3 mg lime/L) into the six leading water mains. The plant, which became fully operational in April 1978, has been described in detail by Pharoah (1980). A monitoring program was set up in December 1977, in advance of the corrective treatment program. The monitoring and sampling schemes were described by Richards et al. (1980).

A total of 780 samples was taken and analyzed from the 15 properties that were sampled frequently, while 1,390 samples were analyzed from the 52 properties that were sampled less frequently. Early results confirmed that the variation in lead concentrations of samples taken from a given property could be pronounced, particularly where high "first-draw" lead concentrations were encountered. It was therefore necessary to obtain relatively large numbers of samples from each property in order to detect trends in lead concentrations in different categories of dwellings. A summary of the treated water supply chemical characteristics is given in Table 4-9.

Richards and co-workers concluded that the treatment program brought about a significant reduction in tap water lead concentrations for both first-draw samples and random daytime samples. They also found considerable galvanic corrosion in premises with "copper → lead → tap" or "lead → copper → lead → tap" mixtures of piping, which caused unsatisfactorily high lead levels in the water. Further, they found that there was poor pH buffering in the distributed water, with many premises having a pH of only 6.8 to 7.0, and that the contact time of the water with the lead pipe was a much more important factor governing the lead level than was the length of the pipe in the building. This latter observation is somewhat contrary to

Table 4-9 Loch Katrine treated water analysis

Characteristic	Concentration*
Color, °*H*	7.5
Turbidity, *ftu*	0.2
Total solids	38.0
pH (units)	7.8
Conductivity, $\mu S/cm$	45.0
Total hardness (as $CaCO_3$)	14.0
Calcium	4.2
Magnesium	0.8
Total alkalinity (as $CaCO_3$)	6.5
Free ammonia	0.05
Nitrite	<0.003
Nitrate	0.57
Chloride	6.0
Total organic carbon	1.4
Aluminum	<0.03
Iron	<0.02
Manganese	<0.02
Lead	<0.005
Copper	0.01
Sodium	2.8
Potassium	0.3
Silica (as SiO_2)	1.3
Phosphate, soluble ortho-	0.001
Free chlorine	0.3
Total chlorine	0.3

Source: Richards et al. (1980).

*All concentrations are in milligrams per litre, unless otherwise specified.

recent theoretical and practical experience with mass-transport modeling, which will be discussed later in this chapter, but may result from other factors.

In a paper published in 1981, Britton and Richards expanded on the conclusions drawn in the 1980 paper of Richards and co-workers, in light of additional data. After April 1980, the pH of the water was continually raised to pH 9, resulting in a pH of 8.5 to 9 in samples taken from consumers' taps.

They estimated that approximately 95 percent of the random daytime samples successfully met the target standard of 0.1 mg Pb/L, up from approximately 80 percent in April 1980. Figure 4-9 shows typical results. Because of the success of the high pH treatment, they suggested that in the few locations where high lead levels were still occurring, it might ". . . be more effective and less costly to encourage flushing, consider limited pipe lining, or undertake the replacement of lead plumbing rather than reinforce the treatment."

They noted a considerable temperature effect on the lead solubility and concluded that because the lead solubility apparently increased with temperature, all water intended for human consumption or cooking should be drawn from the cold-water tap.

When lead lines were partially replaced with copper, some dramatic, short-term increases in water lead concentrations were produced. Obvious causes might have been the dislodging of particulate lead corrosion products or the freshly exposed surface of the pipe undergoing rapid initial corrosion.

Subsequent monitoring by Strathclyde Water revealed that while pH adjustment to the range of 9–9.2 substantially reduced plumbosolvency, the requirements of the new drinking water standard for lead (50 μg/L maximum) could not be met on a continuous basis by pH correction alone (Anon. 1993b). The results from more than 1,000 samples showed that lime dosing to pH 9.2 reduced the number of samples failing the old 100-μg/L standard to between approximately 5 and 16 percent, depending on the area supplied and extent of lead plumbing.

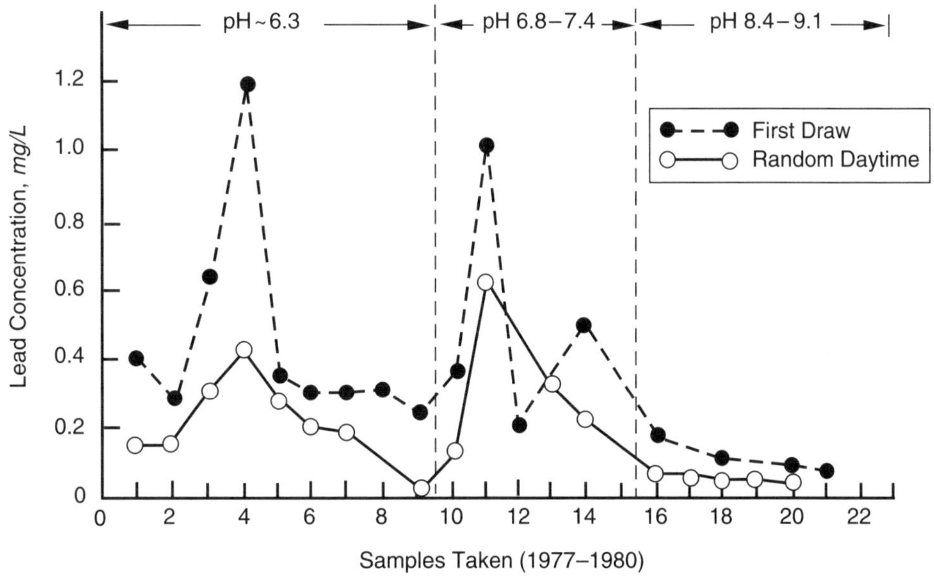

Figure 4-9 Effect of pH on lead levels observed in samples from a lead household system in the Glasgow study of Britton and Richards (1981)

To further reduce the lead levels, orthophosphate dosing was begun throughout the region. Lime dosing was used to achieve a pH of approximately 8 with the phosphate dosing. The usual procedure was to initially dose the orthophosphate (added as food-grade orthophosphoric acid) to a concentration of 3 mg PO_4/L to thoroughly saturate the system. After several weeks, when an equilibrium level of phosphate residual was present throughout the distribution system (showing no depletion through passage), the dosage would be reduced to approximately 1.5 mg PO_4/L.

Within several months of orthophosphate dosing, only 2.5–3 percent of approximately 1,000 samples exceeded a maximum concentration of 50 μg/L. In 1990, 75 of 882 samples collected exceeded 50 μg/L, and in 1991 only 9 of 650 samples exceeded 50 μg/L.

The chemical treatment of the water supplies is considered an interim measure to eliminate lead exposure of the population. Increasingly stringent standards will probably not be met in the future using central water treatment alone. Therefore, Strathclyde Water has instituted a lead pipe replacement program. Since 1986, over 60,000 lead service line pipes have been replaced by copper, polythene, or polyethylene pipework. At this time, if a routine sample or a sample taken at the request of a customer exceeds 50 μg/L, Strathclyde Water immediately replaces the section of service pipe for which it is responsible. The resident of the property is advised of the observed lead levels, and the recommendation is made that all remaining lead plumbing be replaced by the property owner.

Lead solubility studies by the Water Research Centre. Theoretical work at the Water Research Centre (WRc) in England, parallel to that undertaken at USEPA, resulted in a similar comprehensive equilibrium model for lead solubility (Jackson and Sheiham 1980; Hunt and Creasey 1980; Sheiham and Jackson 1981). The model has been useful in predicting treatment approaches in that country.

Sheiham and Jackson (1981) have described pipe tests, performed by the WRc at several field sites, that sought to evaluate the usefulness of several treatment approaches. Results obtained from simple pH and alkalinity adjustment of a low-alkalinity moorland water are given in Table 4-10 and Figure 4-10. Although the observed lead levels were somewhat higher than those predicted by the WRc computer model, the trends were accurately predicted and the usefulness of pH and slight alkalinity enhancement were demonstrated. The differences among the mean levels for the last three columns of "Test 1" in Table 4-10 are not considered to be statistically significant.

Several tests were conducted on a hard water from a chalk aquifer (alkalinity 240 mg $CaCO_3$/L, calcium 124 mg/L, pH 7.1, silica 14 mg SiO_2/L, nitrate 9.0 mg N/L), which yielded some inconsistent results. Good success was achieved when the water was acidified with sulfuric acid, aerated to remove carbon dioxide, and then treated with sodium hydroxide to a pH of 9 with an alkalinity of 60 mg $CaCO_3$ /L. Using this treatment, the resulting finished water would have a reduced DIC content and a lower alkalinity than it would have had without the carbon dioxide gas removal.

Sheiham and Jackson also studied the immediate addition of freshly prepared sodium silicate solution and of aged (48 hours for hydrolysis) solution, using some of their low-alkalinity moorland water. The final concentration was 10 mg SiO_2/L. They concluded that silicate had little or no effect on lead dissolution after 38 days of testing on new pipe at pH 6.5 or after 184 days of testing at pH 7.5 on old pipe.

Table 4-10 Effect of pH and alkalinity on lead concentrations in low-alkalinity moorland water

Characteristic	General Analysis, *mg/L*
Calcium	5.8
Magnesium	0.94
Sodium	4.33
Iron	0.27
Manganese	0.02
Aluminum	0.06
Chloride	8.0
SO_4	3.9
NO_3-N	0.21
SiO_2	1.59
TOC	2.4
Tot-P	0.02

Test 1, 38 Days Duration, 10 Observations, Five pH Conditions

pH	6.5	7.5–8.0	9.0–9.5	8.0–8.5	8.0–8.5
Alkalinity*	10	10	10	60	110
Median µg Pb/L	595	175	81	86	41
Mean µg Pb/L	622	175	78	88	57

Test 2, 28 Days Duration, 7 Observations, Three pH Conditions

pH	6.5	7.5–8.0	8.3–8.7
Alkalinity*	10	10	10
Median µg Pb/L	700	130	100
Mean µg Pb/L	743	175	120

Source: Sheiham and Jackson (1981).

*Alkalinity as mg $CaCO_3$/L.

The median lead levels on the new pipe were approximately 15–30 percent lower than the undosed control, but they were still over 0.5 mg/L.

Numerous experiments with orthophosphate and polyphosphate (pyrophosphate and Calgon SWB [Water Board–quality chips; Albright & Wilson, West Midlands, England]) were also conducted by WRc (Sheiham and Jackson 1981). The experiments used the same low-alkalinity moorland water as was used for the carbonate evaluation (Table 4-10). Table 4-11 summarizes the results. Orthophosphate was more effective in reducing lead on a percentage basis in the new pipes than in the old, but large reductions in the lead level took place in both. The addition of zinc sulfate did not further reduce the lead in the studies. The researchers concluded that orthophosphate was considerably more effective than the types of polyphosphates they tested.

In attempting to understand the orthophosphate's mechanism of action, the investigators analyzed deposits in several pipe specimens. The pyromorphite structure found was considered to be hydroxypyromorphite, even though the observed solubility seemed somewhat high. Plots of log [Pb] versus log [P] for a set of tests produced a straight line of slope –0.61, which was close to the theoretical value of 0.600 for hydroxypyromorphite, $Pb_5(PO_4)_3OH_{(s)}$.

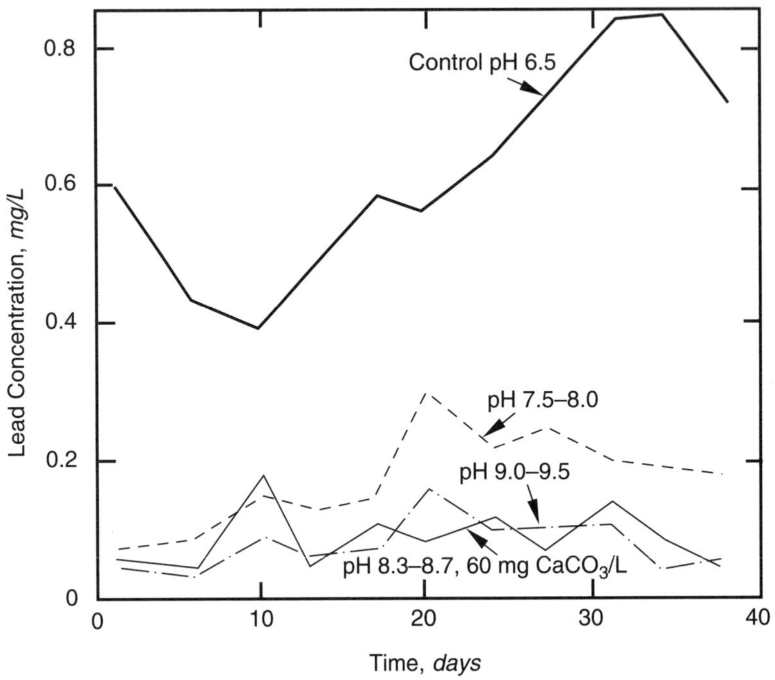

Source: Sheiham and Jackson (1981).

Figure 4-10 Effects of pH and alkalinity on moorland water with initial alkalinity of 20 mg $CaCO_3$/L

Table 4-11 Results of the polyphosphate dosage experiments using moorland water

	Control	Calgon SWB*	$Na_4P_2O_7$*
New Pipes, pH 6.5, 38 Days Duration, 12 Observations			
Median mg Pb/L	0.835	0.500	0.245
Mean mg Pb/L	0.870	0.525	0.261
Old Pipes, pH 7.5, 184 Days Duration, 12 Observations			
Median mg Pb/L	0.266	0.130	0.114
Mean mg Pb/L	0.266	0.133	0.114

Source: Sheiham and Jackson (1981).

*Polyphosphate concentration was 3.1 mg PO_4/L (1 mg P/L).

Another of the WRc tests involved a combination of orthophosphate dosage (3.1 mg PO_4/L) with carbonate and pH adjustment (Table 4-12). The results corroborated what the solubility calculations indicated: increasing the pH and alkalinity (within certain limits) will reduce lead concentrations while orthophosphate is being added. There are no statistically significant differences between the mean concentrations at pH 7.5 of the orthophosphate dosing with and without zinc.

Sheiham and Jackson (1981) described another interesting experiment with a hard groundwater, using both old and new pipe sections. Addition of orthophosphate at 1 mg P/L (3.07 mg PO_4/L) produced very substantial reductions in mean lead concentration to 80 µg Pb/L in the old pipe and 20 µg Pb/L in the new pipe section. Reduction of the orthophosphate dose to 0.5 mg/L caused a noticeable increase in lead in the old but not the new pipe. Increasing the dose from 1 to 2 mg P/L (3 to 6 mg PO_4 /L) had no additional effect. Addition of sodium metasilicate at 5 and 10 mg SiO_2/L produced no effect on lead levels.

The old pipe used for this series of tests provided more sections than were needed for the treatments examined. Three sections left in the "control" sections were used to examine the effect of temperature at 21, 10, and 2°C, which produced mean lead levels of 181, 108, and 44 µg Pb/L, respectively. There was clearly a large effect from temperature under the experimental conditions. Further research is needed to determine if the temperature effect is caused by an increase in the solubility constant of the lead passivation film solid, increasing the formation of soluble lead complexes, increasing the rate of dissolution of the lead material (if not truly an equilibrium situation), inhibiting the formation of the passivation film by other constituents in the water, or a combination of any or all of these factors.

Stagnation curve profiling by WRc. In order to develop a framework for a reliable, practical, and economical method of sampling drinking water for regulatory and public health protection purposes, two important sampling investigations were undertaken by WRc.

Table 4-12 Experimental results for orthophosphate dosage for lead control using a low-alkalinity moorland water

	Control	Orthophosphate*	Zinc + Orthophosphate*
New Pipes, pH 6.5, 38 Days Duration, 12 Observations			
Median mg Pb/L	0.835	0.170	0.200
Mean mg Pb/L	0.870	0.175	0.229
Old Pipes, pH 7.5, 184 Days Duration, 12 Observations			
Median mg Pb/L	0.266	0.080	0.108
Mean mg Pb/L	0.266	0.096	0.109

Source: Sheiham and Jackson (1981).

*Orthophosphate dosage = 3.1 mg PO_4/L.

The first investigation examined stagnation profiles of 36 sites selected to cover several categories of water quality (Bailey et al. 1986). An automatic sampling system was refined and installed on the premises. They reported that the coefficient of variation for 6-hour stagnation samples was less than for fully flushed samples. Thus, it would be a more preferable protocol than estimating standing lead samples from fully flushed samples, although the latter would be a more convenient alternative. They also found a large number of high results that appeared to be aberrations. They were considered to be "outliers" and were discarded from the data analysis.

The researchers analyzed the data using three methods: simple graphical presentation; a theoretical diffusion-based model, such as that of Kuch and Wagner (1983); and an empirical statistical relationship. They found that both expressions fit the data well in terms of curve shape. However, the diffusion-based theory assumes an equilibrium solubility control that should be operative in all houses of a given water quality category. The variability between these sites suggested a discrepancy between that model and actual behavior. To some extent, this may have been caused by the small sample size (150 mL) or nonadhering particulate lead in the samples. They concluded from this investigation that the stagnation curve is a fairly repeatable phenomenon for each given site and that the practical situations appear to be more complicated or variable than diffusion-based model theories currently allow.

The subsequent study covered the development of a defensible rationale for fixed time of stagnation (FTS) as a basis for monitoring (Lacey and Jolly 1986). The authors indicated that 30 minutes would be a reasonable stagnation time for monitoring, under a range of plausible assumptions. People living in lead-plumbed houses should still fully flush the piping before drawing water for human consumption. Periodic repeat sampling is still necessary, especially if water quality fluctuates.

Plumbosolvency studies by Anglian Water. Anglian Water (Cambridge, England) used a pipe rig system with lead when developing an accelerated test for "plumbosolvency" (Colling, Whincup, and Hayes 1987). Colling and co-workers noted that Anglian Water (serving the east midlands of England) had been carrying out phosphate dosing trials for the groundwater supplies to two Lincolnshire cities for five years. Orthophosphate dosing was a nationally recommended treatment for plumbosolvency in high-alkalinity waters, and it represented 75 percent of Anglian Water's entire water production.

New lead pipe was used in their pipe rig system, and scanning electron microscopy (SEM) and energy-dispersive X-ray analysis were used extensively for characterizing the deposits on the pipes. Reviewing historical information from WRc deposit analyses, they pointed out that for many high-alkalinity waters studied, the surface deposit was $Pb_3(CO_3)_2(OH)_2$ (hydrocerussite), rather than the thermodynamically stable $PbCO_3$ (cerussite) mineral. Sometimes both occurred simultaneously.

An approximate relationship existed between the lead values found for the pipe rig test and basic lead carbonate solubility. However, they found that waters appearing to be similar in character can produce different pipe deposits and resultant plumbsolvency. "High-propensity" and "low-propensity" waters were operationally defined on the basis of high (100 ± 20 µg/L) or low (30 ± 10 µg/L) lead levels in the tests.

Their experiments showed that whenever both high- and low-plumbosolvency propensity waters are present in a mixture, the high-propensity component is dominant. When the high-propensity water is dosed with orthophosphate (3.07 mg PO_4/L), it still dominates, because the orthophosphate concentration is essentially diluted when a relatively high fraction of the low-propensity water (without the orthophosphate) is present in the mix.

The test was used to survey all Anglian Water supplies. They found no obvious factors influencing which type of deposit formed (stable cerussite versus metastable hydrocerussite). However, they suggested evidence that a total-hardness-to-total-alkalinity ratio (presumably in units of mg $CaCO_3/L$) of above 1.7 tended to promote plumbosolvency. Additionally, the test confirmed that orthophosphate dosing is an effective treatment for high-alkalinity waters in the absence of factors other than solubility.

In 1992, Colling et al. published further results of investigations at Anglian Water on plumbosolvency control of hard waters (Colling et al. 1992). From a combination of laboratory, pipe rig, and field monitoring experience, they found the following significant results:

- To establish effective plumbosolvency control in hard waters, a minimum dose of approximately 1.8 mg PO_4/L is needed. The orthophosphate concentration was found to affect both the lead solubility level and the rate of lead reduction. Higher orthophosphate doses produced quicker lead reductions. Because of the thick nature of deposits on old lead pipe, a year or more may be required for considerable plumbosolvency reduction to take place. Levels of lead may still continue to drop slowly beyond that time period. The effect of orthophosphate dosage for these waters is shown in Figure 4-11.
- In pipe rig studies, after stabilization, used lead pipes exhumed from sites will ultimately give about the same lead concentrations as new pipe. However,

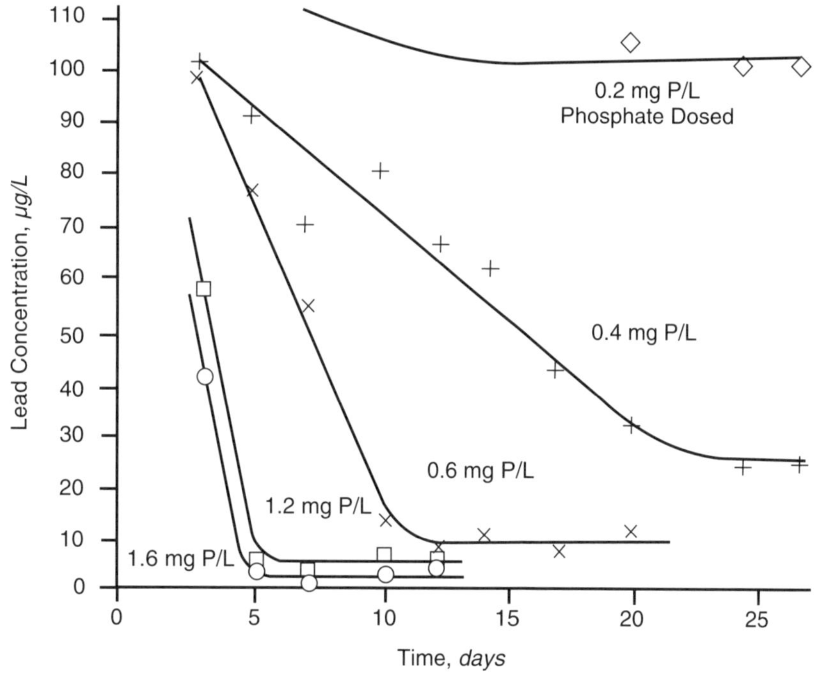

NOTE: Multiply mg P/L concentrations by 3.07 to get mg PO_4/L units.

Source: Colling et al. (1992).

Figure 4-11 Effects of initial orthophosphate concentration on high-plumbosolvency water at 25°C even if the pipes are removed with minimal disturbance and are kept full of water and sealed, the time to reach "stable" lead values may be up to 20 times as long as with new pipe. It may take years to see the ultimate effect of the orthophosphate dosing.

- Short breaks in orthophosphate dosing of one or two days should not affect plumbosolvency control. Effects of longer breaks depend on the orthophosphate dose. High plumbosolvency conditions can return within two weeks at low dosages.
- Where high- and low-plumbosolvency waters are blended before distribution, both waters or the mixture must be properly dosed with orthophosphate.
- Where high- and low-plumbosolvency waters alternate in the distribution system, low-plumbosolvency results are obtained.
- When orthophosphate-dosed high- and low-plumbosolvency waters alternate in the distribution system, a period of erratic behavior occurs. After that, low-plumbosolvency results are seen.
- Orthophosphate-dosed waters tended to show a decrease in plumbosolvency as both the temperature and dosage were raised. The effect was small. Undosed waters showed an increase in solubility as temperature was increased.
- Humic substances at low concentrations in hard waters may be the cause of the higher plumbosolvency propensity.

Studies of lead contamination by particulates. Colloidal hydrous iron oxide and humic acid species were found to be important in creating high lead levels in a system near Glasgow, Scotland (DeMora, Harrison, and Wilson 1987). They used a complex filtration and ion exchange procedure and chemical analyses to gain knowledge of how the completeness of water treatment processes (e.g., coagulation, filtration, disinfection) might impact lead solubility or transport. Four levels of treatment of the same source water were compared in their investigation. They concluded that in low-pH and low-alkalinity source waters prevalent in Scotland, plumbosolvency can be exacerbated by the presence of organic material derived from drainage of upland peaty regions. Slow sand filtration and chlorination were both found to reduce the plumbosolvent property of this particular water. From a variety of chemical arguments and particle size distribution analyses, they concluded that changes in the organic concentration and character, rather than inorganic processes, accounted for the reduction. For the system locations receiving pH-adjusted water (>9), lead levels were lower than predicted purely by solubility equilibria.

Further investigations into the nature and control of high lead levels resulting from particulates were conducted at WRc (Hulsmann 1990). At least two types of particulate problems were identified: adsorption of lead on colloidal particles (mainly iron) and "flaking" lead. The flakes are generated by hydraulic disturbance of an unstable corrosion deposit layer. This problem was thought to be most prevalent in supplies having soft, low-alkalinity water that is high in organics and color and low in pH.

The colloidal–particulate lead problem was reduced by dosing with an unspecified polyphosphate chemical at a level of 5 mg/L as P_2O_5 (3.3 mg/L as PO_4). The polyphosphate chemical reduced the iron concentrations in the water and consequently reduced the lead levels by preventing the formation of the colloidal substrate.

Hulsmann also found that when lead pipes were present, normal maintenance in the distribution system tended to increase the particulate lead and could cause a problem. The mechanism is thought to be either mechanical disturbance or the introduction of dirt. Remedial actions and maintenance programs should aim to reduce the influx of particulate material, particularly iron particles, into lead service line areas.

Related and subsequent studies by Severn Trent (which serves the midlands of England not covered by Anglian Water) provided further elucidation of the particulate lead problem (Breach, Crymble, and Porter 1991). The researchers noted a possible association of the colloidal problem with natural organic material that may chelate the lead from the pipe surface and attach it to the iron particle. The problem was only found in a supply that did not practice coagulation.

Breach and co-workers also pointed out a third type of high and erratic lead phenomenon, which they called "electrochemical lead." This is caused by rapid corrosion by cells set up and activated by the coupling of new copper pipe to old lead pipe or that result from poorly made soldered joints. The problem was found to be particularly prevalent in large institutional-type buildings, such as hospitals, schools, offices, and apartment buildings. These buildings often have long pipe runs, many soldered joints, and relatively low water usage.

To diagnose lead problems, samples are routinely filtered to give estimates of total and dissolved lead. Studies of pipe deposits have also been done using a variety of analytical techniques, such as X-ray diffraction, infrared spectroscopy, and energy-dispersive X-ray analyses.

Fuge, Pearce, and Perkins (1992) reported on an investigation of mobilization of lead and several other metals from existing scale deposits in building plumbing. Significant lead levels in water were found in one instance where no lead plumbing existed. Analysis of the pipe scale material revealed both tin and lead, implicating soldered joints as the original source. Changes in source water treatments, particularly a change in coagulant from $Al_2(SO_4)_3$ to iron salts, were thought to cause remobilization of the original aluminum-containing scale, which had accumulated lead, tin, and other metals over time on the surface deposits on copper pipe.

Studies From Other Areas

Two studies investigating the trace metal levels in some university buildings and distribution water and domestic storage tanks were done in India and Iraq (Ajmal and Uddin 1986; Jawad, Al-Ghazali, and Khorshid 1988). At Aligarh Muslim University (India), lead levels were found to be below 5.5 µg/L in both 24-hour standing and fully flushed water samples from nine different buildings sampled four times in one month. The sample volume was 1 L. It is not clear if the 1-L sample for metals analysis was the first one taken or a second sequential 1-L sample (Ajmal and Uddin 1986). The pH at all sample sites was between 7.00 and 8.00, with most above pH 7.60. Total alkalinities ranged from 128 to 200 mg $CaCO_3$/L, with most between 144 and 176 mg $CaCO_3$/L.

Jawad and co-workers (1988) sampled 220 water storage tanks in 12 main districts in Baghdad. One-litre samples were collected in triplicate from both input and output lines of the tanks. Water was run at full force for about 60 seconds prior to sampling. Mean lead levels of influent water were 28 ± 13 µg/L (1 standard deviation) for the winter samples and 25 ± 15 µg/L for summer samples. The output water averaged 33 ± 16 µg/L for the winter samples and 24 ± 8 µg/L for the summer samples. The relative lack of variability among sites and with seasons indicated lead release in the water distribution system or industrial discharge above the water plant intakes as possible lead sources. Considerable sediment was found to accumulate in the tanks, requiring cleaning every few months. Perhaps the sediment also scavenged lead to some degree.

PARTIAL LEAD SERVICE LINE REPLACEMENT

When attempting to reduce exposure to lead, a critical issue is whether or not service lines need to be replaced if central water treatment does not adequately reduce plumbosolvency. In the United States, the issue is complicated by various definitions of service line ownership and control (Weston and EES 1990; *Federal Register* 1991a).

Scottish Studies

Various studies in Scotland indicated that galvanic corrosion of lead can take place when lead is coupled to copper piping (Britton and Richards 1981; Richards, Britton, and Cochrane 1980). Frequently, when sections of deteriorated lead pipe were replaced by copper, lead levels in the plumbing increased. Britton and Richards (1981) suggested that the lead concentration was increased both by electrochemical corrosion of adjacent surfaces of copper and lead and by the release and deposition of copper onto the lead pipes as water passed through the next pipe section. Each site of copper deposition has the potential to act as an individual galvanic cell, depending on the nature of the lead pipe surface and existing corrosion products on the surface. Many examples are given of increases in the number of erratically high lead results in mixed plumbing systems in the Glasgow area (Britton and Richards 1981; Richards, Britton, and Cochrane 1980).

Britton and Richards (1981), Breach, Crymble, and Porter (1991), and Hulsmann (1990) indicate that any mechanical disturbance of lead plumbing can result in significant elevation of lead levels. This occurs because particulate material dislodges and because relatively fresh and unpassivated lead surfaces come in contact with the water. A study in Glasgow of partial lead service line replacement illustrated that more than four months were required after the replacement for the lead levels to drop and regain consistency (Britton and Richards 1981). Britton and Richards concluded that partial pipe replacement could give rise to unacceptably high lead concentrations in the water, but satisfactory reduction of the lead levels will take place in a matter of weeks or months if the water pH is maintained properly.

US Studies

A short study was conducted in the United States, in which several homes in a very hard water area were to undergo partial lead service line replacement (Wysock, Schock, and Eastman 1991). The utility replaced the lead line from the main to the curb stop with copper. The line from the curb stop to the house connection remained lead. None of the houses tested had lead interior plumbing. Water analyses showed virtual equilibrium with calcium carbonate at a pH of 7.10 and an alkalinity of over 375 mg $CaCO_3$/L during the course of the study. Examination of pipe deposits from removed service lines showed a thick complex-layered structure that included $PbCO_3$, $CaCO_3$, PbO, and PbO_2; some $Pb_3(CO_3)_2(OH)_2$; and additional unidentified solids containing aluminum, SiO_2, iron, and possibly potassium and sodium, in addition to elements present in the compounds previously mentioned.

Only limited prior "baseline" monitoring could be accomplished because the related road construction necessitated a short time schedule for pipe replacement. A statistical analysis of the data, where lead levels were generally below 0.015 mg/L in targeted service line samples, showed essentially no improvement in lead levels after partial replacement in either first-draw or service-line standing samples.

Pilot-plant pipe rig tests from this system, where carefully exhumed service lines were compared to new lead pipes, showed major disturbances and particulate erosion for more than a year after installation in the pipe racks (Wysock et al. 1992; Wysock, Schock, and Eastman 1991). Satisfactory agreement and consistency could not be achieved among replicate removed service-line sections in later water treatment studies.

The studies documented thus far strongly suggest that many benefit and feasibility questions remain concerning partial replacement of lead service lines, and more systematic research is required to answer them. The period of disturbance and temporarily high lead levels caused by partial or full lead service line replacement is almost certainly largely dependent on the water chemistry and the pattern of water use at the individual location. At this time, long-term improvement in tap water lead levels is only likely where the lead levels in flushed samples are below or comparable to those produced by stagnation in contact with brass faucets and soldered joints. Removal of only parts of lead lines will have little impact on first-draw tap water samples unless the water is highly plumbosolvent. The impact on human exposure remains dependent on water usage patterns and the configuration of the plumbing system of the dwelling, as well as water chemistry and flow factors.

CHEMISTRY OF LEAD IN POTABLE WATER

Redox Processes

Taylor (1980) pointed out that the problem of trace metal contamination in drinking water results from both corrosion and solubility processes. In potable water environments, oxidizing agents that transform the metallic lead to one of its oxidized forms are generally present. The most common form encountered is Pb(II), with Pb(IV) only possible under extremely oxidizing conditions. The solids of Pb(II) and Pb(IV) that can form vary with the chemical composition of the water, and they have a wide variety of solubilities that provide the ultimate limitations on dissolved lead input into the water. They can also have considerable differences in crystallinity, compactness, and adherability that will affect oxidation rates and particulate erosion.

When water is passed through lead pipe or pipe containing lead-based solder, the more reactive areas of the surface, such as metal–crystal grain boundaries, become the anodes (Walker and Oliphant 1982). Conversely, the less reactive areas, such as the grains, become cathodic areas. Corrosion occurs at the anodic areas.

The most common oxidant for lead in potable waters is DO. The oxidation, and hence, the corrosion of lead depend on the free lead ion activity, the activity of the DO, and the pH. Manifestations of this reaction are an increase in pH (a decrease in hydrogen ion activity or production of hydroxyl ion), a decrease in the DO content, and an increase in the dissolved lead concentration until it reaches the solubility limit of some solid phase. The reaction is driven by increases in the DO content, by decreases in pH (increases in hydrogen ion activity), and by the complexation of Pb^{2+} by ligands such as carbonate and sulfate. The rate and the extent of the reaction should vary with temperature, but quantitative studies in this regard have not been published.

The oxidation of lead may be expressed as the half-cells:

$$Pb_{(s)} \rightleftharpoons Pb^{2+} + 2e^- \quad E^0 = 0.126\ \text{V} \tag{4-4}$$

$$Pb_{(s)} \rightleftharpoons Pb^{4+} + 4e^- \quad E^0 = 0.784\ \text{V} \tag{4-5}$$

Values for these standard electrode potentials were computed from the internally consistent set of Gibbs free energy of formation values given in Table 4-13. Techniques for computing equilibrium constants and cell potentials from free energies are given in appendix A.

Table 4-13 Internally consistent set of Gibbs free energies of formation at 298.15K and 1 atm

Species	ΔG_f^0 *	Reference†	Species	ΔG_f^0 *	Reference†
		Dissolved Lead Species			
Pb^{2+}	–5.83	a	$Pb(CO_3)_2^{2-}$	–272.4	a
Pb^{4+}	+72.3	b	$PbHPO_4^0$	–270.4	e
$Pb(OH)_2^0$	–96.11	a	$PbHPO_4^+$	–277.7	e
$Pb(OH)_3^-$	–137.6	a	$PbSO_4^0$	–187.5	f
$Pb(OH)_4^{2-}$	–178.4	c	$Pb(SO_4)_2^{2-}$	–366.5	c
Pb_2OH^{3+}	–59.67	a	$PbCl^+$	–39.34	f
$Pb_3(OH)_4^{2+}$	–211.6	a	$PbCl_2^0$	–70.97	f
$Pb_4(OH)_4^{4+}$	–221.6	a	$PbCl_3$	–102.2	f
$Pb_6(OH)_8^{4+}$	–429.0	a	$PbCl_4^{2-}$	–133.1	f
PbO_3^{2-}	–66.34	b	PbF^+	–75.98	g
PbO_4^{4-}	–67.42	b	PbF_2^0	–145.2	g
$PbHCO_3^+$	–150.1	d	$PbNO_3^+$	–33.9	h
$PbCO_3^0$	–141.7	a	$Pb(NO_3)_2^0$	–59.05	h
$PbOH^+$	–52.66	a	$Pb_2CO_3^{2+}$	–149.1	r
			$Pb_3CO_3^{4+}$	–153.50	r
		Associated Dissolved Species			
H^+	0		PO_4^{3-}	–243.48	s
$O_{2(g)}$	0		Cl^-	–31.380	i
$O_{2(aq)}$	3.92	q	F^-	–67.34	i
OH^-	–37.604	a	HSO_4^-	–180.66	i
$H_2CO_3^*$	–148.95	a	SO_4^{2-}	–177.95	i
HCO_3^-	–140.28	a	$H_4SiO_4^0$	–312.60	k
CO_3^{2-}	–126.19	a	NO_3^-	–26.64	i
$CO_{2(g)}$	–94.257	a	$HOCl^0$	–19.1	l
$H_3PO_4^0$	–273.08	j	OCl^-	–8.8	l
$H_2PO_4^-$	–269.80	j	Cl_2^0	+1.65	l
HPO_4^{2-}	–260.33	j			
H_2O	–56.690	i			

Table continued next page

Table 4-13 Internally consistent set of Gibbs free energies of formation at 298.15K and 1 atm (continued)

Species	ΔG_f^o	Reference†	Species	ΔG_f^o	Reference†
		Lead Solids			
Pb	0		$PbSiO_3$	–251.8	h
PbO (red)	–45.33	h	Pb_2SiO_4	–301.5	h
PbO (yellow)	–45.08	m	$PbHPO_4$	–281.8	e
PbO_2 (plattnerite)	–51.98	h	$Pb_3(PO_4)_2$	–565.0	e
Pb_2O_3	–98.42	b	$Pb_4O(PO_4)_2$	–617.0	e
Pb_3O_4	–148.70	h	$Pb_5(PO_4)_3OH$	–902.0	e
Pb(OH) aged	–108.1	a	$Pb_5(PO_4)_3Cl$	–906.1	p
Pb(OH) fresh	–101.3	a	$PbSO_4$	–194.4	k
$PbCO_3$	–149.9	a	$PbSO_4 \bullet PbO$	–246.7	s
$PbCO_3$	–149.5	n	$Pb_{10}O(CO_3)_6(OH)_6$	–1269	t
$Pb_3(CO_3)_2(OH)_2$	–408.9	a			
$Pb_3(CO_3)_2(OH)_2$	–407.8	o			

*Units are kcal/mol.

†References are listed at end of table.

a. Recomputed from: M.R. Schock, *Jour. AWWA*, 72:12:695 and Errata *Jour. AWWA*, 73:3:36 (News), using ΔG_f^o for Pb^{2+} from D.D. Wagman et al. The NBS Tables of Thermodynamic Properties: Selected Values for Inorganic and Chemical Reference Data. *Jour. Phys. Chem. Ref. Data*, vol. 11, Supplement No. 2 (1982). Note that the carbonic acid system species Gibbs free energies of formation are identical to those derived by R.L. Berg and C.E. Vanderzee, *Jour. Chem. Thermodynamics*, 10:1113 (1978).

b. Computed from ΔG_f^o value given by M. Pourbaix, Atlas of Electrochemical Equilibria in Aqueous Solutions. Pergammon Press, Oxford (1966).

c. Computed from log β data of J.D. Hem, *Geochim. Cosmochim. Acta*, 40:599 (1978), using ΔG_f^o for Pb^{2+} and SO_4^{2-} from this list.

d. Computed from the estimated log β_{11} = 2.3 for the reaction $Pb^{2+} + HCO_3^- \leftrightarrows PbHCO_3+$ given by A. Zirino and S. Yamamoto, *Limnol. Oceanogr.*, 17:662 (1972).

e. Computed from data of J.O. Nriagu, *Inorg. Chem.*, 11:10:2499 (1972), using ΔG_f^o values for Pb^{2+}, H_2O_3, OH^-, and phosphate species in this table.

f. Computed from log *K* data of R.M. Smith and A.E. Martell, *Critical Stability Constants: V.4, Inorganic Complexes*, Plenum Press, New York (1976), using ΔG_f^o for relevant species from this table.

g. Computed from log β data of D.R. Turner et al., *Geochim. Cosmochim. Acta*, 45:855 (1981), using ΔG_f^o for Pb^{2+} and F^- from this list.

h. Computed from calculated ΔG_R^o using ΔG_f^o of D.D. Wagman et al. (1968), plus ΔG_f^o for relevant species from this table.

i. V.B. Parker et al. *Selected Thermochemical Data Compatible With the CODATA Recommendations*. NBSIR 75-968, US National Bureau of Standards (1976).

j. Computed from log *K* data in R.M. Smith and A.E. Martell (1976), using ΔG_f^o for PO_4^{3-} from R.A. Robie et al. (1979).

k. R.A. Robie et al. *Thermodynamic Properties of Minerals and Related Substances at 298.15K and 1 Bar (10^5 Pascals) Pressure and at Higher Temperatures*. USGS Bulletin 1452, Reprinted with corrections (1979).

l. D.D. Wagman et al. *Selected Values of Chemical Thermodynamic Properties*. NBS Technical Note 270-3, US National Bureau of Standards (1968).

m. Computed from ΔG_R^o for the reaction PbO (red) $\leftrightarrows$ PbO (yellow) using ΔG_f^o from M.R. Schock (1980), adjusted for the different values of ΔG_f^o PbO (red) from V.B. Parker et al. (1976).

n. Computed from ΔG_f^o value given by R.A. Robie et al. (1979) by adjusting for the different values for ΔG_f^o H_2O used here.

o. Assuming a more soluble log *K* = 18.0 for the reaction $Pb_3(CO_3)_2(OH)_{2(s)} + 2H^{2+} \rightleftarrows 3Pb^{2+} + 2CO_3^{2+} + 2H_2O$ based on evidence presented by M.R. Schock and M.C. Gardels, *Jour. AWWA*, 75:2:87. This agrees closely with ΔG_f^o determined by P. Taylor and V.J. Lopata, *Can. J. Chem.*, 68:395 (1984).

p. Recomputed from J.O. Nriagu, *Geochim. Cosmochim. Acta*, 37:367 (1973), using ΔG_f^o values for Pb^{2+}, Cl^-, PO_4^{3-} from this list.

q. Computed from log *K* = –21.495 for the reaction $\frac{1}{2}\ H_2O_{(aq)} \rightleftarrows \frac{1}{4}\ O_{2(aq)} + H^+ + e^-$ given by L.N. Plummer et al. *USGS WRI* 76-13 (revised and reprinted 1978), using ΔG_f^o for H_2O from this table.

r. Computed from log β values for I=0 from E. Néher-Neumann, *Acta Chem. Scand.*, 46:231–239(1992), using ΔG_f^o values for Pb^{2+}, H_2O, and $CO_{2(g)}$ from this table.

s. From D.D. Wagman et al., The NBS Tables of Thermodynamic Properties: Selected Values for Inorganic and Chemical Reference Data. *Jour. Phys. Chem. Ref. Data*, vol. 11, Supplement No. 2 (1982).

t. From P. Taylor and V.J. Lopata, *Can. J. Chem.*, 62:395–402 (1984).

Differences in oxygen concentration between one place on the pipe surface and another can also produce galvanic concentration cells, which would serve to cause corrosion. The concept of concentration cell corrosion is developed in chapter 1 and has been covered by other authors (Walker and Oliphant 1982; Snoeyink and Jenkins 1980; Butler and Ison 1966).

A second common family of oxidants in most potable waters is chlorine species. For hypochlorous acid or hypochlorite ion, the lead oxidation reaction depends on the pH of the water, the temperature, and the activities of free lead ion, chloride ion, and the hypochlorite ion or hypochlorous acid molecule. Hypochlorous acid species are such strong oxidizing agents that they are unstable or metastable in water, because they are potent enough to oxidize water itself (Butler and Ison 1966). Therefore, quantitative predictions of lead oxidation are difficult.

The chlorine residual could have several possible impacts on lead corrosion, as follows:

- Free chlorine species could act as the primary oxidant, accelerating the conversion of metallic lead to Pb(II) or Pb(IV). This would accelerate the appearance of lead in solution but probably could not affect the equilibrium solubility of lead after sufficient standing time.
- Chloramine species could also act as the primary oxidant and possibly serve to increase lead solubility by dissociating into lead amine complexes.
- Chlorine species could alter lead solubility by oxidizing Pb(II) to Pb(IV) under some conditions. This has not been investigated, so the effect is largely hypothetical (although some PbO_2 has been found in experimental systems [Schock and Gardels 1983]). Notably, the observations by Liverseege and Knapp (1913) of passivation by potassium permanganate solution would support the plausibility of this hypothesis.
- Chlorine species could react with naturally occurring, high-molecular-weight organic molecules, which would alter their ability to mobilize lead through complex formation or to inhibit lead dissolution by forming coatings through adsorption.

Other oxidants from alternate disinfection processes (e.g., O_3, ClO_2) could similarly impact lead dissolution. The impacts of chlorine species and drinking water oxidants have not been widely studied, as was noted previously. Experiments with copper and steel pipe (Singley et al. 1984) indicated that the effects might be additive, and there is little reason at present to think that a similar phenomenon would not be associated with lead.

Only two published studies have directly examined any influence of chlorination on lead corrosion. The study by Schaut (1942), in which he concluded that free chlorine accelerated the attack on lead, was summarized earlier. Because he could not carry out a controlled study to isolate the impact of chlorine, its quantitative significance is uncertain. In the study by James M. Montgomery Consulting Engineers (JMM 1982, 1983), lead solder on copper pipe was believed to be dissolved faster in a soft water with chloramine disinfection (~1.1 mg Cl_2/L) than in one with only a free chlorine residual. In the USEPA studies by Schock and Gardels (1983), differences in rates of free chlorine residual consumption were noted in the pipe loop. At pH 9.2–9.5 in the low-alkalinity water, the overnight (7–8 hours) consumption was approximately 0.1 mg/L as opposed to 0.2 mg/L at pH 8.3–8.5. Direct electrochemical monitoring indicated enhancement of the lead corrosion rate, particularly when chlorine was added to the holding tank. Some anecdotal information from utility pipe loop studies also indicates the corrosion rate or lead level may be increased by the presence of chloramines (Kirmeyer, Sandvig, Pierson, and Neff 1994).

The range of oxidation potentials (E_H) found in drinking water is not precisely known. Even for very low DO levels, there should be enough oxidizing power to promote the conversion of metallic lead to Pb(II). Measurements made during the course of virus inactivation studies using ozone indicated approximately 1.2 V (adjusted to be versus the standard hydrogen electrode) as the upper limit of E_H that would be generated by 1 to 3 mg O_3/L at an unspecified pH (Katzenelson et al. 1973). Measurements from breakpoint chlorination studies at pH 8 also gave a maximum value of approximately 1.2 V (Eilbeck 1984), in good agreement with the ozone study. These high redox potentials are sufficient to at least temporarily move into the hypothetical thermodynamic stability domains for Pb(IV) species, as shown in an example potential–pH diagram (Figure 4-12). The scale analyses of pipe from at least two utilities show the presence of PbO_2 (plattnerite) or similar PbO_2 phase.

By combining the carbonate complexation and hydrolysis chemistry of Pb(II) with Gibbs free energy of formation data for Pb(IV) species (Table 4-13), revised potential–pH diagrams (otherwise known as E_H–pH, Pourbaix, or electromotive force [emf]–pH diagrams) can be generated to show both redox complexation interrelationships simultaneously. In this way, Figure 4-12 was constructed. Here, dissolved carbonate system species (e.g., $H_2CO_3^*$, CO_3^{2-}) had their activities set to 4.8 mg C/L ($10^{-3.40}M$). The thermodynamic activities of the dissolved lead species (e.g., Pb^{2+}, $Pb(OH)_2^0$, $Pb(CO_3)_2^{2-}$) were set at 0.05 mg Pb/L ($10^{-6.62}M$). Areas of passivation or immunity are shaded. Diagrams such as these may give different predictions than those in much of the corrosion literature, because of the importance of carbonate complexation (Schock 1990b).

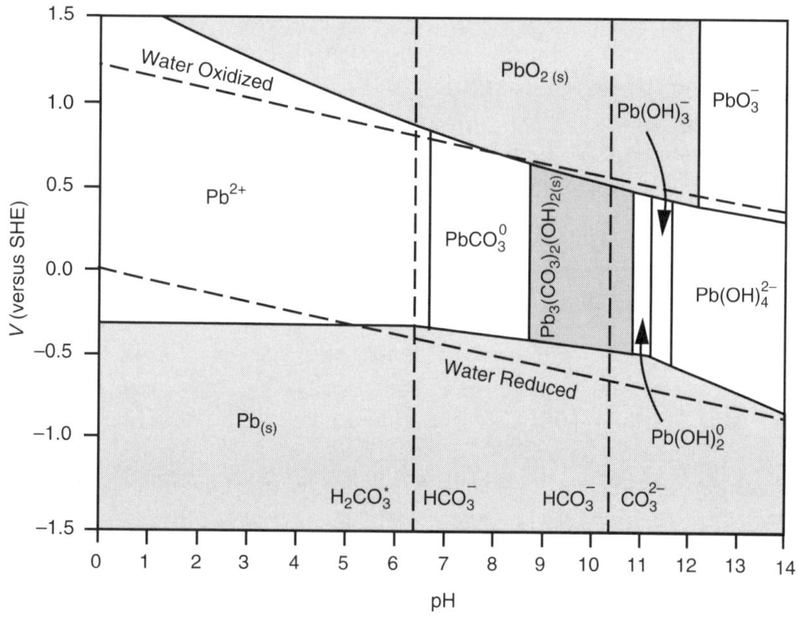

NOTE: Dissolved lead species activities = 0.05 mg/L. Dissolved inorganic carbon species activities = 40 mg $CaCO_3$/L (4×10^{-4} mol/L). Areas of passivation and immunity are unstippled. Ionic strength assumed to be zero.

Figure 4-12 Potential-pH diagram for the Pb-H_2O-CO_3 system at 25°C

Solution Chemistry of Pb(II)

The solution chemistry of Pb(II) is very complicated. This is because of its ability to form a wide variety of complexes under chemical conditions commonly encountered in potable water systems.

Table 4-14 summarizes the most significant Pb(II) reactions in the LEADSOL computer program, which was used in USEPA research (Schock 1980, 1989; Schock and Gardels 1983; AWWA Research Foundation and DVGW, Engler-Bunte Institute 1985) and to compute the values for diagrams in this section. For comparison, equilibrium constants used by the computer program SASIAS in WRc modeling efforts are also given (Jackson and Sheiham 1980). Research into lead complexation by hydroxide and carbonate has also indicated the probable existence of additional complexes, such as $PbHCO_3^+$, $Pb_2CO_3^{2+}$, $Pb_3CO_3^{4+}$, and $Pb(CO_3)OH^-$ (Ferri et al. 1987; Ferri, Salvatore, and Fasca 1989; Fouillac and Criaud 1984; Nèher–Neumann 1992). Of these, $PbHCO_3^+$ and $Pb(CO_3)OH^-$ are most likely to have a noticeable impact on plumbosolvency under some potable water conditions. Current research is under way to refine the solubility models to include these species and determine their practical significance.

Several other solids, particularly lead orthophosphates, could be included in both programs, but these are the solids for which reliable observational confirmation appears to exist (see previous sections). The basic lead carbonate $Pb_{10}(CO_3)_6(OH)_6O$ or $3Pb_3(CO_3)(OH)_2 \cdot PbO$ (plumbonacrite) was not included in earlier modeling efforts, but its existence in lead pipe deposits has now been confirmed (Schock 1989), and it warrants inclusion. Extensive studies into lead dissolution in groundwater relating to underground storage or nuclear wastes has also indicated the possible environmental significance of this solid (Brooker et al. 1983; Haacke and Williams 1981; Shoesmith and Bailey 1988; Shoesmith, Bailey, and Taylor 1988; Taylor and Lopata 1984).

A wide range in experimental values for many lead equilibrium constants has been reported in the literature, which leads to a wide range in derived Gibbs free energies of formation for related aqueous and solid species. Frequently, only a single determination of a complexation or solubility constant has been made, and only at 25°C. Thus, while Table 4-13 represents a critically reviewed set of thermodynamic constants (Gibbs free energies of formation), it cannot be considered to be anymore than tentative until more equilibrium constant studies are done.

In many experimental systems, there is a mutual dependency of solubility and complex formation constants. Some of the problems generated by this situation have been pointed out by Hunt and Creasey (1980), Schock (1980, 1981), and Schock and Gardels (1983) in their reviews of steps taken to select thermodynamic data for their modeling calculations. Additional computations to estimate the solubility constant for hydrocerussite were done by Schock and Gardels using high- and medium-DIC data and resulted in figures that matched those obtained in the published study (Schock and Gardels 1983). The values selected for equilibrium constants incorporated into Tables 4-13 and 4-14 reflect agreement with data generated in USEPA pipe loop studies (Schock 1980, 1981; Schock and Gardels 1983); however, different combinations of hydrolysis, complexation, and solubility constants could generate similar results. Only careful redeterminations of the relevant constants can make the modeling more rigorously correct from a solution chemistry standpoint.

Table 4-14 Summary of equilibrium constants at 25°C and zero ionic strength used in lead solubility modeling computations

Reaction	LEADSOL log K	SASIAS log K
$Pb^{2+} + H_2O \rightleftarrows PbOH^+$	–7.22	–7.25
$Pb^{2+} + 2H_2O \rightleftarrows Pb(OH)_2^0 + 2H^+$	–16.91	–16.91
$Pb^{2+} + 3H_2O \rightleftarrows Pb(OH)_3^- + 3H^+$	–28.08	–28.1
$Pb^{2+} + 4H_2O \rightleftarrows Pb(OH)_4^{2-} + 4H^+$	–39.72	*
$2Pb^{2+} + H_2O \rightleftarrows Pb_2OH^{3+} + H^+$	–6.36	–6.4
$3Pb^{2+} + 4H_2O \rightleftarrows Pb_3(OH)_4^{2+} + 4H^+$	–23.86	–23.9
$4Pb^{2+} + 4H_2O \rightleftarrows Pb_4(OH)_4^{4+} + 4H^+$	–20.88	*
$6Pb^{2+} + 8H_2O \rightleftarrows Pb_6(OH)_8^{4+} + 8H^+$	–43.62	–43.6
$Pb^{2+} + CO_3^{2-} \rightleftarrows PbCO_3$	7.10	7.09
$Pb^{2+} + 2CO_3^{2-} \rightleftarrows Pb\ (CO_3)_2^{2-}$	10.33	10.29
$Pb^{2+} + H^+ + CO_3^{2-} \rightleftarrows PbHCO_3^+$	12.59	*
$Pb^{2+} + H^+ + PO_4^{3-} \rightleftarrows PbHPO_4$	15.41	15.45
$Pb^{2+} + 2H^+ + PO_4^{3-} \rightleftarrows PbH_2PO_4^+$	21.05	21.049
$Pb^{2+} + SO_4^{2-} \rightleftarrows PbSO_4^0$	2.73	2.75
$Pb^{2+} + 2SO_4^{2-} \rightleftarrows Pb(SO_4)_2^{2-}$	3.50	*
$Pb^{2+} + Cl^- \rightleftarrows PbCl^+$	1.59	1.59
$Pb^{2+} + 2Cl^- \rightleftarrows PbCl_2^0$	1.80	1.8
$Pb^{2+} + 3Cl^- \rightleftarrows PbCl_3^-$	1.71	1.7
$Pb^{2+} + 4Cl^- \rightleftarrows PbCl_4^{2-}$	1.43	1.4
$Pb^{2+} + NO_3^- \rightleftarrows PbNO_3^+$	†	1.17‡
$Pb^{2+} + 2NO_3^- \rightleftarrows Pb(NO_3)_2^0$	†	1.4‡
$PbCO_{3(s)} \rightleftarrows Pb^{2+} + CO_3^{2-}$	–13.11	–13.88
$Pb_3(CO_2)_2(OH)_{2(s)} + 2H^+ \rightleftarrows 3Pb^{2+} + 2CO_3^{2-} + 2H_2O$	–18.00	–19.14
$Pb(OH)_{2(s)} + 2H^+ \rightleftarrows Pb^{2+} + 2H_2O$	13.06	12.69
$Pb_5(PO_4)_3OH_{(S)} + H^+ \rightleftarrows 5Pb^{2+} + 3PO_4^{3-} + H_2O$	–62.83	–62.8

Source: Derived from Table 4-16 of Internal Corrosion of Water Distribution Systems *(AWWA Research Foundation and DVGW, Engler-Bunte Institute, 1985, 1st ed.), Gibbs free energy values (LEADSOL), or reported in WRc research (SASIAS).* (Technical Report TR-151).

*Not included in the SASIAS program.

†Not included in the LEADSOL program.

‡Not included in the SASIAS computations.

Effects of Carbonate and pH

The chemical system most comprehensively studied to date, and the system that has probably been the most accurately characterized, is that for lead–water–carbonate. It is the one system that has impact on all water supplies. Many early studies provided useful insight into some of the relationships that exist in this system, but poor methods of pH or alkalinity measurement, if done at all, limited their incorporation into recent modeling efforts.

In a simple system of metallic lead immersed in water containing dissolved carbonate species, the solubility will be controlled by either the simple lead carbonate ($PbCO_3$, cerussite) or one of the two basic lead carbonates (hydrocerussite or plumbonacrite). Above approximately pH 12.5, lead hydroxide ($Pb(OH)_2$) may form. The solubility constants for lead hydroxide reported in the literature vary over about five orders of magnitude (log *K* from –19.85 to –14.9). Samples of deposits from potable water systems and accompanying dissolved lead concentrations do not indicate either lead oxide (PbO, massicot or litharge) or lead hydroxide to be active in governing lead solubility, though they are frequently found.

The pH borders of the region where $PbCO_{3(s)}$, $Pb_{10}(CO_3)_6(OH)_6O_{(s)}$, or $Pb(OH)_{2(s)}$, rather than $Pb_3(CO_3)_2(OH)_{2(s)}$, would be predicted to control the lead solubility are very sensitive to the solubility constant or Gibbs free energy of formation data selected for each of the three solids and the crystallinity of the solid materials themselves.

When only pH and carbonate concentration effects are considered, a picture of lead solubility emerges like that shown in Figure 4-13, computed using the thermodynamic

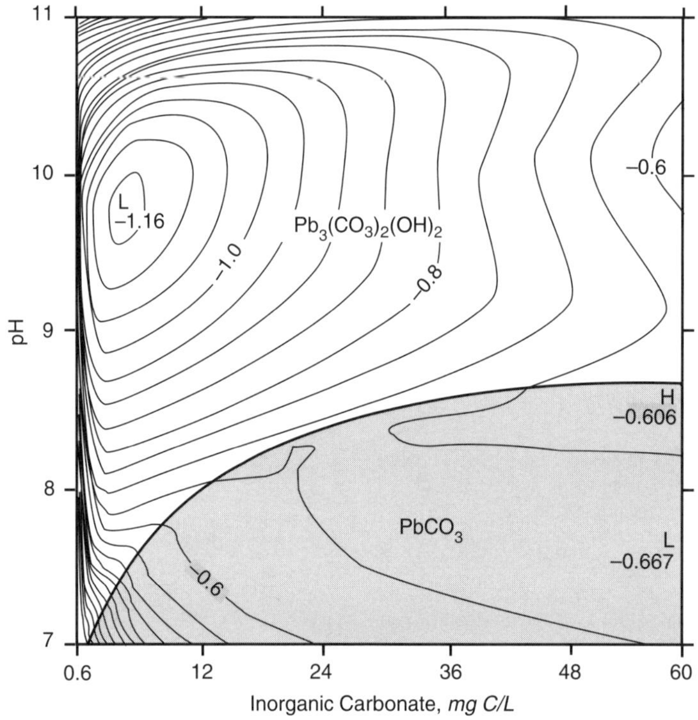

NOTE: The boundary line position between the domains of the lead is approximate. Concentration units are log (mg Pb/L). Local high and low points are marked "L" and "H," respectively.

Figure 4-13 Countour diagram of Pb(II) solubility in a pure system with varying hydrogen ion and carbonate species concentrations, at 25°C and I = 0.005 mol/L data for species in Table 4-14. The contour interval and units were a compromise to show detail with uniform intervals that would be easy to see. A conversion of "log mg Pb/L" to "mg Pb/L" values is provided in Table 4-15, for reference. The minimum lead solubility is clearly indicated to be at a pH of approximately 9.8, with a DIC concentration of between 3.6 and 4.8 mg C/L (0.0003*M* to 0.0004*M*). The symmetry of the contour lines in the basic lead carbonate stability region indicates that both pH and DIC have a significant influence on the lead solubility. In the lead carbonate stability region, above approximately 12 mg $CaCO_3$/L, lead solubility is not very sensitive to either pH or DIC, although the trend is toward slightly lower levels as DIC is increased. A three-dimensional perspective of this relationship is shown in Figure 4-14. Note that if the same complex formation constants were used in the solubility model that were used in the construction of Figures 4-13 and 4-14, but a less-soluble constant was used for $PbCO_{3(s)}$ solubility (such as that chosen by WRc researchers), the simple carbonate solid would be predicted to be stable over a wider pH range.

A "control" experiment was conducted using the recirculation test systems at USEPA to provide a reference for the comparison of effects of silicate and orthophosphate dosing (Schock and Wagner 1985). The "duplicate" test runs involved a pH of 8.2 and a total alkalinity of ~20 mg $CaCO_3$/L. Details of the experimental procedures were given previously (Schock and Wagner 1985). These experiments give useful information on the equilibrium plumbosolvency conditions to a greater degree than the intermittent-flow rigs and, therefore, provide complementary information.

Data from these USEPA experiments (designated 21 and 21A) are shown in Figure 4-15. The generalized saturation index for hydrocerussite (SI_{hc}) may be defined as:

$$SI_{hc} = \log_{10} (IAP / K_{sp}) \tag{4-6}$$

Table 4-15 Unit conversions for concentration units displayed in various solubility diagrams

log mg Pb/L	mg Pb/L	log mg Pb/L	mg Pb/L
–0.4	0.398	–1.4	0.040
–0.5	0.316	–1.5	0.032
–0.6	0.251	–1.6	0.025
–0.7	0.200	–1.7	0.020
–0.8	0.158	–1.8	0.016
–0.9	0.126	–1.9	0.013
–1.0	0.100	–2.0	0.010
–1.1	0.079	–2.1	0.008
–1.2	0.063	–2.2	0.006
–1.3	0.050	–2.3	0.005

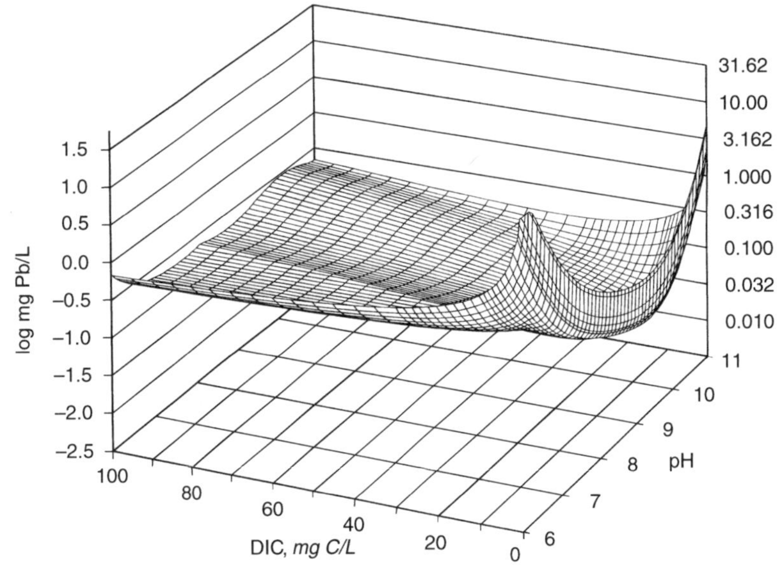

Figure 4-14 Three-dimensional plot of the relationship of lead solubility as log (mg Pb/L) to pH and DIC concentration at 25°C and I = 0.01 mol/L

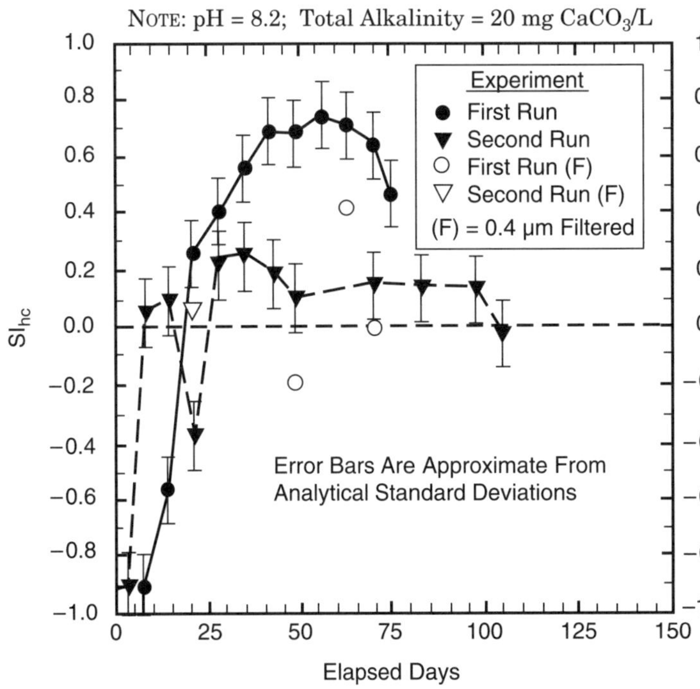

Figure 4-15 Plot of saturation index for hydrocerussite (SI_{hc}) versus time for "control" experiment in USEPA recirculation experiments

Where:

IAP = the ion activity product for the reaction written as:

$$Pb_3(CO_3)_2(OH)_{2(s)} + 2H^+ \rightleftarrows 3Pb^{2+} + 2CO_3^{2-} + 2H_2O \quad (4\text{-}7)$$

K_{sp} = the solubility constant for this reaction form (Schock 1990b) in this case:

$$IAP = \{Pb^{2+}\}^3 \frac{\{CO_3^{2-}\}^2}{\{H^+\}^2} \quad (4\text{-}8)$$

in which {} denote activities rather than concentrations (Snoeyink and Jenkins 1980; Stumm and Morgan 1981). An SI value >0 indicates oversaturation with the mineral phase; an SI value <0 indicates undersaturation with the mineral phase. Thus, analogous to the LI, the SI value <0 represents the thermodynamic tendency toward dissolution or precipitation of that solid under equilibrium conditions. Computations were performed with the WATEQX geochemical equilibrium computer program with a modified thermodynamic database, to make ionic strength and complexation corrections (van Gaans 1989).

After the initial dissolution into the large water volume of the experimental system, both runs indicate fairly good agreement with the predicted equilibrium solubility control by $Pb_3(CO_3)_2(OH)_2$ given by SI = 0. The reference line is drawn for a log K_{sp} value of –18.8 as reported in the literature (Table 4-13). Within experimental error, the two runs are in good agreement with log K_{sp} values in the range of –18.0 to –18.8 as were found experimentally in the USEPA pipe loop systems (Schock and Gardels 1983). The first run may be slightly higher because of the presence of suspended colloidal lead or a fresher and less crystalline basic lead carbonate solid on the pipe surface. The filtered samples indicated significant undissolved lead in the water in the first run.

In mechanistic studies of lead oxidation and solubility in alkaline pH, Shoesmith, Bailey, and Taylor (1988) found that the dissolution rate of Pb2+ from the electrode (pipe) surface was directly proportional to the solubility of the predominant phase present. They observed litharge (PbO) to grow underneath the initially formed basic lead carbonate surface. This unusual presence of the highly soluble solid has been verified in pipe samples from several US cities, including Oakwood (Ohio), Columbus (Ohio), and Lake Forest (Illinois) discussed previously. The stability of this phase near the metallic surface in pipe deposits from water systems not having high lead levels may give an indirect method to estimate the depth of penetration of oxidants through the scale, the permeability to lead ion diffusion, and the difference between corrosion rate and metal release into solution. The researchers hypothesize that the inability of carbonate to pass through the outer basic lead carbonate surface causes the stabilization of PbO below.

Detailed studies by Shoesmith and co-workers (1988) also shed light on the sequence of carbonate-containing solids that may form in the scale. In experiments at pH 12, they found plumbonacrite to be the composition of a compact layer formed at the metal surface mixed with hydrocerussite. A more dispersed layer of hydrocerussite then formed on the plumbonacrite surface. This layer is thought to be the result of precipitation from solution. The underlying hydrocerussite is attributed to a solid-state growth process. The extent of deposition of the outer layer was found to be controlled by the rate of transport of soluble Pb^{2+} species to the bulk solution.

This and related work (e.g., Taylor and Lopata 1984) suggests a sequence of formation and conversion of surface carbonate solids on lead, such as the following:

$$10Pb_{(s)} + 8OH^- + 6CO_3^{2-} \rightleftharpoons Pb_{10}(CO_3)_6(OH)_6O_{(s)} + H_2O + 20e^- \qquad (4\text{-}9)$$

$$Pb_{10}(CO_3)_6(OH)_6O_{(s)} \rightleftharpoons 3Pb_3(CO_3)_2(OH)_{2(s)} + PbO_{(s)} \qquad (4\text{-}10)$$

$$Pb_3(CO_3)_2(OH)_{2(s)} \rightleftharpoons 2PbCO_{3(s)} + H_2O \qquad (4\text{-}11)$$

This sequence would explain the simultaneous presence of $PbCO_3$ and $Pb_3(CO_3)_2(OH)_2$ in many pipe samples and the considerable amount of PbO (litharge) found in many thick deposits. The existence of metastable phases will be a function of the major constituent water chemistry, the oxidation rate of the pipe, and the age of the pipe.

Studies of the precipitation of lead hydroxide, basic lead carbonate, and normal lead carbonate were also reported by Dando and Glasson (1989). They analyzed the precipitate by thermoanalytical techniques, X-ray diffraction, and BET nitrogen gas sorption. By using the BET nitrogen gas sorption technique, they obtained information on the porosity of the precipitates, crystallite sizes, and crystal aging tendencies. A commercial water supply provided a realistic matrix. They found that below pH 8, the basic carbonate was the initial precipitate formed, which later decomposed through a "lead bicarbonate" state to the normal carbonate ($PbCO_3$). The transformation was found to be slow, which they attributed to its being a solid–solid reaction for the most part. They concluded that the small surface areas measured for these solids indicated low porosities and high compactness, which resulted in a highly protective layer that developed as the precipitates aged on the pipe interior. Their studies help support the common observation of the coexistence of hydrocerussite and cerussite in many drinking water supplies and the finding of hydrocerussite when cerussite is the thermodynamically stable phase.

Effects of Sulfate, Chloride, and Nitrate

Several researchers have considered possible effects of simple anions, such as nitrate, chloride, and sulfate, on lead solubility. Beccaria et al. (1981) used a variety of electrochemical and X-ray techniques to study the corrosion of lead in seawater, which provided an extreme case for effects of chloride and sulfate. During the initial passivation stage, $Pb(OH)Cl_{(s)}$ and $Pb_3(CO_3)_2(OH)_{2(s)}$ were found to be constituents in the deposit. After long immersion periods, $PbCO_3 \cdot PbCl_{2(s)}$, $Pb_2O_{3(s)}$, $PbO_{(s)}$, and $PbCl_{2(s)}$ were also found. The presence of sulfate ions did not interfere with the film formation, and sulfate ions did not precipitate to form a lead sulfate solid on the surface. In solutions with extremely high sulfate concentrations, some basic lead sulfate solids were found, notably $Pb_3(SO_4)_2(OH)_2$ and an unidentified compound with a 1:1 oxide-to-sulfate ratio. Beccaria et al. (1981) also noted that the lead corrosion mechanism is affected by the presence of chloride in oxygen-poor environments.

Pb(II) forms somewhat weaker complexes with sulfate and chloride than with carbonate, either of which might occasionally be found in water supplies at a sufficiently high level to impact lead solubility. The greatest impact would be at relatively low pH and low DIC levels, where less lead is complexed by hydroxide and carbonate species. No significant impact of practical proportions is predicted based on solubility. Calculations were performed for chloride, sulfate, and for both sulfate and chloride at DIC levels of 3–30 mg C/L. The ionic strength of the waters for these modeling calculations was set to 0.05 mol/L, to allow higher chloride and sulfate concentrations more plausibly. No significant impact on Pb(II) solubility was indicated. Other calculations have also indicated that a nitrate concentration up to 150 mg/L has no effect on lead solubility (Jackson and Sheiham 1980).

Although orthophosphate complexes of lead have been detected and their stability constants have been measured (Nriagu 1972), they are relatively weak and are not very significant at the concentration levels of orthophosphate and lead normally found in potable waters. However, lead forms several orthophosphate solids of lower solubilities than the carbonates, oxides, and hydroxides, offering a practical opportunity for the control of lead dissolution.

Effects of Orthophosphate

The reduction of lead solubility by dosing orthophosphate or orthophosphate-containing chemicals has received some attention in laboratory studies, although the quantitative accuracy of the predictions has not been refined to the degree that the accuracy for pH and carbonate combinations has. US field studies by Karalekas et al. (1983, 1978) did not find the addition of a commercial zinc orthophosphate chemical to be beneficial; however, several computer modeling and laboratory studies (Schock 1980; Hunt and Creasey 1980; Sheiham and Jackson 1981; Gardels and Schock 1981; Gregory and Jackson 1983; Nriagu 1974) have indicated that orthophosphate can be effective when the important water chemistry variables are properly controlled. Extensive experience in the United Kingdom and Germany corroborate the qualitative predictions of the chemical modeling, as has been noted previously. In order to accurately calibrate the solubility models and to ultimately resolve the issue of the mechanism of orthophosphate inhibitor action, timely analyses of the surface compounds coating the lead pipe interiors must be performed.

A major unknown is the group of factors governing the tendency of the orthophosphate solids to form finely divided precipitates or colloids in some pH conditions, which can cause lead levels to continue above solubility predictions and also to not robustly withstand physical disturbances of the pipe interior.

The lead solubility in a water–carbonate system containing orthophosphate ions will depend on the pH and DIC concentration to a great extent, as well as on the orthophosphate concentration. Most solubility modeling efforts have considered hydroxypyromorphite ($Pb_5(PO_4)_3OH$) to be the most likely lead orthophosphate solid that could control lead solubility (Schock 1980, 1981; Sheiham and Jackson 1981; Schock and Gardels 1983; Gregory and Jackson 1983; Schock 1989). Nriagu (1974) has pointed out that the solubility of $Pb_3(PO_4)_2$ (tertiary lead orthophosphate) is similar to that of hydroxypyromorphite. Although the solubility diagrams presented later in this chapter (Figures 4-18 through 4-22) assume hydroxypyromorphite to be the solubility-controlling phase, the possibility that tertiary lead orthophosphate might control the solubility in some potable water systems should not be discounted for the following reasons:

- A variation of ±1 log unit for each of the log solubility constants of hydroxypyromorphite and tertiary lead orthophosphate can yield a difference of up to about +0.07 mg Pb/L in the pH range of 5–9.5 for the combinations with the largest likely errors. The reliability of the tabulated constants has not

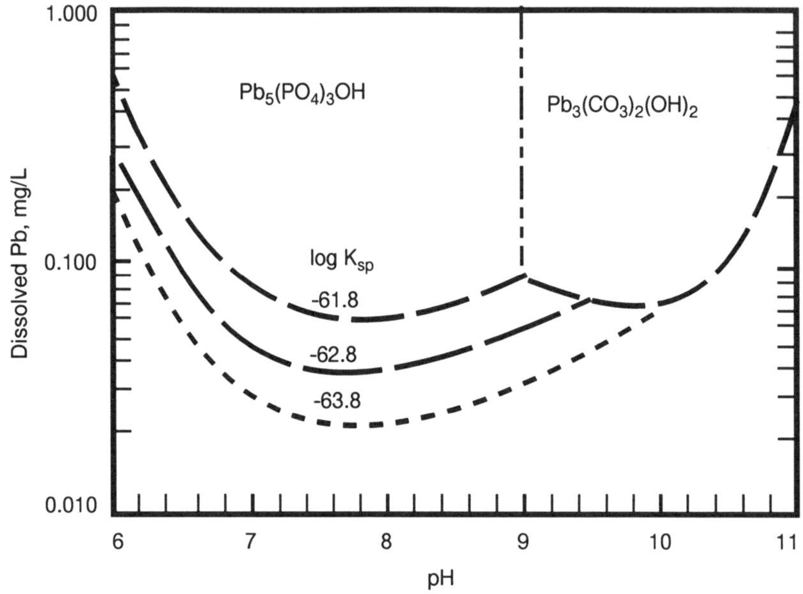

Figure 4-16 Predicted Pb(II) solubility with DIC = 6 mg C/L (0.0005*M*) and PO_4 = 0.5 mg/L, assuming the formation of only $Pb_5(PO_4)_3OH$ (hydropyromorphite) as the orthophosphate phase, with slightly different assumed solubility constant values, temperature = 25°C, I = 0.005 mol/L

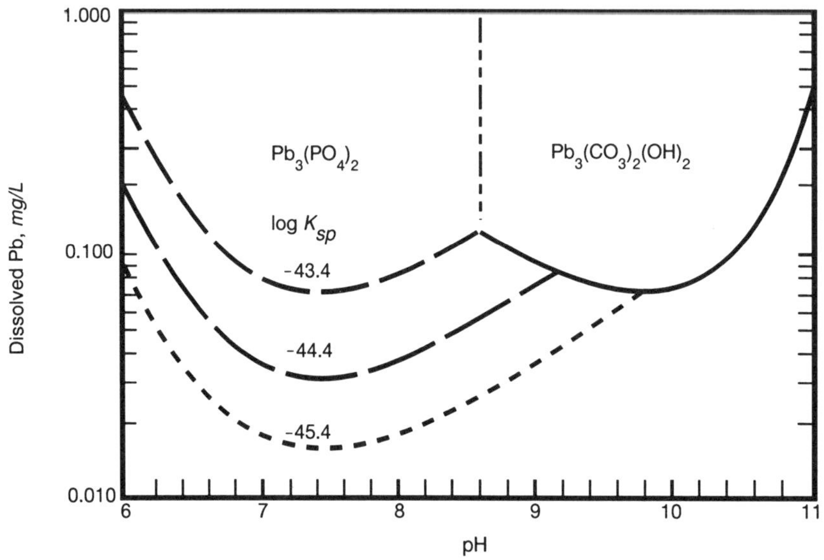

Figure 4-17 Predicted Pb(II) solubility with DIC = 6 mg C/L (0.0005*M*) and PO_4 = 0.5 mg/L, assuming the formation of only $Pb_3(PO_4)_2$ (tertiary lead orthophosphate) as the orthophosphate phase, with slightly different assumed solubility constant values, temperature = 25°C, I = 0.005 mol/L

been well established. Although the trends in predicted solubility are similar (see Figures 4-16 and 4-17), the details of solid-phase dominance could vary significantly depending on the accuracy of the thermochemical data.

- Scale analyses have not frequently been performed in real or experimental systems, and precipitate formation and aging behavior have not been examined in detail. Tertiary lead orthophosphate may or may not have been the primary precipitating solid in the USEPA experiments discussed earlier.
- Lead removal from waste streams as $Pb_3(PO_4)_2$ has been reported (Nriagu 1974).
- By rough analogy with the calcium–orthophosphate system, in which hydroxyapatite is the thermodynamically stable phase but not the initial precipitate (Jenkins et al. 1971; Nancollas 1983; Green and Holmes 1947; Brown 1971), the formation of hydroxypyromorphite may be preceded by a precursor phase of higher solubility.
- Both orthophosphate solids have been found in drinking water pipes from different pipe rigs or water systems in US studies.

Figure 4-18 shows the influence of an orthophosphate dosage of 0.5 mg PO_4/L as pH and DIC are varied (compare with Figure 4-13). Substantial solubility decreases are evident, particularly in the pH range of 7–9. Figure 4-19 illustrates the sensitivity of lead solubility to the orthophosphate concentration at a DIC level of 6 mg C/L (0.0005*M*), and Figure 4-20 illustrates the sensitivity of lead solubility to the orthophosphate concentration at a DIC level of 24 mg C/L (0.002*M*). Note that the contour intervals are on a logarithmic scale and that a relatively small decrease in solubility results from having an orthophosphate dosage of more than about 3 mg PO_4/L in the system above a pH of approximately 7. In the low-DIC system (Figure 4-19), even the presence of 1.0 mg PO_4/L brings about large solubility reductions. Figures 4-19 and 4-20 can be used to help estimate minimum dosages of orthophosphate for the reduction of lead solubility. To illustrate the general trends, Figures 4-21 and 4-22 present solubility three-dimensional surface summaries of the interrelationships between pH and orthophosphate for a high and low DIC level.

These figures easily explain the successful application of approximately 2 mg PO_4/L to control plumbosolvency in many hard groundwaters in the United Kingdom (Gregory and Jackson 1984; Colling, Whincup and Hayes 1987; Colling et al. 1992). They show that DIC plays a major role in controlling the soluble lead level as well as the orthophosphate dosage, particularly between pH 6 and 8.

The sensitivity of lead solubility to orthophosphate dosage lessens at higher orthophosphate dosages, particularly above about 1.5–2 mg PO_4/L in the pH 7–8 range. The higher-pH "trough" in Figure 4-22 is produced by $Pb_3(CO_3)_2(OH)_2$. Clearly, in waters with high DIC, as most hard groundwaters would have, the formation of $Pb_5(PO_4)_3OH$ (the lower trough) or $Pb_3(PO_4)_2$ (not shown but similar to $Pb_5(PO_4)_3OH$) is practical and advantageous. In pipe rig tests using both old and new pipe, Sheiham and Jackson (1981) found very substantial reductions in lead levels by dosing orthophosphate at 1 mg P/L (3.07 mg PO_4/L). The pH was not clearly specified but can be inferred to be about 7.1 from other information. The groundwater used had an initial alkalinity of approximately 240 mg $CaCO_3$/L and a calcium level of approximately 124 mg/L. The lead concentrations were reduced to much lower levels in the new pipe than in the old. The old pipe also showed more of an increase in lead leaching than the new pipe when the orthophosphate dosage was reduced to 0.5 mg P/L (1.5 mg PO_4/L).

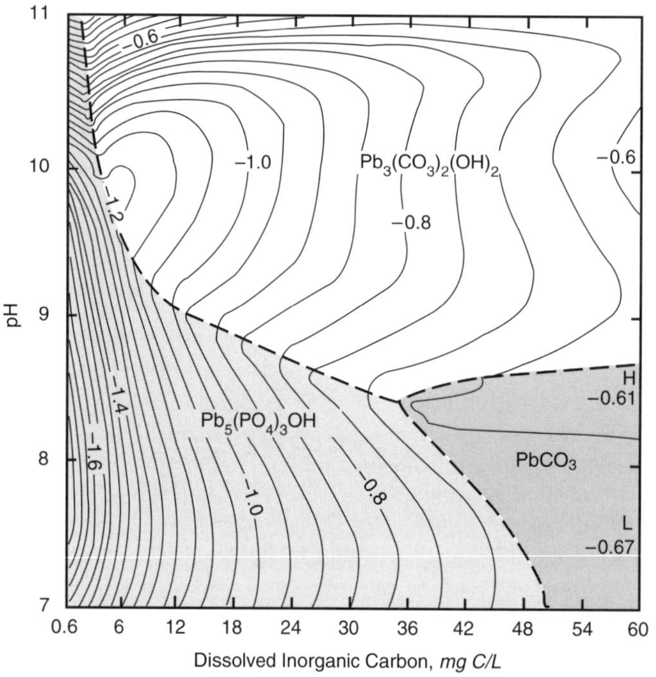

NOTE: Concentration units are log (mg Pb/L). Local high and low points are indicated.

Figure 4-18 Contour diagram of Pb(II) solubility in a system having 0.5 mg PO4/L orthophosphate, with varying levels of pH and DIC, at 25°C and I = 0.005 mol/L

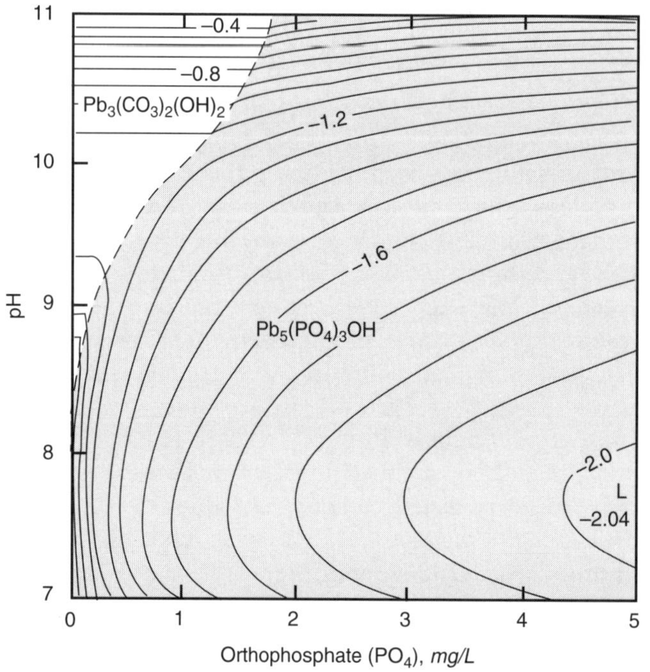

NOTE: Concentration units are log (mg Pb/L). The solid phase boundary lines are approximate.

Figure 4-19 Contour diagram for the impact of varying concentrations of orthophosphate and hydrogen ion, with DIC = 6 mg C/L, temperature = 25°C, and I = 0.005 mol/L

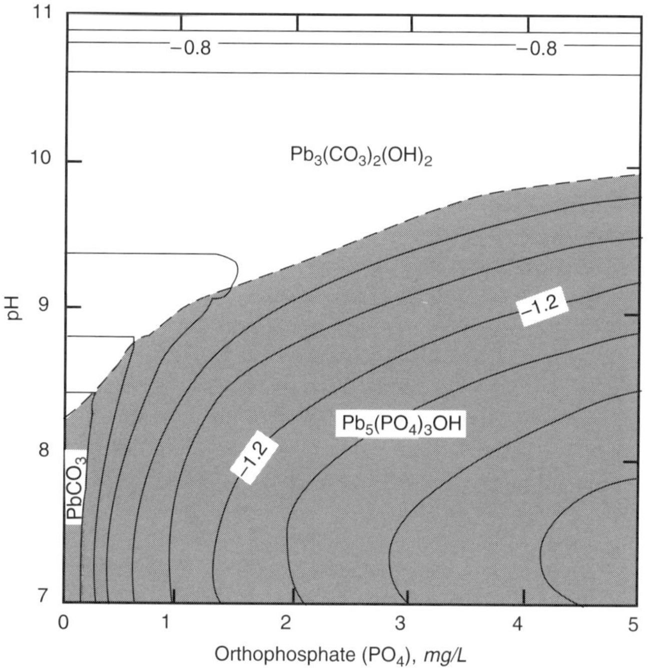

NOTE: Concentration units are log (mg Pb/L). The solid phase boundary lines are approximate.

Figure 4-20 Contour diagram for the impact of varying concentrations of orthophosphate and hydrogen ion, with DIC = 24 mg C/L, (0.002M), temperature = 25°C, and I = 0.005 mol/L

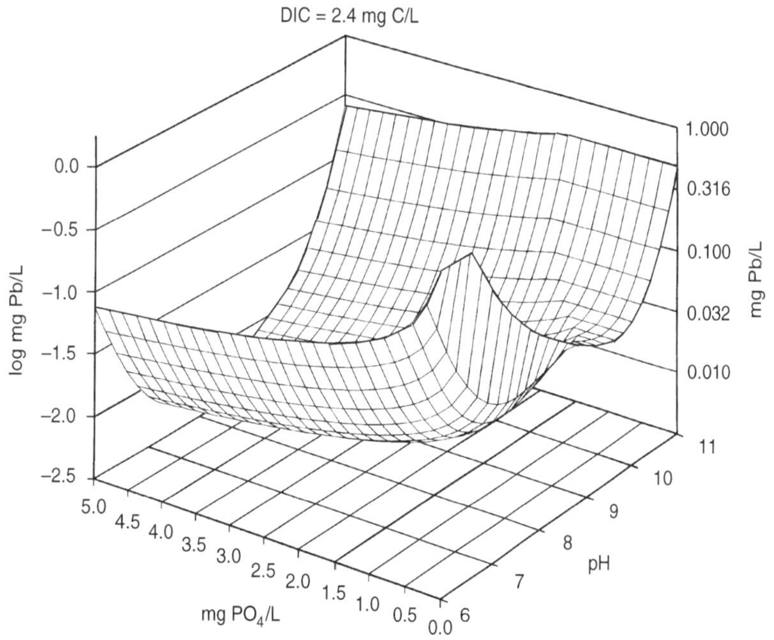

Figure 4-21 Three-dimensional surface plot of lead concentration versus orthophosphate dose for 2.4 mg C/L DIC (DIC = $2 \times 10^{-4}M$, I = 0.01 mol/L, temperature = 25°C)

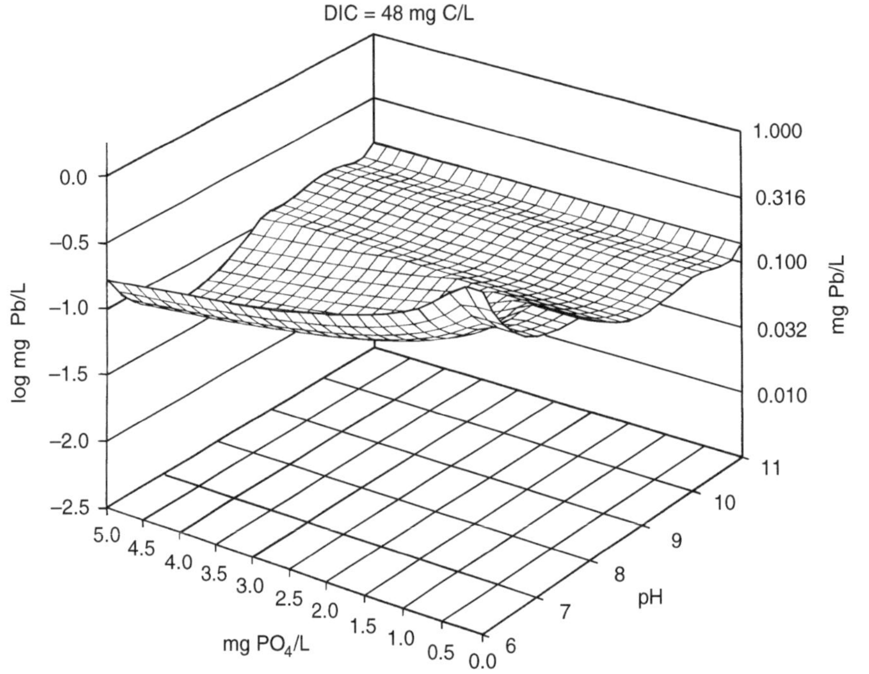

Figure 4-22 Three-dimensional surface plot of lead concentration versus orthophosphate dose for 48 mg C/L DIC (DIC = $4 \times 10^{-3}M$, I = 0.01 mol/L, temperature = 25°C)

The USEPA recirculation experiments at an orthophosphate concentration of ~0.5 mg PO_4/L, alkalinity of ~20 mg $CaCO_3$/L, and pH values of 7.8 and 8.5 provide some additional guidance on the behavior of orthophosphate (Schock and Wagner 1985). Data from these experiments were analyzed using the WATEQX computer program (van Gaans 1989), to look at saturation indices, computed DIC, and speciation under the different experimental conditions. Graphical data summaries of the variation of lead and orthophosphate levels and some other interesting relationships are shown in Figures 4-23 through 4-31. Some samples were filtered through 0.4-μm pore size Nuclepore™ (Nuclepore™ Filtration Products, Pleasanton, California) polycarbonate filters, using all-plastic apparatus and sample rinsing to minimize sorption, to estimate the amount of dissolved lead or zinc present in the water (Schock and Gardels 1983). Summaries of the general water characteristics are given in Table 4-16 for these experiments. Orthophosphate was occasionally added to the test systems, but it was not strictly controlled. The pH, however, was closely controlled, and DIC increased only a little during the runs. Experiments 32A, 33A, and 34A represent "duplicate" runs, consisting of replacing the tank water but not the pipe. Both galvanized and lead pipe were in the system, but zinc levels were not controlled by the orthophosphate and no zinc solids were found on the lead pipe specimens.

The following interesting trends could be extracted from the data:

- Lead concentrations were much lower than those found in the "control" experiment of pH 8.2 for all three pH values tested.
- There was little true difference in the lead levels obtained from the three different pH values, because of the data scatter. The most variability was at pH 8, the least at pH 7.

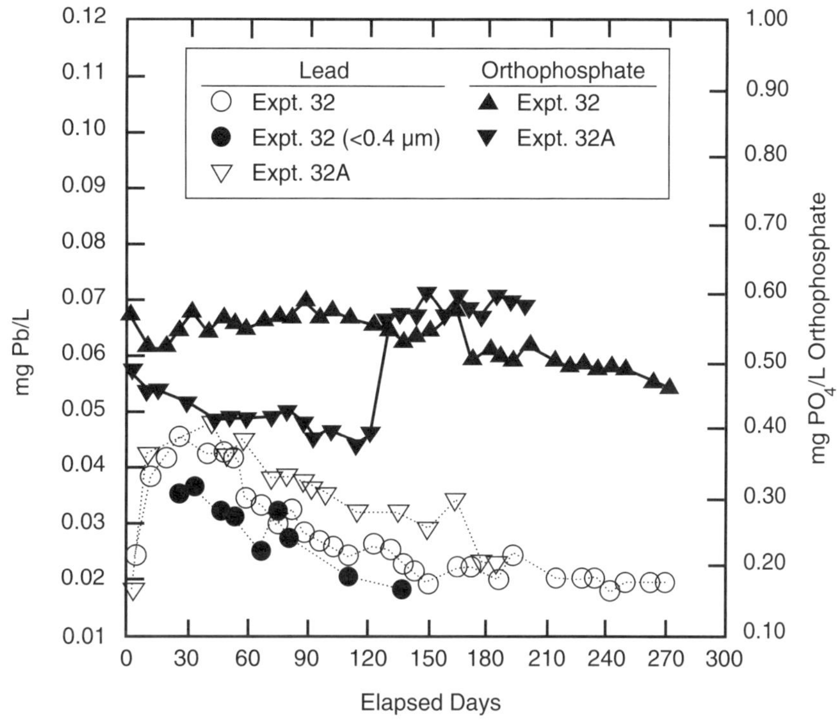

Figure 4-23 Orthophosphate and lead concentration data for low-alkalinity, pH 7 experiments in USEPA recirculation systems

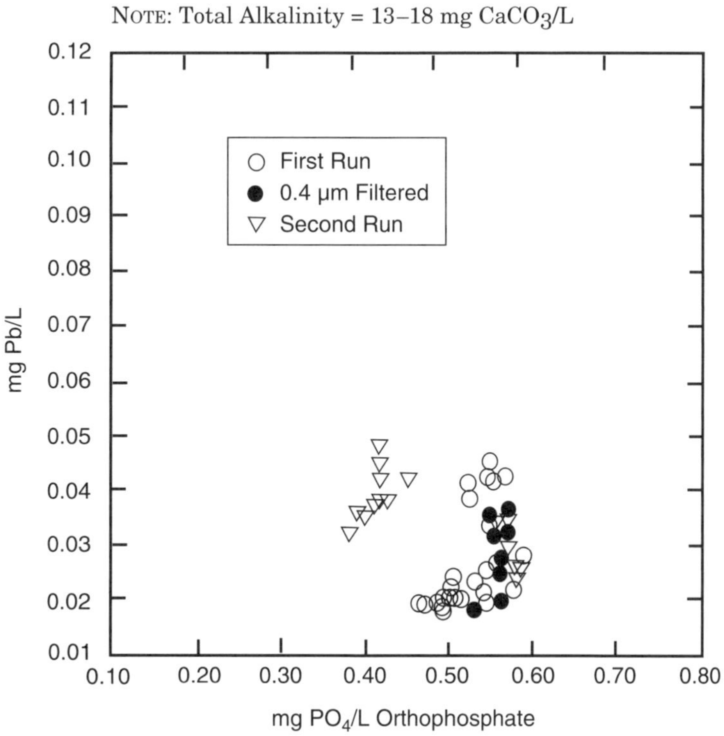

Figure 4-24 Lead concentration versus orthophosphate concentration in USEPA experiments 32 and 32A (pH 7, see Table 4-16)

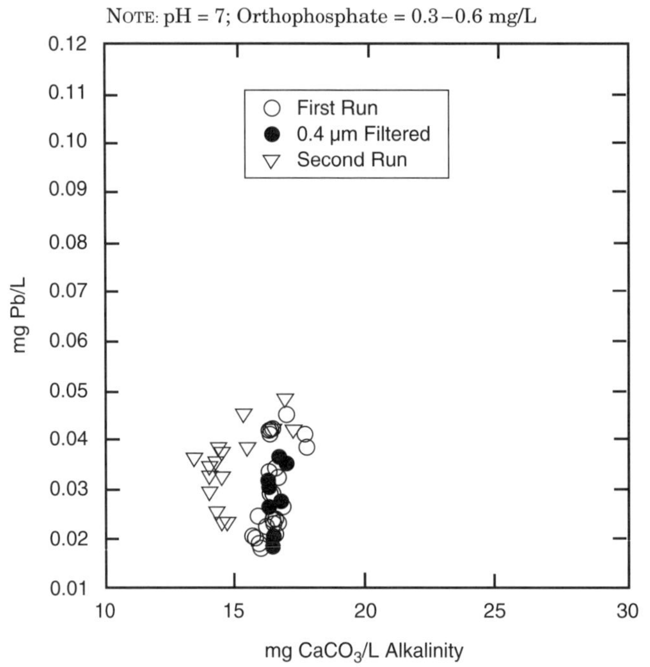

Figure 4-25 Lead concentration versus alkalinity in USEPA experiments 32 and 32A (pH 7, see Table 4-16)

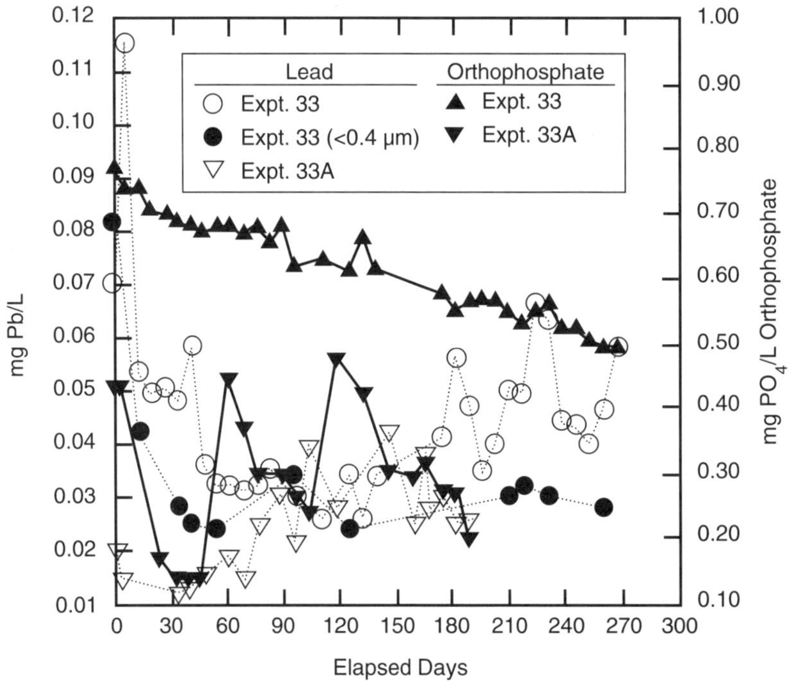

Figure 4-26 Orthophosphate and lead concentration data for low-alkalinity, pH 8 experiments in USEPA recirculation systems

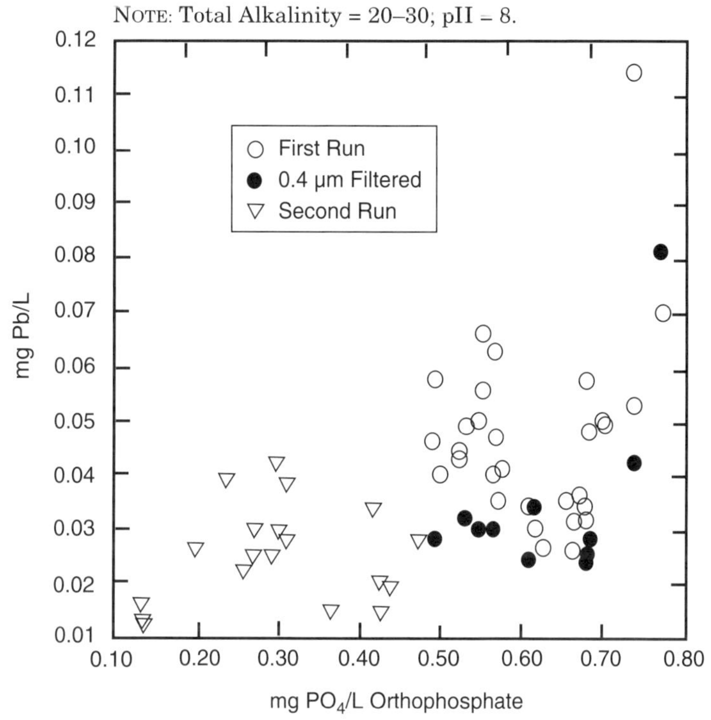

Figure 4-27 Lead concentration versus orthophosphate concentration in USEPA experiments 33 and 33A (pH 8, see Table 4-16)

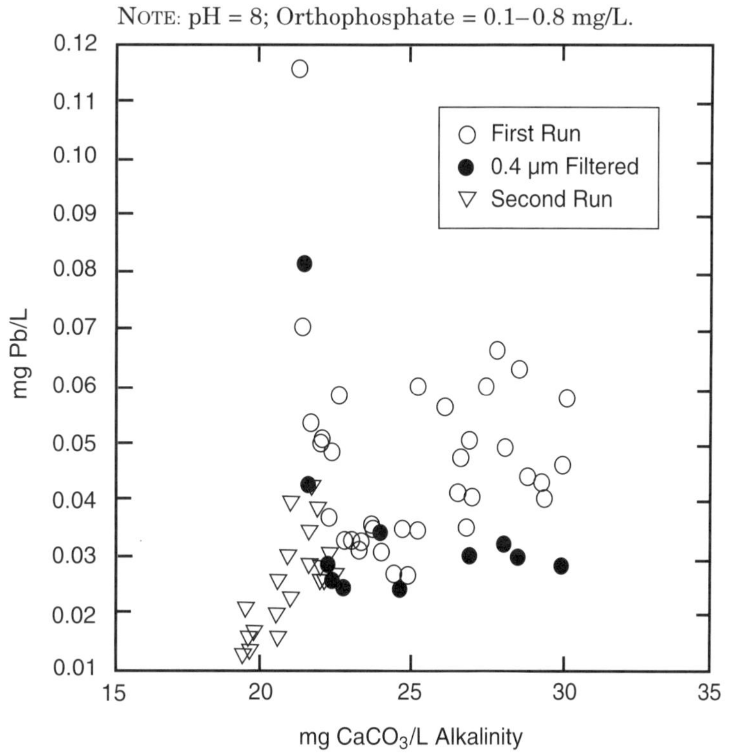

Figure 4-28 Lead concentration versus alkalinity in USEPA experiments 33 and 33A (pH 8, see Table 4-16)

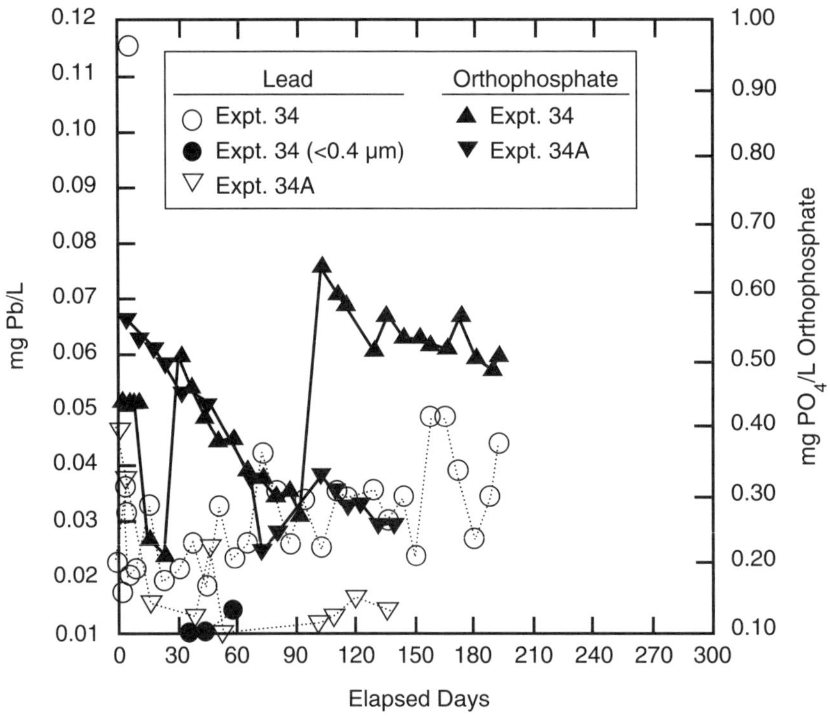

Figure 4-29 Orthophosphate and lead concentration data for low-alkalinity, pH 8.5 experiments in USEPA recirculation systems

NOTE: Total Alkalinity = 20–30; pH = 8.5.

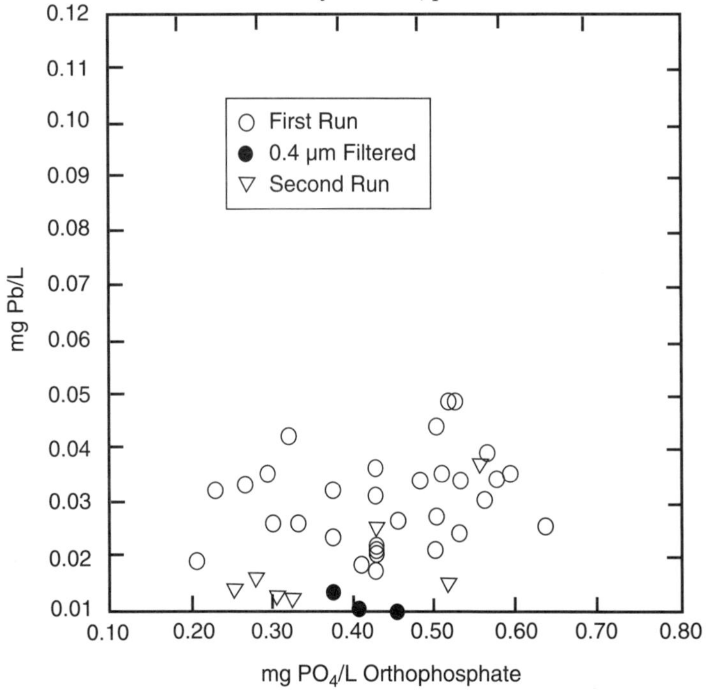

Figure 4-30 Lead concentration versus orthophosphate concentration in USEPA experiments 34 and 34A (pH 8.5, see Table 4-16)

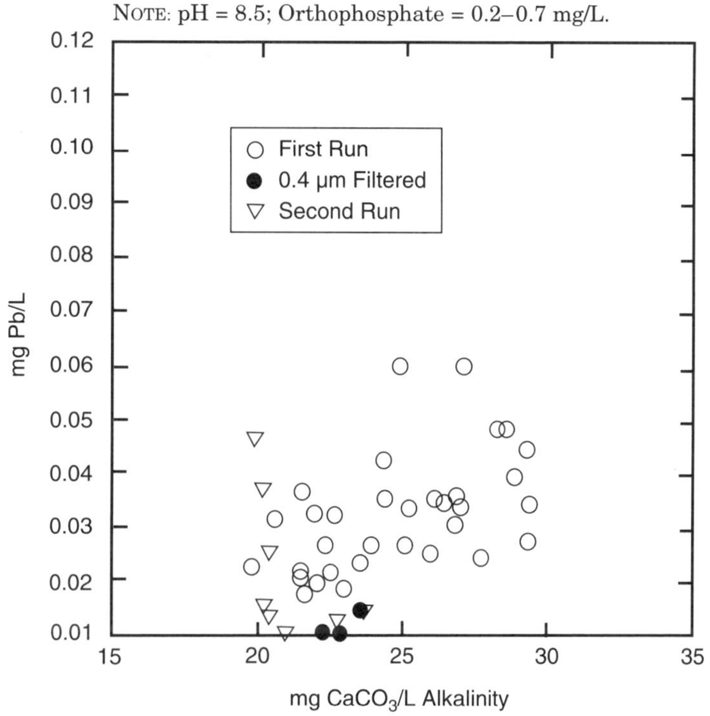

Figure 4-31 Lead concentration versus alkalinity in USEPA experiments 34 and 34A (pH 8.5, see Table 4-16)

Table 4-16 Approximate water qualities for the six USEPA corrosion experiments cited

	Experiment					
Constituent*	32	32A	33	33A	34	34A
Dissolved inorganic carbon, *mg C/L*	4.4–4.8	3.9–4.8	5.0–7.2	4.5–6.8	4.5–6.8	4.6–5.3
pH (units)	7.0–7.1	6.9–7.3	7.9–8.1	7.8–8.2	8.2–8.6	8.2–8.7
Total alkalinity (as $CaCO_3$)	16–18	13–17	21–30	20–22	20–30	20–23
Calcium	4	4	4	3	4	4
Magnesium	1	1	1	1	1	1
Sodium	9	9	11	10	16	11
Potassium	0.6	0.7	0.7	NA†	NA	NA
Chloride	10	10	5	3	9	5
Sulfate	7	7	6	6	8	7
Orthophosphate	0.5–0.7	0.4–0.6	0.5–0.8	0.2–0.5	0.2–0.6	0.3–0.6
Silicate	0.1	<2	1.4	NA	NA	NA
Dissolved oxygen	6.5	NA	7	NA	8	8

*Concentrations are in milligrams per litre for the forms indicated, unless otherwise specified.

†Not analyzed.

- Small trends of lead versus orthophosphate appear in the data, even with the small difference in orthophosphate concentrations during the runs. At pH 7, lead decreases with higher orthophosphate levels. At pH 8, lead levels seem to rise slightly at higher orthophosphate levels. At pH 8.5, not much of a trend is noted.
- At pH 7, there is no trend of lead concentration with alkalinity. At pH 8 and 8.5, there seems to be a small increase in lead apparently associated with higher alkalinity (which correlates with higher DIC).

Most significant, perhaps, is the relative amounts of dissolved versus total lead evident in the six (total) runs. At pH 7, there is clearly little difference between the filtered and unfiltered lead concentrations, indicating essentially all dissolved species present for lead. At pH 8, however, filtered lead samples showed that only about half of the lead appeared to be "dissolved" (i.e., ≤ 0.4 µm). At pH 8.5, the filtered samples uniformly were very low in concentration (<0.015 mg/L) and only comprised about 25–30 percent of the lead present. Also noteworthy is the fact that the second runs for the two pH ≥8 experiments generally produced lower lead levels, which would be consistent with removal of early-formed non-adherent colloids.

Research by Dando and Glasson (1989) into the physical nature of lead orthophosphate precipitates also uncovered some interesting properties. They found much higher surface areas for $Pb_5(PO_4)_3OH$ and $PbHPO_4$ than for cerussite, hydrocerussite, and lead hydroxide. They suggested that these solids did not recrystallize to more compact and protective forms during the time frames tested, unlike the carbonate hydroxide or hydroxy carbonate solids analyzed. Therefore, the lead orthophosphate films would be useful in plumbosolvency reductions but would not form highly stable coatings on pipes. In this regard, prior pipe surface history may also become important.

Equilibrium solubility models predict that at low DIC concentrations some solubility decrease may be obtained by maintaining orthophosphate concentrations considerably above about 0.5 mg PO_4/L above pH 8 (see Figures 4-18 through 4-20). The figures and computations show, however, that the minimum solubility for hydroxypyromorphite generally lies in the range of pH 7.2 to 7.8 for most concentrations of DIC. From a mass-action standpoint and from the results of the field experiments of Colling and co-workers (1987, 1992), higher dosages of orthophosphate at lower pH values are probably preferable to lower dosages at higer pH values. On a mole-per-litre basis, there is little lead present in even corrosive waters, so a major driving force to precipitate a passivating film will be the orthophosphate concentration. Careful examination of the relationship of DIC–pH–orthophosphate concentrations to lead solubility (Schock 1989; EES and ISWS 1990) shows that lower lead levels are usually predicted for the pH 7.2–7.8 range than those for pH >8, even if the solubility of lead orthophosphate is lower than for basic lead carbonate at higher pH.

Effects of Polyphosphates

The aqueous chemistry of lead polyphosphate complexes is largely unknown. Because deposits in potable water systems using polyphosphate chemicals have not been exactly characterized, their influence on the formation of passivating films cannot be clearly stated. Polyphosphate species, particularly simple forms such as pyrophosphate and tripolyphosphate, are known to form strong and significant aqueous complexes with a variety of metals, including lead (Karweik and Huber 1978; van Wazer and Campanella 1950). At least one study has shown the use of a commercial "corrosion inhibitor" polyphosphate to increase copper levels, which would indicate a potential for a similar effect with lead (Bailey 1982). Statistical associations of higher lead levels for utilities using "hexametaphosphates" at pH 6.5–8 have also been reported (Maas et al. 1991), particularly in waters of higher hardness. The phenomenon of "ligand-promoted dissolution" of iron oxides by polyphosphates is well established (Lin and Benjamin 1990). Considerable anecdotal information from utilities on increases in lead levels when sequestering agents are used exists; however, little has been verified because of the lack of a controlled study.

Many early studies referred to the polyphosphate compounds as "metaphosphates," but little, if any, data were presented to confirm their proper structural classification. Further, and importantly, no aqueous species were identified. Many claims have been made with regard to the prowess of polyphosphates as almost universal corrosion inhibitors, so systems containing lead solder or service connections might be exposed to them at the present time.

In the United States, the first laboratory investigation of polyphosphates and lead corrosion was made by Hatch in 1941 (Hatch 1941). His study was designed to look at several factors pertaining to polyphosphate action. One experiment, which looked at (1) the effect of pH in the range of 5.2–8.9, and (2) the release of lead during passage through lead wool, will be discussed here.

In the lead wool experiment, a metaphosphate dosage of 2 mg/L (units unknown) was used. Hydrochloric acid and sodium carbonate were used to adjust the pH of the test water (initially pH 6.0) to pH 5.2, 7.0, and 8.9. Table 4-17 shows the water quality used in the lead wool experiments.

At pH 5.2 and 6.0, the presence of 2 mg/L (units unknown) metaphosphate provided some inhibition of lead dissolution, although the lead levels considerably exceeded 0.5 mg/L (see Figure 4-32). A test of 4 mg/L metaphosphate at pH 6.0 produced results similar to those of the test with 2 mg/L metaphosphate, with only slightly faster lead level reduction. At pH 7.0, the metaphosphate induced additional lead dissolution during exposure of less than approximately 100 hours, possibly indicating interference with the formation of a different kind of passivating film (such as lead carbonate or basic lead carbonate). At pH 8.9, little difference was discernible between

Table 4-17 Average water supply analyses during lead wool corrosion tests of Hatch (1941)

Constituent*	Concentration
Temperature, °*C*	28
pH (units)	6.0
Oxygen	NA†
Chloride	22
Sulfate	162
Calcium	35
Magnesium	8
Total alkalinity (as $CaCO_3$)	3
Total solids	268

*Constituents in milligrams per litre, unless otherwise specified.

†Not analyzed.

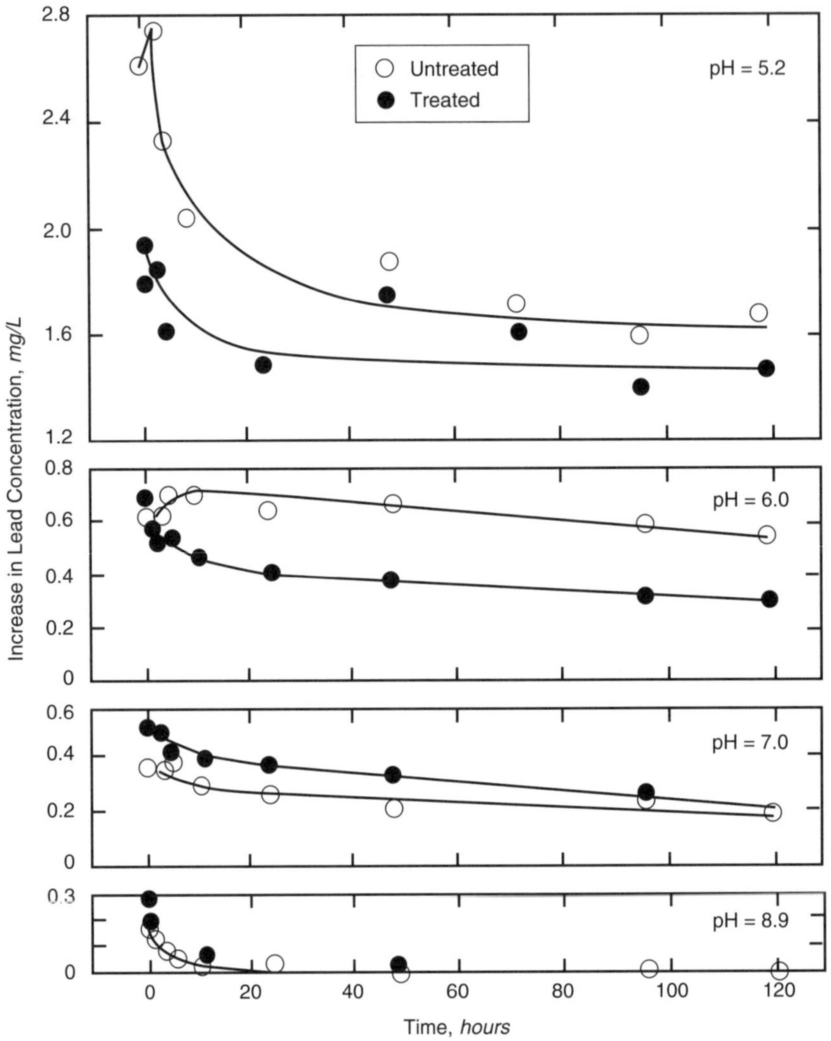

Source: Hatch (1941).

Figure 4-32 Influence of 2 mg/L metaphosphate on the solubility of lead wool at different pH values. (Background water quality is given in Table 4-16.)

the treated and untreated waters. After approximately 10 hours of exposure, the lead level was reduced to 0.05 mg/L or less. Thus, the only experimental condition that effectively reduced the lead to a safe level was pH adjustment to nearly pH 9.

In summarizing the results of his test and the literature available at that time, Hatch noted that the exact concentration of metaphosphate at which corrosion "inhibition" ceased and at which increased lead dissolution commenced had not been determined.

Shortly after the study by Hatch was published, Ruchhoft and Kachmar (1942) performed laboratory experiments on the solubility of two lead salts, basic lead carbonate and lead sulfate, in several different solutions. They included sodium metaphosphate, sodium chloride, calcium hydroxide, sodium hypochlorite, sodium pyrophosphate, aluminum sulfate, and ferrous sulfate. The experimental procedure did not include pH or carbonate control, and the boiling of the test solutions to accelerate the dissolution of the lead salts made the solubility data inapplicable to practical situations, and also probably changed the nature of the polyphosphates through hydrolysis to orthophosphate. Therefore, their study will not be reviewed in detail.

Out of concern for the possible removal of existing deposits in lead pipe by polyphosphate, Moore and Smith (1942) performed additional experiments with lead wool, aged standard lead pipe, three different pH levels, and a slowly dissolving form of polyphosphate called "Micromet." The water sources were: (1) filtered Cambridge water with a turbidity <0.1, approximate hardness of 32 mg/L, total alkalinity of approximately 11.5 mg $CaCO_3$/L, and a pH between 5.9 and 6.3; (2) Cambridge tap water with a pH of approximately 8.8, hardness of 42 mg/L, turbidity ≤0.2, and total alkalinity of approximately 22 mg $CaCO_3$/L; and (3) raw water with a fairly high color, pH of 6.9, 16 mg $CaCO_3$/L alkalinity, and 32 mg/L hardness.

Two series of experiments were conducted with lead wool. The first series involved passing Cambridge tap water through a tube containing lead wool and also through a set of tubes containing "Micromet" followed by lead wool for 39 days. The concentration of metaphosphate was approximately 3 mg/L (presumably by weight of chemical). The second series was the same, except that the Cambridge filtered water (pH ~6.0) was used and the run lasted 35 days. The flow rate was 200 mL/min.

Because of its high pH, only small quantities of lead, generally less than 0.3 mg/L, were released into the tap water. Compared to the untreated water, the metaphosphate solution increased the amount of lead leached from the wool by approximately one third. The filtered water released a considerably larger concentration of lead; generally, the lead level exceeded 0.4 mg/L and rose as high as 1.3 mg/L. Initially, the metaphosphate solution showed a slightly higher lead increase, but after 10 days, the metaphosphate solution exhibited a significant inhibiting action. Additional experiments were conducted with pipe racks and with flowing samples. However, the lead levels in even the most inhibited systems were extremely high in comparison to current standards and known detrimental health effects levels, so they will not be discussed here.

Sheiham and Jackson (1981) tested the application of a polyphosphate to lead pipe in their series of studies under realistic conditions. Two types of polyphosphates were evaluated, the commercial product Calgon SWB and tetrasodium pyrophosphate ($Na_4P_2O_7$). Dosages of 3.1 mg PO_4/L were used with the low-alkalinity moorland water described previously. The results are presented in Table 4-11. Both polyphosphates considerably inhibited lead solubility in comparison to the control but not to as great an extent as that provided by the simple orthophosphate (Table 4-12). Sheiham and Jackson also found that the addition of zinc sulfate, in combination with the polyphosphates, did not enhance the effectiveness of the polyphosphates.

Although iron was studied, the laboratory experiments of Koudelka et al. (1982) may have some important information for understanding the mechanism of action of polyphosphates on metals in general. Electrochemical characteristics and corrosion rates of pure iron in different sodium polyphosphate and sodium polyphosphate–calcium chloride solutions were correlated with surface analyses by X-ray photoelectron spectroscopy (XPS, or electron spectroscopy for chemical analysis [ESCA]). In a 60-mg/L (sodium polyphosphate) solution, a thin and relatively unhydrated film containing iron orthophosphate, probably $FePO_{4(s)}$, was formed. The film was much like an air-formed oxide film in composition and thickness, and indications were that the polyphosphate interfered with the process of film repair, rather than its growth. The following two-step film formation mechanism was proposed:

1. Formation of soluble iron–polyphosphate complex ions, for instance, $Fe^{2+}/P_5O_{16}^{7-}$ and/or $Fe^{3+}/P_2O_7^{4-}$ on the anodic areas of the metal.

2. The hydrolysis (reversion) of the complexes, leading to the precipitation of ferrous or ferric orthophosphate.

Koudelka and co-workers also suggested that the presence of a high concentration of calcium ion accelerates the reversion of polyphosphates to orthophosphate, with the rate of reversion increasing with increasing pH. Thus, orthophosphate films from sodium polyphosphate–calcium chloride solutions would be favored at higher pH values.

An investigation was made to attempt to identify and quantify polyphosphate species of Pb^{2+} (plus four other metals) in water containing Calgon TG and Calgon TG-10 (Calgon Corporation, Pittsburgh, Pennsylvania), which are bimetallic glassy phosphates representative of that class of chemicals (Pagenkopf 1982). Conditional stability determined by differential pulse polarography, for an assumption of 1:1 complex stoichiometry, is generally intermediate between sulfate and carbonate in affinity for Pb^{2+}, so some increase of lead solubility would be expected. The solubility increase might be moderated somewhat by the relatively low concentration of the polyphosphate ligands in most potable water systems and by the competitive complexation of Ca^{2+} (and, presumably, Mg^{2+}) for which some of the polyphosphates have a much greater affinity. Obviously, considerably more work is required in this area of solution chemistry.

In 1991, an experimental determination was made for the conditional stability constants of three commercial proprietary blended and polyphosphate corrosion inhibitors (Holm and Schock 1991a, 1991b). The background included a fixed ionic strength of 0.1, a calcium concentration of 40 mg/L, a pH of 8.0, and total alkalinities of 5, 50, and 250 mg $CaCO_3$/L. Dosages were obtained from the manufacturers' literature, and the methodology used to define the constants was the well-known competing ligand spectroscopy. Experimental complexation data was fitted by assuming a 2-ligand model (Holm and Schock 1991b; Holm and Smothers 1990).

The experimentally determined stability constants were then used to estimate the impact that similar polyphosphate chemicals would have on the equilibrium solubility of lead, assuming either a surface film of hydrocerussite or of hydroxypyromorphite. Figures 4-33 and 4-34 show the predicted solubility enhancement for two of the tested products. When orthophosphate is present, the solubility-enhancing properties of the polyphosphates are offset to some degree, as shown in Figure 4-34 representing 0.4 mg PO_4/L orthophosphate.

The computations using the experimental complexation constants are in qualitative agreement with those of other researchers discussed here. Additional experimentation at pH 7 also revealed significant complexation of lead by the same treatment chemicals (Holm and Schock 1991b). Based on the tendencies found in the experimentation, plus the confusion existing in the water treatment literature on the actual chemistry of these compounds, they concluded the following:

- Many chemical interactions govern when polyphosphate addition could reduce or elevate lead solubility, including pH; DIC; percentage of orthophosphate in the formulation; temperature; calcium, iron, manganese, or magnesium concentration; complexation strength of the polyphosphate ligands included; and the reversion rate.

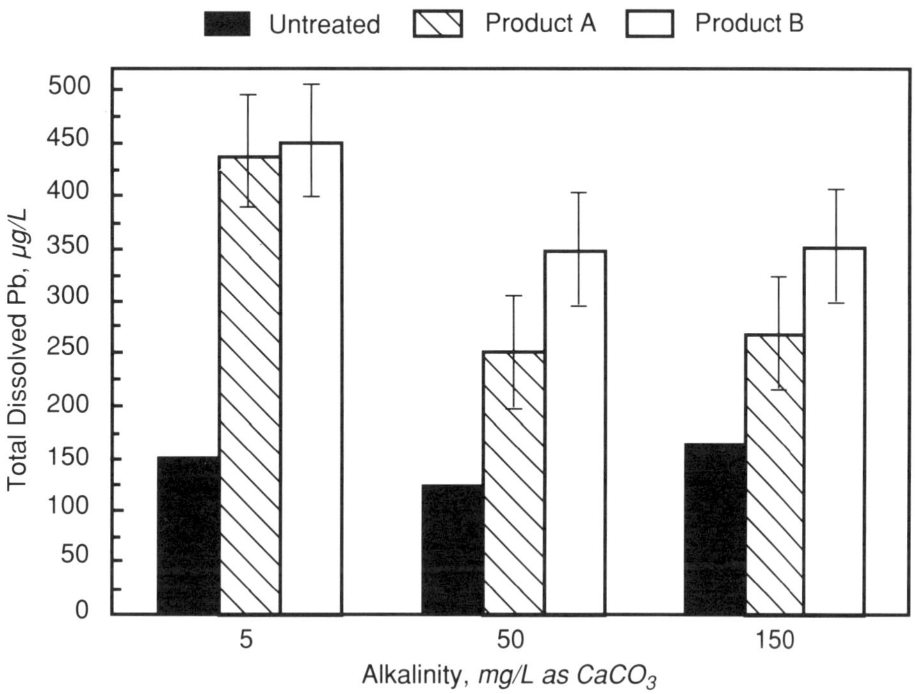

Source: Holm and Schock (1991a, 1991b).

Figure 4-33 Effect of two polyphosphate chemicals on computed lead concentrations for a water of three different alkalinities in equilibrium with $Pb_3(CO_3)_2(OH)_2$ for pH = 8, Ca = 40 mg/L, temperature = 25°C, I = 0.1

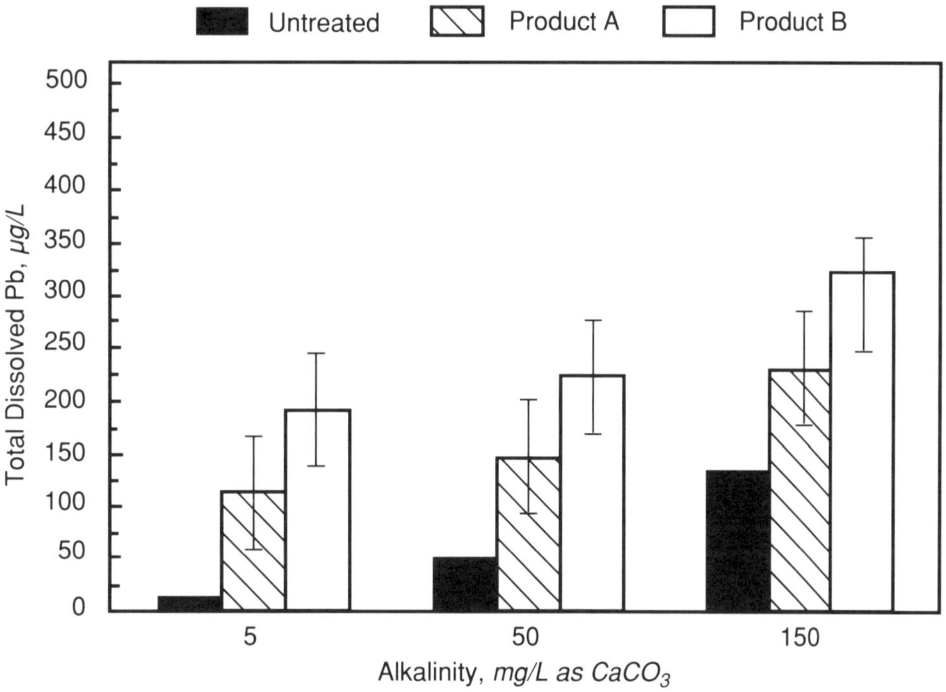

Source: Holm and Schock (1991a, 1991b).

Figure 4-34 Effect of two polyphosphate chemicals on computed lead concentrations for a water of three different alkalinities in equilibrium with $Pb_5(PO_4)_3OH$ for pH = 8, Ca = 40 mg/L, temperature = 25°C, I = 0.1

- The application of polyphosphate chemicals for the specific purpose of controlling plumbosolvency entails considerable uncertainty and a significant risk.
- Much more systematic research is needed on the potential impact of polyphosphate chemicals on lead leaching from a variety of plumbing materials.
- To determine the mechanism of action of the polyphosphate chemicals, detailed compound-specific scale analyses must be performed for a wide range of conditions.

Additional data have recently been presented or published showing that hexametaphosphate chemicals (without blending with orthophosphate) contribute to elevated lead levels. Figure 4-35 shows the average lead levels in standing samples from a system that used sodium hexametaphosphate until the summer of 1990 (Ramaley 1993). The pH in this system ranged from approximately 7 to 8, with an alkalinity of approximately 40 mg $CaCO_3$/L and a total hardness of around 80 mg $CaCO_3$/L. A rapid substantial reduction in the lead levels was apparent when the zinc orthophosphate inhibitor was substituted for the sodium hexametaphosphate.

In a pipe rack test system testing alternative treatments for a very soft, low-pH water in South Carolina, sodium hexametaphosphate at pH 7.5 was found to yield the highest lead levels with the exception of the control water (McAnally et al. 1993). The background water quality and treatment configuration is shown in Table 4-18, and the first-draw lead concentrations are given in Figure 4-36.

Very recent analysis of utility experience under the Lead and Copper Rule also provides strong suggestion that there are potential problems with the use of polyphosphates for lead control. Dodrill and Edwards (1995) pointed out that all 13

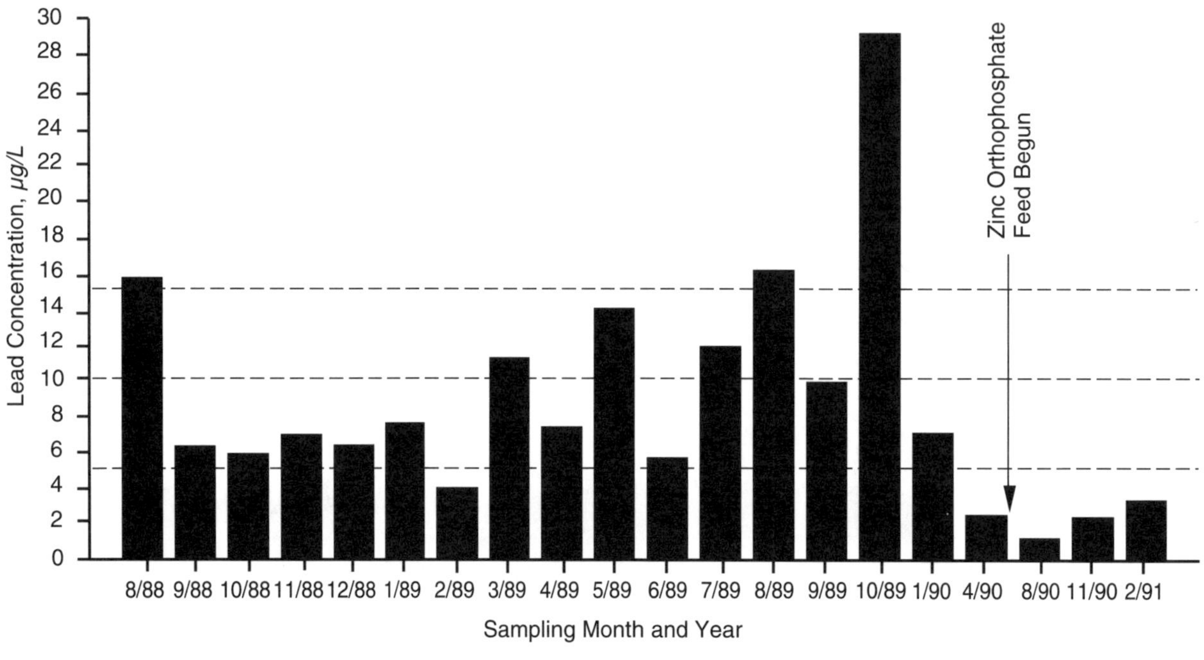

Source: Ramaley (1993).

Figure 4-35 Average tap lead levels in standing samples and the effect of the zinc orthophosphate corrosion inhibition program

Table 4-18 Comparison of water qualities from pipe loop tests for various water treatments on soft, low-alkalinity groundwater

Treatment*	Control	Lime	Soda Ash	Lime/ZOP†	SA/ZOP‡	SA/SHMP§	SA/SHMP/ZOP
pH	5.3 ± 0.2	7.5 ± 0.9	7.7 ± 0.8	7.6 ± 0.8	7.4 ± 0.5	7.6 ± 0.6	7.4 ± 0.4
Alkalinity	5.0 ± 3.7	23.7 ± 7.3	39.8 ± 11.1	25.5 ± 6.2	44.9 ± 0.8	39.5 ± 9.7	42.6 ± 11.6
Hardness	2.5 ± 1.2	20.4 ± 5.7	2.9 ± 1.9	23.5 ±9.1	2.6 ± 1.9	2.7 ± 1.2	2.9 ± 1.9
TDS	20.8 ± 13.9	47.6 ± 11.4	75.4 ± 2.4	66.6 ± 17.7	97.9 ± 26.5	72.6 ± 22.7	97.0 ± 25.0
Conductivity	25.9 ± 5.0	71.5 ± 15.8	122.6 ± 33.6	97.7 ± 23.4	144.9 ± 39.3	110.2 ± 32.5	144.4 ± 38.5
DIC	3.0 ± 0.2	7.9 ± 1.3	10.1 ± 2.3	7.4 ± 1.1	11.9 ± 2.4	10.2 ± 1.8	11.3 ± 2.2
DOP	—	—	—	0.7 ± 0.3	0.8 ± 0.3	—	0.9 ± 0.2
DHP	—	—	—	—	—	0.7 ± 0.3	1.9 ± 0.6

Source: McAnally et al. (1993).
*Alkalinity: Total alkalinity in mg/L as $CaCO_3$.
Hardness: Calcium hardness in mg/L as $CaCO_3$.
TDS: Total dissolved solids in mg/L.
Conductivity: Conductivity in μmhos/cm.
DIC: Dissolved inorganic carbon (mg/L as C).
DOP: Dissolved orthophosphate in mg/L as P.
DHP: Dissolved hydrolyzable phosphate in mg/L as P.
†ZOP is zinc orthophosphate.
‡SA is soda ash.
§SHMP is sodium hexametaphosphate.

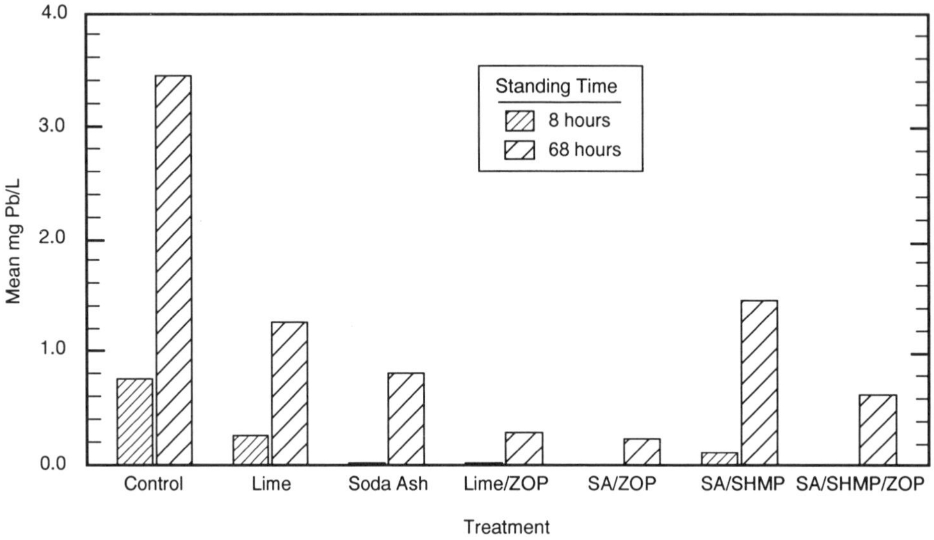

NOTE: ZOP is zinc orthophosphate.
SA is soda ash.
SHMP is sodium hexametaphosphate.

Source: McAnally et al. (1993).

Figure 4-36 Average first-draw lead concentrations from the control and six different treatments in a South Carolina pipe loop study in very soft water

large utilities reporting data with pH >8.4 and alkalinity between 30 and 74 mg $CaCO_3$/L in excess of the lead action level were dosing polyphosphate chemicals.

Effects of Silicate

Several authors have suggested that silicates have the property of reducing lead levels or decreasing the corrosion rate, but few controlled laboratory experiments have been performed. Donaldson (1924) alluded to the use of sodium silicate in England but did not clearly state whether its use was as a specific corrosion inhibitor for lead or for neutralizing acidity in place of lime. Papers published by Thresh in 1922 (Thresh 1922a, 1922b) postulated that alkali silicates are several times more powerful than silicic acid in preventing lead dissolution in water, although the pH effect is not clearly separable from any intrinsic silicate effect.

In 1979 and 1980, USEPA conducted several long-term solubility control experiments that included a brief look at the effect of approximately 10 and 20 mg SiO_2/L (added as sodium silicate) at pH 8.2 in a low-hardness water with an alkalinity of 20 mg $CaCO_3$/L. The recirculation test system described previously was used for these experiments. Plots of the SI for hydrocerussite versus elapsed days are shown in Figures 4-37 and 4-38, and several trends are evident. The lead concentration was not considerably reduced from that in the control experiment (experiments 21 and 21A) when 20 mg SiO_2/L or 10 mg SiO_2/L was present during the first run at each silicate concentration (experiments 27 and 28, respectively). However, in experiments 27A and 28A, after the tanks were refilled with fresh test water and the total pipe exposure had amounted to about eight or nine months (about six months into the new experiment), the amount of lead in solution decreased to a fairly consistent level below that of the control. Possible explanations include the slow formation of a kinetically inhibited lead silicate film and

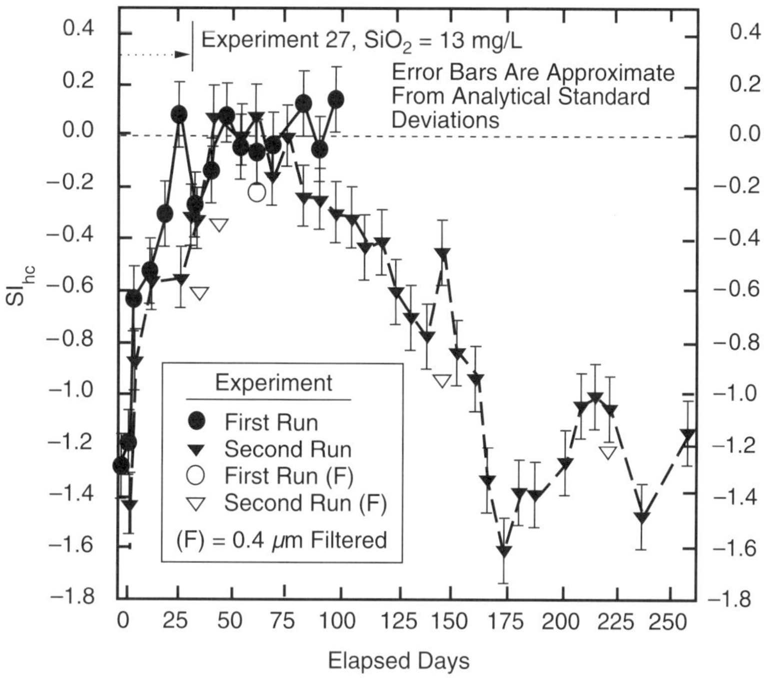

NOTE: SI_{hc} = saturation index for hydrocerussite.

Figure 4-37 Response of the saturation index of $Pb_3(CO_3)_2(OH)_2$ over time to continued contact with 17-20 mg SiO_2/L in USEPA experiments (pH = 8.2, alkalinity ≈ 20 mg $CaCO_3$/L)

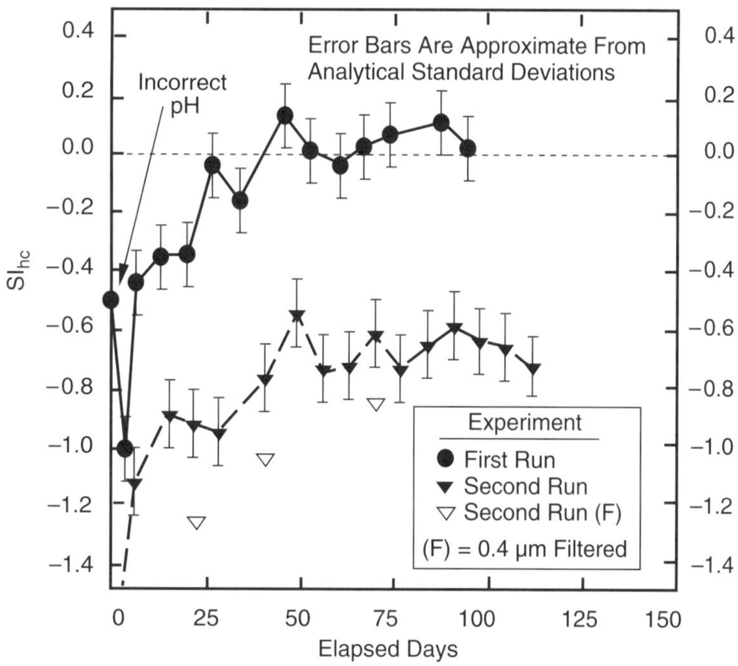

NOTE: SI_{hc} = saturation index for hydrocerussite.

Figure 4-38 Response of the saturation index of $Pb_3(CO_3)_2(OH)_2$ over time to continued contact with 7-10 mg SiO_2/L in USEPA experiments (pH = 8, alkalinity ≈ 20 mg $CaCO_3$/L) adsorption or surface binding of lead to either a silicate film on the inside of the tank or to the inside of the lead pipe. The higher lead levels apparent in the first runs (experiments 27 and 28) could indicate the stabilization of colloidal lead or complexation before the slow film formation process. Slightly lower lead levels were achieved at the 10-mg SiO_2/L dosage for the second run (28A) than for any of the other runs tested. The mechanism behind this observation is not known. A length of galvanized pipe was also in the experiment 28 system, and the result, which was lower than expected, may be influenced by the interaction among the silica in solution, the carbonate ion concentration, the soluble zinc, and the pipe material. Considerably more research is needed into the factors surrounding the deposition rates and both the chemical and physical nature of films formed in the presence of silicate under a variety of pH and background water quality conditions.

Sheiham and Jackson (1981) also studied the immediate addition of freshly prepared sodium silicate solution and of aged (48 hours for hydrolysis) solution, using some of their low-alkalinity moorland water. The final concentration was 10 mg SiO_2/L. Their results are shown in Table 4-19. They concluded that silicate had little or no effect on lead dissolution.

Effects of Organics

In many potable water systems, the behavior of lead is complicated by its interaction with organic ligands, or colloidal material, or both (Harrison and Laxen 1980). Thus, even though the inorganic *dissolved* lead equilibria may be accurately predicted, the accurate prediction of observed levels of *total* lead may not be possible. Incorporation of organic ligands into lead solubility models is handicapped by the fact that the ligands themselves can be metastable and be degraded by both inorganic and organic reactions (Whitfield and Turner 1977) or they are just difficult to isolate and identify. They also may be composed of substances that chemically behave as a mixture of ligands, rather than sample ones like orthophosphate, chloride bicarbonate ion, etc.

Table 4-19 Results of silicate dosage experiments using moorland water (background water quality is given in Table 4-10)

		Additive	
	Control (None)	10 mg SiO_2/L *(No delay)*	10 mg SiO_2/L *(48-hour delay)*
	New Pipes, pH 6.5, 38 Days Duration, 12 Observations		
Median mg Pb/L	0.835	0.680	0.515
Mean mg Pb/L	0.870	0.685	0.503
	Old Pipes, pH 7.5, 184 Days Duration, 12 Observations		
Median mg Pb/L	0.266	0.245	0.300
Mean mg Pb/L	0.266	0.288	0.300

Source: Sheiham and Jackson (1981).

Samuels and Meranger (1984) found concentrated "fulvic acid" solutions of pH 6.2 to be aggressive toward lead in solder and brass, and Thresh (1922b) believed that organic material in the water could prevent the deposition of films on lead surfaces by enhancing the solubility of lead. Miles (1948) found that the organic content of moorland waters appeared to accelerate the initial attack on the lead surface.

Harrison and Laxen (1980) analyzed water from three localities in the United Kingdom and found that significant quantities of dissolved lead (over 50 percent) in some samples appeared to be found in organic complexes. Therefore, simple solubility computations that include only inorganic species may underestimate the levels of dissolved lead attainable in a distribution system. Conversely, organic material that can adsorb or otherwise adhere to the inside surfaces of the pipe could serve to decrease lead solubility. Theoretical interpretations are hindered because the numerous types of organic materials involved have not been characterized. Some simple experiments by Sheiham and Jackson (1981) on two waters with varying degrees of natural color suggested that only highly colored waters might have a significant impact on lead corrosion. They emphasized that this conclusion must be treated with extreme caution because only two waters were tested and because the organic component characteristics might vary from place to place.

The behavior of organic ligands can be modified by their association with or adsorption on particle surfaces, and some natural ligands, such as fulvic acid, may enhance lead removal by precipitation (Saar and Weber 1980). Furthermore, the ligand molecules are difficult to identify and quantify. Therefore, the organic solution chemistry of lead will not be considered here, except to point out that such interactions can take place and that pilot studies may be necessary to take into account reactions that cannot be predicted or quantified by a priori modeling calculations.

Temperature Effects

Little is known about the behavior of the different lead solids and complexes with respect to temperature. Some practical studies indicate that both the equilibrium solubility of lead and its rate of transfer into solution tend to increase as temperature increases when solubility appears to be controlled by carbonate equilibria (Karalekas et al. 1983; Schaut 1942; Moore 1973; Gregory and Jackson 1983; van den Hoven 1987; Colling et al. 1992). However, field data indicate that the reverse trends in hard water containing orthophosphate might exist (Colling et al. 1992). The following important issues must be resolved in future fundamental studies:

- Net solubility trends result from interactions among complex formation, hydrolysis (hydroxide complexation) reactions, and solid precipitation–dissolution reactions. All individual reactions do not necessarily follow the same trend with temperature, so that the new solubility trend is likely a function of water quality. Research on individual complexation and solubility constants is needed, because few data are available for any lead species at a temperature other than 25°C.

- The protective scales on the pipes may be composed of nonlead solids, so the response to temperature will depend on their own solubility trends for the given water quality. This is illustrated by the case of $CaCO_3$ that can either become more or less scale forming in hot water depending on the pH and alkalinity of the water (Schock 1990b).

- The rates of oxidation, dissolution, and diffusion and precipitation will be affected by temperature, and these are largely unknown.
- For systems using blended phosphates, the reversion rates of the polyphosphate components may depend on temperature.

Mass-Transfer Processes

The level of lead found in a sample from a house or building will often reflect that resulting from standing within a pipe system for less time than that required to attain solubility equilibrium (Schock 1990a). In new lead pipe, that state of equilibrium might be attained in as few as 8–24 hours (Schock 1980; Schock and Gardels 1983). In encrusted, older pipe, the time might be longer. Because household taps are typically used for short periods of time, a way to estimate the potential exposure to lead of consumers that takes into account the nonequilibrium situation must be found. The technique should have general applicability and should not require numerous costly sample collections and analyses.

Because many chemical parameters govern equilibrium lead solubility, an obvious approach to this problem would be to fit a polynomial function to experimental or monitoring data with terms relating to different chemical factors (such as pH, DIC, temperature, etc.). For general application, however, different lengths and diameters of pipe must also be considered.

Statistical approach. Bailey and Russell (1981) took a statistical approach in which they fit an empirical log-hyperbolic function of the form:

$$[\mathrm{Pb}] = \exp\left[\alpha - \frac{\beta}{T + \theta}\right] \tag{4-12}$$

to such "stagnation curves." The empirical adjustable parameters α, β, and θ must be estimated for each pipe. T represents the stagnation time. The approach cannot be used to predict stagnation and flushing results for pipes with different combinations of diameters and lengths without direct calibration (Kuch and Wagner 1983; Wagner and Kuch 1981). Furthermore, the function is undefined for the real case at the initiation of a stagnation period, where time is equal to zero.

Model for turbulent flow. To overcome the mathematical difficulties with statistical polynomial stagnation curve models, Kuch and Wagner developed two mass-transfer models, which were derived by analogy to heat-transfer and other well-known hydraulic models. The detailed mathematical derivation will not be repeated here, but the essential assumptions, features, and method of application will be.

The first model to be considered is that pertaining to continuous steady-state turbulent flow. Here, the assumptions are that the flow rate is constant through time; that the flow is fully turbulent, i.e., there is complete radial mixing throughout the pipe bore; and that the pipe is of uniform cross section. The transfer of lead from the pipe wall into the water can then be considered to be controlled by diffusion across a thin layer adjacent to the pipe wall. The transfer is described by the equation:

$$[\mathrm{Pb}]^0 = [\mathrm{Pb}]^\infty - ([\mathrm{Pb}]^\infty - [\mathrm{Pb}]^E)\exp(-kA / \dot{V}) \tag{4-13}$$

Where:

$[Pb]^0$ = lead concentration leaving the pipe
$[Pb]^\infty$ = the equilibrium lead solubility
$[Pb]^E$ = lead concentration of water entering the pipe
A = total area of the internal pipe wall
$\dot{V}$ = volume rate of flow
k = mass-transfer coefficient

In order to perform this computation, the length and diameter of the pipe, the water temperature, the volume rate of flow of the water, the maximum equilibrium lead concentration, and the diffusion coefficient of the dissolved species must be known. The units used must be dimensionally consistent.

To illustrate the effect of pipe length and diameter on the amount of lead transferred into the water upon passage through the pipe, the lead concentration under flow conditions assuming (V = 50 cm/s [1.6 ft/s]) was normalized by the relationship:

$$[Pb]^N = \frac{[Pb]^0 - [Pb]^E}{[Pb]^\infty - [Pb]^E} \qquad (4\text{-}14)$$

and plotted against pipe diameter. Figure 4-39 shows clearly that long pipes with small diameters produce the greatest lead concentrations in the water during flushing.

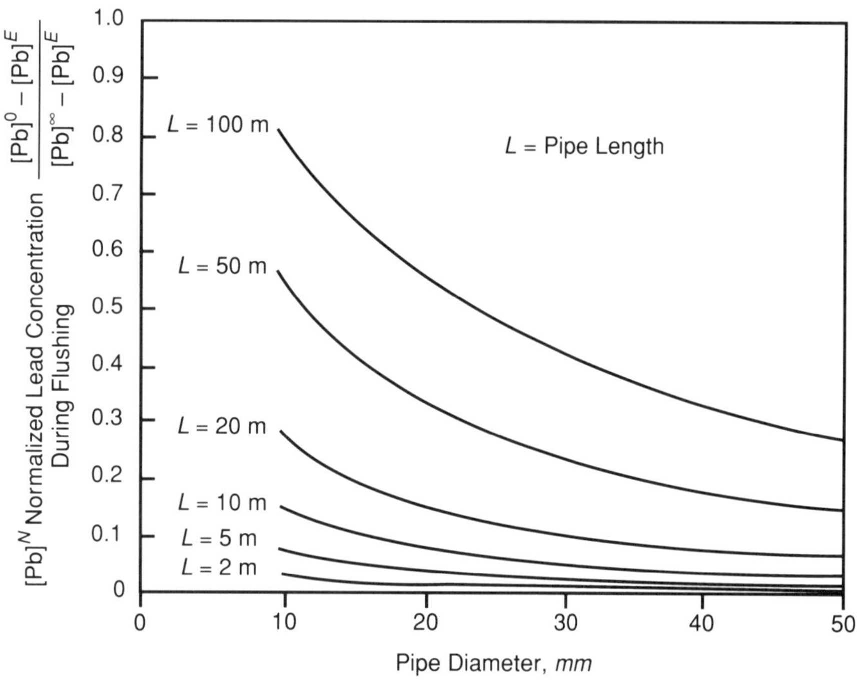

Source: Kuch and Wagner (1983). Reprinted from Water Research, *Volume 17, Mass Transfer Model to Describe Lead Concentrations in Drinking Water, Page 1303, Copyright (1983), with kind permission of Elsevier Science Ltd, The Boulevard, Langford Lane, Kidlington 0X5 1GB, UK.*

Figure 4-39 Plot of normalized lead concentration during flushing ($[Pb]^N$) versus lead pipe diameters for different length pipes, assuming continuous steady-state turbulent flow

Model for stagnation. The second model of interest describes the radial diffusion of lead under conditions of no flow, i.e., stagnation. For practical calculations, the following simplified equation can be used, which is considered to give results within 5 percent of the more rigorous equation (Schlünder 1975).

$$C_t = 1 - \exp\left[\frac{-4Fo'}{(1/Bi') + (33.41 + (4/\pi Fo'))^{-0.5}}\right] \tag{4-15}$$

Where:

$$C_t = \frac{[\mathrm{Pb}]^t - [\mathrm{Pb}]^0}{[\mathrm{Pb}]^\infty - [\mathrm{Pb}]^0}$$

$[\mathrm{Pb}]^t$ = lead concentration after stagnation time t
Fo' = $D \cdot t/4a^2$
Bi' = $\beta_a \cdot 2a/D$
t = time, minutes
D = diffusion coefficient, ~1×10^{-9} m^2/s for Pb^{2+}
a = internal radius of pipe, metres
β_a = mass-transfer coefficient

Generally, β_a is not known a priori, but has been found to be important only in systems where a thick coating of corrosion products, other than lead, or other nonlead scale provides an additional resistance to diffusion. In such cases, stagnation curves are fit to experimental data by trial and error, assuming different values for β_a, until a good fit is obtained.

The data necessary to employ the stagnation curve equation are the pipe diameter, the diffusion coefficient for Pb^{2+}, the maximum equilibrium lead concentration, the lead concentration during flushing, and the mass-transfer coefficient β_a if required. As before, the equilibrium maximum lead concentration can be obtained either by theoretical calculation or by field observation for a particular water type and system. The lead concentration after flushing could also be obtained either through experimental measurement or by calculation using the steady-state turbulent flow equation presented previously (Eq 4-13). The diffusion coefficient for Pb^{2+} is used because of the lack of data for other chemical species that are involved in the solubility reactions. No temperature correction is applied to this term.

The lead concentration at the beginning of the stagnation period is assumed to be equal to $[\mathrm{Pb}]^0$ for the entire length of the pipe. In actuality, the lead concentration along the length of the pipe will vary from $[\mathrm{Pb}]^E$ at the entrance to $[\mathrm{Pb}]^0$ at the end. As samples are taken from the end of the pipe, the lead concentration in sections of the pipe remote from the sampling point will increase proportionally more than in sections near the sampling point. These two refinements tend to offset each other, rendering the assumption of a homogeneous concentration of $[\mathrm{Pb}]^0$ throughout the pipe to be appropriate. Detailed calculations were done to verify this (Kuch and Wagner 1983).

By normalizing the lead concentration as before, one can examine the influence of pipe diameter on lead concentration as a function of time. Figure 4-40 shows that small-diameter pipes produce a rapid increase in lead concentration in the first few minutes of standing.

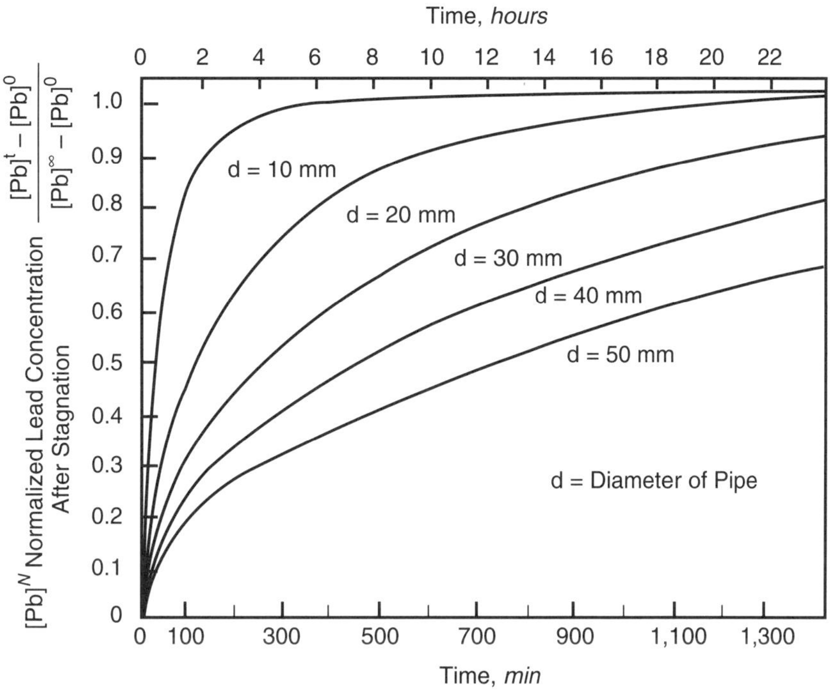

Source: Kuch and Wagner (1983). Reprinted from Water Research, *Volume 17, Mass Transfer Model to Describe Lead Concentrations in Drinking Water, Page 1303, Copyright (1983), with kind permission of Elsevier Science Ltd, The Boulevard, Langford Lane, Kidlington 0X5 1GB, UK.*

Figure 4-40 Plot of normalized lead concentration during stagnation versus time for lead pipes of different diameters, assuming radial diffusion with no flow

Kuch and Wagner (1983) verified the applicability of the mass-transfer model by using data from WRc studies. The stagnation curve model can also be applied to the USEPA flow-through pipe loop studies, although there are relatively few points at each general water quality. As an example, the appropriate terms of the stagnation curve equation are calculated or measured to be as follows, for two of those studies (phases 1 and 2):

$D = 1 \times 10^{-9}$ m^2/s (assumed)
$a = 9.525 \times 10^{-3}$ m
$\beta a = 1$(to negate its effects)
[Pb] = 0.146 mg/L (phase 1)
= 0.076 mg/L (phase 2)
$[Pb]^0$ = 0.030 mg/L (phase 1)
= 0.016 mg/L (phase 2)

Different values for t are inserted into the equation, and $[Pb]^t$ is computed. The lead concentrations $[Pb]^0$ and $[Pb]^\infty$ represent experimental mean values for samples taken at the start of stagnation and subsequently after standing in excess of 20 hours for a given experimental phase (Schock and Gardels 1983). The predicted curves may not exactly pertain to each of the plotted standing time observations (see Figure 4-41); ideally, there should be a family of curves encompassing the range of experimental conditions (pH, alkalinity, etc.). However, the general

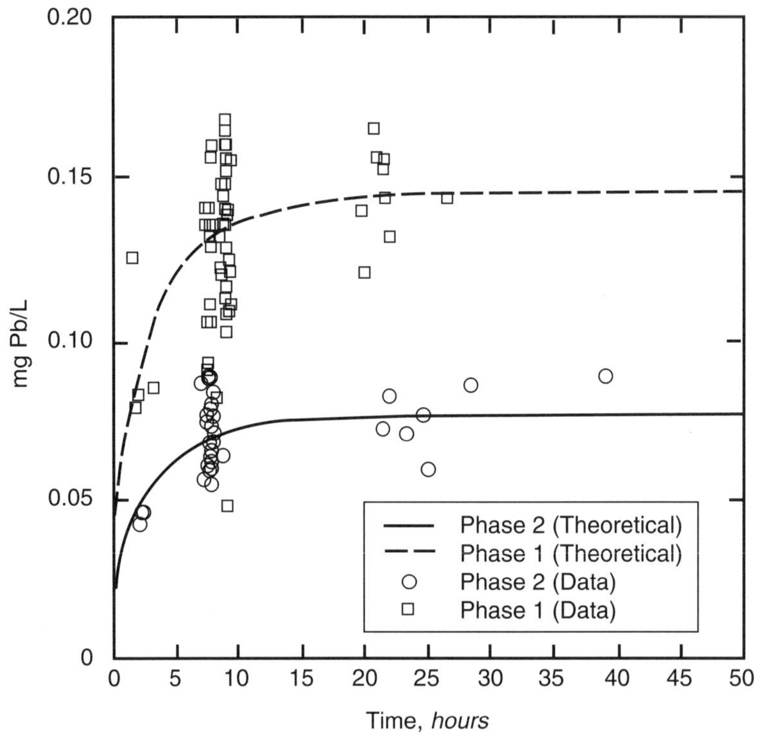

Figure 4-41 Comparison of USEPA pipe loop data for short stagnation periods of the mass-transfer model: phase 1 data (pH 9.5, low alkalinity) and phase 2 data (pH ~9.25, low alkalinity)

predictability of the model is shown to be accurate enough to be useful under practical circumstances, where the equilibrium lead solubility might vary somewhat from location to location. Only in cases where the pipe film is poorly adherent, thus generating occasional spurious high lead values as material sloughs off, does the model fail to closely predict the observed concentrations.

Another interesting aspect of the mass-transfer computation for stagnation conditions is that for the geometry of the USEPA experimental pipe loop, lead levels within approximately 8 percent of equilibrium maximum values are predicted to occur after only 8 hours of standing. This translates into an error of only approximately –0.1 log units in the computed saturation index for hydrocerussite or in the computed solubility constant, which is in close agreement with the computations performed by Schock and Gardels (1983). In their published study and in subsequent unpublished work, they statistically compared data for different periods of stagnation (in excess of 8 hours) and found the inconsistency generally to be on the order of –0.2 log units or less for all experiments, regardless of DIC concentration and pH combination.

Perhaps the most significant practical use of the mass-transfer approach is that a substantial decrease in the size of the monitoring program necessary to assess human exposure is now possible. Calculating the mean lead exposure caused by a lead service pipe (Wagner and Kuch 1981) and assuming a maximum of one long stagnation period and one first draw per day, it was necessary to obtain the probability function of the random daytime draws. The probability function was computed by using the original data from a long-term survey on water consumption patterns in Germany (Wagner and Kuch 1981). The results for three types of houses (i.e., number of inhabitants) are shown in Figure 4-42.

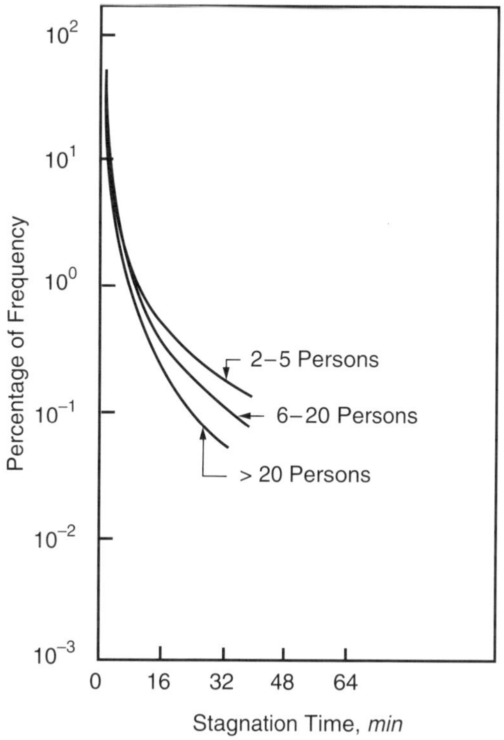

Figure 4-42 Probability function of stagnation time as percentage of frequency (per minutes) for three types of housing

This information was applied for the first time to all lead service pipes installed at Duisburg, Germany. Because Duisburg has three water qualities, the maximum lead levels for each water quality were determined by measuring and recalculating the stagnation curves. These data were then combined with data from the types of houses connected by lead pipes to the water mains and listed in nine tables, i.e., one table for each of the three types of houses and the three water qualities (Wagner 1983). An example is given in Table 4-20.

In this case, the replacement program was conducted according to the following criteria:

Pipe Length	Replacement Priority
>30 m (>100 ft)	Very urgent (4 pipes)
20–30 m (65–99 ft)	Urgent (28 pipes)
10–19 m (32–64 ft)	As soon as possible (181 pipes)
<10 m (<32 ft)	Replace when performing other work in the same vicinity

In addition to the benefits of the model outlined above, the calculations should be applicable to other types of plumbing materials, with substitution of the appropriate constants. As a result of these studies, a lead pipe replacement program was undertaken in Germany. Pipe replacements were prioritized based on water consumption patterns, lead level predictions from efficiently planned monitoring, and the mass-transfer model.

Table 4-20 Summary of data from lead service line study of Duisburg (Germany) Water Works, Distribution Systems West, of single-family houses served by groundwater (the maximum level of lead was 240 μg/L)

	Pipe Length							
	>30 m		20–30 m		10–20 m		<10 m	
Pipe Diameter, *in. (mm)*	Number of Lead Service Pipes	Mean Lead Concentrations, μg/L	Number of Lead Service Pipes	Mean Lead Concentration, μg/L	Number of Lead Service Pipes	Mean Lead Concentration, μg/L	Number of Lead Service Pipes	Mean Lead Concentrations, μg/L
≤ ½ (≤ 15)			19	108	86	78	277	44
½–¾ (15–20)	3	218	7	105	80	76	426	40
¾–1 (20–25)	1	236	2	107	15	75	43	40

Source: Wagner (1983).

Other studies. As was mentioned previously, Bailey et al. (1986) conducted considerable research in the United Kingdom into defining stagnation curves and estimating lead levels after various lengths of standing time. An automatic sampler was installed, and samples were taken weekly after varying standing times up to 360 minutes. Using the same equation form as Bailey (1986), 24 of the 36 sites examined could have curves closely fit to them. Bailey and co-workers concluded their data analysis by observing discrepancies between the lead levels predicted and those observed in their study. These led to several important conclusions:

- From diffusion theory, the asymptotic equilibrium concentration should be dependent on the chemical quality of the water. Houses with lead pipes in the same supply zone should therefore exhibit similar asymptotic levels. This does not appear to have been borne out by these results.
- Genuine differences in the true equilibrium lead concentrations in different pipes may exist because of differences in the physical or chemical properties of their internal surfaces.
- Equilibrium concentrations may be the same but the observed sample concentrations may be affected by mixing. Thus, the hydraulic performance of the real system may be more complicated than assumed in the models.
- The mass-transfer theory may be correct for the dissolved lead but does not apply when significant particulates are present.

Many factors that complicate the interpretation of household data and that contribute to variability have been reviewed (Schock 1990a).

THE NATURE OF CORROSION DEPOSITS ON LEAD PIPE

Several investigations have helped corroborate the solubility models as correctly identifying and largely quantitating plumbosolvency and consequent lead concentrations when equilibrium solubility conditions pertain. Investigators have also uncovered interesting aspects of the interaction of lead pipe with drinking water under different chemistry and flow conditions, and that may lead to more reliable plumbosolvency control techniques.

Studies in the Netherlands

In the systematic investigation of water quality influences on lead levels in many drinking water systems in the Netherlands, Elzenga and Graveland (1980) reported the results of X-ray diffraction analyses of solid corrosion deposits in lead and tin-lined lead pipes from domestic installations. Table 4-21 summarizes the general water quality characteristics of the water supplies investigated and the solids found in the unlined pipes. Interestingly, basic lead carbonate was found in all deposits, regardless of pH and alkalinity. The highly soluble PbO was also frequently found, presumably in the underlayers where it could not readily dissolve into the water. In the tin-lined pipes, SnO and SnO_2 were often found. When lining defects were present, the basic lead carbonates or normal lead carbonate were found.

Studies in Great Britain

Sheiham and Jackson (1981) reported the results of many scale analyses performed on lead pipes from various water supplies in the United Kingdom. Both X-ray diffraction and infrared spectroscopy were used in their study. They noted the widespread existence of hydrocerussite when only cerussite was thought to be thermodynamically stable. Some of their observations and the associated water quality characteristics are summarized in Table 4-22.

Hulsmann (1990) described what was found in cuttings from eight lead service lines exhumed in one area during a study to determine factors affecting the occurrence of particulate lead. She described the innermost layers of the deposits to be dense and have high lead content. They were often creamy white in color. Above these were a number of thin deposit layers of lower lead content, frequently rusty brown or pale tan in color. These outer layers often seemed to be very porous.

Dando and Glasson (1989) found differences in porosity between lead orthophosphate precipitates and the lead carbonate/hydroxycarbonate precipitates. They found the latter to be less porous and possibly more likely to adhere to the surface and provide protection. Colling et al. (1987, 1992) found significant differences in crystal morphology and perhaps porosity, depending on whether the deposit seemed to represent thermodyamic stability or metastability. The latter was thought to produce more erratic lead levels in the water.

Breach, Crymble, and Porter (1991) investigated a variey of lead pipes in the Severn Trent (Birmingham, England) water supply area with X-ray diffraction and energy-dispersive X-ray analysis. They found frequent occurrence of nonlead constituents, such as aluminum, iron, manganese associated with a variey of carbonates, phosphates, and silicates. Scales were found to have a variey of mechanical properties that could influence the lead levels.

Studies in the United States

Numerous samples of lead service line piping and lead gooseneck connections have been analyzed by USEPA, using X-ray diffraction. Several were also qualitatively analyzed for elemental composition by acid digestion and inductively coupled plasma spectroscopy. General summaries of the crystalline composition of many of these samples are given in Table 4-23, and the approximate water quality characteristics for some of the water supplies are given in Table 4-24.

Table 4-21 Chemical composition of drinking water in test installations and deposit phases found using X-ray diffraction

Water Source	pH, (mean)	Alkalinity, *mg $CaCO_3$/L*	Ca^{2+}, *mg/L*	Mg^{2+}, *mg/L*	Na^{+}, *mg/L*	K^{+}, *mg/L*	NO_3^-, *mg/L*	SO_4^{2-}, *mg/L*	Cl^-, mg/L	PO_3^{4-}, *mg/L*	Stable Solids*
Archemerberg	8.2	37	15	1.2	6	1	1.9	7	10	0.05	$Pb(OH)_2$•$2PbCO_3$; PbO; small quantities of PbO•$PbSO_4$
Goor	7.6	257	23	11	23	4	0.8	17	47	0.02	$Pb(OH)_2$•$PbCO_3$; small quantities of PbO
Havelterberg/Hoge	7.9	50	51	5.3	11	1	1.5	13	20	0.12	$Pb(OH)_2$•$2PbCO_3$;
Heksel	7.3	108			12	2	0.8	37	30	0.11	PbO
Nijverdal	8.0	54	22	1.6	6	1	1.1	11	10	0.02	PbO; $Pb(OH)_2$•$2PbCO_3$ a trace of $PbCO_3$
Holten	8.5	86	33	9.6	14	2	5.3	31	28	0.01	$PbCO_3$; PbO; small quantities of $CaCO_3$; trace of $Pb(OH)_2$•$2PbCO_3$
Manderveen	7.6	60	25	1.9	8	2	3.8	14	15	0.10	$Pb(OH)_2$•$2PbCO_3$; small quantities of PbO; traces of $PbCO_3$ and PbO•$PbSO_4$
Leiduin(1)	7.9	157	99	10.8	92	7	26.7	77	164	0.10	$Pb(OH)_2$•$PbCO_3$; $PbCO_3$
Leiduin(2)	8.7	93	26	9.9	119	6	4.6	71	146	0.08	$Pb(OH)_2$•$2PbCO_3$

Source: Elzenga and Graveland (1980). Reprinted with permission of C.H.J. Elzenga

*$Pb(OH)_2 \cdot 2PbCO_3$ = hydrocerussite.

$PbCO_3$ = cerussite.

Table 4-22 Scale analyses of several water systems in WRc studies

System	Treatment*	Deposits
Moorland	NT, pH = 6.5	Amorphous phases† and $Pb_3(CO_3)_2(OH)_2$
Moorland	SSF, pH = 7	Amorphous phases and $Pb_3(CO_3)_2(OH)_2$ + $PbCO_3$
Moorland	Al coag., pH = 8	Amorphous phases and $Pb_3(CO_3)_2(OH)_2$
River (hard)	SSF	$Pb_3(CO_3)_2(OH)_2$
River (soft)	Fe coag., pH = 8	$Pb_3(CO_3)_2(OH)_2$
Chalk groundwater	NT, pH = 7.1	$PbCO_3$
Chalk groundwater and river	Al coag., pH = 7.2	$Pb_3(CO_3)_2(OH)_2$ and $PbCO_3$

Source: Sheiham and Jackson (1981).

*NT = no treatment, SSF = slow sand filtration, Al coag. = alum coagulation, Fe coag. = ferric sulfate coagulation.

†Amorphous phases containing Fe, Ca, and P, typically.

Many of the European researchers' observations are corroborated by this USEPA investigation, which is an ongoing project. Hydrocerussite is frequently found in samples from locations in contact with water qualities where cerussite would be considered to be the thermodynamically stable phase. This may represent the result of slow solid conversion reactions discussed earlier in this chapter. Compounds not containing lead also seem to make up an integral part of the deposits on pipes from several systems, and this could serve to confound attempts to correlate lead levels in lead pipe samples with lead solubility calculations.

Several significant observations can be made from these analyses that go beyond the general corroboration of the European studies:

- Deposits in 1-in. or 2-in. (2.5-cm or 5-cm) pipes often differ somewhat from those in smaller-ID pipes from the same general area of the system, indicating different chemical conditions caused by pipe size.
- The solid PbO_2 has frequently been found in the outermost layers of deposits from several water systems.
- The solid PbO has frequently been found near the point where the deposit contacts with the lead pipe wall, possibly formed by transformation of basic lead carbonates to lower forms, as suggested by Taylor and Lopata (1984).
- Calcium carbonate has been found as a prominent solid in samples from two systems with lime-softened water, but it has rarely been detected in other pipe sections from systems with distributed water near calcite saturation.
- Lead orthophosphate solids (tertiary lead orthophosphate or hydroxypyromorphite) have been found on outer surfaces of pipes from systems dosing "zinc orthophosphate" inhibitors. No crystalline zinc orthophosphate or other zinc solids were found.
- Particularly for systems with pH below 7.5, scales are often highly layered and mineralogically complex. They represent the treatment history of the system, so the response to current water chemistry changes or inhibitor dosages may not be estimable from new pipe surfaces.

Table 4-23 Analyses of lead pipe and pigtail deposits from various US water systems

Location	Inside Diameter, *in. (mm)*	Age, *years*	Major Solids*	Comments
Portland, Oregon	0.5 (13)	40–75	HC, C for most sites; PN, L, also at some sites	Several unidentified solids; eight locations sampled
Providence, Rhode Island	0.5, 0.75 (13, 19)	66–85	PN in most, along with HC; minor C occasionally	Four locations
Oakwood, Ohio	0.75 (19)	>45	C, PbO_2, L, $CaCO_3$, (Al, Fe, K, SiO_2– containing solids)	Complex layer structure; thick deposits; four locations
Durham, North Carolina	0.75 (19)	—	HC, HP	
St. Petersburg, Florida	0.75, 2 (19, 30)	—	HC, L, PbO_2, $CaCO_3$	Two locations
Cleveland, Ohio	0.5 (13)	—	HC, L, PbO_2, PN	
Columbus, Ohio	0.5 (13)	—	HC, TLO, or HP; $CaCO_3$	Changed from lime softening to orthophosphate in 1982
Piqua, Ohio	0.5 (13)	63	HC, SiO_2, $CaCO_3$	Lime softened, total PO_4 ~0.5 mg/L, very uniform hard scale
Los Angeles, California	0.5, 1 (13, 25)	—	HC, L (0.5 in.) C, L, PbO_2 (1 in.)	Two locations; Si, K, Cu also in scale
Chicago, Illinois	0.5 (13)	—	HC, L	Possible trace $CaCO_3$ (vaterite)
Lake Forest, Illinois	0.5 (13)	—	C, L, HC	Very thick L layer at pipe
Gary/Hobart, Indiana	0.5 (13)	—	C, SiO_2, L	Several unidentifieds

Source: USEPA and Illinois State Water Survey 1988–1993, unpublished.

*Solid abbreviations:

C = $PbCO_3$, cerussite

HC = $Pb_3(CO_3)_2(OH)_2$, hydrocerussite

PN = $Pb_{10}(CO_3)_6(OH)_6O$, plumbonacrite

L = PbO, litharge (tetragonal)

HP = $Pb_5(PO_4)_3OH$, hydroxypyromorphite

TLO = $Pb_3(PO_4)_2$, tertiary lead orthophosphate

Considerably more research needs to be done to understand the role natural diffusion barriers play in limiting lead levels in drinking water. An understanding of the role water treatment metals, such as aluminum or iron, may play when they are carried out into the distribution system is also important. The physical nature of the deposits is important in dictating the consistency of the lead levels in the water under operational conditions, as is the speed and extent to which corrosion inhibitors such as orthophosphates or blended phosphates can reduce plumbosolvency.

Table 4-24 Approximate water qualities for several systems with analyzed lead pipe deposits

System	pH	Alkalinity, *mg $CaCO_3$*	Ca, *mg/L*	Mg, *mg/L*	SO_4, *mg/L*	Cl, *mg/L*	SiO_2, *mg/L*	Specific Conductance, *μS*
Cleveland, Ohio	7.5	85	32	8	26	19	1–2	257
Los Angeles, California	7.8	143	58	18	105	28	19	582
Oakwood, Ohio	7.1	375	110	40	64	120	16	—
Providence, Rhode Island	10	15	12	—	12	15	—	—
Durham, North Carolina*	7.0–7.2	13–30	3–19	2–4	16–32	8–12	1–9	100–160
Columbus, Ohio†	7.7	79	28	15	116	—	15	—
Portland, Oregon	6.7–7.1	6–12	1–4	—	—	—	8–10	—
Piqua, Ohio	9.9	64	37	5	—	—	—	—

*Water quality after treatment change in 1979.

†Approximate values analyzed in 1991, representing treatment change starting in 1980.

TREATMENT STRATEGIES FOR CONTROLLING PLUMBOSOLVENCY

General Considerations

Many aspects of corrosion control through water treatment have been widely covered. Some of the more relevant points that apply to corrosion and plumbosolvency from lead pipe were summarized by Schock (1984a, 1990b).

Unlike most distribution system mains, household and building systems frequently have large sections of piping in which the water can be stagnant for hours. During these long periods of standing, corrosion reactions can continue to add by-products to the water until a state of solubility equilibrium is nearly or actually attained. Thus, the metal levels observed in the water are initially governed by the mass-transport of oxidizing or solubilizing agents to the pipe surface and of the corrosion products away from the surface. The upper-limiting case, from a dissolved by-product standpoint, is that ultimately provided by the solubility equilibria of the surface pipe material in the water. Physical factors, such as nonadherence of the corrosion product film or the dislodging of scales by normal turbulent flow, can add particulate metal species to the water, increasing the observed "total" metal concentration. Also, corrosion reactions can continue, even if the equilibrium solubility is attained.

In general, the strategy of corrosion control through water quality adjustment involves interference with at least one link in the chain of interconnected anodic and cathodic corrosion cell reactions (see chapter 1 and chapter 8). Ordinarily, this consists of forming a "barrier" of a "passivating film" that both limits the transport of the metallic species into solution (mostly through solubility limitation) and limits the diffusion of the oxidizing agents (normally dissolved oxygen or chlorine species) to the pipe surface. These complicated couplings of redox and solubility reactions, plus the differences in relative rates of the reactions, prevent simple relationships between corrosion rates determined from pipe coupon or insert data and solubility levels from being developed.

Obviously, extensive monitoring and the establishment of a new treatment to combat lead leaching is a costly endeavor. Though lead levels can often be substantially improved, they cannot necessarily always be lowered to below the 90th percentile 0.015-mg/L action level in the United States or maximum random sample requirements of the United Kingdom or other countries. A program to eliminate lead-containing materials whenever and wherever it is practical is a wise investment in the future. Currently, this would cover complete lead service line or household plumbing and other plumbing devices.

Calcium Carbonate Saturation

Calcium carbonate saturation is the approach traditionally followed by utilities with hard and moderately hard water supplies. Many utilities that use lime softening for clarification also use this approach. Unfortunately, monitoring the calcium carbonate saturation state originally was included as a requirement of the USEPA "corrosivity" regulation (*Federal Register* 1980), even when it was not necessarily applicable to determine the plumbosolvency or lead corrosion potential of a given water system. The monitoring requirement was dropped in 1994 (*Federal Register* 1994).

Many expositions on the general rationale for calcium carbonate saturation are available (Snoeyink and Jenkins 1980; Singley et al. 1984). The primary intent of this approach, with respect to lead, is to deposit a nontoxic barrier film that will "seal" the lead surface from contact and interaction with the water inside the pipe.

The concept has been widely assumed to have been validated; however, virtually no research has been published to define the water quality conditions (i.e., pH, temperature, calcium and carbonate concentrations) or the time necessary to form the film or to examine the metal leaching behavior while film formation is in progress. Intuitively, maintaining a balance between producing a uniform barrier film on pipes through the distribution system and not plugging pipelines is extremely difficult. The approach is unlikely to be effective outside of lime-softened or very hard waters, except where the calcium carbonate deposition is controlled by corrosion processes that affect the pH and alkalinity near the pipe wall. Lead and lead-soldered joints in copper pipe could produce a serious problem if enough lead were exposed (Figures 4-13 and 4-14). At a pH range of 7.5 to 8.5 and a DIC concentration of approximately 24 to 48 mg C/L (carbonate alkalinity of approximately 100 to 200 mg $CaCO_3$/L), the dissolved lead concentration could easily reach 0.16–0.25 mg/L after standing overnight in the absence of a cohesive film. Flowing samples could readily approach the USEPA action level of 0.015 mg/L, given a sufficient length of lead pipe or enough exposed soldered joints.

Special practical considerations when implementing a lead solubility control program based on calcium carbonate saturation should be noted. Numerous researchers have pointed out that two waters with the same LI would not necessarily have the same potential mass of calcium carbonate to deposit (Snoeyink and Jenkins 1980; Singley et al. 1984; Stumm 1960; Rossum and Merrill 1983; Loewenthal and Marais 1976; Schock 1990b; *Standard Methods* 1992) (see chapter 1). The calcium carbonate precipitation potential (CCPP) is really the most enlightening index and would probably be a safer control parameter than the LI (Schock 1990b).

Another consideration is that the calcium carbonate phase that precipitates is frequently the more-soluble aragonite form, rather than calcite. Calcite is the assumed solid phase in almost all calculations. Exactly what pipe, temperature, and water chemical characteristics govern this phenomenon have not been investigated in potable water systems.

To further complicate matters, a third form, vaterite, has been observed in some hot-water systems (Carlson 1983), as well as some cold-water systems (Neff and Schock 1985; Singley et al. 1984). Log solubility constants vary from –7.91 for vaterite to –8.48 for calcite, at 25°C and an ionic strength of zero. The solubility of each different form of calcium carbonate responds to temperature changes somewhat differently, as well.

A third consideration is that the most widely used values for the solubility constant of calcite at different temperatures have been shown to be in considerable error (Plummer and Busenberg 1982; Schock 1984b). Revised values based on the latest research have been presented, and the widely used method of "total dissolved solids" correction originally developed in 1942 has been revisited (Schock 1984b).

pH-DIC-Alkalinity Adjustment

Figures 4-13 and 4-14 show that for a simple metal–water–carbonate system the solubility of Pb(II) can be greatly reduced by increasing pH into the range of 9–10. Frequently, even though a water has a low alkalinity, when the pH is raised into this range, enough DIC is present to yield a carbonate ion concentration high enough to significantly aid in film formation, without additional carbonate or bicarbonate chemical additives. This treatment strategy has been extremely successful in both laboratory and field studies in many US, Dutch, German, and British localities. These studies are covered in an earlier section of this chapter (see Figures 4-1, 4-4, 4-5, 4-6, 4-9, and 4-10).

Because Pb(II) forms strong carbonate complexes, only waters of relatively low carbonate concentrations are amenable to this treatment. Waters with high alkalinities and neutral-to-basic pH values may actually have to have carbonate *removed* or to be softened to enable the maximum reduction in lead solubility and possibly to avoid unwanted calcium carbonate scaling.

Orthophosphate Addition

As discussed earlier, Pb(II) forms at least one orthophosphate solid of low solubility under actual drinking water conditions, which can serve as the basis for corrosion control. USEPA experiments with both lead and galvanized pipes have demonstrated that supersaturation with zinc orthophosphate is not necessary for lead solubility control because the lead forms a passivating lead orthophosphate film.

The passivating action of the orthophosphate depends on pH, DIC concentration, orthophosphate concentration, and temperature. Interrelationships among pH, DIC, and orthophosphate have been widely discussed (Schock 1989, 1990b); only limited data are available on the effect of temperature. On new lead surfaces, a very thin, smooth film is formed. Taylor (1980) suggested that the orthophosphate might not be as effective if the pipe surface is already covered with corrosion products and deposits. There is also little information on the ability of insoluble lead orthophosphate crystallites or colloids to adhere to the pipe surface and withstand mechanical disturbance from water flow or vibration. If the solids are nonadherent, or porous, then lead levels at the tap will not be adequately reduced, even if plumbosolvency is. Information presented previously indicates that the adherability and physical properties of lead orthophosphate passivation films are largely unknown above approximately pH 8 and particularly so at low phosphate dosages. Considerable research is still needed to determine the efficacy of orthophosphate addition in the pH range of 8–10.

Figures 4-19 and 4-20 suggest that orthophosphate additions of approximately 1–2 mg PO_4/L (0.33–0.65 mg P/L) should effectively reduce lead solubility over a much lower pH range than would be possible using carbonate and pH adjustment alone (see also Schock 1989). There are several potentially adverse secondary impacts of lead control that could be reduced in some circumstances by the use of orthophosphate dosing. By using lower pH values, organic halogen formation rates, which are regulated by other US drinking water regulations, should be lower. There is the possibility of reducing lead solubility in high-alkalinity waters without decarbonation through the use of orthophosphate addition. Also, the disinfection efficiency of free chlorine residual species is greater at the lower pH values. Caution must be exercised, however, because upper pH and orthophosphate concentration limits are imposed in hard-water systems, due to the possibility of precipitating calcium carbonate or orthophosphate compounds. Many of these interrelationships were covered by Gregory and Jackson (1983) and Schock (1989, 1990b), and numerous secondary impacts are covered in chapter 10 and other references (e.g., EES and ISWS 1990).

When considering orthophosphate addition as a control strategy for lead, additional benefits in the form of possibly improved corrosion control of other types of plumbing materials in the total distribution system should be considered. Conversely, some detrimental effects may result for other materials. A potential difficulty lies in the destabilization of existing hydroxycarbonate and hydroxide films of lead and other metals when pH must be reduced substantially in order to be in the effective range for orthophosphate passivation. The reestablishment of the "new" films may take a long time (months to years), during which lead levels may become erratic and sometimes high. These issues will be briefly discussed in a subsequent section. There is also some evidence that orthophosphate is ineffective or may produce detrimental results as pH increases above approximately 8, as was shown earlier in the chapter.

Practical experience in the United Kingdom and Germany suggests that systems initiating orthophosphate dosing for lead control may need initial high concentrations (>3 mg PO_4/L) for several weeks to months before a new lead phosphate-based film can be formed on old pipe surfaces. Many months may be required before lead levels become stabilized at lower levels, and the slow formation of lead orthophosphate solids may mean that lead levels continue to decrease for years after the initiation of dosing. Minimum maintenance dosages are usually over 1 mg PO_4/L. US treatment practice has generally been to add minimal amounts of orthophosphate, which may not result in effective film formation or the maintenance of sufficient orthophosphate to the ends of the distribution system. Because of the uncertainty and imprecision of the modeling predictions, general use of higher dosages, as used in European practice, would seem to be a better practice for US utilities to follow, until future research can determine more precisely minimum effective dosages necessary for different water quality and pipe conditions.

Polyphosphate Addition

Polyphosphates are strong chelating agents for calcium, magnesium, iron, and other trace metals. They are useful in reducing scaling, cleaning tuberculation from distribution mains, and reducing "red-water" complaints through sequestration of Fe^{2+} ion and dispersion of ferric hydroxide colloids. However, they have not been proven useful in reducing lead solubility to sufficiently low levels to be safe from a health standpoint. Even though they may be somewhat effective relative to no treatment, particularly in acidic waters, alternative treatment strategies will prove superior in reducing metal levels. Any combined use of polyphosphate chemicals with another treatment (to prevent scaling at high pH or to improve film formation with orthophosphate or sodium silicate) must be accompanied by thorough monitoring of trace metal concentrations. This is necessary because the actual polyphosphate ligands have not been identified, and their complex formation constants have not been determined. Thus, solution modeling calculations cannot even make reliable semiquantitative estimates of solubility behavior in the presence of polyphosphates at the present time.

Silicate Addition

Experimental tests conducted with lead pipe and observations of the inhibitory effect of naturally occurring silica were summarized in a previous section of this chapter. After approximately eight to nine months of exposure to 20 mg SiO_2/L, lead levels were slowly reduced in a soft, low-alkalinity water at pH 8.2. These levels were approximately 0.05 mg Pb/L, but presumably the pH and DIC content of the water would be important variables to consider before extrapolating results to other water qualities. These experiments tend to support the idea of the slow formation of a film on new pipe, which might become effective over the long term as a "cementing" agent for other corrosion products. Corrosion films on lead pipes always tend to be very thin and uniform, unlike those on galvanized pipe. The experiments conducted by Sheiham and Jackson (1981) with low-alkalinity moorland waters led them to conclude that 10 mg SiO_2/L had little effect on lead dissolution.

Studies by Glasser and Lachowski (1980) of the structures of silicate solutions indicated that there is essentially no difference between solutions prepared by the dissolution of anhydrous sodium metasilicate and those prepared from salts with sodium orthosilicate groups. Therefore, either type of additive should provide similar results for corrosion inhibition.

The limited amount of research done thus far with silicate as a lead corrosion inhibitor provides little concrete guidance for anyone considering this option. The experimental data suggest that to show promise, a final concentration of at least 20 mg SiO_2/L should be considered and improvement in lead leaching might take months to become apparent. The role of pH is still in question, but most successful treatments using sodium silicate maintain a pH of 8–9.5.

Potential Destabilization of Passivation Films

The concern about destabilization of existing lead passivation films is illustrated conceptually in Figure 4-43. A solubility diagram is shown, with curves representing theoretical lead solubility in equilibrium with $PbCO_3$ (normal lead carbonate, cerussite), $Pb_3(CO_3)_2(OH)_2$ (basic lead carbonate, hydrocerussite), and $Pb_5(PO_4)_3OH$ (hydroxypyromorphite). To follow the diagram, start by assuming a water utility has been producing a water for many years with a pH of approximately 8.8 and a total alkalinity of 100 mg $CaCO_3$/L (point A). Lead piping in the system would be dominantly protected by hydrocerussite. Suppose, however, that for other treatment objectives, such as optimizing further lead corrosion control or responding to other regulations that would be better met using a water of a lower pH, the utility chooses to add orthophosphate to a concentration of 1 mg PO_4/L. As pH is reduced, the existing basic lead carbonate film will become unstable, while it is being replaced slowly by the new lead orthophosphate. If the dissolution rate of the carbonate solid is faster than the solid conversion reaction rate of the formation of the new

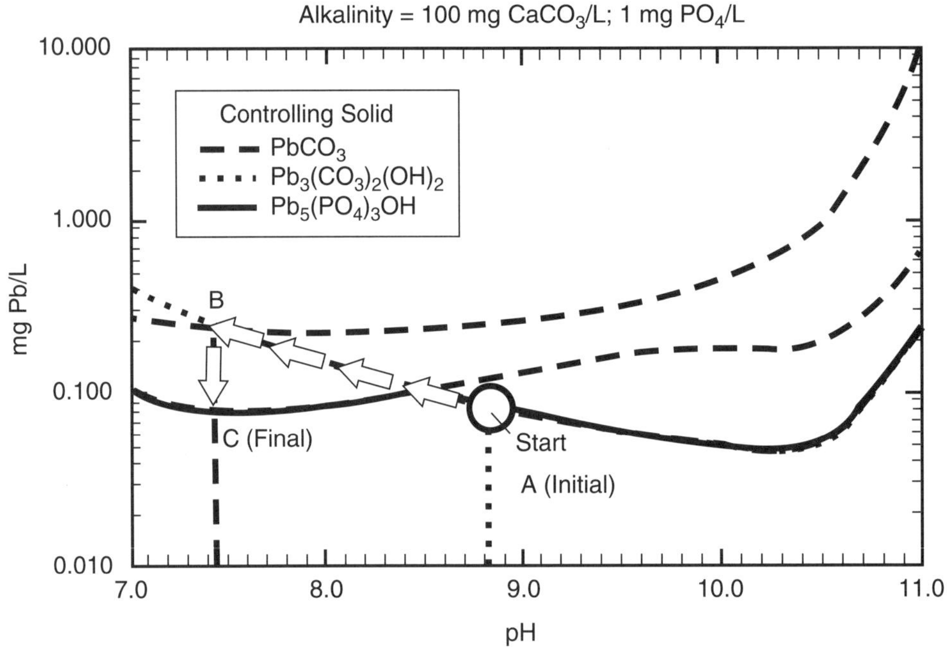

Figure 4-43 Hypothetical path of lead solubility in response to changing stability of passivation films for a water with alkalinity of 100 mg $CaCO_3$/L and orthophosphate dose of 1 mg PO_4/L after pH decrease

hydroxypyromorphite, the solubility will follow along the pathway indicated by the arrows to point B. This would result in a temporary increase of plumbosolvency (difference between solubility at point B versus A). Ultimately, the lead orthophosphate film will form (point C), which would produce a lower lead level.

Unfortunately, very little research has been conducted to examine the relative precipitation, dissolution, and conversion (one solid form to another) rates of important passivating solids in different water qualities and with different doses of orthophosphate. However, field and laboratory evidence cited previously indicate the probability that passivation reactions for lead-containing materials are quickly reversible, depend greatly on the concentration of inhibitor present in the bulk solution (higher film formation rates with higher dosages above the equilibrium solubility levels), and that carbonate reactions are likely faster than those involving orthophosphate. These general principles of responses of films to changing water qualities would also apply to other materials in the distribution system, such as unlined cast-iron mains. Therefore, a utility must take a broad and patient approach to minimize plumbosolvency while maintaining aesthetically pleasing water.

Clearly, determining the chemical and physical nature of existing pipe deposits is a key to ultimately predicting the response of the distribution system to various water treatment change scenarios. Additional research is essential to define the roles that inhibitor dosages, time, and water flow rates play in developing or redeveloping protective pipe deposits. Lead control by central water treatment cannot be done without impact on the rest of the distribution system materials. Therefore, the research must be extended beyond lead-containing materials to cast iron, where important corrosion and microbiological control issues exist when phosphate chemicals are added, and to copper, where the role of orthophosphate and blended phosphates in cuprosolvency reduction is controversial.

REFERENCES

Ainsworth, R.G., et al. 1977. *Lead in Drinking Water*. Tech. Rept. TR-43. Medmenham, England: Water Research Centre.

Ajmal, M., and R. Uddin. 1986. Quality of Drinking Water in the Aligarh Muslim University Campus, Aligarh, U.P. (India), With Respect to Heavy Metals. *Envir. Monit. and Assess.*, 6:195–205.

Akers, C.J., and R. Fellow. 1979. Review of the Problem of Lead in Drinking Water. *Envir. Health,* 87(7):148.

Anon. 1993a. Is There Lead in Your Water? *Consumer Reports,* 2:73–78.

———. 1993b. *Reducing Lead Levels in Water Supplies*. Strathclyde Water Facts Sheet No. 4. Glasgow, Scotland: Strathclyde Regional Council.

Arts, W., H.-J. Bretschneider, C. Michalke, and R. Neemeyer. 1987. Untersuchungen zum Bleieintrag in Trinkwasser durch Bleiinstallationen in Berliner Altbauhäusern. *Wasser Abwasser*, 128(11):591–596.

AWWA (American Water Works Association). 1886. Discussion. In *Proc. of the AWWA Annual Conference. Denver, Colo.: American Water Works Association.*

———. 1991. *Lead and Copper: A Working Explanation of the Lead and Copper Rule*. Denver, Colo.: American Water Works Association.

AWWA Research Foundation and DVGW, Engler-Bunte Institute. 1985. *Internal Corrosion of Water Distribution Systems.* (1st ed.) Denver, Colo.: American Water Works Association Research Foundation.

Bailey, R.J., and P.F. Russell. 1981. Predicting Drinking Water Lead Levels. *Envir. Tech. Letters,* 2:57.

Bailey, R.J., D. Holmes, P.K. Jolly, and R.F. Lacey. 1986. *Lead Concentration and Stagnation Time in Water Drawn Through Lead Domestic Pipes*. Tech. Rept. TR 243. Medmenham, England: Water Research Centre.

Bailey, T.L. 1982. Corrosion Control Experiences at Durham, North Carolina. In *Proc. of the AWWA Water Quality Technology Conference*. Denver, Colo.: American Water Works Association.

Baylis, J.R. 1926. Prevention of Corrosion and "Red Water." *Jour. AWWA,* 15:598.

Beccaria, A.M., et al. 1981. Corrosion of Lead in Sea Water. *Br. Corros. Jour.,* 17(2):87.

Bennett, M.S., O.D. Schneider, J.E. Dyksen, and G.A. Iverson. 1988. Occurrence and Treatment of Lead in Drinking Water at the Tap. In *Proc. of the AWWA Annual Conference*. Denver, Colo.: American Water Works Association.

Breach, R., S. Crymble, and M.J. Porter. 1991. Systematic Approach to Minimizing Lead Levels at Consumers' Taps. In *Proc. of the AWWA Annual Conference*. Denver, Colo.: American Water Works Association.

Britton, A., and W.N. Richards. 1981. Factors Influencing Plumbosolvency in Scotland. *Jour. Inst. Water Engrs. and Scientists,* 35(5):349–364.

Brooker, M.H., S. Sunder, P. Taylor, and V.J. Lopata. 1983. Infrared and Raman Spectra and X-ray Diffraction Studies of Solid Lead(II) Carbonates. *Can. Jour. Chem.*, 61:494–502.

Brown, W.E. 1971. Solubilities of Phosphates and Other Sparingly Soluble Compounds. In *Environmental Phosphorus Handbook*. Edited by E.J. Griffith et al. New York: John Wiley and Sons.

Buelow, R.W., et al. 1980. The Behavior of Asbestos–Cement Pipe Under Various Water Quality Conditions: A Progress Report. *Jour. AWWA,* 72(2):91.

Bunker, G.C. 1921. Lead Poisoning by Water. *Jour. NEWWA,* 35:126.

Butler, G., and H.C.K. Ison. 1966. *Corrosion and Its Prevention in Waters*. London: Leonard-Hill.

Carles, P. 1900. Blei und Trinkwasser. *Jour. Pharm. Chim.,* 6:517–520.

Carlson, W.D. 1983. The Polymorphs of $CaCO_3$ and the Aragonite-Calcite Transformation. In *Carbonates: Mineralogy and Chemistry*. Edited by R.J. Reeder. Mineralogical Society of America.

Chambers, K., and M. Hitchmough. 1992. *Economics of Lead Pipe Replacement.* Report DoE 2956–/1, TMU 9030. London: House of Commons Library.

Chin, D., and P.C. Karalekas, Jr. 1985. Lead Product Use Survey of Public Water Supply Distribution Systems Throughout the United States. In *Plumbing Materials and Drinking Water Quality: Proceedings of a Seminar.* EPA 600/9-85-007. Cincinnati, Ohio: USEPA Water Engineering Research Laboratory.

Clark, H.W. 1927. The Effect of Pipes of Different Metals Upon the Quality of Water Supplies. *Jour. NEWWA,* 41:31.

Colling, J.H., P.A.E. Whincup, and C.R. Hayes. 1987. The Measurement of Plumbosolvency Propensity to Guide the Control of Lead in Tapwaters. *Jour. IWEM,* 1(3):263–269.

Colling, J.H., B.T. Croll, P.A.E. Whincup, and C. Harward. 1992. Plumbosolvency Effects and Control in Hard Waters. *Jour. IWEM,* 6(6):259–268.

Craun, G.F., and L.J. McCabe. 1975. Problems Associated With Metals in Drinking Water. *Jour. AWWA,* 67(11):579.

Dando, K.J., and D.R. Glasson. 1989. Vacuum Microbalance Studies of Lead Deposits From Natural Waters. *Thermochim. Acta,* 152:87–96.

DeMora, S.J., R.M. Harrison, and S.J. Wilson. 1987. The Effect of Water Treatment on the Speciation and Concentration of Lead in Domestic Tap Water Derived From a Soft Upland Source. *Wat. Res.,* 21(1):83–94.

DIN 50900. 1975. Korrosion der Metalle, Begriffe, Allgemeine Begriffe. Berlin: Beuth-Verlag GmbH.

Dodrill, Donna M., and Marc Edwards. 1995. Corrosion Control on the Basis of Utility Experience. *Jour. AWWA,* 87(7):74–85.

DoE (Department of the Environment). 1977. *Lead in Drinking Water: A Survey in Great Britain 1975–1976.* Pollution Paper No. 12. London: Her Majesty's Stationery Office.

Donaldson, W. 1924. The Action of Water on Service Pipes. *Jour. AWWA,* 11:694.

EC (European Community). 1980. Council Directive of 15th July Relating to the Quality of Water Intended for Human Consumption (80/778/EEC). *Off. Jour. Eur. Commun.,* 23(L229/11).

EES and ISWS (Economic and Engineering Services, Inc., and Illinois State Water Survey). 1990. *Lead Control Strategies.* Denver, Colo.: AWWA Research Foundation and American Water Works Association.

Eilbeck, W.J. 1984. Redox Control in Breakpoint Chlorination of Ammonia and Metal Amine Complexes. *Wat. Res.,* 18(1):21.

Elzenga, C.H.J., and A. Graveland. 1980. *Proposal of the European Communities for a Directive Relating to the Quality of Water for Human Consumption, Actual Metal Levels in Drinking Water and Some Possibilities for Reduction by Central Water Conditioning by Waterworks.* Subject 3. 13th IWSA Conference, Paris, 1980. Amsterdam, Netherlands: KIWA and Gemeentewaterleidingen.

Federal Register. 1980. *Interim Primary Drinking Water Regulations, Amendments,* 45(168):57332.

———. 1991a. *Lead and Copper,* 56(110):26460.

———. 1991b. *Lead and Copper,* 56(135):32112.

———. 1992. *Lead and Copper,* 57(125):28785.

———. 1994. *Analytical Methods for Regulated Drinking Water Contaminants; Final Rule,* 59(232): 62456–62471.

Feller, H.G., H. Teepens, W. Arts, and H. Bretschneider. 1984. Untersuchungen zur Korrosion des Bleis in wäßrigen Lösungen und in Trinkwasser. *Z. Metallk.,* 75(8):619–624.

Feller, H.G., H. Zeitner, W. Arts, and J. Bretschneider. 1988. Untersuchungen zur Korrosion von Blei in verschiedenen Wässern. *Z. Metallk.,* 42(9):866–870.

Ferri, D., F. Salvatore, and E. Vasca. 1989. Lead (II) Hydroxide Complexes: On the Formation of Species Higher than $Pb(OH)_3$ in Strongly Alkaline Solutions. *Ann. Chim.,* 79:1–13.

Ferri, D., I. Grenthe, S. Hietanen, and F. Salvatore. 1987. Studies on Metal Carbonate Equilibria. 18. Lead (II) Carbonate Complexes in Alkaline Solutions. *Acta Chem. Scand.,* A41:349–354.

Fouillac, C., and A. Criaud. 1984. Carbonate and Bicarbonate Trace Metal Complexes: Critical Reevaluation of Stability Constants. *Geochem. Jour.,* 18:297–303.

Fuchss, H., H. Bruns, and H. Haupt. 1938. Die Bleivergiftungsgefahr durch Trinkwasser. Verlag Th. Steinkopff, Dresden/u.a.

Fuge, R., N.J.G. Pearce, and W.T. Perkins. 1992. Unusual Sources of Aluminum and Heavy Metals in Potable Waters. *Envir. Geochem. and Health,* 14(1):15–18.

Gardels, M.C., and M.R. Schock. 1981. Corrosion Indices: Invalid or Invaluable? In *Proc. of the AWWA Water Quality Technology Conference.* Denver, Colo.: American Water Works Association.

Gittelman, T.S., C.G. Haiges, P. Luitweiler, and T.L. Yohe. 1992. Evaluation of Lead Corrosion Control Measures for a Multi-source Water Utility. In *Proc. of the AWWA Water Quality Technology Conference.* Denver, Colo.: American Water Works Association.

Glasser, L.S.D., and E.E. Lachowski. 1980. Silicate Species in Solution. Part 1. Experimental Observations. *J.C.S. Dalton Trans.,* 393–398.

Green, J., and J.A. Holmes. 1947. Calculation of the pH of Saturation of Tricalcium Phosphate. *Jour. AWWA,* 39:1090.

Greene, L.A. 1984. Water Supply and Lead in Eastern England. *Effluent Water Treatment Jour.,* 24(7):264.

Gregory, R., and P.J. Jackson. 1983. *Reducing Lead in Drinking Water.* Report 219-S, Water Research Centre, Stevenage Laboratory, Elder Way, Stevenage, Herts.

———. 1984. Central Water Treatment to Reduce Lead Solubility. In Proc. of the AWWA Annual Conference. Denver, Colo.: American Water Works Association.

Haacke, D.F., and P.A. Williams. 1981. Stability of Plumbonacrite. *Jour. Inorg. Nucl. Chem.*, 43:406.

Haring, B.J.A. 1984. Lead in Drinking Water. Ph.D. diss. Universiteit van Amsterdam, Amsterdam.

Harrison, R.M., and D.P.H. Laxen. 1980. Physicochemical Speciation of Lead in Drinking Water. *Nature,* 286:5775.

Hatch, G.B. 1941. Inhibition of Lead Corrosion With Sodium Hexametaphosphate. *Jour. AWWA,* 33(7):1179.

Heap, H. 1913a. A Contribution to the Study of the Actions of Various Waters Upon Lead. *Jour. Soc. Chem. Ind.*, 32(16):811–815.

———. 1913b. A Contribution to the Study of the Actions of Various Waters Upon Lead. *Jour. Soc. Chem. Ind.*, 32(15):771–775.

———. 1913c. A Contribution to the Study of the Actions of Various Waters Upon Lead. Part III. *Jour. Soc. Chem. Ind.*, 32(17):847–856.

Hem, J.D., and W.H. Durum. 1973. Solubility and Occurrence of Lead in Surface Water. *Jour. AWWA,* 65(8):562.

Heumann, D.W. 1987. Los Angeles Department of Water and Power Tap Water Monitoring Program. In *Proc. of the Fifteenth Annual AWWA Water Quality Technology Conference.* Denver, Colo.: American Water Works Association.

———. 1989. Los Angeles Department of Water and Power: Solid Lead Gooseneck Slug Dispersion in Consumer Plumbing System. In *Proc. of the Seventeenth Annual AWWA Water Quality Technology Conference.* Denver, Colo.: American Water Works Association.

Heyer, C. 1888. *Ursache und Beseitigung des Bleiangriffs durch Leitungswasser.* Dessau: Verlagsbuchhandlung Paul Baumann.

Höll, K. 1934. Uber die Trinkwasserversorgung der Insel Helgoland und über das Vorkommen von bleihaltigem Trinkwasser daselbst. *Arch. Hygiene,* 283–295.

Holm, T.R., and M.R. Schock. 1991a. Potential Effects of Polyphosphate Products on Lead Solubility in Plumbing Systems. *Jour. AWWA*, 83(7):74–82.

———. 1991b. Polyphosphate Debate (Reply to Comment). *Jour. AWWA,* 83(12):10–12.

Holm, T.R., and S.H. Smothers. 1990. Characterizing the Lead-Complexing Properties of Polyphosphate Water Treatment Products by Competing-Ligand Spectrophotometry Using 4-(2-Pyridylazo) Resorcinol. *Int. Jour. Envir. Anal. Chem.*, 41:71–82.

Horsford, E.N. 1849. Service-Pipes for Water: An Investigation Made at the Suggestion of the Board of Consulting Physicians of Boston. In *Proc. of the American Academy of Arts and Sciences.*

Howard, C.D. 1923. Lead in Drinking Water. *Amer. Jour. Public Health,* 13:207.

Howard, J.L. 1992. For One NE City: Longer-lasting Pipes and Lower Lead Levels Through Chemistry. *Waterworld News,* 8(2):22.

Hulsmann, A.D. 1990. Particulate Lead in Water Supplies. *Jour. IWEM,* 4(2):19–25.

Hunt, D.T.E., and J.D. Creasey. 1980. *The Calculation of Equilibrium Trace Metal Speciation and Solubility in Aqueous Systems by a Computer Method, With Particular Reference to Lead.* Tech. Rept. 151. Medmenham, England: Water Research Centre.

Jackson, P.J., and I. Sheiham. 1980. *Calculation of Lead Solubility in Water.* Tech. Rept. 152. Medmenham, England: Water Research Centre.

Jawad, I.M., M.R. Al-Ghazali, and M.S.H. Khorshid. 1988. Effect of Domestic Storage Tanks on the Quality of Drinking Waters. Part 1: Chemical Analysis. *Environ. Monit. and Assess.*, 11:79–87.

Jenkins, D., et al. 1971. Chemical Processes for Phosphate Removal. *Wat. Res.,* 5:369.

JMM (J.M. Montgomery Consulting Engineers, Inc.). 1982. Internal Corrosion Mitigation Study, Final Report. Report prepared for the Portland, Oregon, Water Bureau.

———. 1983. Internal Corrosion Mitigation Study, Addendum Report. Report prepared for the Portland, Oregon, Water Bureau.

Jolliff, C.R., P.J. Heidrick, J.A. Cain, and J.E. Ohlsen. 1959. Lead in Domestic Water Supply; Lead Piping as a Source of Toxic Quantities. *Nebraska State Medical Jour.*, 44:156–160.

Karalekas, P.C., Jr., et al. 1976. Lead and Other Trace Metals in Drinking Water in the Boston Metropolitan Area. *Jour. NEWWA,* 90:150.

———. 1978. Alternative Methods for Controlling the Corrosion of Lead Pipe. *Jour. NEWWA,* 92(2):159.

———. 1983. Control of Lead, Copper, and Iron Pipe Corrosion in Boston. *Jour. AWWA,* 75(2):92.

Karweik, D.H., and C.O. Huber. 1978. Polyphosphate Currents at the Lead Dioxide Electrode for Hydrolysis Rate and Instability Constant Measurements. *Anal. Chem.,* 50(8):1209–1212.

Katzenelson, E., et al. 1973. Inactivation of Viruses and Bacteria by Ozone. In *Chemistry of Water Supply, Treatment, and Distribution.* Edited by A.J. Rubin. Ann Arbor, Mich.: Ann Arbor Science.

Kirmeyer, G.J., A.M. Sandvig, G.L. Pierson, and C.H. Neff. 1994. *Development of a Pipe Loop Protocol for Lead Control.* Denver, Colo.: AWWA Research Foundation and American Water Works Association.

Koudelka, M., et al. 1982. On the Nature of Surface Films Formed on Iron in Aggressive and Inhibiting Polyphosphate Solution. *Jour. Electrochem. Soc.,* 129(6):1186.

Kruse, W., and M. Fischer. 1930. Bleivergiftungen durch Trinkwasser im allgemeinen und in Leipzig. *Deutsche med. Wochenschrift* (Leipzig), S. 1814–1818.

Kuch, A., and I. Wagner. 1983. Mass Transfer Model to Describe Lead Concentrations in Drinking Water. *Wat. Res.,* 17(10):1303.

Lacey, R.F., and P.K. Jolly. 1986. *Sampling for Household Water-Lead.* Envir. Rept. TR244. Medmenham, England: Water Research Centre.

Lead Contamination Control Act of 1988. Public Law 100-572. United States Code, Vol. 42, Section 300j-21–300j-26.

Lee, R.G., W.C. Becker, and D.W. Collins. 1989. Lead at the Tap: Sources and Control. *Jour. AWWA,* 81(7):52–62.

Liebrich, A. 1898. Ursachen und Bekämpfungsmethoden des Bleiangriffs durch Leitungswasser. *Zeitschrift für angewandte Chemie,* S. 703–704.

Lin, C-F, and Mark M. Benjamin. 1990. Dissolution Kinetics of Minerals in the Presence of Sorbing and Complexing Ligands. *Envir. Sci. and Technol.,* 24:126–134.

Liverseege, J.F., and A.W. Knapp. 1913. The Action of an Alkaline Natural Water on Lead. *Chemical News,* Oct. 10; p. 176.

Loewenthal, R.E., and G.V.R. Marais. 1976. *Carbonate Chemistry of Aquatic Systems: Theory and Applications.* Ann Arbor, Mich.: Ann Arbor Science.

Maas, R.P., and S.C. Patch. 1990a. Lead Contamination of North Carolina Domestic Tapwater: Prevalence, Risk Factors and Control Measures. UNC–Asheville Environmental Quality Institute. A Technical Report to the Clean Water Fund of North Carolina.

———. 1990b. Lead Contamination of Water in North Carolina Public Schools. UNC–Asheville Environmental Quality Institute. EQI Technical Publication 90-006.

Maas, R.P., S.C. Patch, D.J. Kucken, and B.T. Peek. 1991. A Multi-state Study of the Effectiveness of Various Corrosion Inhibitors in Reducing Residential Lead Levels. In *Proc. of the AWWA Annual Conference.* Denver, Colo.: American Water Works Association.

Mason, W.P. 1904. Notes as to the Action of Water Upon Lead and Zinc. In *Proc. of the AWWA Annual Conference.* Denver, Colo.: American Water Works Association.

McAnally, A.S., S. Bharadwaj, M.E. Meadows, and E.R. Blood. 1993. Lead and Copper Corrosion Mitigation for Several Communities in South Carolina. In *Proc. of the AWWA Annual Conference.* Denver, Colo.: American Water Works Association.

McClelland, N.I., D.A. Gregorka, and B.D. Carlton. 1989. The Drinking-Water Additives Program. *Envir. Sci. and Technol.*, 23(1):14–18.

McFarren, E.F., et al. 1977. Water Quality Deterioration in the Distribution System. In *Proc. of the Fifth Annual AWWA Water Quality Technology Conference.* Denver, Colo.: American Water Works Association.

Miles, G. 1948. The Action of Natural Waters on Lead. *Jour. Soc. Chem. Ind.,* 67:10.

Moore, E.W., and F.E. Smith. 1942. Effect of Sodium Hexametaphosphate on the Solution of Lead. *Jour. AWWA,* 34(9):1415.

Moore, M.R. 1973. Plumbosolvency of Waters. *Nature,* 243:222.

Nancollas, G.H. 1983. Phosphate Precipitation in Corrosion Protection: Reaction Mechanisms. *Corrosion,* 39(3):77.

Neff, C.H., and M.R. Schock. 1985. Corrosion Product—Corrosion Rate—Water Quality Investigation in Building Systems. Final Report on Project EPA CR808566010 Drinking Water Research Division. Cincinnati, Ohio: U.S. Environmental Protection Agency.

Nèher-Neumann, E. 1992. Studies on Metal Carbonate Equilibria 24. The Hydrogen Carbonate and Carbonate Complexes of the Lead(II) and Cadmium(II) Ions in Acid Solutions and a 3 M $(Na)ClO_4$ Ionic Medium at 25°C. *Acta Chem. Scand.*, 46(4):231–239.

Nriagu, J.O. 1972. Lead Orthophosphates—I. Solubility and Hydrolysis of Secondary Lead Orthophosphate. *Inorg. Chem.,* 11(10):2499.

———. 1973. Lead Orthophosphates—II. Stability of Chloropryomorphite at 25°C. *Geochim. Cosmochim. Acta*, 37:367–377.

———. 1974. Lead Orthophosphates—IV. Formation and Stability in the Environment. *Geochim. Cosmochim. Acta,* 38:887.

O'Brien, J.E. 1976. Lead in Boston Water: Its Cause and Prevention. *Jour. NEWWA,* 90(2):173.

Olem, H., and P.M. Berthouex. 1989. Acidic Deposition and Cistern Drinking Water Supplies. *Envir. Sci. and Techn.*, 23(3):333–340.

Pagenkopf, G.K. 1982. *Evaluation of Polyphosphate Complexation Capacity.* Unpublished Report. Drinking Water Research Division. RFQ No. 81-95, Department of Chemistry, Montana State University, Bozeman, Mont. Cincinnati, Ohio: U.S. Environmental Protection Agency.

Patterson, J.W., and J.E. O'Brien. 1979. Control of Lead Corrosion. *Jour. AWWA,* 71(5):264.

Patterson, J.W., et al. 1977. Carbonate Precipitation for Heavy Metal Pollutants. *Jour. WPCF,* 49:2397.

Pharoah, P.L.G. 1980. pH Correction: Technical Aspects of a 500 tcmd Treatment Works. *Jour. Inst. Water Engrs. and Scientists,* 34(4):334.

Plummer, L.N., and E. Busenberg. 1982. The Solubilities of Calcite, Aragonite and Vaterite in CO_2-H_2O Solutions Between 0 and 90°C, and an Evaluation of the Aqueous Model for the System $CaCO_3$-CO_2-H_2O. *Geochim. Cosmochim. Acta,* 46:1011.

Pontius, F.W., and A.E. Warburton. 1992. Congress Debates Lead Legislation. *Jour. AWWA*, 84(1):12.

Pourbaix, M. 1966. *Atlas of Electrochemical Equilibria in Aqueous Solutions.* Oxford, England: Pergamon Press.

Proskauer, B. 1893. Beitrag zur Bleiaufnahme durch Quellwasser. *Zeitschrift für Hygiene,* 14:292–298.

Pullmann. 1887. Zur Frage der Verunreinigung des Wassers durch bleierne Leitungsröhren. *Deutsche Vierteljahresschrift für öffentliche Gesundheitspflege,* 19:265.

Ramaley, B.L. 1993. Monitoring and Control Experience Under the Lead and Copper Rule. *Jour. AWWA*, 85(2):64–67.

Rapmund, O. 1889. General-Sanitätsbericht für die Jahre 1886–1888 des Regierungsbezirks Aurich, S. 111–116.

Rees, O.W., and A.L. Elder. 1928. The Effect of Certain Illinois Waters on Lead. *Jour. AWWA,* 19:714.

Richards, W.N., M. Britton, and L. Cochrane. 1980. Reducing Plumbosolvency—The Effect of Added Lime on the Loch Katrine Supply to Glasgow. *Jour. Inst. Water Engrs. and Scientists*, 35(4):315–345.

Rossum, J.R., and D.T. Merrill. 1983. An Evaluation of the Calcium Carbonate Saturation Indexes. *Jour. AWWA,* 75(2):95.

Ruchhoft, C.C., and J.F. Kachmar. 1942. Study of Solubility of Two Lead Salts in Dilute Solutions With Special Reference to the Lead Hazard in Drinking Water. *Jour. AWWA,* 34(1):85.

Saar, R.A., and J.H. Weber. 1980. Lead(II) Complexation by Fulvic Acid: How It Differs From Fulvic Acid Complexation of Copper(II) and Cadmium(II). *Geochim. Cosmochim. Acta,* 44:1381.

Samuels, E.R., and J.C. Meranger. 1984. Preliminary Studies on the Leaching of Some Trace Metals From Kitchen Faucets. *Wat. Res.,* 18(1):75.

Schaut, G.C. 1942. The Action of a Chlorinated Water Supply Upon Lead Pipe. *Amer. Jour. Pharm.,* 241–249.

Schlünder, E.U. 1975. *Einführung in die Wärme-und Stoffübertragung.* Braunschweig: Verlag Friedrich Vieweg and Sohn.

Schock, M.R. 1980. Response of Lead Solubility to Dissolved Carbonate in Drinking Water. *Jour. AWWA,* 72(12):695.

———. 1981. Response of Lead Solubility to Dissolved Carbonate in Drinking Water. *Jour. AWWA,* 73(3):36 (errata).

———. 1984a. Treatment or Water Quality Adjustment to Attain MCL's in Metallic Potable Water Plumbing Systems. In *Proc. of the Seminar on Plumbing Materials and Drinking Water Quality.* Cincinnati, Ohio: U.S. Environmental Protection Agency.

———. 1984b. Temperature and Ionic Strength Corrections to the Langelier Index—Revisited. *Jour. AWWA,* 76(8):72.

———. 1989. Understanding Corrosion Control Strategies for Lead. *Jour. AWWA,* 81(7):88–100.

———. 1990a. Causes of Temporal Variability of Lead in Domestic Plumbing Systems. *Envir. Monit. and Assess.,* 15:59–82.

———. 1990b. Internal Corrosion and Deposition Control. In *Water Quality and Treatment: A Handbook of Community Water Supplies.* New York: McGraw-Hill, pp. 997–1111.

Schock, M.R., and R.W. Buelow. 1981. The Behavior of Asbestos–Cement Pipe Under Various Water Quality Conditions: Part 2, Theoretical Considerations. *Jour. AWWA,* 73(11):609.

Schock, M.R., and M.C. Gardels. 1983. Plumbosolvency Reduction by High pH and Low Carbonate–Solubility Relationships. *Jour. AWWA,* 75(2):87.

Schock, M.R., and I. Wagner. 1985. The Corrosion and Solubility of Lead in Drinking Water. In *Internal Corrosion of Water Distribution Systems.* 1st ed. Denver, Colo.: American Water Works Association Research Foundation.

Schock, M.R., R. Levin, and D.D. Cox. 1988. The Significance of Sources of Temporal Variability of Lead in Corrosion Evaluation and Monitoring Program Design. In *Proc. of the Annual AWWA Water Quality Technology Conference.* Denver, Colo.: American Water Works Association.

Sheiham, I., and P.J. Jackson. 1981. The Scientific Basis for Control of Lead in Drinking Water by Water Treatment. *Jour. Inst. Water Engr. and Scientists,* 35(6):491.

Shoesmith, D.W., and M.G. Bailey. 1988. Anodic Oxidation of Lead in Aqueous Carbonate Solutions. I. Film Formation and Dissolution at pH 12. *Can. Jour. Chem.,* 66:2652–2657.

Shoesmith, D.W., M.G. Bailey, and P. Taylor. 1988. Anodic Oxidation of Lead in Aqueous Carbonate Solutions. II. Film Formation and Dissolution in the pH Range 9 to 14. *Can. Jour. Chem.,* 66:2941–2946.

Singley, J.E., et al. 1984. *Corrosion and Calcium Carbonate Saturation Index in Water Distribution Systems.* EPA R805400-01-1. Washington, D.C.: U.S. Environmental Protection Agency.

Snoeyink, V.L., and D. Jenkins. 1980. *Water Chemistry.* New York: John Wiley and Sons.

Standard Methods for the Examination of Water and Wastewater. 18th ed. 1992. (A.E. Greenberg, L.S. Clesceri, and A.D. Eaton, eds.). Washington, D.C.: American Public Health Association, American Water Works Association, and WEF.

Stich. 1930. Einige Untersuchungen über die Löslichkeit des Bleies im Wasser mit Bezug auf das Leipziger Leitungswasser. *Pharmazeutische Zeitung,* 75:1262–1263.

Stumm, W. 1960. Investigation on the Corrosive Behavior of Waters. *Jour. San. Engrg. Div.—ASCE,* 86:27.

Stumm, W., and J.J. Morgan. 1981. *Aquatic Chemistry, an Introduction Emphasizing Chemical Equilibria in Natural Waters.* 2nd ed. New York: Wiley-Interscience.

Taylor, F.B. 1980. Metals—Corrosion or Dissolution? In *Proc. of the NEWWA Seminar on Corrosion Control in Drinking Water Systems.* Randolph, Mass.: New England Water Works Association.

Taylor, P., and V.J. Lopata. 1984. Stability and Solubility Relationships Between Some Solids in the System PbO-CO_2-H_2O. *Can. Jour. Chem.,* 62:395–402.

Tergast. 1899. Ursache und Verhütung des Bleiangriffs durch Leitungswasser nach Erfahrungen bei der städtischen Wasserleitung in Emden. *Zeitschrift für Medizinalbeamte,* 12:165–180.

Thresh, J.C. 1922a. The Action of Natural Waters on Lead, Part I. Effect of Various Saline Constituents. *Analyst,* 47(560):459.

———. 1922b. The Action of Natural Waters on Lead, Part II. The Action of Moorland Waters on Lead. *Analyst,* 47(560):500.

Townsend, T.L., and M.R. Pollen. 1988. Get the Lead Out. In *Alaska Water Management Association Conference.* Anchorage, Alaska: Alaska Water Management Association.

TrinkwV. 1990. Verordnung über Trinkwasser und über Wasser für Lebensmittelbetriebe (Trinkwasserverordnung–TrinkwV.) vom 12. December 1990. BGBl. I, 2613–2629.

USEPA (U.S. Environmental Protection Agency). 1978. *National Interim Primary Drinking Water Regulations.* EPA-570/9-76-003. Washington, D.C.: Government Printing Office.

———. 1991. *Lead and Copper Rule Guidance Manual. Volume I: Monitoring.* Denver, Colo.: American Water Works Association.

———. 1992. *Lead and Copper Rule Guidance Manual. Volume 2: Corrosion Control and Treatment.* Denver, Colo.: American Water Works Association.

Vaccari, D.A., and R. Mueller. 1991. Factors Affecting Lead Corrosion Byproduct Formation; Field and Laboratory Studies. In *Proc. of the AWWA Annual Conference.* Denver, Colo.: American Water Works Association.

van den Hoven, T.J.J. 1985. How to Determine and Control Lead Levels in Tap Water: The Dutch Approach. In *Proc. of the Thirteenth Annual AWWA Water Quality Technology Conference.* Denver, Colo.: American Water Works Association.

———. 1987. A New Method to Determine and Control Lead Levels in Tap Water. *Aqua,* 6:315–322.

van den Hoven, T.J.J., and M.W.M. van Eekern. 1988. *Optimal Composition of Drinking Water.* KIWA (Netherlands Waterworks Testing and Research Institute). N. V. KIWA. Report No. 100.

van Gaans, P.F.M. 1989. WATEQX—A Restructured, Generalized, and Extended FORTRAN 77 Computer Code and Database Format for the WATEQ Aqueous Chemical Model for Trace Element Speciation and Mineral Saturation, for Use on Personal Computers or Mainframes. *Comp. and Geosci.*, 15(6):843–887.

van Wazer, J.R., and D.A. Campanella. 1950. Structure and Properties of the Condensed Phosphates. IV. Complex Ion Formation in Polyphosphate Solutions. *Jour. Amer. Chem. Soc.*, 72(2):655–663.

Wagner, I. 1983. Beurteilung der Hausanschlussleitungen aus Blei bei den Stadtwerken Duisburg AG.

———. 1989. Effect of Inhibitors on Corrosion Rate and Metal Uptake in Drinking Water Systems. In *Sem. Proc.: Internal Corrosion Control Developments and Research Needs,* AWWA Annual Conference. Denver, Colo.: American Water Works Association.

Wagner, I., and A. Kuch. 1981. *Trinkwasser und Blei.* Karlsruhe: DVGW-Forschungsstelle Berichte Band 18.

Walker, R., and R. Oliphant. 1982. Corrosion of Lead in Drinking Water. *Anti-Corrosion,* 29(4):8–11.

Watkins, W.H. 1886. *Lead Poisoning.* New York: American Water Works Association, p. 47.

Weston, Roy F., Inc., and EES (Economic and Engineering Services, Inc.). 1990. *Lead Service Line Replacement: A Benefit-to-Cost Analysis.* Sponsored by The Water Industry Technical Fund. Denver, Colo.: American Water Works Association.

Weston, R.S. 1920. Lead Poisoning by Water and Its Prevention. *Jour. NEWWA,* 34:239.

Whitfield, M., and D.R. Turner. 1977. The Theoretical Studies of the Chemical Speciation of Lead in Seawater. In *Lead in the Marine Environment.* Edited by M. Branica and Z. Konrad. Oxford, England: Pergamon Press.

WHO (World Health Organization). 1993. *Guidelines for Drinking-water Quality. Volume 1. Recommendations.* 2nd ed. Geneva, Switzerland: World Health Organization.

Wolffhügel, G. 1887. Wasserversorgung und Bleivergiftung. Arbeiten am Kaiserlichen Gesundheitsamte II, S. 484.

Worth, D.J., et al. 1975. Relationship of Blood Lead Levels to Lead in Household Drinking Water. In *Proc. of the AWWA Annual Conference.* Denver, Colo.: American Water Works Association.

Württemberg. 1793. Handschrift von Gottes Gnaden Carl Herzog zu Württemberg und Teck etc. Württ. Staatsarchiv CVI—Bd. 1.

———. 1878. Amtsblatt des Königlich Württembergischen Ministeriums des Inneren, Stuttgart, Nr. 7, S. 101–106.

Wysock, B.M., M.R. Schock, and J.A. Eastman. 1991. A Study of the Effect of Municipal Ion-Exchange Softening on the Corrosion of Lead, Copper and Iron in Water Systems. In *Proc. of the AWWA Annual Conference.* Denver, Colo.: American Water Works Association.

Wysock, B.M., A.M. Sandvig, M.R. Schock, C.P. Frebis, and B. Prokop. 1992. Statistical Procedures for Corrosion Studies. In *Proc. of the Twentieth Annual AWWA Water Quality Technology Conference.* Denver, Colo.: American Water Works Association.

Yaziz, M.I., H. Gunting, N. Sapari, and A.W. Ghazali. 1989. Variations in Rainwater Quality From Roof Catchments. *Wat. Res.,* 23(6):761–765.

Young, E.S., Jr., and W.E. Sharpe. 1984. Atmospheric Deposition and Roof-Catchment Cistern Water Quality. *Jour. Envir. Qual.,* 13(1):38–43.

5

Corrosion of Copper in Potable Water Systems

John F. Ferguson, Otto van Franqué, and Michael R. Schock

Since World War II copper has been the most common material for consumer premise plumbing because of its excellent characteristics, including ease of installation, low cost, and corrosion resistance. Corrosion problems, although infrequent, can be severe for the affected consumers and systems. Failure of copper tubing by pitting, blue or green water problems, and, more recently, failure to meet the US Environmental Protection Agency (USEPA) action level for copper in tap water samples are major problems when they do occur.

In this chapter, the chemistry of copper and its corrosion chemistry are presented to explain copper scale formation, copper solubility, and the major types of copper corrosion. Uniform corrosion and copper release are reviewed along with erosion corrosion. Pitting problems with copper are categorized into two groups, cold water pitting and hot water pitting, in order to review their characteristics, causes, and mitigation. Microbiologically induced copper corrosion is also discussed.

Information is developed on the role of water quality in affecting copper corrosion and release, but corrosion mitigation, monitoring, and control are addressed in more detail in chapters 8, 9, and 10 for copper, along with other materials.

NOTE: The first edition of this chapter was authored by Henry Cruse, Otto van Franqué, and Richard D. Pomeroy.

COPPER AS A PLUMBING MATERIAL

Copper came into use as a plumbing material after World War II. While use of copper tubes for drinking water applications dates back to the late 19th century, a major breakthrough came with the introduction of the capillary solder fitting for jointing pipes in the 1920s (Winkler 1990; Franqué 1987). By virtue of its corrosion resistance and ease of installation, copper has progressively displaced alternative materials for pipe sizes at which it is competitively priced (up to about 2 in. [50 mm]). Today copper accounts for between 50 and 90 percent of all tube installed in drinking water service in industrialized countries (Cohen and Myers 1984). In the United States, this amounts to well over 500 million lin ft (160 million m) of copper water tube installed each year.

Phosphorus deoxidized copper is used for copper tubes and capillary fittings. Soft tempered copper tubes (for outside diameters up to ⅞ in. [22 mm]) are commonly used for service lines, while both soft and hard tempered copper are used in interior plumbing. Wall thickness of copper tubes is 0.04 in. (1 mm) or greater in the lower range of diameters and sometimes depends on the temper of the tubes.

Copper is also an essential element in brass and bronze. Fittings and fixtures from both groups of alloys are an integral part of any plumbing installation.

Failure of Copper Tubing

Though failures of copper tubing are relatively rare, some do occur, often because of pitting corrosion in cold water. Statistics from the US Copper Development Association (CDA) (Cohen and Myers 1984) and the Deutsches Kupfer-Institut (DKI) (Franqué 1983) reflect situations in the early 1980s and 1990s and show the relative distribution of the different types of tube failure that occur (Table 5-1). Erosion corrosion has the second largest incidence of occurrence. A number of reports reflect the experiences in other countries (Japan CDA 1982).

The situation has improved due to improved production techniques, which give better cleanliness of the inner bore, removing carbon films that are said to initiate pitting (Campbell 1954). In the United States, production techniques have not required modification. In Europe and elsewhere, fabricators have followed two different approaches in order to improve quality. Some used blasting techniques to clean the inner bore as a

Table 5-1 Service experience with copper plumbing systems in the United States and Germany

	Frequency, %		
Cause of Failure	United States 1983	Germany 1983	Germany 1990
Pitting corrosion (all types)	58	85	60
Erosion corrosion	24	8	15
Faulty workmanship*	5		
Corrosion of outer surface of tubes	7	3	8
Fatigue	2	3	5
Other	4	1	11

*For Germany this is included in "Other" and "Fatigue" categories.

final production stage. Water–sand blast or water–air blast were used initially. Today the predominant practice is to use iron grit as a blasting material. The introduction of this process has reduced the frequency of severe cases of pitting by 90 percent or more (Cornwell, Wildsmith, and Gilbert 1973). The other choice is to preoxidize the inner bore (Baukloh and Triquet 1982). Preoxidizing includes the removal of any carbon present and produces an oxide scale, which is said to improve corrosion resistance.

Regulations

In recent years, release of dissolved and particulate copper corrosion products has emerged as a concern in the water industry. The Lead and Copper Rule (USEPA 1991a) has established an action level of 1.3 mg/L for copper. If this level is exceeded in 10 percent of selected tap water samples, corrosion control treatment and source water treatment requirements are triggered. Tap water monitoring in the early 1990s has revealed that the copper criteria are exceeded at least as often as the lead criteria in US drinking water. The European Community has recommended that a level of 3 mg/L in pipe after 12 hours of stagnation be introduced into the national regulations of the European countries. To date (1993), few other countries have followed the US regulation, and health-based copper standards in stagnant tap water are not widely adopted. However, occasional occurrences of green or blue water from copper plumbing have long been seen as an aesthetic problem.

COPPER CHEMISTRY

Introduction

Copper, a light red metal, is soft, malleable, and ductile. It is highly conductive toward electricity and heat, and its resistance to oxidation gives it a high degree of corrosion resistance. In the galvanic series, copper is more noble than any of the other metals commonly used in water distribution systems (see chapter 1), with the exception of passivated stainless steels. Hence, it is rarely expected to corrode due to galvanic action. Among the more common metals, only titanium, silver, gold, and platinum are less readily oxidized. However, copper may not be noble when compared to silver solders or carbon.

Copper, when it corrodes, is oxidized to Cu(I) (cuprous) and Cu(II) (cupric) states and forms insoluble precipitates and soluble complexes in each. The precipitates are, or closely resemble, copper minerals and are often identified by mineral names; they may form passivating scales or they may be found in water, often as blue or green turbid suspensions. The soluble complexes influence the solubility of passivating solids and contribute to copper concentrations in drinking water.

The thermodynamic stability of copper is depicted in Figure 5-1, which is a Pourbaix diagram for the Cu–H_2O system for $10^{-4}M$ soluble copper. The diagram, using Pourbaix' classical terms, indicates that copper is immune to corrosion at low pE values, that cuprite (Cu_2O) is stable over a narrow range of pE values, and that tenorite (CuO) is stable at alkaline pH values for water at equilibrium with atmospheric oxygen. These solids are expected to form during corrosion, especially during periods of stagnation, and may be passivating. At lower pH values, Cu^{2+} is stable, which should allow corrosion of copper to take place at nearly all pE values. Under strong oxidizing conditions, copper can exist in Cu(III) and Cu(IV) oxidation states (not shown in Figure 5-1). These forms are not known to be important in plumbing corrosion.

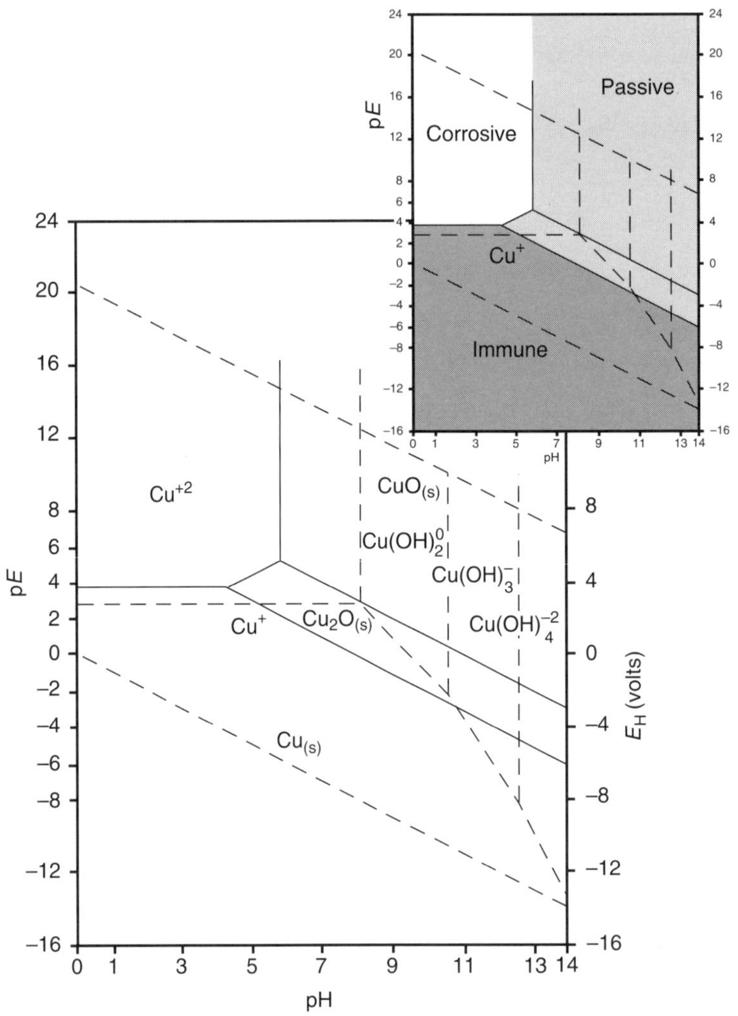

NOTE: Solid lines represent solids and predominant species; dashed lines represent aqueous species

Figure 5-1 Thermodynamic stability of copper solids and aqueous species for total copper $10^{-4}M$

For copper in contact with drinking water, the corrosion process can be interpreted using the Pourbaix diagram. If dissolved oxygen (DO) and other oxidized substances were completely absent, copper would be immune to oxidation because it is quite noble. In drinking water, the oxidizing agents (electron acceptors) that will cause the corrosion of metallic copper are predominantly DO and aqueous chlorine species (Atlas and Zajicek 1982; Stone et al. 1987; Reiber 1989). The anode of the corrosion cell is the location where the oxidized copper species (Cu^+ or Cu^{2+}) is produced, while the cathode is the place where a reacting species is reduced. The oxidation of Cu(m) in drinking water is thought to involve a one-electron transfer, forming Cu^+ as the initial product, followed by precipitation of Cu_2O. However, cuprite (Cu_2O) may also form by a solid-state nucleation process, resulting in a scale on the metal surface.

$$Cu(m) \rightarrow Cu^+ + e^- \tag{5-1}$$

$$2Cu^{+} + H_2O \rightarrow Cu_2O_{(s)} + 2H^{+} \qquad (5\text{-}2)$$

Cuprite may be further oxidized to Cu(II), which can precipitate as tenorite (CuO) at alkaline pH values (Eqs 5-3 and 5-4). (Chemical reactions and thermodynamic data are included in Table 5-2, and methods for using these data to compute equilibrium constants are presented in appendix A.)

$$\tfrac{1}{2}Cu_2O_{(s)} + \tfrac{1}{2}H_2O \rightarrow Cu^{2+} + OH^{-} + e^{-} \qquad (5\text{-}3)$$

$$Cu^{2+} + H_2O \rightarrow CuO_{(s)} + 2H^{+} \qquad (5\text{-}4)$$

These reactions require an electron acceptor, usually DO or residual chlorine. Many papers and texts have documented the impact of DO on copper oxidation and dissolution. Several studies have also shown that aqueous chlorine species have a significant impact on copper oxidation and corrosion rates (Atlas and Zajicek 1982; Suzuki et al. 1983; Reiber, Ferguson, and Benjamin 1987; Reiber 1989; Singley et al. 1986). Chlorine and DO affect the oxidation rate of the copper metal and may also alter the crystalline characteristics and porosity of the oxide corrosion product film produced at the pipe surface.

Copper oxides include several forms in addition to the stable forms, cuprite and tenorite. Cuprous oxide is usually brownish red, although it may vary from yellow to red depending on particle size. It occurs in nature and corrosion products as cuprite, which is the stable form, and often occurs as a thin, nearly continuous layer. Cupric oxide is black and occurs as tenorite; however, cupric hydroxides or oxyhydroxides, which are more soluble but also form more readily, are often present during corrosion.

Cuprous ion is unstable with respect to disproportionation to cupric ion and copper (Eq 5-5), but the reverse reaction is implicated in copper corrosion processes in pitting and during stagnation.

$$2Cu^{+} \rightleftarrows Cu(m) + Cu^{2+} \qquad (5\text{-}5)$$

Table 5-2 lists a selected set of internally consistent Gibbs free energies of formation for copper and related chemical species, taken from a variety of compilations or computed from equilibrium constants as noted in the table. Adjustments were made for differences in ΔG_f^o values among the data sources for species used in deriving the tabulated values. In subsequent equilibrium calculations, equilibrium constants for 25°C were derived from ΔG_f^o values (see appendix A). Although some researchers have made predictions of copper corrosion behavior at elevated temperature (Adeloju and Hughes 1986), adequate information is generally not available to characterize stability constants for many important copper compounds and complexes at temperatures above or below 25°C.

Corrosion Chemistry

Copper corrosion may proceed as a uniform process or, to varying degrees, it may be patchy or localized, resulting in pitting attack that sometimes causes perforation of copper pipe. Localized corrosion has been further distinguished based on the water type in which it occurs and the morphology of pits on the copper surface under aggressive attack. All these processes involve oxidation of metallic copper to

Table 5-2 Internally consistent Gibbs free energies of formation—copper and related species at 25°C

Species	ΔG_f^o kcal/mol	ΔG_f^o kJ/mol	Reference
Cu(I) Aqueous Species			
Cu^+	11.95	49.98	1
$CuCl^0$	–23.12	–96.72	2
$CuCl_2^-$	–58.29	–243.89	3
$CuCl_3^-$	–88.76	–371.35	3
$Cu_2Cl_4^{2-}$	–115.71	–484.12	3
$CuNH_3^+$	–2.44	–10.20	4
$Cu(NH_3)_2^+$	–15.46	–64.67	4
Cu(I) Solid Species			
Cu(OH)	–43.50	–182.00	1
Cu_2O	–35.01	–146.47	5
Cu(II) Aqueous Species			
Cu^{2+}	15.65	65.48	1
$Cu(OH)^+$	–30.18	–126.28	6
$Cu(OH)_2^0$	–75.57	–316.21	6
$Cu(OH)_3^-$	–117.72	–492.55	7
$Cu(OH)_4^{2-}$	–157.14	–657.47	8
$Cu_2(OH)_2^{2+}$	–67.65	–283.03	8
$Cu_3(OH)_4^{2+}$	–151.49	–633.82	8
$CuHCO_3^+$	–127.09	–531.73	9
$CuCO_3^0$	–119.84	–501.43	9
$Cu(CO_3)_2^{2-}$	–251.19	–1050.98	9
$CuCO_3(OH)_2^{2-}$	–206.00	–861.90	10
$Cu(OH)CO_3^-$	–161.43	–675.43	11
$CuCl^+$	–16.28	–68.10	8
$CuCl_2^0$	–46.95	–196.42	12
$CuCl_3^-$	–76.35	–319.44	12
$CuHPO_4^+$	–250.14	–1046.57	13
$CuH_2PO_4^+$	–256.61	–1073.64	13
$CuSO_4^0$	–162.52	–692.53	13
$CuNH_3^{2+}$	3.80	15.92	13
$Cu(NH_3)_2^{2+}$	–7.21	–30.16	13
$Cu(NH_3)_3^{2+}$	–17.36	–72.65	13
$Cu(NH_3)_4^{2+}$	–25.72	–107.59	13
$Cu(NH_3)_5^{2+}$	–32.06	–134.14	13
$CuNO_3^+$	14.97	62.63	8
$Cu(NO_3)_2^0$	16.20	67.76	8
$CuH_2P_2O_7^-$	–468.47	–1960.08	12
$CuHP_2O_7^0$	–463.16	–1937.88	12
$Cu_2P_2O_7$	–427.31	–1787.86	12
$CuP_2O_7^{2-}$	–451.96	–1891.00	1
$Cu(P_2O_7)_2^{6-}$	–913.81	–3823.40	1

Table continued next page

Table 5-2 Internally consistent Gibbs free energies of formation—copper and related species at 25°C (continued)

Species	ΔG_f^o kcal/mol	ΔG_f^o kJ/mol	Reference
Cu(II) Solid Species			
CuO	–30.15	–126.16	14
CuO	–30.62	–128.10	14
CuO	–30.32	–126.84	15
$Cu(OH)_2$	–85.60	–358.16	14
$Cu(OH)_2$	–85.90	–359.41	14
$Cu(OH)_2$	–85.22	–356.56	15
$CuCl_2$	–56.29	–232.52	13
$Cu_2(OH)_3Cl$	–160.09	–669.83	13
$Cu_4(OH)_6Cl_2$ (atacamite)	–320.51	–1341.00	10
$Cu_4(OH)_6Cl_2$ (paratacamite)	–320.27	–1340.00	10
$Cu_2(OH)_2CO_3$ (malachite)	–215.74	–902.68	16
$Cu_2(OH)_2CO_3$ (malachite)	–215.34	–900.96	13
$CuCO_3$	–123.67	–517.45	12
$Cu_3(CO_3)_2(OH)_2$ (azurite)	–341.63	–1429.36	12
$CuSO_4$ (chalcocyanite)	–158.17	–661.80	8
$Cu_3(SO_4)(OH)_4$ (antlerite)	–345.75	–1446.60	8
$Cu_4(OH)_6(SO_4)\cdot H_2O$ (langite)	–488.53	–2044.00	8
$CuSO_4\cdot 5H_2O$	–271.36	–1135.37	12
$CuO\cdot CuSO_4$	5.95	24.89	12
$Cu_4(OH)_6SO_4$ (brochanite)	–434.51	–1818.00	10
CuP_2O_7	–448.11	–1874.91	12
$Cu_2P_2O_7$	–447.97	–1874.30	8
$Cu_3(PO_4)_2\cdot 2H_2O$	–606.27	–2536.62	12
$Cu_3(PO_4)_2$	–490.30	–2051.40	12
$CuSiO_3\cdot H_2O$		–1207.50	10
$CuSiO_3\cdot 2H_2O$		–1443.90	10
Associated Dissolved Species			
H^+	0		
$O_{2(g)}$	0		
$O_{2(aq)}$	3.92	16.40	17
OH^-	–37.604	–157.34	17
$H_2CO_3^*$	–148.95	–623.21	17
HCO_3^{2-}	–140.28	–586.93	17
CO_3^{2-}	–126.19	–527.98	17
$CO_2(g)$	–94.257	–394.37	17
H_3PO_4	–273.08	–1,142.57	13
$H_2PO_4^-$	–269.80	–1,128.84	13
HPO_4^{2-}	–260.313	–1,089.15	13
H_2O	–56.690	–237.19	17
PO_4^{3-}	–243.48	–1,018.72	13
Cl^-	–31.380	–131.29	17

Table continued next page

Table 5-2 Internally consistent Gibbs free energies of formation—copper and related species at 25°C (continued)

Species	ΔG_f^o kcal/mol	ΔG_f^o kJ/mol	Reference
F^-	–67.34	–281.75	17
HSO_4^-	–180.66	–755.88	17
SO_4^{2-}	–177.95	–744.54	17
$H_4SiO_4^0$	–312.6	–1,307.92	17
NO_3^-	–26.64	–111.46	17
$HOCl^0$	–19.1	–79.91	17
OCl^-	–8.8	–36.82	17
$Cl_{2(aq)}$	1.65	6.90	17
$NH_{3(aq)}$	–6.33	–26.48	13

References

1. Taken directly from: Wagman, D.D. et al. The NBS Tables of Chemical Thermodynamic Properties. *Jour. Phys. & Chem. Ref. Data*, 1982. vol. 11.

2. Cited in: Moffett, J.W., and R.G. Zika, Oxidation Kinetics of Cu(I) in Seawater: Implications for its Existence in the Marine Environment. *Marine Chem.*, 1983. vol. 13, pp. 239–251.

3. Computed from log β values given in: Fritz, J.J., Chloride Complexes of CuCl in Aqueous Solution. *Jour. Phys. Chem.*, 1980. vol. 84, pp. 2241–2246.

4. Computed from conditional stability constants for I = 2 mol/L tabulated on p. 301 of: Ringbom, A., *Complexation in Analytical Chemistry*. 1963, New York: Wiley-Interscience, using ΔG_f^o values from this table. Although not rigorously correct, there should be little dependence of log β on ionic strength for these complexes because there is no change in charge. Hence, these should provide reasonable estimates, given other data uncertainties and difficulties in correcting to I = 0 from I = 2 mol/L.

5. Computed from log β values given in: Baes, C.F., Jr., and R.E. Mesmer, *The Hydrolysis of Cations*. 1976, New York: Wiley-Interscience.

6. Computed from log β values given in: Paulson, A.J., and D.R. Kester, Copper (II) Ion Hydrolysis in Aqueous Solution. *Jour. Solution Chem.*, 1980. vol. 9(4), pp. 269–277.

7. Computed from log β values given in: Vuceta, J., and J.J. Morgan, *Hydrolysis of Cu(II). Limnol. & Oceanog.*, 1977. vol. 22, pp. 742–746.

8. Computed from log β values given in: Martell, A.E., and R.M. Smith, *Critical Stability Constants*. Vol. 5: First Supplement. 1980, New York: Plenum Press.

9. Computed from log β values given in: Byrne, R.H., and W.L. Miller, Copper (II) Carbonate Complexation in Seawater. *Geochim. Cosmochim. Acta*, 1985. vol. 49, pp. 1837–1844.

10. Either the only value tabulated or the average of values tabulated in: Woods, T.L., and R.M. Garrels, *Thermodynamic Values at Low Temperature for Natural Inorganic Materials: An Uncritical Summary*. 1987, New York: Oxford University Press.

11. Experimental log β data determined at the ionic strength of seawater of: Symes, J.L., and D.R. Kester, Copper(II) Interaction with Carbonate Species Based on Malachite Solubility in Perchlorate Medium at the Ionic Strength of Seawater. *Marine Chem.*, 1985. vol. 16, pp. 189–211, extrapolated to infinite dilution (I = 0) using activity coefficient data of Byrne and Miller [9] for seawater, and estimating γ for the complex as 0.65.

12. Derived from data given in: Lindsay, W.L., *Chemical Equilibria in Soils*. 1979, New York: John Wiley & Sons, p. 449, through accepting log β values, log K values, or computed ΔG_f^o and rederiving ΔG_f^o by substituting values for other reactant or product species from this table.

Table continued next page

Table 5-2 Internally consistent Gibbs free energies of formation—copper and related species at 25°C (continued)

13. Computed from log β values given in: Smith, R.M., and A.E. Martell, *Critical Stability Constants*. Vol. 4, Inorganic Ligands. 1976, New York: Plenum Press, using the Davies equation when necessary to extrapolate from values determined at $I \leq 0.5$ mol/L to $I = 0$.

14. Computed from log K or log β values given in: Schindler, P.W., Heterogeneous Equilibria Involving Oxides, Hydroxides, Carbonates and Hydroxide Carbonates, In *Equilibrium Concepts in Natural Water Systems*. 1967, American Chemical Society: Washington, DC, after correcting to $I = 0$ by the Davies equation.

15. Computed from log K data given by: de Zoubov, N., C. Vanleugenhaghe, and M. Pourbaix, Copper. In *Atlas of Electrochemical Equilibria in Aqueous Solutions*. 1974, National Association of Corrosion Engineers: Houston, TX, pp. 384–392.

16. Computed from log K data in: Symes, J.L., and D.R. Kester, Thermodynamic Stability Studies of the Basic Copper Carbonate Mineral, Malachite. *Geochim. Cosmochim. Acta*, 1984. vol. 48, pp. 2219–2229.

17. From the table of ΔG_f^o values in: Schock, M.R., and I. Wagner, *The Corrosion and Solubility of Lead in Drinking Water. In Internal Corrosion of Water Distribution Systems*, 1985, AWWA Research Foundation/DVGW (German Gas and Waterworks Association) Forschungsstelle: Denver, CO.

cuprous and cupric forms, usually including some scale-forming precipitates. The oxidation is supported by DO, residual chlorine, and possibly by other electron acceptors, but is not thermodynamically possible with reduction of water to hydrogen.

The classic view of electrochemical corrosion involves separation of anodic sites from cathodic sites, with transport of electrons through the metal and transport of ionic charges through the solution. If such a separation is stable, a pit can result (Figure 1-9). There are many mechanisms invoked to establish stable anodic pits in complex metallic systems, such as found with copper plumbing. These include the presence of different metals; deposits on or disruptions of the copper surface; heterogeneity of the copper crystal structure; differences in water composition, especially oxygen concentration; and velocity differences affecting the accumulation of scales. With copper and some other metals, corrosion may be nearly uniform, occurring over large parts of the metal surface at relatively uniform rates. This may result from the random, dynamic movement of anodic and cathodic sites over the surface of a homogeneous metal surface.

However, with copper a considerable amount of evidence supports corrosion mechanisms that do not involve electron transport through the metal. Copper immersed in an oxygenated, aqueous solution rapidly corrodes to form a compact, adherent film of cuprous oxide (cuprite) in good electrical contact with the underlying metal. This cuprite film layer may function as an electrical semiconductor with migration of lattice defects (Cu^{2+} or electron holes) that support oxidation and reduction reactions at the film boundaries. In uniform corrosion (Ives and Rawson 1962a, 1962b, 1962c, 1962d), copper is oxidized, forming a cuprite adherent layer at the metal surface, while oxygen is reduced at the exterior of the adherent cuprite film (Figure 5-2). Further oxidation of cuprite and accumulation of cupric precipitates may occur in an overlying porous film layer. In pitting corrosion of copper (Lucey 1967), a break in the cuprite layer allows diffusion of cuprous ions away from the surface (Figure 5-3). Lattice defects in the cuprite layer support oxidation of Cu^+ to Cu^{2+}, and corrosion of copper occurs by reaction of Cu^{2+} with Cu(m) (Eq 5-5). The inner surface of the cuprite layer and the metal become separated as the pit deepens. Oxidized corrosion products accumulate over the pit, and copper oxidation is supported by reduction of oxygen.

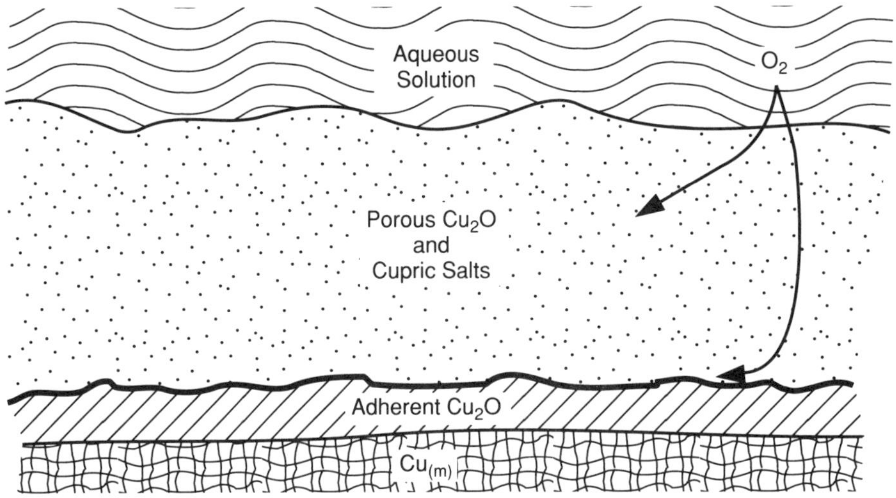

Source: After Ives and Rawson (1964c).

Figure 5-2 Duplex film model of copper corrosion

These mechanisms involve transfer of electrons away from the metal surface in reactions involving an electronically conducting cuprite layer. The site of copper oxidation is underneath the site of the oxygen reduction. They do not involve movement of electrons through the metal from an oxidation site (anode) to a reduction site (cathode).

Copper Scale-Forming Solids

The identity of insoluble scales that form on copper is strongly dependent on water chemistry, and the type of scale that forms will influence the extent of protection offered copper and the magnitude of the associated corrosion by-product release. Solids that might form in systems containing chloride, sulfate, nitrate, and bicarbonate are listed in Table 5-3, along with corresponding data related to their solubility and enthalpy (Edwards, Ferguson, and Reiber 1994). Solids that have been observed in copper pitting corrosion are also noted, using Edwards' categorization of cold-water (I), hot-water (II), and soft-water (III) pitting.

Cu(I) Chemistry

Carefully deaerated water does not corrode copper. With the addition of electron acceptors to the solution, Cu_2O (cuprite), or CuOH, or both may form. There are considerable discrepancies in the literature on the solubility constants for these two solids, and the differences are major enough to cause prediction uncertainties for cuprous solubility of roughly 10 to 100. Additionally, the conditions under which protective oxide layers form are poorly known (de Zoubov, Vanleugenhaghe, and Pourbaix 1974). The transformation from cuprous hydroxide to cuprous oxide is likely to be affected by the nature and concentrations of ions in solution and hydrodynamic factors. Some research has indicated that Cu_2O may also be formed below the surface under oxidizing conditions through rapid $Cu(OH)_2$ dissolution, followed by disproportionation and redeposition of Cu_2O (Shoesmith and Lee 1977). This is most favored in fractures and crevices of the surface films and helps explain the formation

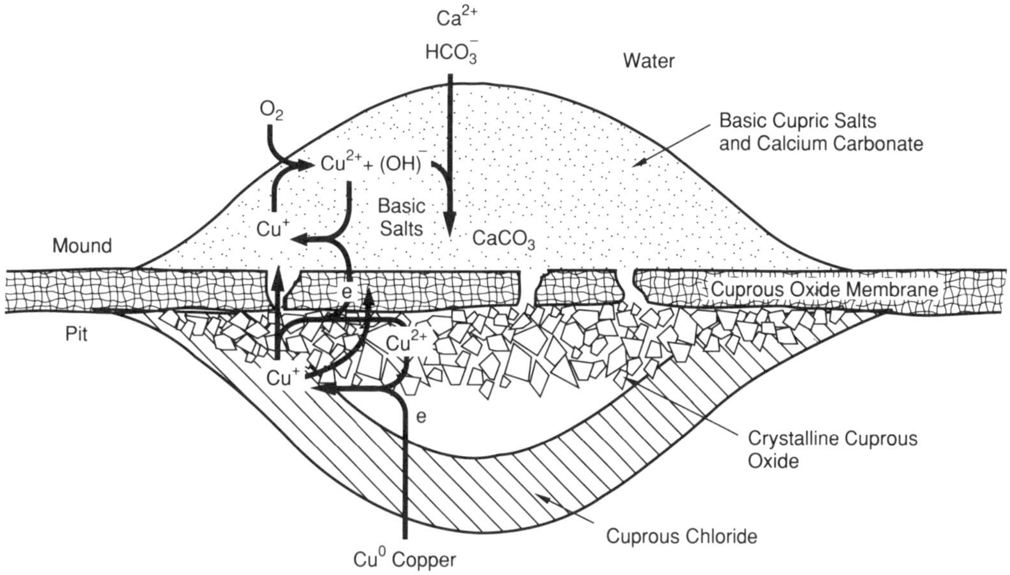

Source: Lucey (1967).

Figure 5-3 Pitting corrosion of copper

of layered copper corrosion deposits that have cupric solids on the surface. Le Gal La Salle et al. (1992) have also done electrochemical studies that indicate a cyclic transformation of rapidly formed Cu^{+} and Cu^{2+} cations, a compact Cu_2O layer, and then the reoxidation and formation of a porous external layer of CuO or $Cu(OH)_2$.

Cuprous ion forms very weak hydrolysis species (i.e., $CuOH^0$), if any. No species of this type is generally reported in relevant reviews of copper chemistry (de Zoubov, Vanleugenhaghe, and Pourbaix 1974; Baes and Mesmer 1976). The formation of bicarbonate and carbonate complexes $CuHCO_3^0$ and $CuCO_3^-$, respectively) have been hypothesized to explain some trends in oxidation rates by dissolved oxygen, but no thermodynamic values are available (Millero, Izaguirre, and Sharma 1987). The aqueous chemistry in drinking water is dominated by the formation of several stabilizing complexes in drinking water besides the carbonate complexes, notably by NH_3 and Cl^-.

Cuprous ammine complexes (see Table 5-2) can be formed either directly or by reduction of cupric ammine complexes by a route such as the one indicated in Eq 5-6 (Cotton and Wilkinson 1988).

$$Cu(m) + Cu(NH_3)_4^{2+} \rightarrow 2Cu(NH_3)_2^{+} \qquad (5\text{-}6)$$

Also, complexes of Cu(I) are generally formed by direct interaction of ligands with Cu(I) halides, and reduction of Cu(II) compounds, or reduction of Cu^{2+} in presence of the ligand (Cotton and Wilkinson 1988). In the presence of ammonium salts, the domain of corrosion of copper expands considerably (de Zoubov, Vanleugenhaghe, and Pourbaix 1974). Severe attack of copper pipe attributed to the presence of high concentrations of ammonia in the soldering flux has also been noted (Akkaya and Ambrose 1987).

Table 5-3 Copper solids that might form in water distribution systems, including their solubility and enthalpy of formation

Solid Foundation	Solid Name	Color	log (K)	ΔH^o, *kcal/m*	Pitting Type
$2Cu^+ + H_2O - 2H^+ \rightarrow Cu_2O$	Cuprite	Yellow, red, brown	1.55	–6.25	I, II, III
$Cu^+ + H_2O - H^+ \rightarrow CuOH$	Cuprous hydroxide		–0.91		
$Cu^+ + Cl^- \rightarrow CuCl$	Nantokite (cuprous chloride)	White, grey	6.76	–9.98	I
$Cu^+ + CO_3^{-2} \rightarrow Cu_2CO_3$	Cuprous carbonate	Yellow	—	—	
$2Cu^+ + SO_4^{-2} \rightarrow Cu_2SO_4$	Cuprous sulfate	White–grey	1.95	4.56	
$2Cu^+ + HS^- - H^+ \rightarrow Cu_2S$	Chalocite	Black, grey, blue	34.62	–49.4	
$4Cu^{+2} - 6H^+ + 6H_2O + SO_4^{-2} \rightarrow Cu_4(OH)_6(SO_4)$	Bronchantite	Light blue	–15.3		I, II, III
$4Cu^{+2} - 6H^+ + 7H_2O + SO_4^{-2} \rightarrow Cu_4(OH)_6(SO_4) \cdot H_2O$	Langite	Light blue	–16.8	39.61	
$3Cu^{+2} - 4H^+ 4H_2O + SO_4^{-2} \rightarrow Cu_3OH_4SO_4$	Antlerite	Blue–green	–8.29		
$2Cu^{+2} - 2H^+ + H_2O + SO_4^{-2} \rightarrow CuO \cdot CuSO_4$	Dolerophane		–11.5	35.58	
$2Cu^{+2} - 2H^+ + 2H_2O \rightarrow Cu(OH)_2$	Cupric hydroxide	Light blue/ blue–green	–8.64	15.25	II
$2Cu^{+2} - 2H^+ + H_2O \rightarrow Cu(O)$	Tenorite	Black	–7.62	15.24	II
$2Cu^{+2} - 2H^+ + 2H_2O + CO_3^{-2} \rightarrow CuCO_3 \cdot Cu(OH)_2$	Malachite	Blue–green	5.48	15.61	I, II, III
$3Cu^{+2} - 2H^+ + 2H_2O + 2CO_3^{-2} \rightarrow 2CuCO_3 \cdot Cu(OH)_2$	Azurite	Blue–green	16.92	23.77	
$2Cu^{+2} - 3H^+ + 3H_2O + Cl^- \rightarrow Cu_2OH_3Cl$	Atacamite		–7.34	18.69	II
$2Cu^{+2} - 3H^+ + 3H_2O + NO_3^- \rightarrow Cu_2OH_3NO_3$	Basic copper nitrate		–9.24	17.35	
Other Solids of Potential Interest in Distribution Systems					
$3Cu^{+2} + 2PO_4^{-3} \rightarrow Cu_3(PO_4)_2$	Cupric phosphate	Blue	36.85		
$H_4SiO_4 + CU^{+2} - 2H^+ \rightarrow CuH_2SiO_4$	Dioptase		–6.5	8.96	II, III
$Ca^{+2} + CO_3^{-2} \rightarrow CaCO_3$	Calcite	White	8.475	2.59	I

Source: Edwards, Ferguson, and Reiber (1994).

Cuprous chloride complexes (see Table 5-2) are not as strong as the ammine complexes, but they are significant in many cases because chloride concentrations in most drinking water are many times those of ammonia. They have been shown to play an important role in retarding the rate of cuprous ion oxidation in natural waters, particularly at neutral to alkaline pH values (Millero, Izaguirre, and Sharma 1987; Millero 1990a, 1990b; Eary and Schramke 1990).

Figures 5-4 and 5-5 show the effect of chloride and ammonia complexation on Cu(I) solubility at 25°C. For these calculations, CuOH was assumed to be the solubility-controlling solid. Cu_2O would be 60- to 100-fold less soluble, but it would not affect the relative positions of the solubility curves. Clearly, when Cu(I) is present in corrosion products, levels of chloride and ammonia found in many water supplies could significantly increase soluble copper concentrations. Some research has also indicated that the presence of organic ligands in solution may tend to complex Cu^+, thus increasing the dissolution rate of copper metal (Brown and Fernando 1991). Whether or not these observations apply to the natural organic matter that is present in drinking water remains to be seen.

Cu(II) Chemistry

Considerable research has been devoted to the study of the aqueous speciation of copper in natural water and seawater (Mann and Deutscher 1977; Lindsay 1979; Paulson and Kester 1980; Symes and Kester 1984; Byrne and Miller 1985; Byrne, Kump, and Cantrell 1988). The selection of the aqueous species to be incorporated follows the selections of Byrne and Miller (1985) and Stiff (1971). Initially, these calculations were performed without the inclusion of the $CuCO_3(OH)_2^{2-}$ complex. However, exploratory calculations showed that this species may have an important impact on copper solubility at pH levels found in many lime-softened or

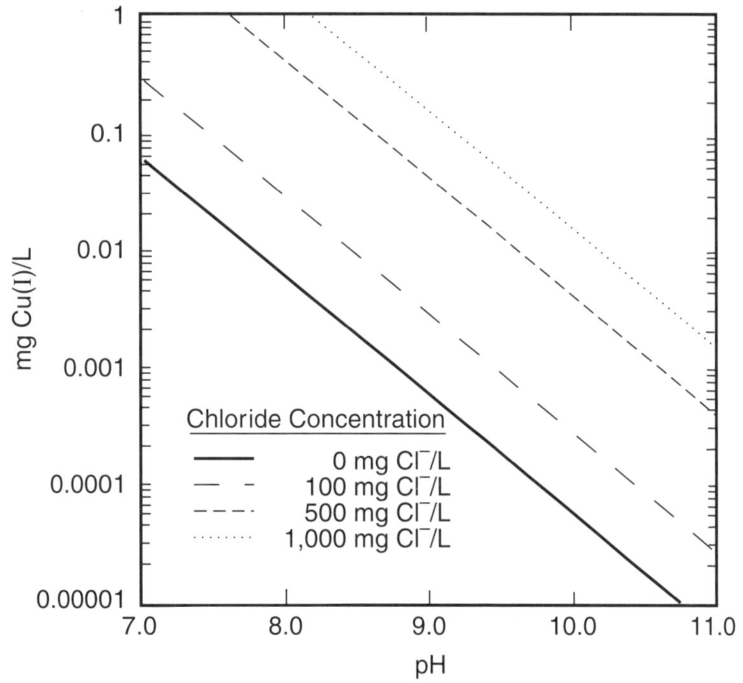

Figure 5-4 Effect of chloride complexation on CuOH solubility; I = 0.02, temperature = 25°C

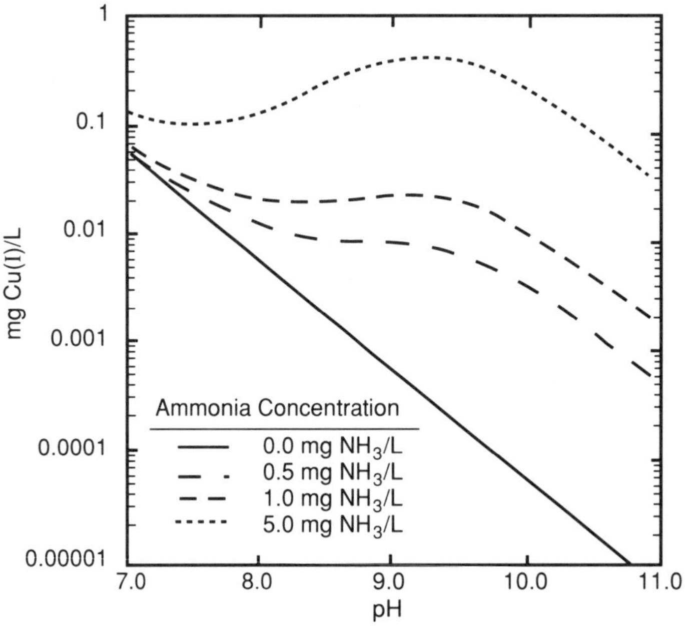

Figure 5-5 Effect of ammonia complexation on CuOH solubility; I = 0.01, temperature = 25°C

other high-pH water supplies and also in water with lower pH values, and it should generally be included for completeness.

Major inconsistencies exist, particularly in the hydrolysis constant for $Cu(OH)_2^0$, with values for log K ranging from 13.7 to "less than 17.3" for the reaction:

$$Cu^{2+} + 2H_2O = Cu(OH)_2^0 + 2H^+ \qquad (5\text{-}7)$$

There is also considerable spread in the reported constants for $Cu(CO_3)_2^{2-}$ and $CuHCO_3^+$.

Figures 5-6 and 5-7 show the predicted aqueous speciation for two concentrations of dissolved inorganic carbon (DIC), assuming the formation of $Cu_2(OH)_2CO_3$ (malachite) and $Cu(OH)_2$ (cupric hydroxide) solids. In the pH and carbonate concentration ranges of most drinking water, complexes with carbonate species are the most important ones. $CuCO_3^0$ and $Cu(CO_3)_2^{2-}$ are predominant at alkaline pH values. At the higher DIC (Figure 5-7) at a pH of approximately 9.5, $CuCO_3(OH)_2^{2-}$ becomes important in enhancing the solubility of $Cu(OH)_2$. While the general effect of complex formation on solubility is shown with malachite and cupric hydroxide, these solids may form very slowly over time periods much longer than normal stagnation. If metastable solids occur, the copper solubility would be increased at all pH values. The prediction of increased solubility by waters of increasing DIC is supported by the observations of many research and field studies (Tronstad and Veimo 1940; Shalaby, Al-Kharafi, and Said 1990). In tap water monitoring in the United States, a strong correlation between alkalinity and copper in stagnant tap water samples has been reported with copper concentration increasing as alkalinities increase above 75 mg/L as $CaCO_3$ (Montgomery Watson, EES, and Karalekas 1993).

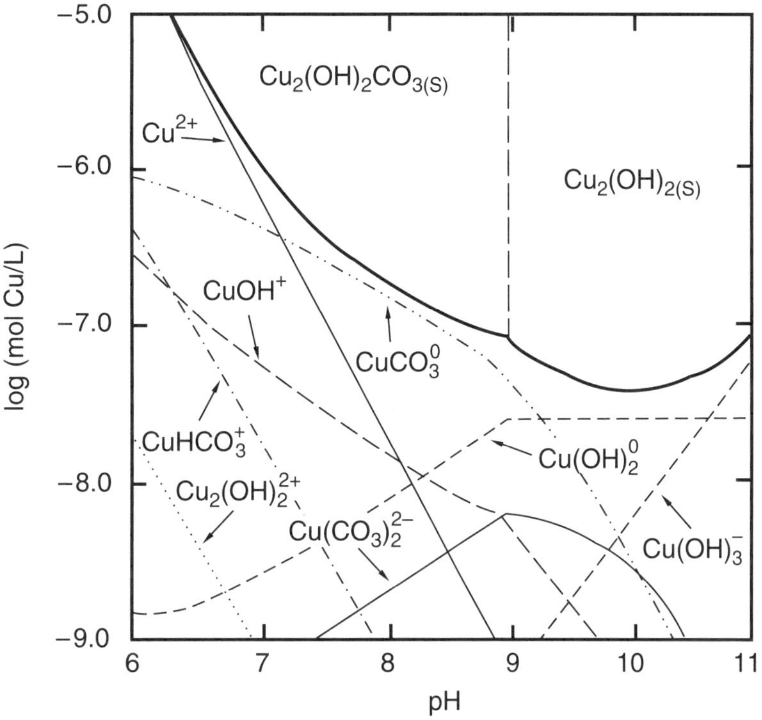

Figure 5-6 Cu(II) speciation for DIC = 4.8 mg C/L, I = 0.005, temperature = 25°C

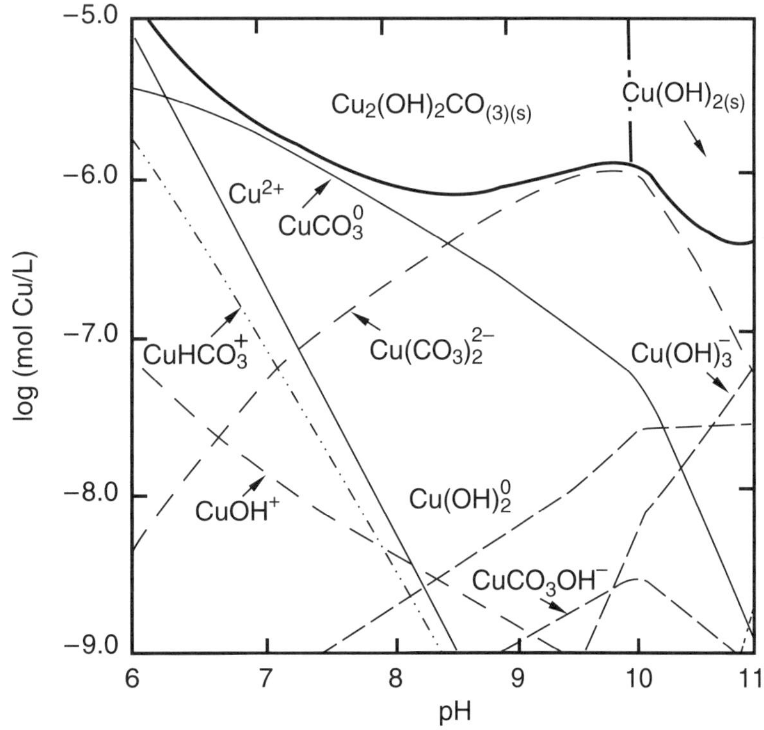

Figure 5-7 Cu(II) speciation for DIC = 96 mg C/L, including the $Cu(OH)_2CO_3^{2-}$ complex

Several investigators have looked at interactions between Cu^{2+} and natural organic matter (NOM). NOM is diverse, mainly consisting of humic or fulvic substances. In chemical equilibrium models, it is generally characterized empirically by statistical models that describe binding of metals by the formation of one or several theoretical ligands under a specific set of experimental conditions (such as pH and ionic strength). These are numerical constructs that cannot always be used for comparison of data for different water supplies. Nonetheless, some features of interaction with copper can be examined.

Some studies suggest that while copper can form complexes with NOM ligands at concentrations typical of drinking waters, copper speciation is more likely dominated by hydroxide or carbonate complexes (Cabaniss and Shuman 1988a, 1988b; Holm and Curtiss 1990). Cu^{2+} ion will also bind with adsorbed organic material containing appropriate functional groups. The binding with adsorbed organic matter seems to be stronger than direct binding with surface sites on several materials tested (Davis 1984). Copper present as an organic complex may bind preferentially with adsorbed organic material (Hirose 1990), and adsorbed organic material may mobilize colloidal copper corrosion products (Korshin et al. 1994). The significance of these opposite tendencies of NOM on copper concentrations in standing water is not known.

The potential impact of relatively high concentrations of sulfate, orthophosphate, ammonia, and chloride on Cu(II) equilibrium solubility of $Cu_2CO_3(OH)_2$ and $Cu(OH)_2$ have also been computed (Figure 5-8) (low DIC). At low DIC levels, there is marginal reduction of copper solubility at pH below 7 by the formation of $Cu_3(PO_4)_2 \cdot 2H_2O$. Some enhancement of solubility is produced by the formation of the $CuSO_4^0$ aqueous complex. Ammonia (5 mg NH_3/L) and chloride (250 mg/L) have

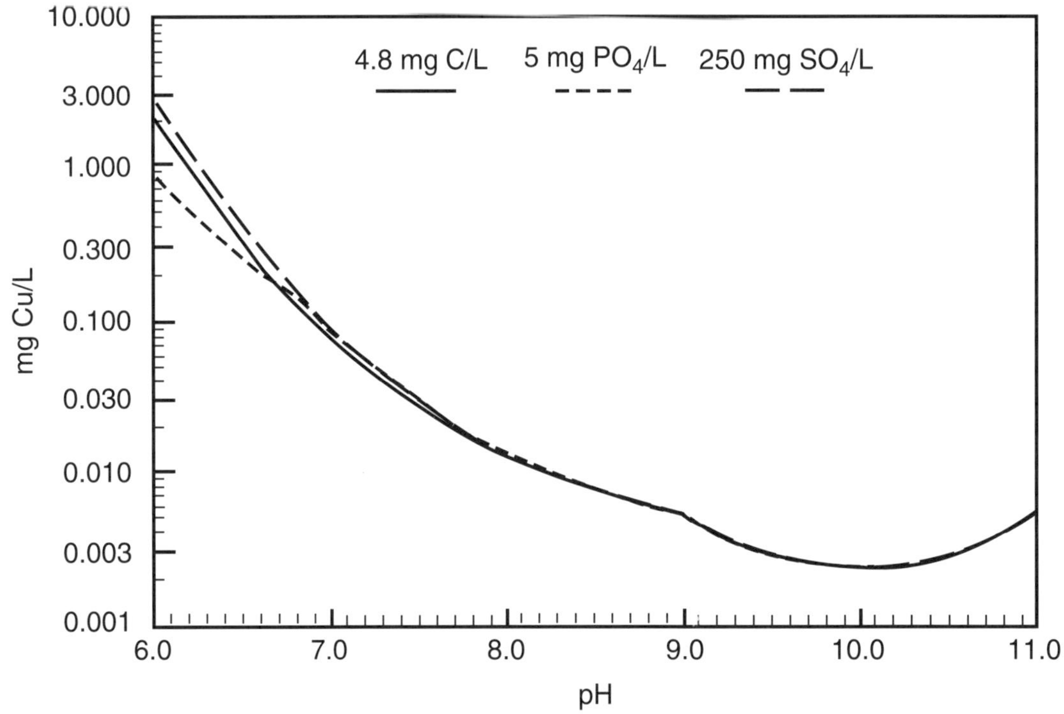

Figure 5-8 Effects of phosphate and sulfate on Cu(II) solubility for DIC = 4.8 mg C/L, I = 0.005, temperature = 25°C

insignificant effects on solubility. Chloride complexes of Cu(II) are weak and have little direct impact on the solubility of the oxidized copper form. Similarly, the sulfate and orthophosphate complexes are not very strong, and little change in copper solubility is predicted. Little information on the solubility of copper orthophosphate solids is available. From tabulations of ΔG_f° data (see Table 5-2), both $Cu_3(PO_4)_2$ and $Cu_3(PO_4)_2 \cdot 2H_2O$ are too soluble to make a major impact on copper solubility at orthophosphate dosages normally applied to drinking water (0–5 mg PO_4/L). Two studies (Reiber 1989; Benjamin et al. 1990) have shown reduction in copper corrosion rates by the addition of orthophosphate. Another study reported reductions in copper concentrations, but the orthophosphate level was approximately 950 mg PO_4 /L (Drogowska et al. 1992).

Ammonia complexation constants are much higher than those for cupric ion with either sulfate or chloride. However, it also has little predicted effect on overall Cu(II) solubility. Enhancement of copper solubility during periods of chloramination with excess ammonia present, followed by a decrease in solubility with breakpoint chlorination, was observed (EES 1990). However, others have seen no difference in copper corrosion rates or copper release when free chlorine or combined chlorine were used in parallel tests (Reiber 1993). Calculations indicate that cuprous ion ammine complexes and Cu(I) complexes may be involved in any effects of ammonia on copper corrosion. Carbonate and hydroxide complexation dominates the solubility of copper, so similar calculations at high DIC levels show even less effect from the weaker ligands.

While some species, such as chloride, NOM, and sulfate, may not directly enhance Cu(II) solubility, they may indirectly influence it by modifying the identity, crystal structure, morphology, and compactness of the solids formed at the pipe surface (Campbell 1954; Shalaby et al. 1990; Edwards, Ferguson, and Reiber 1994).

In drinking waters, the solubility-controlling thermodynamically stable solid is most likely to be CuO (tenorite), $Cu(OH)_2$, or $Cu_2(OH)_2CO_3$ (malachite). Although basic copper chlorides and sulfates such as bronchantite and atacamite (see Table 5-3) are found in corrosion scales in pitting attack (Edwards, Ferguson, and Reiber 1994), they are generally not as stable as the oxide, hydroxide, or basic carbonate solids in the pH range above approximately 6 under almost all drinking water conditions.

The quantitative prediction of copper solubility in drinking water, made much more important by recent US regulations, relies heavily on the properties of cupric oxide, hydroxide, and basic carbonate solids. Because of considerable uncertainty in the values for the solubility constants of all of these minerals, the tendency for prolonged existence of thermodynamically metastable phases such as $Cu(OH)_2$, the possibly slow formation and dissolution rates, and the effects of particle size (Schindler et al. 1965; Schindler 1967), quantitation of copper levels becomes highly problematic. Furthermore, particulate copper often is a significant fraction of copper in drinking water supplies.

For example, Schindler and co-workers (1965, 1967) found the log equilibrium constant for CuO (Eq 5-8) to vary from 8.92 to 9.13 at I = 0.2 as the molar surface area increased.

$$CuO_{(s)} + 2H^+ \rightarrow Cu^{2+} + H_2O \tag{5-8}$$

The reported values (log K) for the solubility constant of $Cu_2(OH)_2CO_3$ (malachite) range from –5.16 to –6.20. The most recent experimental work by Symes and Kester (1984) leads to a value of log K = –5.48, which has been used for most of the modeling reported here. Little compelling evidence currently exists for the selection of this value over that of Schindler, Reinert, and Gamsjäger (1968) of –5.18. Thus, the uncertainty in the computed copper concentration in equilibrium with malachite is at least a factor of 2, aside from uncertainties in aqueous speciation.

COPPER CORROSION

Uniform Corrosion and Copper Release

Under some circumstances, copper tubing corrodes uniformly, releasing copper compounds into solution as well as accumulating copper-containing scale. In relative terms, copper is much more resistant to generalized corrosion than the other metals typically found in potable water systems. The work of Bengough and Wormwell as cited in Evans (1982) showed that in aerated 0.5*M* sodium chloride solution, the scale of the relative corrosion rates after 100 days of exposure was Cu < Fe < Zn.

Corrosion increases rapidly as the pH drops below 6, and at low pH values corrosion is almost always of a generalized nature. Excess metal release is commonly observed. Often little scale deposit occurs because of the high solubility of copper minerals at slightly acidic pH values. General corrosion seldom causes rapid failure of tubing but can cause significant thinning and reduced service life. "Green water" from dislodged copper precipitates is a common manifestation of high corrosion rates; however, soluble copper concentrations exceeding the 1.3-mg/L USEPA action level may also occur without evidence of green water. Related to green water is green staining; even a few parts per million of copper in water can react with soap scums and cause green staining of plumbing fixtures. Staining has largely disappeared because detergents have replaced soaps.

At higher pH values, copper corrosion problems are almost always associated with nonuniform corrosion processes. If corrosion occurs uniformly at pH values above 7, it usually occurs slowly; the uniformity arises from an oxide film and slow copper oxidation. However, the effect of pH on copper corrosion is very marked and often is described as generalized corrosion if the problem is associated with low pH values.

Shull and Becker (1960) investigated the effects of pH on copper release in a typical service application. Figure 5-9 gives their results and indicates low release at pH above 7. Similar work in Germany (Meyer 1981) found the same general effect over a wider range of water compositions (Figure 5-10). In practice, concentrations of up to 5 mg/L have been observed in particularly susceptible environments. Experimental studies have shown that soft potable waters develop rather high concentrations of copper when the system is not operating and that repeated flushing of the system may be required to bring the copper level down to a palatable level. The problem is less apparent in waters with moderate alkalinity in which there is sufficient bicarbonate to assist in formation of a protection film. At high alkalinities, carbonate forms predominant complexes with Cu(II) and results in higher copper release.

Calcium carbonate scale formation has been found to protect copper and reduce copper release under favorable conditions. However, if no scale forms or if it is discontinuous, generalized corrosion may occur. The influence of other major ions on the corrosion rate is usually subtle and is not completely understood. Generally, it may be stated that cations (calcium, magnesium, sodium, and potassium) exert little effect on the rate of corrosion. However, anions (chloride, sulfate, bicarbonate) do exert some influence on the rate of corrosion and on pitting (Fischer, Fußinger, and

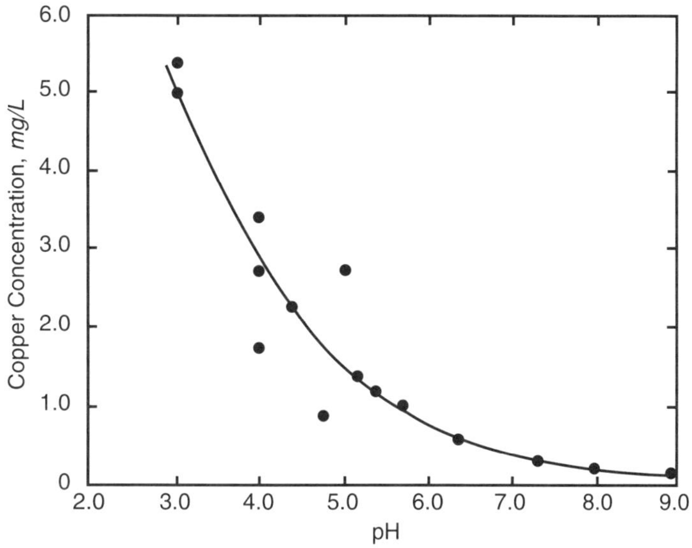

Conditions: Water at the desired pH was allowed to flow through 60 ft (18.5 m) of new 0.75-in. (1.9-cm) copper tubing at a rate of 0.5 gpm (2 L/min) for 1 hour and then stopped for 16 hours to simulate overnight conditions.

Figure 5-9 Effect of pH on the release of copper into solution

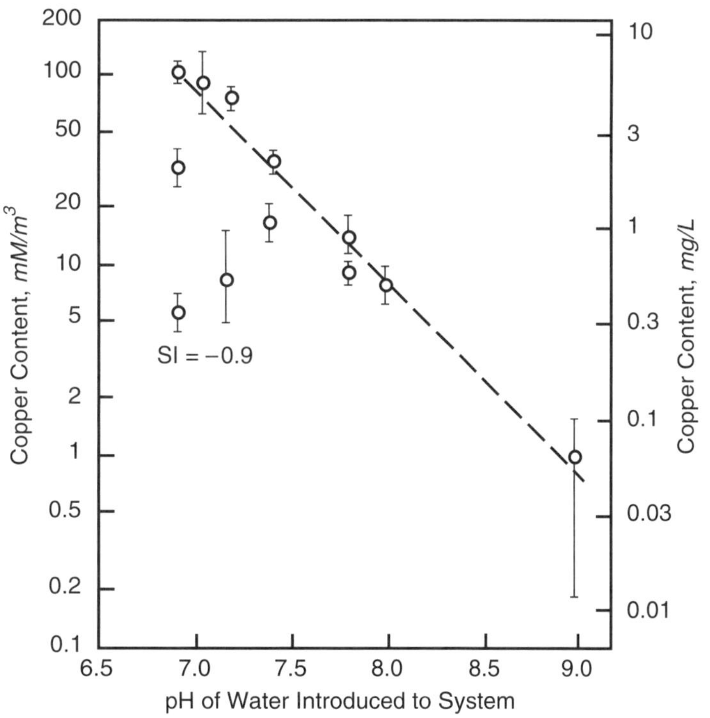

NOTE: Average copper content in different drinking waters after 8 hours' stagnation. The values were measured after a service period of eight months.

Figure 5-10 Effect of pH on copper release in different waters

Hänßel 1992). Sulfate is a strong corrosion catalyst implicated in pitting corrosion, while chloride is much less corrosive. Bicarbonate tends to increase copper solubility and copper release in standing water; however, in low-alkalinity waters, alkalinity additions buffer the pH and may help form passivating scales. It is difficult to quantify these observations, but the chloride, sulfate, and bicarbonate concentrations are significant in corrosion and copper release.

Generalized corrosion and green-water complaints typically arise when soft waters of slightly acid pH (6 to 7) are involved. The phenomenon clearly occurs below pH 6, and potable water should not be served at acid pH values. Waters in the pH range of 6 to 7 typically contain high levels of carbon dioxide. There is considerable disagreement whether carbon dioxide per se causes corrosion, whether pH is the cause, or whether the factors interact. In Germany and elsewhere, carbon dioxide concentration is often cited as a causative factor subject to control by removal or pH adjustment. In the United States, where much experience is with very low-alkalinity surface water, pH has been more often emphasized as the cause of high corrosion and as the parameter requiring adjustment.

Water velocity and temperature also play roles in copper corrosion. A protective scale can be eroded by the fast-flowing water, thereby leaving the underlying copper surface unprotected. In hot-water systems, increased temperature increases the rate of attack. Figure 5-11 schematically shows the combined effects of water velocity and increasing temperature on a softened water. The corrosion rates are satisfactorily low when the velocity is less than about 3 ft/s (1 m/s) but increase rapidly as velocity and temperature increase. These data show only relative effects and should not be generalized.

Nonstationary Copper Release

In household plumbing short periods of flow are usually separated by longer periods of stagnation. During the stagnation, corrosion reactions take place, leading to low oxygen concentrations near the metal. For iron, Kuch showed that corrosion continues after oxygen is depleted (see chapter 2). Instead of oxygen, ferric compounds are reduced and dissolved ferrous iron is formed. Similar reactions have been found to occur in copper tubing. Cu(II) may be reduced after oxygen is depleted. Corrosion reactions under stagnation result in time-varying patterns of copper, oxygen, and corrosion rates. These phenomena are termed nonstationary, or Kuch-type, corrosion. They may lead to high or variable metal concentrations in the standing water.

Werner (1992) studied this problem for copper installations using a well water. Experiments were conducted at pH values of 5.8, 6.5, 7.2, and 8.0 and also with 3 mg/L orthophosphate, using 0.3-in. (8-mm) copper tubing in 16-ft (5-m) lengths. Three different operating conditions were used: continuous flow; 12 hours flow and 12 hours standing; and 1 hour flow and 23 hours standing. It was found that Kuch-type corrosion needs some time to start and becomes more pronounced in pipes after several months of operation.

In varying degrees, a similar pattern was seen in nearly all of the 18 conditions tested. DO usually declined over 5–15 hours to values near zero, while dissolved copper initially increased and then decreased, as illustrated with results for pH 7.2 and 8.0 (Figure 5-12).

Dissolved copper increased during the first 5 to 10 hours to maximum values as high as 20 mg/L. These maximum values are higher at the lower pH values. The values, which decreased from about 20 mg/L at pH 5.8 to 3 mg/L at pH 8.0, also depend on the flow pattern and the age of the pipe. The high soluble copper concentrations occur

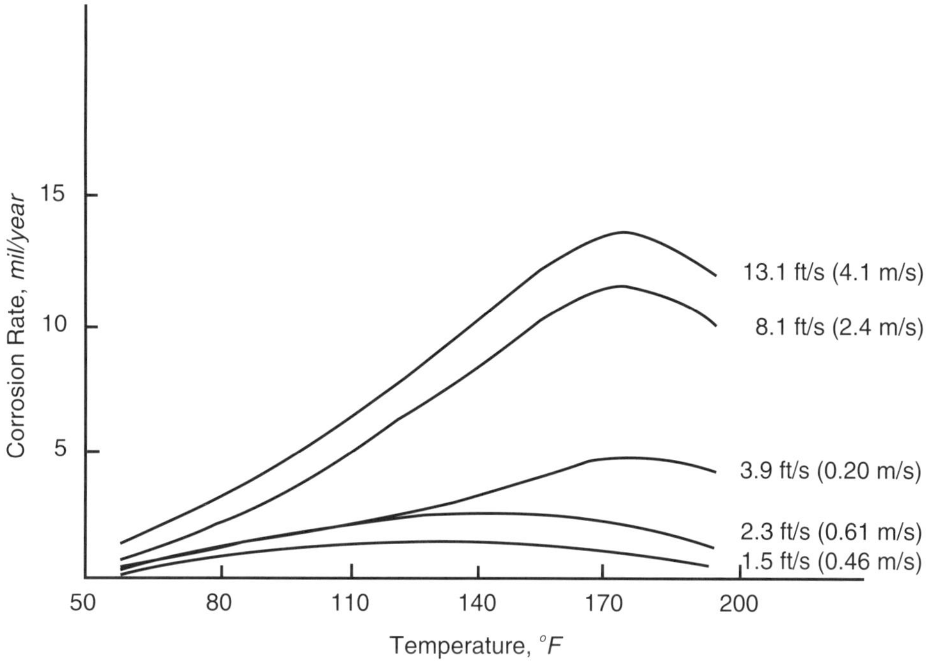

Source: Obrecht and Quill (1960).

Figure 5-11 Effect of temperature and velocity on corrosion of copper tubing by a sodium-zeolite softened water

under conditions where $CuCO_3^0$ predominates, which results in high solubilities of minerals such as malachite or metastable solids such as $Cu(OH)_2$. After about 5 hours the oxygen concentration reaches low values. From this time on, the concentration of dissolved copper decreases sometimes nearly to 0 after about 24 hours.

While the increase in copper concentration may result from corrosion and oxidation reactions, the decrease results from the reaction of the dissolved Cu(II) constituents with the metal, leading to a layer of cuprous oxide. This layer tends to slow down further cathodic Cu(II) reduction. Orthophosphate was ineffective in halting this phenomenon.

This Kuch-type corrosion of copper has only recently been discovered, but it may help explain the considerable scatter in standing water concentrations that is found in many surveys. The kinetics of corrosion and precipitation reactions may be important parameters affecting copper in standing water.

Blue- or green-water problems are often associated with high concentrations of copper in standing water. However, in parts of the service area of the East Bay Municipal Utilities District (Oakland, California), excessive release of blue, green, and occasionally red or brown precipitates developed in newly constructed homes over a period of one to two years. Extensive investigation eventually revealed that the blue-water phenomena could be treated by chlorination or heat disinfection of the affected homes.

Erosion Corrosion

Erosion corrosion is readily recognized by the characteristic roughening of the pipe interior (Figure 5-13). Very often, typical U-shaped pits that are undercut in the direction of water flow in the tube are present.

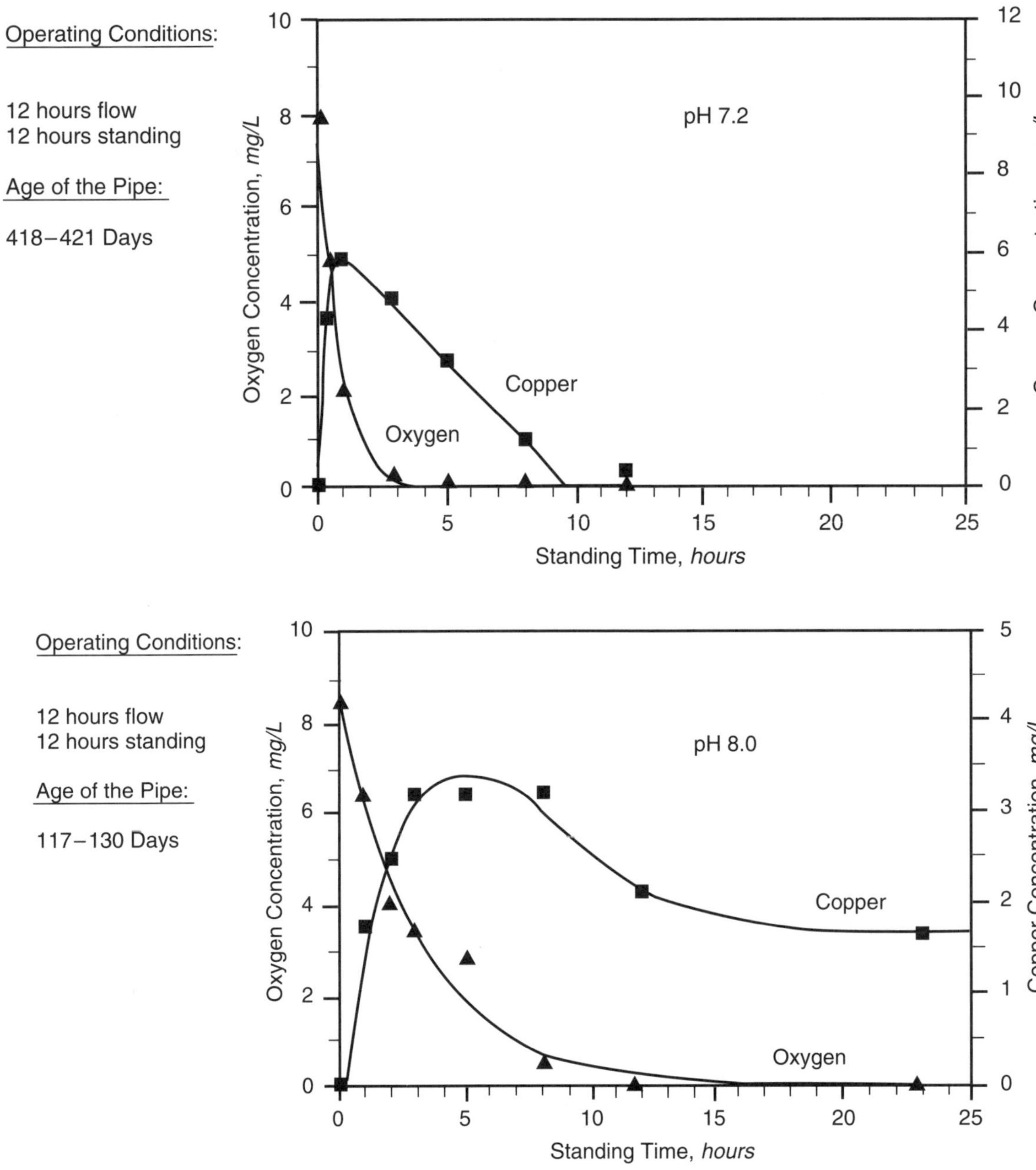

Figure 5-12 Kuch-type copper release during stagnation periods for pH 7.0 and pH 8.0 (well water, 12°C, alkalinity 5.3 meq/L, calcium 125 mg/L, magnesium 11 mg/L, sodium 22 mg/L, chloride 50 mg/L, nitrate 19 mg/L, sulfate 54 mg/L, organic carbon 1.8 mg/L)

The primary cause of impingement attack is high water velocity. The presence of entrained gases and entrained solids increases the corrosion rate and reduces the velocity threshold at which the corrosion starts. The underlying cause of the attack is the inability of copper to form protective scales under the abrasive effects of water and/or entrained material. Temperature also interacts with velocity, increasing corrosion attack, especially at temperatures exceeding 60°C (Obrecht and Quill 1960).

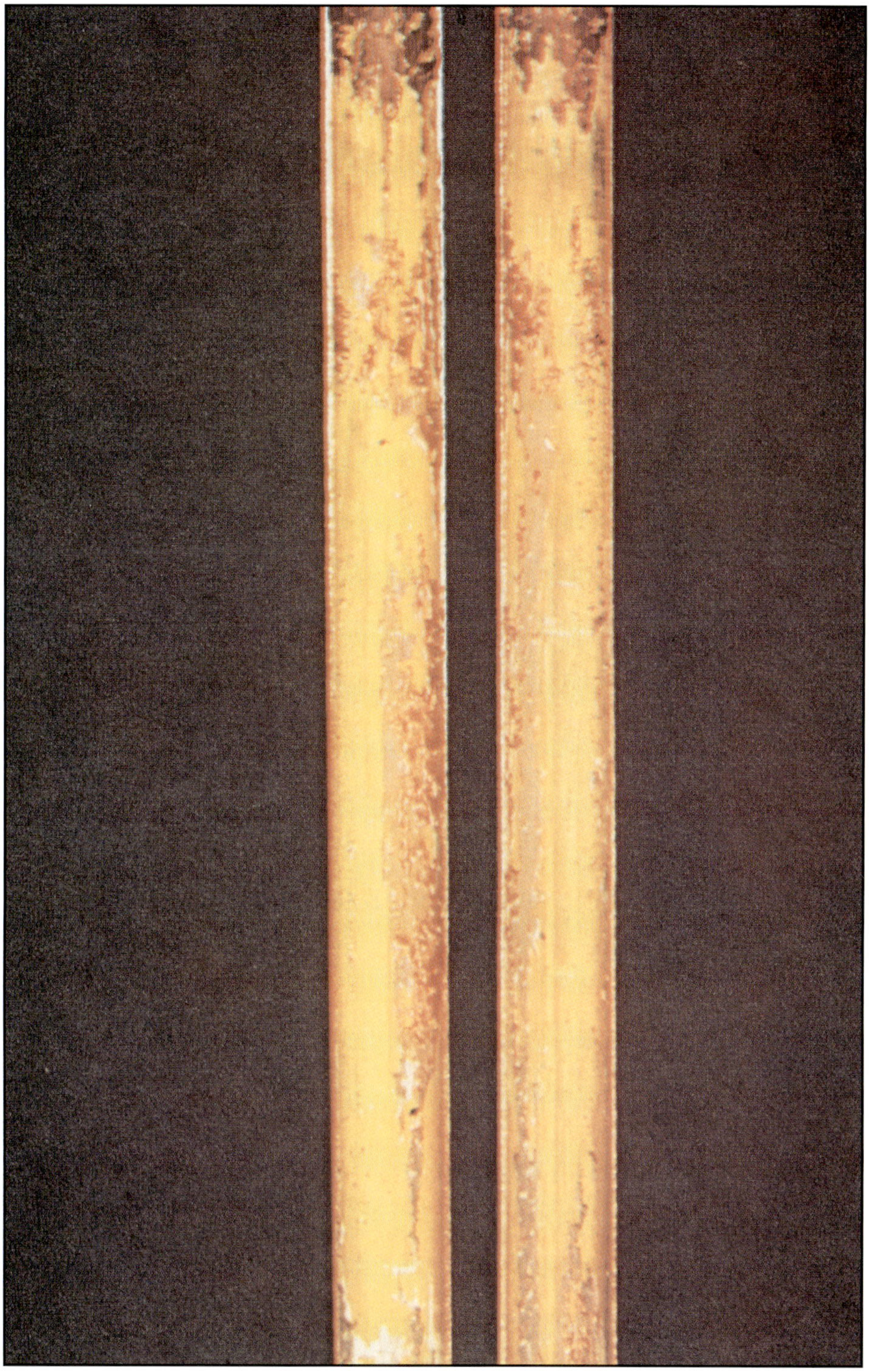

Figure 5-13 Impingement attack

The surface films are not eroded uniformly, and when openings develop they give rise to anodes. Cuprous ions form rapidly, because the small anode is surrounded by a large cathode. These ions are swept away by the impinging water; thus the copper ion concentration at the anode remains low. This, together with depolarization of the cathode by the rapid delivery of DO, enables the pit to maintain a high rate of corrosion.

Erosion corrosion is primarily the result of faulty design and workmanship that allows continuous high velocities. It is important to recognize that the calculated design velocity may be much less than an actual local velocity, which may be strongly influenced by burrs or solder beads, etc. Guidelines and codes of practice invariably refer to the calculated velocity.

National standards and codes of practice give guidance for proper design. They often include an upper limit in the range of 6 ft/s (1.8 m/s) for normal service and 15 ft/s (4.6 m/s) for lines with continuous flow or a table that considers the different modes of service. The US CDA has recommended a nominal ½-in. (12.7-mm) tube to provide service to a tap at the kitchen sink. The National Association of Corrosion Engineers has recommended a maximum design velocity of 4 ft/s (1.2 m/s) for type K copper tube.

Many of the reported cases of erosion corrosion occur in hot-water tubing where the water is continuously circulated by a pump. Under such conditions, it is recommended that the water velocity not exceed 1.5 ft/s (0.5 m/s). Although there is little information on actual velocities within failed systems, this limit seems to be safe (Landgrebe 1980) and is slightly below the values given in the international literature.

For cold water lines, the usual recommendation is to restrict the velocity to less than 6 ft/s (2 m/s). No failures have been reported with velocities of 6 ft/s (2 m/s) or less, and tests have shown that copper tubes can endure even higher velocities under favorable conditions (Sick 1972).

Pitting Corrosion (or Localized Attack)

Copper corrosion processes range from those that appear macroscopically to be uniform over broad areas of the surface to those of localized or pitting phenomena. There is much debate and confusion in the literature classifying and characterizing the pitting that is observed. A recent review (Edwards, Ferguson, and Reiber 1994) has included three categories: type I (cold water), type II (hot water), and type III (soft water), along with uniform corrosion. The characteristics for this taxonomy are summarized in Table 5-4. In the following sections, simplified designations of cold-water pitting and hot-water pitting are used. Soft-water pitting, which is less widely recognized or understood, results in broad, shallow pits and high corrosion by-product release. It is not addressed in this chapter.

Cold-water pitting. Several authors (Campbell 1968; Cruse and Pomeroy 1974; Mattson and Fredericksson 1968) have described the occurrence of cold-water pitting. Typically, the tube fails in a short period of time (in as little as a few months or as long as a few years). It occurs sporadically (i.e., in only some houses in a development, only for short periods in a given water system, with some waters but not others of apparently the same composition), and it occurs primarily with groundwaters free from organic matter.

Table 5-4 Guide to identification of copper corrosion problems, including a summary of exacerbating and ameliorating factors

Characteristic	Uniform Corrosion	Type I Pitting (Cold Water)	Type II Pitting (Hot Water)	Type III Pitting (Soft Water)
Pit shape	No pits	Deep and narrow	Narrower than type I	Wide and shallow
Problem	Blue or green water, high by-product release	Pipe failure	Pipe failure	Blue water, voluminous by-product releases, pipe blockage
Scale morphology on attacked surface	Tarnished copper surface or loose powdery scale	Underlying Cu_2O with overlying malachite, calcite, or other basic copper salts; often CuCl underlies Cu_2O	Underlying Cu_2O with overlying bronchantite, some malachite	Underlying Cu_2O with overlying bronchantite, some malachite
Water qualities	Soft waters of low pH	Hard, cold, well waters between pH 7 and 7.8; high sulfate relative to chlorides and bicarbonate	Hot waters, pH below 7.2; high sulfate relative to bicarbonate	Soft waters, pH above 8.0
Initiating factors	None noted	Stagnation early in pipe life; deposits within pipe, including dirt or carbon films, high chlorine residuals, water softeners, alum coagulation	Higher temperatures, high chlorine residuals, alum coagulation, particles	Stagnation early in pipe life, pH values above 8.0, alum coagulation
Ameliorating factors/treatments	Raise pH or increase bicarbonate	NOM/increase bicarbonate and pH	Lower temperatures, high pH values/increase bicarbonate and pH	NOM/avoid stagnation early in pipe life, add $Ca(HCO_3)_2$

Source: Edwards, Ferguson, and Reiber (1994).

The characteristics of the pitting of copper water pipes are described in several places; the German standard DIN* 50930, Teil 5, gives a good summary. The main characteristics are as follows:

- Pits penetrate the pipe wall in a relatively short time, usually between three and four years, sometimes within a few months. In typical cases, the frequency of leaks is much less after 5 years, although there are a few cases in which pits have penetrated the wall after 15 years or more.

*Deutsches Institut für Normung (German Institute of Standardization), Burggrafenstr. 6, Berlin, Germany.

- Type I pitting is observed in cold-water systems. Associated hot-water systems rarely have the problem.
- Type I pitting is usually found only in pipes carrying water from wells and rarely in pipes carrying water from untreated rivers or lakes.
- If there are multiple pits in a pipe, they often tend to lie along longitudinal lines; sometimes the tubercles over a line of small pits form a continuous longitudinal mound (see Figure 5-14).
- The interior exhibits a characteristic green malachite scale, which may contain traces of paratacamite ($Cu_2(OH)_3Cl$), atacamite ($Cu_2(OH)_3Cl$), and nantokite (CuCl) (see Figure 5-15).
- Beneath the malachite scale, the entire inner surface is covered with a smooth, shiny film of dark, reddish-brown cuprite, sometimes in amounts of up to 400 mg/dm^2. In amounts exceeding 100 mg/dm^2, the cuprite can be spalled by bending the pipe. The underside is not shiny but is crystalline. Sometimes, tubercles can be slid off the surface to reveal a continuous cuprite film with one or a few pinholes in it. Then, when the film is broken, a pit is revealed. Copper ions from the corrosion reaction have migrated out of the cavity through the pinholes. The cavities are loosely filled with crystals of cuprite or sometimes a mass of fine crystals of nantokite.

Observations of cold-water pitting strongly indicate that at least two or more factors are involved. Early literature describes many observations about assumed reasons for pitting corrosions of copper, such as the water supplied, service conditions, or the temper of the tubes affected.

"Cold-water type pitting" of copper tubing has predominantly been a cold-water phenomenon. If hot-water pipes were affected, it was often found that these pipes had remained cold for long periods or were pipes supplying remote and rarely used taps. Pitting is almost always associated with hard well waters that usually have pH values in the range of 7 to 8.2. It seems that cold-water pitting is not observed below pH 7, since more rapid, generalized corrosion will occur. Above pH 8.2 and in hot-water systems, most of these waters deposit hardness scales that are usually protective and free CO_2 is absent.

A significant observation, which was corroborated by most European investigators in the 1960s and 1970s, is that most of the failures occurred with soft-annealed tubes. The statistics of the DKI show that in 1982, 88 percent of the pitted tubes were in the soft-annealed temper (Franqué 1983). Similar results were found in laboratory tests by Devroey and Depommier (1965), which showed that soft-temper tubes are more susceptible than half-hard tubes, which in turn are more susceptible than hard-temper tubes (see Table 5-5).

North American reports do not quite agree with these observations. CDA identifies about the same number of failures in annealed and hard-drawn tubes. However, since the tightening of quality standards in Europe, the picture there has changed, and DKI reports the portion of soft-annealed tubes of all tubes affected has dropped to 42 percent in 1992. This means that the temper might play a role; however, the overriding factor had been the quality of the inner bore.

If hard-drawn tubes are affected, there are obvious differences in the distribution of pits compared to affected soft-annealed tubes. Soft-annealed tubes show pitting along the length of the tubes, whereas hard-drawn tubes are mainly affected near

Figure 5-14 Type I or cold-water pitting

Figure 5-15 Cross section of a pit

Table 5-5 Effects of temper on pitting corrosion susceptibility in tests by Visseries et Trefileries Réunies (VTR) in Belgium

		Maximum Depth of Pitting After One Year		
Temper	Number of Tubes	<0.1 *mm*	0.1–0.45 *mm*	>0.45 *mm*
Soft	13	3 (23.1%)	2 (15.4%)	8 (61.5%)
Half hard	62	41 (66%)	13 (21%)	8 (13%)
Hard	33	33 (100%)	0	0

joints or places where the tube has been annealed by the plumber in order to make it bendable.

Regardless of what temper is used, most pits and leaks are found in horizontally installed tubes. The German experience with incidence of pitting as a function of orientation shows 85 percent of leaks occurring in horizontal tubes. Since the average ratio of horizontally to vertically installed tube is estimated to be 4:3 in Germany, there is a strong association of pitting with horizontal runs.

Causes of cold-water pitting. There is general agreement that type I pitting is caused by the combined effects of several factors. Apart from the influence of water composition, these factors can be divided into three categories: those due to the fabrication of tubes, those due to faulty craftsmanship, and those due to operating conditions. The highest incidence must be ascribed to the first factor; fortunately, the situation has improved because these factors have been addressed (Eckardt, Klingler, and Sick 1990).

In order to understand the discussion that follows, it is important to note that the conditions that cause a pit to start (inception of pitting) are different from those that allow the pit to continue to develop (propagation of pitting). As a general guide, the inception of pitting seems associated with the condition of the metal before it is exposed to water, whereas propagation of the pit is a function of water quality. Also, it should be noted that current knowledge of pitting is not complete and differences of opinion still exist as to the causes of pitting inception.

Factors due to the condition of the inner bore (mainly manufacturing defects). The original systematic investigations into the inception of pitting were carried out by Campbell (1950). He noted that pipes that had failed due to pitting often contained a wrinkled film of carbon on the interior surface, and he measured the carbon layer on the inner bore. The carbon was attributed to the breakdown of drawing lubricant residues during the bright annealing process of tube manufacture. Cruse and Pomeroy (1974) attempted to confirm these quantitative data in a study of pitted copper tubes in the United States. Though carbon was found, the quantities did not correlate well with the occurrence of pitting. It appears that quantity is not important, but the presence of carbon as a film is.

The importance of carbon films in causing cold-water pipe pitting is now widely recognized. The national standards for copper tubing in the United Kingdom, the United States, and New Zealand require tubes to be "clean and free from deleterious films in the inner bore." Other countries are adopting similar standards and specifying test methods for the presence of carbon. The DVGW (German Gas and Waterworks Association) had specified limited carbon content in the inner bore in its DVGW standard GW 392 (DVGW 1992) until 1982; at that time, the additional requirement "free from films" was introduced. This measure resulted in a considerable decrease of pitting failures in Germany. Methods used to measure carbon films are described in appendix B.

As already mentioned, some hard-drawn tubes fail near brazed joints or in places where the tube has been annealed by the plumber. Brazing requires a working temperature of 700°C. If the tube contains a certain amount of lubricant residue, it can apparently develop carbonaceous films as in the industrial annealing process. Such films were detected near joints in failed hard-drawn tubes. Consequently, German and French quality specifications now limit the amount of residual lubricant (measured as carbon) on the inner surface of copper tubes to 0.2 mg/dm^2 (DVGW 1992).

This development resulted in a reduction of pitting in hard-drawn tubes (Eckardt, Klingler, and Sick 1990), but in contrast to soft-annealed tubes, failures still occurred in which no carbon films were detected. The possibility that scale as such could cause pitting corrosion was again discussed. Baukloh and co-workers (Baukloh et al. 1989) showed in field tests that a particular oxide scale was able to initiate pitting without the presence of carbon.

Factors due to faulty craftsmanship. The main factor that can be clearly defined as faulty craftsmanship is the excessive use of fluxes. Fluxes, which are corrosive by their nature, are used for soldering and brazing copper tubes. Brazing fluxes have not caused any problems up to now, so discussion can be restricted to soldering fluxes. If substantial amounts of flux are introduced into the tube bore and allowed to remain, pitting may occur in areas adjacent to the flux residues or on the edge of such residues (Figure 5-16). There are indications that the use of solder–paste (a mixture of solder and flux) has increased the number of pitting cases connected with flux contamination during the late 1970s and early 1980s.

Even the failures connected to fluxes reflect the general picture of cold-water pitting. Most of these cases are reported to have occurred with soft-annealed tubes and, in most cases, carbon films were also present. But even with carbon films present, pitting was primarily restricted to the contaminated zones.

Observations of the influence of fluxes prompted further study into the role of flux and the development of a standard test procedure for evaluating the corrosivity of fluxes to copper in the United States as well as Germany. Extensive research by the CDA (Cohen and Costas 1982) led the CDA to recommend to the American Society for Testing and Materials (ASTM) that a standard test for fluxes be adopted. This standard test is ASTM B 813-91.

Parallel to this development, a new DVGW standard GW 7 was developed and became effective in 1987. This specification restricts the amount of corrosive agents in the flux and requires the flux to be water-soluble and removable. Although the specification does not adopt US testing procedures (Cohen and Costas 1982), the requirements produce results similar to those in the United States (Franqué and Zimmerman 1988).

Faulty craftsmanship may result in dead ends or faulty joints where the water remains stagnant, a condition that increases the risk of pitting.

Influence of service conditions. Once pipes are filled with water for pressure testing, they become wet and will probably never be dried. If the system is not used, the stagnant conditions in the system are favorable for pitting. If the system is drained, the effect is more pronounced. These conditions are significant if there

Figure 5-16 Flux-induced corrosion

is a prolonged period of stagnation between the completion of the plumbing installation and occupancy of the building and use of water. Stagnation periods of any length are undesirable, and periods over three months are clearly detrimental. It should be emphasized that the stagnation does not cause pitting but accelerates it if other conditions are unfavorable. If installations must be left without normal use, the water should be run intermittently to mitigate this condition. Similar effects will be evident in systems that are used infrequently.

In general, conditions that lead to irregularities on the inner surface and that cause defects in the oxide layer are likely to induce pitting. Deposits are a principal cause of irregularities and should be removed by flushing the plumbing installation before it is used or should be prevented by placing filters at the point where water enters the installation. Deposits are seldom the sole cause of pitting; deposits may explain why horizontally installed tubes are more affected by cold-water pitting than vertical ones.

However, this fact might also be explained by Lucey's (1967) observation that the settling out of corrosion products can have adverse effects on copper surfaces. No matter which explanation one prefers, both factors are removed once water is flowing in the tubes. A relatively high velocity of water apparently yields an improved tube life with regard to cold-water pitting.

Hot-water pitting. Type II pitting occurs only in certain soft water areas and rarely occurs at a water temperature less than 60°C (140°F). It is characterized by deep pits of small cross section, which contain very hard crystalline cuprous oxide and are capped by small black or greenish-black mounds of cuprous oxide and basic copper sulfate. The surface between the pits is usually covered by a nearly black layer of silt deposited by the water. Some hot soft waters that contain small amounts of manganese can produce a special kind of type II pitting in which the pits are slightly larger and the black deposit on the surrounding areas of the copper is mostly manganese dioxide.

This type of pitting has been reported primarily from Sweden and Germany, but it has also occurred in the United Kingdom. A few cases have been reported each year in the United States, and it has occurred in the Canadian North Pacific Coast area.

Type II pitting is much slower than type I, rarely producing a perforation in less than eight years.

Effects of water quality. There is some consensus about the composition of type II pitting waters. Levelton (1965) gave a typical analysis for Canadian waters as follows: pH 6.3, total solids 25.2 mg/L, DO 10.9 mg/L, total hardness 4.2 mg/L, total alkalinity 3.5 mg/L, and free carbon dioxide 1.5 to 4.5 mg/L. Mattson and Fredericksson (1968) reported that in Sweden, type II pitting only occurs in waters with a pH below 7.4 and a ratio of bicarbonate to sulfate of less than unity. Their observation agrees with that of Franqué (1968) in Germany, and similar observations have been reported in the United Kingdom.

The variety of type II pitting that involves deposition of manganese dioxide requires, of course, the presence of manganese in the water, and problems have arisen with waters containing as little as 0.03 mg/L. Such waters are soft moorland-derived supplies that have usually been treated to raise the pH above 7.4. Type II pitting has not been reported in seawater, even for seawater that has been acidified and used at high temperature, as in desalinization plants.

In the United States, flocculation of snowmelt by treatment with aluminum sulfate to overcome turbidity has caused type II pitting. Type II pitting is a minor problem in Germany. It is not connected with the quality of the tube but is attributed to the water composition. Hence, water treatment is the most common remedy recommended, consisting mainly of raising the pH. Germany, following Swedish experience, has included the following formula in its standard DIN 50930): Type II pitting is unlikely if the water has a pH greater than 6.5 and if the ratio of bicarbonate to sulfate is greater than 2:1.

Microbiologically Induced Corrosion of Copper

Bacteria, often growing in biofilms, have long been known to occur on interior pipe surfaces, often with scales and corrosion products, and often associated with corrosion problems. However, the association has usually been thought to represent an innocuous colonization of an available surface, rather than a causative or exacerbating factor in copper corrosion. In recent years, several instances of corrosion have been strongly linked to microbial processes, either because biofilms were present or because the problems could be mitigated by effective disinfection. There has also been a significant amount of research on the properties and effects of bacterial colonization of copper surfaces.

In the mid-1980s, new hospitals in Germany and Scotland developed pitting corrosion shortly after their opening (Fischer et al. 1992; Wagner, Fischer, and Paradies 1992). In a county hospital in Germany, cold- and hot-water pipes were damaged within months of opening. Construction had been completed nearly four years before the hospital was opened to the public. The pits occurred in both soft and hard copper tubes in cold- and hot-water lines. The pits were atypical in that the copper tubercles were overlain with black precipitate; there was also general corrosion. Copper was released sporadically into drinking water and an insoluble film was present in the corrosion deposits that was inferred to be of bacterial origin. The water was soft, adjusted to pH 9–9.5, and disinfected with chlorine and chlorine dioxide. Cleaning with citric acid, followed by hot water at 60°C, was at least partially effective in removing the corrosion products, producing a clean surface and temporarily stopping further damage.

In Scotland, pitting attack has been observed mainly in hot-water pipes used in soft, low-alkalinity, high-NOM waters (Keevil et al. 1988). Biofilms have been detected in the affected piping, and further study has associated copper corrosion and biofilm presence to hot-water systems in which temperatures were not always maintained above 130–140°F (55–60°C). Chemostat studies have confirmed that biofilms are not generated at temperatures of 140°F (60°C) or higher and that biofilm viability is reduced greatly by exposure to 140°F (60°C) water (Walker, Dennis, and Keevil 1992).

The biofilms have been described as thin and continuous, containing natural polysaccharides, or as thick, copious slime, creating aerobic–anaerobic environments and promoting growth of sulfate-reducing bacteria. Research is addressing the bacteria involved in these biofilms, the mechanism of effecting pitting corrosion, the role of water quality, and the mitigation of the problems. Bremer and Geesey (1991a) have isolated and grown bacteria from biofilms on copper and, using advanced surface examination techniques, have shown the production of exopolymers and the initiation of attack on copper by some strains of bacteria and not by others (Geesey and Bremer 1990; Bremer and Geesey 1991b). Paradies, Fischer and co-workers have investigated the role of polymer films, using model films, to simulate the biofilm and natural polysaccharides (Paradies et al. 1992).

An overview of pitting corrosion of copper associated with microbial growths is coauthored by researchers from the American, German, and British groups (Geesey et al. 1993).

SUMMARY

Copper corrosion problems continue to challenge the drinking water community. Stringent requirements for monitoring copper release in tap water have rekindled interest in corrosion and mobilization of corrosion products. Understanding of copper corrosion continues to develop, with recent emphasis being placed on the role of microorganisms and natural organic matter, as well as inorganic anions, pH, and alkalinity effects. Copper release has also been shown to be a dynamic process in laboratory studies, with concentrations passing through a maximum concentration and then declining during a stagnation period. Pitting and tubing failures continue to occur infrequently, but at high cost when they do occur. Mitigation of copper corrosion, especially copper release, has become a major treatment objective in many US systems because of the requirements of the Lead and Copper Rule or because of needs to minimize wastewater discharges to sensitive receiving waters.

REFERENCES

Adeloju, S.B., and H.C. Hughes. 1986. The Corrosion of Copper Pipes in High Chloride–Low Carbonate Mains Water. *Corros. Sci.,* 26(10):851.

Akkaya, M., and J.R. Ambrose. 1987. An Electrochemical Technique for the Prediction of Long-Term Corrosion Resistance of Copper Plumbing Systems. *Mat. Perf.,* 26(3):9.

ASTM (American Society for Testing and Materials). 1991. *Standard Specifications for Liquid and Paste Fluxes for Soldering Application of Copper and Copper Alloy Tube,* Designation B813-91. Philadelphia: ASTM.

Atlas, D.C., J. Zajicek, and O.T. Zajicek. 1982. The Corrosion of Copper by Chlorinated Drinking Waters. *Wat. Res.,* 16:693.

Baes, C.F., Jr., and R.E. Mesmer. 1976. *The Hydrolysis of Cations.* New York: John Wiley and Sons.

Baukloh, A., and C. Triquet. 1982. Internally Surface Treated Copper Sanitary Tubes With a High Resistance to Pitting in Cold Waters. In *Proc. of International Symposium on Corrosion of Copper and Copper Alloys in Building.* Tokyo: Japan Copper Development Association.

Baukloh, A., H. Protzer, U. Reiter, and B. Winkler. 1989. Kupferrohre in der Hausinstallation. Einfluss von Produktqualität, Verarbeitungs- und Installations-bedingungen auf die Beständigkeit gegen Lochfrass Typ I, METALL, Heft 1, Metall-Verlag GmbH, Berlin/Heidelberg.

Benjamin, M.M., S.H. Reiber, J.F. Ferguson, E.A. Vanderwerff, and M.W. Miller. 1990. *Chemistry of Corrosion Inhibitors in Potable Water.* Denver, Colo.: AWWA Research Foundation and American Water Works Association.

Bremer, P.J., and G.G. Geesey. 1991a. Laboratory-Based Model of Microbiologically Induced Corrosion of Copper. *App. Envir. Micro.,* 57:1956–1962.

———. 1991b. Properties of the Exopolymer of a Bacterium That Causes Pitting Corrosion of Copper. In *Short Communications of the 1991 International Marine Biotechnology Conference.* Dubuque, Iowa: W.C. Brown.

Brown, F.R., and Q. Fernando. 1991. Kinetics of the Dissolution of Copper Metal in Aqueous Solutions Containing Unsaturated Organic Ligands and Copper(II). *Talanta,* 38(3):309.

Byrne, R.H., and W.L. Miller. 1985. Copper (II) Carbonate Complexation in Seawater. *Geochim. Cosmochim. Acta,* 49:1837.

Byrne, R.H., L.R. Kump, and K.J. Cantrell. 1988. The Influence of Temperature and pH on Trace Metal Speciation in Seawater. *Mar. Chem.*, 25:163–181.

Cabaniss, S.E., and M.S. Shuman. 1988a. Copper Binding by Dissolved Organic Matter: I. Suwannee River Fulvic Acid Equilibria. *Geochim. Cosmochim. Acta,* 52:185.

———. 1988b. Copper Binding by Dissolved Organic Matter: II. Variation in Type and Source of Organic Matter. *Geochim. Cosmochim. Acta,* 52:195.

Campbell, H.S. 1950. Pitting Corrosion in Copper Water Pipes Caused by Films of Carbonaceous Material. *Jour. Inst. Metals,* 77:345.

———. 1954. A Natural Inhibitor of Pitting Corrosion of Copper in Tap Waters. *Jour. App. Chem.,* 4:633–647.

———. 1968. *Corrosion of Metals in the Water Supply Industry.* British Non-Ferrous Metals Research Association, Miscellaneous Publication 574. London: BNF.

———. 1972. *A Review: Pitting Corrosion of Copper and Its Alloys.* British Non-Ferrous Metals Research Association, Miscellaneous Publication 574. London: BNF.

Cohen, A., and L.P. Costas. 1982. Flux Corrosion of Copper Plumbing Systems. In *Proceedings of the International Symposium on Copper and Copper Alloys in Building.* Tokyo: Japan Copper Development Association.

Cohen, A., and J.R. Myers. 1984. Mitigation of Copper Tube Cold Water Pitting by Water Treatment. Paper presented at the National Association of Corrosion Engineers, Corrosion '84, at New Orleans, April.

Cornwell, F.J., G. Wildsmith, and P.T. Gilbert. 1973. Pitting Corrosion in Copper Tubes in Cold Water Services. *Br. Corros. Jour.,* 8:202.

Cotton, F.A., and G. Wilkinson. 1988. *Advanced Inorganic Chemistry.* New York: John Wiley and Sons.

Cruse, H., and R.D. Pomeroy. 1974. Corrosion of Copper Pipes. *Jour. AWWA,* 67(8):479.

Davis, J.A. 1984. Complexation of Trace Metals by Adsorbed Organic Matter. *Geochim. Cosmochim. Acta,* 48:679.

Devroey, P., and C. Depommier. 1965. Visseries et Trefileries Réunies, S.A. Unpublished Report. (Reported in H.S. Campbell. 1972. British Non-Ferrous Metals Research Association, Miscellaneous Publication 574.)

de Zoubov, N., C. Vanlengenhaghe, and M. Pourbaix. 1974. Copper. In *Atlas of Electrochemical Equilibria in Aqueous Solutions.* Houston, Texas: National Association of Corrosion Engineers/Centre Belge d'Etude de la Corrosion (CEBELCOR).

DIN 50930, Teil 5, Ausgabe Februar 1993. Korrosion der Metalle; Korrosion metallischer Werkstoffe im Innern von Rohrleitungen, Behältern und Apparaten bei Korrosionsbelastung durch Wässer; Beurteilung der Korrosionswahr-scheinlichkeit von Kupfer und Kupferlegierungen. Berlin: Beuth-Verlag GmbH.

DIN EN 1057. 1993. Kupfer und Kupferlegierungen: Nahtlose Rundrohre aus Kupfer für Wasser- und Gasleitungen für Sanitärinstallationen und Heizaungsanlagen. Berlin: Beuth-Verlag GmbH.

Drogowska, M., L. Brossard, et al. 1992. Effects of Phosphate Ions on Copper Dissolution and Passivation. *Jour. Electrochem. Soc.,* 139(10):2787.

DVGW. 1992. Nahtlosgezogene Rohre aus Kupfer für Gas- und Wasserinstallationen-Anforderungen und Prüfbestimmungen. Bonn: Wirtschafts- und Verlagsges. Gas und Wasser, mBH.

Eary, L.E., and J.A. Schramke. 1990. Rates of Inorganic Oxidation Reactions Involving Dissolved Oxygen. In *Chemical Modeling of Aqueous Systems II.* Washington, D. C.: American Chemical Society.

Eckhardt, K., E. Klingler, and H. Sick. 1990. Verbindungstechnik und Korrosion im Sanitärbereich. *SBZ*, 45:316–320.

Economic and Engineering Services (EES). 1990. *Lead Control Strategies.* Denver, Colo.: AWWA Research Foundation and American Water Works Association.

Edwards, M., J.F. Ferguson, and S.H. Reiber. 1994. The Pitting Corrosion of Copper. *Jour. AWWA*, 86(7):74–90.

Evans, U.R. 1982. *An Introduction to Metallic Corrosion*. 3rd ed. London: Arnold.

Fischer, W., B. Füßinger, and I. Hänßel. 1992. Influence of Anions on the Pitting Behaviour of Copper in Potable Water. In *Proc. of Twelfth Scandinavian Corrosion Congress and Eurocorr 92*, Fincorr, Vantaa: Corrosion Society of Finland.

Fischer, W., H.H. Paradies, D. Wagner, and I. Hänßel. 1992. Copper Deterioration in a Water Distribution System of a County Hospital in Germany Caused by Microbially Induced Corrosion—I. Description of the Problem. *Werkst. u. Korr.*, 43:56–62.

Franqué, O. von. 1968. Über Bedentung, Umfang und Stand der Untersuchungen des Lochfrasses bei Kupferrohren. *Werkst. u. Korr.*, 5:377.

———. 1983. Lochkorrosion Type I bei Kupferrohren. Symposium Korrosion in Kalt- und Warmwassersystemen der Hausinstallation, Bad Nauheim, Germany.

———. 1987. Normung und Qualitätsprüfung von Kupferrohren. Tagungsband "Rohre aus Kupfer und Kupferlegierungen—Anwendungen in Wärmeaustauschern, Industrie und Haustechnik." Deutsches Kupfer-Institut, Berlin. Deutsche Gesellschaft für Metallkunde e. v. Oberursel, Germany.

Franqué, O. von, and K.-F. Zimmermann. 1988. Korrosionsschäden bei der Kupferrohrinstallation verringern. *SBZ*, 12: 778–783.

Geesey, G.G., and P.J. Bremer. 1990. Applications of Fourier Transform Infrared Spectrometry to Studies of Copper Corrosion Under Bacterial Biofilms. *Mar. Tech. Soc. Jour.*, 24(3):36–43.

Geesey, G.G., P.J. Bremer, W.R. Fischer, D. Wagner, C.W. Keevil., J.T. Walker, and A.H.L. Chamberlain. 1993. Unusual Types of Pitting Corrosion of Copper Tubes in Potable Water Systems. Edited by G.G. Geesey, Z. Lewandowski, and H.-C. Flemming. In *Biofouling/Biocorrosion in Industrial Water Systems*. Chelsea, Mich.: Lewis Publishing.

Hirose, K. 1990. Chemical Speciation of Trace Metals in Seawater: Implication of Particulate Trace Metals. *Mar. Chem.*, 28:267.

Holm, T.R., and C.D. Curtiss III. 1990. Copper Complexation by Natural Organic Matter in Ground Water. In *Chemical Modeling of Aqueous Systems II,* Los Angeles: American Chemical Society.

Ives, D.J.B., and A.E. Rawson. 1962a. Copper Corrosion I: Thermodynamic Aspects. *Jour. Electrochem. Soc.,* 109:447.

———. 1962b. Copper Corrosion II: Kinetic Studies. *Jour. Electrochem. Soc.,* 109:452.

———. 1962c. Copper Corrosion III: Electrochemical Theory of General Corrosion. *Jour. Electrochem. Soc.,* 109:458.

———. 1962d. Copper Corrosion IV: The Effect of Saline Additions. *Jour. Electrochem. Soc.,* 109:462.

Japan CDA (Copper Development Association). 1982. *Proceedings of the International Symposium on Corrosion of Copper and Copper Alloys in Building.* Tokyo: Japan Copper Development Association.

Keevil, C.W., J.T. Walker, J. McEvoy, and J.S. Colbourne. 1988. Detection of Biofilms Associated with Pitting Corrosion of Copper Pipework in Scottish Hospitals. In *Proc. of the Biodeterioration Society Meeting on Biocorrosion.* Paris.

Korshin, G.V., S.A.L. Perry, M.A. Edwards, and J.F. Ferguson. 1994. Influence of Natural Organic Matter on Corrosion in Potable Water. In *Proc. of the 1994 Annual Meeting of the American Water Works Association.* Denver, Colo.: American Water Works Association.

Landgrebe, H. 1980. Wasserfluss Stoppt Korrosionverdruss. *IKZ,* 35(4):86–94.

Le Gal La Salle, A., A. Jardy, R. Rosset, M. Keddam, A. Caramel, and D. Noel. 1992. Copper Corrosion, Passivation and Protection in Aqueous Solutions. II. Proposition of Copper Dissolution Mechanism in Acidic Sulphate Medium Through the Use of a Quartz Microbalance. *Mémoires et Etudes Scientifiques Revue de Métallurgie,* 89(4):238–243.

Levelton, B.H. 1965. Service Life of Pipes Exposed to Domestic Waters. NACE Publication 4E 160. Houston: NACE.

Lindsay, W.L. 1979. *Chemical Equilibria in Soils.* New York: John Wiley and Sons.

Lucey, V.F. 1967. Mechanism of Pitting Corrosion of Copper in Supply Waters. *Br. Corros. Jour.*, 2:175–187.

Mann, A.W., and R.L. Deutscher. 1977. Solution Geochemistry of Copper in Water Containing Carbonate, Sulphate and Chloride Ions. *Chem. Geol.*, 19:253–265.

Mattson, E., and A.M. Fredericksson. 1968. Pitting Corrosion in Copper Tubes: Causes of Corrosion and Countermeasures. *Brit. Corros. Jour.*, 3:246.

Meyer, E. 1981. Gesetzmäßigkeiten des Eintrags von Schwermetallen in das Trinkwasser durch Rohrwerkstoffe bei unterschiedlicher Wasserbeschaffenheit. *Schr.-Reihe Verein WaBoLu*, 52:9–30.

Millero, F.J. 1990a. Effect of Ionic Interactions on the Oxidation Rates of Metals in Natural Waters. In *Chemical Modeling of Aqueous Systems II*. Washington, D.C.: American Chemical Society.

———. 1990b. Marine Solution Chemistry and Ionic Interactions. *Mar. Chem.*, 30:205.

Millero, F.J., M. Izaguirre, and V.K. Sharma. 1987. The Effect of Ionic Interaction on the Rates of Oxidation in Natural Waters. *Mar. Chem.*, 22:179–191.

Montgomery Watson EES (Economic Engineering Services), and P. Karalekas. 1993. Initial Monitoring Experiences of Large Water Utilities Under USEPA's Lead and Copper Rule. Denver, Colo.: American Water Works Association.

Obrecht, M.F., and L. Quill. 1960. How Temperature and Velocity of Potable Water Affect Corrosion of Copper and Copper Alloys. A series of seven articles published in *Heating, Plumbing and Air Conditioning*, Jan., 165–169; March, 109–116; April, 131–137; May, 105–113; July, 115–122; Sept., 115–133; and April, 1961, 129–134.

Paradies, H.H., W.R. Fischer, I. Hänßel, and D. Wagner. 1992. Characterization of Metal Biofilm Interactions by Extended Absorption Fine Structure Spectroscopy. In *Microbiol Corrosion—2*. Edited by C.A.C. Sequeira, and A.K. Tiller. London and New York: Elsevier Applied Science.

Paulson, A.J., and D.R. Kester. 1980. Copper(II) Ion Hydrolysis in Aqueous Solution. *Jour. Solution Chem.*, 9(4):269.

Reiber, S. 1989. Copper Plumbing Surfaces: An Electrochemical Study. *Jour. AWWA*, 81(7):114.

———. 1993. *Chloramine Effects on Distribution Systems Materials*. Denver, Colo.: AWWA Research Foundation and American Water Works Association.

Reiber, S.H., J.F. Ferguson, and M.M. Benjamin. 1987. Corrosion Monitoring and Control in the Pacific Northwest. *Jour. AWWA*, 79(2):71.

Schindler, P., M. Reinert, and H. Gamsjäger. 1968. "Löslichkeitskonstanten und Freie Bildungsenthalpien von $Cu_2(OH)CO_3$ (Malachit) und $Cu_3(OH)_2(CO_3)_2$ (Azurit) bei 25°C. *Helv. Chim. Acta*, 51(2):1845–1856.

Schindler, P., H. Althaus, F. Hofer, and W. Minder. 1965. "Löslichkeitsprodukte von Zinkoxid, Kupferhydroxid und Kupferoxid in Abhängigkeit von Teilchengröße und molarer Oberfläche. Ein Beitrag zur Thermodynamik von Grenzflächen fest-flüssig. *Helv. Chim. Acta*, 48(5):1204–1215.

Schindler, P.W. 1967. Heterogeneous Equilibria Involving Oxides, Hydroxides, Carbonates and Hydroxide Carbonates. In *Equilibrium Concepts in Natural Water Systems*. Washington, D.C.: American Chemical Society.

Shalaby, H.M., F.M. Al-Kharafi, and A.J. Said. 1990. Corrosion Morphology of Copper in Dilute Sulphate, Chloride, and Bicarbonate Solutions. *Br. Corros. Jour.*, 25(4):292.

Shoesmith, D.W., and W. Lee. 1977. The Dissolution of Cupric Hydroxide Films From Copper Surfaces. *Electrochimica Acta*, 22:1411.

Shull, K.E., and R.J. Becker. 1960. Cold Water Corrosion of Copper Tubing. *Jour. AWWA*, 59(3):401.

Sick, H.H. 1972. Die Erosionbeständigkeit von Kupferwerkstoffen gegenüber strömendem Wasser. *Werkst. u. Korr.*, 23:12–18.

Singley, J.E., R.E. Pisigan, A. Ahmadi, P.O. Pisigan, and T.-Y. Lee. 1986. *Corrosion and Calcium Carbonate Saturation Index in Water Distribution Systems, Project Summer Water Engineering Research Lab.* Cincinnati, Ohio: U.S. Environmental Protection Agency.

Stiff, M.J. 1971. Copper/Bicarbonate Equilibria in Solutions of Bicarbonate Ion at Concentrations Similar to Those Found in Natural Water. *Wat. Res.*, 5:171.

Stone, A., D. Spyridakis, M. Benjamin, J. Ferguson, and S. Osterhus. 1987. The Effect of Short-term Change in Water Quality on Copper and Zinc Corrosion Rates. *Jour. AWWA,* 79(2):75–82.

Suzuki, I., Y. Ishikawa, et al. 1983. The Pitting Corrosion of Copper Tubes in Hot Water. *Corros. Sci.,* 23(10):1095.

Symes, J.L., and D.R. Kester. 1984. Thermodynamic Stability Studies of the Basic Copper Carbonate Mineral, Malachite. *Geochim. Cosmochim. Acta,* 48:2219.

Tronstad, V.R., and R. Veimo. 1940. Preliminary Researches on the Action of Water on Copper Pipes. *Jour. Inst. Metals,* 66:17.

USEPA (U.S. Environmental Protection Agency). 1991a. *Lead and Copper,* Final Rule Correction. *Fed. Reg.* 56:135:32112.

———. 1991b. *Lead and Copper,* Final Rule Correction. *Fed. Reg.* 56:110:26460.

———. 1992. *Lead and Copper,* Final Rule Correction. *Fed. Reg.* 57:125:28785.

Wagner, D., W. Fischer, and H.H. Paradies. 1992. Copper Deterioration in a Water Distribution System of a County Hospital in Germany Caused by Microbially Induced Corrosion—II. Simulation of the Corrosion Process in Two Test Rigs Installed in This Hospital. *Werkstoffe und Korrosion*, 43:496–502.

Walker, J.T., P.J. Dennis, and C.W. Keevil. 1992. Simulation and Control of Copper Pipework Corrosion Using a Laboratory Chemostat Model. In *Microbial Corrosion—2.* Edited by C.A.C. Sequeira and A.K. Tiller. London and New York: Elsevier Applied Science.

Werner, W. 1992. Untersuchungen zum Mechanismus der Flächenkorrosion in Trinkwassen leitungen aus Kupfer. Ph.D. diss. University of Karlsruhe, Germany.

Winkler, B., ed. 1990. *The Use of Copper as a Plumbing Material.* Johannesburg: The Penrose Press.

Woodward, J. 1975. Copper Corrosion in Potable Water Service. Western States Corrosion Seminar.

6

Copper Alloys and Solders

Roger J. Oliphant and Michael R. Schock

Nonferrous alloys are commonly used in domestic plumbing systems, either as fixtures, fittings, or in the making of joints. These alloys have been identified as a source of lead contamination in drinking water. In light of changing regulations, the use of these alloys may be regulated in order to limit the amount of lead found in potable water systems today.

This chapter discusses both copper alloys and solders. For each, the composition of the various alloys commonly used in potable water systems is outlined. This is followed by a discussion of the implications of recent changes in the regulatory requirements. Also covered are the common forms of corrosion for copper alloys and solders, followed by a detailed treatment of the corrosion processes for each. The procedures used to study these corrosion problems are covered as well as the methods used to counter the corrosion.

COPPER ALLOYS USED IN WATER SUPPLY SYSTEMS

The two copper alloy types most frequently used in potable water systems are brasses and gunmetals.

Brasses

Brasses are basically alloys of copper and zinc, with other minor constituents added to enhance certain properties. The proportion of zinc can vary between about 10 and 45 percent. Up to a level of 37 percent, the alloys are single-phase α brasses; above

NOTE: This chapter did not appear in the first edition.

this value and up to 45 percent zinc, a two-phase duplex structure is the most stable—these are the α–β brasses.

Two compositions of brass are commonly used as fixtures or fittings in water systems. An α brass containing about 30 percent zinc is usually used for tube or sand-cast fittings. However, fittings are commonly manufactured from duplex brasses containing about 40 percent zinc, because these are more suitable for hot stamping. Many fittings can be made more quickly and more cheaply by forging (alternatively called hot stamping) than by casting, and hot-stamped fittings have the additional advantages of freedom from porosity and a superior surface finish. Some brazing alloys are also duplex brasses and may contain as much as 50 percent zinc with, consequently, a high proportion of β phase.

Lead is added to brasses, typically up to about 6 percent, to improve machinability or reduce porosity in castings. For some applications where increased resistance to erosion attack is required, alloys with 1 percent tin are used.

There are several US manufacturing standards for brass alloys. The American Society for Testing and Materials (ASTM) has published B62-93, Standard Specification for Composition Bronze or Ounce Metal Castings, and B5-89, Standard Specification for Electrolytic Tough-Pitch Copper Refinery Shapes.

Gunmetals

Gunmetals are alloys of copper, tin, and zinc with or without lead; they are casting alloys. Virtually all model building and plumbing codes in the United States (drawing on the Unified Numbering System [ASTM and SAE 1986]) recognize alloy C83600 (85% Cu:5% Sn:5% Pb:5% Zn) and alloy C84400 (81% Cu:3% Sn:7% Pb:8.5% Zn) for domestic plumbing valves. The American Water Works Association (AWWA) Standards committee on valves, however, only recognizes alloy C83600. Although the majority of valve bodies in domestic use are composed of one of these alloys, red brass components, such as cap nuts, cartridges, etc., may be of other alloys such as C32000. Laboratory-type goosenecks are often of alloy C26000. In the United Kingdom, the two most common gunmetals used for water fittings are leaded gunmetals (LG) LG2 (84% Cu:5% Sn:5% Pb:5% Zn) and LG4 (88% Cu:5% Sn:4% Pb:2% Zn).

EFFECT OF REGULATORY REQUIREMENTS ON THE USE OF COPPER ALLOYS

US Regulations

In the United States, only one federal regulation specifically affects the use of copper alloys and solders for drinking water applications. The Lead Contamination Control Act (LCCA) of 1988 (USEPA 1988), an amendment of the Safe Drinking Water Act, imposed a prohibition on the use of plumbing materials containing more than 8 percent lead for drinking water systems. Materials (except solders) having lower percentages of lead are deemed by the act to be "lead free." In addition to this specific regulation, the US Department of Housing and Urban Development amended its Minimum Property Standards to reflect these requirements of the LCCA.

Because the Lead and Copper Rule (USEPA 1991) requires sampling of first-draw 1-L samples at the consumers' taps, it indirectly provides a reason for reconsidering the use of the traditional brasses for household faucets and other auxiliary features. The rule promulgated action levels for lead of ≥15 µg/L and for copper of ≥1,300 µg/L at the consumers' taps. Thus, the contribution of lead and copper from plumbing fixtures and soldered joints became a real concern to utilities. If detectable levels of lead and undesirable levels of copper are found, utilities must attempt to reduce that contribution through central water treatment.

Thus far, no regulations have been set in the United States for tin, antimony, zinc, or bismuth levels at the consumers' taps.

European Regulations

In the countries of the European Community (EC), the minimum required quality of potable water is defined by the Drinking Water Directive (EC 1980). Each member state of the EC has enacted this directive, with stricter modifications where thought necessary, into their respective domestic legislation.

The concentrations currently allowed by the directive at the consumers' taps, after the water has stood for 12 hours, are 3,000 μg/L for copper and 5,000 μg/L for zinc. For lead, the definition of the limit is ambiguous but is usually interpreted as 50 μg/L in a flushed sample and up to 100 μg/L in a standing sample. In Great Britain, the legislation enacting the EC directive, the 1989 Water Act for England and Wales and the 1990 Water Act for Scotland, has made the lead standard more stringent and less ambiguous, being set at a maximum of 50 μg/L for any water sample. This may change in the future as a result of the new guideline value, an average value of 10 μg/L, suggested by the World Health Organization (WHO 1993). In anticipation of a significant reduction in the allowable lead level in drinking water, research is currently in progress to develop alternatives to the leaded copper alloys.

There is no limit set for tin in the directive, but there is a maximum admissible concentration value for antimony of 10 μg/L.

TYPES OF COPPER ALLOY CORROSION IN POTABLE WATERS

The main forms of copper alloy corrosion in potable waters are dezincification of the brasses, general dissolution, and impingement attack. Dezincification was the major problem in the past because fittings were being blocked by the corrosion product. This form of attack was named "meringue" dezincification after the physical appearance of the corrosion product. The subject has been reviewed by Jester (1985). Little work has been done on the tendency of dezincification to promote lead contamination.

General dissolution of the alloy is becoming more significant because levels of contamination deemed acceptable in drinking water are being reduced. Impingement attack has mainly been a problem with brass condenser tubes. In potable water systems, it is primarily associated with copper pipes downstream of fittings or at bends.

Dezincification

The main corrosion problem associated with brasses is dezincification, which is the selective removal of zinc. Any simple brass containing more than about 15 percent zinc is susceptible to this form of corrosion. However, it can be prevented completely in α brasses by addition of a very small amount of arsenic (0.02–0.06 percent) to the alloy. When duplex brasses dezincify, it is primarily the β phase that is attacked, though the attack often extends subsequently into the α phase. Unfortunately, introduction of arsenic into the alloy does not prevent dezincification of the β phase, and dezincification of duplex brasses cannot therefore be prevented by this means (BNF 1982).

The phase suffering dezincification becomes porous and mechanically very weak, although the overall dimensions of the piece do not change very much. If a piece of equipment is covered with dirt or deposits, it is possible to have sudden, unexpected failure because of the poor strength of some undetected dezincified area. The loss of zinc from brass is obvious to the naked eye because the corroded alloy changes from the normal yellow to a red or copper color.

When duplex brazing alloys were used to braze copper hot-water tanks (cylinders), severe dezincification of the seams was often experienced, in many cases in waters in which duplex brass fittings generally gave no trouble. This is because the rate of attack on the brazed joint is greatly accelerated by the large area of copper adjacent to it. It has therefore become the standard practice to braze copper hot-water tanks with special alloys, such as copper–silver–phosphorus, which do not contain zinc. The effect of a large area of copper adjacent to an α–β brass in encouraging dezincification is also shown by a tendency for the bosses of copper tanks to suffer more severe attack than other fittings in the hot-water circulation system. The British standards for direct and indirect cylinders (BS 699 1984 and BS 1566 1984, respectively) require all screwed connections to be made of copper, gunmetal, or α brass inhibited against dezincification.

General Dissolution

Dissolution is the gradual dissolving of components of the alloy into the water being carried. With copper alloys, the main concern with this form of attack is the lead contamination that can occur with certain compositions.

Impingement Attack

Impingement attack can affect the majority of copper alloys to a greater or lesser extent. It occurs under conditions of high velocities with waters that form thin or otherwise poorly protective corrosion product layers and that contain large amounts of dissolved or entrained gas. In potable water systems, the attack most typically affects copper pipe downstream of fittings with hard borehole waters that have low pH values because of the presence of dissolved carbon dioxide.

CHEMISTRY OF COPPER ALLOY CORROSION

Electrochemistry

The mechanism of dezincification has been extensively studied from an electrochemical point of view. The following three mechanisms have been proposed:

- general dissolution of the material followed by deposition of copper
- selective dissolution of zinc leaving copper as a porous structure
- a combination of the first two mechanisms

The first mechanism is proposed in papers by Sugawara and Ebiko (1967) and by Horton (1970). The second mechanism has been supported by Langenegger and Robinson (1969). Heidersbach and Verink (1972) showed that both mechanisms could operate under specific conditions of pH and electrochemical potential. They attempted to explain this model by superimposing the Pourbaix diagrams of the 70:30 Cu:Zn alloy and those for the two elements in a 0.1*M* chloride solution. The resulting composite diagram (Figure 6-1) has regions where only the zinc forms stable ions

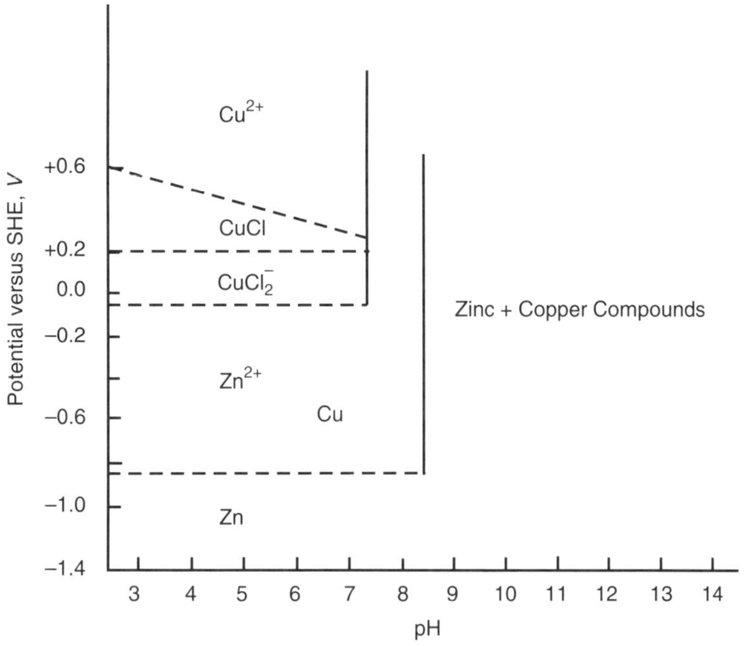

Source: After Heidesbach and Verink (1972).

Figure 6-1 70:30 Cu-Zn alloy in 0.1M chloride solution with superimposition of copper-zinc equilibrium diagrams

(second mechanism) and other regions, at more positive potentials, where the copper as well as the zinc produces ions, although these are anionic (negative) species. The latter species were believed to be unstable with respect to the metal so that in this region of the diagram the first mechanism predominates. At yet more positive potentials, both elements produced cationic species and general dissolution of the alloy occurred (i.e., dezincification as such ceases).

In service, the corrosion rate is believed to be controlled by the cathodic reaction, i.e., the reduction of an electron accepting species (whether or not mediated by the reduction of copper ions). This is mainly due to the low solubility of oxygen in water or the low concentration of any other likely electron-accepting species. However, the mechanism that predominates is determined by the anode potential. The anode potential is controlled by the composition of the solid solutions of the different phases of the alloy as well as the presence of metallurgical defects, impurities at grain boundaries, partial shielding by deposits, etc.

The ability of arsenic to prevent dezincification in single-phase brasses has been known for many years. An electrochemical mechanism explaining its action and why it fails for two-phase α–β brasses has been proposed by Lucey (1965). This assumes the first mechanism is operative and produces positive copper species; any process that blocks the reduction of copper ions back to the metal will then stifle the corrosion.

The potential of film-free α brass and β brass in the type of solutions believed to exist at the advancing corrosion front have been measured as –0.38 V and –0.56 V, respectively, versus a standard hydrogen electrode (SHE). At the corrosion front, the copper ions produced are precipitated as the sparingly soluble Cu_2Cl_2. Taking into account the activities of copper ions that exist in equilibrium with this salt, the Cu^{2+} can be reduced to the metal at –0.16 V SHE, whereas the reduction of Cu^{+} requires potentials more negative than –0.41 V SHE. And so, while the corrosion of the β phase of the alloy can use both Cu^{+} and Cu^{2+} as the electron acceptors, the α phase can only reduce Cu^{2+}.

The arsenic is believed to function by reducing the Cu^{2+} at the corrosion front to Cu^{+}, after which it is regenerated by the corrosion of more copper. The overall reaction can be represented by the equations:

$$3Cu^{2+} + As = 3Cu^{+} + As^{3+} \tag{6-1}$$

$$3Cu + As^{3+} = 3Cu^{+} + As \tag{6-2}$$

Consequently, with no Cu^{2+} at the corrosion front, the dezincification of α brass is inhibited; dezincification of the β phase, on the other hand, can continue via the reduction of Cu^{+}. In the absence of arsenic, α brass corrodes via the Cu^{2+} ions produced by the disproportionation reaction:

$$2Cu^{+} \rightleftharpoons Cu^{2+} + Cu \tag{6-3}$$

Of course, this explanation is not valid if the second dezincification mechanism operates.

Water Composition Factors and Dezincification

Over the years, various rules of thumb have been developed to explain the conditions under which dezincification will occur. As with many other corrosion phenomena, certain conditions are required to "initiate" attack and different conditions are required to ensure its continuation or "propagation." Both sets of conditions have to be satisfied for the corrosion to be a significant problem.

A water's chloride-to-carbonate (temporary) hardness ratio is the primary constituent determining its ability to propagate dezincification. This relationship is summarized in Figure 6-2, the Turner diagram. This figure was derived from a study of meringue dezincification and was based on numerous laboratory tests and the service behavior of a large number of actual water supplies. However, further experience has shown that it is generally applicable as an indicator of a water's ability to support dezincification.

Waters with compositions defined by points within the shaded area of the Turner diagram are liable to support dezincification. The boundary between the shaded area and the "safe" area in the diagram cannot be defined very exactly. For waters with compositions near the boundary, the sulfate content also has to be considered, as do the natural variations in composition of water from any one supply.

Unfortunately, the effect of sulfate is not straightforward; at some concentrations it increases the value of the chloride-to-carbonate-hardness ratios required for a water to support dezincification, while at yet higher values, in what is otherwise the same water, it has the reverse effect. Nitrate has been found to reduce the dezincification propensity of waters, especially at high sulfate levels (>150 mg SO_4/L). Orthophosphate, even at low concentrations, significantly increases the rate of attack (Oliphant 1975).

For dezincification to be initiated by the water, its pH has to be above 7.6. Thus, waters with a composition within the shaded area of the Turner diagram, but with a pH of less than 7.6, will not generally cause problems. Corrosion will occur occasionally in fittings that have become individually activated, say by mechanical abrasion of the normally protective air-formed oxide film.

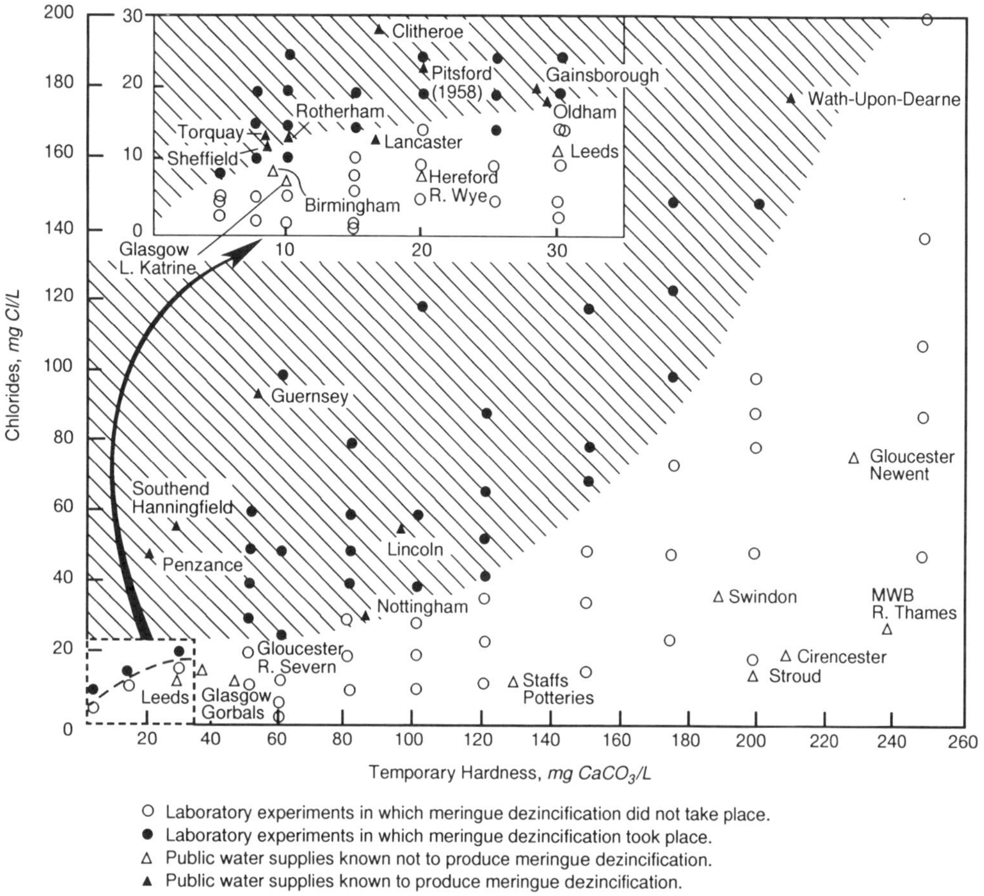

Source: BNF Metals Technology Centre, Oxon, England.

Figure 6-2 Influence of water composition on meringue dezincification

At pH values between 7.6 and 8.2, a water in the shaded area of the Turner diagram will both initiate and propagate the attack, which will take the form of a slow penetration through the wall, producing leakage from a fitting after approximately 15 years. When the pH rises above 8.2, the corroded zinc actually precipitates inside the fitting, usually as some form of basic salt. It is the physical form of this deposit, a series of hollow shells, that gives it its name, meringue dezincification. This is the most troublesome form of the corrosion, leading to substantial restriction of water flows within three to five years.

Dezincification occurs in both hot- and cold-water systems, but the rate of attack is greater in hot-water systems. Waters that have a chloride-to-carbonate-hardness ratio that will support dezincification but that are borderline in pH may give no problems in cold-water systems. However, these waters could cause problems in hot-water systems, as heating the water normally increases the pH.

Contamination Problems From Alloy Dissolution

The most significant element in copper alloys from the metal dissolution point of view is lead, its maximum allowed concentration in drinking water being very much less than the other components. The lead occurs in the alloys as a separate, presumably less noble, phase.

Nielsen (1975) was one of the earliest researchers to identify the possibility of significant contamination from leaded copper alloys. She conducted field tests for up to six months using two hard waters and one slightly hard water in the pH range 7.3–7.7. The experiments showed that lead levels generally decreased quickly to below 50 μg/L in galvanized iron tubes but to a level above this limit in brass and gunmetal tubes. She concluded that the differences in salt content and hardness had little impact on the leaching from these metals. A standard test method to evaluate the potential for lead and cadmium leaching from metallic components in drinking water was presented.

Further evidence that leaded copper alloys could cause significant contamination problems arose from work in Fife, Scotland (Purdie 1985). Samples from a series of schools, taken as part of a national lead survey, were found to have high lead levels. These were attributed to badly made soldered joints. As part of the remedial work to overcome this problem, the joints were cut out and replaced by compression fittings. After resurveying, little improvement was found, and in one case, where the whole system had been replaced, the results were actually worse!

This led to a detailed examination to track down the source of this new contamination. Lead corrosion products were found in the copper tube remote from the old joint areas. These probably had their origin in solder runs and indicated that not enough of the pipe had been replaced when the joints were renewed. These deposits were easily dislodged and could produce particulate lead contamination on an intermittent basis. No lead was leached from new samples of the copper or stainless steel piping used in the different systems.

Some lead was found in the jointing compound originally used (this linseed-oil-based material is no longer considered acceptable in the United Kingdom for potable waters because of its ability to support microbiological growths). A small amount of lead could also be leached from the chrome-plated taps being used.

Eventually, the main lead source was found to be the brass compression fittings. Different components of the fittings produced different degrees of contamination; the compression olive and the nut usually gave higher values than the body. After a few months, the level of contamination gradually dropped to acceptable values.

Historically, few sample programs have been designed in a manner that would have enabled identification of brass faucets as a significant potential source of contamination. Studies at the Illinois State Water Survey were among the first in the United States to identify this source (Neff 1985; Neff, Schock, and Marden 1987; Schock and Neff 1988). The studies showed that the existence of relatively new brass sampling faucets produced significant contamination of both standing and flushed 125-mL samples. The contamination from the faucets largely confounded all statistical efforts to relate leaching and corrosion rates of galvanized iron and copper pipe to major water chemistry variables.

At approximately the same time, Samuels and Meranger (1984) were investigating leaching of trace metals from kitchen tap assemblies. Eight common types of kitchen faucet were purchased from two local plumbing suppliers. All were chrome-plated brass and had brass-to-copper connectors. In general, those containing copper connecting tubes also had visible lead solder. Five water qualities were studied in duplicate for 24-hour leaching.

Samuels and Meranger found sizable quantities of lead in the water, even with faucet assemblies that had no visible lead solder. The faucet that produced the highest lead levels in most waters had an excessive amount of solder visible at the joints. The lead levels, sometimes reaching 50–100 mg/L, were all of a magnitude that should cause concern, even with the understanding that the tests could not reflect actual service conditions.

Further confirmation of this problem has come from the work of Lee, Becker, and Collins (1989) and Gardels and Sorg (1989). The first study attempted to correlate lead levels in drinking supplies with various water quality parameters and plumbing characteristics. They concluded that leaded brass faucets contributed an average of one third of the total lead at sites with leaded soldered joints and all the lead at homes with only galvanized and plastic systems.

Although water quality and site characteristics could not be used to predict lead levels, certain factors were associated with higher values. Other than the presence of corrosion inhibitors, pH was the only water quality factor that appeared to influence tap lead levels. Sites with elevated pH values (≥8.0) had significantly lower lead values. Zinc orthophosphate produced the lowest levels, with polyphosphates being as effective as elevated pH values. Alkalinity, chlorine residual, calcium, and hardness did not affect lead levels. Site factors were also important, with age being the most significant.

In the work of Gardels and Sorg, 12 faucets representing various designs, materials of construction, and manufacturers were studied in the laboratory. The fittings were tested for one year using distilled water (representing an aggressive supply) and Cincinnati, Ohio, tap water (representing a nonaggressive supply). Retention times were from 30 minutes to 34 days. They concluded that, depending on water quality, new cast-brass faucets could contribute lead in excess of the then-proposed action level of 10 µg/L. They also concluded that 60–75 percent of the lead leached appeared in the first 125 mL of water collected and 95 percent appeared after 200–250 mL of water was collected. Determining the maximum lead concentration from a faucet required both the cold- and hot-water sides to be sampled using a sample volume no larger than 100–125 mL.

TEST METHODS FOR COPPER ALLOYS

Resistance to Dezincification

British Standards BS 2872 (1989) and BS 2874 (1986) specify a test for determining the resistance of brass alloys to dezincification. This test involves exposing samples to a 1 percent solution of cupric chloride at 75°C for 24 hours followed by examination of sections to establish the maximum depth of any dezincification that has occurred. The sample passes this test if the maximum depth of dezincification in a forging, or in the transverse direction of extruded material, does not exceed 100 µm. A maximum depth of 200 µm is permitted in the longitudinal direction of extruded material. The International Organization for Standardization method ISO 6509 (ISO 1981) is the same as the British Standard, except it does not give acceptance limits.

Currently, a European standard is being developed that will define two classes of dezincification-resistant alloys. Grade A material, after being tested by the ISO 6509 test, will not show a depth of dezincification greater than 200 µm in any direction. With Grade B material, the mean depth of dezincification will not be more than 200 µm and the maximum depth not more than 400 µm in any direction.

Galvanic Test Cell to Investigate Water Composition and Dezincification

Where brass fittings are used to join copper pipe, a galvanic couple is automatically created. Oliphant (1978) developed the galvanic test cell, originally devised by Lucey, to quantify a water's ability to support dezincification. The cell consists of a duplex brass electrode (the anode) (area 2 cm^2 [0.31 $in.^2$]) held in close contact but physically separate from a copper electrode (the cathode) (area 13.5 cm^2 [2.1 $in.^2$]). The electrode assembly is contained in a boiling tube that is closed by a bung. The bung contains inlet and outlet ports for the test water and channels for the external circuit wires, which join the electrodes (Figure 6-3). The whole assembly is held in a 60°C water bath; test water is pumped once through at a minimum rate of 35 mL/h using a peristaltic dosing pump.

The galvanic current is determined by measuring the potential drop across a 10-kΩ resistance in the external circuit and applying Ohm's law. The change in galvanic current over time can be plotted graphically. The area under such curves can be related by Faraday's laws to a mass of corroded metal (see the section, Relationship Between Galvanic Current and Contamination Levels, later in this chapter for a more detailed discussion of this). A trace typical of waters that support dezincification is shown in Figure 6-4. The current gradually increases as the attack penetrates into the metal, presumably because changes occur in the solution chemistry at the corrosion front as the exchange with the bulk water becomes restricted.

Effects of Alloys on Water Quality

UK Study. A standard test for assessing the potential of metallic materials to contaminate potable waters was developed in the United Kingdom. As part of the development, the lead leached from a range of the common copper alloys was examined. The test, which was developed primarily to test materials used in fittings, has been published by the British Standards Institution as BS 7766 (1994).

The method involves preparing coupons of the material being tested to defined dimensions and surface finish. After a degreasing and washing procedure, the coupons are suspended in 100 mL of a defined test water for 14 days. The test water is replaced every 24 hours, except over weekend periods, for a total of 10 changes. The temperature is maintained at 24 ±1°C. In the test, the lead concentration in the last day's extract is measured.

Figure 6-5 shows the decline of lead contamination during one of the development tests. From initially very high values during the first few days (not shown in the figure), the contamination quickly falls to quasi-equilibrium levels. It should be noted that the test conditions were chosen for practical convenience, rather than to accurately reflect exposure conditions of a fitting in service. Thus, the results, as presented, only provide a ranking test for the different alloys.

Interestingly, there is only an approximate correlation between lead levels in the alloys (see Table 6-1 for their compositions) and their potential to contaminate supplies. Thus, the nominally 5 percent leaded gunmetal LG2 consistently gave lower contamination after 14 days than the nominally 4 percent LG4 gunmetal (see Figure 6-6).

US Study. In 1991, the US Environmental Protection Agency (USEPA) initiated experiments to investigate the potential for lead leaching from standardized coupons of different plumbing materials, including brasses (Lytle and Schock 1993). The coupons, which measured 1 in. × 2 in. × 0.125 in., were contained in an all-Teflon™

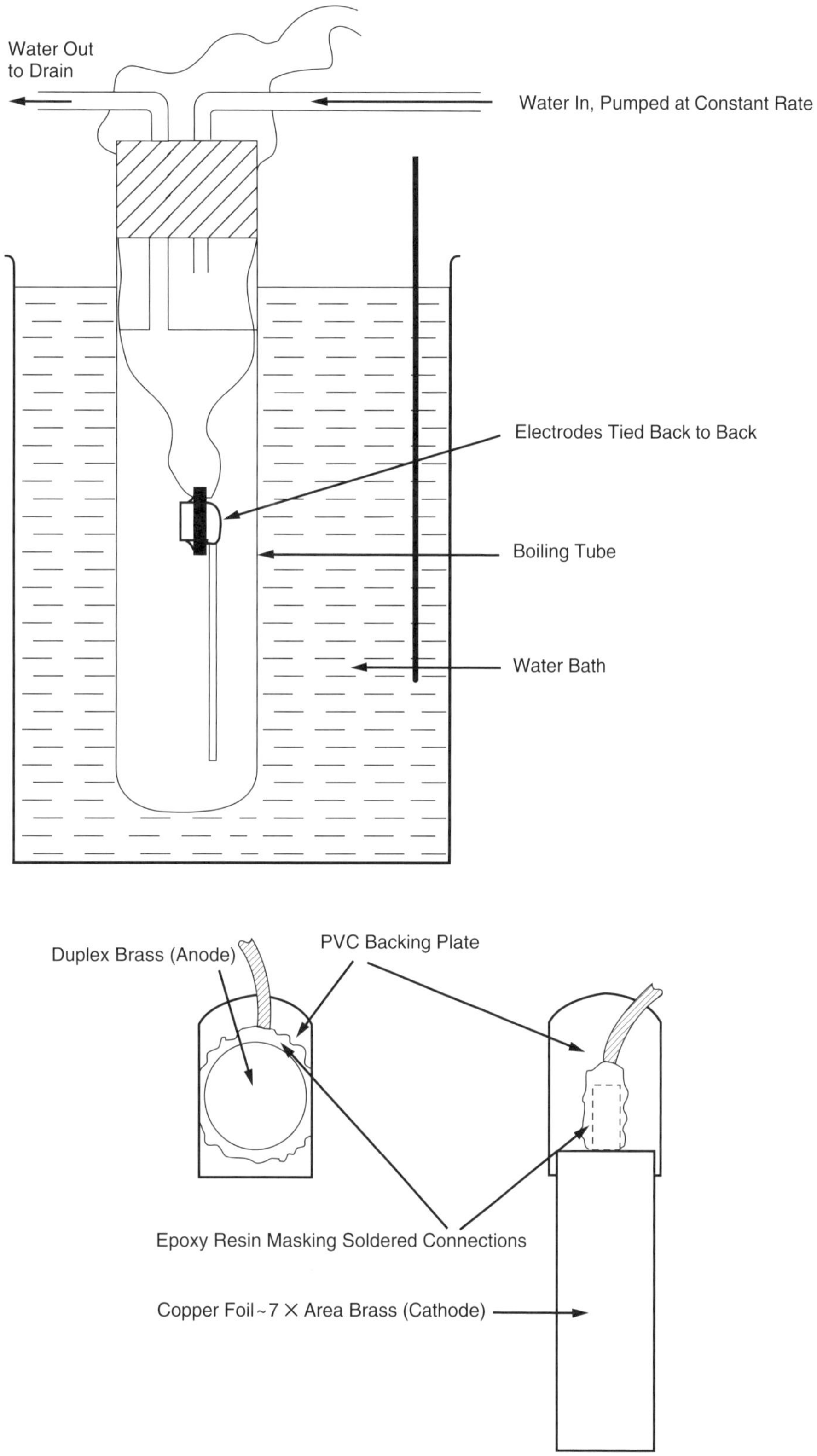

Figure 6-3 Dezincification corrosion cell

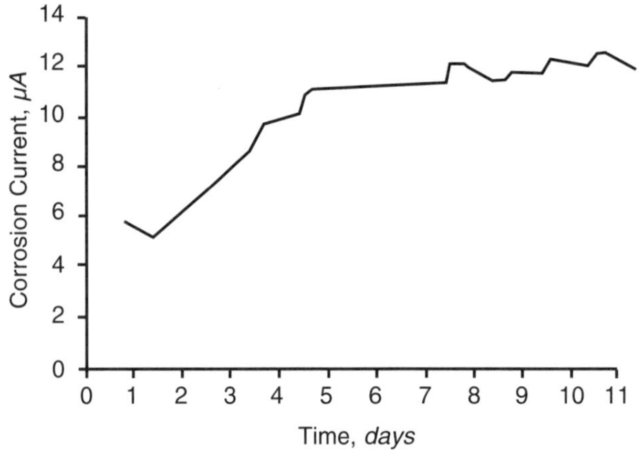

Figure 6-4 Typical corrosion trace of a dezincifying water

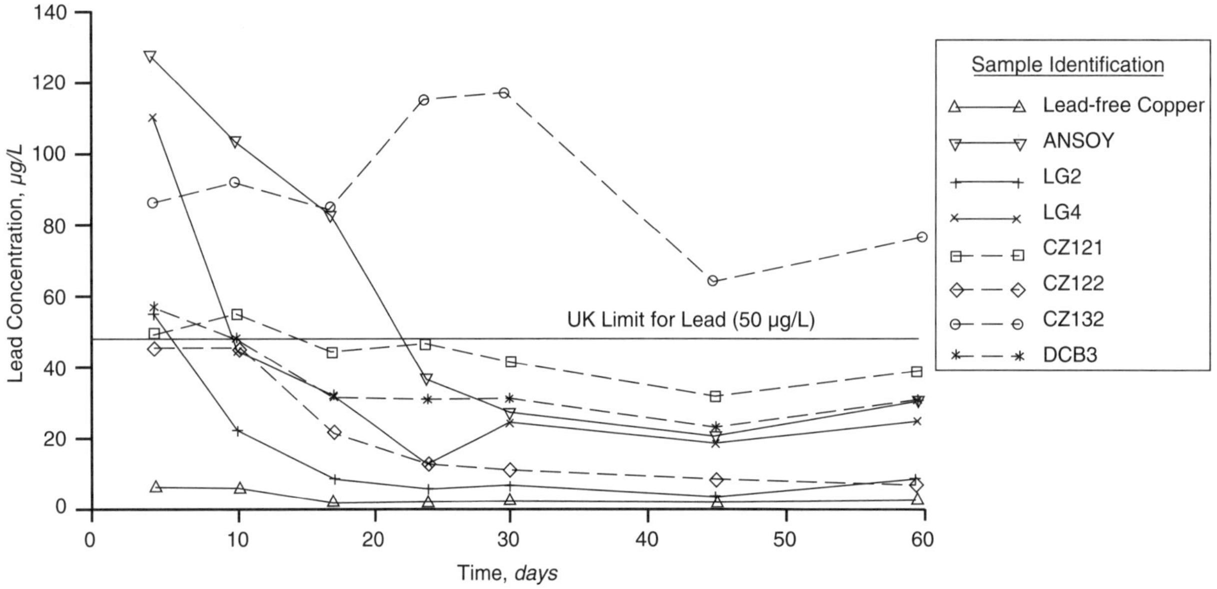

Source: Amos, unpublished report, Water Research Centre, Swindon, England.

Figure 6-5 UK study of lead release (average of three results) versus time (days 4–59) (see Table 6-1 for alloy analyses)

cell system that was fed with water from a high-density polyethylene tank with a floating lid. No metal materials were used in the plumbing system or were in contact with the feed waters to the test cells. The cells were approximately 30 mL in volume, and fresh water was flushed daily through all of them. Samples were taken after 24 hours of stagnation.

Table 6-1 Chemical composition of test coupons used in UK leaching studies (see Figures 6-5 and 6-6 for the results)

Sample	Composition, % W/W										
	Copper	Tin	Lead	Iron	Nickel	Aluminum	Manganese	Silicon	Zinc	Arsenic	Antimony
Lead-free copper	Rem*	<0.001	<0.005	<0.001	<0.001	<0.001	<0.001	<0.001	<0.005	<0.001	
CZ132	61.1	0.09	2.43	0.05	0.01		<0.01		Rem	0.11	
CZ121Pb4	58.3	0.17	3.64	0.15	0.04				Rem		
CZ122	57.9	0.12	2.18	0.12	0.04				Rem		
DCB3	61.0	0.24	1.81	0.16	0.16	0.54	0.02		Rem		
LG2	Rem	4.9	5.38	0.19	0.42	<0.004			5.18		
LG4	Rem	5.32	3.88	0.01	0.78	<0.01	<0.01	<0.01	2.05	0.07	0.09
ANSOY	Rem	3.7	8.3		0.40				9.7		

Source: Amos, unpublished report, Water Research Centre, Swindon, England.
*Rem = Remainder

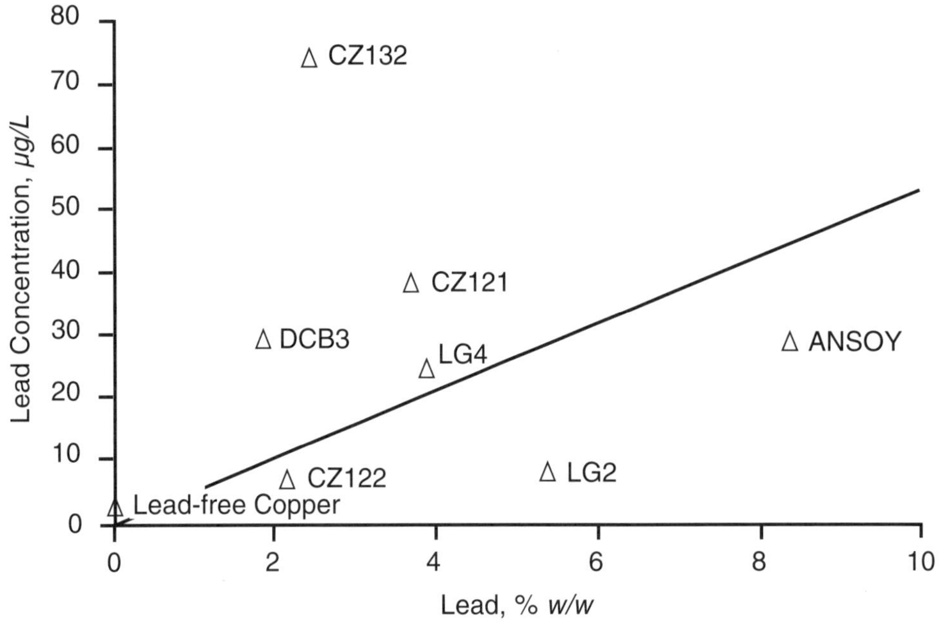

Source: Amos, unpublished report, Water Research Centre, Swindon, England.

Figure 6-6 Lead release on day 59 versus lead content of alloys

Three water qualities were tested with Cincinnati, Ohio, tap water as the basis (see Table 6-2). The major water qualities examined were

- Cincinnati tap water, pH ~ 8.5
- Cincinnati tap water, pH ~ 7.0
- Cincinnati tap water, pH ~ 7.5, orthophosphate ~ 3 mg PO_4/L

The materials tested are summarized in Table 6-3 and include the two alloys (C36000 and C84400) most commonly used by US valve and faucet manufacturers. The results of these experiments for the brasses are shown in Figures 6-7 through 6-12 and may be summarized as follows:

- The amount of lead leached from the coupons during the first four months of exposure generally paralleled the percentage of lead in the material.
- The levels of leaching in the three waters tested dropped substantially within several days; however, steady-state leaching rates were not reached even after four months.
- The differences in results from duplicate coupons of C36000 exceeded the differences between several of the other alloys with different lead percentages.
- The orthophosphate addition at pH 7.5, using disodium orthophosphate, produced rapid and substantial reductions in lead leaching from all materials.
- The decrease in lead leaching from the C83600 alloy produced by the addition of orthophosphate diminished over time, eventually becoming comparable with the effect produced by increasing the pH.

Any attempt to extrapolate these results to results obtained from real valves or faucets under normal usage conditions is beset by several problems. Commercial

Table 6-2 Composition of waters used in the US leaching studies

Analyte	Run 1					Run 2					Run 3				
	Min.	Max.	Mean	Std. Dev.	Median	Min.	Max.	Mean	Std. Dev.	Median	Min.	Max.	Mean	Std. Dev.	Median
Lead, µg / *L*	<0.002	11.9	1.4	2.4	0.48	<0.002	14.1	0.68	2.12	0.50	<0.002	2.0	0.29	0.49	0.10
Calcium, *mg / L*	34.7	54.8	42.4	4.6	40.9	25.6	51.2	33.5	6.3	33.3	34.5	44.5	40.0	2.6	40.3
Copper, *mg / L*	<0.02	0.03	0.00	<0.02	<0.02	<0.02	2.3	0.05	0.37	<0.02	<0.003	0.02	0.003	0.01	<0.003
Iron, *mg / L*	<0.05	0.07	<0.05	<0.05	<0.05	<0.05	0.66	<0.05	0.13	<0.05	<0.002	0.17	0.01	0.03	0.01
Potassium, *mg / L*	2.3	4.4	3.5	0.45	3.6	1.6	3.2	2.0	0.39	1.96	<2.00	3.4	3.0	0.42	3.1
Magnesium, *mg / L*	9.1	16.6	13.5	1.8	13.5	6.0	12.8	8.1	1.7	8.05	<0.025	11.1	9.7	1.4	9.9
Manganese, *mg / L*	<0.01	0.03	<0.01	0.01	<0.01	<0.01	0.03	<0.01	0.02	<0.01	<0.0004	0.01	<0.0004	<0.0004	<0.0004
Sodium, *mg / L*	14.8	44.5	33.4	6.8	34.7	3.8	53.9	15.7	7.4	14.0	16.3	26.2	21.3	1.1	22.8
Zinc, *mg / L*	<0.01	0.39	0.03	0.06	0.01	<0.01	0.07	<0.01	0.03	<0.01	<0.001	0.01	<0.001	<0.001	<0.001
Alkalinity, *mg* $CaCO_3$ */ L*	43.5	74.1	60.2	5.6	60.4	21.7	52.2	34.9	6.4	33.9	43.1	58.4	52.2	3.0	52.6
Sulfate, *mg* SO_4 */ L*	104.0	130.0	116.3	7.9	115.0	52.8	108.9	68.2	10.7	66.2	64.6	93.6	78.6	7.5	78.8
Chloride, *mg / L*	39.0	43.8	40.3	1.2	39.9	21.4	46.0	29.6	5.6	28.0	0.04	36.5	30.8	5.0	31.7
Silica, *mg* SiO_2 */ L*	2.4	2.8	2.6	0.2	2.8	4.3	7.7	6.0	0.73	6.1	3.7	8.5	6.1	1.1	5.6
Nitrate, *mg N / L*	<0.02	1.1	0.77	0.17	0.73	<0.02	2.0	1.1	0.30	1.2	<0.02	1.1	0.85	0.33	0.96
Ammonia, *mg N / L*	<0.03	<0.03	<0.03	0.00	<0.03	<0.03	<0.03	<0.03	0.11	<0.03	<0.03	<0.03	<0.03	<0.03	<0.03
Orthophosphate, *mg* PO_4 */ L*	<0.02	0.37	<0.02	0.08	<0.02	<0.02	0.31	0.02	0.05	<0.02	2.5	3.3	2.8	0.16	2.8
Dissolved oxygen, *mg / L*	na*	na*	na*	na*	na*	6.0	10.2	8.9	0.72	9.0	5.5	10.8	8.7	0.76	8.8
Total inorganic carbon, *mg C / L*	na*	na*	na*	na*	na*	10.9	14.1	12.5	1.21	12.4	11.2	14.8	13.7	0.76	13.7
Free chlorine, *mg* Cl_2 */ L*	0.65	3.1	2.0	0.56	2.00	1.0	2.2	1.4	0.28	1.3	0.69	2.3	1.4	0.49	1.2
pH, *pH units*	8.04	8.79	8.53	0.18	8.54	6.9	7.1	7.0	0.05	7.0	7.4	7.6	7.5	0.05	7.5

*na = not analyzed

Table 6-3 Chemical composition of alloys used in US leaching studies (see Figures 6-7 through 6-12 for the results)

Loop Number	Alloy Number	Metal Type	Copper, %	Zinc, %	Lead,* %	Tin, %	Iron, %
1	C36000	FC† brass	60.49 ± 1.87§	37.56 ± 0.89§	2.73 ± 0.34§	na‡	0.16 ± 0.04§
2	C36000	FC brass	61.26 ± 4.14	37.38 ± 1.24	2.78 ± 0.33	na	0.17 ± 0.03
3	C83600	Red brass	84.56 ± 3.47	5.20 ± 1.84	5.06 ± 0.69	na	0.10 ± 0.06
4	C84400	Red brass	77.25 ± 11.63	10.61 ± 3.55	6.61 ± 0.92	na	0.17 ± 0.04
5	C84500	Red brass	78.89 ± 2.26	11.17 ± 1.13	5.97 ± 1.37	na	0.22 ± 0.10
6	C85200	Yellow brass	69.30 ± 0.36	25.60 ± 2.75	1.98 ± 0.37	na	0.62 ± 0.05
7	C85400	Yellow brass	64.82 ± 3.45	30.16 ± 6.24	1.59 ± 0.19	na	0.19 ± 0.26
8	C122	Copper pipe	97.54 ± 2.93	0.00 ± 0.35	0.01 ± 0.08	na	0.03 ± 0.00
9	Pure lead	Pure lead	0.20 ± 2.80	0.05 ± 0.56	91.23 ± 6.85	na	0.02 ± 0.03
10	Pure lead	Pure lead	na	0.00 ± 0.43	91 ± 19.42	na	0.03 ± 0.02
11	Pure zinc	Pure zinc	na	99.66 ± 0.89	0.07 ± 0.06	na	0.03 ± 0.01
12	Lead/zinc	Solder	na	na	44.22 ± 5.55	58.29 ± 1.73	0.06 ± 0.11

*Lead analyzed by flame atomic adsorption.
†FC = free cutting.
‡na = not analyzed.
§ ± = 95% confidence interval.

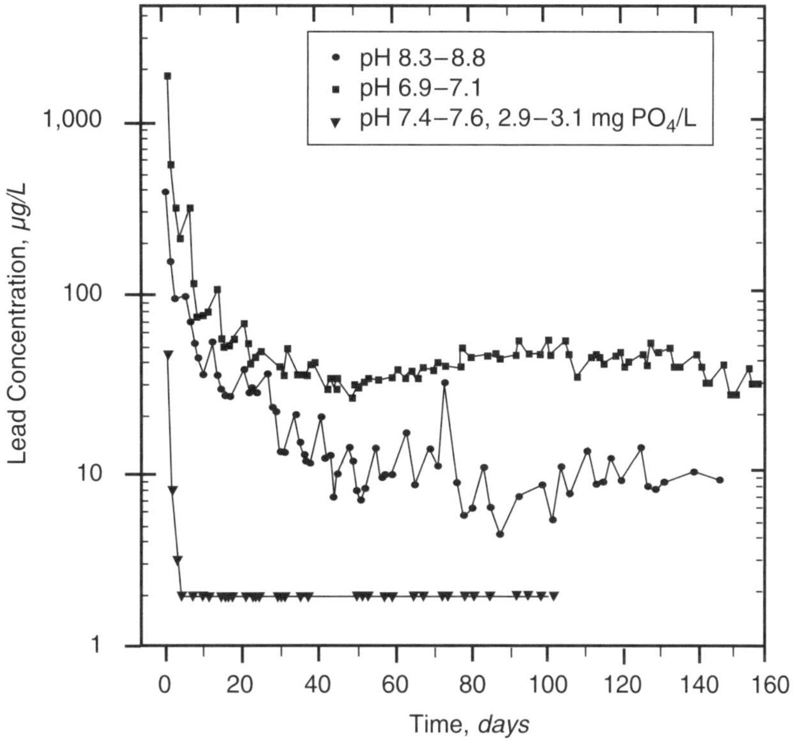

Figure 6-7 Lead release from brass alloy CDA85400, yellow brass, 1.6 percent lead, versus time in three test waters

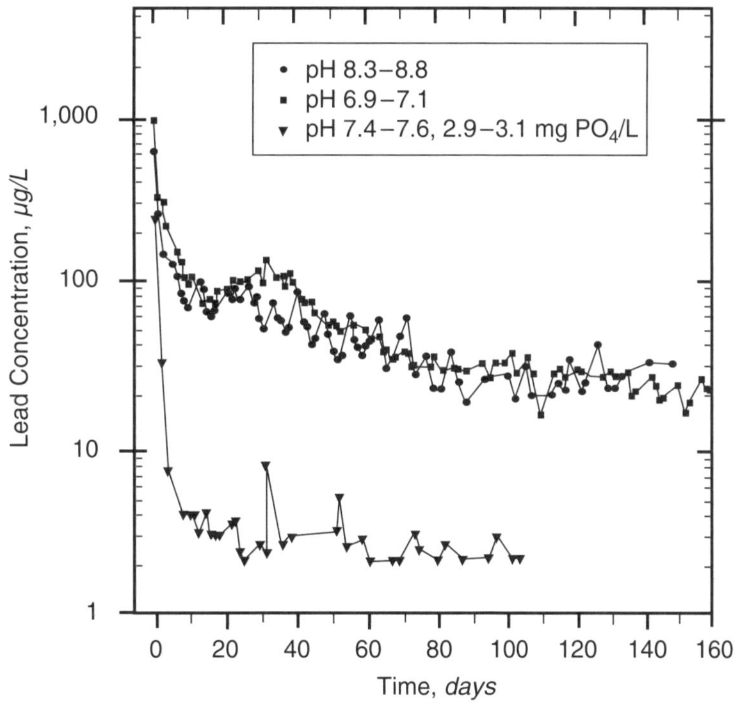

Figure 6-8 Lead release from brass alloy CDA85200, yellow brass, 2.0 percent lead, versus time in three test waters

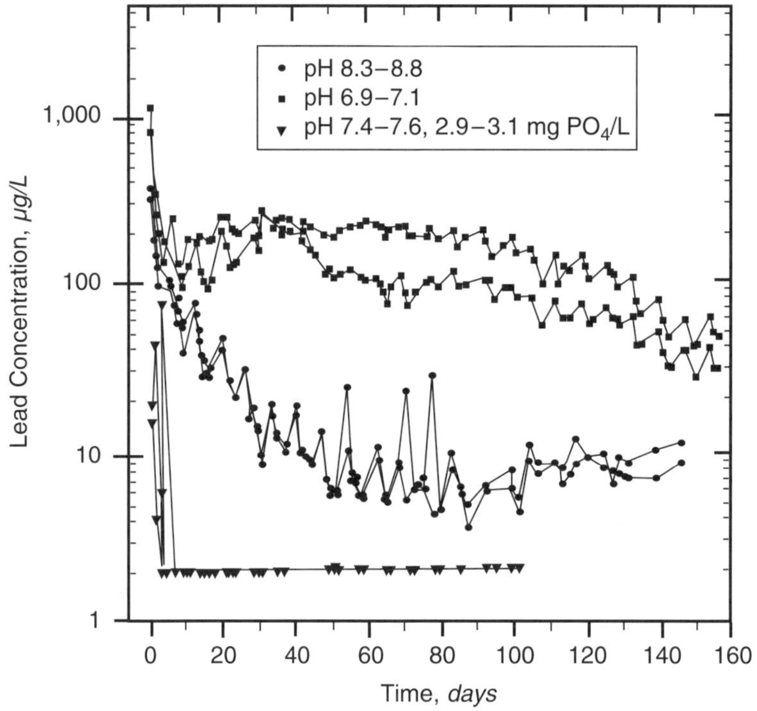

Figure 6-9 Lead release from brass alloy CDA36000, free cutting brass, 2.75 percent lead, versus time in three test waters

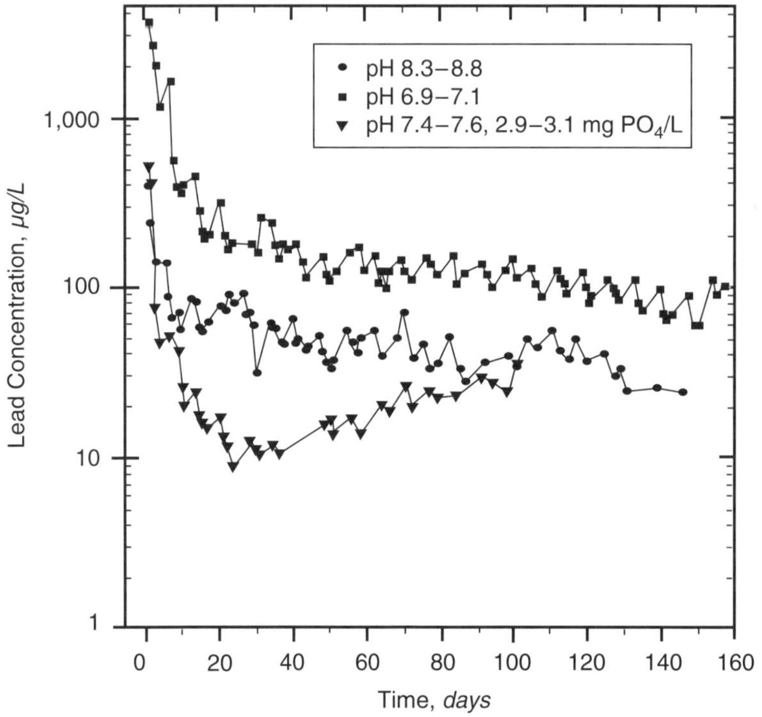

Figure 6-10 Lead release from brass alloy CDA83600, red brass, 5.1 percent lead, versus time in three test waters

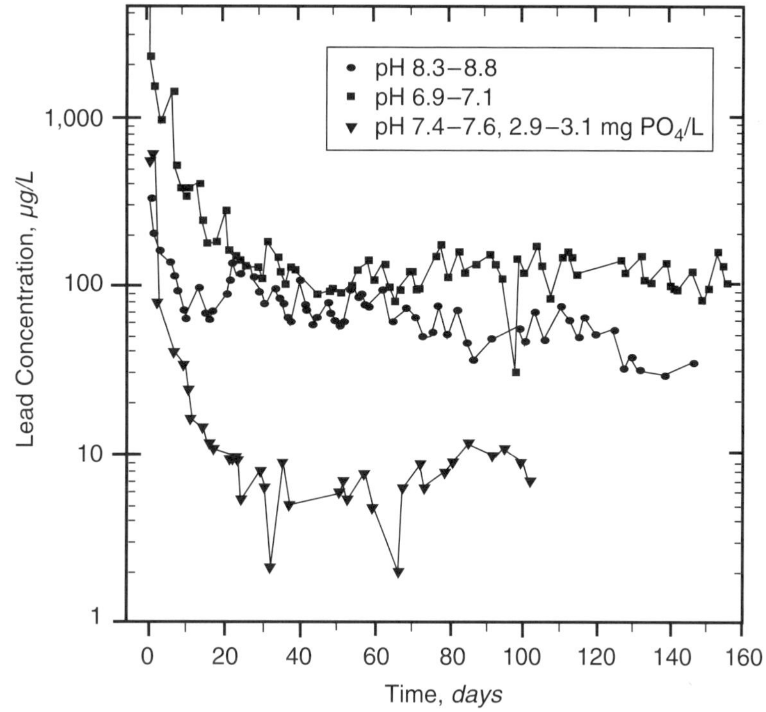

Figure 6-11 Lead release from brass alloy CDA84500, red brass, 6.0 percent lead, versus time in three test waters

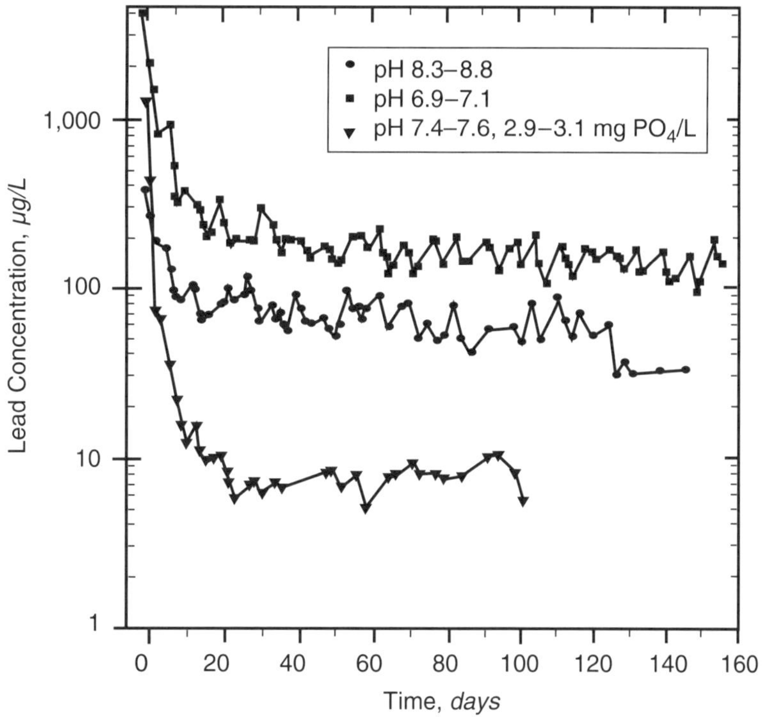

Figure 6-12 Lead release from brass alloy CDA84400, red brass, 6.6 percent lead, versus time in three test waters

products are usually composed of parts made from a variety of alloys; perhaps three or four are used in a single device. Also, the internal surface-area-to-volume ratios differ for different models of the same article. The method of manufacture can be different as well, e.g., sand cast or forged. All of these factors will influence lead leaching profiles in ways that are not quantitatively predictable.

It is difficult to predict the extent of human exposure to the alloys found in plumbing materials. Difficulties arise for two primary reasons: (1) the complexity of the chemical reactions taking place over time in different water qualities, and (2) operation of system valves and the resultant changes in water velocities, which create physical disturbances of the corrosion product formed.

Other studies. Other work (Oliphant 1993) has indicated that an increase in the pH of a soft water, to the range 8.5–9, could have the same beneficial effect as that found for contamination from the lead pipes; a mean lead contamination level from a 3 percent leaded brass of 13 μg/L was found in a solution with a carbonate hardness of 16 mg/L at pH 9 versus 39 μg/L for the same solution at pH 7. The addition of orthophosphate was also examined using the same alloy. At a level of 1 mg P/L, lead levels of 3 μg/L (standard deviation [SD] 0.5) versus a control of 46.4 μg/L (SD 1.7) were obtained (the test procedure was that of the BS 7766 method [BS 7766 1994] apart from solution composition). This would suggest that contamination from leaded copper alloys is mainly controlled by the solubility of the corrosion product.

When the coupons were tested while electrically coupled to a large area of copper, the corresponding results with the orthophosphate solution were mean 6.6 μg/L (SD 2.4) versus a control mean of 61.2 μg/L (SD 21.2). Galvanic coupling with copper has the effect of increasing the scatter in the results, making it unclear if a statistically significant increase has been produced. This area obviously requires further examination. However, orthophosphate addition as a method of avoiding contamination from leaded copper alloys could cause problems in waters that support dezincification (see the section, Water Composition Factors and Dezincification, earlier in this chapter).

STRATEGIES TO COUNTER CORROSION PROBLEMS WITH COPPER ALLOYS

Development of a Dezincification-Resistant Alloy

Following the difficulties that arose from the dezincification of duplex brass fittings in some areas in the late 1950s, manufacturers developed new ranges of capillary soldered fittings made from copper tube. The new copper fittings were generally cheaper than the brass fittings that they replaced; however, it should be noted that it is not practical to produce copper compression fittings or to use copper for complicated shapes such as valve or tap bodies.

An increased use of gunmetal, which is immune to dezincification, was another way of avoiding the problem. Although gunmetal is completely satisfactory for these purposes, cast gunmetal fittings are considerably more expensive than the same fittings made from duplex brass by hot stamping or gravity die casting. Cast gunmetal fittings also have a surface finish that is inferior.

Consequently, research was undertaken in the United Kingdom by the BNF Metals Technology Centre and others to develop a new type of brass. This new brass would contain sufficient β phase at the hot-stamping temperature to forge satisfactorily and, by subsequent heat treatment, could be converted to an all-α structure at room temperature that could be protected against dezincification by incorporating arsenic in the alloy.

The work was successful (Bower, Oseland, and Davies 1978) and resulted, in 1980, in the inclusion of a new dezincification-resistant brass, designated as CZ132 in British Standards (typical composition 61% Cu:36.4% Zn:2.5% Pb:0.11% As). CZ132 is a leaded brass. Its machinability is comparable with the leaded duplex brass, CZ122 (typical composition 58% Cu:40% Zn:2% Pb), commonly used in production of water fittings. CZ132 rods and bars for machining are heat treated by the materials supplier. CZ132 forging stock is supplied untreated and must be heat treated in the temperature range of 450 to 550°C after forging to ensure resistance to dezincification. This is done by the fittings manufacturer.

Use of Corrosion Inhibitors to Prevent Dezincification

Work using the galvanic corrosion cell described earlier identified a potable water that, although it had an unfavorable chloride-to-carbonate-hardness ratio and a pH above 8.2, appeared naturally inhibited against dezincification (Oliphant 1978). That this was due to the presence of an inhibitor, rather than the absence of some aggressive agent, was demonstrated by its ability to pass on this property when blended with supplies that otherwise supported corrosion.

This discovery led to a search for other agents that might be used as inhibitors. Of course, for those waters that are only borderline corrosive, according to the Turner diagram, blending with nonaggressive supplies or adjustment of their carbonate hardness levels might be an option.

Health constraints and costs limit potential inhibitors to the usual silicates, ortho- and polyphosphates, and zinc. The effect of silicates was found to be very sensitive to the time delay between dosing the water and its coming into contact with the duplex brass. While freshly dissolved silicate was a very effective inhibitor, its efficacy rapidly declined with time (see Figure 6-13). This time sensitivity was reduced somewhat by the simultaneous addition of zinc.

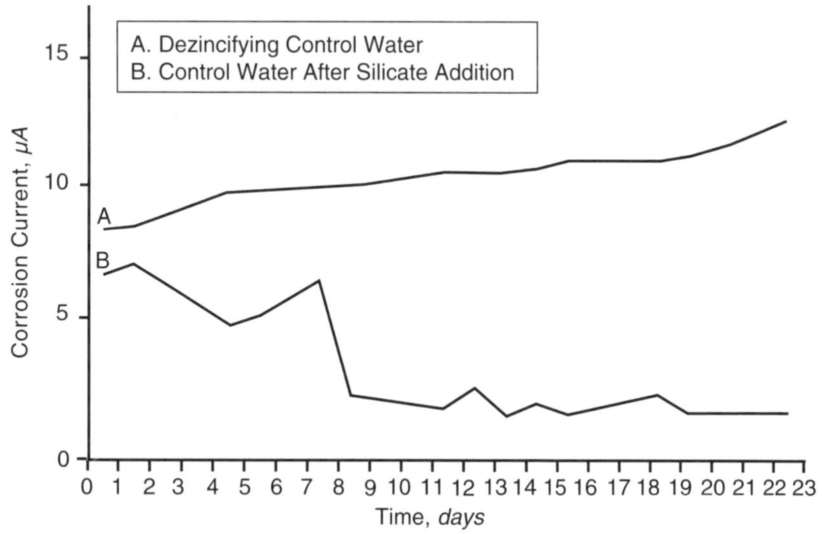

Figure 6-13 Effect of 10 mg SiO_2/L on dezincification; silicate added using a column of "neutral" sodium silicate glass, with minimum delay between addition and entry to test cell

The explanation for this activity loss may be the hydrolysis of the silicate ion or polymer to silicic acid. Certainly, by following the recommended procedure for preparing a silicic acid sol, which is used as a flocculating aid in water treatment, i.e., lowering the pH of a 2 percent silicate solution to between 8.8–9.2 and allowing it to age for 1 hour before dosing, the reagent had lost all ability to inhibit.

The previous work by Lucey using a galvanic corrosion cell (see Oliphant 1975) established that orthophosphate stimulated dezincification. Thus, dezincification problems presumably could occur in waters that support the corrosion and that are being dosed with orthophosphate to prevent lead contamination. This possibility was not investigated because at that time orthophosphate was not being widely used to control lead contamination.

Polyphosphates, either as pyro- or metaphosphates, proved to be completely effective in suppressing the corrosion, as judged by the reduction in the flow of galvanic current at a level of 5 mg P/L, provided zinc was also present at a concentration of 1 mg Zn/L (see Figure 6-14). Even after several days in contact with the test solution, the bright metallic luster on the brass anode was maintained. Any change in the type or amount of metal leaching was not examined, however, because this was not an objective of the study. Because there was little interest in treating waters in this way to avoid dezincification problems, determination of the minimum level at which significant inhibition was still achieved was not pursued.

Avoidance of Lead Contamination

As mentioned previously, the main problem with dissolution of copper alloys is the lead contamination this can produce. A lead-free alloy containing bismuth is already on the market, but it is only suitable for cast products. Currently, no authoritative sources have considered it necessary to issue a water quality standard for bismuth.

Other alloys, in which the lead is replaced by silicon or sulfur, have been investigated for both cast and wrought fixtures and fittings. Unfortunately, many of these show a significant loss of strength at water temperatures typical of those

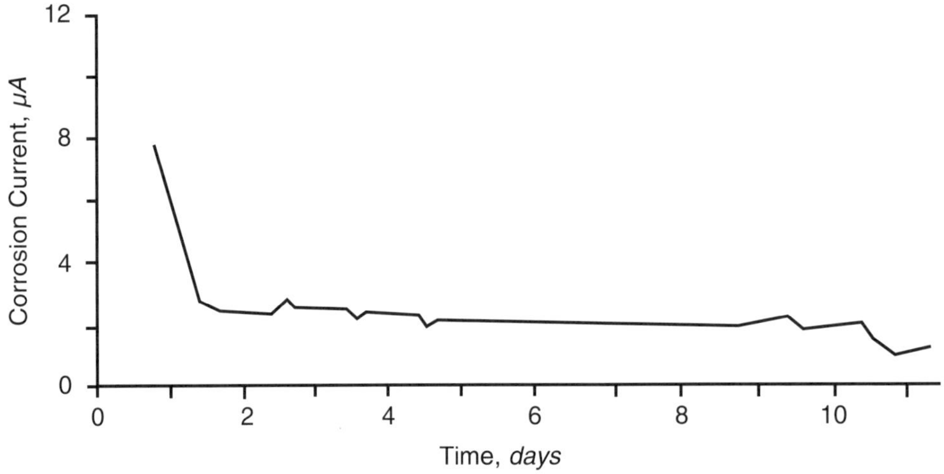

Figure 6-14 Effect of 5 mg P/L polyphosphate addition with 1 mg Zn/L dezincification (addition is to same control water described in Figure 6-13)

found in hot-water plumbing systems so that currently there are still no lead-free copper alloys suitable for wrought products.

SOLDER ALLOYS USED IN WATER SUPPLY SYSTEMS

Solders have been used in water supply systems since Roman times. Soft solders have melting points below 800°F (427°C). Alloys with melting points above 1,000°F (538°C) are known as brazing alloys; the low-temperature brazing alloys are often termed "hard solders." For soldering and brazing, the alloy used as the filler metal in the joint has a lower melting point than that of the metal sections being joined.

Tin is the most important constituent in most soft solders. This is because of its low melting point and its favorable atomic size, which allows it to bond with most common engineering metals. The other constituents in soft solders produce a melting point range for convenience of working, e.g., in the plumber's wiped joint, and may lower the basic cost of the alloy.

Traditionally, the solders used in European domestic plumbing systems have contained about 50–60 percent lead, added to this level purely to lower the cost of the alloy. However, the European standards currently being prepared will ban the use of all leaded solders in potable water systems. The specified replacements for this application are tin : silver (95% Sn:5% Ag) and tin:copper (97% Sn:3–0.4% Cu) alloys, as already sanctioned by German Standard DVGW GW2 (1980). In the United States, where solders are specified in ASTM B32-93 (1993), a tin : antimony alloy (95% Sn:5% Sb) has been favored.

Most information on the corrosion of solders in potable waters concerns leaded alloys, the ones most commonly used for the last 50 years. Because of changes in legislation, primarily to avoid health problems, this has become mainly of historical interest, certainly as far as new building construction is concerned. These considerations highlight the need for equivalent, in-depth studies of the new alloys being specified for potable water systems.

IMPACT OF REGULATORY REQUIREMENTS ON THE USE OF SOLDER ALLOYS

The potential contamination problem with leaded solders is now generally recognized both in Europe and the United States. The US LCCA of 1988 prohibits the use of solders with more than 0.2 percent lead, and a similar ban was introduced in the United Kingdom by a modification of the Model Byelaws (UK Model Water Byelaws 1986) in 1989.

The specification of alternative nonleaded alloys (discussed in the previous section) has solved this contamination problem for new installations. While there is currently no limit in the United States for antimony in drinking water, the EC Directive specifies a maximum admissible concentration of 10 μg/L, and the new WHO guidelines specify a mean of 5 μg/L.

TYPES OF SOLDER ALLOY CORROSION IN POTABLE WATERS

The main form of corrosion of solders is the slow dissolution of the solder components. This is particularly important in light of the health implications related to leaded solders.

Dissolution of Lead From Leaded Solders

The fact that tin–lead soldered joints can be a significant source of contamination gradually emerged during investigations of consumer exposure to lead from drinking waters. In a British survey (DoE 1977), many houses were found to have high lead levels in spite of the fact that no lead pipes were present in the drinking water systems. Often, the levels were as high as those found in lead-plumbed houses. It was concluded that in these copper-plumbed systems, lead was leaching from the tin–lead soldered joints.

Other studies (Dangle 1975; Karalekas et al. 1976; Britton and Richards 1981; JMM 1982, 1983; Samuels and Meranger 1984; Neff and Schock 1985; Gregory 1990) have implicated solder as the source of lead, which was often found to exceed the maximum concentration limit. Relatively high lead levels that tend to decrease with exposure time in standing samples from new copper lines in old systems have also been observed (Neff and Schock 1985; Gregory 1990).

CHEMISTRY OF SOLDER CORROSION

The corrosion of solders at joints is strongly affected by the joint's geometry and the metal being joined. Normally, the area of solder exposed is small, so that when the metal being joined is more electropositive, and the solution has sufficient conductivity, the situation of small anode–large cathode applies and the corrosion damage can be severe.

Tin–lead solders are anodic to copper and Monel metal but cathodic to steel, zinc, and cadmium. Where these alloys are used to join aluminum, destructive corrosion can occur in the presence of water. This arises from a grain boundary effect in the tin, produced by the diffusion of aluminum into the solder during the fusion process, such that the water produces a severe embrittling intercrystalline attack.

The corrosion resistance of tin–lead solders, in the absence of conditions particularly corrosive to tin, decreases as the proportion of lead increases. The correlation is not usually completely linear. Thus, in potable waters the decrease is fairly gradual up to a lead content of 70 percent; above this value the resistance falls much more quickly.

A problem specific to soldered joints is the corrosion that can occur because of the use of flux. The function of the flux is to dissolve oxide layers on the two surfaces to be joined, to maintain these surfaces oxide-free at the elevated temperatures at which the joint is made, and to lower the viscosity of the molten solder so that it runs freely into the joint space. If excessive flux is used, there is an increased risk of producing a run of solder along the bore of the tube. This increased area of exposed metal, plus the presence of any acidic residues, can increase the severity of any contamination problem.

With the active varieties of flux, e.g., those based on acid halides, it is recommended that the workpiece be washed after assembly, first in acidified water and then in neutral water to remove the last traces of flux. So-called nonaggressive fluxes, which are based on organic resins, and others that are only active when hot, are now available to minimize this problem. In the United States, fluxes are specified in ASTM B813-93 (1993). German standard DVGW GW7 recommends fluxes that are completely eliminated after the first flushing procedure with drinking water.

LABORATORY STUDIES OF WATER CHEMISTRY EFFECTS

The usual method of studying a corrosion process in which contamination is the main outcome is to construct a test rig and monitor the contaminant concentration. While this has the advantage of measuring the parameter of immediate interest, it provides little insight into the corrosion reaction itself, which is necessary if generalizations are to be made from a limited amount of empirical data.

Three notable studies have specifically observed lead leaching from solder in some detail: Wong and Berrang (1976), Lyon and Lenihan (1977), and Oliphant (1983). The first two studies primarily used the corrosion rig approach; the last study used a galvanic test cell. More recent work in this area using electrochemical techniques has been carried out by Reiber (1991).

Wong and Berrang Study

Wong and Berrang reported on the lead contamination of the water supply to Victoria, British Columbia. Although they did not find any unacceptable levels arising from solder in their survey of domestic systems, they did obtain high results from infrequently used taps in their own laboratory.

To confirm that the lead present was coming from the soldered joints, they fabricated a series of laboratory rigs 50 ft (15.25 m) long in ½-in. (15-mm) copper tube using 20 soldered joints. The joints were made with either 50:50, 60:40, or 95:5 tin–lead alloys. Results were also obtained from a copper-only rig and another in which silver was possibly used as a solder, perhaps to test the effect of the soldering process itself, although no explanation is given in the text.

Their results supported the findings of many other authors, namely that high initial lead levels gradually fall as the cumulative volume of water passed through such systems increases. All results were for water samples that had been static in the rigs for 1 hour. It was not until a cumulative volume of 12,000 L (3,170 gal) had passed through each rig that a significant difference was observed between the rig constructed using 50:50 solder and the one with the 60:40 alloy (tin–lead). Lead levels of 96 μg Pb/L for the 50:50 solder and 49 μg Pb/L for the 60:40 solder were then obtained after the 1-hour static period; no indication of the sample volume is given. As the internal volume of their rig can be calculated to have been 2 L (0.53 gal), it is probable that they used its entire contents for their sample. The other rigs, including the one with 95:5 soldered joints, all had lead concentrations as background levels of 1–3 μg Pb/L from the start.

Wong and Berrang reported dissolution rates of 400 μg Pb/soldered joint/hour for the first hour, falling to 100 μg Pb/soldered joint/hour over 24 hours, after a cumulative volume of 150,000 L (39,630 gal) had been flushed through each rig. They suggested that this volume of water was equivalent to one year of normal usage. Tests performed on a real one-year-old house gave a rate of 0.4 μg Pb/soldered joint/hour. They concluded that such systems, with a volume of about 2 L (0.53 gal), would exceed the 50-μg Pb/L Canadian limit for lead in 4–20 hours assuming a rate of 0.1–0.5 μg Pb/soldered joint/hour. It would appear, although not stated, that the total number of joints in such a system was taken to be between 10 and 50.

Their overall conclusions were as follows: the normal tin–lead solder should be replaced by the 95:5 alloy, a new system should be flushed with a year's normal water usage, and the first 2 L (0.53 gal) drawn from a tap should not be used for human consumption if it has been static in the system for more than 24 hours.

Lyon and Lenihan Study

Lyon and Lenihan conducted an extensive study on lead contamination in Glasgow, Scotland, tap water arising from solder used to join thin-walled copper tubing. A sampling program in a large, modern building revealed that of 216 samples 44 percent had lead levels greater than 0.1 mg/L, with one sample showing 4.3 mg Pb/L. No lead plumbing was present in the system, so the only plausible lead source was the lead-based solder used to join the capillary fittings.

Laboratory tests were performed on both newly made joints and joints that had been in use for several years. Tests were also performed on joints under service conditions. Experimental circuits were constructed to evaluate the leaching of lead by recirculating deionized water from new capillary fittings joining thick-wall (half-hard) tube, thin-wall (hard-drawn) tube, and thin-wall (hard-drawn) tube with poorly made joints using excessive amounts of flux. Three runs were necessary to clean the superficial dirt from compression-fitted "blank" circuits, so only the results of runs 4–6 were reported.

Even with circulation times of 1 hour or less, lead levels in excess of 0.1 mg Pb/L were reached in the approximately 20-L (5.3-gal) water volume. The removal of flux through a degreasing process resulted in a substantial reduction of the lead leaching, although it approached 0.1 mg Pb/L after 3 hours of circulation. However, the leaching of copper was more than doubled after the cleaning process, reaching almost 1 mg Cu/L after 3 hours.

Lyon and Lenihan described further tests on static deionized water and, more important, on static Glasgow tap water. The values of lead release are shown in Figure 6-15. In each experiment, the trend of the lead release as a function of time was the same; initial high values dropped to a steady value of about 0.8 mg/L after four to five weeks.

An examination of the homogeneity of water samples taken from three large buildings in the Glasgow area showed that mixing in pipe sections was relatively poor and that the distribution of lead in the water was very inhomogeneous. Thus, direct comparisons of field and laboratory data are difficult, and trends in real systems may be somewhat obscured.

Ten lengths of 15-mm (ca. ½-in.) hard-drawn tube containing one capillary fitting were taken from a Glasgow building. The copper tubing had been in use five years. Lead leaching by deionized water after 16 hours' contact was measured five times for each tubing section. The tubes were then degreased with hot water and detergent and the experiment repeated. The results are shown in Figure 6-16.

The effect of adjusting the pH of Glasgow tap water on lead leaching was estimated by using a 0.75-m (28.5-in.) length of copper tubing containing five capillary fittings. Each experiment was carried out in triplicate, with an exposure time of 16 hours. The results are plotted in Figure 6-17.

An electrochemical study was performed to look at the influence of remaining flux on the corrosion reaction. Lyon and Lenihan concluded that the flux was an excellent electrolyte and that the lead released from capillary-jointed service lines

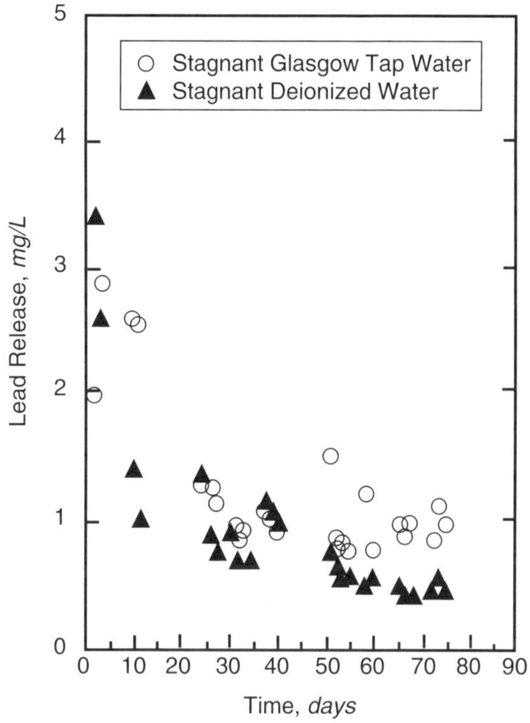

Source: Lyon and Lenihan (1977). Reprinted courtesy of the Institute of Materials.

Figure 6-15 Lead leaching from hard-drawn copper tube with well-made joints by a 16-hour exposure to Glasgow tap water

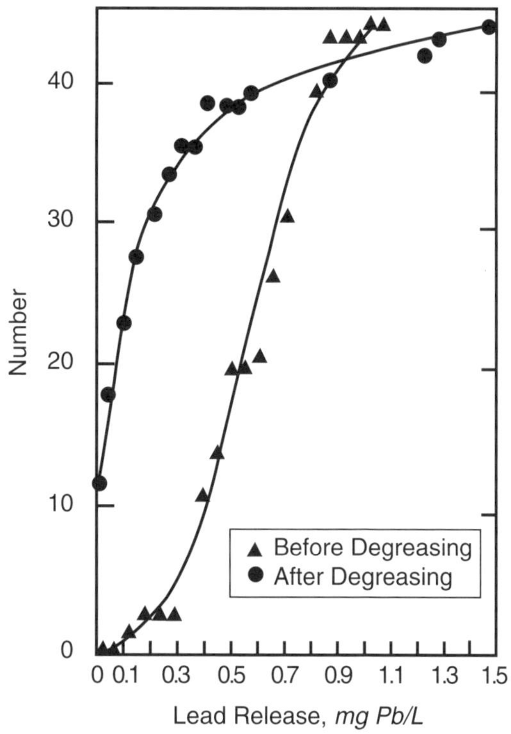

Figure 6-16 Cumulative frequency diagram of lead leaching from five-year-old capillary joints by deionized water after 16-hour exposures

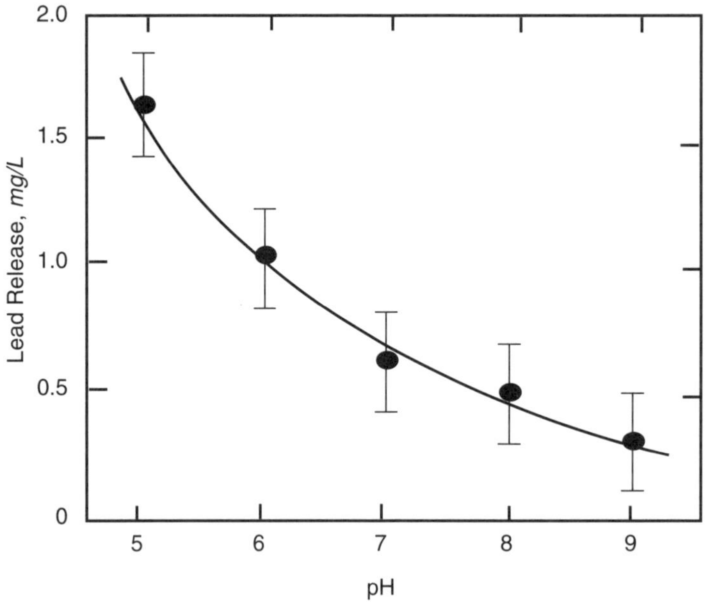

Figure 6-17 Effect of adjusting the pH of Glasgow tap water on lead dissolution from copper tubing containing five capillary fittings exposed for 16 hours

resulted primarily from this type of electrochemical corrosion of the "solder/Glasgow water/copper" cell. They further concluded that

- In the circulatory experiments, no significant difference was observed between the release from the thin-wall (hard-drawn) tube and that from the thick-wall (half-hard) tube.
- Differences arising as a result of the jointing being well or poorly made are difficult to interpret. Although it seems not unreasonable to expect that the release from a poorly made joint would be greater than that from a well-made joint, unfortunately, there is no way to control the amount and distribution of flux and solder in the tube, and the terms "well-made" and "poorly made" are necessarily subjective.
- The mean lead release in micrograms/fitting/16 hours for run 3 of the circulatory experiment and reading 3 of the static experiment was 322 μg and 216 μg, respectively; the latter figure fell to 20 μg after four to five weeks, where it remained constant for at least three months. Capillary joints taken from building B, which had been in service for five years, gave a mean release of 22 μg/fitting/16 hours. It appears then that the initial release can be in the range of 200–300 μg/fitting/16 hours and fall by a factor of 10 in four to five weeks to around a level of 20 μg/fitting/16 hours, which can be maintained for a period of many years.

Oliphant and the Effect of Water Composition

Oliphant (1983) further developed his galvanic cell approach in a comprehensive study of the lead contamination from soldered joints. The anode consisted of a 2-cm^2 (0.31-$in.^2$) solder electrode fabricated by dipping a copper template smeared with flux into a bath of molten solder. The rest of the cell was as described earlier in this chapter. Tests were run at 20°C for a 28-day period.

Validity of the test cell. Initially, it was necessary to determine the extent to which the differences between the cell design and actual soldered joints affected the results. The main differences were the presence of the large resistance in the cell's external circuit impeding the galvanic interaction between the copper and solder and the larger solution resistance caused by the physical separation of the electrodes. The significance of these factors was examined using an Evans diagram.

An Evans diagram (Figure 6-18) was constructed by changing the resistance in the external circuit of the cell through a series of values. For each resistance, after a 5-minute equilibration period, the galvanic current was determined and the potential of each electrode measured separately against a saturated calomel electrode. This information allowed the two lines, the upper one for the cathode and the lower one for the anode, to be drawn on the diagram. In this way, the different reactions occurring at the two electrodes could be studied independently.

The slope of the two lines on the Evans diagram gives the activation resistances of the oxidation (anodic) and reduction (cathodic) reactions concerned. In waters of low aggressivity, but without actual inhibitors, both electrodes were found to have an activation resistance of around 20 kΩ. The internal solution resistance, which was measured directly in the galvanic cells to take account of the electrode sizes and spacing, usually varied between 2 and 3 kΩ. Thus, with the external circuit resistance of 10 kΩ, the total cell resistance was approximately 52 kΩ. With the combined activation resistances constituting nearly 77 percent of this total circuit resistance, it was the rates of the reactions at the electrodes that were the main factors determining the value of the galvanic current in the cell, as is the case at a soldered joint.

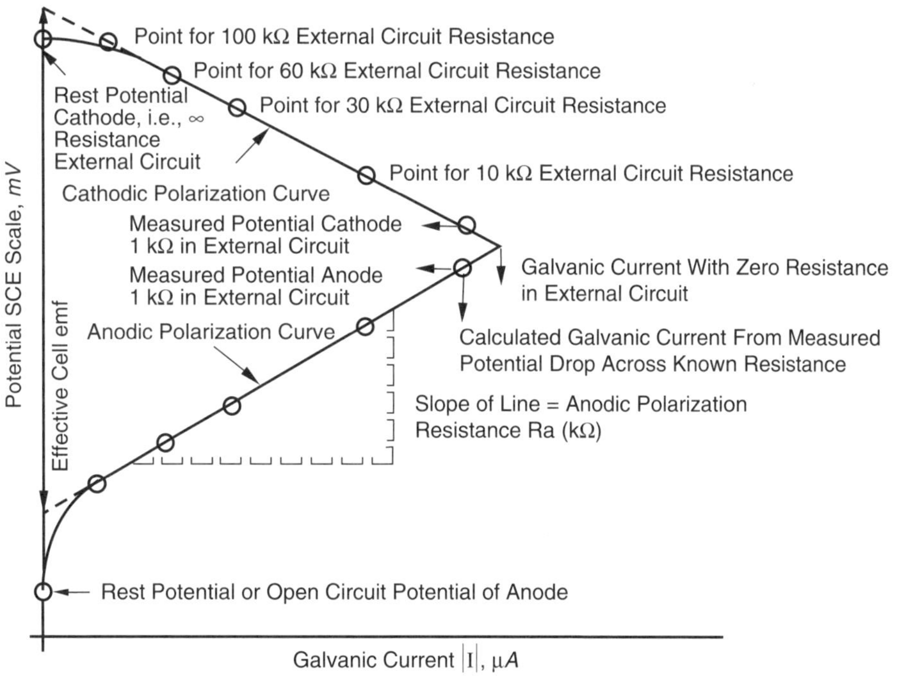

Figure 6-18 Evans diagram (schematic)

This can be taken further by extending the lines on the Evans diagram until they cross. This is done to determine the galvanic current in the absence of any resistance in the external circuit, i.e., the situation at a joint where the copper and solder are in physical contact. This can also be done by using the following formula:

$$\frac{1}{I} = \frac{1}{I'} - \frac{Re}{E} \tag{6-4}$$

Where:

I	=	galvanic current with no resistance in the external circuit
I'	=	galvanic current with a resistance in the external circuit
Re	=	the value of the external resistance
E	=	effective electromotive force (emf) of solder–copper cell obtained from the Evans diagram

Theoretically, the Evans diagram should obey the Tafel equation, which requires the logarithm of the current to obtain a straight line. However, with the usually small currents found with naturally occurring corrosion reactions, a linear dependency seems to be a good approximation.

Relationship between galvanic current and contamination levels. Of course, the galvanic current may constitute only a portion of the total corrosion current if the anode (solder) is also corroding using internal cathodes. (Any tendency to do this would be exacerbated by the resistance in the external circuit of the galvanic test cell.) Matters are further complicated by the fact that solder is an alloy of tin and lead, and this raises the question as to whether corrosion is concentrated on the lead-rich phase.

Although the corrosion current provides a direct measure of the rate of attack, there is no necessary connection between this and the level of contamination in the water, which is the question to be considered when discussing the corrosion of soldered joints. With the contamination from lead pipe, for example, it is the solubility of the corrosion product that is the controlling factor, not the rate of corrosion.

To address these questions, a series of cells were corroded under static conditions so that all the corroded material could be collected and correlated with the total coulombs of charge generated by the galvanic cell. The latter was determined by calculating the area under a plot of galvanic current against time. Where it was necessary to include the corrosion product adhering to the solder surface, the product was removed by submerging the electrode into 1 L (0.26 gal) of 1 percent boiling acetic acid for 5, 10, or 15 minutes, depending on the amount of material that had to be removed. Uncorroded electrodes were exposed to the same conditions in order to obtain a blank correction for the dissolution by the acetic acid of uncorroded metal.

This work established that for all solder compositions except the eutectic alloy (62% Sn:38% Pb), the total amount of lead corroded could be predicted from Faraday's laws of electrolysis, after allowing for the proportion of lead in the alloy (Figure 6-19). This meant that the galvanic current was the total corrosion current; i.e., the coupling with copper totally suppressed the operation of internal cathodes on the solder, even with the 10-kΩ resistance in the external circuit. The results also meant that both phases of the alloy corroded equally.

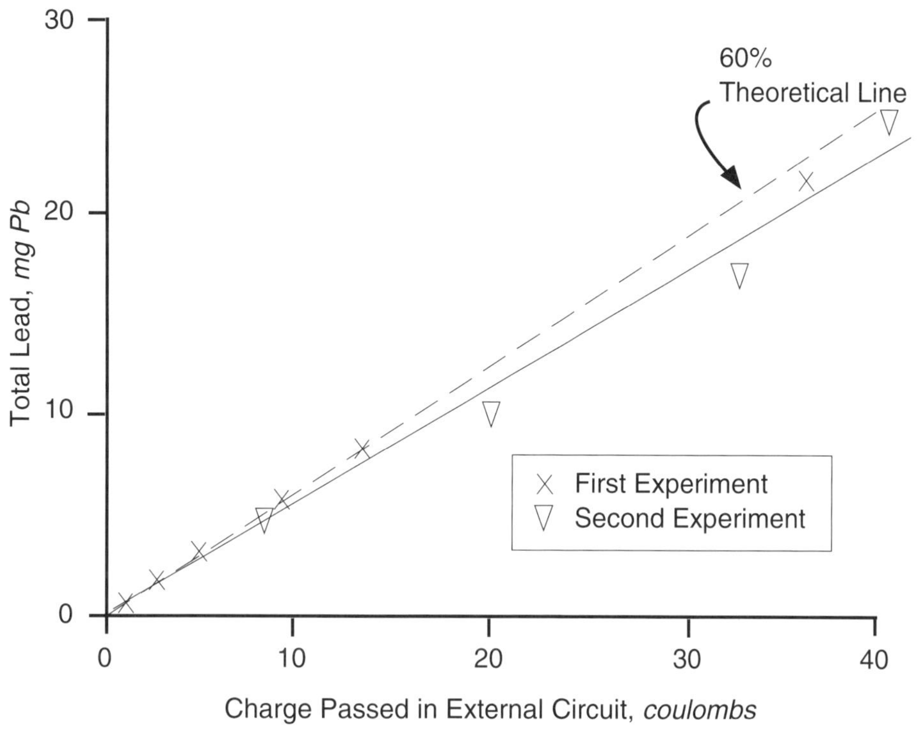

Figure 6-19 Total lead oxidized versus charge passed through the external circuit for a 40% Sn:60% Pb alloy

The exception to this was the near-eutectic alloy, which only has a single phase. Figure 6-20 shows that for this alloy the lead results fall significantly below the alloy's corresponding theoretical line calculated from Faraday's law; compare this with the results given in Figure 6-19. This suggested that for the near-eutectic alloy there is preferential corrosion of the tin. This was confirmed by a second experiment that looked at the amount of tin oxidized; these values exceeded their corresponding theoretical line by an amount equivalent to the underoxidization of the lead.

To determine how the corrosion product splits between the electrode surface and the aqueous phase, only the amount of lead in the water was determined after acidification with nitric acid. This was done in static experiments run for different periods to give a range of coulombic charge values. From the previous results, it was known that the total amount of lead corroded was that predicted by Faraday's laws. Thus, comparison between the Faraday line and the levels of lead actually appearing in the aqueous phase gave the extent of contamination arising from a given level of corrosion (Figure 6-21). The results indicated that initially very little of the corroded lead appeared in the water. However, as the corrosion product layer on the electrode surface became more complete, the level of contamination gradually increased. Eventually, the curve on the graph can be expected to become parallel with the Faraday line. This occurs when the corrosion product layer on the electrode has reached the thickness that is at the limit of its mechanical stability. After this point, all the corroded lead would enter the water, either in true solution or as particles of corrosion product, and the galvanic corrosion current would then be a measure of the average rate of contamination of the water.

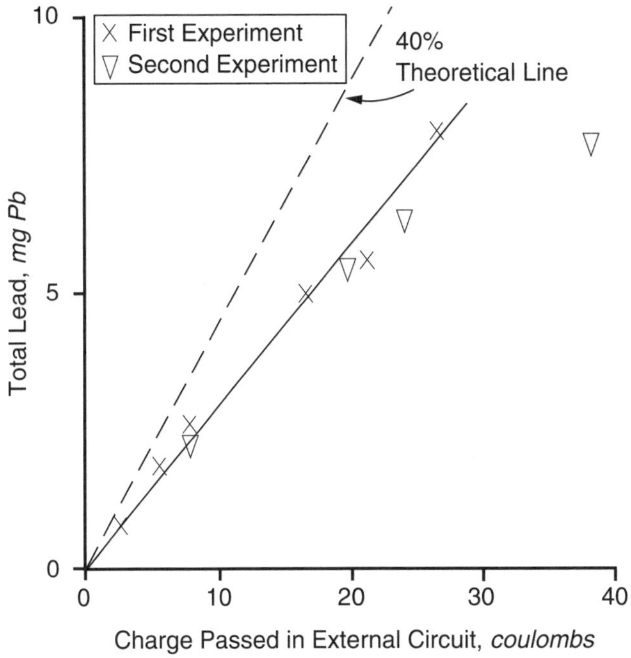

Figure 6-20 Total lead oxidized versus charge passed through the external circuit for near-eutectic alloy, 60% Sn:40% Pb alloy

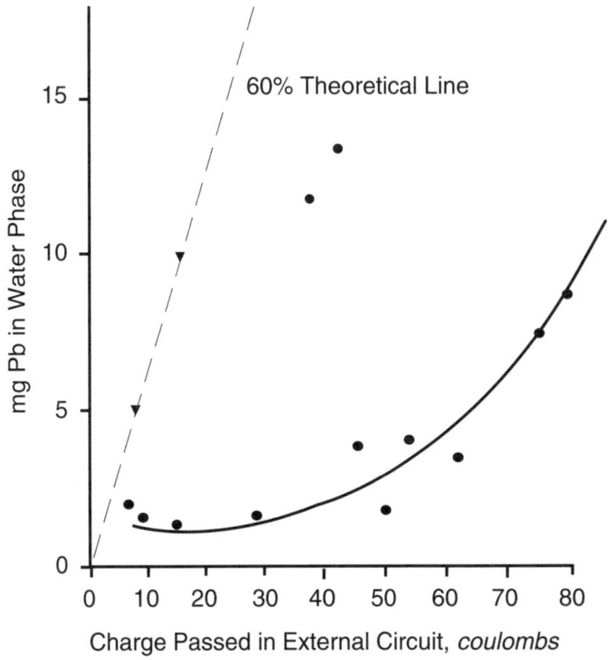

Figure 6-21 Oxidized lead in liquid phase only versus charge passed through the external circuit for a 40% Sn:60% Pb alloy

Because these experiments were run under static conditions, there was by necessity a buildup of corrosion product in the aqueous phase. In one particular case, a cell was exposed for 15 months under static conditions to 1.2 L (0.32 gal) of laboratory tap water, which was not particularly aggressive. After this period, the galvanic current had decreased to 10 percent of its four-week value and there was a white precipitate in the bottom of the container. After the precipitate was taken into solution with nitric acid, it produced a lead concentration of 2.56 g/L. The observed reduction in corrosion rate was of the same order as would be expected under flowing conditions with such a nonaggressive water. (For more aggressive waters, e.g., high-chloride waters, the long-term reduction in corrosion rate was nearer 50 percent of the fourth week value.)

These results suggest that the increased concentration of corrosion product in solution has only a limited effect on the corrosion rate, thus indicating that, unlike the case with lead pipe, reducing solubility is not effective in controlling contamination. Presumably, it is the buildup of corrosion product on the solder surface that produces the observed long-term reduction in the corrosion rate. However, before the film can achieve sufficient thickness to reduce the corrosion rate to the rate of dissolution, at which point the solubility of the corrosion product is the controlling mechanism, the film exceeds the limit of its mechanical stability. Consequently, the rate of corrosion remains the factor controlling the level of contamination where there is galvanically stimulated attack. See the last section of this chapter for further discussion of these points.

Effect of water composition. The test cell was used to determine the effect of water composition on corrosion rates. These experiments were run for 28 days and carried out under flowing conditions. The average galvanic current during the fourth week was used as the measure of corrosivity of the solution under test. These solutions were made using laboratory tap water, a hard borehole supply with virtually no organic material. The water was diluted with deionized water to give the required carbonate hardness and then various inorganic ions were added from concentrated stock solutions.

Problems were experienced with reproducibility of results. It appeared that the reactivity of the individual electrode surfaces was often a major factor determining corrosion rates; standard deviations of 40 percent were not uncommon. In some instances, all cells were started on the same solution to identify high and low electrode reactivities. The cells with different reactivities were then uniformly distributed among the solutions under test in order to eliminate electrode bias in the results. In these experiments, the corrosion currents were expressed as the difference from the mean of the first week's results for each cell. The scatter was much reduced in highly aggressive or strongly inhibited solutions, i.e., when the water compositions became the overriding factor.

Parameters found to increase the rate of corrosion were decreases in pH (Figure 6-22) and increases in chloride (Figure 6-23) and nitrate. Nitrate produced some very high but erratic results, which may have been influenced by the corresponding chloride concentration in the test solution.

Sulfate was found to stifle corrosion when present in sufficient concentration to overcome the effect of the chloride. In fact, as predictors of corrosivity, these two factors were best combined as a sulfate-to-chloride ratio (Figure 6-24).

The effect of the sulfate was not due to the formation of a more insoluble corrosion product layer; even at concentrations of 2,000 mg SO_4/L no traces of sulfate

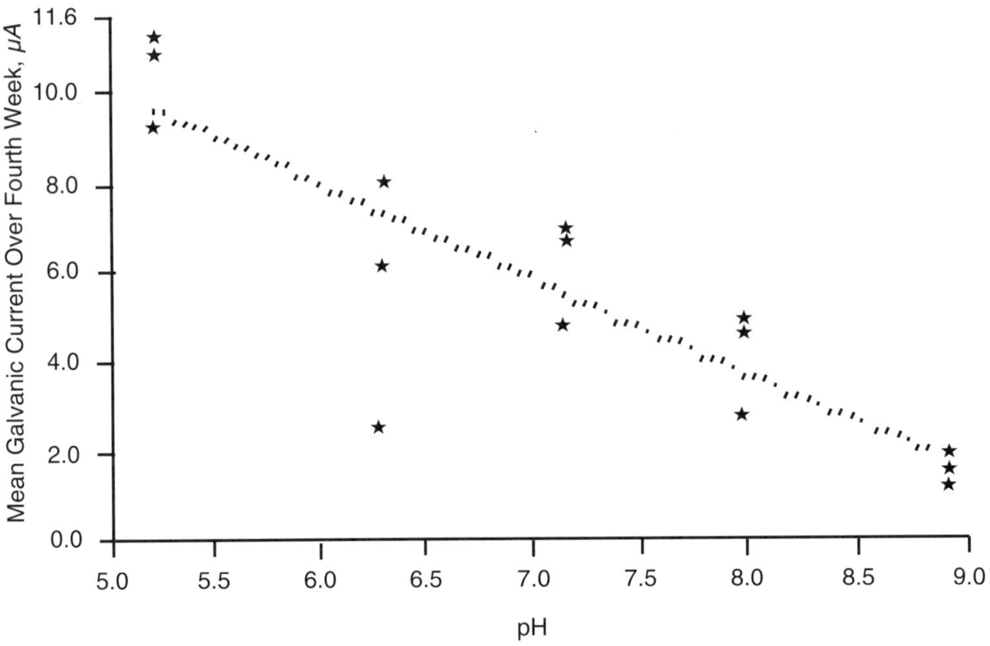

Figure 6-22 Effect of pH on the galvanic corrosion of solders

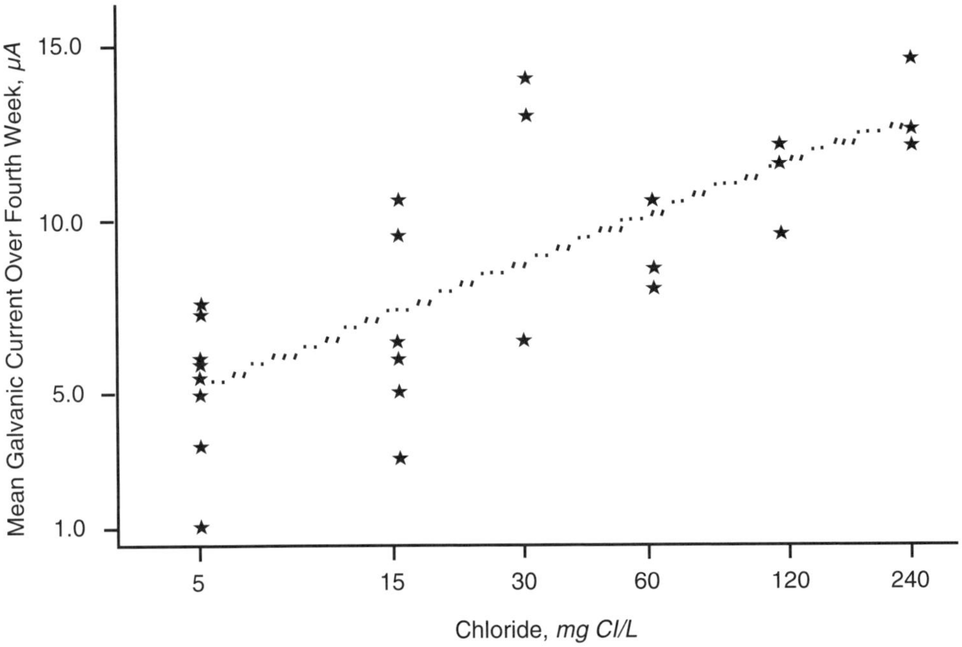

Figure 6-23 Effect of chloride on the galvanic corrosion of solders

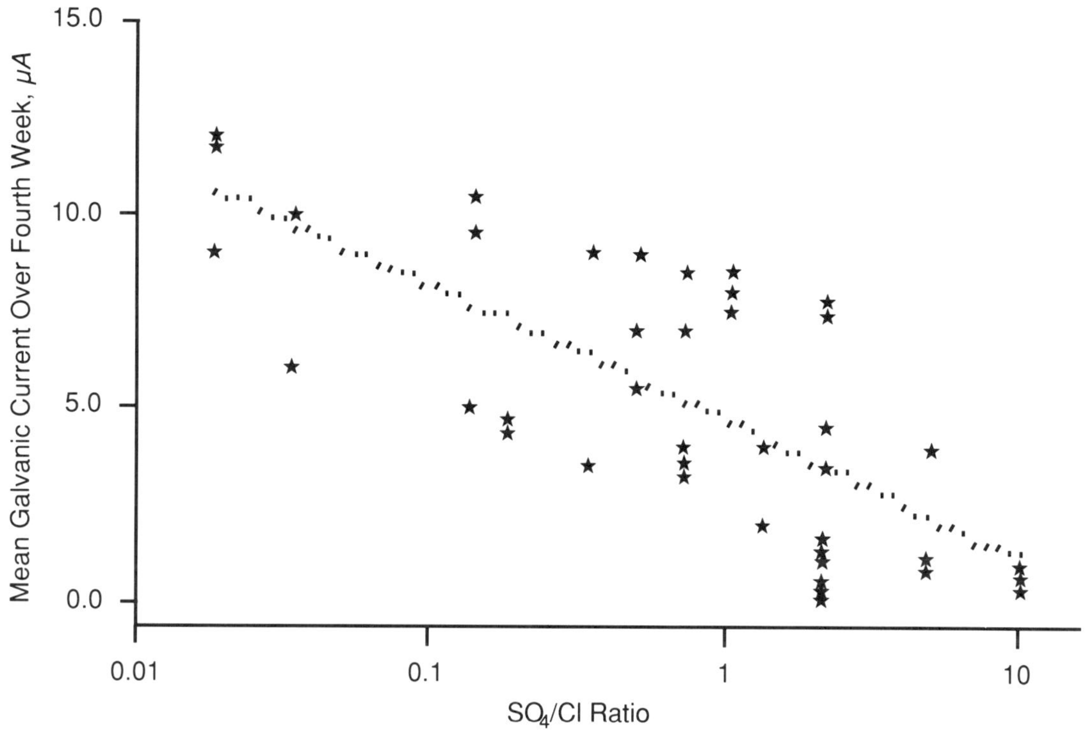

Figure 6-24 Effect of sulfate, expressed as a ratio with chloride, on the galvanic corrosion of solders

were found in the corrosion product. Its effect was to change the physical form of the normal corrosion product to crystalline plates that were more protective.

In the study carried out by Reiber (1991) using a galvanic test cell, preliminary work suggested that sulfate was without effect. As a consequence, in the main part of the project, he adjusted some of the parameters of his baseline test solution, i.e., hardness, conductivity, and pH, using compounds containing sulfate as the anion. His results showed very low corrosion rates with the solder anode assuming high positive potentials.

It appears that the test solution contained no aggressive anions, only bicarbonate, which is without effect, and sulfate, which is strongly inhibiting under the right circumstances. It was for this reason that a highly protective film developed and raised the corrosion potential of the solder in the way described. It is not to be expected that this would be a general result, because in most waters there are aggressive ions, e.g., chloride and nitrate. Carbonate hardness was found to have no effect on the corrosion rate.

The galvanic test cell was used on actual supply waters by Gregory (1990). Thirty-eight different waters from throughout the United Kingdom were investigated. These waters represented the typical extremes of water quality parameters found by laboratory work to have significant effects on galvanic corrosion.

In general, the results corroborated those found in the laboratory. However, because the water parameters could not be varied at will and, as a result, there were strong correlations between them, the trends were not as well defined. Thus, chloride was found to increase attack and sulfate to reduce it. However, statistical examination established that the chloride-to-sulfate ratio was a better predictor of corrosion than the inverse ratio used by Oliphant.

As observed previously, the presence of nitrate produced large and often erratic currents that swamped the effects of the other ions. With the sample set examined, pH was not found to have the well-defined effect observed in the laboratory.

Investigations of Nonleaded Solders

Two studies have investigated the corrosion behavior of tin–antimony (95%Sn:5%Sb) solders. In the first, Herrera, Ferguson, and Benjamin (1982) conducted a theoretical analysis of the behavior of the solders, including the application of chemical equilibrium concepts and E_H–pH Pourbaix diagrams. In addition, they performed two experimental studies. These involved coating small copper sheet specimens with the solder, so that the exposed surface area was 9.7 cm^2 (1.5 $in.^2$) of solder and 24.5 cm^2 (3.8 $in.^2$) of copper. They also used pure antimony coupons. Exposure experiments to a treated water from a Seattle, Washington, reservoir were carried out for 0.5 to 98 hours for soldered copper and 70 hours for the pure antimony. Field tests consisted of overnight standing samples of 0.2-L (0.05-gal) volume from a variety of taps in buildings at the University of Washington, where the tin–antimony solder had been in use for up to 10 years.

In the coupon tests, much less antimony was leached from the soldered specimens than from the pure antimony coupons. Antimony corrosion appeared to be stifled by the passivation of the tin in the solder. Antimony concentrations were 2–3 μg/L after standing exposures of 98 hours for the soldered coupons. The volume of the solution was not given. Interestingly, 12–23 μg/L of lead were also found in the samples that stood for 9 to 27 hours.

The tap-water samples from all but one of the sites had <0.6 μg/L of antimony. Lead contents ranged from <1 to 11.5 μg/L.

In the second study, Isaacs (1988) conducted tests on the galvanic corrosion of soldered coupons using a scanning vibrating electrode technique. He investigated both tin–antimony and tin–lead (60%Sn:40%Pb) solders. While his tests did not provide leaching data, he showed that the principal bonding of the solder to copper was predominantly through the antimony–copper phase for the tin–antimony solder. The core of antimony-depleted tin near the edges of the solder underwent increased localized corrosion.

TREATMENTS TO REDUCE CORROSION OF LEADED SOLDERS

Laboratory Studies

The laboratory studies by Oliphant (1983) established that silicate and zinc act as corrosion inhibitors for galvanically coupled leaded solders. Silicate, if present at high enough concentrations (>20 mg SiO_2/L), would maintain the bright metallic appearance of the solder surface.

The favorable effect of zinc was discovered when examining the effect of flux residues. Contrary to expectations and previous work, no increase in corrosion rates was observed in the presence of flux, even though it was lowering the pH of the test water. The flux used was a mineral grease, which was found to contain 4 percent zinc chloride. It was found that the zinc entering solution more than compensated for the known effect on the corrosion produced by the decrease in pH and increase in chloride.

Polarization measurements established that the zinc was affecting the ability of the copper electrode to function as a cathode, presumably by reacting with the hydroxyl ions being produced there and forming a precipitate of zinc hydroxide. In the experiments on zinc, it was not possible to use zinc sulfate as the salt to add the cation, because of the known effect of the anion. Consequently, zinc acetate had to be used, after first demonstrating that this anion was neutral in the corrosion of solders (Figure 6-25).

Somewhat surprisingly, Oliphant did not find orthophosphate to be an effective inhibitor of leaded solders, even at concentrations of 60 mg PO_4/L. Although it is by no means clear why these results do not concur with other work (see below), one difference is the fact that he investigated synthetic waters whereas the other studies investigated real supplies.

Field Studies

Sulfate, silicate, zinc, and orthophosphate (the last two separately and together) additions were examined as potential inhibitors in the work by Gregory (1990). The details are summarized in Tables 6-4 and 6-5. The principal observations were

- all treatments reduced the corrosion rate by half in the low-alkalinity water tested
- in the three high-alkalinity waters tested, silicate and sulfate dosing were ineffective; the effectiveness of orthophosphate dosing seemed to increase with time
- zinc dosing, especially in combination with orthophosphate, was the most effective treatment for all waters

The optimum dose of zinc was 1.0 mg Zn/L, with orthophosphate also present at about 0.4 mg P/L, but this depended on the water quality and the solubility of zinc. Doses much greater than the saturation solubility could lead to dirty-water problems.

The overall conclusion of this work was that, despite the possibility of water treatments, the use of nonleaded solders for potable water systems was still the most effective, convenient, and relatively inexpensive option.

The results obtained by Oliphant for orthophosphate are not inconsistent with those from Gregory's work. Oliphant's test solutions were made by diluting laboratory tap water, an organically pure borehole supply, with distilled water. This produced a "baseline" water, containing 50 mg $CaCO_3$/L carbonate hardness and little else. Such a water is very rare in the natural world because soft waters tend to come from upland impounded sources and contain all sorts of organic materials, as well as inorganic constituents. Organic materials can have a significant effect on galvanic corrosion processes. This was illustrated by the discovery in the Oliphant study of a water naturally inhibited against dezincification even though its inorganic makeup ought to have supported the corrosion.

In Gregory's results (Table 6-4), the phosphate added had no inhibitory effect on the borehole water tested (water 1). Inhibition was only observed with the surface-derived sources in which organics would be present. Thus, a simple solubility model to explain the observed effect of phosphate may be misleading.

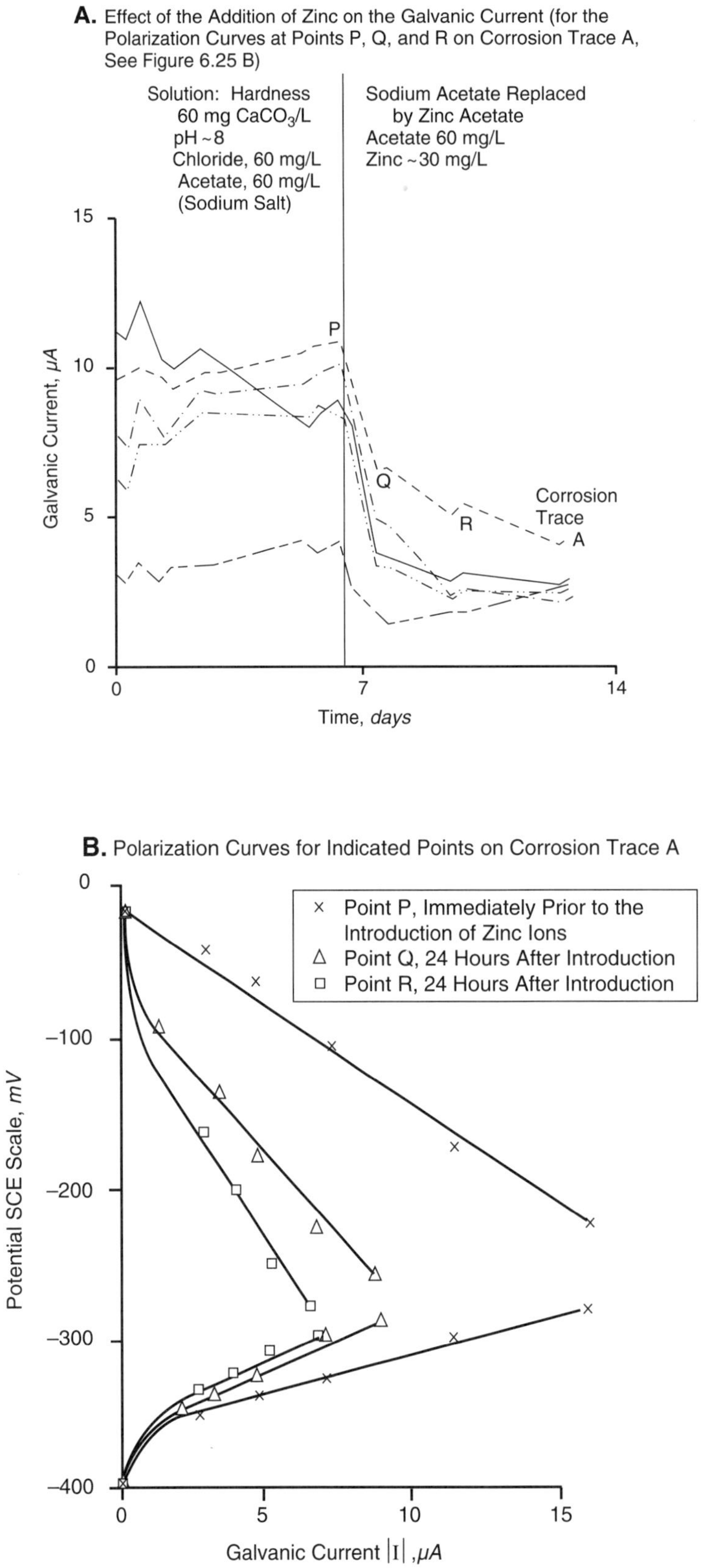

Figure 6-25 Effects of dosing zinc on the galvanic corrosion of solders

Table 6-4 Field study results on the effect of water treatment on the galvanic corrosion of solders*

Reagent	Approximate Concentration, μg / L	Reduction in Corrosion Potential Compared to Control, % Water 1	Water 2	Water 3	Water 4	Average
SiO_2	15 as SiO2	4	4	73	–6	19
SiO_2	10 as SiO_2	0	–13	52	—	13
$MgSO_4$	10 as SO_4	– 34	—	47	—	7
Na_2HPO_4	2 as P	–1	16	57	34	27
$ZnSO_4$	5 as Zn	55	14	65	78	53
$ZnSO_4$	0.5 as Zn	—	—	—	40	40
$ZnSO_4$ + Na_2HPO_4	0.5 as Zn	—	28†	—	75	52
$ZnSO_4$ + Na_2HPO_4	0.2 as P					

Source: Adapted from Gregory (1990).

*See Table 6-5 for average qualities of waters before test period.

†Two-week test period only.

Table 6-5 Average qualities of water during test period, before addition of reagents

Characteristic	Water 1	2	3	4
pH	7.4	7.6	8.3	8.5
Alkalinity, *mg $CaCO_3$/L*	264	185	10	116
Nitrate, *mg N/L*	6.9	13	0.28	5.9
Chloride, *mg Cl/L*	19.6	44.7	8.5	69
Sulfate, *mg SO_4/L*	26.9	121	6.3	133
Chloride to sulfate ratio	0.73	0.37	1.35	0.52
Silicate, *mg SiO_2/L*	15.3	3.8	2.6	1.5
General description	Chalk groundwater, no chemical treatment	Lowland river water, full chemical treatment	Impounded upland water, no coagulation, pH adjusted	Stored lowland river water, coagulated and softened

Source: Adapted from Gregory (1990).

SOLUBILITY AND CORROSION MODELS AS THE CONTROLLING MECHANISMS DETERMINING CONTAMINATION LEVELS

General Considerations

A metal's tendency to contaminate water is determined by the rates of two independent processes—the rate of corrosion and the rate of dissolution of the corrosion product. Initially, with bare metal, the corrosion rate far exceeds the dissolution rate so a corrosion product layer builds over the metal's surface. As this layer tends to stifle the corrosion, the corrosion rate falls toward that of the dissolution rate. If the corrosion product layer can achieve sufficient thickness, the corrosion rate can be reduced to the dissolution rate. At this point, equilibrium is achieved and the level of contamination is controlled by the dissolution rate. The dissolution rate, other things being equal, e.g., the residence time of the water in the pipe, etc., is controlled by the solubility of the corrosion product in the supply.

The thickness of the corrosion product actually required to reduce the corrosion rate to the dissolution rate is determined by various factors. Chief among these are the magnitude of the initial corrosion rate and the effectiveness of the product layer in stifling the corrosion. The former is determined by such factors as the aggressivity of the water and the latter is often determined by the physical form of the corrosion product. In situations where the corrosion rate is very fast and/or the corrosion product is not very effective at stifling the reaction, the thickness of the layer required may be beyond its mechanical strength. In these circumstances, a stable layer of sufficient thickness will never be formed and the contamination will remain in the control of the corrosion rate. Consequently, under these circumstances, changing the solubility of the corrosion product in the water will not be an effective treatment.

The Mechanism Controlling Contamination From Lead Pipe

The corrosion rate of lead pipe in potable water is sufficiently low for the corrosion product layer to be able to stifle the rate of attack so that it is equal to that of the dissolution rate within the limits of the layer's mechanical stability. Water treatment to reduce the solubility of the corrosion product causes the layer to thicken, which in turn reduces the corrosion rate further. As a result, a new equilibrium is achieved at a lower level of contamination.

However, in certain waters, e.g., those with high particulate iron levels or high color, the mechanical stability of the corrosion product formed in a lead pipe can become compromised. Under these circumstances, even the thin layer normally required to bring the process under solubility control cannot be attained and high particulate lead levels appear in the supply.

The Controlling Mechanism in the Presence of Galvanic Coupling

For galvanically coupled lead or solder, the corrosion rate in water is significantly higher than it is for lead pipe. Also, the corrosion deposit tends to have a less compact, more friable structure. The first of these factors means that, other things being equal, the corrosion product layer will have to be thicker in order to reduce the corrosion rate to the dissolution rate. The second implies that the mechanical strength of the layer is unlikely to be sufficient to sustain such a layer. It is thus hardly surprising that the contamination from soldered joints, or that from galvanically stimulated lead corrosion, remains under corrosion control and is largely unaffected by reductions in the solubility of the corrosion product.

The favorable effect of sulfate on galvanically coupled lead and solder arises because of the changed physical form of the corrosion product. This layer is made far more effective at stifling the corrosion at a thickness well within its mechanical stability. The effect of sulfate illustrates that the solubility of the corresponding lead salt is not a reliable indicator of which ions will have a favorable affect on the corrosion rate. Chloride shows behavior similar to that of sulfate in protecting freely corroding lead. However, when the corrosion potential is raised by galvanic coupling, the effect disappears and chloride becomes an aggressive ion. Thus, potential is also a factor, in a more subtle way than is normally recognized, in determining contamination levels.

The mechanism whereby orthophosphate reduces contamination from galvanically coupled solders is not clear. The temptation is to ascribe it to a simple solubility mechanism, but neither Oliphant in the laboratory nor Gregory in his field studies could get it to function as an inhibitor with organically pure borehole waters. If the presence of organic material is also required, this implies a more subtle mechanism than simple solubility.

Apparently favorable results were observed for the effect of orthophosphate on the contamination from a galvanically coupled leaded copper alloy in an organically pure test water (see the section, Effects of Alloys on Water Quality, earlier in this chapter). This may be an illustration of how misleading it can be to only determine contamination levels without looking at the corresponding corrosion rates. The situation may be like that illustrated in Figure 6-18 where, because the solubility has been significantly reduced by the addition of orthophosphate, initially little contamination appears in the water phase. It is only when the buildup of the corrosion product has progressed to the point where the limit of its mechanical stability has been reached that significant (particulate) contamination will be observed in the aqueous phase. The tests in question were not run long enough (only 14 days) for this point in the process to have been reached. Certainly, the scatter observed in all the results from the galvanically coupled leaded copper alloy is suggestive of increased instability in the corrosion product resulting in an intermittent particulate contamination problem.

From the above considerations, it will be apparent that contamination from metals in contact with water is a far more subtle and varied process than a simple solubility mechanism would allow. Although this represents an inconvenience to the end user of these materials, it provides the subject with its interest and intellectual fascination for the researcher.

SUMMARY

Copper alloys and solders are a rather disparate group of metals used in drinking water systems mainly to join copper pipe. Many of the alloys are potential sources of lead contamination, the significance of which has increased with the reduction in the maximum concentration of lead acceptable in drinking water. For the solders, suitable nonleaded compositions are already available for potable water applications. However, this is not yet the case for all the fabrication methods in which copper alloys are currently used. There are also significant cost implications in not using leaded copper alloys.

Although each has its individual modes of failure, the copper alloys and solders are generally resistant to corrosion in potable waters. Because of their use with copper pipes, the possibility of galvanic interactions is built in to their normal applications. This makes the mechanism of contamination by lead from leaded solders quite distinct from that from lead pipe, which in turn changes the effectiveness of the potential remedial measures. The mechanism controlling lead contamination from leaded copper alloys remains to be established.

REFERENCES

ASTM (American Society for Testing and Materials). B 5-89. 1989. *Standard Specification for Electrolytic Tough-Pitch Copper Refinery Shapes. Philadelphia, Pa.: American* Society for Testing and Materials.

ASTM B 32-93. 1993. *Standard Specification for Solder Metal.* Philadelphia, Pa.: American Society for Testing and Materials.

ASTM B 62-93. 1993. *Standard Specification for Composition Bronze or Ounce Metal Castings.* Philadelphia, Pa.: American Society for Testing and Materials.

ASTM B 813-93. 1993. *Standard Specification for Liquid and Paste Fluxes for Soldering Applications of Copper and Copper Alloy Tube.* Philadelphia, Pa.: American Society for Testing and Materials.

ASTM DS-58C and SAE (Society of Automotive Engineers, Inc.) HSJ 1086. 1986. *Metals and Alloys in the Unified Numbering System.* 4th ed. Philadelphia, Pa.: American Society of Testing and Materials and Society of Automotive Engineers.

BNF Metals Tech. Ctr. 1982. *Dezincification in Supply Waters.* Miscellaneous Publication 491.

Bower, J.E., P.W.R. Oseland, and G.C. Davies. 1978. Development of a Hot-Stamping Brass Resistant to Dezincification. *Br. Corros. Jour.,* 13:4.

Britton, A., and W.N. Richards. 1981. Factors Influencing Plumbosolvency in Scotland. *Jour. Inst. Water Engrs. and Scientists,* 35(4):349.

BS 699. 1984. *Copper Direct Cylinders for Domestic Purposes.* London: British Standards Institution.

BS 1566. 1984. *Copper Indirect Cylinders for Domestic Purposes.* London: British Standards Institution.

BS 2872. 1989. *Copper and Copper Alloy Forging Stock and Forgings.* London: British Standards Institution.

BS 2874. 1986. *Copper and Copper Alloy Rod and Sections (Other Than Forging Stock).* London: British Standards Institution.

BS 7766. 1994. *Specification for Assessment of the Potential for Metallic Materials to Affect Adversely the Quality of Water Intended for Human Consumption.* London: British Standards Institution.

Dangle, R.A. 1975. Study of Corrosion Products in the Seattle Water Department Tolt Distribution System. EPA 670/2-75-036. National Environmental Research Center. Cincinnati, Ohio: U.S. Environmental Protection Agency.

DoE (Department of the Environment). 1977. Lead in Drinking Water: A Survey in Great Britain 1975–1976. Pollution Paper No. 12. London: Her Majesty's Stationery Office.

DVGW GW2. 1983. *Verbinden von Kupferrohren für die Gas- und Wasserinstallation innerhalb von Grundstücken und Gebäuden.* Bonn: Wirtschafts- und Verlagsges, mbH.

DVGW GW7. 1992. *Flux for Soldering of Copper Pipes for Gas and Water Installation; Specifications and Test Requirements for Suitability Test.* Bonn: Wirtschafts- und Verlagsgesellschaft Gas und Wasser, mbH.

EC (European Community). 1980. Council Directive of 15th July, 1980 Relating to the Quality of Water Intended for Human Consumption (80/778/EEC). *Off. Jour. Eur. Comm.,* 23 (L 229/11).

Gardels, M.C., and J. Sorg. 1989. A Laboratory Study of the Leaching of Lead From Water Faucets. *Jour. AWWA,* 81(7):101.

Gregory, R. 1990. Galvanic Corrosion of Lead Solder in Copper Pipework. *Jour. Inst. Wat. and Envir. Managers,* 4(2):112.

Heidersbach, R.H., and E.D. Verink. 1972. The Dezincification of Alpha and Beta Brasses. *Corrosion—NACE,* 28(11):397–418.

Herrera, C.E., J.F. Ferguson, and M.M. Benjamin. 1982. Evaluating the Potential for Contaminating Drinking Water From the Corrosion of Tin-Antimony Solder. *Jour. AWWA,* 74(7):368.

Horton, R.M. 1970. New Metallographic Evidence for Dezincification of Brass by Redeposition of Copper. *Corrosion—NACE,* 26(6):160–163.

Isaacs, H.S. 1988. The Measurement of the Galvanic Corrosion of Soldered Copper Using the Scanning Vibrating Electrode Technique. *Corros. Sci.,* 28(6):547.

ISO (International Organization for Standardization) 6509. 1981. *Corrosion of Metals and Alloys—Determination of Dezincification Resistance of Brass.* Geneva, Switzerland: International Organization for Standardization.

Jester, T.C. 1985. Dezincification Update. *Jour. AWWA,* 77(10):67.

JMM (J.M. Montgomery Consulting Engineers, Inc.). 1982. Internal Corrosion Mitigation Study, Final Report. Prepared for Portland, Oregon, Water Bureau.

JMM (J.M. Montgomery Consulting Engineers, Inc.). 1983. Internal Corrosion Mitigation Study, Addendum Report. Prepared for Portland, Oregon, Water Bureau.

Karalekas, P.A., G.F. Craun, A.F. Hammonds, C.R. Ryan, and D.J. Worth. 1976. Lead and Other Trace Metals in Drinking Water in the Boston Metropolitan Area. *Jour. NEWWA,* 90:150.

Langenegger, E.E., and F.P.A. Robinson. 1969. A Study of the Mechanism of Dezincification of Brasses. *Corrosion—NACE,* 25(2):59–66.

Lee, R.G., W.C. Becker, and D.W. Collins. 1989. Lead at the Tap: Sources and Control. *Jour. AWWA,* 81(7):52.

Lucey, V.F. 1965. The Mechanism of Dezincification and the Effect of Arsenic. *Br. Corros. Jour.,* 1(1):9–14, 53–59.

Lyon, T.D.B., and J.M.A. Lenihan. 1977. Corrosion in Solder Jointed Copper Tubes Resulting in Lead Contamination of Drinking Water. *Br. Corros. Jour.,* 12(1):41.

Lytle, D.A., and M.R. Schock. 1993. Impact of pH and Lead Composition on Metal Leached From Brass Coupons. In *Proc. of the AWWA Annual Conference.* Denver, Colo.: American Water Works Association.

Neff, C.H. 1985. Impact of Copper Galvanised Pipe and Fittings on Water Quality. In *Plumbing Materials and Drinking Water Quality: Proceedings of a Seminar.* EPA/600/9-85/007. Cincinnati, Ohio: U.S. Environmental Protection Agency.

Neff, C.H., and M.R. Schock. 1985. Corrosion Product—Corrosion Rate—Water Quality Investigation in Building Systems. Final Report on Project EPA CR808566010. Drinking Water Research Division. Cincinnati, Ohio: U.S. Environmental Protection Agency.

Neff, C.H., M.R. Schock, and J.I. Marden. 1987. Relationships Between Water Quality and Corrosion of Plumbing Materials in Buildings. EPA/600/S2-87/036. Cincinnati, Ohio: U.S. Environmental Protection Agency.

Nielsen, K. 1975. Measurement and Test Procedure for Cadmium and Lead Pick Up in the Water From Drinking Water Installations. *Scand. Corros. Consr.,* 7:36.

Oliphant, R.J. 1975. *Dezincification of Domestic Plumbing Fittings—Limitations of the Turner Diagram.* TM 112. Medmenham, England: Water Research Centre.

———. 1978. *Dezincification by Potable Water of Domestic Plumbing Fittings: Measurement and Control.* TR88. Medmenham, England: Water Research Centre.

———. 1983. *Summary Report on the Contamination of Potable Water by Lead From Soldered Joints.* ER 125E. Medmenham, England: Water Research Centre.

———. 1993. Changing Perceptions of the Significance of Potential Sources of Lead Contamination in Domestic Water Systems. *Water Supply,* 11(3/4):399.

Purdie, L. 1985. Fife Engineering Department, Fife House, North Street, Glenrothes, Fife, Scotland. Personal communication.

Reiber, S. 1991. Galvanic Stimulation of Corrosion on Lead-Tin Solder-Sweated Joints. *Jour. AWWA,* 83(7):83.

Samuels, E.R., and J.C. Meranger. 1984. Preliminary Studies on the Leaching of Some Trace Metals From Kitchen Faucets. *Wat. Res.,* 18(1):75.

Schock, M.R., and C.H. Neff. 1988. Trace Metal Contamination From Brass Fittings. *Jour. AWWA,* 80(11):47.

Sugawara, H., and H. Ebiko. 1967. Dezincification of Brass. *Corros. Sci.,* 7:512–523.

UK Model Water Byelaws. 1986. *Guide to the Application and Interpretation of the Model Water Byelaws.* Chichester, England: Ellis-Horwood.

USEPA (U.S. Environmental Protection Agency). 1988. The Lead Contamination Control Act. 24 CFR Part 200. *Federal Register,* April 6, 53(66):11270–11271.

———. 1991. Drinking Water Regulations; Maximum Contamination Level Goals and National Primary Drinking Water Regulations for Lead and Copper. *Federal Register,* June 7, 56(135):32112.

WHO (World Health Organization). 1993. *Guidelines for Drinking-water Quality.* Geneva, Switzerland: World Health Organization.

Wong, C.S., and P. Berrang. 1976. Contamination of Tap Water by Lead Pipe and Solder. *Bull. Envir. Contam. Toxicol.,* 15(5):530.

7

Cement-Based Materials

Pierre Leroy, Michael R. Schock,
Ivo Wagner, and Heinrich Holtschulte

The corrosion of cement-based materials is of concern for several reasons. Though there are fewer health concerns about waterborne asbestos fibers, there are still standards governing the number and size of fibers allowed in the water. Further, there are still concerns about inhalation of airborne asbestos from showers, dehumidifiers, etc., and the clogging of filter systems.

The degradation of water quality by the leaching of hydroxide ion and hardness is a major water treatment concern. Stabilization of the chemical and structural integrity of asbestos–cement, cement, and cement–mortar-lined distribution pipes should be an important economic and infrastructure maintenance concern for all utilities.

This chapter begins with an introduction to the chemistry and physical nature of cementitious materials used for the manufacture or rehabilitation of distribution system piping. That is followed by a discussion of the chemistry of the corrosion of these materials and strategies for stabilization and corrosion control for the cement matrix and asbestos fibers. Several sections include information on methodologies for assessing the condition of the pipe surfaces in contact with the water.

The last part of the chapter gives detailed information on techniques available for the rehabilitation of distribution system pipes by cement–mortar lining and related processes, as well as relevant engineering and economic considerations.

NOTE: The first edition of this chapter was authored by the late Heinrich Holtschulte and by Michael R. Schock. This chapter was chapter 6 in the first edition.

MATERIAL PROPERTIES

The cement-based materials discussed in this chapter include two general components:

- The aggregates—these constitute the basic structure of the material, e.g., sand in the case of cement–mortar lining, sand and gravel in the case of reinforced or prestressed concrete pipes, asbestos fibers in the case of asbestos–cement pipes.
- The binder—responsible for the cohesion and mechanical properties of the material.

Figure 7-1 shows the basic structure of cement-based material, exemplified by a cement–mortar lining.

Types of Cement Used

The binder is made of hydrated cement obtained by reacting water with powdered cement. The powdered cement consists of calcium silicates and calcium aluminates in various proportions depending on the type of the cement.

Cement is made of a calcinated mixture of calcareous and argillaceous rocks with siliceous or aluminous compounds. The basic mixture, the clinker, is made from calcareous and argillaceous rock. Other types of cement can be made with pozzolane, blast furnace slag, fly ash, or bauxite ($Al_2O_3{\cdot}2H_2O$). They contain calcium oxide, aluminum oxide, or silica. These cements can be described using a triangular graph as shown in Figure 7-2.

Generally, the concrete used for water treatment plant pipes, water tanks, and filters is made of portland cement, whereas cement–mortar linings can be made of different types of cements, i.e., portland cement, high-alumina cement, blast furnace cement, or super-sulfated cement. The standard composition of these types of cements are given in Table 7-1.

The different basic elements indicated in the table are present as compounds that vary with the basic chemicals used. Portland cement contains calcium silicates (tricalcium silicate, abbreviated as C_3S, and dicalcium silicate, abbreviated as C_2S), calcium aluminate (abbreviated as C_3A), iron calcium aluminate (abbreviated as C_4AF),

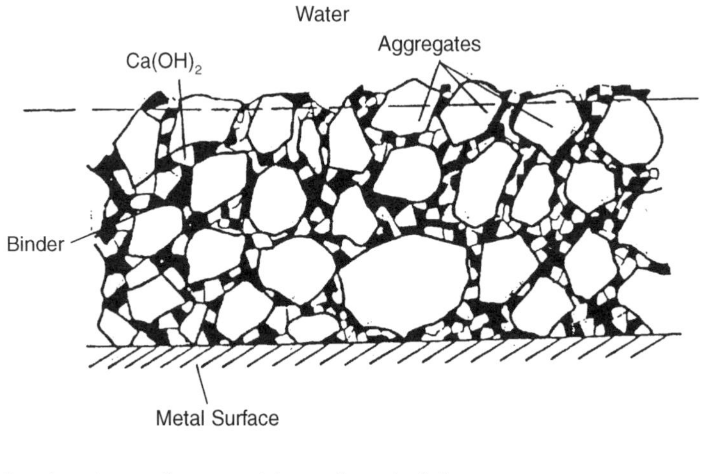

Figure 7-1 Schematic structure of cement-based material

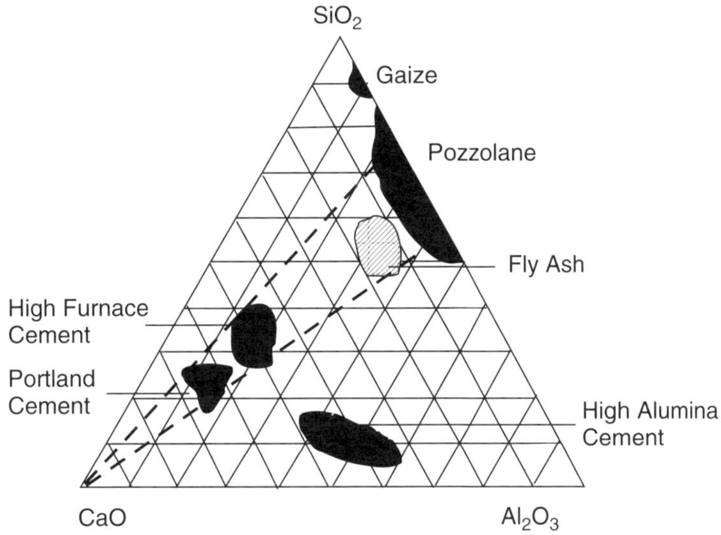

Source: Coanda (1987). Reprinted with permission of Serban Coanda.

Figure 7-2 Triangular graph of cements, slag, and pozzolane

Table 7-1 Standard composition of the primary types of cements used for pipes manufacturing

Parameter	Blast Furnace Cement	Pozzolanic Metallurgical Cement	High-Alumina Cement	Portland Cement
SiO_2, %	27	29	3.5	21
CaO, %	48	44	38	65
Free CaO, %	0.5	1.5	0.5	2.5
Al_2O_3, %	13	13	36	5
Fe_2O_3, %	1.5	3	18	2.5
SO_3, %	3.8	2.2	0.1	1.9
Na_2O, %	0.4	0.3	0.1	0.1
K_2O, %	0.8	0.7–1.7	0.05	0.8
MgO, %	2.8	3	0.4	0.5
Ignition loss, %	0.2	1.2–2.4		2
Density, g/cm^3	2.9	2.8	3.2	3.1
Specific surface, cm^2/g	3,350–3,450	3,650	2,750	3,870

and gypsum (abbreviated as $CaSO_4 \cdot 2H_2O$) to delay hardening. The other types of cement contain blast furnace slag, pozzolane, silica ash, etc. The shorthand notations for the major chemical compounds approximately correspond to the following compositions (Highway Research Board 1972): $C_2S \simeq Ca_2SiO_4$; $C_3S \simeq Ca_3SiO_5$; $C_3A \simeq Ca_3Al_2O_6$; $C_4AF \simeq Ca_4Fe_2Al_2O_{10}$.

Each type of cement has a standardized composition that can vary slightly from one country to another. The reference standards for some European countries and the United States are given in Table 7-2.

Table 7-2 International standards for cement compositions

Country	Portland Cement	High-Alumina Cement	Pozzolanic Cement	Blast Furnace Slag Cement
United States	ASTM C150-92	ASTM C618-93	ASTM C618-93	ASTM C618-93
Europe	EN 197-1-1*	EN 197-1-VA/B*	EN 197-1-IVA/B*	EN 197-1-IIIA/B/C*

*Source for all European standards listed here is CEN 1992.

Hydration of Cement

Reactions of portland cements. The binder of cement-based materials is the hardened and hydrated cement. Its composition and chemical and mechanical resistance depend on the type of cement and the conditions of hydration used. After mixing with water, the cement undergoes various hydrolysis, hydration, and precipitation reactions. These lead to the hardening of the material.

During mixing and until hardening, the various reactions of hydration only affect the superficial layer of the cement particles. This reaction then significantly slows down, and the various solid phases formed through the layer become in thermodynamic equilibrium with the interstitial water, which is rich in calcium, sodium, and potassium hydroxide phases.

These reactions continue until, over a period of years, the cement particles are completely hydrated. Other reactions can also take place, such as the change of crystal form of some compounds initially formed. Thus, calcium trisulfoaluminate is transformed into calcium monosulfoaluminate. In this way, the hardened materials have a heterogeneous structure that is porous on both micro and macro scales. The texture of the binder, like the porosity of the material, results from the formation of crystals of hydrated compounds, which fix a part of the added water.

The addition to the basic clinker of alkaline blast furnace slag or ashes, containing primarily some CaO and SiO_2 with a CaO/SiO_2 ratio of about 1.7, does not modify the composition of the formed products but modifies their structure. On the other hand, the addition to the clinker of acidic compounds like fly ashes or pozzolane, containing primarily SiO_2 and Al_2O_3 with a CaO/SiO_2 ratio of about 1.3, fundamentally modifies the composition of the formed products and the porosity of the material. The most important effect is the fixation by such compounds of the lime formed during hydration of the cement.

Reactions with alumina cement. The setting of alumina cement is very different from the behavior of any other type of cement. The compounds formed are not the same as in the preceding cases and the nature of the process depends on the temperature. The reactions do not lead, as in the case of portland cement, to free calcium hydroxide, but to an excess of aluminum hydroxide.

Characteristics of the Medium in the Pores

As indicated earlier, the water in the pores is in thermodynamic equilibrium with all the crystallized compounds forming the structure, such as calcium hydroxide, calcium silicates, calcium aluminates, and calcium sulphoaluminate. The water in the pores also contains other components coming from the cement itself, such as sodium hydroxide and potassium hydroxide.

The presence of these alkaline and earth-alkaline hydroxides raises the pH value of the medium in the pores to about 13 to 13.5. Table 7-3 gives the composition of pore medium for two types of cement—portland and blast furnace cement.

The Aggregates

For cement–mortar linings, the aggregate is sand. The effective size of the sand differs depending on the type of process being used. A siliceous sand with hard, clean, rounded particles is generally selected for cement–mortar linings. For concrete pipes, which are mainly steel-reinforced, sand and gravel are used as aggregates.

The stability of a mineral in alkaline medium is very important. Some minerals, such as hornfels, opal, feldspar, mica, and flint, may cause the cement properties to degrade.

The aggregate in asbestos–cement materials is the asbestos fiber, of which there are several mineral varieties. The use of the asbestos fibers significantly increases the stress and pressure resistance of the hardened cement.

DEGRADATION OF CEMENT MATERIALS

Types of Degradation

Several types of degradation of cement materials can occur. The most important and widely known is degradation caused by aggressivity of the water (acid waters or waters aggressive to calcium carbonate). Other types of attacks can be caused by other ions present in the water, e.g., sulfate and magnesium. In certain cases, expansive by-products can form between the cement medium and the aggregates. The increase in volume of these by-products over those originally present causes weakening of the material. Finally, degradation can take place if the total inorganic carbon (TIC) content is too low to form precipitates within the cement matrix (see the section, Influence of Very Soft Water With Low TIC, later in this chapter).

Effect of Water Contact on Cementitious Materials

Contact of cement materials with the transported water produces changes in the cement matrix. The interactions between the cement material and the water result

Table 7-3 Composition of the interstitial medium of two types of cement

Parameter	Portland Cement	Blast Furnace Slag Cement
pH	13.4	13.2
Calcium, *mg Ca/L*	65	37
Magnesium, *mg Mg/L*	0.25	0.30
Sodium, *mg Na/L*	2,200	1,600
Potassium, *mg K/L*	11,500	8,500
Aluminum, *mg Al/L*	4.5	25
Sulfate, $mg\ SO_4^{2-}/L$	550	160
Silica, $mg\ SiO_2/L$	12	32

from the great surface of contact between these two phases, especially in the pores. The ions (such as OH^-, Ca^{2+}, Na^+, K^+, and SO_4^{2-}) in the concentrated medium of the pores, diffuse into the transported water. At the same time, some ions in the water (e.g., CO_3^{2-}, HCO_3^-, Cl^-) can diffuse into the pores. At this point, several interactions are possible depending on the characteristics of the transported water.

The first and primary type of interaction with water concerns the calcium carbonate saturation state and carbonate speciation of the water, which can lead to the rapid degradation of the cement material. This is generally called carbonic aggressivity. Another type of interaction concerns the chemical equilibria with the various calcium aluminate or calcium silicate phases.

The calcium and hydroxide ions present in the pore water tend to escape into the transported water, which contains smaller concentrations of these constituents. Conversely, the carbonate species and other ions present in the transported water tend to enter into the pore water. The reaction between the hydroxide ions from the pore water and the dissolved aqueous carbon dioxide and bicarbonate ion forms carbonate ion (CO_3^{2-}) (see chapter 1).

As a result, conditions suitable for spontaneous calcium carbonate precipitation may exist in the pores, near their openings. These calcium carbonate deposits reduce the interactions between transported water and the medium inside the pores, thus protecting the cement material. However, such a deposit can only be effective if it is stable and can fill the pores to eliminate permeability to the transported water.

The pores become blocked if, in general, all calcium ions from the calcium hydroxide of the cement contribute to the formation of calcium carbonate deposits without any "leakage" into the transported water. Similarly, the calcium ions from the transported water also contribute to the formation of the calcium carbonate deposits during this first step. Thus, in this ideal case, the transported water loses calcium ions to the formation of the deposits (Figure 7-3).

Blockage of the pores cannot occur if a significant fraction of the calcium ions from the calcium hydroxide leaches into the transported water, because the solubility product

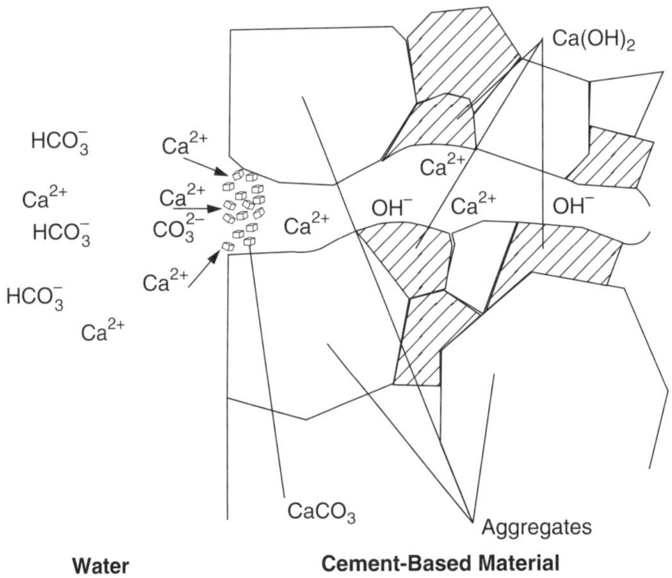

Figure 7-3 Ion migration giving an efficient blockage of pores

for calcium carbonate is not sufficiently exceeded. Then the Ca^{2+} ions present in the transported water do not contribute to the formation of interstitial calcium carbonate deposits, and consequently, the transported water gains calcium ions (Figure 7-4).

If the calcium carbonate deposits that are formed do not dissolve upon interaction with the transported water, the deposit is "stable." To achieve this, the transported water must not be "aggressive" to calcium carbonate. If the transported water is aggressive, the pores can only be obstructed by calcium carbonate for a limited period of time, until these deposits are finally redissolved into the transported water. Thus, when the transported water's chemical properties do not allow the "sealing" of the cement pores, the deterioration of the cement material is inevitable.

Evolution of Cement Material Depending on Composition of Transported Water

The composition of the water in the pores is influenced first by the equilibrium with the solid phase and by the composition of the water used for manufacturing. When the pipeline system is operated, the chemical environment of the cementitious material is then influenced by the composition of the transported water. In order to foresee the behavior of a pipe, it is important to know if the transported water is or is not aggressive to calcium carbonate. However, it is not the *only* important consideration.

Legrand et al. (1981) have suggested criteria to show the effect of water composition on cement materials. These criteria are described in more detail in chapter 8.

Alternatively, a calculation, based on pH, alkalinity, calcium ion concentration, carbonate ion concentration, temperature, and ionic strength, can be made to determine the calcium carbonate precipitation potential (CCPP) of the water (see chapter 1). A negative value for CCPP would indicate a water quality that would tend to dissolve or be unable to precipitate calcium carbonate.

Influence of moderate and well-buffered waters. In a water highly aggressive toward calcium carbonate, the dissolution of the locally precipitated

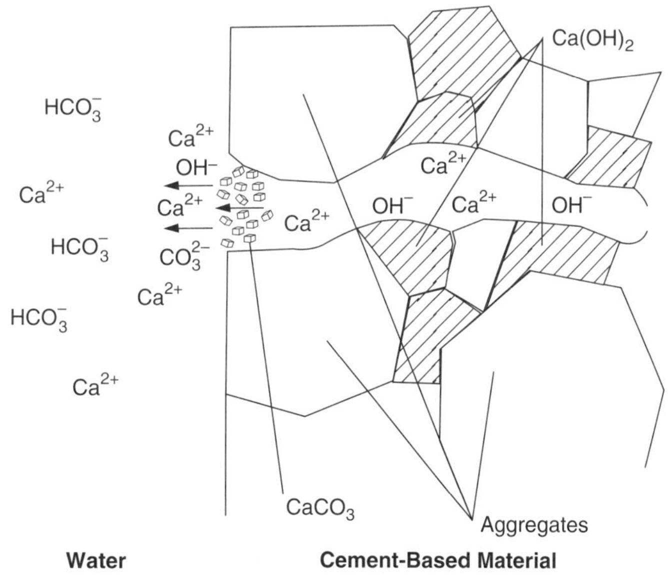

Figure 7-4 Migration of ions when protection is not possible

$CaCO_3$ proceeds continuously into the cementitious material. The alkaline properties of the cement are inadequate to neutralize the $CaCO_3$-dissolving properties of the water.

In waters slightly aggressive toward calcium carbonate, basically the same process happens; however, the velocity of the attack is sufficiently low with respect to the lifespan of the network so that significant deterioration does not take place.

If the transported water is in equilibrium or oversaturated with respect to calcium carbonate, the evolution leads to blocking of the pores. The calcium carbonate deposit is then stable and the cement material is protected for as long as the transported water remains under these conditions.

A limit can be approximately defined between very aggressive waters that degrade cement material rapidly and slightly aggressive waters. The limit, which is a quantity of CO_2 aggressive to calcium carbonate of about 10 mg $CaCO_3$/L, is based on practical experience. This value is only an approximate value, because waters of very high alkalinity may allow higher values, and in very soft water, the leaching effects may override the kind of deterioration just described.

Influence of very soft water with low TIC. If the transported water has low TIC, the evolution may not lead to blocking of the pores, regardless of whether the water transported is oversaturated with $CaCO_3$. In waters of low TIC, the amount of $CaCO_3$ precipitation possible may not be sufficient to protect the material if it is smaller than the amount of lime dissolved. This is analogous to the concept of CCPP discussed in chapter 1 and elsewhere, for example, Schock 1990. Hard waters of high TIC will potentially produce a greater mass of $CaCO_3$ precipitate than waters of lower TIC but the same hardness and would therefore be more protective.

With inadequate precipitation of calcium carbonate the water can dissolve a significant quantity of lime, because lime solubility is about 20 m*M*/L (1,120 mg CaO/L). Consequently, the cement material will be degraded. The lime dissolved into the transported water also results in an increase of pH.

In soft water, the calcium hydroxide dissolves into the water at a rate that decreases over time with the depth of the attack. But this evolution also depends on the evolution of the material's porosity. With portland cement, the high calcium hydroxide content of the hydrated material leads to a significant increase in the porosity, thereby increasing the degradation rate.

This type of degradation of cement material has been studied by Scandinavian researchers (Kristiansen 1978; Vik and Weideborg 1991). The water in Scandinavia is very soft and aggressive to calcium carbonate. These studies were mainly concerned with the effect of cementitious materials on the quality of transported water. A study by Vik and Weideborg (1991) related the amount of calcium extracted from the cement material to the calcium concentration of the water introduced into the pipes. These results are reported in Figure 7-5.

A 14-year study on soft water is reported by Soukatchoff (1990). Samples of cement mortar made with different types of cement were exposed to a very soft water. The characteristics of the water are shown in the following:

pH, 5.00
Alkalinity, 0.5 mg/L $CaCO_3$
Hardness, 2.6 mg/L $CaCO_3$
Dissolved oxygen, 12 mg/L
Free CO_2, 1.7 mg/L
Calcium, 0.45 mg/L

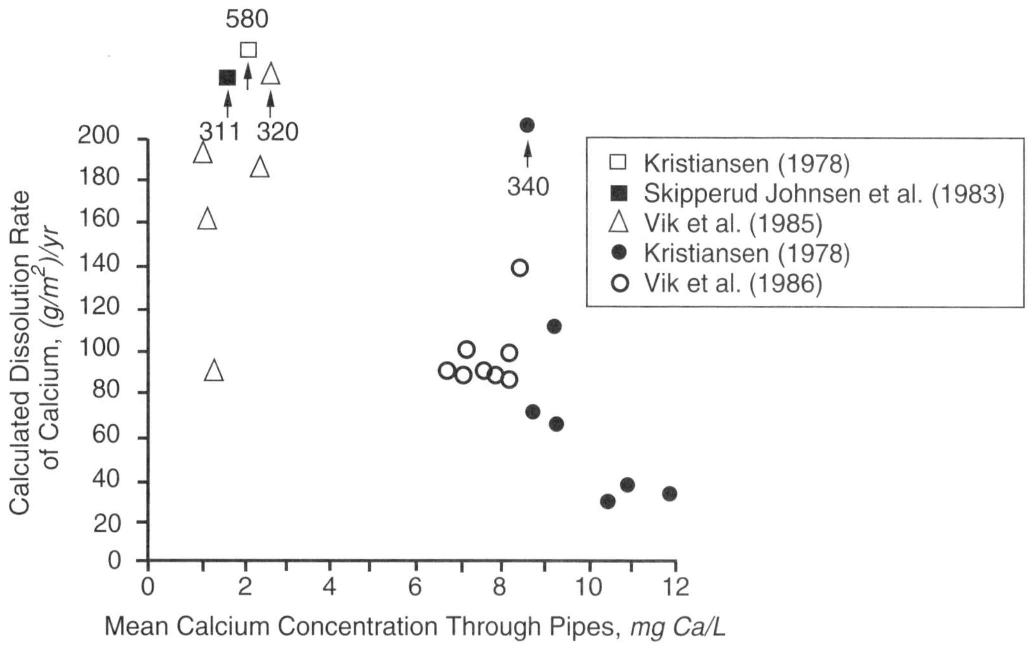

Source: Vik and Weideborg (1991). © Copyright by NACE International. All Rights Reserved by NACE; reprinted with permission.

Figure 7-5 Calculated dissolution rate of calcium from cement material as a function of the mean Ca concentration

The types of cement tested were portland cement, blast furnace slag cement, and high-alumina cement. The parameters studied were the variation of weight, porosity, and compressive strength. Figure 7-6 and Table 7-4 give the principal results.

Influence of Other Ions

Rapid degradation of cement materials can occur when the material is in contact with water that has a high sulfate concentration. In such cases, sulfate reacts with calcium aluminates present in the hydrated cement, giving a highly hydrated calcium sulfo-aluminate (ettringite). The mechanism of ettringite formation is as follows:

$$3CaO \cdot Al_2O_3 \cdot 12H_2O + 3CaSO_4 \cdot 2H_2O + 13H_2O \rightarrow 3CaO \cdot Al_2O_3 \cdot 3CaSO_4 \cdot 31H_2O \quad (7\text{-}1)$$

This compound has a significantly larger volume than the initial aluminate. Under the internal pressure caused by the formation of this product, cracks appear that reduce the material's mechanical strength.

Sulfate concentrations as low as 400 mg/L can lead to ettringite formation. Such a level is higher than the typical sulfate concentrations found in most potable waters. However, in certain gypsum-containing areas, such a level can be reached or even exceeded.

The effect of sulfate is reduced if chloride is also present in high concentrations. This is the case with seawater in which no ettringite formation is observed even with a sulfate content higher than 1,000 mg/L.

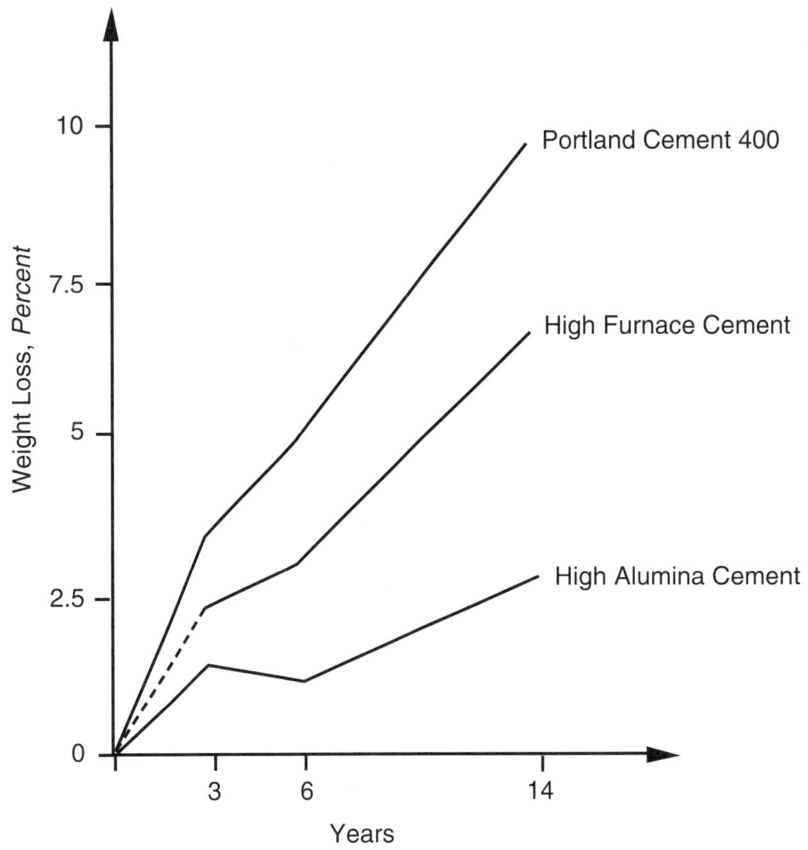

Source: Soukatchoff (1990). Reprinted with permission of Pont-à-Mousson S.A.

Figure 7-6 Weight loss of different types of cement mortar after 14 years in very soft water

Table 7-4 Change in porosity and compressive strength of cement-based pipes after 14 years in soft water

Cement Type	Change in Porosity, %	Change in Compressive Strength, %
Portland	+32.8	–62
Blast furnace slag	+21.5	–32
High alumina	+10.3	+134

Some other ions may also lead to deterioration of cement materials. For example, polyphosphate ions can also accelerate calcium removal and consequent cement degradation through sequestration (Weston Water Utility 1983).

Alkali-Granulate Reactions

Some degradations have an endogenic origin, meaning that the material properties may cause the degradation themselves. This is the case for the alkali reaction of granulates and is observed mainly with concrete pipes. This phenomenon appears as swelling of the material along with significant cracking. It results from the reaction of alkaline bases present in the cement with certain specific granulate materials. The reactive granulate materials are mainly dolomite, schist, opal, caledonia, micaceous feldspar, and flint.

EFFECTS OF CEMENT MATERIALS ON WATER QUALITY

Primary Interactions Between Water and Cement Materials

As indicated earlier, the primary interaction between water and cement materials leads to lime leaching, which induces a substantial pH increase. Such leaching can have important consequences on water quality. The amount of impact on the increase of the pH of a water depends on the total alkalinity and buffer capacity of the transported water, the type of cement used, the contact time between the water and the cement material, and the pipe diameter.

The result of the primary effects are

- increase in pH
- increase of calcium hardness
- increase in alkalinity
- migration of aluminum

Studies

A field case in France. A significant change in water pH caused by cement materials occurred in France near La Roche sur Yon in 1986. A new main made of steel pipes with a portland cement lining, 40 km (25 mi) long with a decreasing diameter from 400 mm (approximately 16 in.) at the beginning to 100 mm (approximately 4 in.) at the end, had been put in operation. It conveyed a moderately soft water with an alkalinity of 30 mg/L $CaCO_3$ and a pH of 8.1. The flow rate was 300 m^3/d (79,260 gpd) and outlets existed along the pipeline. The longest contact time between water and the cement lining was about 30 hours. The pH value of the water rose to 10 after the first 10 km (62 mi) and reached 12.6 at the end of the system. This pH represents essentially the pH in equilibrium with $Ca(OH)_2$, similar to that of the commonly used pH buffer solution.

In order to maintain the pH value of the distributed water under the maximum value of pH 9.5 required by the European Community (EC), it was necessary to increase the flow rate of the water to 600 m^3/d (158,520 gpd). This was done by discharging water using hydrants near the end of the line. The flow was progressively reduced in correspondence with the reduced rate of pH increase as the cement lining aged. The last hydrant was shut nine months after the start of operation.

Other European cases. As reported by Collins and Smith (1978), a 9.7-km (6-mi) long pipeline made of cast-iron pipe with factory-applied cement–mortar lining conveying a soft water showed a clear dependency of increased pH with water throughflow. The original pH of 7 increased with decreasing flow rate to nearly pH 10, even after years of normal operation (Figure 7-7).

A similar instance occurred in an in situ cement–mortar-relined pipeline in Aachen, Germany, where the last 1 km (0.6 mi) of the pipeline served only three houses. Even after three years of operation, the low discharge on this line led to a pH increase from 9 to 12.3 in the low-buffered water (alkalinity 25 mg/L $CaCO_3$). The problem was partially resolved by increasing the TIC content of the water, resulting in a pH of the water at the starting point of the relined section to pH 6.8–7.0. In this case, the increase of TIC, which ironically increased the level of calcium carbonate aggressive CO_2 improved the properties of the very low-buffered water such that calcium carbonate was formed in the lining and the $Ca(OH)_2$ leaching was suppressed. After sufficient time, under close monitoring, the addition of CO_2 was stopped to avoid redissolving of the $CaCO_3$ layer.

Investigations of the Water Research Centre. Investigations performed by the Water Research Centre (WRc) of Swindon, England, concerned cement–mortar lining applied in situ and in the factory. The investigations showed and quantified the influence of the water's original pH and alkalinity on the amount of pH increase observed in pipe use.

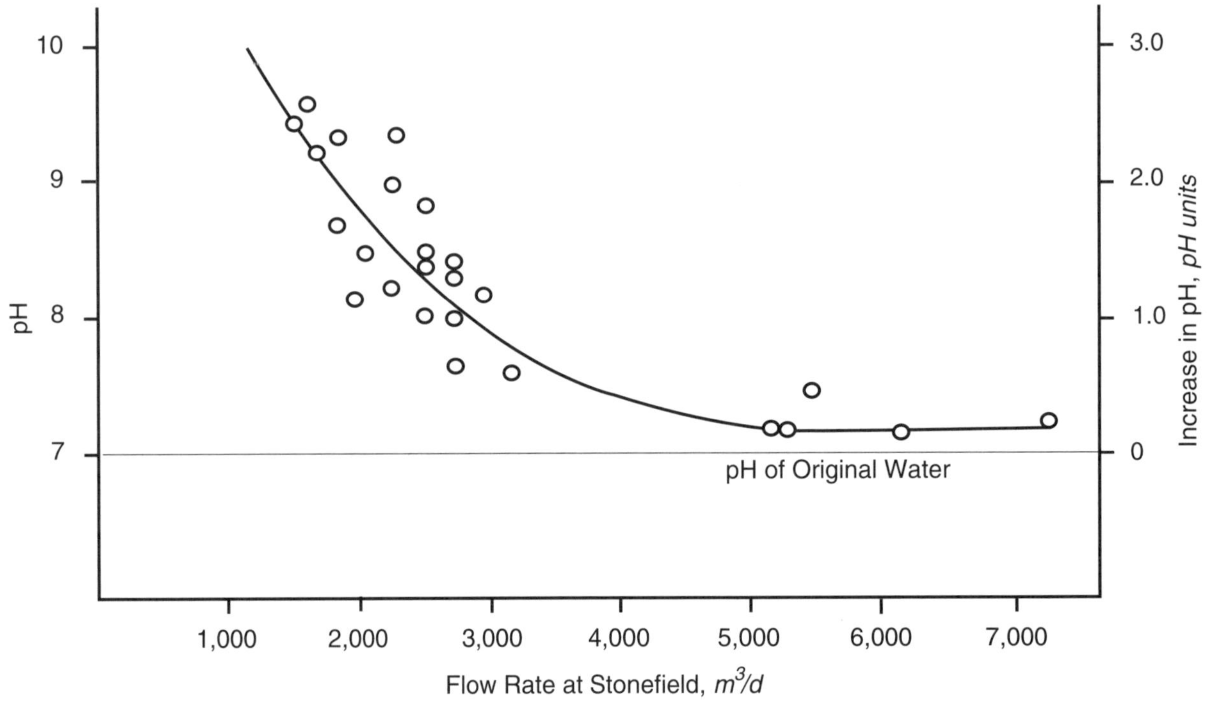

Figure 7-7 Effect of flow rate on rise in pH; Gorthy Wood, United Kingdom, November 1965-November 1966

A study of in situ linings was performed at four sites and using a laboratory test rig (Conroy and Oliphant 1991). An ordinary portland cement lining prepared in accordance with the Water Research Centre in situ cement–mortar lining operational guidelines was used. The water quality for each of the test series is given in Table 7-5. The waters cover a wide range of alkalinity, from very soft to hard. Table 7-6 gives the pH levels measured after a contact time of 1 hour and Table 7-7 gives the aluminum levels measured after contact times of 1.5 and 6 hours.

The conclusions of this first report concern the pH increases and the variations of calcium, sodium, potassium, and aluminum concentrations after various contact times. It is shown that the extent and duration of the pH increase appear to be related primarily to the alkalinity of the supplied water. pH increases were observed to varying degrees in all cases except for site D, which had a hard water supply.

In all areas, except site D (high-alkalinity water), the initial calcium, sodium, and potassium levels were all increased after relining.

Aluminum was also monitored at all sites because laboratory experiments showed it to leach from cement mortar into deionized water. The tests indicated that only sites in areas of low-alkalinity water showed any increase in aluminum levels after contact with cement mortar.

Based on these data, it was suggested that the portland cement mortar should not be used for in situ relining where the water alkalinity is below 55 mg/L as $CaCO_3$. This study also indicates that the use of cement containing blast furnace slag can reduce significantly the increase of pH. A similar result could be obtained with high-alumina cement, but in very soft water it led to an unacceptable level of aluminum.

A second WRc study of factory-applied cement–mortar linings (Conroy et al. 1991) was performed with similar water qualities. The cement mortars used were made of a mixture of portland cement and pulverized fuel ash in variable ratios from 34 to 42 percent of fuel ash. Different diameters were used, from 80 to 150 mm (3 to 6 in.).

This study concluded that the effect of factory-applied linings on water quality is less important than the effect of in situ linings. It is suggested that factory-applied linings could be used when the supply water alkalinity is greater than 25 mg/L as $CaCO_3$.

Table 7-5 Water quality analysis for each of the test series

Trial Site	pH	Alkalinity, *mg/L CaCO₃*	Aluminum, *mg/L Al*	Calcium, *mg/L Ca*	Sodium, *mg/L Na*	Potassium, *mg/L K*
Test rig	9.2	8	0.04	3.5	3.5	0.3
Site A	7.8	10	0.1	4.6	6.0	1.0
Site B	8.7	35	<0.4	26	13.2	3.4
Site C	7.6	55	<0.4	37	5.8	1.9
Site D	8.00	160	<0.4	92	2.2	5.5

Table 7-6 pH levels after a contact time of 1 hour

Time, *months*	pH Units				
	Test Rig	Site A	Site B	Site C	Site D
0	9.2	7.8	8.7	7.6	8.0
0.03	11.7	11.9	11.3	11.6	8.3
0.25	11.0	11.2	9.9	7.0	8.0
0.5	10.9				
0.75	10.6				
1	10.5	10.6	9.4	7.6	8.0
2	10.4				
3	10.2	10.4	9.1		
4	10.2				
6	10.2	10.5			
12	10.2				
24	10.0				

Table 7-7 Aluminum levels after contact times of 1.5 and 6 hours

Time, *months*	Aluminum, *mg/L*	
	1.5-Hour Contact Time	6-Hour Contact Time
0.03	0.080	
0.25	0.120	0.260
0.5	0.080	0.300
0.75	0.100	0.300
1	0.100	0.250
2	0.060	0.190
3	0.070	0.190
6	0.080	0.140
12	0.056	
24	0.046	0.063

ASBESTOS-CEMENT PIPES FOR DRINKING WATER SYSTEMS

Introduction

Since the late 1970s, the use of asbestos has been under considerable discussion in the United States and in Europe. The original discussion was evoked by health concerns mainly relating to lung cancer (asbestosis) but also included fears of adverse health effects from ingested fibers (Buelow et al. 1980; AWWARF 1974; NAS 1982). Public concern caused authorities in many countries to ban asbestos for nearly any use.

Although the health question has not been clearly resolved, the potential of "aggressive" water to cause structural deterioration, impairment of water quality through pipe dissolution, or degradation of hydraulic efficiency remains an important justification for continuing to develop a better basis for operational guidance in controlling corrosion and extending the longevity of these materials.

Currently in the United States, asbestos in drinking water is limited to a concentration of 7 MFL (million fibers per litre) of longer than 10 μm. The sampling and analysis requirements are specified under the regulation promulgated in 1991 (USEPA 1991). The standard is based on reducing possible human cancer risks from drinking water exposure.

Some research has also raised the question of potential risk to asbestos exposure by the inhalation of fibers mobilized from drinking water by household humidifiers (Hardy et al. 1992).

Originally, the US Environmental Protection Agency (USEPA) proposed a ban on the further production and new installation of asbestos–cement (A–C) pipes for drinking water distribution as did agencies in some European countries. This proposal was promulgated in 1989 under the Toxic Substances Control Act (USEPA 1989). The basis of the regulation was primarily the unavoidable health risks during the production of asbestos-related products, but also a not quite clearly described health risk by potential deterioration of A–C pipes and the ingestion of drinking water containing asbestos fibers. The A–C pipe production ban was challenged in court, and the rule was invalidated by the US Court of Appeals of the Fifth Circuit in 1991 (Pontius 1992).

Hundreds of thousands of miles of A–C pipes remain in use worldwide for conveying drinking water. An immediate replacement is neither possible nor necessary in many cases. Therefore, it is worthwhile to consider corrosion behavior of this pipe material to prolong the service life of existing pipe installations. Additionally, as pointed out by Pontius (1992), controlling exposure to asbestos from existing uses as much as possible should be a prudent and necessary public health objective.

Historical Background

A–C pipe was developed in Italy approximately 70 years ago (Olson 1974). In the years 1906–1913, Società Anonima Eternit Pietra Artificiale of Génova, Italy, combined asbestos fibers with cement to produce a reinforced pipe that would take the high pressure necessary to pump salt water up into the city of Génova for its street flushing system. By 1938, there were already 40,000 mi. (64,400 km) of A–C pipe installed. There were an estimated 1.5 million mi (2.4 million km) of A–C pipe in service as of 1974, 200,000 mi (322,000 km) of which were in the United States. Logsdon (1983) reported studies showing that in 1980 13.9 percent of US water mains were made of A–C pipe. In Germany, A–C pipes were first produced in Berlin in 1930 by the Deutsche Asbestzement AG. The increased use of A–C pipes in Germany has resulted in large part from economic advantages over other types of piping materials.

Many of the factors apparently causing the worldwide acceptance, and the rapid growth in use of A–C pipe in the United States, were summarized by Olson (1974). Noteworthy among these are the following:

- A–C pipe is resistant to corrosion, particularly structural deterioration, both internally and externally. In the United States, the expectation of good corrosion resistance and the elimination of red-water complaints have played an important role in the widespread acceptance of A–C pipe.

- A–C pipe has enough strength to withstand the high internal forces imposed by water hammer and shock earth loads from earthquakes.
- A–C pipe has contributed to high water quality. Manufacturers of the pipe have pioneered the concept of rubber ring joints—joints containing no jute, which offers a focal point for bacterial growth. A–C pipe does not rust and cause discoloration of the water.
- A–C pipe is lightweight, economical, and easy for the contractor to install. These qualities result in lower installation costs.
- A–C pipe has a permanently smooth interior wall, so pumping costs are low.

Standards Governing Asbestos-Cement Pipe Selection

German standards. In the Federal Republic of Germany, A–C pipes were produced in accordance with DIN* 19800, up to an internal diameter of 2,000 mm (80 in.). From 50 mm (2 in.) ID to 600 mm (24 in.) ID, four classes of wall thickness are listed. Their dimensions are in accordance with DIN 19800, which requires that the pressure that leads to breakdown of the pipe in an inside pressure test is not less than 3 times the nominal pressure. A–C pipes with diameters larger than 600 mm (24 in.) have special requirements concerning wall thicknesses, and under pressure tests they have to withstand a minimum of 2½ times the nominal pressure.

The factory tests are carried out with a pressure that is twice the nominal and that is held for 30 seconds. During the pressure tests, no visible water drops or moist parts are allowed on the outside of the pipe being examined.

The following minimum strengths of A–C pipes are defined in DIN 19800:

Circular tensile strength, 22–25 N/mm^2 (approx. 15–17 kPa)
Circular bending strength, 45–51 N/mm^2 (approx. 31–35 kPa)
Longitudinal bending strength, 25 N/mm^2 (approx. 17 kPa)

As it ages, the material becomes stronger, with its strength increasing up to 25 percent beyond its level of original strength.

US standards. In the United States, several American National Standards Institute/American Water Works Association (ANSI/AWWA) standards apply to the selection and use of A–C pipe. In numerical order, these are described in the following paragraphs.

ANSI/AWWA C400-93 (1993), AWWA Standard for Asbestos–Cement Pressure Pipe, 4 in. through 16 in. (100 mm through 400 mm), for Water Distribution Systems. This standard covers three pressure classes of type I or type II A–C pipe intended for private rights-of-way. This pipe is to be used in systems having relatively unpredictable flows and many appurtenances that do not permit good hydraulic analyses. The standard includes sections on general information, definitions, references, affidavit of compliance, materials specifications of composition, physical and chemical requirements, design, pipe classes, pipe diameters, pipe lengths, couplings, joints and wall thickness, workmanship and finish including imperfections, inspection,

*Deutsches Institut für Normung (German Institute of Standardization), Burggrafenstr. 6, Berlin, Germany.

physical test requirements, retests and basis for rejection, test for uncombined calcium hydroxide, test records, marking and delivery, and preparation for delivery.

ANSI/AWWA C401-93 (1993), AWWA Standard for the Selection of Asbestos–Cement Pressure Pipe, 4 in. through 16 in. (100 mm through 400 mm), for Water Distribution Systems. This standard was prepared with the aim that design engineers can quickly determine the proper class of A–C pipe to use under various combinations of internal pressure and external loading. It includes references, pipe design and selection guides including strength and design factors, combined loading theory, three-edge loads, and selection curves.

ANSI/AWWA C402-89 (1989), AWWA Standard for Asbestos–Cement Transmission Pipe, 18 in. through 42 in. (450 mm through 1,050 mm), for Potable Water and Other Liquids. This standard covers nine pressure classifications of type I and type II A–C pipe to be used in both public and private rights-of-way. The transmission pipe is to be used in systems with reasonably predictable flows and few appurtenances. Sections are included on general information, definitions, affidavit of compliance, materials specifications of composition, physical and chemical requirements, design pipe classification, pipe diameters and length, couplings, joints, wall thickness, inspection, physical test requirements, retests and basis for rejection, test for uncombined calcium hydroxide, test records, marking and delivery, and preparation for shipment.

ANSI/AWWA C403-89 (1989), AWWA Standard Practice for the Selection of Asbestos–Cement Transmission and Feeder Main Pipe, Sizes 18 in. through 42 in. (450 mm through 1,050 mm). This standard was designed to provide design engineers with information so that they can quickly determine the correct strength classification of pipe to use under various combinations of internal pressure and external load. The standard includes general information, general design including strength and design factors, combined loading theory, three-edged bearing load, external loads, earth loads, superimposed loads, hydrostatic pressure, operating pressure, surge pressure, pipe selection, combined loading curves, safety factors, use of selection charts for economical surge pressure analysis, frictional power requirements, and a friction head loss chart.

ANSI/AWWA C603-90 (1990), AWWA Standard for Installation of Asbestos–Cement Pressure Pipe. This document covers the installation of A–C pressure pipe with fittings or appurtenances of asbestos–cement, cast iron, other materials, or combinations of materials. Included is coverage of definitions, references, material acceptance, storage and handling, general and detailed work to be performed including alignment and grade, excavation and preparation of trench, laying of pipe, jointing of pipe to valves, hydrants and fittings, setting of hydrants, valves and fittings, leakage, backfilling procedure before tests and after tests, and disinfection.

Several important aspects of work practice in handling, cutting, and tapping A–C pipe are given in the AWWA handbook, titled *Work Practices for Asbestos–Cement Pipe* (1995).

Internal Corrosion of Asbestos-Cement Pipes

In natural waters, the presence of the different species of carbonic acid (CO_2, HCO_3^{3-}, CO_3^{2-}) in combination with calcium hydroxide from the cement material can reduce free lime leaching by forming calcium carbonate within the pipe wall or the cement–mortar lining. The formation of calcium carbonate depends on the water supply characteristics, particularly pH, hardness, and alkalinity. Sufficiently high buffer capacity and alkalinity favor calcium carbonate formation, although even in hard water supplies, lime leaching will occur (Muller 1978), mainly in new pipes and during initial operation. Afterwards, the calcium carbonate seals the inner surfaces more or less completely against further reactions, as described earlier.

Groundwaters with high buffer capacity but low pH values contain remarkable amounts of calcite-dissolving carbon dioxide. Classically, these waters are said to possess "aggressive CO_2," meaning that the actual dissolved carbon dioxide concentrations exceed those that would be in equilibrium with solid calcium carbonate at the observed pH, calcium, alkalinity, mineral contents, and temperature. Thus, in a second step, the $CaCO_3$ in the layer is redissolved, increasing the water's hardness, pH, and alkalinity. The process of "carbonation" (here meaning precipitation of $CaCO_3$) and dissolution continues into the pipe wall, deteriorating the structure of the "cement stone" until the system consists of only the insoluble products and has lost its stability.

The mechanism of attack by soft waters of low mineral content is generally different. First, the calcium hydroxide is removed by leaching. Then, if the water still possesses sufficient acid content, calcium carbonate is either dissolved or not formed. Because alkaline conditions cannot be maintained within the cement after the initial calcium hydroxide is removed, attack begins on the hydrated calcium silicates. These convert to more calcium hydroxide, and the cycle of reactions can continue either until sufficient calcium carbonate is formed and the water is neutralized to the point at which the cycle stops or until no cementitious material remains to bind the aggregates together (Muller 1978). Some of the reactions have also been described by Heinrich et al. (1978).

Flentje and Sweitzer (1955) observed a chalky brownish layer below the surface of pipe exposed to water. This layer is a zone of high iron and manganese salts, which can act as a kinetic barrier to further calcium removal, as had been observed by others (Carson 1927; Chappell 1930), and reduce the speed of dissolution significantly (see the section, Primary Interactions Between Water and Cement Materials, earlier in this chapter).

Laboratory studies. In 1950, Tracy published a study of observations in some small Vermont systems having naturally soft waters. Over a period of approximately four years, sampling was done at different locations in Brattleboro, North Bennington, South Shaftsbury, Proctorsville, South Troy, and South Burlington. In many instances, large increases in pH and hardness were observed for short travel distances in the pipe (approximately 0.5 to 1.5 mi [0.8 to 2.4 km]). Increases were most severe in newly installed A–C lines, and the water quality changes were noted to decrease after the pipe was in service for several years. The water quality changes were also found to be most severe when the flow rate was low. Tracy concluded that the A–C pipe tended to stabilize after several years, but at that time no quantitative relationship could be developed to predict the stabilization period, with the study being so preliminary.

In 1960, Eick studied the corrosion behavior of A–C pipe under several extreme conditions, which were not comparable to drinking waters. The results showed that the attack by mineral acids used was not linearly related to the contact time. It began with a high rate of attack that continuously decreased. Eick attributed this behavior to an A–C layer that reduced the velocity by partially blocking the transport of the aggressive liquid to the surface of the pipe.

In 1977, Kristiansen reported on a Norwegian A–C pipe research project, whereby water from the Oslo waterworks was put through two experimental pipe loops. A description of the water quality used in the experiments is found in Table 7-8. The colloidal humic matter contained in the water was found to form a deposit in the pipes that slowed the rate of attack on the pipes somewhat.

Table 7-8 Quality of Oslo water used in experiments of Kristiansen (1977)

Parameter	Concentration*
pH (units)	6.6
Calcium	3.4
Magnesium	0.50
Sodium	1.4
Potassium	0.45
Chloride	1.7
Sulfate	7.0
Nitrate	0.9
Alkalinity, *meq / L*	0.08
Organic carbon	3.0

Source: Kristiansen (1977). Reprinted with permission.

*All concentrations are in milligrams per litre, unless otherwise noted.

Kristiansen (1977) made the following important observations on the chemical behavior of the pipe material:

- The water had a higher pH value and calcium content after passing through the pipes. This was a result of the leaching of free lime from the cement matrix. Except for the first analytical data, the extraction of calcium was constant. During the first year of operation, the water-cured pipe showed less calcium leaching. This must be explained by the fact that the pipe had been kept in storage long enough after manufacture that a deposit of calcium carbonate had formed on the surface; this deposit had to dissolve before the extraction of free lime, and calcium carbonate dissolves very slowly.
- An analysis of the humic material deposit in the pipes showed that it mainly consisted of organic matter, ferrous hydroxide, and silicic acid. After removal of the deposits, the analytical results for 1974 showed an increase in the release of calcium for the water-cured pipe. This indicated that the deposit, to a certain extent, acted as a protection against release from the pipe material.
- The formation of a deposit of calcium carbonate on the surface prevented the release of lime from the pipe material. The results of analysis of the pipe material showed that calcium carbonate was only present in a thin layer nearest the water. The results also indicated that there was a marked difference in extraction of calcium from the water-cured and autoclave-cured material. The release from the latter seemed to be more extensive. The dissolution of calcium from the water-cured and autoclave-cured pipe material did not exceed 50 percent and 75 percent, respectively. For the water-cured material, however, the loss of calcium was extracted from deeper below the surface of the encrusted material.
- The greatest lime extraction occurred in that part of the pipe where the water enters. Thus, the change in the chemical composition of the test water caused by the lime extraction had, as to be expected, a diminishing effect on leaching from the pipe material downstream.
- The magnesium content in the water decreased as it passed through the pipes, indicating that a precipitation of magnesium hydroxide had taken

place. Some of this magnesium was found in the precipitated deposit, but magnesium also entered into the outermost layers of the pipe material and replaced calcium. An earlier study indicated magnesium silicates were formed on the inner surface layer of concrete pipes.

Hünerberg and Tessendorff (1977) described tests performed with A–C pipes in several different solutions. Their results in systems most appropriate to drinking waters can be summarized in the following way.

They found that by increasing the buffer capacity of the water, for a given amount of "aggressive CO_2," the attack on the cement would be reduced. Unlined A–C pipe was found unsuitable under aggressive conditions.

In waters containing high sulfate concentrations, the composition of the cement becomes important. In order to provide good resistance to sulfate attack, a tricalcium aluminate (C_3A) concentration of less than 3 percent is required. They considered the cement with low C_3A content to retain its structural integrity when exposed to sulfate concentrations of up to 1,000 mg/L.

Buried pipes from Berlin, Germany, excavated after 42 years of service without failures, were tested for mechanical strength, resistance to abrasion, and the influence of time on corrosion processes. Some analytical data for Berlin A–C pipe is presented in Table 7-9, and some of the results of the mechanical testing are summarized in Table 7-10.

Hünerberg and Tessendorff concluded that the experience with long-term operation of A–C pipes and the results of laboratory experiments proved that these pipes maintained their structural integrity at pH values greater than or equal to pH 6 and sulfate concentrations less than 1,000 mg/L. If "aggressive CO_2" is in the water, the influence of buffer capacity has to be taken into account.

A study carried out by the Committee of European Organizations on Corrosion (CEOCOR) from 1968 through 1976 generally confirmed the other experimental results with regard to the structural integrity of A–C pipe in aggressive media (CEOCOR 1979). Many of their tests did not reflect the circumstances or concerns of drinking water systems, however.

In 1978, Hallenbeck and co-workers published a study of 15 hard or moderately hard water systems in northeastern Illinois. In that study, they attempted to ascertain whether asbestos fibers were released into the water by the A–C distribution pipe. No statistically significant increases in fiber levels were found. The authors concluded that there could have been fiber increases at levels below the detection limit of their asbestos analytical methodology or at levels that were obscured by laboratory contamination or naturally occurring chrysotile in some of the water supplies investigated.

Table 7-9 Relevant water quality data for Berlin, Germany, at pH 7.45

Parameter	Concentration
Sulfate	125
Total hardness (as $CaCO_3$)	1,700
Carbonate hardness (as $CaCO_3$)	400
Free CO_2 (as CO_3)	19

Source: *Hünerberg and Tessendorff (1977). Reprinted with permission of Springer-Verlag from* Handbuch für Asbestzementrohre, *K. Hünerberg and H. Tessendorff, 1977, © Springer-Verlag.*

*In milligrams per litre.

Table 7-10 Mechanical testing results from Hünerberg and Tessendorff (1977) for Berlin A-C pipe compared with the German Standard (DIN) values

	Mean Breaking Stress,		Minimum Values From DIN 19800,	
	N/mm^2	*(kPa)*	N/mm^2	*(kPa)*
Longitudinal binding strength	32.2	(~21.9)	25.0	(~17.0)
Circular binding strength	70.0	(~47.6)	45.0	(~30.6)
Circular tensile strength	25.1	(~17.1)	22.0	(~15.0)

Source: Hünerberg and Tessendorff (1977). Reprinted with permission of Springer-Verlag from Handbuch für Asbestzementrohre, *K. Hünerberg and H. Tessendorff, 1977, © Springer-Verlag.*

The National Academy of Sciences review (NAS 1982) reported on a 1978 pilot study in Seattle, Washington, performed by Kennedy Engineers, who evaluated the aggressiveness of two water supplies, the Tolt and Cedar rivers. Various treatments or metal corrosion inhibitors were investigated for their ability to retard attack on the A–C pipe by the aggressive waters. Both supplies had low pH, calcium, and alkalinity levels. For the Cedar River, addition of lime and soda ash appeared to be promising. For the system with the Tolt River as the source, lime and zinc orthophosphate additions were suggested.

Figure 7-8 is a scanning electron microscope (SEM) photograph of the interior surface of a section of A–C distribution main taken by the USEPA before the corrosion control program was initiated. Note how the cement binder has greatly dissolved, exposing numerous bundles of fibers for erosion, in comparison with an SEM photograph of unused A–C pipe for the same system (Figure 7-9).

Kirmeyer et al. (1980) reported on a USEPA survey of asbestos fiber input to the drinking water from A–C mains in the Seattle, Washington, area. Considerable increases in calcium content and pH in the water systems were found, which was evidence of pipe deterioration. These data are shown in Table 7-11. In addition, asbestos fiber counts increased from levels that were not statistically significant in the source water to levels as high as 1.59×10^6 fibers/L. Most of the test section contained US type II, autoclaved A–C pipe; only a very small amount of pipe was manufactured in Belgium. Samples were collected between February and November 1975. Since the pipe had been installed in the 1965–1967 time frame, this was not an initial sloughing of loose or uncombined material.

An interesting study was published by Mah and Boatman (1978) on evidence of A–C pipe corrosion collected by SEM, transmission electron microscopy (TEM), and energy-dispersive X-ray elemental analysis (EDXA). The surface of new A–C pipe was found to be relatively smooth and free from exposed asbestos fibers. Several different types of experiments were carried out, and of particular interest is the inhibitor study. For the inhibitor study, the test units of A–C pipe consisted of eight 3.3-m (11-ft) lengths of standard 100-mm (4-in.) diameter, class 150 pipe arranged in a loop system, in a continuous once-through water mode.

The only inhibitor of the six tested for which data were reported and that was believed to be beneficial was prepared by a mixture of lime and zinc orthophosphate. This mixture consisted of lime and orthophosphate (5.0 mg PO_4/L) plus zinc (0.3 mg/L). Zinc was found to deposit on the surface of the A–C pipe, whereas no

Figure 7-8 Scanning electron microscope photograph of A-C pipe from the Seattle, Washington, distribution system (magnification ~ 100×)

Figure 7-9 Scanning electron microscope photograph of unused A-C pipe from the Seattle, Washington, distribution system (magnification ~ 100 ×)

Table 7-11 Water quality changes resulting from dissolution of A-C pipe

Parameter	Tolt River	Cedar River	Location Control Works (Baseline)	1,860 m (6,100 ft) From Control	9,790 m (32,119 ft) From Control
pH	5.8–6.2	6.8–7.4	7.2	7.6	9.0
Calcium, *mg $CaCO_3$ / L*	6–7	16–18	5.8	8.0	11.0
Alkalinity, *mg $CaCO_3$ / L*	2	15–18	14.0	18.8	24.6

Source: Kirmeyer et al. (1980).

peak for phosphorus was found after 218 days of exposure. At that time, "zinc orthophosphate" corrosion inhibitors were believed to operate by depositing a film of a zinc solid such as $Zn_3(PO_4)_2$ or $Zn_3(PO_4)_2 \cdot 4H_2O$ that would seal the surface from further water attack, which made the results surprising to many workers. However, subsequent detailed computations of the chemical equilibria by Schock and Buelow (1981) for many systems showed that the results of Mah and Boatman were reasonable and predictable and that orthophosphate was not necessarily involved in A–C pipe protection mechanisms.

Kanarek et al. (1981) studied nine water systems in the San Francisco Bay area to determine if asbestos fibers were released into the water during passage through the system, as part of a larger epidemiological study. The systems contained calcium levels of approximately 9 to 35 mg/L, alkalinities from approximately 50 to 130 mg $CaCO_3$ / L, and pH values of approximately 7.7 to 9.2. Eight of the systems showed increases in asbestos fibers after passage through the monitored part of the system. On the average, a 100-fold increase in the number of fibers was found. Significantly, they found that the system with the highest pH and aluminum value, but the lowest calcium and alkalinity, showed the largest fiber count increase.

DVGW (the German Gas and Waterworks Association) ordered a study on the presence of asbestos fibers in drinking waters of the Federal Republic of Germany. The study investigated both the number of fibers from natural sources and the number input to the water from A–C pipe. This study was conducted from 1977 to 1980 and was reported by Meyer (1982). Meyer found the use of TEM to identify and count the fibers was essential. A summary of his results is given in Table 7-12.

The table shows results before and after distribution in A–C pipes for water of different qualities. An approximate Langelier index (LI) value is given for each water. The results show no remarkable number of fibers leaching from A–C pipes in water qualities near calcium carbonate saturation.

Pipelines 1–10 had no lining. Pipes from lines 11–14 had a bituminous lining that did not prevent an attack on the wall and the leaching of bitumen and asbestos particles.

In pipeline 14, even after 15 years in operation and a high negative LI, no fibers were found. A test pipe showed defects and pores in the bituminous layer, but the surface was protected by an additional layer of iron hydroxides from the water that reduced corrosion attack.

Table 7-12 Results of analyses of asbestos fibers in some German distribution systems with A-C pipes

Pipeline Number	Number of Samples Taken	Langelier Index	Asbestos Fibers* at Waterworks	Asbestos Fibers* After Distribution
1	4	> –0.2	<0.005	<0.005
2	2	> –0.2	<0.005	<0.005
3	2	> –0.2		<0.005
4	4	> –0.2	0.04	0.02–0.04
5	5	> –0.2	<0.005	<0.005
6	3	> –0.2	<0.005	<0.005
7	3	> –0.2	<0.005	<0.005
8	2	> –0.2	<0.005	<0.005
9	2	> –0.2	0.01	<0.005
10	3	> –0.2		<0.005
11	7	–0.4	<0.005	0.01–0.04
12	4	–0.8		0.18–0.39
13	7	–1.2	0.01	0.005–1.0
14†	4	–2.0	<0.005	<0.005
15‡	4	–2.5	<0.005	0.01

Source: Meyer (1982).

*Fiber counts are in units of 10^6 fibers/L for asbestos fibers greater than 5 μm in length.

†Raw water pipelines, bituminous layer, water with iron content.

‡Length of pipeline only 300 m (985 ft).

The data from fiber counting were compared to data gathered in an examination of the inner surface of the pipe walls. Pipe samples from lines subjected to aggressive waters showed an attack on the cement and a residue of a fibrous fleece. Pipes with a single bituminous coating showed only partial deterioration of the surface. Samples taken from lines transporting waters that were approximately saturated with calcium carbonate did not show evidence of either attack on the surface or softening of the surface, even after 30 years in operation.

USEPA laboratory studies. In this section, the historical development of the investigations performed by the USEPA will be reviewed. Much of this summary is taken directly from the overview provided by Logsdon and Millette (1981).

In late 1973, the USEPA Drinking Water Research Division (DWRD) began investigating the performance of A–C pipe exposed to a variety of water quality conditions. A recirculating pilot-scale experimental pipe loop was built to accommodate several full lengths of 10-cm-diameter (4 in.) and 15-cm-diameter (6 in.) type II A–C pipe. This system included a holding tank that was open to the atmosphere and equipment for feeding chemicals to maintain the water quality desired for the experiments. Details of the system and experiments have been given by Buelow et al. (1980).

Two of the experiments performed were with waters having "aggressiveness index" (AI) values of 11.7 and 10.4. The AI 11.7 conditions were achieved by maintaining a pH of 7.5, a calcium concentration of 150 mg $CaCO_3$/L, and an alkalinity of 100 mg $CaCO_3$ /L. Continuous feed of sodium bicarbonate to maintain the desired alkalinity was opposed by continuous feed of HCl to maintain the pH at 7.5. The difficulties encountered in maintaining the pH and the alkalinity at the desired concentrations were caused by the system's design—the open reservoir allowed contact between the water and carbon dioxide in the atmosphere. Buelow et al. (1980)

concluded that a recirculating pipe loop in which water passed through a holding tank open to the atmosphere several times per hour was not representative of a water distribution system.

The unsatisfactory performance of the pilot-scale A–C pipe loop experiments led to the development of a bench-scale experimental apparatus described by Buelow et al. (1980) and Buelow (1979). This experimental arrangement was designed to allow ease in operation and magnification of effects of water attack on pipe specimens by the extended use of recirculated water between changes. The 380-L (100-gal) volume of the recirculating water acted on a small piece of A–C coupon (see Figure 7-10). The water flow was 0.2 L/s (3 gpm), providing a velocity of about 0.35 m/s (1 ft/s) past the coupon, a reasonable flow. An important feature of the test equipment was the cover and Tygon™ tubing ring gasket that prevented atmospheric contact with the water at the small gap between the cover and the tank wall. Because the water in the tank and piping system was isolated from the atmosphere to a large degree, water quality could be held much closer to the desired value, and with substantially reduced effort.

100-Gal Recirculation System for Corrosion Testing

Part Number	Description
1	100-gal Stainless Steel Tank
2	1-in. Plastic Ball Valve
3	1-in. Plastic Tee
4	Plastic Boiler Drain
5	1-in. Plastic Union
6	1 1/2-in. by 1-in. Reducer Coupling
7	1 1/2-in. by 6-in. PVC Nipple
8	1 1/2-in. PVC Coupling
9	1/2-in. PVC 90° Ell
10	Magnetic Drive Pump With Plastic Impeller
11	1/2-in. PVC 45° Ell
12	Plastic Water Meter
13	1/2-in. PVC Tee
14	1/2-in. Pipe to Tube Ell
15	1/2-in. ID Tygon™ Vacuum-Type Tubing
16	Floating Cover (The floating cover is made from 1/8-in. sheet PVC. A 1-in. strip of PVC cemented around the edge of the disc forms the floating pan. The threaded fittings in the cover require a buildup of PVC.)
17	Tygon™ Tube Ring Gasket Under Floating Lid
18	Tank Legs
19	1/2-in. PVC Pipe
20	Sample Valve
21	3/4-in. PVC Nipple
22	Rubber Stopper

NOTE: All pipe bushings and nipples required to make connections are not numbered and described. Water flow is indicated by arrows.

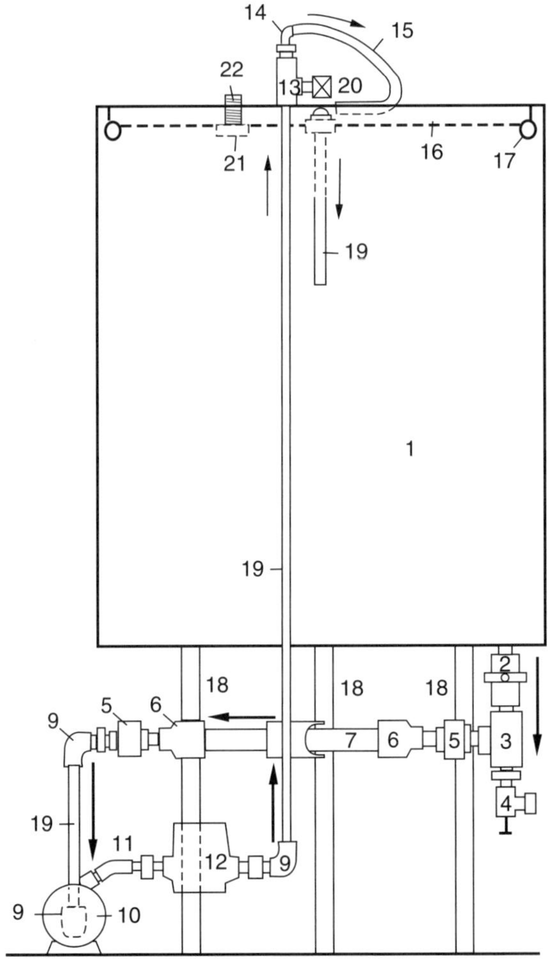

Figure 7-10 A-C pipe small-scale corrosion experimental setup used in USEPA studies

Results of the recirculating tank coupon tests were evaluated in several ways. Water quality was monitored, and changes in pH, calcium, and alkalinity were noted. When an additive such as zinc chloride, ferric chloride, or zinc orthophosphate was used, important components such as zinc, orthophosphate, or total iron in the water were measured. Pipe coupons were inspected visually, and many were photographed with SEM and analyzed by EDXA. In addition, Buelow noted that some coupons had been softened in the six-month-long test, so that they could be gouged with a fingernail. New A–C pipe coupons are hard and cannot be gouged or scratched with a substance as soft as a fingernail. This finding was pursued in subsequent projects, in which quantitative methods of hardness testing were developed.

From 1978 through 1982, 32 different experiments were conducted with uncoated A–C pipe coupons by using the apparatus shown in Figure 7-10. The results of most of these experiments have been reported in detail by Buelow et al. (1980) and by Schock and Buelow (1981) and will not be repeated here except in terms of general conclusions. Unless the specific impact of high calcium or alkalinity levels were the object of investigation, the general water qualities used were soft waters (calcium levels of 5–20 mg $CaCO_3$/L), with alkalinities of approximately 20 mg $CaCO_3$/L and pH values of 7 to 8.5. All pipe coupons were cut from 4-in.-diameter (10 cm) type II A–C pipe.

Buelow et al. (1980) and Schock and Buelow (1981) were able to draw several significant conclusions from their laboratory studies, in conjunction with their field observations. These conclusions were the result of combining straightforward observation with computer calculations of the chemical equilibria of the water systems. These conclusions are

- Small-scale experiments in systems closed to the atmosphere can be used to evaluate the performance of A–C pipe under realistic field conditions.
- Calcium carbonate saturation indices, such as the AI or LI, are insufficient in themselves to predict whether A–C pipe will be attacked by a given water.
- Calcium carbonate deposition can be used to prevent attack on A–C pipe.
- The AI, as traditionally conceived, is not theoretically sound from a chemical point of view for use as a predictor of fiber release and pipe interior degradation in the absence of calcium carbonate deposition. An accurate index must take into account the presence of protective constituents such as iron, manganese, silica, and zinc, as well as more commonly measured parameters such as pH, calcium concentration, carbonate concentration, temperature, etc. These constituents have been observed to form protective deposits in many field investigations, along with unidentified organic material.
- Some coating mechanisms serve to bind the asbestos fibers against release but not to seal the cement pipe matrix from dissolution. Thus, calcium and pH increases during flow through a distribution system must be interpreted with caution, as they do not necessarily indicate fiber release. Other chemical reactions in the water may also affect the pH.
- The chloride, sulfate, and orthophosphate salts of zinc have been tested and found to provide substantial protection of A–C pipe when they are added to the water at the proper concentration and pH ranges and when those chemical conditions are maintained throughout the distribution system. Zinc is the active agent. It acts to coat the pipe and protects the pipe against fiber release and water attack. An example of protection by a zinc coating is given by the SEM photograph shown in Figure 7-11. The reaction mechanisms

Figure 7-11 Scanning electron microscope photograph of A-C pipe specimen from USEPA laboratory experiment, showing the surface covered by a zinc-containing solid (magnification ~ 300×)

have not been delineated. One possibility is that initially zinc reacts with the water to form a zinc–hydroxycarbonate precipitate such as $Zn_5(CO_3)_2(OH)_6$, hydrozincite. The zinc solid may then react with the pipe surface itself, possibly converting some or all of the coating to a harder zinc–silicate solid phase. The chemical dosages required for a given general water quality are closely estimable by diagrams constructed from computer-assisted calculation of simultaneous aqueous equilibria (Schock and Buelow 1981). Examples of the diagrams developed to predict areas of solid deposition and zinc dosages are given in Figures 7-12 and 7-13, for a water with the composition summarized in Table 7-13.

- When systems involve both metal and A–C pipe, the zinc chloride or sulfate salts provide little or no protection to the metal pipe. However, DWRD tests at pH 8.2 with lead and A–C pipe in the same recirculation system showed that the orthophosphate salt containing zinc provides corrosion inhibition for both types of piping material. The lead solubility in such systems is governed by the formation of lead orthophosphate (rather than zinc orthophosphate) compounds. Cement linings should behave essentially the same or be even more amenable to zinc coating than is the A–C pipe matrix. Presumably other metals forming orthophosphate scales of low solubility (such as zinc or iron) will have their dissolution reduced more effectively by the zinc orthophosphate salts. Their protection would also at least depend on the pH and dissolved carbon concentrations in the water.

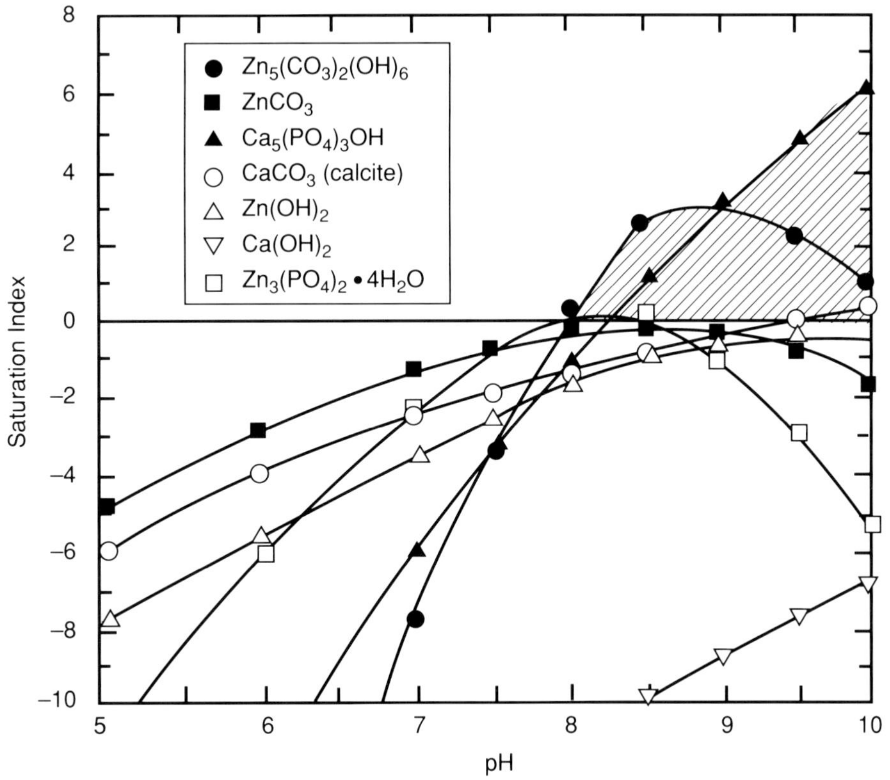

Figure 7-12 Saturation index diagram for model system M10P (see Table 7-13)

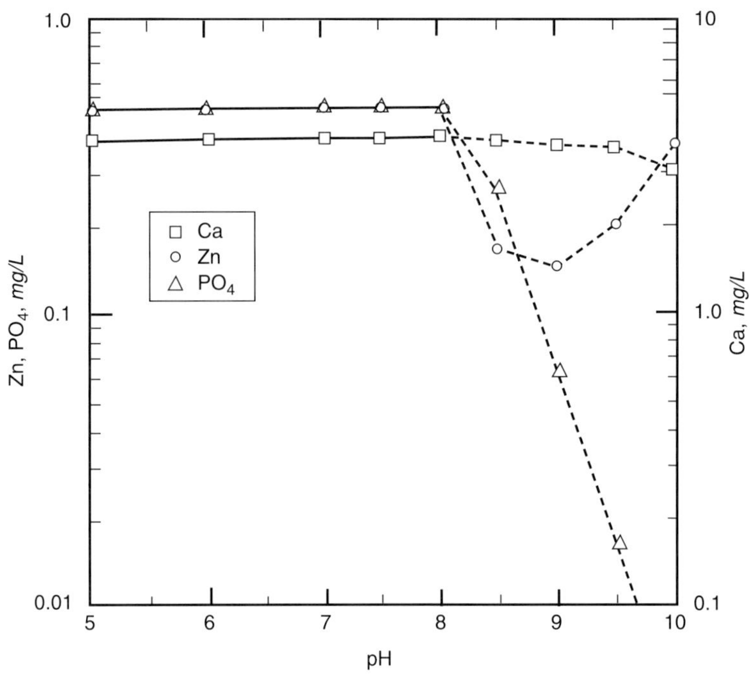

Figure 7-13 Precipitation diagram for zinc, calcium, and orthophosphate in model system M1OP (see Table 7-13)

Table 7-13 Initial composition of water for system M1OP, used to construct Figures 7-12 and 7-13

Parameter	Concentration, *mg/L*
Calcium	4.0
Magnesium	1.0
Dissolved inorganic carbon (as C)	2.4
Sodium	12.0
Chloride	15–29
Zinc	0.5
Orthophosphate (as PO_4^{-3})	0.5

Source: Schock and Buelow (1981).

- Strong sequestering or complexing agents, such as polyphosphates, can attack the pipe by enhancing calcium, aluminum, iron, and magnesium leaching from the cement matrix. They could also prevent the formation of protective coatings by metals such as zinc, manganese, and iron.

Additional USEPA experiments reported previously (Holtschulte and Schock 1985) were evaluated with the assistance of some quantitative hardness testing and cross-sectional EDXA elemental scanning techniques (Millette 1983; Tate et al. 1984; Millette et al. 1984). The hardness testing was accomplished with a modification of a standard instrument (Durometer) that should be of wide applicability to the evaluation of degradation of cementitious materials.

The Durometer was a model 307L Type D Scale instrument (Pacific Transducer Company, Los Angeles, California), with the flat base replaced with one machined to match the curvature of the A–C pipe specimens in order to obtain accurate measurements (Millette 1983; Millette et al. 1984). The Durometer scale ranges from 100 (no penetration) to 0 (0.25 cm [0.1 in.] penetration) (Millette 1983).

The additional USEPA experiments concentrated on the effect of high levels of silica (35 mg SiO_2/L) at different pHs (Holtschulte and Schock 1985). None of the nine systems investigated showed major deterioration, but there was a trend toward better performance at pH 8. The results at pH 6 were the most erratic. EDXA of the surfaces of pipe specimens from experiments with iron content in the test water did not reveal substantial iron coatings, as would be expected from a simple consideration of ferric iron hydrolysis and hydroxide solubility. A likely explanation would be the complexation or colloidal dispersion of Fe(III) by silica, as was suggested by Schock and Buelow (1981) based on their literature review. If verified by additional experimentation, this finding would be very important. Silica would have a complicated role in the stabilization of the cement pipe matrix. If absent, or at low concentration, significant pipe softening and calcium leaching could take place (providing no other deposit-forming inhibitors are present). Above this level, some support to maintain pipe hardness would be available, although calcium leaching would not be prevented by silica alone.

Above some threshold level, perhaps 25–30 mg SiO_2/L, depending on pH and other interaction properties (such as charge or complex stability constants), silica might interfere with the formation of protective ferric iron films that might slow calcium leaching and bind fibers to the surface. The effect of pH alone (independent of alkalinity) on the rate of cement matrix dissolution has not been studied. Such an investigation would be important in elucidating A–C pipe dissolution mechanisms.

Millette (1983) extended the experimental bench-scale approach of the earlier USEPA studies to emphasize the use of quantitative hardness testing techniques and X-ray elemental mapping and scanning. He used SEM on systems containing ferric iron, a natural inhibitor for A–C pipe dissolution and fiber release. The feasibility of using sacrificial corroding iron rings or pipe sections was investigated by using all-plastic tank assemblies and pipe loops developed by JM Montgomery Consulting Engineers at the suggestion of USEPA (Tate et al. 1984). Millette also analyzed many pipe samples collected as part of various USEPA field studies. The laboratory studies produced the following important findings:

- Coatings containing iron were found on many A–C pipe specimens exposed to a variety of water qualities considered to be aggressive to nonaggressive based on their calcium carbonate saturation state.
- Under water conditions considered to be very aggressive, the iron coatings formed had a granular, porous structure that did not necessarily retard calcium leaching from the cement matrix. The iron did help in preventing the exposure of asbestos fibers at the pipe surface.
- The iron coatings were not formed as a continuous film, but rather as crystals of an unidentified iron compound that accumulated on the surface of the pipe. Under the conditions of the experiments, the iron accumulated as a linear function of time.

- When the water flow velocity was in the "turbulent" range, it had no measurable effect on the rate or quality of the iron deposit on the pipe surface. However, the pipe coupons from the systems with higher flow velocities had more asbestos fibers exposed on their surfaces. Thus, competing mechanisms of dissolution and coating are probable, and both of them depend on the velocity and turbulence of the water.
- Both the Rockwell "L" and the modified Durometer "D" scale tests were found to be promising methods to measure A–C pipe condition. The test values were related to calcium leaching from the pipe, not to fiber exposure.
- Cast-iron sacrificial rings would have to be of an impractically large size to be of benefit for in situ coating of A–C pipe.
- The silica level is a critical parameter to be considered in the protection of A–C pipe with iron.

The results of this study are in close agreement with the observations of Flentje and Sweitzer (1955), who found iron deposits and coatings to be beneficial in slowing down the rate of calcium leaching from cement-lined cast-iron mains.

Field Studies of In Situ Protection and Rehabilitation of A-C Pipe

USEPA-sponsored studies. Logsdon (1983) and Logsdon and Millette (1981) have summarized many aspects of the important field investigations undertaken by USEPA to determine the feasibility of in situ rehabilitation of A–C pipe, and methods appropriate for that task.

At Greenwood, South Carolina, storage and chemical feed tanks were set up at each of the water treatment plants to enable the dosage of zinc orthophosphate (0.3 mg Zn/L) to control pipe softening and fiber leaching in the distribution system. Two sections of new A–C pipe, removable for testing, were used to determine the effect of the treatment under low and high flow conditions. During the test period, routine tests such as pH and alkalinity showed no significant changes (Logsdon 1983). On the other hand, asbestos fiber concentrations were reduced substantially. Also, SEM and EDXA photographs showed the development of a partial zinc coating on the pipe specimens from the two locations (Logsdon 1983; Schock et al. 1981). Although the zinc dosage was lower than would be desirable for the pH and alkalinity of the water, as well as the length of pipe in the system, corrosion of A–C pipe was greatly reduced, if not arrested. The hardness of the pipe was not restored, however.

Extremely softened and deteriorated pipe in Weston, Wisconsin, led to a cooperative agreement with USEPA to study rehabilitation methods. The study, which commenced in 1980, investigated three treatments to alleviate the problem: zinc chloride addition to the water supply, water main flushing to reduce fiber counts, and cement–mortar lining of the pipe after the use of polypigs or steel scrapers (USEPA 1984; Weston Water Utility 1983).

Water quality analyses, hardness tests, weight gain/loss determinations, and electron microscopic analyses (SEM, EDXA, and fiber counts) were all used in this study to evaluate the different treatment options. The results showed that scraping or pigging A–C mains could effectively remove the soft, deteriorated layers of pipe, but it greatly increased the fiber counts in the water when normal operation resumed.

Deposition of a zinc coating as a result of the zinc chloride addition did not materialize. Sequestration of the zinc by the polyphosphate chemical added for iron control is the probable cause for the inhibition of zinc precipitation as well as being a significant factor in the initial calcium removal and pipe softening. The action of polyphosphates as agents that could enhance calcium leaching by complexation, as well as increase the solubility of other metal compounds, was pointed out by Schock and Buelow (1981).

Weight measurements of pipe coupons were generally unsatisfactory. The difficulty in obtaining meaningful constant weights because of the complicated hydration–dehydration chemistry of the cement materials makes this approach unsuitable to a great degree (USEPA 1984; Weston Water Utility 1983).

The cement–mortar lining was essentially successful, even though contamination from the tank truck supplying water for the test section of pipe generated detectable fiber counts after the lining process was completed. Also, poor precision for the counting technique was evident at low fiber concentrations. Electron microscopic examination indicated excellent surface coverage and sealing against fiber erosion.

The main flushing experiment revealed that chrysotile asbestos fiber concentrations were somewhat elevated in comparison with undisturbed water, even after 90 minutes of flushing (USEPA 1984; Weston Water Utility 1983). Further study was suggested.

Following the pilot-scale tests reported in an earlier section, a field investigation was implemented in Bellevue, Washington, to determine the effect of a corrosion control program featuring lime addition to raise the pH to 8.0, plus the addition of 0.6 mg Zn/L added as zinc chloride (Tate et al. 1984).

The USEPA-sponsored field study in Bellevue, Washington, began shortly after Seattle started its corrosion control program, which consisted mainly of pH and alkalinity adjustment of the Tolt River supply. Monitoring sites were selected to sample the influent water, effluent from the treatment plant, a point after zinc chloride addition was made (to monitor the applied dosage), dead ends, and sewers draining the area. The dead ends enabled the monitoring of possible particle accumulation, and the sewers provided the opportunity to find out how much zinc was actually leaving the system. A control area was selected for parallel testing, where zinc addition was not used. During the study, the pH varied from 7.9 to 8.7, and alkalinity ranged from 8 to 22 mg $CaCO_3$/L.

Analysis of the zinc mass-balance results and the samplings of dissolved and particulate zinc indicated that precipitation of a zinc solid (or solids) was occurring and that deposition was taking place in the test system. The exact fraction of zinc removed to the pipe surface could not be accurately quantified. This was verified by EDXA of pipe samples, showing the formation of a zinc-containing deposit. The interior surfaces of pipe samples removed from two of the test sites were harder to scratch than the interior surfaces by specimens from the control section and did not have exposed fibers (as did the control specimens).

Among the conclusions derived from the field study were

- No confident determination was possible as to whether the zinc chloride reduced the asbestos fiber release from the test sections. Initial tests indicated that pH, calcium, and asbestos fiber concentration increases were less than those in the control sections. However, water quality differences between the control and test sections were less pronounced after a year of the Seattle corrosion control program.

- The Seattle corrosion control program in itself had a significant impact on slowing the chrysotile fiber release from A–C pipe.
- Hardness testing was inconclusive in determining the effectiveness of zinc chloride addition in arresting deterioration of A–C pipes that had been in service.
- Calcium loss measurements showed no difference between A–C pipe behavior in test and control sections; however, visual and manual examination suggested the control sections showed more pipe softening.
- Zinc precipitation and deposition occur in the full-scale application of zinc chloride (given the proper pH, alkalinity, and zinc levels).
- A brief cost analysis suggests that chemical treatment is the lowest cost alternative for dealing with A–C pipe deterioration, when it is appropriate.

Confirmation of the ability of zinc, iron, and manganese to provide inhibition of fiber leaching and lessening of pipe deterioration has been discussed for several distribution systems by Buelow et al. (1980), Schock and Buelow (1981), and Schock et al. (1981). An interesting case is shown by Figures 7-12 and 7-13, which provide a basis to understand the condition of a specimen from an A–C pipe in service for 21 years in a water with a pH of approximately 7.4, an alkalinity of approximately 25 mg $CaCO_3$/L, and a calcium hardness of 40 mg $CaCO_3$/L (Schock 1980). The A–C pipe section was located after a considerable length of cast-iron pipe. In order to reduce iron corrosion and red-water complaints, a liquid zinc orthophosphate formulation had been added, at a concentration of 0.4–0.5 mg Zn/L. The A–C pipe surface is shown in Figure 7-14 to be covered by numerous platy crystals, which cover virtually all of the chrysotile fibers. The EDXA (Figure 7-15) reveals naturally occurring manganese, zinc from the metal corrosion control treatment, and iron from natural sources and from the cast iron pipe dissolution, acting together to help "preserve" the A–C pipe surface (the calcium and silicon are from the A–C pipe itself). A water sample taken when the pipe specimen was removed had undetectable chrysotile fibers (<250,000 fibers/L). Thus, the compatibility of a corrosion control program for both iron and A–C pipe is shown, even though the program was not instituted with A–C pipe in mind.

Other studies on lined A–C pipe. The CEOCOR (1979) study gave some proposals for useful liners or coatings to be applied to the inside of A–C pipes. Materials used frequently in the past are

- bitumen
- coal tar
- coal tar–epoxy
- epoxy
- chlorinated rubber
- natural rubber
- butadiene
- other synthetic, nonsaponifiable resins

Because of potential taste, odor, or health problems, linings or coatings by any material other than cement mortar should only be considered after careful investigation

Figure 7-14 Scanning electron microscope photograph of an A-C pipe specimen from a New England water system that uses zinc orthophosphate for iron corrosion control (magnification ~ 300 ×)

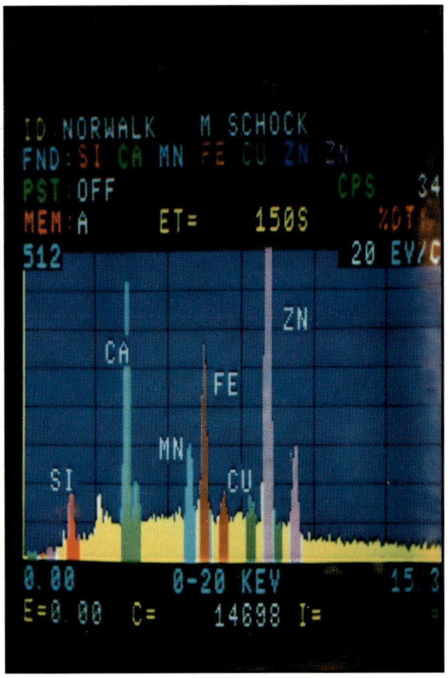

Figure 7-15 EDXA of the surface of the A-C pipe in Figure 7-14

of local, state, and federal regulations. For example, numerous problems with leaching of tetrachloroethylene and other contaminants from pipe coated with a vinyl toluene liner have been reported in the United States (Larson, Love, and Reynolds 1983; Gardner 1981; Anderson 1981; McCallie 1981; Woodhull 1981). Although that particular type of coating has been discontinued for several years (Larson, Love, and Reynolds 1983), continued caution must be exercised when considering new lined pipe and in monitoring water from previously installed lined pipe. Meyer (1982) found that deterioration (lime and fiber leaching) took place in many bitumen-lined pipe systems in Germany.

A detailed discussion of cement–mortar lining of pipes will be covered in later sections of this chapter.

Monitoring Asbestos-Cement Pipe Condition

The question of the quality of already-buried distribution systems needs continued attention, to provide water distributors with information concerning the potential risk of fiber release. This may take any of several approaches, which are discussed in the following paragraphs.

Chemical analyses of distributed water. Traditionally, the aggressivity of water toward A–C pipe has been equated with the calcium carbonate saturation index, in one form or another. The limitations of this approach, with respect to the prediction of chrysotile fiber release and calcium loss (cement binder dissolution and breakdown), has been discussed in detail (Schock and Buelow 1981; Schock et al. 1981).

Simple monitoring of calcium and pH increases as water passes through a system containing A–C pipe can certainly give an indication that further investigation might be necessary. To predict the deposition of naturally occurring films other than calcium carbonate, an analysis program should include temperature, ferric iron, silica, and manganese, in addition to pH and calcium, plus enough major constituents (total alkalinity, magnesium, sulfate, chloride, etc.) and an estimate of redox conditions (chlorine residual, dissolved oxygen) to enable the computation of saturation indices or precipitation potentials of other potentially protective solids (Schock and Buelow 1981; Schock et al. 1981). Numerous computer programs are available to do such calculations, and many are now available for use on microcomputers.

Some of the solids that might be useful as predictors of naturally occurring films are amorphous ferric hydroxide, lepidocrocite (a ferric oxyhydroxide mineral), manganese dioxide (perhaps birnessite form), and chalcedony (cryptocrystalline quartz). Other solids may be protective, and continued investigation into the identities of the solids depositing on A–C pipe surfaces should be an important area of investigation. No single index has been developed, or is likely to be developed in the near future, to take into account the many solids that could be protective, plus organic matter and complexation interactions.

Evaluation of the state of pipelines in use. Historically, drinking water distribution systems containing A–C pipes have often been put into service without considering the possible corrosive properties of the distributed water. Obtaining data on the impact of the water on the pipe material and the possible release of fibers into the water led to the evaluation of methods to assess the physical state of the pipe. Two approaches are used to describe the actual situation: fiber control in the distributed water and the evaluation of the state of the pipe itself.

Measurement of fiber release. Three analytical and sampling problems relating to the concentrations and sizes of fibers released during the conveyance of drinking water require careful attention. They are

1. The significance of proposed or existing maximum contaminant levels (MCLs).
2. The analytical method used for fiber counting.
3. The sampling protocol.

Two major standards concerning fiber content of drinking water exist: The USEPA regulation of 7×10^6 fibers/L of a length of 10 µm and longer (USEPA 1991) and the German Federal Health Office (BGA) recommendation, not related to health risk data, which says 10^4 fibers/L of 5 µm and longer is to be the figure for a "non asbestos polluted" water (BGA 1981; DVGW 1993). The difference between the two standards reflects the uncertainty of the toxicological impact of asbestos fibers in drinking water.

The BGA standard is strictly linked to analysis by TEM with a fixed sample preparation method, fiber counting procedure, and the identification of asbestos fibers beside any other possible fibrous materials. If the method is performed correctly, the analytical detection limit is approximately 2,000 fibers/L.

For the US drinking water standard, a strict TEM-based methodology is also specified (USEPA 1991). The procedure is essentially that developed by Chatfield and Dillon (1983), with additional methodological clarification and guidance (Feige, Clark, and Brackett 1993). Although the relationship to compliance-monitoring legal validity has not been fully established, investigations of procedural improvements have continued to be published (*Standard Methods* supplement 1994; Brackett and Clark 1993).

The complete procedural methodology covers filtration materials and processing, sample containers and treatment, detection limit, counting rules, and fiber identification (chrysotile or amphibole) by a combination of electron diffraction pattern and quantitative X-ray analysis. Any other analytical methods, like SEM, provide too poor resolution for the procedure of fiber identification.

Aside from any analytical uncertainties inherent in even the most rigorous TEM-based methodology, a major interpretive uncertainty remains because the observed fiber count may depend strongly on the sampling procedure. The leaching of fibers into the water is strongly related to the variable mechanical forces that are present at the sampling point, as well as the chemical history of the water. Also the operational conditions such as throughflow, pipe geometry, sampling tap geometry, and similar local effects greatly influence the asbestos fiber level in an individual sample. Finally, a third complication is that the water itself may contain naturally occurring asbestos fibers, and a comparison of water quality before and after transport through A–C pipes is necessary. This calculation then involves determining the precise and accurate difference between two somewhat variable and imprecise numbers.

All this means that, in practice, a single sample or duplicate samples for asbestos fiber content can only give ambiguous information about the physical status of a distribution network. However, collecting a statistically sufficient number of samples is so time- and money-consuming that such a sampling and analytical endeavor becomes practically unaffordable.

Measurement of the physical state of the pipe. The use of more objective physical test methods has been investigated primarily by Logsdon and Millette (1981), Millette et al. (1981), and Millette et al. (1984) at USEPA and by researchers at JM Montgomery Consulting Engineers (Tate et al. 1984). An interesting consequence of this line of investigation is that standard "weight loss" coupon tests, commonly relied on in metallic systems, are generally inappropriate for an A–C pipe, because of hydration–dehydration reactions and complicated phase changes within the material itself.

The use of reflected-light optical microscopy with A–C pipe cross sections mounted and polished in Bakelite mounts has also been suggested as a possibly useful technique (Logsdon and Millette 1981; Schock et al. 1981). Obvious visual differences between the outer pipe layers attacked by water and the underlying zone beyond the solution penetration are visible. For a better understanding, these differences need to be coupled with mineralogical identifications and further understanding of A–C solution reactions and solid-phase transitions.

Because of the problems connected to fiber counting in Germany, a threefold method was developed to be able to give useful information to the water utilities on the physical status of A–C pipes in their drinking water distribution networks. This method is based on three procedures for the physical and chemical examination of the pipes.

For the test, pieces of 0.5-m to 1-m (20-in. to 40-in.) pipe must be excavated. These are used for physical inspection and the analytical (chemical) investigations. Additionally, the corresponding information on the distributed water quality should be available, if possible, for the whole period the A–C pipe network was installed and operating.

In the first step, the pieces of pipe are broken to get fresh surfaces of the pipe wall. The surfaces are moistened with a 0.01 percent phenolphthalein solution. As shown in Figures 7-16 through 7-21, the parts of the pipe walls that kept their original condition produced a violet color caused by the alkaline (hydroxide) content of the cement. The parts where $CaCO_3$ deposition and/or deterioration of the cement occurred—through either contact with the conveyed water or by attack on the pipe exterior by the surrounding material—showed no coloration by the phenolphthalein.

Figure 7-16 Original A-C pipe sample (fresh surface) with bituminous lining

Figure 7-17 Sample as shown in Figure 7-16 after staining with phenolphthalein (no calcium carbonate cementation)

Figure 7-18 Freshly broken A-C pipe sample without lining, phenolphthalein-stained, exposed to nonaggressive water

Figure 7-19 A-C pipe sample with attack from the inside by transport of water aggressive to calcium carbonate

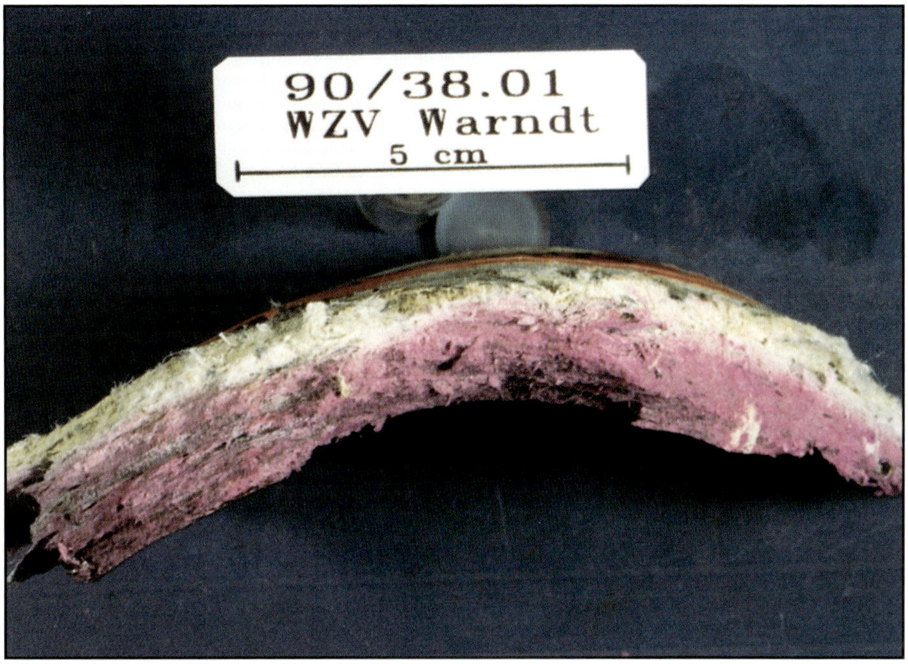

Figure 7-20 A-C pipe sample with attack from the outside by environment that is aggressive to calcium carbonate

Figure 7-21 Inside and outside attack of an A-C pipe

To complement the phenolphthalein staining procedure, a scratching test is performed, comparable to the "fingernail test," proposed by Buelow et al. (1981) or Schock and Buelow (1981). For this purpose, the original inner surface of the pipes, which is often covered with reddish or brownish layers from iron and/or manganese, is carefully cleaned by a smooth brush and water. After the cleaning, a blunt metal scraper is used to scratch on the surface.

Depending on the mechanical stability of the scratched surface, the procedure either shows no effect or rips asbestos fibers from the surface. Using the magnifications of an optical microscope, a decision can be made on whether asbestos fibers can be mechanically abraded into the water. The result of this "scratch test" can be documented better by a photograph, as demonstrated in Figures 7-22 through 7-24.

Finally, samples from different parts of the cross section of the pipe wall are taken. To analytically distinguish $CaCO_3$ scaling from deterioration by dissolution and leaching, in the uncolored sections of the wall, the quantity of calcium is determined. For this purpose, the samples are extracted in boiling HCl (3 percent) for 1 hour. The reduced liquid volume is adjusted afterward with 3 percent HCl solution to a fixed level and the calcium concentration in the sample is determined.

The result of this kind of analysis is shown in Figures 7-25 through 7-29, which represent the different stages of deterioration found on real samples.

The results of the three testing methods are evaluated and combined to provide a recommendation for the respective water utilities on water treatment modifications, such as pH correction. Recommendations are also made for physical maintenance measures at the network, e.g., rehabilitation or replacement of the pipes, depending on the degree of degradation.

Conclusions

No simple index exists that can predict the behavior and service life of A–C pipe when it is exposed to waters of varying chemistries. In general, calcium carbonate saturation has been presumed to be an effective treatment strategy, and there is reasonable field evidence that the idea is basically sound. Some of the success of this approach may be attributed to the mass-action effect of waters with high calcium concentrations and pH values that would tend to stabilize the cement matrix. The primary drawback to calcium carbonate saturation is that in the absence of actual scale formation, or the formation of protective coatings by naturally occurring substances such as iron and manganese, fibers are left exposed at the surface of the pipe, where they could be physically eroded.

In addition to the protection afforded by naturally occurring iron (as a ferric hydroxide or oxyhydroxide deposit) and manganese (as a Mn(IV) oxide deposit), silica may also be beneficial as an agent that can help maintain the hardness of the pipe. This effect has also not been studied in detail. Laboratory experiments indicate that high concentrations of silica will inhibit iron hydroxide precipitation, possibly through complexation or by colloid stabilization and dispersion. If the pipe deterioration can be stabilized without needing the iron precipitate to bind the fibers and help coat the surface, then the usable silica level can be increased substantially. More research is obviously needed in this area.

Orthophosphate, sulfate, and chloride salts of zinc have been found to be useful in preventing softening of A–C pipe specimens and in slowing calcium leaching in numerous laboratory experiments. The dosage of zinc necessary depends at least on the pH and dissolved inorganic carbon concentration of the system. How useful zinc addition is in extending the potential lifetime of newly installed A–C lines or

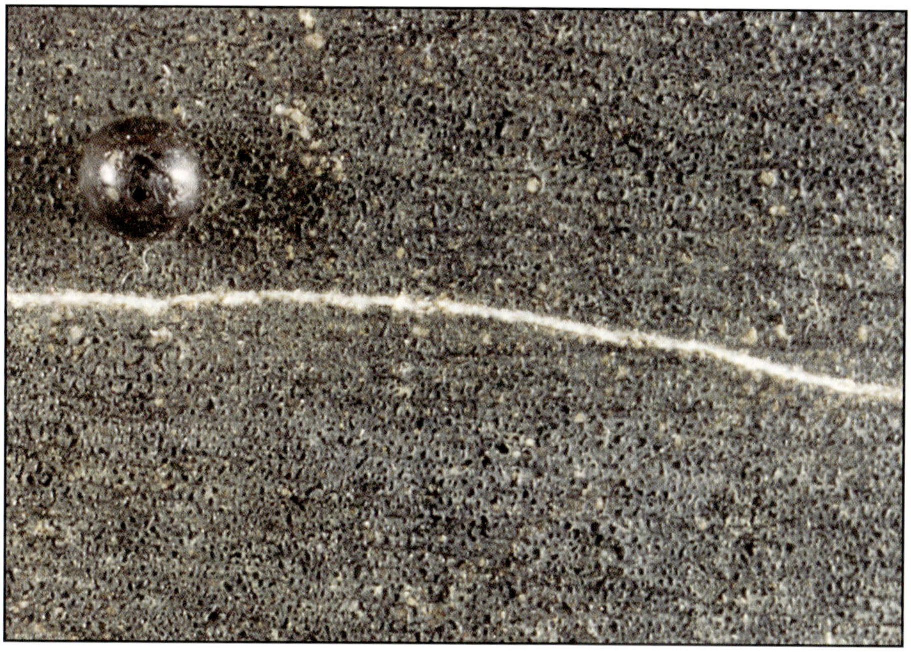

Figure 7-22 Scratch test on an undisturbed bitumen-lined inner surface of an A-C pipe

Figure 7-23 Scratch test on an unlined inner A-C pipe surface exposed to a water aggressive to calcium carbonate without rust deposit

Figure 7-24 Scratch test on an unlined inner A-C pipe surface exposed to a water aggressive to calcium carbonate with rust deposit (brownish surface)

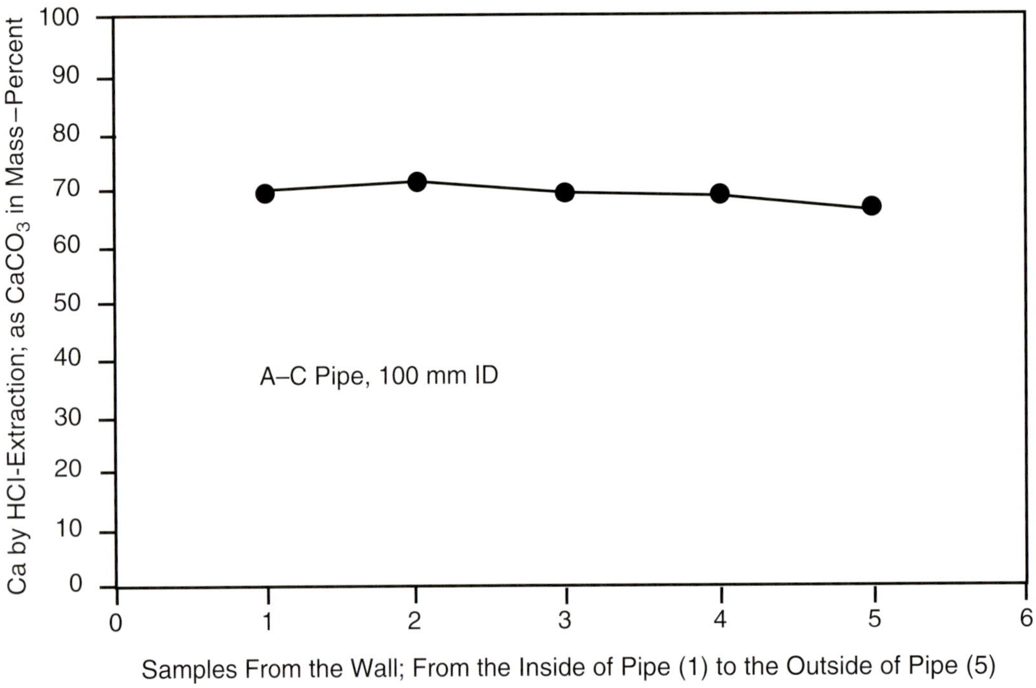

Figure 7-25 Calcium contents in an unaffected A-C pipe

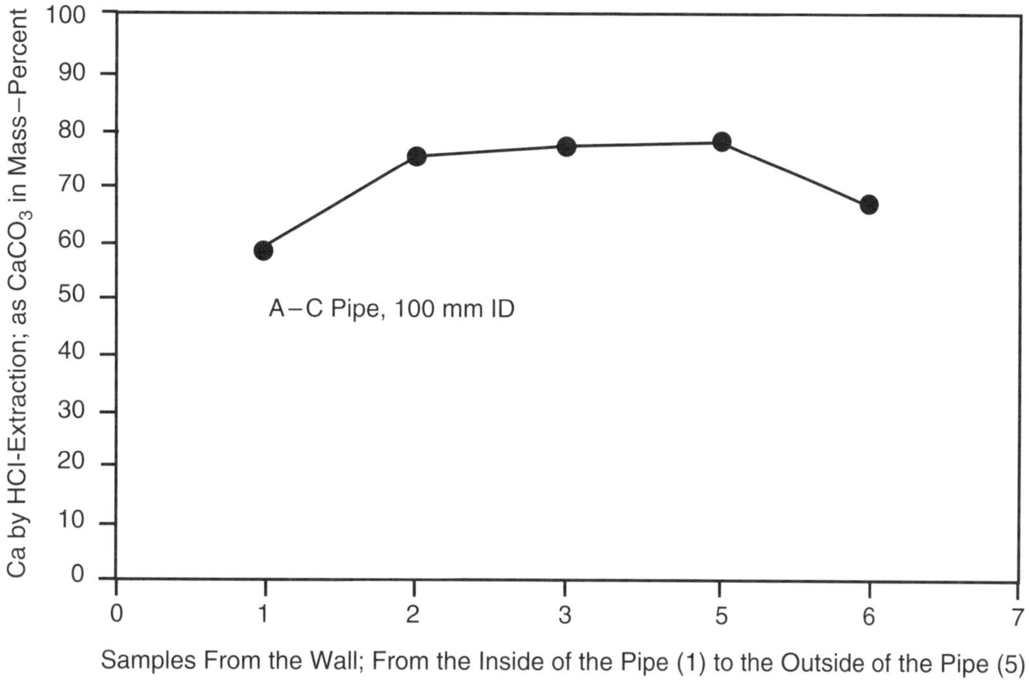

Figure 7-26 Calcium contents in a slightly affected A-C pipe (inside and outside)

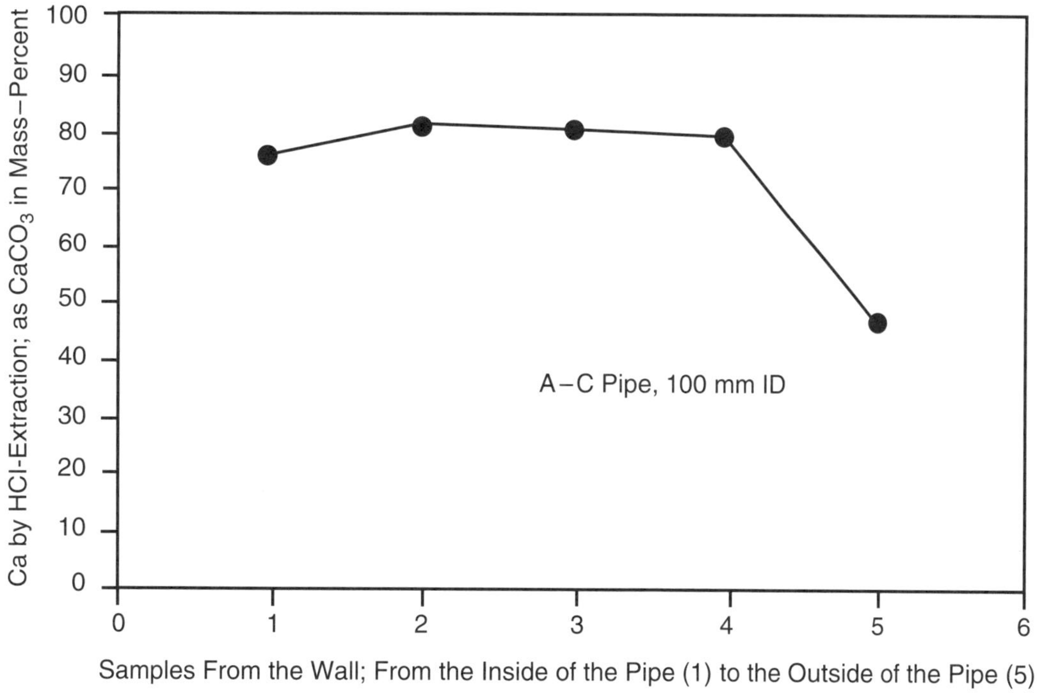

Figure 7-27 Calcium contents in an A-C pipe undergoing outside attack

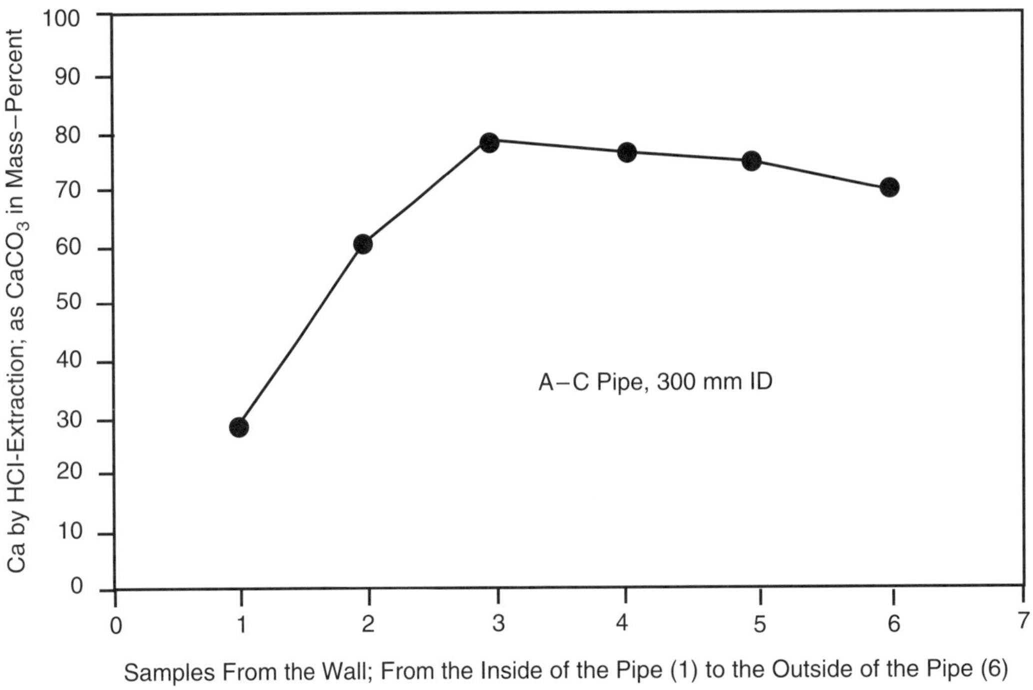

Figure 7-28 Calcium contents in an A-C pipe undergoing inside attack

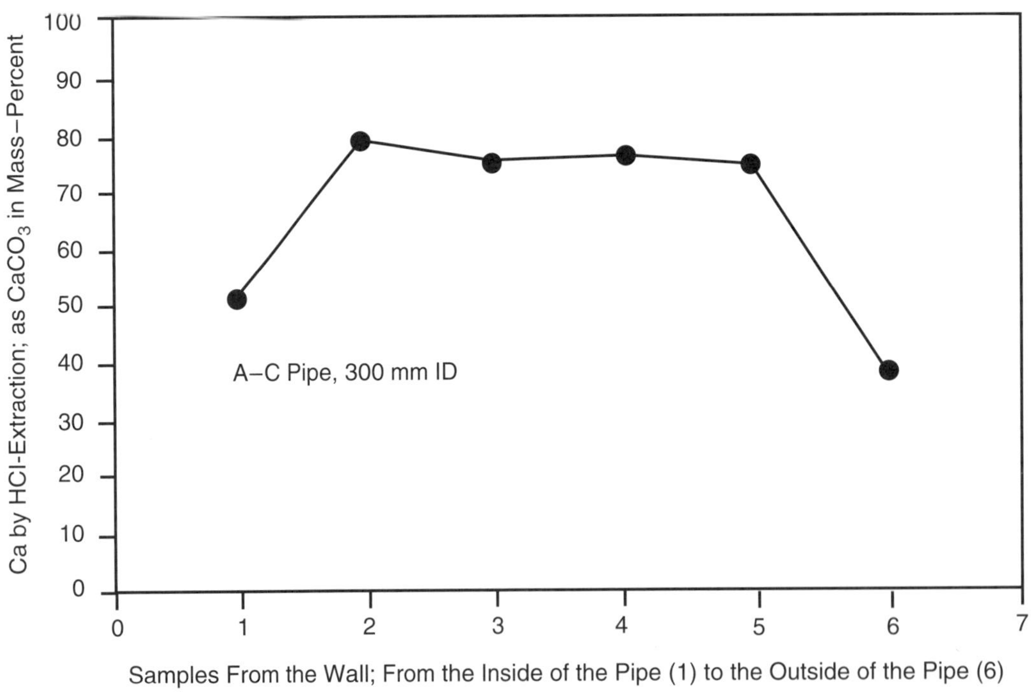

Figure 7-29 Calcium contents in an A-C pipe undergoing inside and outside attack

in increasing the service life of existing mains has not been established in the short field tests performed thus far. In these experiments, zinc has been found to be the active agent, not its anionic associate. Thus, the choice of salt can primarily be based on economic considerations, impact on other metallic compounds in the water system, or whether the system water would ultimately contribute to nutrient loading in waste streams. If the system consists of metallic pipe as well as asbestos–cement, the orthophosphate salt might be preferable to cut down on metallic corrosion. In such cases, the pH of the water would need to be adjusted or maintained in a range appropriate for orthophosphate passivation of the metal.

Polyphosphates strongly complex calcium. This property will accelerate the leaching of calcium from the pipe, resulting in rapid additional softening and fiber release from the pipe surface.

Thus far, the only successful technique used for the rehabilitation of A–C pipe is cement–mortar lining. Whether scraping and removing the softened surface layers in contact with the water are essential before the lining is applied has not been established. The technique of in situ cement–mortar lining will be covered in later sections of this chapter.

SEM has been found to be very useful to observe the extent of exposed fibers at the surface that could be subject to erosion and as an indicator of the effectiveness of treatment chemicals in coating fibers and sealing the pipe surface against leaching (Mah and Boatman 1978; Millette et al. 1984; Schock et al. 1981; Clark et al. 1980). As an adjunct to SEM, the EDXA technique has been used to gain important information on elemental distributions in the films and the pipe itself. Originally, only qualitative information could be obtained; however, correction techniques, particularly those using computerized calculations of chemical equilibria, sometimes can enable actual quantitation of the elements. This was found particularly useful by Millette (1983), Millette et al. (1984), and Clark et al. (1980) in showing the penetration of water into the pipe and the effectiveness of the pipe coating in preventing leaching, through the use of cross-sectioned pipe specimens carefully polished and mounted in Bakelite. EDXA is also useful in indirectly identifying exposed fibers by mapping of the element magnesium. High concentrations of magnesium on the surface are likely caused by exposed fibers, because there are few magnesium solids that will precipitate on A–C pipes under normal water quality conditions.

CEMENT-MORTAR-LINED AND CONCRETE PIPES

The method of manufacturing cement materials depends on the type of cement and the type of product. Cement mortar is used as a lining applied in the factory or in the field, while concrete is used to manufacture pipes and to build water tanks or other water installations (i.e., water treatment plants, pumping stations).

Cement-Mortar Linings

Linings applied in manufacturing plants. There are two main processes for producing cement–mortar-lined pipe. They are

- the centrifugal casting method in which fresh fluid mortar is poured into the pipe. The pipe is then spun at very high speed, thus compacting the mortar lining and expelling the excess water.
- the sprayed mortar process in which fresh mortar is sprayed onto the pipe wall by a fast revolving head, which is moved along inside the pipe. The

pipe is then rotated at low speed to smooth the lining surface, but no water is expelled. The mix therefore must contain no excess water, which necessitates a higher cement content than for centrifugal process mixes.

Once applied, the lining is cured for sufficient time in a controlled atmosphere so that it gains enough strength to allow other fabrication operations to proceed. Controlled curing prevents loss of water by evaporation from the fairly thin lining; it can take place at room temperature in a fog room or at elevated temperature by steam curing. The curing time is determined by the type of cement, the curing temperature, and the desired strength level. The mean curing time at ambient temperature is about one week; at elevated temperatures it is a few hours. The maximum strength is reached after about four weeks of hardening at ambient temperature.

General requirements for cement–mortar linings, such as materials specification, thickness, and surface condition, can be found in various standards, including

International:	ISO 4179 (1985)
France:	NF A48 901 (AFNOR 1985)
United States:	ANSI/AWWA C104/A21.4 (1990)
United Kingdom:	BS 4772 (BSI 1988), Appendix C
	WRc, *In Situ Cement–Mortar Lining, Operational Guidelines* (1990)
Germany:	DVGW W 342, DVGW W 344, DIN 2614, and DVGW W 343

The composition of the cement mortar is defined by two ratios: the sand/cement ratio (S/C) and the water/cement ratio (W/C). Generally, centrifugally cast cement–mortar linings have S/C ratios between 1.5 and 2, depending on pipe diameter and lining thickness. For mortar applied by the spray process, the S/C ratio does not exceed 1.2. In all cases, the W/C ratio is about 0.35 to 0.45.

Linings applied in situ. The following two methods are used to rehabilitate pipes by in situ cement–mortar lining:

- the drag trowel method used for pipes with diameter less than 200 mm (8 in.)
- the centrifugal method used for all diameters

According to DVGW standards W 343 and W 344 (DVGW W 343, 1981; DVGW W 344, 1985), the characteristics of the cement mortar are as follows (Hirner 1990):

- the W/C ratio is 0.35 (DVGW) and can rise to 0.39 (BS) according to pumping conditions during application
- the S/C ratio is about 1.7 for the drag trowel method and 1.0 for the centrifugal method
- the compressive strength must be 50 N/mm^2 (34 kPa) for the drag trowel method and 64 N/mm^2 (43.5 kPa) for the centrifugal method
- the bending/tensile strength must be about 7 N/mm^2 (4.8 kPa) for the drag trowel method and 8 N/mm^2 (5.4 kPa) for the centrifugal method
- the thickness of the lining must be about 3 mm (0.12 in.) for small diameters with a tolerance of 2–3 mm (0.08–0.12 in.) with increasing thickness for pipes with diameters up to 900 mm (36 in.) for centrifugal applications

Figure 7-30 shows the pipe-cleaning phase and the application of the cement–mortar lining by drag trowel method (Jaryd and Ljunggren 1990). Figure 7-31 shows the centrifugal method for large pipes with diameters up to 600 mm (24 in.).

Several other methods of in situ lining have been proposed. For example, some entail using glass-fiber-reinforced cement lining (Sakurai et al. 1990). Others involve rebuilding a new pipe inside the old one using a cement mortar applied by pressure injection between the pipe wall and an internal air-pressurized plastic former (Conpipe method; Lannblad 1990). The principles of this method are explained in Figure 7-32.

Experiences With Cement-Mortar Lining

In the past, pipes transporting drinking water were either unlined or coated with coal tar or bituminous enamel. However, neither coating could prevent corrosion. Incrustation took place, red water was delivered to the consumer, and a decrease in the flow capacity of pipes followed. Even special coatings of bituminous enamels could not prevent problems that arose after long periods of service.

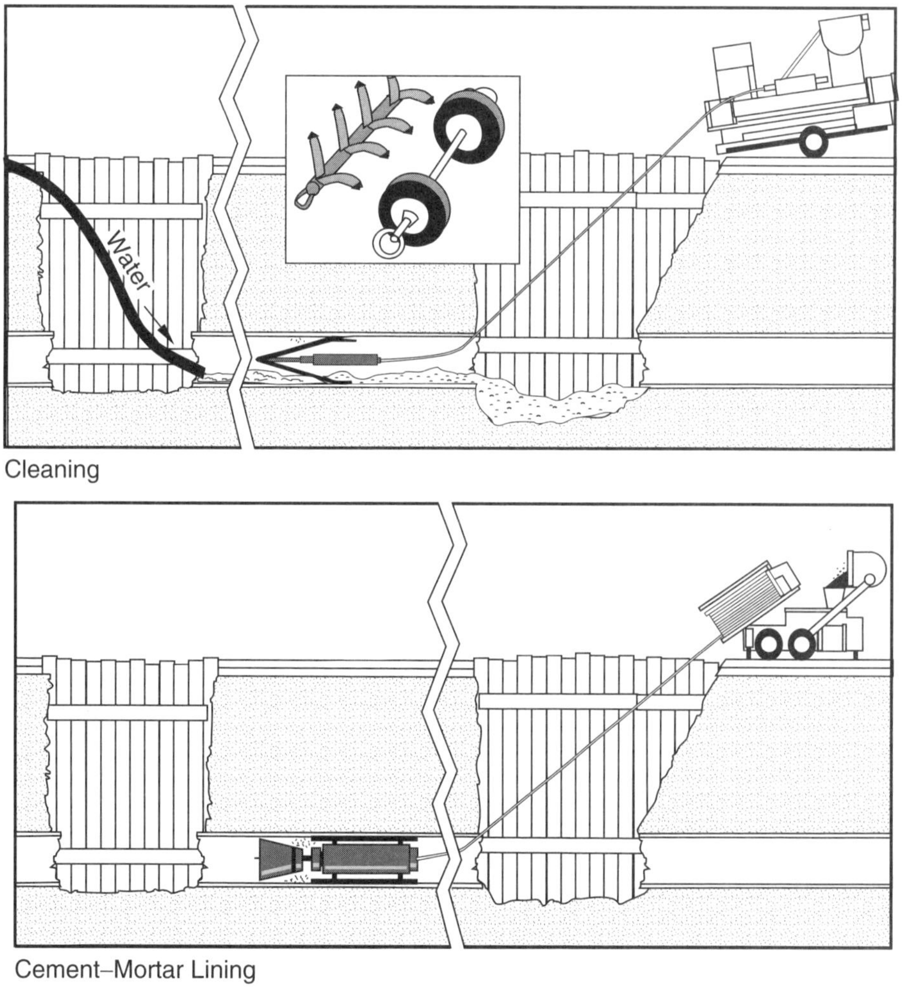

Source: Jaryd and Ljunggren (1990). Reprinted with permission of Water Supply.

Figure 7-30 Pipe cleaning and application of cement-mortar lining by drag trowel method

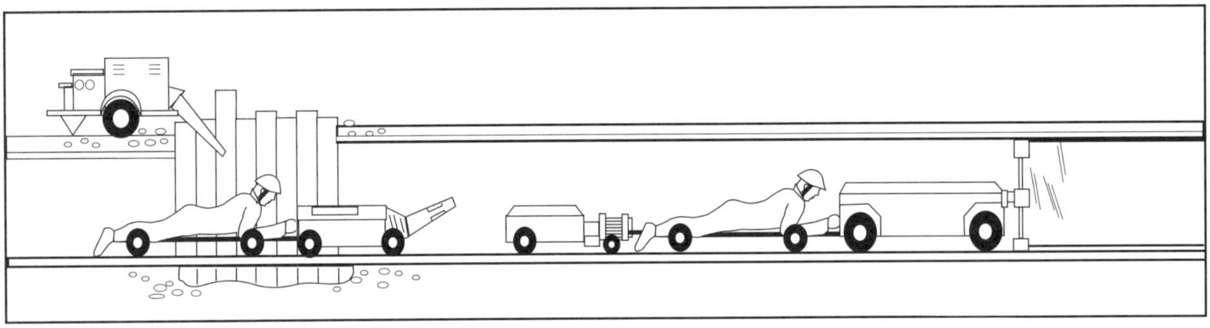

Figure 7-31 Centrifugation method for in situ cement-mortar lining (Tate method)

Compared to coal tar or bituminous enamel linings, cement–mortar lining has been successful in preventing corrosion. For example, in Dortmund, Germany, cement–mortar lining was applied especially for reducing the problems with red water. The results of the lining project in one district are shown in Figure 7-33. The cement–mortar lining in this area started in 1969. By the time the project was finished in 1973, the number of red-water reports was reduced to zero, although the total amount of water delivered had increased.

All literature on cement–mortar lining discusses the protection cement mortar provides against corrosion, thus ensuring a high flow coefficient for many years. Regarding corrosion attack on and durability of cement–mortar lining, many references cite the work of Chappell (1930) and Miller (1965). Because the purpose of cement–mortar lining is to guarantee uninterrupted operation during the lifetime of pipelines, it is important that leaching does not affect durability of the lining, that softness of the cement–mortar layer does not occur or is limited, and that sand loss does not occur.

Experience with cement–mortar lining in Dortmund, Germany (Holtschulte 1972, 1974, 1978a, 1978b, 1980a, 1980b). In 1972, a factory-lined pipeline (1,000 mm [40 in.] ID) was taken out of service. During inspection, a brittle layer was found on the surface of the cement–mortar lining. The layer (1–1.5 mm [0.04–0.06 in.] thick) could be rubbed off easily by hand. The pipeline, laid in 1966, was operated for three years with a water flow velocity of about 1 m/s (3.3 ft/s). From 1969 to 1972, the water in the pipeline, which was not operational, was only changed twice a week.

Because of the composition of Dortmund's water (see Table 7-14), systematic tests were undertaken. Sections of factory-lined pipe (80, 100, 200, 500, and 1,000 mm [3, 4, 8, 20, and 40 in.] ID) laid since 1961 and sections of in situ–lined pipes of the same diameter were examined. The following tests were made:

- visual inspection of the lining
- determination of the mechanical stability of the lining
- determination of the total thickness of the lining and the thickness of the hard portion of the lining

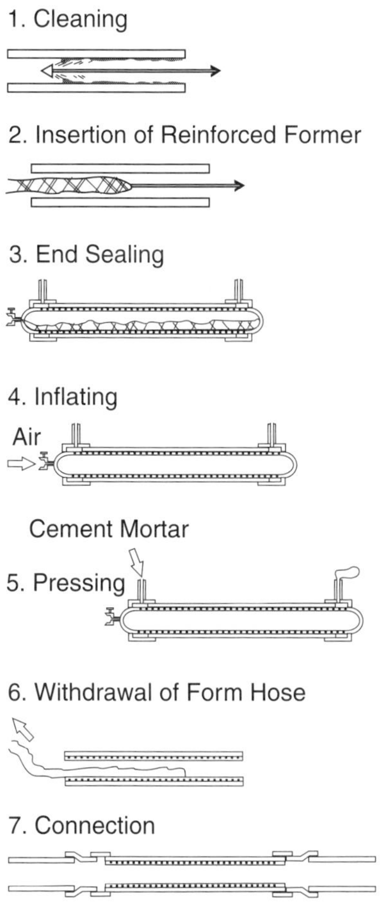

Main Pipes

1. Cleaning of existing pipe by means of a high-pressure sluice.
2. Insertion of reinforced former.
3. End sealing.
4. Inflating.
5. Injection of cement mortar. The thickness and strength of the lining can be amended to suit loading requirements.
6. Withdrawal of former.
7. Connection.

Laterals

Lateral pipes can be renovated in many different ways. For example,

- Conpipe renovation with cement mortar.
- Pipecracker, exchange of lateral pipes.
- Slip lining.

Source: Lannblad (1990). Reprinted with permission of Water Supply.

Figure 7-32 Conpipe method for cement-mortar relining

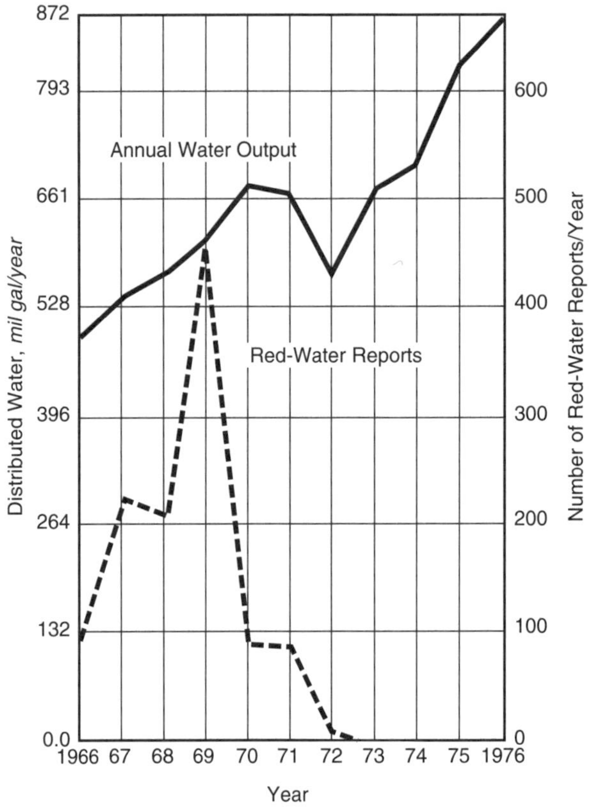

Figure 7-33 Reports of red water in a Dortmund, Germany, district from 1966 to 1976 (lining with cement mortar began in 1969)

Table 7-14 Quality of Dortmund drinking water

	1968–1972	
Extreme Values	Minimum	Maximum
pH	6.4	6.9
Sulfate, *mg / L*	33.6	67.2
Hardness, *mg / L* $CaCO_3$		
Total	115	170
Temporary	60	100
Iron, *mg / L*		0.5
Manganese, *mg / L*		0.93
Carbon dioxide, *mg / L*		
Free	6.2	27.2
$CaCO_3$ aggressive	4.4	24.4
Conductivity, μS	248	386

- analysis of the layer on the lining (in each case layers 0.5 mm [0.02 in.] thick), with regard to the CaO concentration and to the iron and manganese in the layer on the water side
- determination of the proportion (by weight) of the sand and mortar in the lining in the present condition and in the original condition
- determination of the grain-size distribution of the sand

Factory lining and in situ lining. The linings of factory-lined pipes were soft in places to a depth of about 0.5 to 1.5 mm (0.02 to 0.06 in.), depending on the time of operation and the flow rate. The softened part was brittle and crummy. The rest of the cement–mortar layer was hard and stable and could not be scraped off by a knife. Leaching increased under the same conditions with time of operation.

Pipelines with low flow were still in excellent condition after 8, 11, or 14 years of operation. Neither softening nor leaching occurred. These examples demonstrate clearly that the leaching and softening of the cement–mortar lining are reduced at low flow rates. If the water stagnates, equilibrium occurs. If the CO_2 concentration increases during intensive flow, leaching is accelerated.

The decrease in CaO content resulting from leaching of factory cement–mortar linings after different years of service is shown in Figure 7-34, where zero on the x-axis represents the surface exposed to the water and the x-scale shows the composition into the interior of the lining.

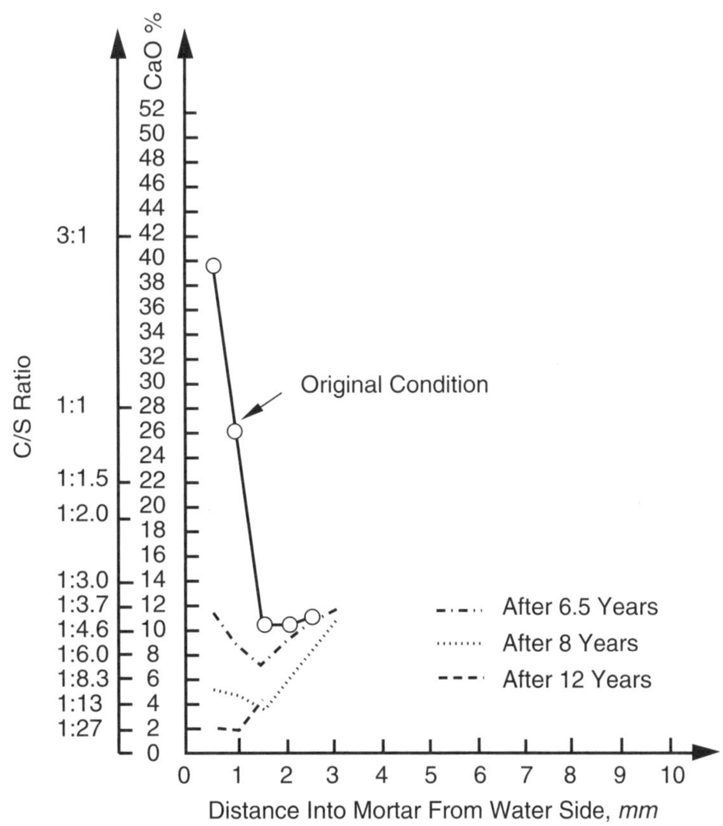

Figure 7-34 CaO content in relation to the C/S ratio of factory cement-mortar linings after different durations of service (100 mm ID)

Pipes (1,000 mm [40 in.] ID) lined in situ by different firms using the process first introduced by the Australian Centriline Company were in excellent condition (very stable and hard linings) after as few as 3 and as many as 12, 13, or 18 years of operation. The visual inspection of the cement–mortar linings showed no differences. Interestingly, a surface layer about 1 mm (0.04 in.) thick was present on the water side of the lining in all pipes. The cement–mortar lining could therefore be separated into two layers during its removal. This separation of the layers demonstrated the hardness of the thin layer nearer the water side, which prevented any possibility of sand loss, even after many years of operation. Thus, it is not surprising that the rest of the layer, with a thickness of about 7 mm (0.3 in.), was very hard. The reason for this separation of layers is unclear. Perhaps it is the influence of trowels used to smooth large pipes after the lining process; separation of the layer does not occur in small-diameter pipes.

Calcium oxide leaching after three years of service is shown in Figure 7-35.

Both factory-applied linings and in situ linings had a brown coating on the water side, which consisted mostly of silicon and hydrous oxides of aluminum, iron, and manganese.

In summary, the tests in 1972 gave the following results:

- A layer was found on the water side of all cement–mortar linings.
- Whether of cast iron or steel, all factory-lined pipes exhibited softening, the extent of which depended on the time of operation. Leaching of CaO and a deposition of iron hydroxide occurred. Chappell's (1930) and Miller's (1965) observations were confirmed.

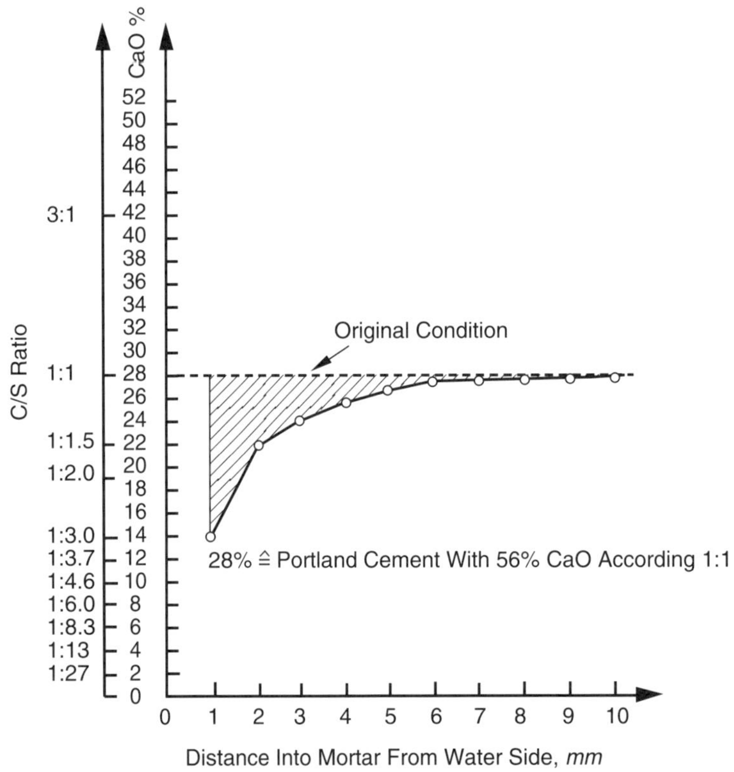

Figure 7-35 Decrease of CaO content because of leaching, after three years of service (Centriline process, 1,000 mm ID [40 in.], mix 1:1)

- The pipes lined in situ with the Centriline process were all in good condition, even after 18 years in operation.

In situ versus factory lining. On the basis of the test results obtained in 1972, the question arose whether the in situ lining process is inherently better than the factory lining process. In 1972, the differences between the methods were explained in the following way.

The leaching resistance of the lining itself is due to the mortar density. According to Biczôk (1968), the simplest method of protecting concrete against corrosion is the production of a high-quality concrete under the following conditions:

- selection of a proper cement
- determination of the correct content of cement on the basis of granulate structure
- selection of suitable aggregates
- selection of a corresponding graduation of grain size
- quantity of water
- production of a dense concrete
- curing of fresh concrete

Factory linings are a mixture of cement and sand in a ratio of 1:2, 1:2.5, or 1:3, whereas the in situ lining of the Centriline process is a 1:1 mixture.

Based on the experiences in Dortmund, it was concluded that a cement mortar with the 1:1 mix is more resistant to concrete corrosion. The cement–mortar lining of a factory-lined pipe, part of which was lined with a 1:1 mix and part of which was lined with a 1:2 mix, was tested. During the time in operation, more CaO was removed from the portion lined with the 1:1 mix, but a high concentration of CaO was still available (see Figure 7-36).

The production of a dense concrete depends on the quantity of water and on the curing of the fresh concrete.

The advantages of in situ lining are:

- a low water-to-cement value
- a high density caused by a homogeneous structure over the whole layer
- a hardening of the mortar under unchangeable conditions

Buried pipes do not experience a change in temperature. In most cases, especially for small-diameter pipe, the buried pipe is soon filled with water after it is lined, and hardening occurs under water.

The consistency of the mortar and the time and speed of spinning should be so adjusted as to reduce as much as possible the segregation of the sand from the cement. However, the polished sections of factory linings and analysis of layers show that in practice segregation does occur. During this process, a layer of cement is formed on the water side. The effects are shown in Figure 7-37. The lining is divided in layers 0.5 to 1 mm (0.02 to 0.04 in.) thick. As a result, the original mixing proportion changes. Based on the concentration of CaO, the mixing proportion ranges from 3:1 to 1:3.7 C/S. The in situ lining shows no separation of grain sizes. The cross section shows a constant mixture of 1:1 or 1:2 (see Figure 7-38).

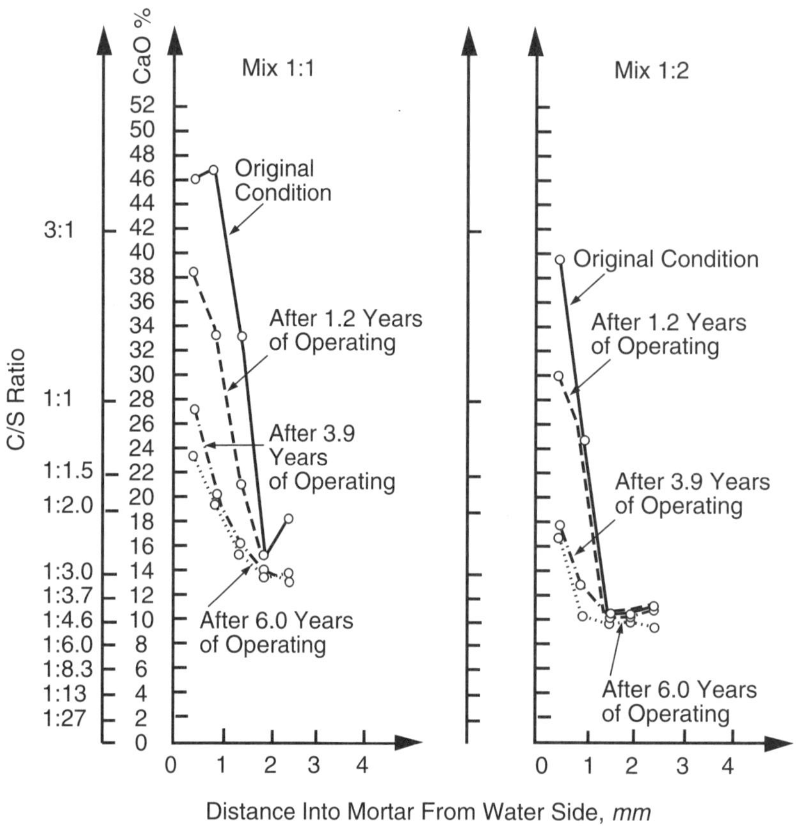

Figure 7-36 Analysis of CaO in the layer on a factory-lined pipe as a function of mix ratio (100 mm ID [4 in.])

Behavior of the inner surface in practice. Other questions have also been raised. For example, does the softening of the cement–mortar lining impair its service life, and is sand loss, for instance, possible?

The composition of the Dortmund water does not reduce the service life of the lining. However, the original composition of the lining changes because the CaO is partly leached out and replaced by iron hydroxide.

As a result of the deposition of iron and manganese compounds, a gelatinous sediment is formed that makes the cement–mortar layer very tight and leaching more difficult.

Tests have shown that this sediment layer exists after 18 years of operation. Although water hammers occur, the fine sand that is a principal component of the slime layer is still present. According to the experiences gained up to 1972 and also between 1972 and 1984, sand loss from the softened cement–mortar lining does not occur.

The softened layer is preserved under normal conditions of operation, but sand loss occurs if the cement–mortar lining dries up. Therefore, if cement–mortar pipes that transport water containing "aggressive CO_2" are taken out of operation, drying of the lining must be avoided by using suitable precautions to prevent temporary desication of the mortar lining and consequent sand loss.

The original 1972 technology of lining water pipelines with cement mortar could protect pipes from corrosion. This technology was improved by the field test described later in this chapter. The lining is unaffected by nonaggressive water;

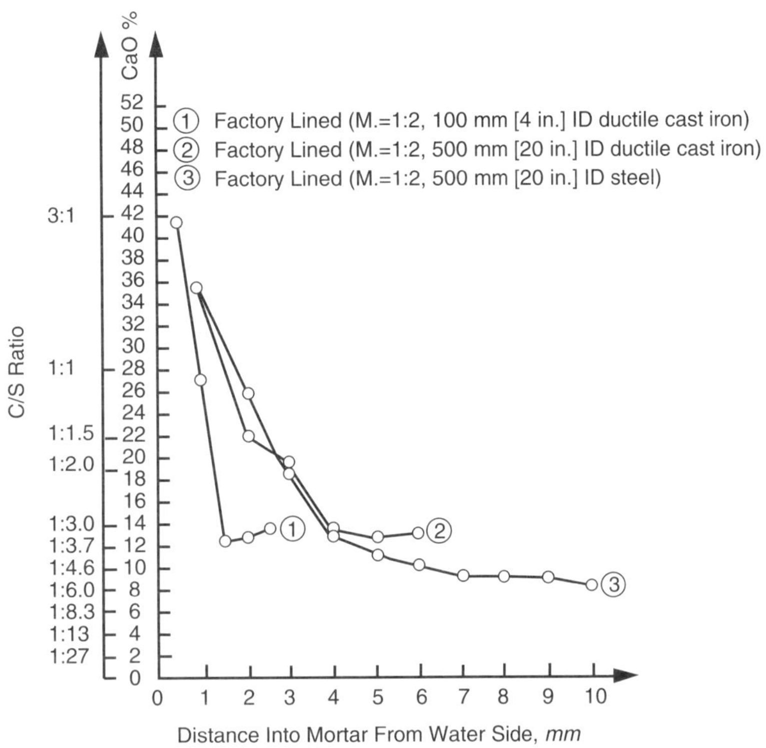

Figure 7-37 Analysis of CaO in the layers on cement-mortar linings in their basic condition

aggressive water, like that in Dortmund, leaches the cement out of the lining. But because of the development of a protective layer, sand loss has not occurred and is not expected to on the basis of the tests performed.

Rehabilitation of a water pipeline network by cement–mortar lining. A water pipeline's longevity is limited. During operation, a pipeline must be inspected and maintained. After a certain period of operation, it may be necessary to recondition or replace the pipeline.

The assessment of pipeline rehabilitation requires planned and systematic work on water pipelines. The major objective of rehabilitation is to obtain a fully restored pipeline that can compete against a replaced pipeline of the same dimension, hydraulic efficiency, and technical valuation.

Cement–mortar lining of water pipelines. From the technical and economical points of view, only cement–mortar lining, which is used on cast-iron and steel pipelines, is considered for rehabilitation of water mains. A permanent internal protection is guaranteed with a cement–mortar lining, which also eliminates limited leaks of pipes and pipe connections as a result of the high resistance of cement mortar to pressure. Cement–mortar lining also enhances the hydraulic characteristics of the mains and prevents internal corrosion damages.

German pipe rehabilitation procedure must be in accordance with DVGW Standard W 343, which covers such issues as crew size. Table 7-15 shows the weekly efficiency of a six-to-eight-person crew.

As mentioned earlier, cement–mortar lining closes small leaks (up to 1 in. [25 mm]). Experiments have shown that drilled steel pipe, with a 60-mm-diameter (2.4-in.) hole and a 5-mm (0.2-in.) lining, can withstand a pressure as great as 20 bar (290 lb/in.^{2}).

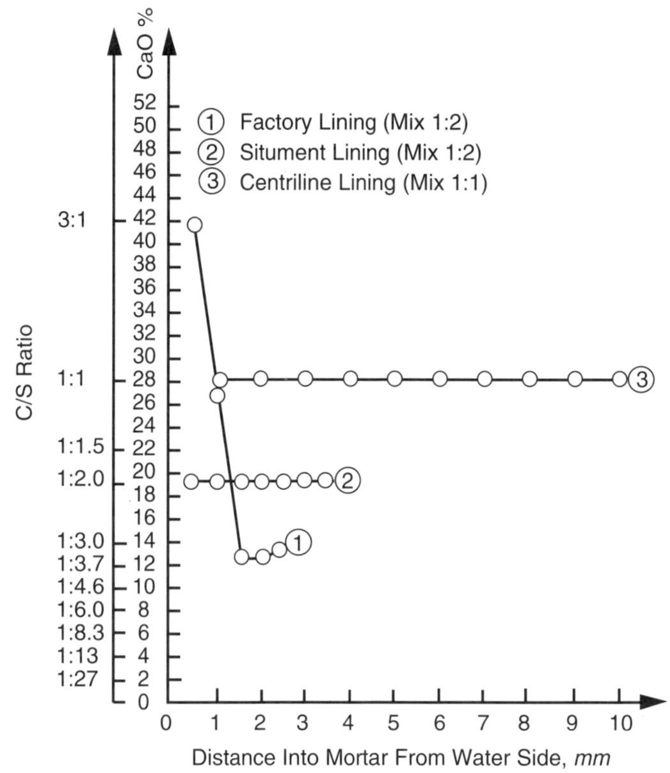

Figure 7-38 Analysis of CaO in layers—principal differences between two cement-mortar lining processes and original condition

Table 7-15 Weekly* efficiency of one crew lining pipelines with cement mortar

Workers per Crew			
Excavation and Pipe Fitting	Cleaning and Lining	Internal Diameter, *mm (in.)*	Efficiency
6–8†		<400 (<16)	400–500 m (1,300–1,600 ft) = 4–5 sections
6–8	6–8	500 (20)	400–500 m (1,300–1,600 ft) = 2 sections
6–8	6–8	>500 (>20)	600–700 m (2,000–2,300 ft) = 2 sections

*Five days per week, nine hours per day.

†For smaller pipes, a six-to-eight-person crew is sufficient to handle both types of work.

All armatures and built-in equipment must be overhauled before pipeline renovation begins.

Costs for rehabilitation in relation to pipe replacement. Table 7-16 shows the proportional costs in Germany for cement–mortar lining a pipeline compared to those for installing a new pipeline. These costs include the necessary bypass as well as the exchangeable armatures, etc.

Table 7-16 Example comparison of costs for rehabilitating and replacing pipes in relation to pipe size

	Costs for Rehabilitation				
Internal Diameter, *mm (in.)**	Cleaning and Cement–Mortar Lining, %	Excavation, Pipe Fitting, Restoration of Road Surface, %	Material, %	Quota of City Work Laborer and Remaining Costs, %	Costs for Rehabilitation in Relation to Pipe Replacement, %
80–150 (30–6)	33	49	7	11	39.5
200–300 (8–12)	33	48	7.5	11.5	41
500 (20)	22	55	13	10	33.2
600–800 (24–32)	37	47	8	8	19.4
1,000–1,200 (40–48)	30	57	5	5	13

*Equivalent inch measurements are calculated in rounded inches.

Economy of rehabilitation by cement–mortar lining. The economics of a renovation method are expressed by comparing amortization of invested capital for the rehabilitation costs with costs for replacement.

Mathematical expressions covering the economics of pipe rehabilitation and replacement may be developed and summarized as follows:

$$(E–S) \cdot p^{n_s} = E \tag{7-2}$$

$$p^{n_s} = \frac{E}{E-S} = \left(1-\frac{S}{E}\right)^{-1} \tag{7-3}$$

$$n_S \times \log p = \left(1-\frac{S}{E}\right)^{-1} \tag{7-4}$$

$$n^S = \frac{\log\left(1-\frac{S}{E}\right)^{-1}}{\log p} \tag{7-5}$$

Where:

n_S = time necessary to operate rehabilitated pipe in order to reach break-even point with the replacement
S = rehabilitation costs
E = replacement costs
p = $1 + i$
i = rate of interest

Expressed in another way, the replacement of a pipe requires investment costs of E. The reconditioning of the same pipe requires costs of S. Choosing the appropriate reconditioning process ought to result in a difference between E and S, the amount of money saved.

The minimum time of operation for a reconditioned pipeline has to be long enough so that the interest earned for the amount of money saved is equal to E. Therefore, the minimum time of using the invested capital is equal to the minimum time of operation for the reconditioned pipeline. Table 7-17 shows short periods of 2.3 to 6.6 years for n_s, depending on the diameter.

If a reconditioned pipeline is still in service after the calculated time, reconditioning is an economical success. Furthermore, after reconditioning, the costs for maintenance and repair decrease and the requirements for personnel are reduced.

Hydraulic characteristics after cement–mortar lining. Incrustations in pipelines cause decreases in flow capacity. Figure 7-39 shows a comparison between a new pipeline with a roughness coefficient (c_{WH}) of c = 133 and incrusted pipelines of different c_{WH} values, illustrating that incrusted pipelines have greatly reduced capacity for constant head loss.

A relining of the pipes leads to full efficiency, despite the smaller inner diameter (Figure 7-39). For example, a c_{WH} value of c = 140.7 increases the hydraulic efficiency of a lined pipe, even in comparison with a new pipeline, because the lining provides a continuous internal surface, without interruptions by the appurtenances.

Compared to the capacity of an old (100-mm-ID [4 in.]) incrusted pipe, the lining process may increase the capacity up to about 180 percent of the original. This assumes a final diameter of 94 mm (3.7 in.), with a lining thickness of 3.0 mm (0.12 in.).

The thickness of the lining has, because of improved hydraulic quality of the inner surface, little influence on efficiency even in small pipes. Similarly, with a lining thickness of 4.5 mm (0.18 in.) and a diameter of 91 mm (3.6 in.), the increase in flow capacity was still 70, 120, or 150 percent, depending on the prior extent of incrustation (Figure 7-40).

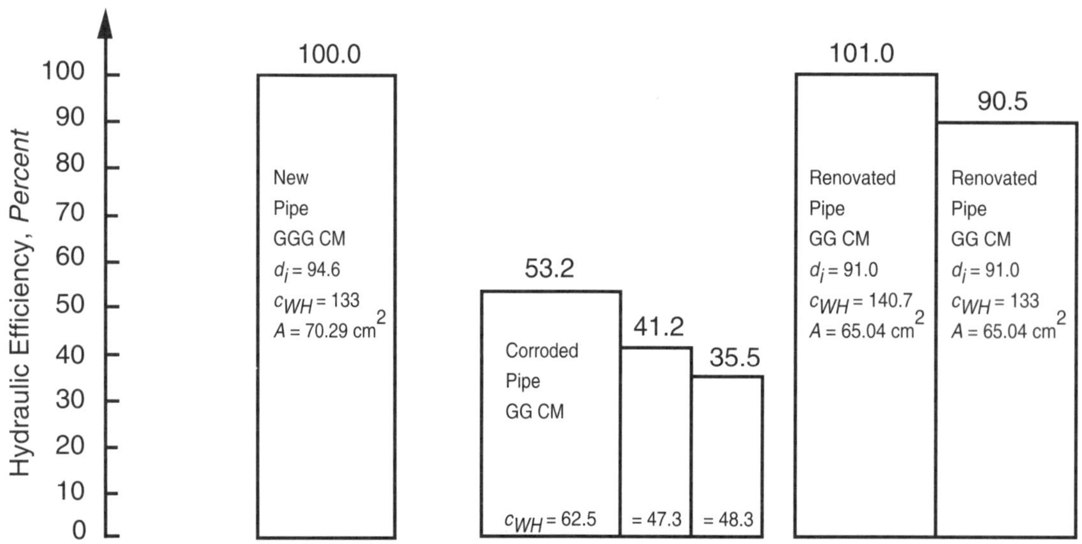

NOTE: A = cross-sectional area; CM = cement–mortar lined; d_i = internal diameter; GG = ductile iron; GGG = gray cast iron; c_{WH} = roughness coefficient.

Figure 7-39 Flow rate independence of internal diameter and c_{WH} value with Δp constant (100 mm ID)

Table 7-17 Calculation of break-even time for rehabilitated pipes

Relation of Cost for Renovation S to Renewal E	Internal Diameter, *mm (in.)*	n_S = minimum line of operation in years to break even for a reconditioned pipe*
0.40	<500 (20)	6.6
0.332	500 (20)	5.2
0.162	>500 (20)	2.3

$$* \, ns = \frac{\log \left(1 - \frac{S}{E}\right)^{-1}}{\log p}$$

rate of interest i = 8 percent

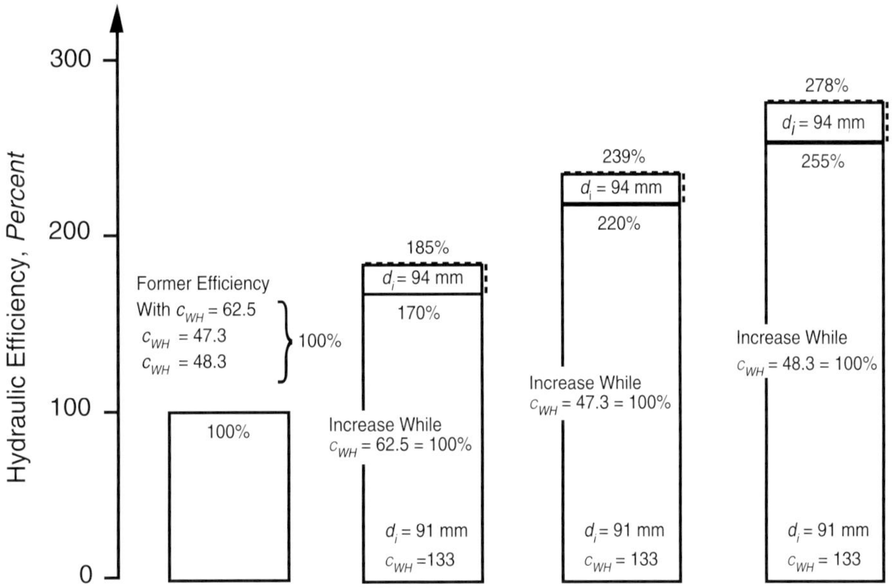

NOTE: c_{WH} = roughness coefficient; d_i = internal diameter.

Figure 7-40 Increase in hydraulic efficiency of a pipeline after applying cement-mortar lining (Δp constant, 100 mm ID)

Mechanical Characteristics

Cement–mortar linings. The mechanical characteristics of cement–mortar linings are covered in the standards for manufacturing already referenced.

As reported by Grandpierre (1986), the mix proportions affect, to a large extent, the characteristics of fresh mortars as well as hardened mortars. Of great importance are the W/C, the S/C, and the percentage of fine elements (cement + finer sand grains).

The rheological characteristics of fresh mortars can be controlled as follows:

- increasing W/C renders the mortar more fluid, but too much water will cause segregation when the mix is pumped and cast in the centrifugal process
- on the contrary, increasing S/C results in a less mobile mix

- having a certain amount of fine particles in the mortar for the mix to be pumpable and to obtain a smooth and regular lining

The properties of cement mortars in the hardened state are also affected by these parameters, which may lead to conflicts, as follows:

- increasing W/C produces at least two undesirable effects: a loss of strength and higher shrinkage
- for S/C between approximately 0 and 2, S/C is rather positive as it decreases shrinkage and gives higher flexural and compressive strengths

Figures 7-41 and 7-42 give examples of the correlation between the W/C and S/C ratios and the resulting compressive and bending strengths of cement mortars. The W/C ratio in the vicinity of the 0.35 generally used corresponds to strength in excess of 6 MPa (9 N/mm^2) in bending and 50 MPa (73 N/mm^2) in compression.

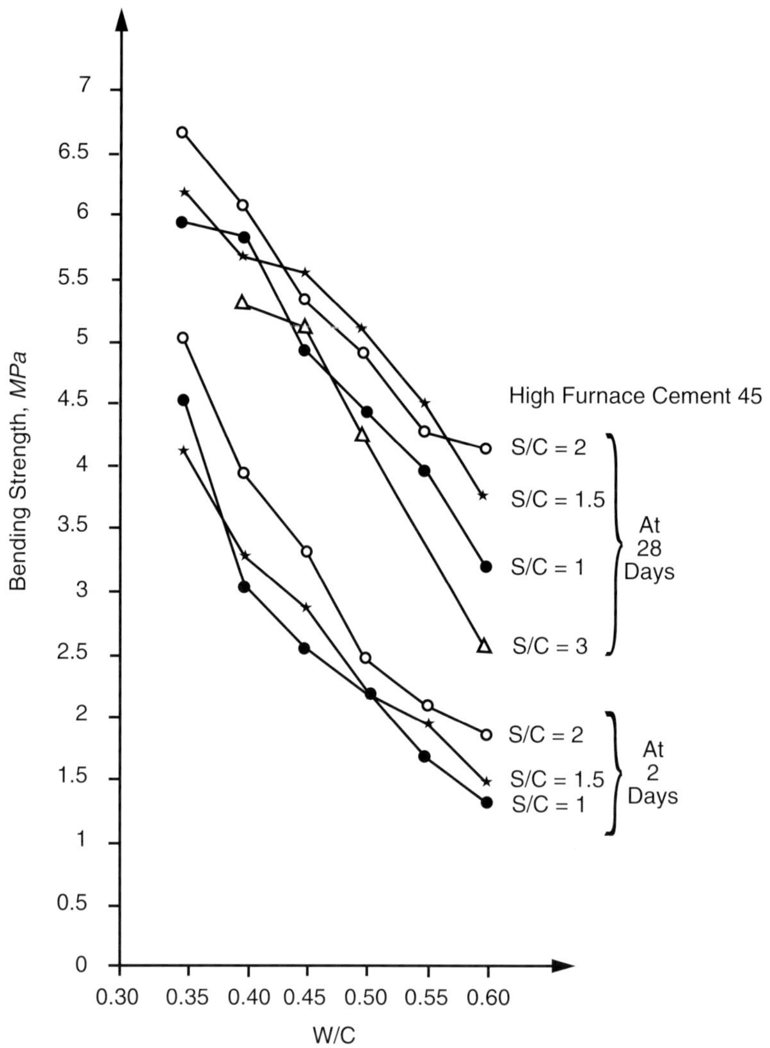

Source: Grandpierre (1986). Reprinted with permission of Pont-à-Mousson S.A.

Figure 7-41 Influence of W/C value and S/C ratio on bending strength of cement mortars

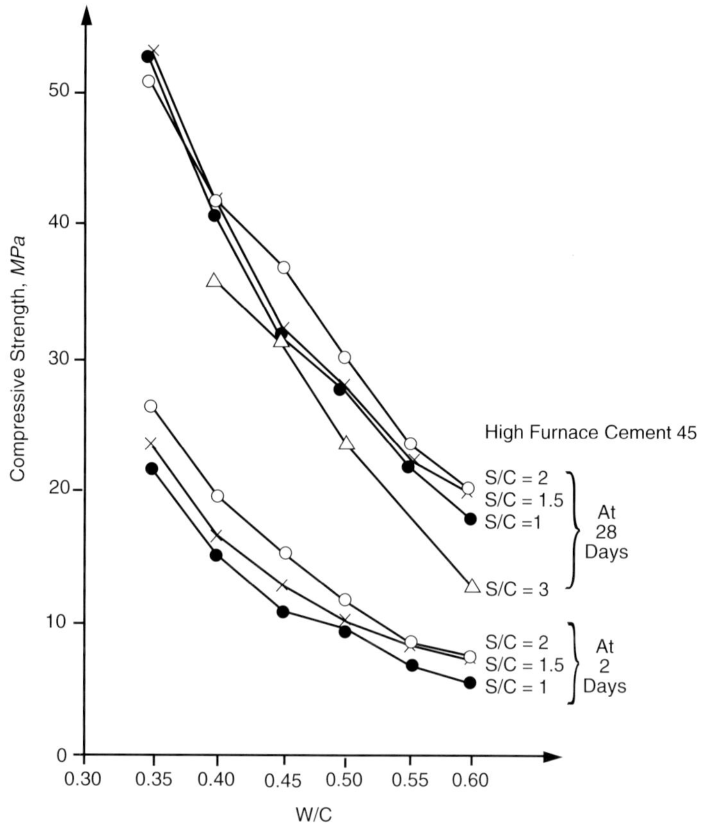

Source: Grandpierre (1986). Reprinted with permission of Pont-à-Mousson S.A.

Figure 7-42 Influence of W/C value and S/C ratio on compressive strength of cement mortars

Concrete pipes. The two kinds of concrete pipes most often used in potable water supply are

- reinforced concrete pipe with a steel sheet for water tightness
- prestressed concrete pipe

Pipes with steel sheet can be jointed with a rubber ring or by welding the steel sheets.

The production of concrete pipe is based on vibro-press-spun concrete. The process can produce prestressed pipes using longitudinal and radial prestressing systems and reinforced concrete with the inner steel sheet. These products are made by specific- or variable-size molds. Ordinary concrete pipes are produced by standard spinning machines, each of which has a fixed range of inner diameters. In contrast, a universal spinning machine can produce all ranges from 600 mm (24 in.) up to 2,500 mm (100 in.) by hydraulically adjusting the moving mobile axle. Reinforced concrete pipes can be produced by a vibrating process for certain special uses.

Molds are used for producing reinforced concrete pipes. Before concreting, each mold is properly prepared with a release oil sprayed on the inner face. Prestressing wires are then tensioned to the required load by means of a special jack. The molds are then positioned on the spinning machine or in the vibrating chambers for concreting.

Case Study of Remedies for Reinforced Concrete Pipes, Al Khobar

To improve water supply conditions for several towns near Dammam, Saudi Arabia, new installations were built between 1979 and 1981. The installations include

- A desalination plant located in Al Khobar, which uses the process of multi-stage flash distillation of seawater at a nominal capacity of 250,000 m^3/d (approximately 66 mgd).
- A treatment plant in which distillated water is mineralized by adding carbon dioxide and lime. The water is then disinfected with chlorine. The chlorine is manufactured by electrolysis of the seawater. The disinfected water is thus enriched with dissolved salts found in seawater, particularly sodium and chloride.
- A water transport system that connects the treatment plant to surrounding cities. The system comprises 240 km (150 mi) of water mains, with diameters of 1,200 mm (48 in.) at the output of the plant and 500 mm (20 in.) at the end of the system. The maximum operating pressure does not exceed 8 bar (116 lb/in.^{2}). The mains consist of reinforced concrete pipes, which have a steel cylinder to ensure watertightness and strength. Pipes are jointed using rubber seals covered on the internal face with cement mortar. The transport system includes a main pumping station and, for each city served, a delivery station to regulate the water pressure. In each of these stations, the mains are made of steel with an epoxy lining.

When first commissioned, the new network was connected to only some of the cities, and the flow introduced in the system was about 20,000 m^3/d (approximately 5 mgd). Under such conditions, it appears the water entering the secondary stations had a pH of up to 12.

Also, some important cracks appeared along the joints formed by a slight displacement of the pipes after laying. Some cracks were about 5 mm (0.2 in.) wide and the steel plate had started to corrode (Figure 7-43). During this period, the mineral composition of the water introduced into the pipes was as indicated in Table 7-18.

In Figure 7-44, the condition of the water entering the pipe is represented by point M_0 and the condition of the water at the output of the pipe by point M_1.

In order to reduce the pH drop and protect the pipe in the area around the cracks, the mineral reconstitution process was modified by increasing the calcium hardness of the water to a value higher than 60 mg/L. The lime and CO_2 balance was adjusted to obtain a slightly calcifying water. Table 7-19 gives the composition of the water introduced in the network after this modification.

Immediately after introducing this treatment, the pH of the water at the output of the network decreased to below 10 and within a short time the pH value was lower than 9.5 everywhere. After one year of operation, the cracks that were initially corroded were protected by a significant deposit of interstitial $CaCO_3$.

Since 1983, the transported water has been slightly oversaturated with $CaCO_3$. The new treatment proved to be perfectly effective in preventing internal corrosion on the inner walls of concrete reinforced pipes (Figure 7-45). After about one year, as soon as the cracks in the concrete reinforced pipes had filled, the levels of carbon dioxide and lime could be gradually reduced. The calcium hardness has been decreased to the initial value, but the pH has been kept high enough to ensure that the water remains oversaturated with respect to calcium carbonate.

Table 7-18 Characteristics of the water introduced in the network of Al Khobar when first commissioned

Characteristic	Value
Temperature, °*C*	35
pH	7.75
Calcium, *mg* / *L*	12.4
Magnesium, *mg* / *L*	1.2
Sodium, *mg* / *L*	8.1
Potassium, *mg* / *L*	2.0
Bicarbonate, *mg* / *L*	41.1
Sulfate, *mg* / *L*	3.8
Chloride, *mg* / *L*	12.8
Nitrate, *mg* / *L*	<1
Saturation index after Legrand and Poirier	–0.8

Figure 7-43 View of the cracks on the joints of concrete pipe shortly after commissioning

Table 7-19 Characteristics of the water after modification of treatment

Characteristic	Value
Temperature, °*C*	35
pH	8.39
Calcium, *mg / L*	25.4
Magnesium, *mg / L*	3.4
Sodium, *mg / L*	8.1
Potassium, *mg / L*	2.0
Bicarbonate, *mg / L*	78.7
Sulfate, *mg / L*	12.5
Chloride, *mg / L*	12.8
Nitrate, *mg / L*	<1
Saturation index after Legrand and Poirier	+0.4

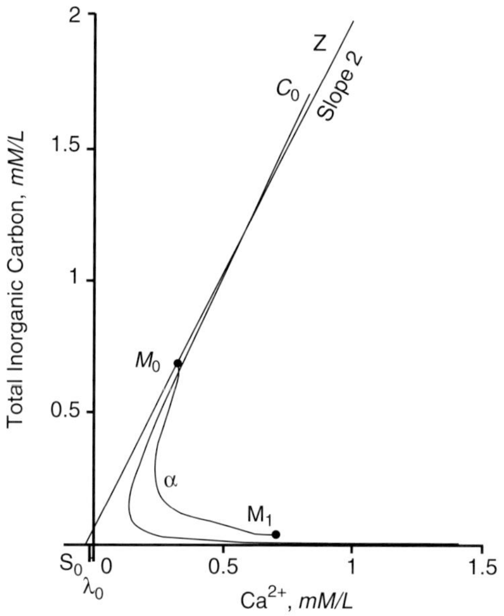

Figure 7-44 Legrand and Poirier style representation of the water entering and at the output of the network. Also shown is the calcium carbonate equilibrium curve C_0 and a possible evolution α of the water during its transport

Case Studies of Rehabilitation of Pipe by Cement-Mortar Lining, Germany

The case of the Hamburg, Germany, network, in which the pipeline was rehabilitated by addition of a cement–mortar lining, has been described by Hoppe (1990). The main conclusions are reported here.

The water supply network in Hamburg, which is 5,400 km (3,400 mi) long, is one of the oldest pipe networks in Germany. About 90 percent of the supply pipelines consist of cast iron or steel. Initially lined with bituminous material, the cast-iron

Figure 7-45 View of the internal wall of the concrete pipes after one year of operation with oversaturated water and a calcium hardness of 60 mg/L $CaCO_3$

pipes were strongly incrusted by corrosion products. Disruptions to supply, lack of pressure, and coloration of the water were frequently observed.

In 1958, 1,210 m (4,000 ft) of 350 mm (14 in.) ID and 206 m (680 ft) of 915 mm (35 in.) ID old, gray cast-iron pipe were cleaned and relined with cement mortar applied by centrifugation. Supply pipes (80 mm [3 in.] ID to 250 mm [10 in.] ID) have been similarly coated since 1971, at first using a troweling process and since 1979 using a centrifugal technique similar to the technique used in 1958. To date, about 156 km (97 mi) of pipe have been relined.

A statistical study has been performed on the number of failures registered on the relined pipes. In Figure 7-46, defects observed on relined pipes are shown according to the year of lining. The number of defects was normalized by length of the lining applied each year. This figure shows that the change of lining technique induced an increase in defects in the early years, when the centrifugal technique had still not been optimized.

To estimate the efficiency of protection by the cement–mortar lining and the extent of erosion, a sample was taken from the first lined pipeline after 27 years of operation. The sample was examined physically and chemically. The protective layer of cement lining remained intact, without any chipping. After removal of the cement lining, no corrosion of the pipe surface was found, although in some areas (lower part of the pipe) it had become visibly rust colored due to ferric hydroxide. Due to the flow, the estimated erosion of the cement lining was 1 mm (0.04 in.)—the cement binding agent was washed out and the aggregate grains were exposed. The surface became softer. The surface hardness according to the Mohs scale was

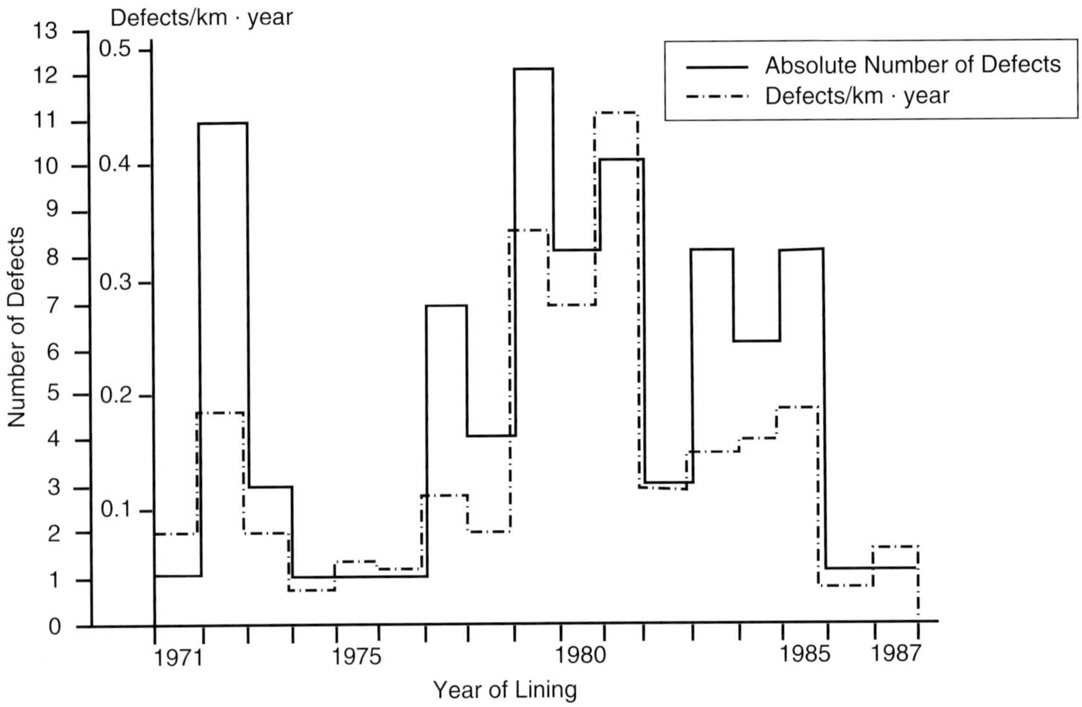

Source: Hoppe (1990). Reprinted with permission of Water Supply.

Figure 7-46 Defects in pipe after subsequent lining in the Hamburg network

<1 on the surface and 1–2 after removal of the loose layer. The surface hardness was 4 at a depth of 1–2 mm (0.04–0.08 in.).

The observed adhesive strength of 3.10 N/mm^2 (~2 kPa) is high and indicative of a well-applied lining. The adhesive strength between the cement lining and the metal is higher than the inherent adhesive strength of the cement. Due to chemical reactions—cement / metal—the higher adhesive strength was to be expected.

Chemical analyses of the coating were made at each 0.5-mm (0.02-in.) depth of the lining and showed

- the content of calcium oxide (CaO), ferric oxide (Fe_2O_3), and manganese dioxide (MnO_2) increased from the water side toward the metal side
- an accumulation on the water side of magnesium oxide (MgO), silica (SiO_2), and aluminum oxide (Al_2O_3), which decreased toward the metal side
- significant decalcification had taken place in the cement lining nearest the water, causing increased sand erosion and a decline in surface hardness

In summary, the cement lining in the examined pipeline was shown to prevent the formation of deposits and corrosion of the metal during 27 years of operation.

The water introduced into the distribution network of Dortmund, Germany, was aggressive to $CaCO_3$ and has an alkalinity of about 50 mg/L as $CaCO_3$, a calcium hardness of 115 mg/L of $CaCO_3$, and a pH value of about 6.8.

The network's gray unlined cast-iron pipes were heavily corroded, and the quality of the water deteriorated as it passed through the network. Since 1969, a part of the network has been relined each year. A rapid improvement in the quality of the distributed water was then observed as was a decrease in red-water complaints.

The behavior of the cement lining was studied by examining the depth of attack, the calcium leaching, and sand erosion. After 18 years of operation, the depth of attack was found to be between 1 and 4 mm (0.04 and 0.2 in.). The influence of the cement–mortar composition of the corroded lining on the amount of sand loss is shown in Figure 7-47. It appears that the linings made with high alumina cement and with acrylic resin were resistant to corrosion.

Remedies for Potable Water Tanks or Filters

For reinforced or prestressed concrete tanks, as for filters in water treatment plants, there are some specific cases of corrosion that must be examined. The degradation of cement materials in tanks involves the roof and the wall sections that are alternatively in contact with air and water. For filters, the degradation of cement materials involves those parts in contact with water.

Degradation by condensation water. When the air temperature outside a tank is lower than that of the water in the tank, which is generally the case in winter in cold climates, the moisture in the air of the tank condenses on the ceiling, forming droplets. These droplets grow with time and eventually fall into the water in the tank.

The condensation water is almost unmineralized, containing only dissolved gases such as carbon dioxide and oxygen. This water is extremely aggressive to the concrete surface of the tank and dissolves the lime found in the pores of the cement

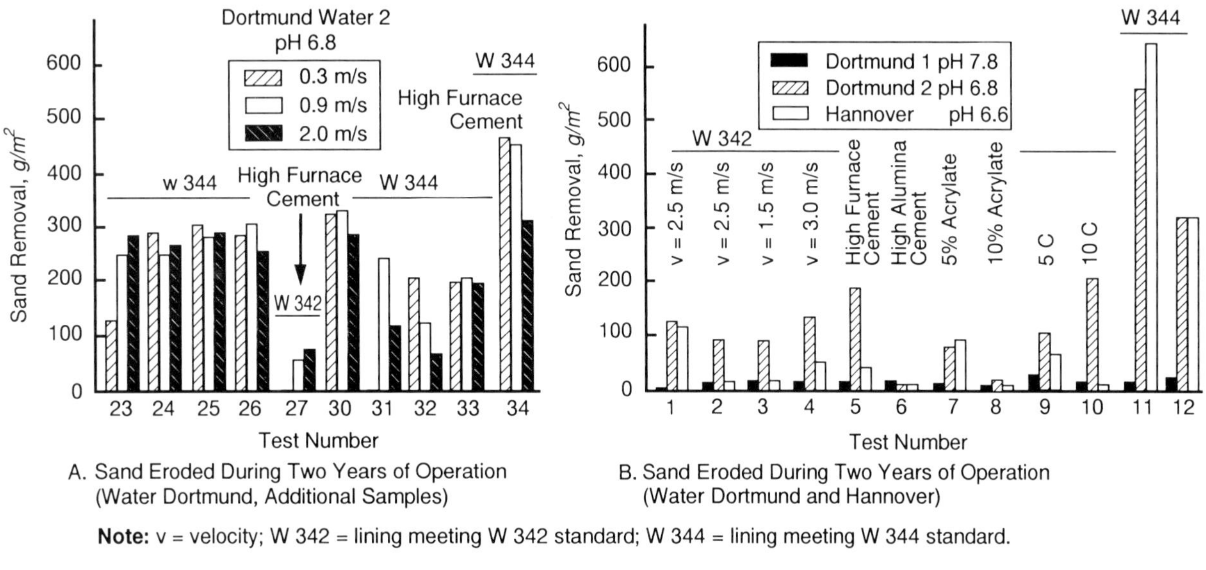

Source: Holtschulte et al. (1984). Reprinted with permission of Das Gas- und Wasserfach Wasser. Abwasser gwf.

Figure 7-47 Sand loss in relation to the quality of cement used material. Such attack can be very fast if the difference in temperature between the external air and the water is large. This attack will lead to a fast deterioration of the tank ceiling because no calcium carbonate can deposit in the pores of the material. And as the lime dissolves in the condensation water, the porosity of the surface layer increases, leading to a fast penetration of the aggressive water into the material. Then, the pH value of the medium in the material near the frame of the tank decreases to values lower than 9. Under such conditions, the tank's steel frame starts to corrode. Under pressure of the ferrous corrosion products, cracks appear along the armatures and the material is rapidly degraded. The exposed armatures will also corrode quickly in contact with moist air containing chlorine gas or ozone.

Such corrosion may be slowed or avoided by use of a different quality of cement having a hydrated phase with a low lime content. High alumina or pozzolanic cements can be used for building the tank roof (Conio 1988).

If a tank is built with portland cement materials, it is possible to protect the parts of the tank wall that are not regularly in contact with water by applying a coating made of cement–epoxy or cement with polyvinyl chloride. In each case, the surface must be cleaned and the deteriorated layer must be removed. The Les Lilas water tank feeding Paris, France, was lined with cement–epoxy 10 years ago. This particular coating appears to be very effective.

Degradation due to water level variations. Degradation of tank surfaces due to variations in water level involves the upper part of the tank walls. This wall section is alternately in contact with water or air depending on the variations in water level.

Rapid leaching of lime from the pores of the cement material can be observed at a rate that is dependent on the porosity of the cement material and the type of cement used. The highest rate is observed with portland cement.

When the water level in the tank is high, the pores of the concrete fill with water. Lime in the concrete dissolves into the water inside the pores. As the water level drops, the water from the pores, which has a high lime content, mixes with the stored water. As a result, calcium carbonate precipitates near the wall surface. These calcium carbonate deposits, which look like agglomerate sand, can be observed on the vertical wall surfaces (Figure 7-48).

Such leaching leads to a rapid degradation of the cement material with a decrease in the pH value inside the material near the armatures. The upper tank wall can deteriorate rapidly, while the lower part of the wall, forever in contact with stored water, has a normal evolution with a slow protection of the surface by interstitial calcium carbonate precipitation.

In the L'Haye les Roses reservoir, which serves Paris, France, the upper part of the reservoir walls was deteriorated. The deteriorated layer was 15 mm (0.6 in.) thick from the surface of the material to near the armatures; the reservoir had been in operation for eight years. In contrast, the lower part of the walls, the part in permanent contact with water, was carbonated to only a 3-mm (0.12-in.) thickness. The stored water was always slightly oversaturated, but this characteristic appears to be not sufficient enough to give good protection to the upper part of the walls. However, it does explain the good condition of the lower part of the walls.

Such degradation can be avoided by applying an appropriate coating to the walls. A cement–epoxy or polyvinyl chloride–cement coating can be used with good success.

Figure 7-48 View of calcium carbonate deposits formed on a sample of the vertical walls of the Les Lilas water tank in Paris, France

Coating a tank with a fiber-reinforced bitumen layer covered with an aluminum sheet can lead to rapid degradation of the coating itself and degradation of the cement-based wall. In practice, it is quite impossible to have a continuous coating without any defect. Consequently, water can go through the coating to the cement material. Once in contact with the cement, water dissolves the lime, and the water's pH increases to a value high enough to attack the aluminum sheet when it comes back through the coating. In this way, the coating is rapidly deteriorated. This leads to circulation of water between the tank and the pores of the cement material and a direct contact between the stored water and the bituminous coatings, resulting in bad tastes in the stored water. These events took place in the Les Justices tank feeding a small network near Paris, France. Figure 7-49 shows the appearance of such a coating after few years of operation. This type of material is not suitable for protecting the walls of water tanks.

Degradation of the concrete walls in settling tanks and filters. Deterioration of the concrete walls in settling tanks and filters can occur in water treatment plants that use prechlorination and clarification by flocculation, settling, and filtration. This has been observed by Giacasso (1988) and in the Orly Water Treatment Plant, which serves Paris, France. In these two cases, the raw water was slightly hard and calcifying but required high chlorine and flocculant dosages. The flocculant used was aluminum sulfate.

Figure 7-49 View of the appearance of the coating made of fiber-reinforced bitumen covered with an aluminum sheet after a few years of operation in the Les Justices water tank in a suburb of Paris, France

After chemical addition the water becomes aggressive to calcium carbonate. This water in contact with the concrete walls slowly dissolves the lime present in the wall material. Certainly, the calcium carbonate does precipitate but it is rapidly dissolved in contact with aggressive water. This leads to a loss of aggregates on the surface of the material. Giacasso (1988) reported a loss of thickness of about 3 mm (0.12 in.) after 10 years.

As in the case of water tanks discussed earlier, the pH of the cement material near the armatures decreases and can reach a value low enough to start corrosion of the metal.

It is possible to resolve this problem by adding lime or caustic soda to the water at the same time the other acidic chemicals are added. When using aluminum sulfate, the pH may be high enough to obtain excessive levels of aluminum from the flocculant in the potable water. In such cases, it is possible to use ferric chloride or sulfate instead of aluminum sulfate, these two chemicals being less soluble in the water for a wider range of pH.

REFERENCES

AFNOR (Association Française de Normalisation). 1985. NF—A48 901. *Produits de Fonderie—Tuyaux en Fonte à graphite Sphéoïdal Pour Canalizations Avec Pression—Revêtement Interne au Mortier de Ciment Centrifugé—Préscriptions Générales.* Paris: Association Française de Normalisation.

Anderson, P.T. 1981. Vinyl Lined Asbestos Cement Pipe Problems—Health Effects. *Jour. NEWWA,* 217–219.

ANSI/AWWA (American National Standards Institute/American Water Works Association). C104/A21.4. 1990. *American National Standard for Cement–Mortar Lining for Ductile-Iron Pipe and Fittings for Water.* Denver, Colo.: American Water Works Association.

ANSI/AWWA C400-93. 1993. *Standard for Asbestos–Cement Pressure Pipe, 4-in. Through 16-in. (100-mm Through 400-mm), for Water Distribution Systems.* Denver, Colo.: American Water Works Association.

ANSI/AWWA C401-93. 1993. *Standard for the Selection of Asbestos–Cement Pressure Pipe, 4-in. Through 16-in. (100-mm Through 400-mm), for Water Distribution Systems.* Denver, Colo.: American Water Works Association.

ANSI/AWWA C402-89. 1989. *Standard for Asbestos–Cement Transmission Pipe, 18-in. Through 42-in. (450-mm Through 1050-mm), for Potable Water and Other Liquids.* Denver, Colo.: American Water Works Association.

ANSI/AWWA C403-89. 1989. *Standard Practice for the Selection of Asbestos–Cement Transmission and Feeder Main Pipe, Sizes 18-in. Through 42-in. (450-mm Through 1050-mm).* Denver, Colo.: American Water Works Association.

ANSI/AWWA C603-90. 1990. *Standard for Installation of Asbestos–Cement Pressure Pipe.* Denver, Colo.: American Water Works Association.

ASTM C150 92. 1992. *Standard Specification of Portland Cement.* Philadelphia, Pa.: American Society for Testing and Materials.

ASTM C618 93. 1993. *Standard Specification for Fly Ash and Raw or Calcinic Natural Pozzolane for Use as a Mineral Admixture in Portland Cement Concrete.* Philadelphia, Pa.: American Society for Testing and Materials.

AWWA. 1995. *Work Practices for Asbestos–Cement Pipe.* Denver, Colo.: American Water Works Association.

AWWA C400-77. 1977. *Standard for Asbestos–Cement Pressure Pipe, 4-in. Through 24-in., for Water and Other Liquids.* Denver, Colo.: American Water Works Association.

AWWARF Committee Report. 1974. A Study of the Problems of Asbestos in Water (Part 2). *Jour. AWWA,* 66:9.

BGA, Bundesgesundheitsamt. 1981. Gesundheitliche Risiken von Asbest; Eine Stellungnahme des Bundesgesundheitsamts; BGA-Berichte 4/1981, Berlin.

Biczôk, I. 1968. Beton Korrosion—Betonschutz, ed. Bauverlag, Wiesbaden–Berlin.

Brackett, K.A., and P.J. Clark. 1993. A Comparison of the Efficiency of Polycarbonate and Mixed Cellulose Ester Filters for Use in the Filtration of Water Samples. *Environmental Choices, Technical Supplement,* 1(2):25.

British Standard Institute (BSI). BS 4772. 1988. Ductile Iron Pipes and Fittings. Section 3.2. London: British Standards Institute.

Buelow, R.W. 1979. Laboratory Techniques for Determining Corrosivity of Water to Asbestos–Cement Pipe. In *Proc. of the Seventh Annual AWWA Water Quality Technical Conference.* Denver, Colo.: American Water Works Association.

Buelow, R.W., et al. 1980. The Behavior of Asbestos–Cement Pipe Under Various Water Quality Conditions: A Progress Report. *Jour. AWWA,* 72(2):91.

Carson, H.Y. 1927. Cement-lined Water Mains. *Ind. Eng. Chem.,* 19:781.

CEN (Comité Européen de Normalisation). 1992. PrENV 197-1. *Cement Composition, Specifications, and Conformity Criteria. Part 1—Common Cements.* Brussels, Belgium: Comité Européen de Normalisation.

CEOCOR (Committee of European Organizations on Corrosion). 1979. Asbestzementrohre in aggressiven Medien—CEOCOR-Studie. *gwf Wasser/Abwasser,* pp. 62–68.

Chappell, E.L. 1930. Chemical Characteristics of Cement Pipe Linings. *Ind. Eng. Chem.,* 22:1203.

Chatfield, E.J., and M.J. Dillon. 1983. *Analytical Method for the Determination of Asbestos in Water.* EPA 600/4-83-043. Athens, Ga.: U.S. Environmental Protection Agency.

Clark, P.J., et al. 1980. Asbestos–Cement Products in Contact with Drinking Water: CEM Observations. *Scanning Electron Microscopy,* 1:341.

Coanda, S. 1987. La Dégradation des Canalisations en Matériaux à Base de Ciment. In *Les Canalisations d'Eau et de Gaz.* Paris: Technique et Documentation.

Collins, H.H., and T.R. Smith. 1978. Bore Coatings for Spun Iron Pipes in Water Distribution Systems. In *Papers and Proceedings, WRc Conference, Oxford, September. 1978.* Medmenham, England: Water Research Centre.

Conio, O. 1988. La Corrosione Nell'Installatione Coperto di Filtrazione del Brugneto. Journées Italo-Franco-Suisses AGHTM, SVGW, Federgasacqua. Avignon, Mai.

Conroy, P.J., and R. Oliphant. 1991. *Deterioration of Water Quality in Distribution Systems. The Effects of Water Quality Arising From in Situ Cement Lining.* WRc report N DoE 2435-SW. Medmenham, England: Water Research Centre.

Conroy, P.J., S. Canfer, T. Olliffe, and R. Oliphant. 1991. *Deterioration of Water Quality. The Effects Arising from the Use of Factory Applied Cement Mortar Linings.* WRc report N DoE 2723-SW. Medmenham, England: Water Research Centre.

DIN (Deutsches Institut für Normung) 19800. 1973. Asbestzementrohre und Formstücke für Druckrohrleitungen. Berlin: Beuth-Verlag GmbH.

DIN 2614. Ausgabe 2/1980. Zementmörtelauskleidungen für Gußrohre, Stahlrohre und Formstücke; Verfahren, Anforderungen, Prüfungen.

DVGW (German Gas and Waterworks Association). 1993. Untersuchung von Asbestfasern in Trinkwasser; Wasser-Information Nr. 40. 9/93. Bonn: Wirtschafts und Verlagsges, MBH.

DVGW W 342. 1978. Werksseitig hergestellte Zementmörtelauskleidungen für Guss- und Stahlrohre—Anforderungen und Prüfungen, Einsatzbereiche. Bonn: Wirtschafts und Verlagsges, MBH.

DVGW W 343. 1981. Zementmörtelauskleidung von erdverlegten Guss- und Stahlrohrleitungen—Einsatzbereiche, Anforderungen und Prüfungen. Bonn: Wirtschafts und Verlagsges, MBH.

DVGW W 344. 1985. Zementmörtelauskleidung auf dem Lagerplatz. Bonn: Wirtschafts und Verlagsges, MBH.

Eick, H. 1960. Korrosionsfragen aus dem Transportwasser bei Asbestzement-Druckrohren. *Vom Wasser,* 27.

Feige, M.A., P.J. Clark, and K.A. Brackett. 1993. Guidance and Clarification for the Current EPA Test Method for Asbestos in Drinking Water. *Environmental Choices, Technical Supplement,* 1(5): 13–14.

Flentje, M.E., and R.J. Sweitzer. 1955. Solution Effects of Water on Cement and Concrete in Pipe. *Jour. AWWA,* 47(12):1173.

Gardner, R.L. 1981. Vinyl Lined Asbestos–Cement Pipe Problems—Operator Aspects. *Jour. NEWWA,* 220–223.

Giacasso, P. 1988. Corrosion du béton dans une station de traîtement d'eau de boisson d'eau déséquilibré calco-carbonique. Journées Italo-Franco-Suisses AGHTM, SVGW, Ferdergasacqua. Avignon, Mai.

Grandpierre, M. 1986. *Ductile Iron Pipe Compendium.* Nancy, France: Pont-à-Mousson S.A.

Hallenbeck, W.H., et al. 1978. Is Chrysotile Asbestos Released From Asbestos–Cement Pipe Into Drinking Water? *Jour. AWWA,* 70:97.

Hardy, R.J., et al. 1992. Indoor Asbestos Concentrations Associated With the Use of Asbestos-Contaminated Tap Water in Portable Home Humidifiers. *Envir. Sci. and Technol.,* 26(4):680.

Heinrich, B., et al. 1978. Untersuchungen der Korrosionseigenschaften von Zementmörtelauskleidungen für Stahlrohre. 3R International 17, 448–458.

Highway Research Board. 1972. *Guide to Compounds of Interest in Cement and Concrete Research.* Special Report 127. NRC, NAS, NAE.

Hirner, W. 1990. Rehabilitation Methods: A National Overview. Rehabilitation Methods in Germany. *Water Supply,* 8:199–202.

Holtschulte, H. 1972. Zementmörtelauskleidung von erdverlegten Wasserrohrleitungen kleiner Nennweiten. *Neue DELIWA-Zeitschrift,* 10:449.

———. 1974. Praktische Erfahrungen mit der Zementmörtelauskleidung in Wasserrohrleitungen. DVGW-Schriftenreihe Wasser Nr. 4, S. 26:35.

———. 1978a. Investitions- und Unterhaltungskosten im Rohrleitungsbau. *gwf—Gas/Erdgas,* 119(10):429.

———. 1978b. Practical Experiences With Cement–Mortar Lining of Water Pipelines. Water Research Centre Conference.

———. 1980a. Ergebnisse des Forschungs- und Entwicklungsprogramms Korrosion und Korrosionsschutz, Band 1: 1974–1977. FE-KKs-Project No. 0-5.2/4 B and 5.2/4 B.

———. 1980b. Sanierung von Wasserrohrnetzen 5.Wassertechnisches Seminar "Wasserverteilung." Bericht aus Wassergütewirtschaft und Gesundheitsingenieurwesen, Technische Universität München.

Holtschulte, H., and M.R. Schock. 1985. Asbestos–Cement and Cement–Mortar-Lined Pipes. In *Internal Corrosion of Water Distribution Systems*, 1st ed. Edited by AWWA Research Foundation and DVGW, Engler Bunte Institute. Denver, Colo.: AWWA Research Foundation.

Holtschulte, H., et al. 1984. Feldversuche zum Korrosionsschutz von Wasserleitungen durch Zementmörtel- und Zementauskleidungen.*gwf—Wasser /Abwasser, Heft 12.*

Hoppe, P. 1990. Rehabilitation Methods: Linings. Water Mains Rehabilitation of the Hamburger Wasserwerk: A Comparison and Evaluation of Replacement and Cement Lining and Establishment of an Evaluation System. *Water Supply,* 8:283–290.

Hünerberg, K., and H. Tessendorff. 1977. Handbuch für Asbestzementrohre. 2. Aufl. Berlin: Springer Verlag.

ISO (International Standard Organization). 1985. ISO 4179. *Tuyaux en Fonte Ductile Pour Canalisations Avec et Sans Pression. Revêtement Interne au Mortier de Ciment Centrifugé—Préscriptions Générales.* Geneva, Switzerland: International Standard Organization.

Jaryd, A., and O. Ljunggren. 1990. Rehabilitation Methods: A National Overview. Rehabilitation of Water Mains in Sweden. *Water Supply,* 8:231–236.

Kanarek, M.S., et al. 1981. Chrysotile Asbestos Fibres in Drinking Water from Asbestos–Cement Pipe. *Envir. Sci. and Technol.,* 15(8):923.

Kirmeyer, G.J., et al. 1980. *Seattle's Corrosion Control Plan—Chemical Treatment Coupled With Selection of Corrosion Resistant Materials.* Municipal Environmental Research Laboratory. Cincinnati, Ohio: U.S. Environmental Protection Agency.

Kristiansen, H. 1977. Corrosion of Asbestos–Cement Pipes. *Vatten,* 1:53.

———. 1978. Degradation of Asbestos–Cement and Concrete Pipes. In *Papers and Proceedings, WRc Conference, Oxford, September 1978.* Medmenham, England: Water Research Centre.

Lannblad, B. 1990. Rehabilitation Methods: Linings. Cement Mortar Lining of Water Mains by Mortar Injection Principle. *Water Supply,* 8:279–282.

Larson, C.D., O.T. Love, Jr., and G. Reynolds, III. 1983. Tetrachloroethylene Leached From Lined Asbestos–Cement Pipe Into Drinking Water. *Jour. AWWA,* 75:184.

Legrand, L., G. Poirier, and P. Leroy. 1981. Les Equilibres Carboniques et l'Equilibre Calcocarbonique dans les Eaux Naturelles. Ed. Eyrolles, 61 Boulevard Saint-Germain, F-75005 Paris.

Logsdon, G.S. 1983. Engineering and Operating Approaches for Controlling Asbestos Fibres in Drinking Waters. Envir. Health Perspect., 53:169.

Logsdon, G.S., and J.R. Millette. 1981. Monitoring for Corrosion of Asbestos–Cement Pipe. In *Proc. of the Ninth Annual AWWA Water Quality Technology Conference.* Denver, Colo.: American Water Works Association.

Mah, M., and E.S. Boatman. 1978. Scanning and Transmission Electron Microscopy of New and Used Asbestos–Cement Pipe Utilized in the Distribution of Water Supplies. *Scanning Electron Microscopy,* 1:85.

McCallie, M. 1981. Personal communication.

Meyer, E. 1982. Untersuchungen zum Vorkommen von Asbestfasern im Trinkwasser in der Bundesrepublik Deutschland und gesundheitliche Bewertung der Ergebnisse. *gwf—Das Gas- und Wasserfach,* 123:85–96.

Miller, W.T. 1965. Durability of Cement–Mortar Linings in Cast-Iron Pipe. *Jour. AWWA,* 57:773.

Millette, J.R. 1983. Investigation of the Protective Effect of Iron Containing Coatings on Asbestos–Cement Pipe Exposed to Aggressive Water. Ph.D. diss., University of Cincinnati, Cincinnati, Ohio.

Millette, J.R., et al. 1980. Aggressive Water: Assessing the Extent of the Problem. *Jour. AWWA,* 72(5):262.

———. 1981. Asbestos Analytical Technique. In *Proc. of the Eighth Annual AWWA Water Quality Technology Conference.* Denver, Colo.: American Water Works Association.

———. 1984. Evaluating the Condition of Asbestos–Cement Pipe. *NACE Materials Performance,* 14–20.

Muller, J.R. 1978. Deterioration of Portland Cement Concrete in Natural Waters. *Corrosion and Coatings,* 11–23.

NAS (National Academy of Sciences). 1982. *Drinking Water and Health, Volume 4.* Washington, D.C.: National Academy Press.

Olson, H.L. 1974. Asbestos in Potable-Water Supplies. *Jour. AWWA,* 66:515.

Pontius, F. W. 1992. USEPA's Ban on Asbestos-Containing Products Invalidated. *Jour. AWWA,* 84(92):22.

Sakurai, Y., S. Inoue, T. Nakajima, and A. Myamoto. 1990. New Methods for Pipeline Replacement With Ductile Iron Pipes. *Water Supply,* 8:253–257.

Schock, M.R. 1980. Unpublished Data. DWRD, Cincinnati, Ohio: U.S. Environmental Protection Agency.

———. 1990. Internal Corrosion and Deposition Control. In *Water Quality and Treatment,* 4th ed. New York: McGraw-Hill and American Water Works Association.

Schock, M.R., and R.W. Buelow. 1981. The Behavior of Asbestos–Cement Pipe Under Various Water Quality Conditions: Part 2, Theoretical Considerations. *Jour. AWWA,* 73:609.

Schock, M.R., et al. 1981. *Evaluation and Control of Asbestos–Cement Pipe Corrosion.* NACE Corrosion/81 and EPA 600/D-81-067.

Soukatchoff, P. 1990. Résistance des Mortiers aux Eaux Aggressives. *TSM,* 85(4):197–201.

Standard Methods for the Examination of Water and Wastewater. 18th ed. suppl. 1993. Washington, D.C.: American Public Health Association, American Water Works Association, and Water Environment Federation.

Tate, C., et al. 1984. *Control of Asbestos Loss From Asbestos–Cement Pipe in Aggressive Waters.* EPA Cooperative Agreement CR807789010, Municipal Environmental Research Laboratory. Cincinnati, Ohio: U.S. Environmental Protection Agency.

Tracy, E.L. 1950. Observations in Water Samples From Cement-Asbestos Pipe Systems in Vermont. *Jour. NEWWA,* 64:164.

USEPA (U.S. Environmental Protection Agency). 1984. *Control of Asbestos Fibre Loss From Asbestos–Cement Watermain.* Project Summary EPA 600/S2-84-014. Washington, D.C.: Government Printing Office.

———. 1989. Asbestos; Manufacture, Importation, Processing, and Distribution in Commerce Prohibitions. Final Rule. *Federal Register,* 54(132):29460.

———. 1991. National Primary Drinking Water Regulations—Synthetic Organic Chemicals and Organic Chemicals; Monitoring for Unregulated Contaminants; National Primary Drinking Water Regulations Implementation; National Secondary Drinking Water Regulations. Final Rule. *Federal Register,* 56(20):3526.

Vik, E.A., and M. Weideborg. 1991. Corrosion Control and Corrosion Monitoring in Low Alkalinity Waters—Norwegian Experiences. In *Corrosion and Corrosion Control in Drinking Water Systems.* Houston, Texas: NACE.

Weston Water Utility. 1983. *Control of Asbestos Fibre Loss From Asbestos–Cement Watermains.* EPA M2361. Washington, D.C.: Government Printing Office.

Woodhull, R.S. 1981. Use of Vinyl-Lined Asbestos–Cement Pipe—A Case History. In *Proc. of the Ninth Annual AWWA Water Quality Technology Conference.* Denver, Colo.: American Water Works Association.

WRc Engineering. 1990. *In Situ Cement Mortar Lining—Operational Guidelines.* Swindon, Wiltshire (U.K.): Water Research Centre.

8

Mitigation of Corrosion Effects

Eilen A. Vik, Robert A. Ryder, Ivo Wagner, and John F. Ferguson

Water utilities have several ways to overcome the difficulties that arise from internal corrosion of piping and plumbing systems. First and most important among them is careful attention to the properties and performance of all materials used in contact with the water. Materials are available that will function for reasonable service lifetimes for any potable water quality without significant loss of capacity, excessive failure, release of corrosion by-products, or other damages commonly associated with corrosion. However, water systems are old, complex, and costly investments of the water utility and its customers. Most often, the problem faced is mitigation of corrosion effects on materials already in use. The strategy most widely used is to change the water quality through measures like pH and alkalinity adjustment, use of chemical corrosion inhibitors, or mixing of different waters to change quality and reduce corrosivity. Materials specification through building codes and utility policies and, when necessary, replacement with more resistant materials are essential parts of corrosion mitigation. Good workmanship in installing pipe is also an important factor. However, corrosion-control water treatments are frequently also necessary.

In this chapter, water qualities and their modification are emphasized as critical factors in corrosion control. Low-alkalinity, low-mineral-content waters arise naturally in many parts of the world; they may be exacerbated by acid precipitation or they may result from the processing of water by softening, desalination, or chemical treatment. In each of these cases, adjustment of pH, alkalinity, and sometimes hardness is

NOTE: Portions of this chapter appeared in chapters 7 and 8 of the first edition.

needed for corrosion mitigation. In many other waters, hardness and alkalinity limit the pH values that can be attained without calcium carbonate precipitation, which under favorable circumstances may form a thin protective film, but under other circumstances may form thick deposits or increased turbidity. Calcium carbonate saturation, as discussed in chapter 1, is an important criteria in understanding corrosion mitigation, but often is incorrectly interpreted in mitigation treatment. In some waters, naturally occurring substances, including natural organic matter (NOM) or silica, may minimize corrosion. As water treatment processes are implemented that change NOM and other water qualities, natural inhibition may be reduced.

Corrosion inhibitors have a major role in many aspects of corrosion control. While the range of chemical products is broad, the constituents are restricted by regulation, as well as by efficacy. Ortho- and polyphosphates, zinc salts, and sodium silicates are the major chemicals used in corrosion inhibitors, either as generic chemicals or in proprietary mixtures. The properties and use of these are described in the latter part of this chapter.

BASIC CORRELATIONS BETWEEN WATER QUALITY AND MATERIALS

The corrosiveness of the water passing through the distribution system, service lines, and home plumbing systems can either be due to the natural water quality, to water treatment processes, or to chemicals used in the drinking water treatment plant.

Table 8-1 lists some of the chemical factors that are of importance. Many of these factors are closely related, and a change in one often changes another. The most important example of this is the relationship between pH, carbon dioxide (CO_2), dissolved inorganic carbon (DIC), and alkalinity.

Natural Low-Alkalinity Drinking Water

Low-alkalinity, soft waters with low pH naturally occur in humid regions where there is significant runoff and little opportunity for dissolution of soil minerals. These are usually surface waters, but occasionally low-alkalinity groundwaters are also found. Geological and geographical factors must be favorable for a low-alkalinity water. The soil or bedrock must be poorly water soluble and have low amounts of bases that produce alkalinity when dissolved. These water qualities are found in parts of North America, Nordic countries, the United Kingdom, and central Europe.

Surface Water Acidification

Acidification of surface water and groundwater, primarily by airborne pollutants, has played a significant role in production of low-pH and low-alkalinity waters. The geological susceptibility for acidification in North America as well as European surface waters has been assessed by several groups of researchers (Galloway and Cowling 1978; Henriksen et al. 1989). Waters susceptible to acidification typically occur in eastern and southern regions of the United States and in areas of the Pacific coast, including the western mountain chains. The same water qualities are found in Canadian surface waters. In Europe, surface waters in the Scandinavian countries, including Finland, Norway, and Sweden, are very susceptible. The same problem is also seen in parts of Scotland, in certain regions in east central Europe, and probably elsewhere.

Table 8-1 Chemical factors influencing corrosion and corrosion control

Factor	Effect
pH	Low pH may increase corrosion rate; high pH may protect pipes and decrease corrosion rates or could cause dezincification of brasses or soft blue hydroxide deposits in copper pipe.
Alkalinity, dissolved inorganic carbon (DIC)	May help form protective coating; helps control pH changes. Low to moderate alkalinity reduces corrosion of most materials. High alkalinities increase corrosion of copper, lead, and zinc.
Dissolved oxygen (DO)	Increases rate of many corrosion reactions.
Chlorine residual	Increases metallic corrosion, particularly for copper, iron, and steel. Reduces microbial-induced corrosion.
Total dissolved solids (TDS)	High TDS tends to increase conductivity and corrosion rate.
Calcium hardness	Calcium may precipitate as $CaCO_3$ and may provide protection and reduce corrosion rates, but may cause turbidity or excessive scaling
Chloride, sulfate	High levels increase corrosion of iron, copper, and galvanized steel.
Hydrogen sulfide	Increases corrosion rates.
Ammonia	May increase solubility of some metals, such as copper and lead.
Polyphosphates	May reduce tuberculation of iron and steel, and provide smooth pipe interior. May increase iron and steel corrosion at low dosages. May attack and soften cement linings and asbestos–cement pipe. Increase the solubility of lead and copper. Prevent $CaCO_3$ formation and deposition. Sequester ferrous iron and minimize occurrence of red water. Can revert as orthophosphate and form protective films.
Orthophosphate	May form protective films.
Silicate	Sequesters metal ions and forms protective films. Silicate solutions increase pH and alkalinity.
Natural color, organic matter	May decrease corrosion by coating pipe surfaces. Some organics can complex metals and accelerate corrosion or metal release.
Iron or manganese	May react on interior of asbestos–cement pipe to form protective coating. May precipitate film on interior of metallic pipes, which may be protective or may cause pitting in copper.
Zinc	May react on interior of asbestos–cement pipe to form protective coating. May react on interior of pipe to form protective coating. May tend to precipitate at cathodic sites and form protective deposits.
Copper	Causes pitting in galvanized pipe.
Magnesium (and possibly other trace metals)	May inhibit the precipitation of calcite form of $CaCO_3$ on pipe surfaces and favor the deposition of the more soluble aragonite form of $CaCO_3$.

Modified from Schock (1990).

Abundant evidence of acidification now exists. Schindler et al. (1989) used the US Environmental Protection Agency's (USEPA) Eastern Lake Survey database to predict a 0.6 median pH unit decline for lakes in the Adirondacks and a 0.7 unit decline for lakes in the Poconos and Catskills. Bolze and Beyea (1989) presented results from the Citizens Acid Rain Monitoring Network started in the United States. Volunteers monitor pH of precipitation during a minimum one-year period. Data for June 1988 are presented in Figure 8-1, showing that acid rain fell in 34 states. These are the states shown in the figure with a monthly average pH in rainwater of <5.0. Wright (1988) compared the sensitivity of lakes to acidification in the eastern United States and southern Norway and found the latter to be inherently more sensitive, with a higher percentage of acidic lakes. At present, about 70 percent of the lakes sampled in the southern part of Norway are estimated to have lost their bicarbonate buffering capacity (Henriksen et al. 1988). In central Europe, for example, Germany, the typical composition of soil and bedrock has made the problems of acidification less obvious than in northern Europe. However, in certain areas of Germany, acidification is causing serious problems. Hantge (1992) addresses especially the drinking water quality problems occurring in both groundwater and surface water in the Rheinland–Pfalz region. An investigation in this region has shown that more than 200 water utilities have problems meeting the German drinking water regulations with respect to pH for soft water due to acidification.

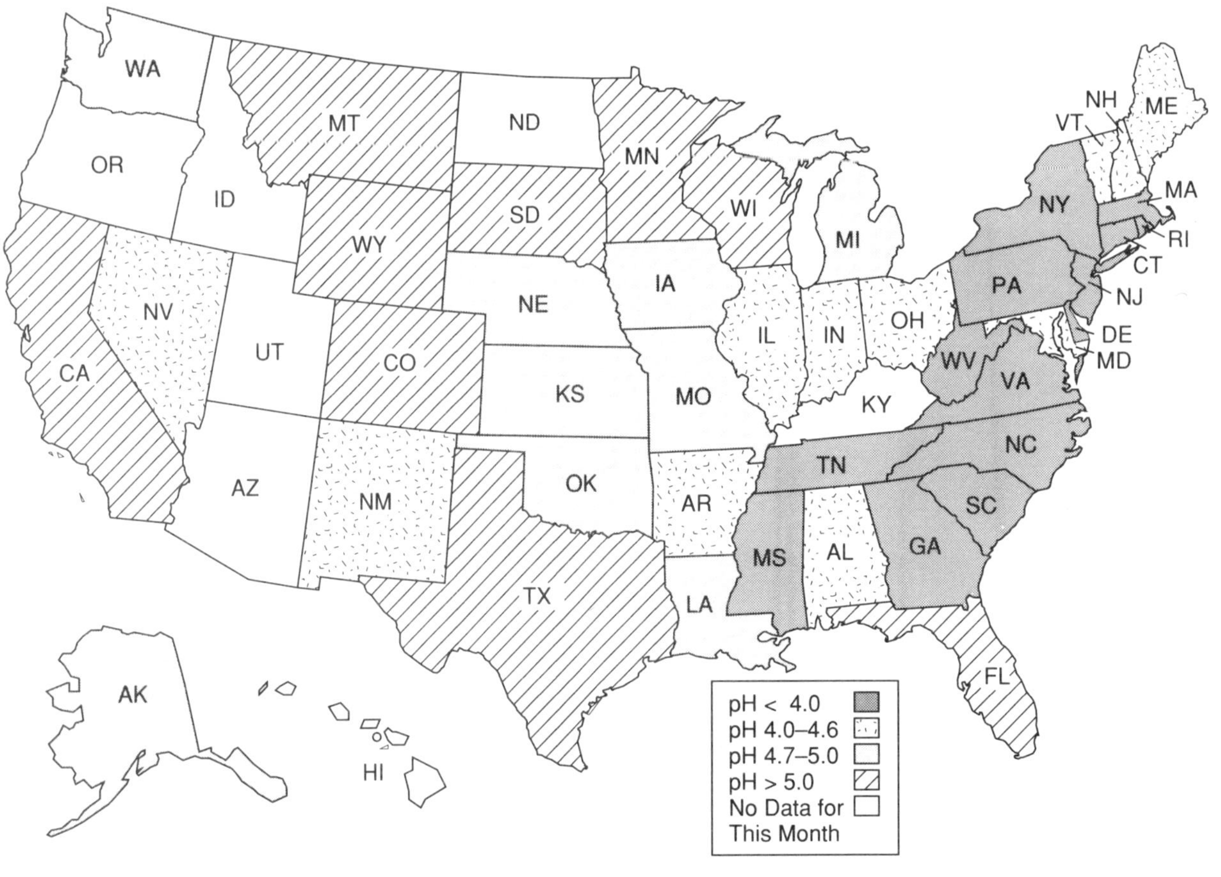

Source: Bolze and Beyea (1989). Reprinted with permission from Environmental Science and Technology © *1989 American Chemical Society.*

Figure 8-1 Average pH values in rainwater in June 1988

Effects of acid precipitation extend beyond losses of alkalinity and pH decreases. Anion concentrations are increased as well. There are changes in the amounts and functional properties of NOM in surface waters undergoing acidification. Gjessing, Taud, and Peterson (1992), in summarizing preliminary results of ongoing research, observed that humic substances can contribute directly to acidification, interact with other dissolved materials, influence microbial enzyme processes, and directly influence aquatic biota. Acidification can also cause extremely high aluminum concentrations due to dissolution of aluminum from the soil (Boman and Vik 1985). The correlation between aluminum and pH in the drinking water is shown in Figure 8-2.

Mitigation of acid precipitation is an important international problem, requiring long-term cooperation to reduce acid air emissions. The local negative environmental effects of acidification can be addressed by liming programs for lakes and forests. pH and alkalinity can be supplemented in water treatment, and water treatment can address other water quality changes in the water source. Increased corrosion in drinking water systems is a likely effect of acidification that should be assessed and mitigated.

Water Treatment Effects on Water Corrosivity

Natural, low-mineralized waters are usually not very corrosive to steel, galvanized, or copper pipes. The natural inhibitors are site specific, but they usually include NOM, relatively high concentrations of silicate and bicarbonate concentrations (meq/L) that exceed the chloride and sulfate anions (meq/L). Water treatment often increases the corrosive tendency of the water by removing or changing the NOM, decreasing the alkalinity, and increasing the chloride and sulfate concentrations when inorganic aluminum or ferric salts are used for coagulation, and when high chlorine dosages are used (Singley et al. 1981). Fluoridation with hydrofluosilicic acid can also lower the alkalinity and the pH in these kinds of water. Softening or membrane filtration processes often can increase the corrosiveness of the treated

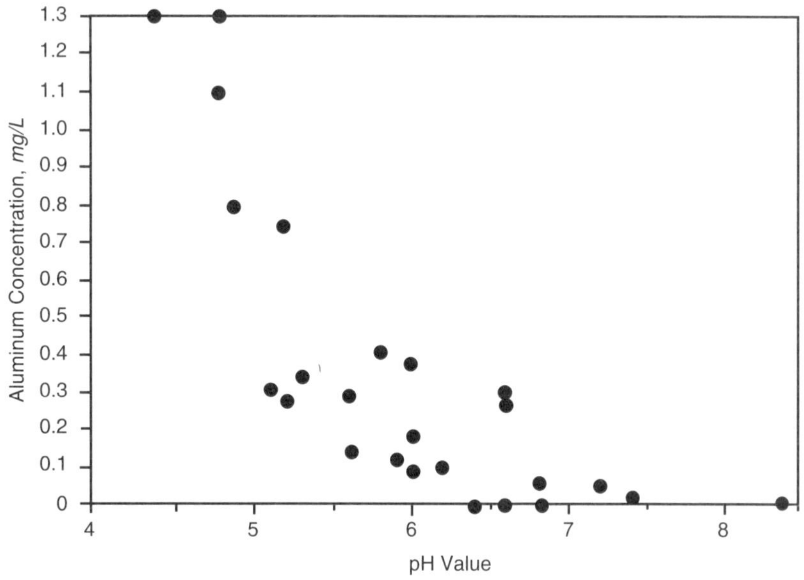

Figure 8-2 Relationship between aluminum and pH in acidified lakes of southern Norway

water. The observed corrosion effect depends on the raw water and the treatment processes used and can only be predicted in a general way.

Water treatment processes play an important role in increasing the corrosiveness of drinking water, which needs to be taken into consideration when the water treatment plant is being designed. Typical examples are discussed below.

Chemically treated water. The use of coagulant chemicals as an aid to sedimentation of particulate matter has been practiced from ancient times. The predominant chemical agent has been aluminum sulfate, or alum. Other coagulants such as ferric sulfate ($Fe_2(SO_4)_3$) or ferric chloride ($FeCl_3$) are also widely used. These inorganic chemicals are available commercially for coagulation. Recently, polymeric and partially hydrolyzed coagulants such as polyaluminum chloride and sulfate are being used as coagulants with less reduction of pH and alkalinity, and as a consequence, with less of an increase in corrosivity than the traditional alum and iron coagulants. Table 8-2 gives an overview of the changes in quality that can relate to corrosivity caused by chemical treatment, disinfection, and fluoridation processes.

When alum is used, alkalinity is reduced as shown in the following reaction when bicarbonate and pH are decreased.

$$Al_2(SO_4)_3 \cdot 14H_2O + 6HCO_3^- = 2Al(OH)_{3(s)} + 3SO_4^{2-} + 6CO_2 + 14H_2O \quad (8\text{-}1)$$

When ferric chloride is used, the same stoichiometric reaction is seen, except that the chloride concentration instead of the sulfate concentration is increased.

$$2FeCl_3 \cdot 6H_2O + 6HCO_3^- = 2Fe(OH)_{3(s)} + 6Cl^- + 6CO_2 + 6H_2O \quad (8\text{-}2)$$

Table 8-2 Alkalinity reduction and aggressive anion increases associated with the common water treatment chemicals

Chemical	Formula	Basis for Comparison	Alkalinity Change, *eq.*	Cl^-, SO_4^{2-}, or F^- Change, *eq.*
Alum	$Al_2(SO_4)_3 \cdot 14H_2O$*	Al^{3+}	3	3
Polymerized aluminum chloride/sulfate		Al^{3+}	1–2	3
Ferric chloride	$FeCl_3 \cdot 6H_2O$†	Fe^{3+}	3	3
Ferric sulfate	$Fe_2(SO_4)_3 \cdot 6H_2O$‡	Fe^{3+}	3	3
Cl_2 disproportionation		Cl_2	1	1
Cl_2 oxidation		Cl_2	1–2	1
Cl_2 plus ammonia		Cl_2	0	1
Chloramine oxidation		Cl_2	0–1	1
Fluoridation	Hydrofluosilicic acid	F	1	1

Source: Ferguson (1985).

*$Al_2(SO_4)_3 \cdot 18H_2O$ is commonly used. In the calculations, alum with molecular weight 600 and 0.09 mg Al^{3+}/mg alum has been used.

†For ferric chloride: 0.34 mg Fe^{3+}/mg ferric chloride has been used.

‡For ferric sulfate: 0.26 mg Fe^{3+}/mg ferric sulfate has been used.

The most important variable for the proper use of chemicals as coagulants is the maintenance of an effective pH value. The normal operational pH range for alum is between 5.5 and 7.3.

As an example, if an alum dosage in the range of 3 to 100 mg/L is used, the changes in water quality illustrated in Table 8-3 would occur. Alum can significantly change the corrosiveness of the water.

Disinfection with chlorine. When gaseous chlorine is used, disproportionation of gaseous chlorine occurs according to the following reaction:

$$Cl_2 + H_2O = HOCl + H^+ + Cl^- \tag{8-3}$$

The result is a pH and alkalinity decrease and a Cl^- increase (see Table 8-2). HOCl is a strong oxidizing agent. If organic matter is present, HOCl will oxidize some of these compounds, and the result is a further decrease in alkalinity with formation of CO_2 or other oxidized organics.

$$HOCl + \tfrac{1}{2}CH_2O = \tfrac{1}{2}CO_2 + Cl^- + H^+ + \tfrac{1}{2}H_2O \tag{8-4}$$

When chlorination takes place with ammonia to form monochloramine, ammonia (which is a base) is added in proportion to chlorine. Monochloramine is a less reactive oxidant, and chloramination has less effect on alkalinity and pH than free chlorine.

Fluoridation. Fluoridation of water for dental prophylaxis commonly occurs in North America. If fluosilicic acid is used, the alkalinity will decrease according to the following equation:

$$H_2SiF_6 + 4H_2O = H_4SiO_4 + 6H^+ + 6F^- \tag{8-5}$$

Table 8-2 gives the net result of chemical dosages with fluosilicic acid. For a low alkalinity, the effect on pH and alkalinity can be dramatic. Fluoride also probably enhances corrosion much as do chloride and sulfate ions. Sodium fluoride or sodium silicafluoride can be used instead of fluosilicic acid and will not reduce pH or alkalinity.

Softening. Hardness is removed in the lime or lime-soda softening process by precipitation, ion exchange processes, or reverse osmosis (RO). Water quality changes are significantly different, depending on which process is used. The chemical

Table 8-3 Water quality changes from chemical treatment with alum and calcium bicarbonate

	Water Quality Changes			
Alum Dosage*	Final pH	Alkalinity Decrease, *meq*	Alkalinity Decrease, $CaCO_3$, *mg / L*	Increase SO_4, *mg/L*
3 mg/L	5.5–7.3	0.03	1.4	0.5
50 mg/L	5.5–7.3	0.45	22.5	8.0
100 mg/L	5.5–7.3	0.9	45	16

*$Al_2(SO_4)_3 \cdot 14.6H_2O$, molecular weight 600.

precipitation caused by the lime softening process can be illustrated by the following two equations:

$$Ca^{2+} + 2HCO_3^- + Ca(OH)_2 = 2CaCO_{3(s)} + 2H_2O \tag{8-6}$$

$$Mg^{2+} + 2HCO_3^- + 2Ca(OH)_2 = 2CaCO_{3(s)} + Mg(OH)_{2(s)} + 2\ H_2O \tag{8-7}$$

The lime softening process reduces the alkalinity, but the excess lime will increase the pH of the water, as illustrated in a case study in Florida (Table 8-4) comparing water quality of lime and membrane softening processes. Ion exchange softening uses a strong acid resin and exchanges two Na^+ ions for each Ca^{2+} ion removed. The resin is normally regenerated with salt brine. Ion exchangers can remove nearly all the hardness from the water and produce a soft, more corrosive water. It is, therefore, frequently necessary to provide some bypass of untreated water to reduce corrosivity. Since regeneration is performed with brine, leakage of chloride to softened water also occurs and can aggravate corrosion.

When membrane processes are used for softening, the removal of calcium and alkalinity can be more effective than when lime softening is used. The membrane processes are also efficient in removing other inorganic ions, as well as NOM. The American Water Works Association's (AWWA's) Membrane Technology Committee (1992) reviewed the use of membrane processes in potable water treatment and found that membrane processes have been most widely used in RO applications. In 1989, more than 4,000 land-based RO units were in operation with a desalting capacity of approximately 1 bil gal/d (3.8×10^9 L/d). Membrane processes are becoming more frequently used for softening. Taylor et al. (1989) compared the cost and effectiveness of membrane processes for softening and trihalomethane (THM) control with a conventional lime softening plant at Flagler Beach, Florida. Results from pilot studies showed that THMs could be controlled and a finished water with superior water quality could be obtained with the membrane process (see Table 8-4). The membrane process was also observed to be less labor intensive and easier to operate than a traditional softening plant. A major advantage of the newer

Table 8-4 Predicted water quality for lime softening and membrane processes at Flagler Beach, Florida

Parameter	Raw Water	Finished Water	
		Softening Plant	Membrane Plant
pH	7.4	8.9	9.0
Alkalinity, *mg* $CaCO_3$ */ L*	301	52	35
Ca, *mg / L*	100	13	12
Cl, *mg / L*	111	111	30
Na, *mg / L*	60	60	20
SO_4, *mg / L*	2	2	0
Color, *cfu*	37	5	0
Dissolved organic carbon, *mg / L*	11	17	1
Trihalomethane formation potential, *mg / L*		350	20

Source: Taylor et al. (1989).

lower-pressure RO process or nanofiltration membrane systems relative to earlier high-pressure RO systems is that enough alkalinity and calcium remains in the product water to permit stabilization with little or no addition of chemicals.

Duranceau, Taylor, and Mulford (1992) studied removal of synthetic organic chemicals using a nanofiltration process. Very good results were obtained, and nearly all the organics were removed with only 6 mg Ca/L and an alkalinity of 20 mg $CaCO_3$/L remaining after treatment. Depending on the efficiency required for organics removal, the alkalinity and calcium content in the permeate seems to be controllable by the choice of membranes. However, pH will most likely have to be regulated for corrosion control purposes.

Desalination. Desalination of drinking water can be performed by various processes. Most common are evaporation processes and RO membranes. Evaporation can be done by distillation at high pressure and low temperature (<40°C) or at low pressure and high temperature. The first process is the most frequently used because of less energy use. An elevated bacteria concentration in the finished water is often encountered, and disinfection needs to be included in the system. Electrolytic production of sodium hypochlorite is then often used, since seawater or brine is usually available. Treated water generally has a low pH, low alkalinity, low calcium concentration, and relatively high temperature. Leroy (1983) and Legrand and Leroy (1990) studied the corrosivity of desalinated water in Saudi Arabia and showed the need for corrosion control, even when steel pipe with internal epoxy protection was used in the distribution system. In the Yanbu–Medina water utility (Saudi Arabia), lime and carbon dioxide were added to prevent corrosion of the distribution lines, which consisted partly of steel pipes with internal epoxy lining and partly of asbestos–cement pipes. Prior to corrosion control treatment, after only three months of operation, severe corrosion problems were observed in the steel pipes. Peeling of the internal epoxy lining was observed in places, and the steel was severely corroding. The pipe system was operated at high pressure, and corrosion control countermeasures were needed immediately.

The following countermeasures against corrosion were chosen: (1) reduce seawater volume through the brine electrolysis unit used for NaOCl production in order to increase the concentration of hypochlorite and thus reduce the dosage volume and (2) increase the dosage of lime plus CO_2 in order to increase pH, alkalinity, and calcium concentration of the treated water.

The suggested corrosion control methods proved to be very effective. Table 8-5 presents the water quality before and after the corrosion control treatment.

Table 8-5 Composition of potable water before and after modifications of treatment to prevent corrosion in the Yanbu-Medina water supply system in Saudi Arabia

Treated Water Process	Temperature, *°C*	pH	Alkalinity, *mg $CaCO_3$ /L*	Calcium, *mg /L*	Chloride, *mg /L*
Before modifications	36	7.5–7.7	30–55	12–14	55–60
After corrosion control modifications	36	8.1	60–65	26–29	13–18

Source: Leroy (1983).

pH, ALKALINITY, AND HARDNESS CONTROL

Water treatment practices to reduce corrosiveness of drinking water vary from country to country. In a workshop in Oslo, Norway, in 1990 (Vik and Hedberg 1991), a group of scientists from Europe (Denmark, Finland, the Netherlands, Norway, Germany, Sweden, and the United Kingdom) and the United States tried to agree on common countermeasures for corrosion problems in water distribution systems. The group agreed on some aspects, although the different water treatment practices and raw water qualities in each country have resulted in rather different approaches to corrosion mitigation. Table 8-6 summarizes corrosion and water quality problems that all participants agreed were caused by materials in contact with drinking water, and Table 8-7 summarizes the consensus view of the desirable water quality for the various materials.

Details relating types of corrosion of various materials and methods for monitoring are discussed in other chapters.

In practice, a distribution system consists of many materials in contact with water that varies somewhat in quality, both temporally and from one part of the system to another. Corrosion problems can vary extensively in a single system. To minimize internal corrosion problems, the optimum quality is, more realistically, a compromise to sustain a suitable water quality for each pipe material in each location and each time. The first approach to achieving such water quality is often establishing guidelines or standards for pH, alkalinity, and related parameters.

Different organizations and countries have proposed different water qualities to avoid internal corrosion problems in various pipe and house installation materials. These are summarized in Table 8-8. One of the main points of corrosion control is pH adjustment; the considerable differences in pH guidelines are illustrated in Figure 8-3. As an example, a European Community (EC) directive for drinking water states that the appropriate pH range is between pH 6.5 and pH 8.5, and for extremely soft waters pH 9.5 (EC 1980). In the German national regulation, if the calcium carbonate saturation index is <0, the pH must be adjusted to 8.0, or so that pH is not more than 0.2 pH units lower than pH_s if $pH_s \leq 8.0$. This regulation is intended to protect cement materials and to reduce water quality deterioration, especially with regard to heavy-metal release.

As seen from Table 8-8 and Figure 8-3, all countries and organizations consider pH, alkalinity, and hardness to be very important characteristics. Several ways to control these characteristics are presented in the next section.

Low-Alkalinity Water

Corrosion problems arising from low-alkalinity, soft, low-pH waters are often treated by increasing the buffer capacity and the concentrations of divalent cations like calcium and magnesium. The pH is also increased in these treatments. The treatment strategies usually are intended to adjust all three characteristics.

The available technologies that can be used, depending on the water to be treated, are

- dosing alkaline chemicals such as sodium bicarbonate, sodium hydroxide, calcium hydroxide, or soda ash. The latter three can be used together with carbon dioxide to adjust pH.
- deep bed filtration over alkaline media filters with $CaCO_3$ or half-burned dolomite limestone. CO_2 may be added before alkaline media filters to increase total carbonate content and to adjust the pH.

Table 8-6 Corrosion and water quality problems caused by materials in contact with drinking water

Material	Corrosion Type	Tap Water Quality Deterioration
Cement Based		
Asbestos–cement* Concrete Cement mortar†	Uniform corrosion	Calcium dissolution. Increased pH values (up to 12). For asbestos–cement pipes, in unstable waters, pH increases and asbestos fibers can be found in the water. Surface roughening, strength reduction, and pipe failure.
Iron		
Cast	Uniform corrosion	Rust tubercles (blockage of pipe).
Ductile	Graphitization Pitting under unprotective scale	Iron and suspended particles release.
Steel	Pitting	Rust tubercles (blockage of pipe). Iron and suspended particles release.
Galvanized Steel	General pitting corrosion	Excessive zinc, lead, cadmium, iron release, and blockage of pipe.
Copper	Uniform corrosion	Copper release.
	Localized attack	
	Cold-water pitting (type I) Hot-water pitting (type II) Other types of localized attack	Pipe perforation and subsequent leakage from pipes.
	Microbial-induced corrosion Corrosion fatigue	Leakage from pipes and sporadic blue deposit release. Rupture of pipes and fittings and consequently leakage.
	Erosion corrosion	Leakage from pipes.
Lead		
Lead pipe	Uniform corrosion	Lead release.
Lead–tin solder		Lead and cadmium release.
Brass	Erosion and impingement attack Dezincification Stress corrosion	Penetration failures of piping. Blockage of pipes and fittings. Lead and zinc release.
Plastic	Unknown Degraded by sunlight and microorganisms?	Taste and odor.

Modified from Vik and Hedberg (1991).

*No internal lining (e.g., tar).

†Used as internal lining of iron and steel materials.

Table 8-7 General water quality conditions to avoid internal corrosion in various pipe and house installation materials

Type of Material	Corrosion Type	Water Quality Parameters: pH	Alkalinity, *meq/L*	Calcium, *mg/L*	SO_4, *mg/L*	Chlorine, *mg/L*	$\frac{HCO_3}{SO_4 + Cl}$‡, $\frac{meq/L}{mg/L}$	$\frac{HCO_3}{SO_4}$, $\frac{meq/L}{mg/L}$	O_2, *mg/L*	Organic Materials	Aggressive CO_2, *mg/L*	Remarks
Cement-Based Pipe												The most important parameter is aggressive CO_2.
Asbestos cement Concrete Cement mortar	Uniform	>7	>0.3*	>10	<200						<5	
Iron												
Cast Ductile Steel	Uniform	>7	>0.2 to 0.5*	Desirable			≥1.5			Important†		
	Pitting								>2			Tends to occur under deposits.
Galvanized Steel							>1.0					After some months, see iron pipes.
Copper												
	Uniform	>7	Can be low							Might act as an inhibitor	As low as possible	Gives high copper release to the water.
	Cold-water pitting type (I)				As low as possible	As low as possible		>0.8				Type I can be avoided by use of qualified pipe and soldering.
	Hot-water pitting type (II)	>7	1 to 2									
	Other types of localized pitting											
	Microbial-induced corrosion											Chlorine residual >0.4 mg/L.

Table continued next page

Source: Modified from Vik and Hedberg (1991).

*Scandinavian experience with low-alkalinity waters suggests higher alkalinity, >0.5 meq/L.

†High molecular weight acidic groups.

‡The inverse of this ratio is also known as the Larson index (see chapter 1).

Table 8-7 General water quality conditions to avoid internal corrosion in various pipe and house installation materials (continued)

		Water Quality Parameters										
Type of Material	Corrosion Type	pH	Alkalinity, *meq/L*	Calcium, *mg/L*	SO_4, *mg/L*	Chlorine, *mg/L*	$\frac{HCO_3}{SO_4 + Cl}$ ‡, $\frac{meq/L}{mg/L}$	$\frac{HCO_3}{SO_4}$, $\frac{meq/L}{mg/L}$	O_2, *mg/L*	Organic Materials	Aggressive CO_2, *mg/L*	Remarks
Copper (continued)	Corrosion fatigue											
	Erosion corrosion											Velocity >1.1 m/s.
Lead												Should be replaced in low-alkalinity waters.
Lead pipe Lead-tin solder		>8										Orthophosphate dosing at high alkalinity. Flush before drinking.

Source: Modified from Vik and Hedbery (1991).

*Scandinavian experience with low-alkalinity waters suggests higher alkalinity, >0.5 meq/L.

†High molecular weight acidic groups.

‡The inverse of this ratio is also known as the Larson index (see chapter 1).

Table 8-8 Recommended strategy and water quality versus internal corrosion problems from various organizations and countries

Parameters Affecting Corrosion	Country*†								
	N	DK	SF	S	NL	UK	US	CDN	G
pH	7.5–8.5	8.0	7–9	7.5–9	7.8–8.3	5.5–9.5	6.5–8.5	6.5–6.8	6.5–9.5
Alkalinity, *meq /L*	0.6–1	>1.6	>0.6	>1	>2	>0.5			
Hardness, *mg Ca / L*	15–25	20–200	20–30	20–60					
Chloride, *mg Cl / L*	<100	<50	<100	<100			<250	<250	<250
Sulfate, *mg* SO_4 */L*	<100	<50		<100			<250	<250	<240
Copper, *mg Cu / L*	<0.1			<0.05			<1.3	<1.0	≤3.0 H
Iron, *mg Fe / L*	<0.1	0.05		0.05			<0.3	<0.3	≤0.2 A
Manganese, *mg Mn / L*	0.05	0.02	0.1	0.02			0.05	0.05	≤0.05
Aluminum, *mg Al / L*	0.1	0.05	0.3	0.1			0.1	0.1	≤0.2 H
Langelier index (LI)					-0.2–0.3		0		0
$\frac{[Cl^-] + [SO_4^{2-}]}{[HCO_3^-]}$, all concentrations in *meq/L*					<1				<1
Cadmium, *mg Cd/ L*	0.005 H	0.005 H	0.005 H	0.001 H	0.2 H	0.005 H	0.005 H	0.01 H	0.005 H
Lead, *mg Pb / L*	0.05 H	0.05 H	0.05 H	0.01 H			0.015 H		0.04 H
Zinc, *mg Zn / L*	1 H	5 H	1	0.3 T		5 H	5 A		5 A

Modified from Vik and Hedberg (1991).

*N = Norway, DK = Denmark, SF = Finland, S = Sweden, NL = the Netherlands, UK = United Kingdom, US = United States, CDN = Canada, G = Germany.

†A = aesthetic, H = health, T = taste.

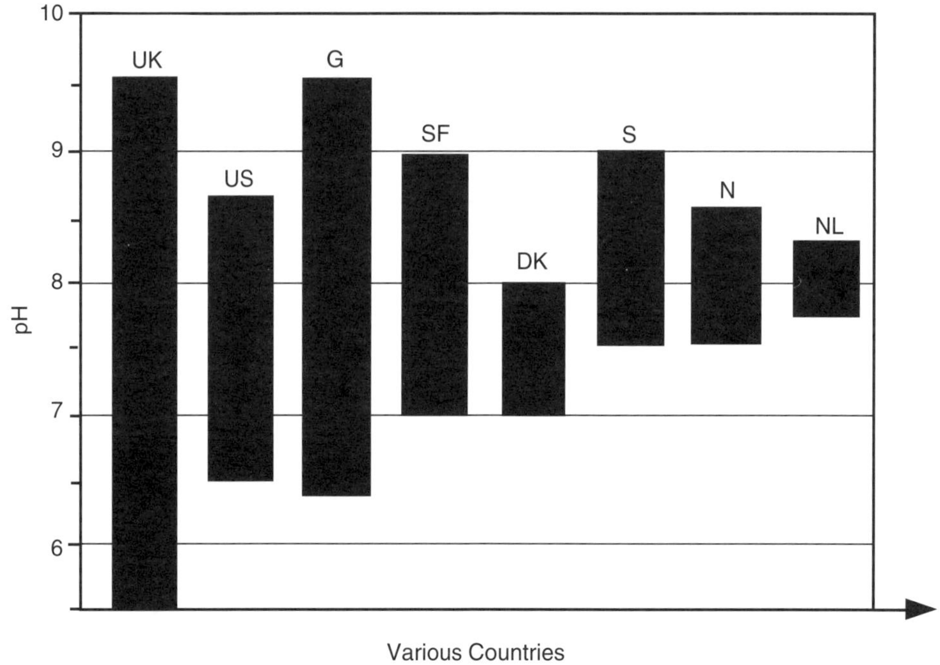

Figure 8-3 Guideline for acceptable pH values in different countries

The different pH-control techniques are discussed and advantages and disadvantages pointed out in the following sections.

Alkalinity and pH control. In a water system that requires both pH and alkalinity control, several options exist. Table 8-9 gives information about the various chemicals used. Table 8-10 gives the effect of various corrosion control methods on water quality, while Table 8-11 lists recommended criteria for choosing methods. Both tables are from German recommendations (DVGW 1992, 1993). The following chemicals and combination of chemicals are most commonly used in water utilities:

- soda ash (Na_2CO_3)
- sodium bicarbonate ($NaHCO_3$)
- sodium hydroxide (NaOH) and carbon dioxide (CO_2)
- lime ($CaO/Ca(OH)_2$) or lime with CO_2
- lime and soda ash ($Ca(OH)_2/CaO + Na_2CO_3$)
- alkaline media filters, sometimes with carbon dioxide ($CaCO_3$, $CaCO_3 \cdot MgO + CO_2$)
- limestone slurry in combination with carbon dioxide and sodium hydroxide ($CaCO_3 + CO_2 + NaOH$)

Table 8-9 Chemicals used for pH or alkalinity adjustment

pH Adjustment Chemical	Alkalinity, *mg $CaCO_3$ /L added by 1 mg/L dose**	Equipment Needed	Chemical Standards to Be Followed
Lime milk ($Ca(OH)_2$)	1.35	Quicklime slaker,† and/or hydrated lime solution tank and feed pump with erosion-resistant lining	AWWA B202
Lime water ($Ca(OH)_2$)	1.35	Additional settling tank	AWWA B202
Caustic soda (NaOH)(50% solution)‡	1.25	Proportioning pump or rotameter	AWWA B501 CEN Proposed Standard 1992c
Soda ash (Na_2CO_3)‡	0.94	Solution tank, proportioning pump, or rotameter	AWWA B201 CEN Proposed Standard 1992c
Sodium bicarbonate ($NaHCO_3$)	0.59	As for soda ash	
Carbon dioxide (CO_2)			CEN Proposed Standard 1992a

Source: Modified from Schock (1990).

*Caustic soda and lime only add hydroxide alkalinity. Soda ash and sodium bicarbonate add carbonate or bicarbonate alkalinity depending on pH.

†If bought as CaO.

‡Limits for Na^+ addition are given, depending on state regulations.

Table 8-10 Effect of various corrosion control methods on drinking water quality

	Water Quality Changes*†					
Corrosion Control Method	pH	m	p	DIC	Ca	Ca+Mg
Physical stripping of free CO_2	+	0	+	–	0	0
Chemical addition of						
$Ca(OH)_2$	+	+	+	0	+	+
NaOH	+	+	+	0	0	0
$NaHCO_3$	+	+	0	+	0	0
Na_2CO_3	+	+	+	+	0	0
Filtration through						
$CaCO_3$	+	+	+	+	+	+
$CaCO_3$•MgO	+	+	+	+	+	+

Source: DVGW (1992a).

*m = total alkalinity = $[HCO_3^-] + 2[CO_3^{2-}] + [OH^-] - [H^+]$

p = carbonate alkalinity = $[CO_3^{2-}] - [CO_2] + [OH^-] - [H^+]$

DIC = dissolved inorganic carbon

Ca = calcium

Ca+Mg = total hardness = calcium + magnesium

†Symbols: + means increase; – means decrease; 0 means no significant change.

Table 8-11 Criteria for choosing corrosion control method to reach wanted pH value and calcium carbonate saturation

	Side Effects		
Objective	Wanted	Unwanted	Corrosion Control Method
Increasing pH value	Increasing Ca^{2+} concentration and total alkalinity		Lime dosing, filtration through $CaCO_3$
Increasing pH value	Increasing Ca^{2+} concentration and total alkalinity	Increasing the Mg^{2+} concentration	Filtration through half-burnt dolomite ($CaCO_3$•MgO)
Increasing pH value	Increasing the total alkalinity	Increasing the Na^+ concentration	Dosage of NaOH $NaHCO_3$ Na_2CO_3
Increasing pH value		Dissolved oxygen increase Possible scaling	Stripping of CO_2
Increasing Ca^{2+} concentration and total alkalinity	Changing the pH value		Addition of CO_2 plus filtration through $CaCO_3$ or with dosage of $Ca(OH)_2$, addition of $Ca(OH)_2$ with Na_2CO_3 or $NaHCO_3$
Increasing Ca^{2+} concentration and total alkalinity	Changing the pH value	Increasing the Mg^{2+} concentration	Addition of CO_2 plus filtration through $CaCO_3$•MgO

Modified from DVGW (1992a).

Soda ash, Na_2CO_3. Sodium carbonate, or soda ash, is considered relatively easy to work with. It dissolves to form sodium and carbonate ions.

$$Na_2CO_3 = 2Na^+ + CO_3^{2-} \tag{8-8}$$

Soda ash is mainly used in powder form, delivered in bulk or in sacks. Soda ash hydrolyzes by absorbing water from air, which can make it cake. Sacks should therefore not be stored longer than six months. The chemical readily dissolves, is considered nonhazardous, and, relatively easy to handle and to dose; it is often used in smaller water facilities. The same type of dosing equipment as is used for NaOH (Figure 8-4) can be used. This includes a solution tank and a proportioning pump.

Soda ash provides an increase in total alkalinity and DIC in addition to a pH increase. A 1-mg/L dosage increases the alkalinity by 0.94 mg $CaCO_3$/L.

Cohen and Myers (1987) reported on a successful water treatment application of soda ash to prevent cold-water copper pitting corrosion failures in residential plumbing systems of private homes and condominiums in the Highland Greens subdivision of Fort Shawnee, Ohio. It was believed that the pitting was facilitated by a low pH and elevated carbon dioxide and by the presence of appreciable amounts of sulfates and chlorides in the water. Sodium bicarbonate was successfully used to increase pH and to eliminate dissolved carbon dioxide, which seemed to support the pitting. Table 8-12 presents qualities of the raw water, which comes from wells, the aerated and blended water at the time of the pitting corrosion failures, and the corrosion-controlled treated water that successfully reduced the incidence of pitting failures. Sodium hydroxide or any pH-adjusting chemicals, such as caustic soda or lime, might have been an acceptable solution to the problem, but Cohen and Myers (1987) selected soda ash to raise pH above 8.3 and to remove excess CO_2. The treatment plant was not continuously staffed, and there was concern that low winter temperatures in an unheated building could lead to freezing or caking of the alternative chemicals, caustic soda, or lime.

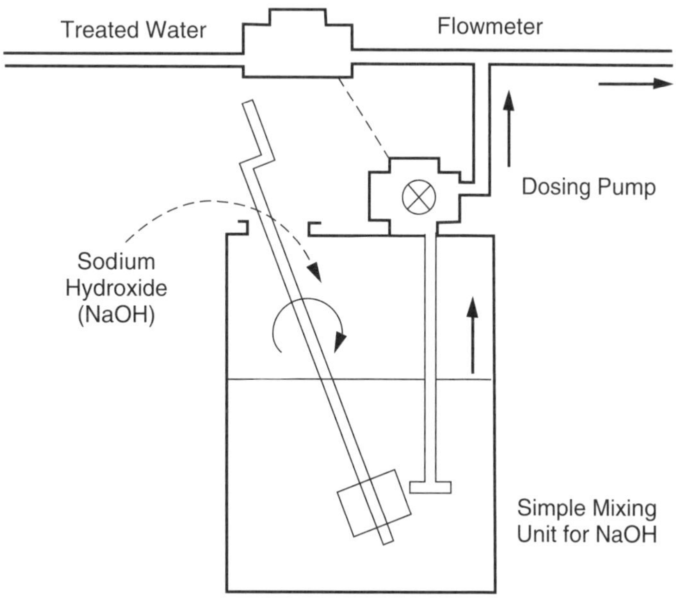

Figure 8-4 Single-step mixing unit used for chemical addition of NaOH

Table 8-12 Representative analyses of raw and treated water from the Highland Greens subdivision, Fort Shawnee, Ohio

Parameter	Raw Well Water	Aerated and Blended Water	Soda Ash Treated Water
pH	7.1	7.3	8.4
Dissolved oxygen, *mg O_2 / L*	0.7	4.4	8.0
Free carbon dioxide, *mg / L*	54	27.5	0
Total alkalinity, *mg $CaCO_3$ / L*	298	296	360
Chlorides, *mg / L*	28	13.5	18
Sulfates, *mg / L*	362	342	333
Calcium, *mg / L*	145	135	124

Source: Cohen and Meyers (1987).

Sodium bicarbonate, $NaHCO_3$ (baking soda). Sodium bicarbonate is a nonhazardous chemical that is easy to work with. The following dissolution reaction takes place:

$$NaHCO_3 = Na^+ + HCO_3^- \tag{8-9}$$

When sodium bicarbonate is used, both pH and alkalinity of the water are increased, but not as much as when the same dosage of soda ash is used. The pH cannot be increased above about 8.3. It is as easy to handle as soda ash. The cost of this chemical per unit of alkalinity increase is usually higher than others.

Sodium hydroxide (NaOH) and carbon dioxide (CO_2). When NaOH is being used, the following chemical reaction takes place:

$$NaOH = Na^+ + OH^- \tag{8-10}$$

Even a small surplus (overdosage) of NaOH (caustic soda) in very soft waters with low CO_2 increases pH to excessively high values. Therefore, strict control of its dosage is necessary. For soft, low-alkalinity water, NaOH is recommended only when used in conjunction with CO_2 or when free CO_2 is present in the water. Then the following reaction takes place:

$$NaOH + CO_2 = Na^+ + HCO_3^- \tag{8-11}$$

NaOH is normally delivered in 50 percent solution, but it can also be delivered as or diluted to 20 percent solutions that have a much lower freezing temperature, or it can be purchased as a crystalline bead chemical. The solution can be delivered in drums or in bulk. The solution is a hazardous chemical. Severe burns can occur to skin or eyes on exposure; this requires special precautions to be taken. Safe handling of tanks, pumps, piping, and dosage equipment must be carefully implemented as specified by building and fire codes.

A simple system for dosing NaOH solution from a dry chemical addition is shown in Figure 8-4. Storage of 50 percent NaOH solutions requires temperatures higher than 15°C, because the freezing point of the 50 percent solution is 12 to 13°C, and the freezing point of a 30 percent solution is still 10°C. The following comments can be made about the use of NaOH:

- easy to dose, but needs precise dosing control
- small capital and investment costs
- requires special safety precautions
- causes no HCO_3^- or CO_3^{2-} increases unless free CO_2 is present in the water or CO_2 can be added to the system
- adds sodium to the water

An example of the results obtained in reducing copper and iron corrosion with the use of NaOH + CO_2 at the Munkfors (Sweden) water facility (Vik and Hedberg 1991) is shown in Table 8-13.

Lime and carbon dioxide. When lime is used, the following reaction takes place:

$$Ca(OH)_2 = Ca^{2+} + 2OH^- \tag{8-12}$$

Overdosage of lime in very soft waters with low free CO_2 will increase the pH above the acceptable range. Addition of lime to soft waters is sometimes accompanied with carbon dioxide, and the following overall reaction takes place:

$$Ca(OH)_2 + 2CO_2 = Ca^{2+} + 2HCO_3^- \tag{8-13}$$

Lime can be used either as hydrated lime ($Ca(OH)_2$) or quicklime (CaO). The use of quicklime requires a lime slaker to produce hydrated lime. CaO is usually used in larger water facilities due to lower chemical costs that override the higher capital equipment and operational costs to maintain clean slaking equipment. Because of its low solubility and high hygroscopy, which increase the risks of caking and blockages, lime is difficult to use for operational and maintenance reasons. Newer dosing equipment is acceptable even from an operational standpoint, but it requires fairly large investments. Also the quality of lime varies, and some types produce fewer lime clinkers than others.

Quicklime must be stored in moisture-proof areas free of CO_2, as quicklime has a great affinity for absorbing water and CO_2 to form calcium carbonate cake. Consequently, quicklime is normally used within a few weeks after manufacture. Dry hydrated lime is less hygroscopic and much easier to store than quicklime, but carbonation can cause deterioration. Lime is shipped in sacks or as a dry bulk

Table 8-13 Typical metal-release changes at consumers' taps after the use of NaOH + CO_2 at Munkfors, Sweden, water utility

Water Quality Parameter	Treated Water Before NaOH + CO_2 Treatment	Treated Water After NaOH + CO_2 Treatment
pH value	<7.7	8–9
Calcium, *mg / L*	3–4	3–4
Alkalinity, *mM / L*	0.15	0.75–0.85
Copper, *mg / L*	0.15–0.29	<0.01
Iron, *mg / L*	0.4–0.6	0.08–0.23

Source: Vik and Hedberg (1991). © Copyright by NACE International. All Rights Reserved by NACE; reprinted with permission.

chemical that is stored in silos. More detailed consideration for handling of lime is given in AWWA Standard B202-83 for Quicklime and Hydrated Lime.

Lime can contain contaminants, such as heavy metals and high concentrations of aluminum. This can be due to the natural quality of the limestone or contaminants incorporated during mining and calcining or due to poor quality control in the calcining process. AWWA Standard B202-83 provides a level of purity necessary to meet drinking water requirements.

Typical compositions of commercial quicklimes are given in Table 8-14, and typical compositions of hydrated limes are given in Table 8-15. Lime can be used as limemilk (~10 percent $Ca(OH)_2$) or limewater (~1 percent or less $Ca(OH)_2$). Two typical limewater-preparation systems—a European and a Japanese technology—are shown in Figures 8-5 and 8-6.

The addition of lime as the final treatment process frequently increases the turbidity of the treated drinking water. Many water utilities requiring large dosages of lime to increase the pH value have therefore hesitated to use lime unless they have an efficient sedimentation tank or saturator to prevent lime particles from going into treated water. This is shown in Figures 8-5 and 8-6. In other cases, deep bed filtration is used after lime addition.

Since lime often contains trace impurities of aluminum, a higher residual aluminum concentration will often follow the final pH adjustment with lime. The water quality regulations in various countries often require that after a chemical treatment process using alum, residual aluminum concentration should not exceed 0.1 or 0.2 mg/L. If large concentrations of lime are needed, the increase in aluminum concentration due to lime addition can be as high as 0.5 mg/L. The contribution of aluminum from lime can be minimized by good turbidity control (Vik et al. 1992).

Table 8-14 Typical compositions of quicklimes

Component	High-Calcium Quicklimes, *range percent of total composition*	Dolomitic Quicklimes, *range percent of total composition*
CaO	93.25–98.00	55.50–57.50
MgO	0.30–2.50	37.60–40.80
SiO_2	0.20–1.50	0.10–1.50
Fe_2O_3	0.10–0.40	0.05–0.40
Al_2O_3	0.10–0.50	0.05–0.50
H_2O	0.10–0.90	0.10–0.90
CO_2	0.40–1.50	0.40–1.50

Source: Parsons (1965). Reprinted with permission of National Lime Association.

Table 8-15 Typical compositions of hydrated limes

Component	High Calcium, *range percent*	Dolomitic, *range percent*
CaO	72–74	46–48
MgO	—	33–34
H_2O	23–24	15–17

Source: Parsons (1965). Reprinted with permission of National Lime Association.

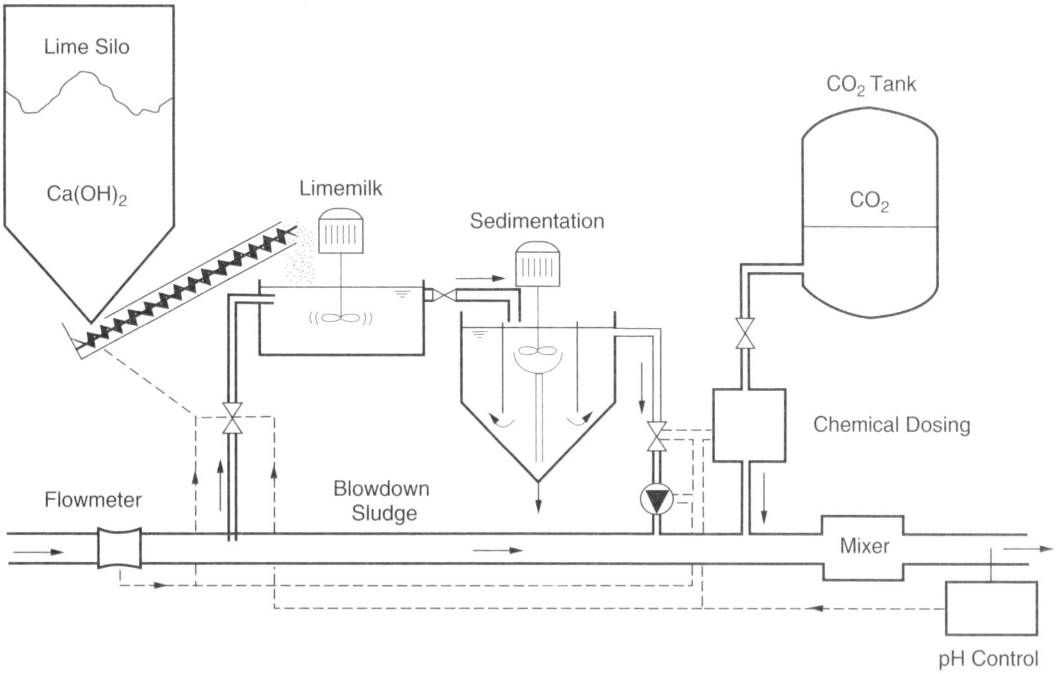

Figure 8-5 Lime-carbon dioxide dosing system used in Europe

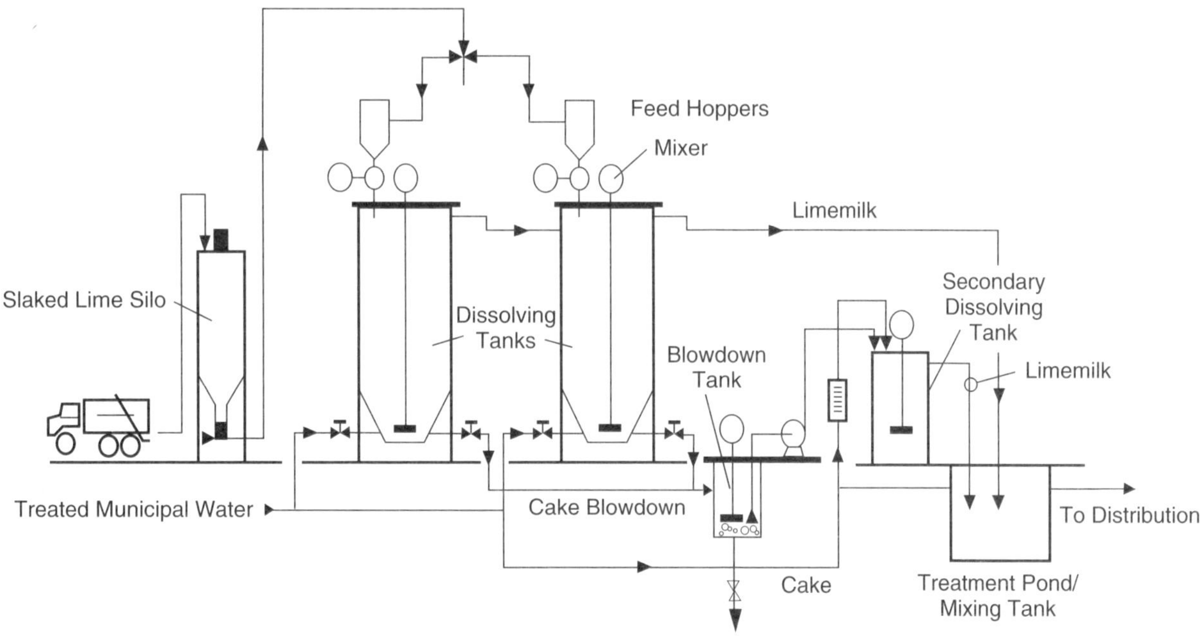

Source: Kojima (1989).

Figure 8-6 Typical continuous-feed lime saturation system used in Japan

The experiences from the use of lime for final pH control in water utilities can be summarized as follows:

- a relatively inexpensive chemical
- increases pH and calcium for soft water
- requires large investments in chemical dosing equipment
- can increase turbidity and aluminum in treated water unless there is good control of particulates
- can cause dust problems if special precautions are not taken
- can clog dosing equipment

Lime and carbon dioxide are used in many conventional water utilities in Finland, Sweden, and Norway having soft, low-alkalinity water. The same treatment is also being used in desalination plants in Saudi Arabia (Leroy 1983; Legrand and Leroy 1990). When lime and carbon dioxide are used, the desired concentrations of bicarbonate, pH, and calcium can easily be controlled by selecting the right sequence and proportion of chemicals.

Leroy (Leroy 1983; Legrand and Leroy 1990) describes several examples from desalination plants in Saudi Arabia. Where water is desalinated by distillation, good protection of cement mortar and iron can be obtained by adjusting the treatment to a final calcium hardness and an alkalinity of not less 60 mg $CaCO_3$/L during an initial period of operation. The water initially must be slightly calcifying with a Langelier index (LI) of 0.3. Leroy states that a protective calcium carbonate layer can be obtained in less than six months.

After the initial period for protective layer formation, the treatment can be reduced to achieve an alkalinity and a calcium hardness of about 30 to 40 mg $CaCO_3$/L. The saturation ratio can then be reduced to about 1.5 (LI – 0.1 to 0.2). Such treatment enables stabilization of the deposit without any increase in scale. An example from the distribution system of Yanbu–Medina in Saudi Arabia is described by Leroy (1983). Leroy emphasizes the importance of adding the carbon dioxide to the system before lime. If lime is added first, lime will not stay in solution long enough to form a protective layer.

Legrand and Leroy (1990) give recommendations for the mode of operation of a lime and carbon dioxide treatment system for naturally soft water. When natural organic substances are present, a higher $CaCO_3$ saturation ratio is needed. French experience is that the ratio of $[Ca^{2+}][CO_3^{2-}]/K_{so}$ should be between 2 and 4 or an LI between 0.3 and 0.6 should be present. If the temperature of the water is lower than 10°C, the time to form an effective $CaCO_3$ protective layer can be up to one year.

Hedberg (1983) presents experiences from the Arvika water utility in Sweden. This treatment plant includes the processes shown in Figure 8-7 and has chemical dosages of 25 mg aluminum sulfate/L, 35 mg Ca(OH)/L, and 32 mg CO_2/L. Water qualities before and after an increase in alkalinity and hardness are presented in Table 8-16. The pH values and iron and copper concentrations were monitored at several points in the distribution system for one year before (1981) and one year after the corrosion control treatment took place. The pH values in the distribution system were more stable than before, within ±0.5 pH units as compared to >1 before treatment. The iron concentrations were slowly reduced, but still showed occasional high values. The copper concentrations were reduced to below 0.1 mg/L, but immediately increased if the pH was reduced by 0.5 pH units. The finished

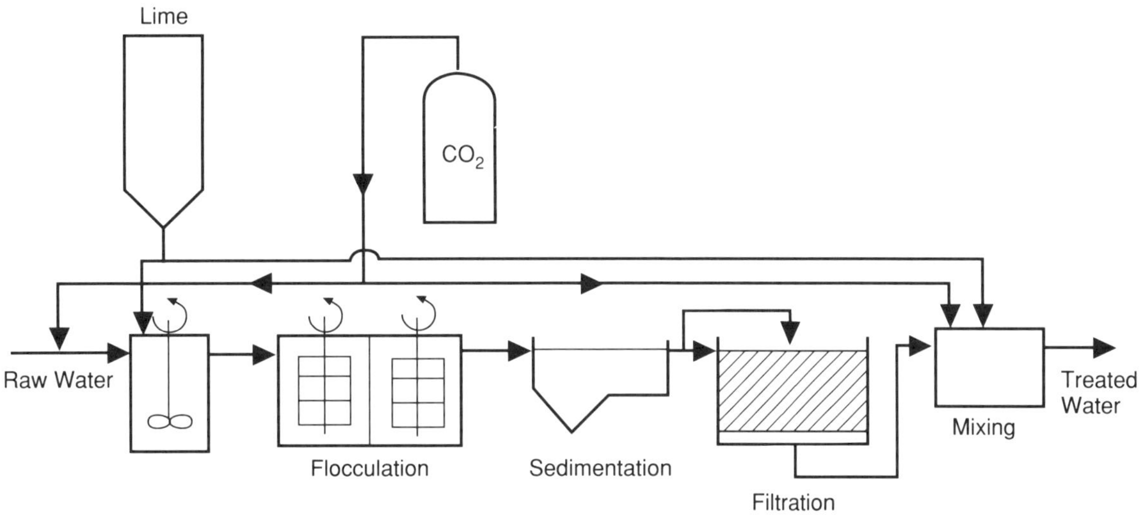

Figure 8-7 Corrosion control with lime and carbon dioxide included in a conventional water treatment plant

Table 8-16 Water quality data from Arvika, Sweden, water utility

		Finished Water	
	Raw Water	Before Hardness Increase	After Hardness Increase
Alkalinity, *meq / L*	0.07–0.10	0.07–0.10	0.6–0.8
pH	6.3–6.5	7.5–8.5	7.8–8.5
Hardness, *mg Ca / L*	2–4	2–4	18–25
Color, *mg Pt / L*	20	<5	<5
Chloride, *mg Cl / L*	5	5	5
Sulfate, *mg SO_4 / L*	9	23	23

Source: Vik and Hedberg (1991).

water quality in Arvika is, however, lower in calcium and alkalinity than recommended by Legrand and Leroy (1990), Baldauf and Henkel (1991), and Klein (1992) as well as by the German Gas and Waterworks Association (DVGW 1992, 1993).

Lime and soda ash. The Seattle Corrosion Study (Ryder 1978b) is an example of the effective use of lime and soda ash for pH, alkalinity, and hardness control to reduce the corrosivity of a soft, low-mineralized water. In Seattle, Washington, lime and soda ash were used to increase pH from ~6.0 to ~8.0, with an increase in alkalinity from ~5 to ~17 mg $CaCO_3$/L. Seattle's program has reduced the overall rate of corrosion on aged copper surfaces by 70 percent as well as reduced the release of copper, lead, zinc, and iron in standing water samples from homes. The results from an 18-month monitoring program evaluating the relationship between copper plumbing corrosion and variations in delivered water quality in several communities in the Pacific Northwest have been presented by Reiber, Ferguson, and Benjamin (1987). Significant relationships were found for the copper corrosion rate, which decreased as the pH increased, but increased with the free chlorine residual.

Alkaline media filters ($CaCO_3$, $CaCO_3 \cdot MgO + CO_2$). A limestone reactor or filter is a bed or layer of granular limestone through which water passes. It has principally been used in smaller water treatment systems to increase the hardness, pH, and alkalinity of soft drinking water. The process simulates the natural processes that occur when water passes through limestone or other alkaline minerals. The process is well suited to smaller water systems because it eliminates the need to feed chemicals in liquid or powder form and cannot result in overdosing. There are numerous forms and types of limestone, varying in chemical composition, mineralogy, crystallinity, color, texture, and hardness. Next to sand and gravel, limestone is the second greatest tonnage material produced in the United States. The two fundamental types of limestone that are used in alkaline contactors are high-calcium and dolomitic limestone. Pure high-calcium limestone is 100 percent calcium limestone (calcite or aragonite). Pure dolomite is ~54 percent $CaCO_3$ and ~46 percent $MgCO_3$. High-quality, high-calcium limestone is 97 to 99 percent $CaCO_3$ (54 to 56 percent CaO). Some examples of limestone composition are given in Table 8-17.

Chemically, high-calcium limestone is predominantly $CaCO_3$ and is divided into the following forms: limestone, marble, and chalk. Dolomite limestone is $MgCO_3 \cdot CaCO_3$ and half-burnt dolomite $MgO \cdot CaCO_3$.

Alkaline media filters are mainly used in small water utilities. In many areas of North America and in sparsely populated areas of Europe with low-alkalinity surface waters or groundwaters, which may be affected by acid precipitation, alkaline media filters are an important corrosion control technique for individual homeowners and small public and private water users. It is also an important corrosion control procedure on offshore production platforms in the North Sea and in other areas, where desalination using evaporation or RO produces potable water. Alkaline media filters are used to improve taste and to control corrosion.

Table 8-17 Composition of different types of limestone

	Various Types of Limestones,* *percent composition*										
Parameter	No. 1	No. 2	No. 3	No. 4	No. 5	No. 6	No. 7	No. 8	No. 9	No. 10	No. 11
CaO	54.5	28.9	41.8	31.2	29.5	45.7	55.3	52.5	29.4	40.1	40.4
MgO	0.6	2.7	1.9	20.5	21.1	7.1	0.5	0.6	21	25.8	26.0
CO_2	42.9	33.1	32.9	47.9	46.9	43.6	43.7	41.9	—	33.2	32.6
SiO_2	0.7	19.8	13.4	0.11	0.14	2.6	0.42	2.4	2.9	0.3	0.3
Al_2O_3	0.7	5.4	5.6	0.3	0.04	0.23	0.13	1.6	0.3	n.a.	n.a.
Fe_2O_3	0.08	1.6	0.56	0.19	0.1	0.2	0.05	0.56	0.4		
P_2O_5	—	—	0.22	—	0.05	0.04	—	—	0.01		
SO_3	0.31	—	0.33	—	—	0.33	0.01	—			
Na_2O	0.16	—	0.31	0.06	0.01	0.04	—	—			
K_2O	—	—	0.72	—	0.01	0.03	—	—			
H_2O	—	—	0.16	—	0.16	0.23	—	n.d.			
Other	—	—	0.29	—	0.01	0.06	0.08	0.2	0.13		

Source: Boynton (1980) and Hedberg (1983).

*No. 1 = Indiana high-calcium stone; No. 2 = Lehigh Valley, Pennsylvania, "cement rock"; No. 3 = Pennsylvania "cement rock"; No. 4 = Illinois Niagaran dolomitic stone; No. 5 = Northwestern Ohio Niagaran dolomitic stone; No. 6 = New York magnesium stone; No. 7 = Virginia high-calcium stone; No. 8 = Kansas Cretaceous high-calcium stone (chalk); No. 9 = Dolomite, Glanshammar, Sweden; No. 10 = Akdolit-Gran; No. 11 = Magno dol.

n.a. = not available

n.d. = not determined

Low cost and simple operation are important criteria in establishing the feasibility of a corrosion control technique for small water-supply systems. Alkaline media filters do, in most cases, meet these criteria. The chemistry of the water is altered as the limestone dissolves. The chemical reaction taking place when calcium–limestone is used is illustrated in the following equation:

$$CaCO_{3(s)} + CO_2 + H_2O = Ca^{2+} + 2HCO_3^- \tag{8-14}$$

Dolomitic materials dissolve slowly. The half-burnt media is more frequently used. When water passes through half-burnt dolomite materials, the following reaction takes place:

$$MgO \cdot CaCO_3 + 3CO_2 + 2H_2O = Ca^{2+} + Mg^{2+} + 4HCO_3^- \tag{8-15}$$

Experiments on the kinetics of calcium carbonate (calcite) dissolution, done by Rickard and Sjöberg (1983), have shown that the $CaCO_3$ dissolution rate in slightly acid to alkaline solutions is controlled by an interfacial calcium-ion mass-transfer resistance and a first-order surface reaction acting in series. Letterman and co-workers (Letterman et al. 1987; Letterman, Haddad, and Driscoll 1991; Letterman and Kothari 1992) used this information and experimental results from laboratory-scale experiments, pilot-scale experiments, and full-scale small water treatment plants to develop design criteria with nomographs and a computer program for designing limestone contactors.

To evaluate the usefulness of an alkaline media filter using German guidelines (DVGW 1993), the following water quality parameters should be known:

- alkalinity, to pH = 4.3
- acidity, to pH = 8.2
- calcium concentration
- iron concentration
- manganese concentration
- aluminum concentration

Table 8-18 lists criteria for evaluating the applicability of alkaline media filters, and Table 8-19 lists physical and chemical characteristics of the recommended alkaline media.

The German Gas and Waterworks Association (DVGW 1993) has made recommendations for design of alkaline media filters. The recommended minimum volume of alkaline material is:

$$V_M = Q \times t_F \times f \times 1/60 \tag{8-16}$$

Where:

V_M = minimum volume of alkaline media, m^3
Q = filtration rate, m^3/h
t_F = theoretical contact time, min (see Figure 8-8)
f = temperature correction factor, $f = 0.5$ at 15°C, 1 at 10°C, 1.5 at 5°C

Table 8-18 Recommended water quality criteria for using alkaline media filters

Water Quality Parameter	Comments	$CaCO_3$ Filter*	$CaCO_3 \cdot MgO$ Filter*
Fe	Will produce excessive sludge in the filter.	≤ 0.5 mg/L	≤ 0.5 mg/L
Mn	Will produce excessive sludge in the filter.	≤ 0.1 mg/L	≤ 0.1 mg/L
Al	Will produce excessive sludge in the filter.	≤ 0.2 mg/L	≤ 0.2 mg/L
Treated water pH = 8.0		$ANC_{4.3} + 2BNC_{8.2} \leq 1.5$ m*M*/L $[Ca^{2+}] \leq 0.75$ m*M*/L	$ANC_{4.3} + 2BNC_{8.2} \leq 1.5$ m*M*/L $[Ca^{2+}] \leq 0.75$ m*M*/L
Treated water pH = pH_s (at $CaCO_3$ equilibrium)		$ANC_{4.3} + 2BNC_{8.2} \leq 1.5$ m*M*/L $[Ca^{2+}] \leq 0.75$ m*M*/L	$ANC_{4.3} + 2BNC_{8.2} \leq 2.5$ m*M*/L

Source: Data from DVGW (1993).

*ANC = acid neutralizing capacity, alkalinity.
BNC = base neutralizing capacity, acidity.

Table 8-19 Physical and chemical characteristics of typical alkaline media

	$CaCO_3$		
Parameters	Dense $CaCO_3$*	Porous $CaCO_3$	$CaCO_3 \cdot MgO$ Half-Burnt Dolomite
Media size, *mm*	0.71–1.25 1.0–2.0 1.6–2.5	1.0–3.0	0.5–1.2 0.5–2.5 2.0–4.5 4.0–7.0
Color	Gray-white-yellow	Gray	White-gray-brownish
Density, g / cm^3	2.700	2.400	2.750
Media shape	Angular	Angular-round	Angular-round
Composition, *g / kg DS*†	~990	~940	~700
$CaCO_3$, *g / kg DS*	(>980)	(>850)	
$MgCO_3$, *g / kg DS*	2–4	50	
Free MgO, *g / kg DS*	—	—	270 (>230)
Free CaO, *g / kg DS*	—	—	10
SiO_2, g/kg DS	—	—	20

Source: Data from DVGW (1993).

*Available as Devonite and Jura lime.

†DS = dry solids.

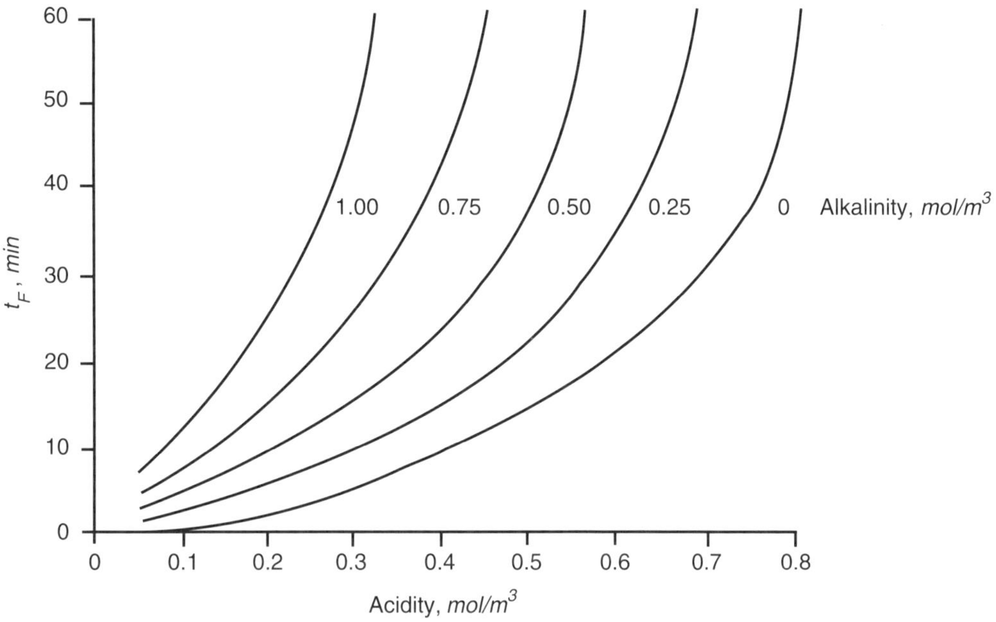

Source: DVGW (1993). Reprinted with permission of DVGW.

Figure 8-8 Dependency of contact time t_F on alkalinity and acidity of a water in a $CaCO_3$ filter (dense $CaCO_3$, granular size 1-2 mm) to achieve an effluent pH of 8 at 10°C (50°F)

The filter volume, V_F , can be calculated as$V_F = V_M + V_B$, where V_B = volume of alkaline media used. The volume of media used depends on operational aspects such as the maximum allowable time before new medium is used.

The filter volume depends on the difference in $CaCO_3$ dissolution capacity of the raw and treated water. If the change in base capacity (base neutralizing capacity [$BNC_{8.2}$]) is 1 m*M*/L, then 100 g $CaCO_3$/L will be used, and one should assume that an additional 10 percent of the limestone is lost during backwashing. DVGW (1992b) recommends the following backwashing steps when clogging reduces the flow capacity:

1. 5 to 10 min air, v_a ~60 m/h
2. 5 to 10 min air–water mixture, v_a = 60 m/h, v_w = 10 m/h
3. backwashing until clean water is obtained, v_w = 25 m/h

To increase the rate of $CaCO_3$ dissolution and/or to stabilize the pH from $MgO \cdot CaCO_3$ media filters, carbon dioxide gas is frequently used (Figure 8-9). This combination has given very good results for low-pH and low-alkalinity waters that have minimal free CO_2. Addition of CO_2 is often necessary to prevent excessive pH values in low-alkalinity water treated with half-burnt dolomite filters.

Table 8-20 gives examples of typical water quality changes obtained in alkaline media filters and observed changes in corrosion rates or metal release.

Limestone slurry ($CaCO_3$ + CO_2 + NaOH). Limestone slurry has been used successfully for corrosion control in Norwegian water utilities (Eikebrokk, Østerhus, and Brattebø 1990). A sketch of a limestone slurry system is shown in Figure 8-10. The slurry is prepared by the chemical supplier in very fine particles. The physical and chemical characteristics of a slurry produced by a Norwegian supplier, Hustadmarmor A/S, is given below. The particles' sizes are 0.2 to 3 µm, and impurities are <1 percent

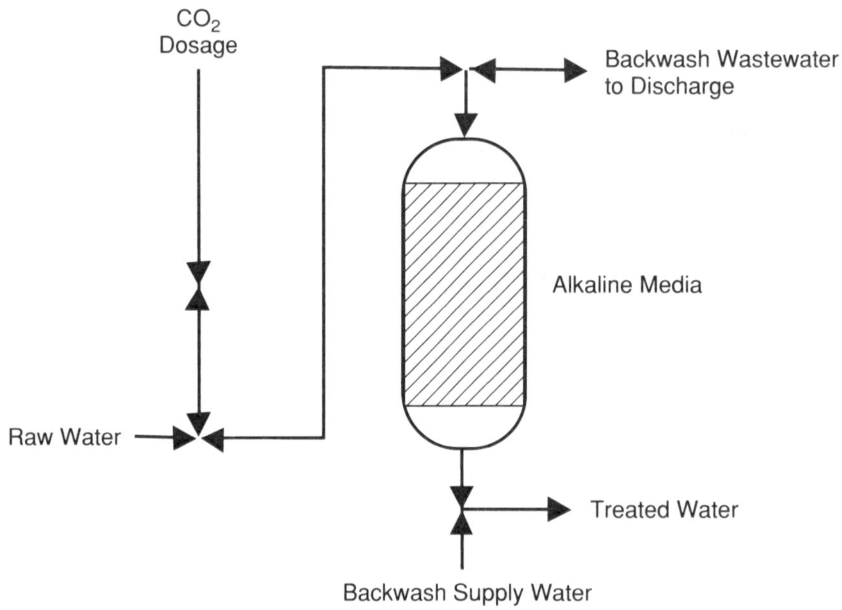

Figure 8-9 Alkaline media-CO_2 treatment system for water with free CO_2

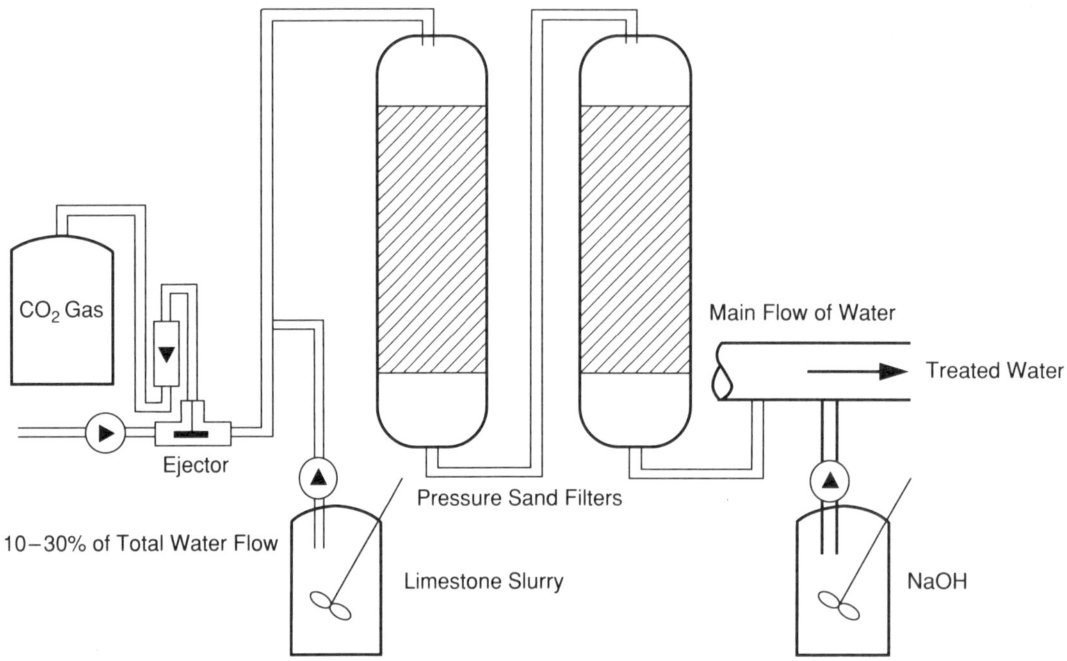

Source: Eikebrokk et al. (1990). Reprinted with permission of Norwegian Water Works Association.

Figure 8-10 Schematic of a limestone slurry system used for corrosion control of soft, low-alkalinity drinking water in Norway

Table 8-20 Typical water quality changes from calcium carbonate water treatment and the effects on corrosion

Case Studies With Water Quality Parameters and Treatment Process	Raw Water	Untreated Tap Water	Treated Water	Treated Tap Water
Mandal Water Utility, Norway*				
Calcium, *mg / L*				
Alkalinity, *mM / L*	1.5	1.5	14.0	14.0
pH	0.02	0.03	0.7	0.7
Iron, *mg / L*	5.0	5.2	7.2	8.0
	0.05	0.6	0.05	0.1–0.2
Treatment process, open limestone filters				
Bay Side Cottage, New York†				
Calcium, *mg / L*	4.1	4.1	11.3	11.3
Alkalinity, *mM / L*	0.15	0.15	0.57	0.57
pH	6.3	6.3	7.3	7.3
Copper, *mg / L*, standing water	<0.0005	1.9		0.5
Lead, *mg / L*, standing water	0.003	0.033		0.015
Treatment process, limestone filters				
Malvik Water Utility, Norway‡				
Calcium, *mg / L*	7.5–8.5		17.5–18.5	
Alkalinity, *mM / L*	0.3–0.45		1.75–1.9	
pH	6.9–7.1		7.5–7.7	
TOC, *mg C / L*	6–7		6–7	
Corrosion rate, *μm / year*				
Copper	6–6.5		~4	
Ductile iron	~90		45–70	
Steel	~63		50–55	
Treatment process, limestone slurry plus CO_2				

**Skagen (1992).*

†Letterman et al. (1987, 1991).

‡Eikebrokk, Østerhus, and Brattebø (1990).

$MgCO_3$, <0.2 percent SiO_2, <0.02 percent Fe_2O_3, and <0.05 percent Al_2O_3. To keep a high dissolution rate and low particle loss to the treated water, high CO_2 pressure, low pH, high concentrations of $CaCO_3$ particles, small particle size, and a filter system to capture undissolved particles are favored. Final pH adjustment is needed to achieve the desired pH value of 8.0 to 8.5 by adding caustic soda.

Chemical Dosages for pH and Alkalinity Increases

Several computer programs and diagrams have been developed to help compute dosages of treatment chemicals to achieve desired pH, alkalinity, and calcium carbonate saturation levels. Figure 8-11 shows the Deffeyes (1965) diagram, which includes an explanation of how it is used for addition or removal of CO_2; for addition of $CaCO_3$, $NaHCO_3$, and Na_2CO_3; for dilution; and for adding an acid (C_A) or a base (C_B). This diagram has C_T = total carbonate carbon = $[HCO_3^-] + [CO_3^{2-}] + [H_2CO_3]$ on the x-axis and total alkalinity = $[HCO_3^-] + 2[CO_3^{2-}] + [OH^-] - [H^+]$ on the y-axis. It does not take into account $CaCO_3$ solubility equilibria.

The use of the diagram to compute chemicals for alkalinity increase is illustrated in the following example calculation.

SAMPLE WORK PROBLEM

A water identified as A (marked in Figure 8-11) has the following composition:

Alkalinity (total)	=	0.5 meq/L
pH	=	6.6
Calcium	=	5 mg/L

The water utility wants to increase the alkalinity to >1 meq/L and the pH to ~8.0 (B in the diagram) and wants suggestions on which chemicals are suitable. The following chemicals can be used:

Alternative 1. CO_2 can be added to C_T = 1 mM/L and NaOH or $Ca(OH)_2$ can be added to pH = 8.0. When NaOH is used, this will give an increase in sodium concentration. When $Ca(OH)_2$ is used, an increase in calcium concentration occurs.

With $Ca(OH)_2 + CO_2$:

$Ca(OH)_2$ dosage: 0.5 meq/L • 37 mg/meq = 18.5 mg $Ca(OH)_2$/L
CO_2 dosage: 0.3 mM/L • 44 = 13.2 mg CO_2/L
Calcium concentration in water: (5 + 10) mg/L = 15 mg/L

Alternative 2. $CaCO_3$ or Na_2CO_3 can be used. If 0.5 mM/LC_T is added, the pH value will increase to approximately 8.5. $CaCO_3$ will require very high detention time in the contactors to produce a pH of 8.5, because it dissolves slowly. Note that calcium carbonate solubility is not addressed by the Deffeyes diagram. The required dosages for the two chemicals are

using $CaCO_3$ = 28 mg/L: increase in calcium concentration = 11 mg/L
using Na_2CO_3 = 30 mg/L: increase in sodium concentration = 13 mg/L

In a number of European countries (for example, Germany, Denmark, and some parts of France), the authorities require that drinking water should be calcifying to protect against corrosion. In these countries, the LI, or the difference in pH values between $CaCO_3$ equilibrium and the actual water quality, is regularly calculated. In France, for example, Legrand and his associates have developed a computer program that is being used widely (Legrand and Leroy 1984, 1990).

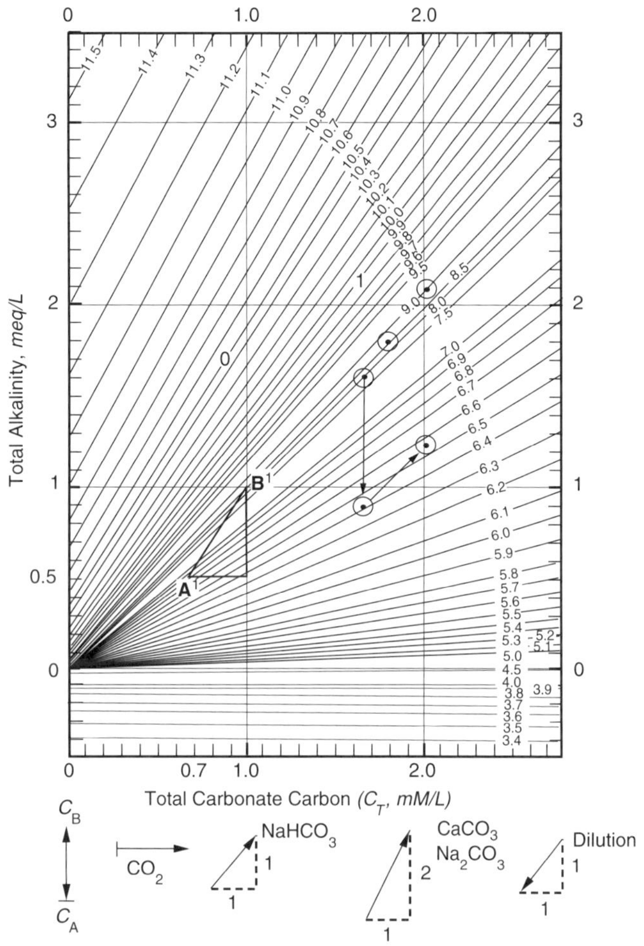

NOTE: The point defining solution composition moves as a vector in the diagram as a result of the addition (or removal) of CO_2, $NaHCO_3$ (Na_2CO_3), or C_B and C_A.

Figure 8-11 Diagram of pH contours in alkalinity versus C_T (carbonate carbon)

An example of this model as it is being used in France is given in Figure 8-12, which illustrates $CaCO_3$ saturation and the importance of adding CO_2 before lime when these chemicals are chosen. The figure uses calcium in millimoles per litre on the x-axis and total inorganic carbon in millimoles per litre on the y-axis. A desalinated water contains no calcium and no inorganic carbon and is illustrated as M0 in the figure. This figure is constructed for a specific water with a given concentration of ions and temperature. The computer program will construct a curve for each water of interest. The line C is the $CaCO_3$ equilibrium line for the water presented in this figure. For all points to the right of C, $CaCO_3$ is supersaturated. The curve C′ is the spontaneous precipitation curve defined as $[Ca^{2+}][CO_3^{2-}] = 40 \cdot K_{so}$, where K_{so} = solubility constant for $CaCO_3$. Addition of M1 mM CO_2/L, followed by M2 mM Ca/L (lime), produces a water just saturated, while addition of lime then CO_2 would exceed the spontaneous precipitation curve and cause precipitation during chemical dosing.

Other widely used approaches to incorporating calcium carbonate solubility into chemical addition calculations have been developed by Caldwell and Lawrence (1953), Loewenthal and Marais (1974), and by AWWA (1978).

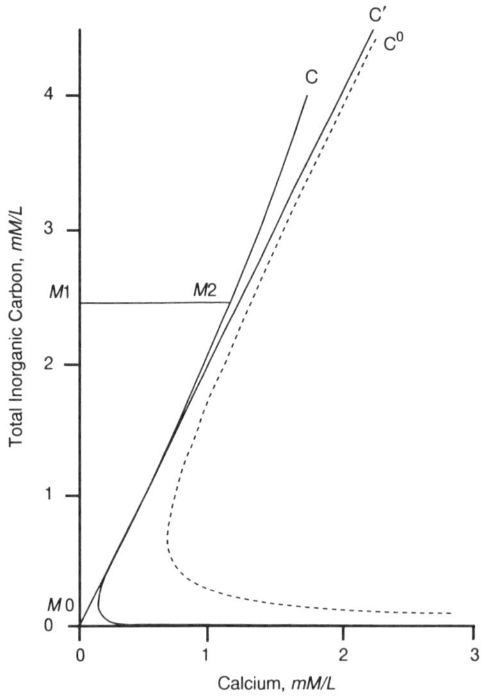

NOTE: M0 is the representative point for desalinated water, M1 is the water after CO_2 addition, and M2 is the water after lime injection. *C* is the equilibrium curve and *C′* is the spontaneous precipitation curve ($[Ca^{2+}]\,[CO_3^{2-}] = 40K'_s$.

Source: After Legrand and Leroy (1984 and 1990).

Figure 8-12 Evolution of water during lime and CO_2 addition, viewed in Legrand and Poirier graph

Moderate- and High-Alkalinity Water

Corrosion caused by a low-pH water and/or a high free CO_2 concentration may be effectively curtailed in many potable water systems by addition of an alkaline chemical for simple neutralization of pH. This can be done with several different chemicals. If pH and aggressive CO_2 are the only parameters to control, the following methods can be used:

- aeration for stripping of CO_2
- addition of NaOH or Na_2CO_3
- addition of lime

Methods for chemical additions for pH adjustment have been described earlier. A major consideration is not exceeding calcium carbonate saturation to avoid excessive scaling in the distribution system or turbidity in the water.

Aeration for stripping of CO_2. In waters with high carbon dioxide concentrations, aeration for stripping carbon dioxide reduces the ratio of $[H_2CO_3]/[HCO_3^-]$, which results in an increased pH. For groundwater and spring waters, contact with air in the water tanks may remove enough CO_2 to produce scaling.

Despite an appreciable increase in dissolved oxygen, coupon tests have indicated in some instances a significant reduction in corrosivity (Mullen, Johansen, and Ubdelraman 1989). Typical water quality changes at various water utilities where this corrosion control approach has been used are presented in Table 8-21.

Table 8-21 The effect of CO_2 stripping of different water qualities

Water Quality	pH	Alkalinity, m*M*/*L*	Free CO_2, *mg*/*L*	O_2 Concentration, *mg*/*L*	Corrosion Rate,* *mil*/*yr*
Gaupne Water Utility, Norway†					
Raw water	6.0–6.4	~0.6	61.6	3.5	—
Treated water	7.7	~0.6	2.0	—	
Fort Lewis, Washington‡					
Raw water	6.2	0.83	25	2.0	—
Treated water	7.0–7.8	0.83	6.3	6.0	
Charles Street Plant, Elizabethtown, New Jersey§					
Raw water	7.1	1.7	26	4.9	—
Treated water	8.0	1.6	4	11.7	
Netherwood Plant, Elizabethtown, New Jersey§					
Raw water	7.3	1.9	26	10.4	7.9
Treated water	8.3	1.9	<1	12.1	2.4

*LPR, Steel.
†Hem (1993).
‡Ballantyne (1986).
§Mullen et al. (1989).
— indicates information not available.

Stripping can be performed with a number of different stripping technologies. Typical residual CO_2 concentration can be reduced to ≤5 mg CO_2/L, depending on technology and raw water quality.

CORROSION INHIBITORS

As described previously in this chapter, one of the main options of corrosion control is pH adjustment. However, in many instances pH control or pH and alkalinity adjustment are not sufficient to suppress corrosion problems. This is because of the complex impact of water quality on different materials and stringent limits concerning lead and copper or other materials in drinking water (Wagner 1989). To overcome corrosion problems that cannot be resolved by pH control, water systems and private households often require addition of corrosion inhibitors, often along with pH control.

Natural Inhibitors

Natural inhibitors provide corrosion protection in many surface water systems. Natural inhibitors identified in different locales and with different materials have included natural organic matter (NOM), or humic and fulvic materials, dissolved silica, and phosphate. The materials that have received some degree of corrosion protection from natural inhibitors include iron, zinc coatings, lead, copper, and asbestos cement or cement mortars.

In the past, drinking water from the Lake of Constance distributed in the Stuttgart, Germany, area caused little corrosion with respect to all materials. On the other hand, some serious corrosion and red-water problems, especially in galvanized steel pipes, were observed with most groundwaters in the Stuttgart area. Neither pH, alkalinity, saturation index, buffer intensity, or neutral salt concentration adequately explained the difference in corrosivity of the water sources.

During intensive research on these problems (Werner, Wurster, and Sontheimer 1973, 1976; Werner 1976), it was found that the NOM (humic and fulvic acids) in water from the Lake of Constance was the major factor in the low corrosivity. Figure 8-13, for example, shows how the corrosion rate of zinc depends on the humic acid (from Lake of Constance) concentration.

Studies with humic substances and iron have shown that the molecular weight and the polarity (acidity) are important parameters. Both change the adsorbability of the organics on the metal and on the corrosion product layer. They delay carbonate precipitation and retard oxidation to the ferric state (Rudek 1975, 1979; EBI 1980). Both effects result in a more protective, denser layer of corrosion products, which results in reduced corrosion. Organic matter also has significant influence on unprotected steel or unlined cast- or ductile-iron pipes and on copper (Edwards and Ferguson 1993).

Red-water problems often occur in waters with a low concentration of organic matter. Removal or modification of organic matter during drinking water treatment is a treatment objective for color removal or control of disinfection by-products and occurs coincidentally with turbidity removal or in use of disinfectants. In these cases, addition of a corrosion inhibitor may be necessary. Further research is needed to determine which treatment processes may retain the effects of natural organic inhibitors without causing other problems.

An important process affecting NOM is ozonation. While ozonation increases the acidic groups in the humic and fulvic acid molecules, it may also reduce the molecular weight and aromaticity of the substances, which leads to lower adsorbability and also increases complexation with metal ions. Therefore, implementation of ozonation or other oxidative treatments should include evaluation of changes in corrosion protection and additional corrosion mitigation, if necessary.

Drinking Water Inhibitors

In contrast to the chemicals that are added to water for pH control, corrosion inhibitors are chemicals that are dosed in small quantities to obtain a passivating film to block the electrochemical corrosion reactions and, thus, provide a significant reduction in corrosion, assessed as corrosion rate or metal release or both.

To avoid negative effects on drinking water quality, the inhibitor composition and dosage are quite limited by toxicity and purity regulations. For example, in the United States, specific types of phosphates, including zinc phosphates and different silicates, are tested and certified by NSF International.* In Germany, the regulations are restricted to allowing only "food-grade" sodium or potassium phosphates and silicates at concentrations not higher than 5 mg phosphates/L as P_2O_5 and 40 mg silicates/L as SiO_2. Similar restrictions, even more stringent, exist in other European countries. Because of toxicity

*NSF International, National Sanitation Foundation, P.O. Box 130140, Ann Arbor, Michigan 48113-0140.

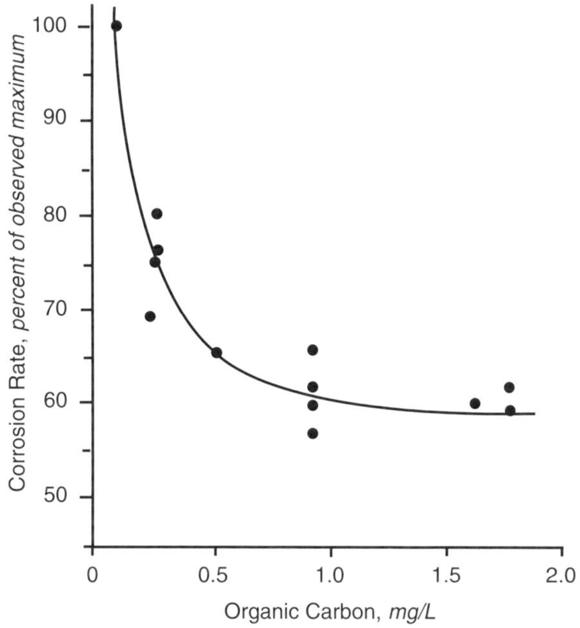

Figure 8-13 Effect of adding humic acids from Lake of Constance to a model water on the corrosion of zinc

regulations, only phosphate and silicate inhibitors, and blends thereof, may be used.

Orthophosphate and silicate chemicals form passivating films on anodic sites to suppress the electrochemical corrosion reactions. There may be zinc, calcium, or iron deposition that further enhances the suppression of corrosion. A discussion of the various types of corrosion inhibitors commonly used in potable water systems follows.

Phosphates. There are three principal types of phosphates used as corrosion control inhibitors: orthophosphates, molecularly dehydrated polyphosphates, and bimetallic (zinc-containing) phosphates.

Orthophosphates. Typical orthophosphates are phosphoric acid (H_3PO_4), monobasic sodium phosphate (NaH_2PO_4), dibasic sodium phosphate (Na_2HPO_4), and tribasic sodium phosphate (Na_3PO_4).

The sodium orthophosphate compounds, also called crystalline phosphates, are now extensively used in North America and Europe as individual products and in proprietary blends, or as a result of the hydration and reversion of polyphosphates. There are now AWWA standards for the purchase of monobasic and dibasic sodium phosphate (AWWA Standard B504-88; AWWA Standard B505-88). They can be purchased in the United States from chemical suppliers that specify these products either as food-grade chemicals or NSF approved.

Molecularly dehydrated polyphosphates. Typical compounds in this group are sodium tripolyphosphate ($Na_5P_3O_{10}$), sodium hexametaphosphate (polyphosphate mixture) $(NaPO_3)_6$, and other sodium and potassium polyphosphates.

There are AWWA standards for purchase of sodium tripolyphosphate and sodium hexametaphosphate (AWWA Standard B503-89). These are all formed from the combination of soda ash or caustic soda or potassium hydroxide with phosphoric acid. Many different formulations of polyphosphates are available; their differences stem from the heating and crystallization conditions selected by the manufacturers.

Sodium tripolyphosphate ($Na_5P_3O_{10}$) contains Na_2O and P_2O_5 in a molar ratio of approximately 1.67:1, respectively. An amorphous mixture of long-chain polyphosphates is formed by differing crystallization conditions. The polyphosphates have a form of $Na_{n+2}P_nO_{3n+1}$. The average values of n are between 10 and 16: typically $Na_{12}P_{10}O_{31}$. These are called glassy phosphates and have no definite molecular structure. Varying the ratio of sodium oxide and P_2O_5 produces species with different chain lengths and physicochemical properties, although the inhibitive action of these chemicals is not thought to be significantly altered (Boffardi 1984). Polyphosphates have a record of successful corrosion suppression in some waters, but in others they can increase corrosivity. The primary use of the dehydrated phosphates has been to sequester metals and prevent red water or suppress calcium carbonate scales. Testing conducted by the Illinois State Water Survey (ISWS) found an acceleration of copper and lead corrosion by polyphosphates (Neff, Schock, and Marden 1987). If there is effective metal corrosion protection, it may be due to reversion to orthophosphate.

The pH of the water has been found to have an effect on the protective properties of phosphates. Orthophosphates are effective at slightly alkaline conditions. The phosphates are relatively insensitive to dissolved solids concentrations; however, a sufficient, protective phosphate dosage must be used since insufficient dosage can accelerate corrosion.

The effectiveness of polyphosphates in drinking water has been observed, but studies have seldom demonstrated conclusively how and why the inhibitors work. The polyphosphates have been used as inhibitors of galvanic corrosion attack between two dissimilar metals. Divalent ions, calcium in particular, seem to be needed with the polyphosphates for effective inhibition of steel. The ratio of calcium ion concentration to polyphosphate concentration should be at least 0.2:1 and preferably 0.5:1 or more. Formation of many different minerals in deposits or protective scales is possible during the corrosion of iron, and undoubtedly is affected by the inhibitor formulations.

Polyphosphates prevent formation of slightly soluble scales of calcium carbonate and calcium sulfate and sequester and stabilize iron and manganese to prevent red water. The reduction in scale formation can be attributed to the adsorption of the polyphosphate on crystal faces, thus arresting the growth of the crystal.

The glassy phosphates were initially used as sequestering agents to minimize scaling. Later, it was found that they also could minimize red water from natural iron or corroding pipe. These glassy phosphates first found use as corrosion inhibitors about 1940 (Illig 1957). Their continued use for corrosion protection should be evaluated since in many instances a cosmetic suppression of discoloration occurs, while the real corrosion rate is aggravated. However, their use in blends with orthophosphate to penetrate corrosion scales to aid in orthophosphate filming on the metal surface can be beneficial.

In the United States (Hatch 1973), high polyphosphate dosages were recommended after the cleaning of iron pipelines. In these cases, corrosion control primarily meant masking the iron in the water so that it was not visible to the consumer.

Bimetallic (zinc-containing) phosphates. The bimetallic phosphates usually combine zinc in concentrations of 10 to 30 percent zinc with ortho- or polyphosphates. They were developed and first used as corrosion inhibitors about 1950 (Kleber 1965). These are usually proprietary chemicals and can be purchased from many producers in the United States or Europe.

Generally, bimetallic phosphates are less soluble and dissolve less readily than sodium orthophosphates or polyphosphates. They usually are provided in sulfuric or hydrochloric acid solutions to maintain the zinc in solution and require stainless steel or plastic tanks, pumps, and valves. Many suppliers offer several grades of bimetallic ortho- or polyphosphate, which differ with respect to the zinc-to-phosphate ratio and solubility. Bimetallic phosphates often perform better than individual phosphates at lower dosages, particularly in more mineralized and harder waters. Bimetallic phosphates can be blended with silicates and sodium phosphates by some suppliers.

The zinc orthophosphate inhibitor treatment was originally developed by Murray (1970) in Long Beach, California, for well water. It was originally derived from a blend of zinc sulfate, monobasic sodium phosphate, and sulfonic acid in the following proportions: $ZnSO_4 \cdot H_2O$, 56 percent; NaH_2PO_4, 24 percent; and H_3NO_3S, 20 percent.

It has been found that the protective film consists of about equal portions of zinc and phosphate that display a synergistic effect (Kelly et al. 1978). It has also been postulated that zinc forms carbonate compounds and blocks the cathodic sites, while phosphate forms compounds at the anodic sites. This inhibitor was initially provided in a dry granule form, with a zinc content of 19.4 percent.

There are other products with phosphate-to-zinc mass ratios that vary from 3:1 to 1:5 and zinc concentrations of 5 to 25 percent. The higher zinc concentrations usually reduce the time to establish corrosion control by rapid film formation at the cathode. However, primarily because of environmental regulation of zinc in wastewater effluents and sludges, it has been found by long-term testing in the San Jose, California, and the Philadelphia, Pennsylvania, water systems that with elevated pH the zinc concentration may be as low as 1 in 5 parts to 1 in 10 parts of phosphate (Ryder et al. 1994).

Phosphate blends. Many commercial products used in practice contain blends of sodium orthophosphate or polyphosphate like tripolyphosphate, hexametaphosphate, or more condensed products. Their concentration (as P_2O_5) is limited in Germany to a maximum of 5 mg/L in the treated water. The choice of product and dosage depends on its effectiveness as an inhibitor. In many cases, mixtures of ortho- and polyphosphates are used to combine corrosion protection with hardness stabilization or red-water suppression. The same effect may occur if only polyphosphates are used because a partial hydrolysis occurs within the piping systems, producing a certain amount of orthophosphate.

The fraction of orthophosphate typically ranges from 5 to 40 percent. High proportions of orthophosphate provide more corrosion protection, while higher proportions of polyphosphate enhance sequestering. The dosage requirements are primarily dependent on the pH and the calcium concentration in the water. Insufficient dosages can form an inadequate film and can aggravate pitting corrosion at fewer anodic sites.

In moderately hard water, these blended phosphates have been shown to be quite effective for lead and copper suppression. Initiation of corrosion control with blended polyphosphates is often slow; low dosages are used to avoid soft scale formation and discolored water. Positive corrosion control may require three to six months of dosing.

Silicates. Sodium silicate comes in several forms: dry chemicals or liquid silicate solutions. In dry form, the silicates are Na_2SiO_3, Na_6SiO_7, and $Na_2Si_3O_7$, with variable waters of hydration. They are produced by fusing sodium carbonate with silica sand. These are generally poorly soluble in cold water. Under heat and pressure,

Table 8-22 Sodium silicates commonly used as corrosion inhibitors

	Chemical Properties				
Type	Na_2O, %	SiO_2, %	Na_2O:SiO_2 Ratio	Silicate Solids, %	Density, g/cm^3
Neutral silicate "N" liquid*	8.9	28.7	1:3.22	37.6	1.38
Disilicate "D" liquid*	14.5	29.0	1:2	43.5	1.53
Metasilicate, granular	20.5	28.7	1:1	58.2	Solid

*PQ Corporation (Philadelphia, Pa.) trade names.

soluble silicates are produced and these vary in their proportions of sodium oxide (Na_2O) and silica (SiO_2). Both dry solid and soluble sodium silicates have been used as corrosion inhibitors and sequestering chemicals since the 1930s, and the more common types used in drinking water are listed in Table 8-22.

The "N" soluble silicates with Na_2O:SiO_2 ratios of 1:3.22 are the most commonly used in water with a pH range of 7 to 9 for corrosion inhibition of cold- or hot-water systems up to 180°F (82°C). The more alkaline "D" soluble silicates are used when the pH of the water is 7 or less. AWWA Standard B404-92 is for "N" liquid sodium silicate.

The dry granular form is frequently used in pot feeders on individual hot-water systems. It is also blended with phosphates and available in many proprietary compounds.

Soft waters of low pH and high oxygen content are most likely to be treated by silicates, although successful use of silicates as corrosion inhibitors has been demonstrated with a relatively wide variety of water qualities. Protection is achieved at silicate dosages varying from 4 to 30 mg/L, with higher dosages used for waters with higher concentrations of hardness, chloride, and dissolved solids, as well as at higher temperatures (Lane, Larson, and Schilsky 1977). Frequently, high silicate dosages are used for low pH conditions, but it is often more economical to increase the pH to between 7.5 and 8 by adding caustic soda or soda ash to lower the silicate requirement and the overall cost of inhibitor treatment. Although silicates are predominantly anodic inhibitors at low concentrations, there is a propensity similar to the phosphates for pit formation if the silicate dosage is insufficient.

The dosage of silicates in distribution systems or in individual (house) installations is increasing in North America. In practice, the maximum concentration of silicate (as SiO_2) after dosage is limited to 40 mg/L, and a major benefit of using silicates appears to be the accompanying pH increase. Because silicate corrosion control is relatively expensive, either caustic soda or slaked lime should be used in larger water systems if pH increase is the objective.

The corrosion protection offered by pure silicates appears to be caused mainly by the pH increase and film formation. Due to metal ion sequestering properties, silicates can also reduce corollary red-water problems, more so than pH or pH and alkalinity adjustment alone.

Surface analysis indicates that protection by the silicates occurs by forming a thin layer over the corroded metal layer, as contrasted to phosphates that form protective films on the bare metal surface by chemical reaction with either the corrosion products or the bare metal.

A proposed mechanism for inhibition by silicate is the formation of a protective film that consists of ferric or another metal oxide and silicate. Corrosion must occur before the metal surface can be protected. The films are self-limiting and do not build up in thick layers. The protective films will gradually break down and protection will cease within a short time if the dosage is stopped. Larger dosages of silica during pretreatment speed up the formation of the protective film, but also increase the chances of accumulated corrosion products being removed from old piping systems.

Sodium silicate corrosion-inhibitor treatment has been extensively used in point-of-use applications at individual buildings and for industrial processes. Silicate solutions have also been used in municipal supplies in the eastern United States and Canada. Test applications at Baltimore, Maryland (Ebner 1975), and Seattle, Washington (Ryder 1978a), have shown a generally excellent protection of copper and steel with silicate inhibitors. However, in these cases most of the inhibiting effects may have been due to the pH increase.

The cost of the silicate inhibitors is higher than that of phosphates or bimetallic inhibitors. Silicates can be combined with zinc, to provide bimetallic inhibitors, or with phosphates. These combinations are used in many of the proprietary inhibitor formulations.

Phosphate–silica mixtures. The use of phosphate–silicate products has increased in the past few years, and they have been shown to provide corrosion protection as good as or better than that of phosphate alone, but at reduced phosphate concentrations (Ryder 1990).

The products are orthophosphate–polyphosphate–silicate mixtures, with various phosphate-to-silica ratios. The positive effect of silicates in combination with reduced quantities of phosphates is a synergistic mechanism (Wehle 1979). Phosphates combine with metals forming precipitates that protect the surface layers, while the silicates may improve this effect by clogging pores and covering the more acidic anodic parts of the surfaces.

Inhibition Effects of Phosphates or Silicates on Different Materials

Phosphate and silicate corrosion inhibitors have been used with or without pH control. There are now data on their use to reduce metal release and to prolong the service life of a distribution system or a domestic installation. Inhibitor dosages, if concentrations are limited, may not avoid pitting or other localized corrosion caused by material or installation faults. Nor are corrosion rates of the usual materials—galvanized steel, steel, cast iron, copper, or lead—sufficiently reduced to extend the life span of a system to beyond 75 to 100 years.

Corrosion inhibitors can be very beneficial and necessary, if concerns about water quality deterioration have to be resolved. Unfortunately, there is no simple answer if a problem such as red water, high copper release, or increased lead release shows up, especially if two or three problems are present at the same time. The choice of the optimal inhibitor and of the necessary concentration depends primarily on the system, its operational conditions, the distributed water quality, and the particular quality concerns.

Lead. Concern about lead piping is rarely associated with corrosion of the material but rather with the health effect of ingesting dissolved and particulate lead. Inhibition measures are often intended to reduce the plumbosolvency of the water. Schock (1980b) and Sheiham and Jackson (1981) showed that pH and alkalinity of water have the most important influence on plumbosolvency. But only under very specific conditions, i.e., low alkalinity combined with very high pH, will the lead content in water be reduced to a value less than the limits of USEPA, EC, and the World Health Organization (WHO) standards at equilibrium conditions. In the United States, a maximum contaminant level goal of zero and an action level of 15 μg/L for the first 1-L kitchen-tap samples, following 6 to 8 hours of stagnation of water in lead-containing pipe or soldered pipe systems, was prescribed by the Lead and Copper Rule (USEPA 1991a, 1991b, 1992).

Alkalinity and pH adjustment are not sufficient to reduce lead solubility to meet these stringent limits, nor do they reliably control release of particulate lead in corrosion products. Since elimination of lead-containing materials from contact with drinking water is not practical in the short term, reducing lead solubility and corrosion rates by adding appropriate inhibitors can be an effective temporary measure (see chapter 4).

Some work done in the early 1950s at the German Federal Health Office indicated that polyphosphates and silicates had an inhibiting effect on lead levels in drinking water (Gad and Naumann 1953). Further and more detailed research at the British Water Research Centre showed, however, that polyphosphates and silicates did not reduce plumbsolvency. Polyphosphates may increase the lead levels by complexation (Gregory and Jackson 1983; Neff, Schock, and Marden 1987).

In contrast, orthophosphate dosages can reduce plumbosolvency by building up a lead phosphate layer that has a significantly low solubility. An example of test results for lead solubility with increasing orthophosphate dosage is shown in Figure 8-14. Though the thermodynamic solubility of lead phosphate is very low (Schock 1980a), measurements of lead concentrations in water exposed to new lead pipes exceeded theoretical solubility and depended on the orthophosphate concentration (Figure 8-15). Similar results found with old lead pipes in Britain show that the reduction of plumbosolvency is also determined by the kinetics of formation of the less soluble lead hydroxyphosphates (Gregory and Jackson 1983; Wagner 1989).

Copper. Copper may be affected by either general corrosion or pitting. With respect to general corrosion in cold and hot potable waters, Wehle (1979, 1983) showed that orthophosphate somewhat reduced the corrosion rate in cold waters, whereas polyphosphates increase the corrosion rate and silica corrosion suppression seems to be related mostly to an increase in pH. The combination of ortho- and polyphosphates can reduce the corrosion rate more than orthophosphate alone. Tests carried out by Stichel (1981) compared effects for orthophosphates and silicates in hot water. Orthophosphates reduced general corrosion, but silicates showed no significant influence. However, polyphosphates and orthophosphate–polyphosphate mixtures in hot water significantly reduced general corrosion. The apparent contradiction in polyphosphate behavior as a function of water temperature can be explained by the partial hydrolysis of polyphosphates to orthophosphate in hot water.

Results on copper corrosion performed with test units comparable to the ISWS Machined Nipple Test and using Seattle (Washington) Tolt River supply (Ryder 1978b) showed that pH increase to 8.0 by dosing lime and soda ash reduced the corrosion rate by 75 percent. Additional phosphate or silicate dosage had insignificant benefit compared to the increased pH. However, subsequent tests in Seattle found that orthophosphate caused a very pronounced reduction, compared to pH and alkalinity adjustment alone (Benjamin et al. 1990). Orthophosphate protection was also shown to be very pH dependent with protection lost at slightly acidic pH values. Similar experience in Germany indicated that this effect was mostly due to poor mechanical properties of corrosion product scales under these conditions.

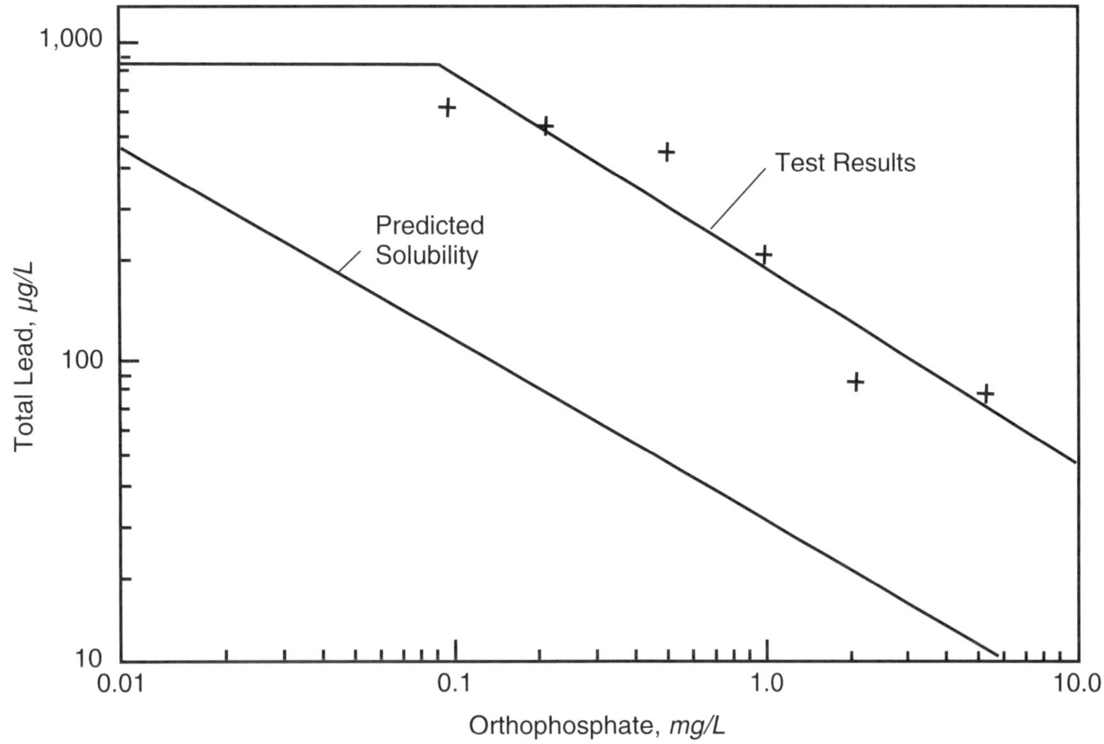

Source: Sheiham and Jackson (1981).

Figure 8-14 Lead solubility as a function of orthophosphate dosage: Theoretical and experimental results (water quality: pH—6.5-6.8; alkalinity—50 mg $CaCO_3$/L; maximum lead concentration in untreated water after long-term stagnation—830 μg/L)

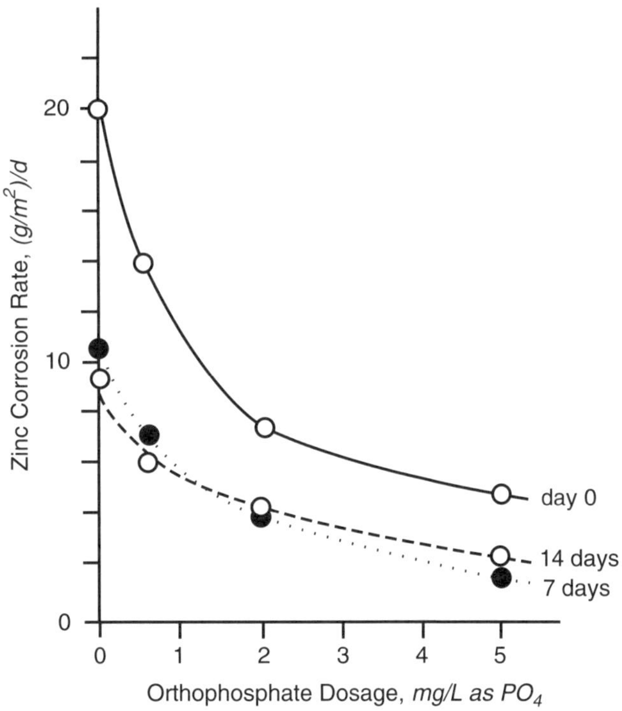

Figure 8-15 Effect of orthophosphate on the corrosion of zinc (flow velocity—1.6 m/s [5 ft/s]; pH of water—7.15)

Tests on copper and copper alloys showed a reduction in copper leaching after dosage of orthophosphate–polyphosphate mixtures (Borankovich 1973). This finding may be important in solving copper-induced pitting problems in houses with galvanized steel pipes and brass fittings and valves.

The influence of phosphates and silicates on pitting corrosion has not been reproducible under many test conditions. Experiments conducted by Franqué (1982) corroborated field experience that pitting corrosion was not diminished by phosphates and silicates. However, these tests may have used insufficient inhibitor addition or there was not an increase in pH, since more recent data in Florida have shown a reduction in copper pitting.

Galvanized steel. General corrosion of galvanized steel pipe depends mainly on the pH of the water and the flow velocity. If the zinc layer is corroded, corrosion continues in the zinc–iron alloy phase and finally reaches the basic iron material. Orthophosphates effectively reduce the general corrosion of the zinc layer. The effect increases if the zinc–iron alloy phase is reached. Tests performed by Werner (1976) showed that the corrosion rate of the pure zinc layer of galvanized steel depends on the orthophosphate concentration (Figure 8-15). The relative magnitude of the inhibition effect is dependent on the flow velocity, at least in the test range of 0.5 to 5 ft/s (0.2 to 1.6 m/s). Analogous tests with polyphosphates have shown that they do not reduce the corrosion rate of the pure zinc layer. An increase in corrosion rate can occur with the use of polyphosphates (Wagner 1977).

Benjamin et al. (1990) found that orthophosphate dosages of 1 and 5 mg P/L had only minimal effects in reducing corrosion of zinc from Seattle tap water at pH 8. Indeed, penetration of the galvanized layer to iron occurred under all conditions tested with Seattle water in one to two years.

Silicates have not been shown to inhibit zinc corrosion beyond the effect of increased pH. Test results with silica dosages up to 20 mg/L and parallel tests with NaOH dosage to increase pH to the same value showed corrosion rates comparable to those of galvanized pipe.

Iron. Orthophosphates can reduce the corrosion rate of steel, and they strongly impact the iron-release rate. Figure 8-16 shows corrosion and iron-release rates as a function of orthophosphate concentration. Small amounts of orthophosphate greatly reduced iron-release rates, increasing the concentration from 0.5 to 1 mg/L reduced iron-release by two-thirds. Orthophosphate also reduced the corrosion rate, but a concentration of 3 mg/L was necessary to reduce the corrosion rate by two-thirds. Benjamin et al. (1990) also showed that orthophosphate (1 mg P/L) effectively reduced corrosion by approximately two-thirds in Seattle tap water at pH 8.0.

Polyphosphates and blends of ortho- and polyphosphates with small proportions of orthophosphates have been shown to actually increase corrosion rates and iron release.

Tests carried out on a low-mineralized water from the Aggerverband (Germany) water utility reservoir compared results obtained with orthophosphate and silicate blends and with varying increases in pH to results with a phosphate–silicate mixture (Table 8-23). A large reduction in the corrosion rate was achieved by dosage of a phosphate–silicate mixture alone at a pH of 8.7. Even in the case of very high pH values, when the iron-release rate decreases to very low values, the corrosion rate remains relatively high. The result of this high corrosion rate and the low iron-release rate is that a large quantity of ferric precipitates are deposited in the pipes. The high corrosion rate also leads to rapid depletion of oxygen in water during longer transport or stagnation periods. This creates a high risk of non-steady-state (Kuch-type) corrosion (see chapter 2). Fe(III) deposits become reduced to Fe(II) compounds, causing ongoing corrosion. The reduction reaction is responsible for conspicuous red-water problems as well.

Summarizing the results of the different inhibitors on the metals, one can conclude that polyphosphates usually have little or no effect on either corrosion inhibition or the reduction of metal release. Increased corrosion and metal-release rates often occur because of the complexing and sequestering properties of polyphosphates, leading to a higher solubility of metal salts or to the formation of less protective layers.

Orthophosphates normally have an inhibiting effect if the corrosion products or the corroding metal can form insoluble phosphates and improve the protective scale quality by increasing its impermeability and adherence. The corrosion products have a low solubility and reduce the metal release of the water.

Silicates are generally less effective. Often the observed effects seem to depend on the pH increase caused by the alkaline character of the silicate compound. It is possible that silicate can maintain a corrosion inhibitor film, following film initiation by a phosphate–silicate blend. Variable corrosion inihibition has been found for mixtures of ortho- and polyphosphates. A wide range of effects was found from good inhibition on copper, to mixed effects on lead and zinc, to an increased influence on corrosion and iron release on unalloyed steel. However, phosphate–silicate mixtures often show good corrosion reduction in steel pipe.

The influence of water quality, insufficient dosage, and overfeeding is site specific. Further research will be necessary to discern the mechanisms of the different phenomena.

In operational practice, the minimum concentration necessary for long-term inhibitor dosage has to be determined to minimize the cost and adverse environmental aspects of excessive phosphate concentrations. Experience shows that stopping

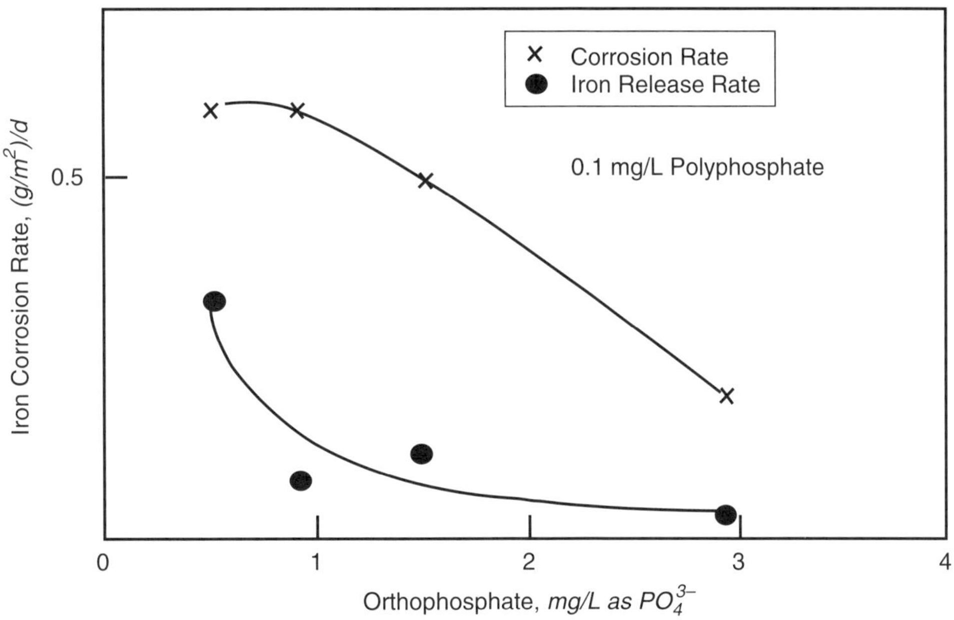

Figure 8-16 Corrosion rate and iron release rate versus orthophosphate concentration in ¾-in. diameter (20 mm) steel pipe (water flow—0.5 m/s [1.6 ft/s]; water quality: pH 7.3-7.4; alkalinity—275 mg $CaCO_3$/L; polyphosphate concentration—0.1 mg/L)

Table 8-23 Corrosion test results* on steel pipe at Aggerverband (Germany) water utility

Water Quality	pH	Iron-Release Rate (v_{Fe}), $g\ Fe/M^2$	Corrosion Rate (v_{corr}), $(g\ Fe/m)/d$	v_{Fe}/v_{corr}, %
Raw water	7.6	0.3	0.7	43
Drinking water (lime addition)	8.4	0.14	0.77	18
Carbonate-silicate (soda ash and silicate)	9.7	0.04	0.72	6
Carbonate-silicate-phosphate (soda ash and silicate and mixed phosphates)	9.7	0.025	0.6	4
Orthophosphate	7.6	0.2	0.5	40
Silicate-phosphate	8.7	0.08	0.25	32

*After 90–150 days.

v = rate

v_{Fe} = iron release rate

v_{corr} = iron corrosion rate

inhibitor dosage usually in time revives the corrosion problems. If good corrosion protection was provided by the inhibitor, a certain time without dosing can be accepted without risk, as shown in Figures 8-17 and 8-18.

It is necessary to control a reduction in inhibitor dosage by a monitoring program.

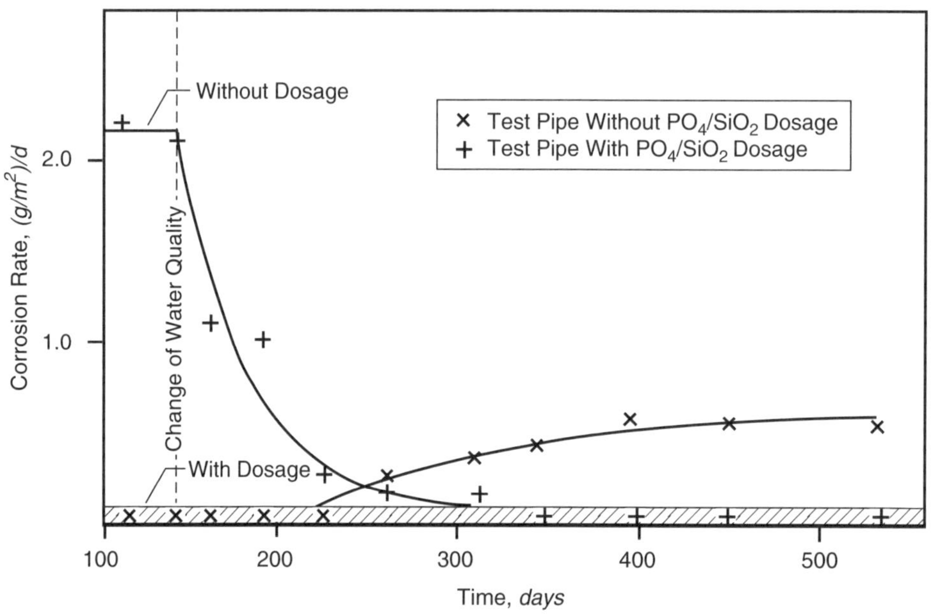

Figure 8-17 Development of corrosion rate with exposure time on two steel pipe test loops with and without dosage (changing water quality during the test; permanent flow conditions—0.5 m/s; water quality: pH 7.6; alkalinity—150 mg $CaCO_3$/L; 160 mg Cl^-/L; 80 mg SO_4^{2-}/L; water source: Düsseldorf Water Works)

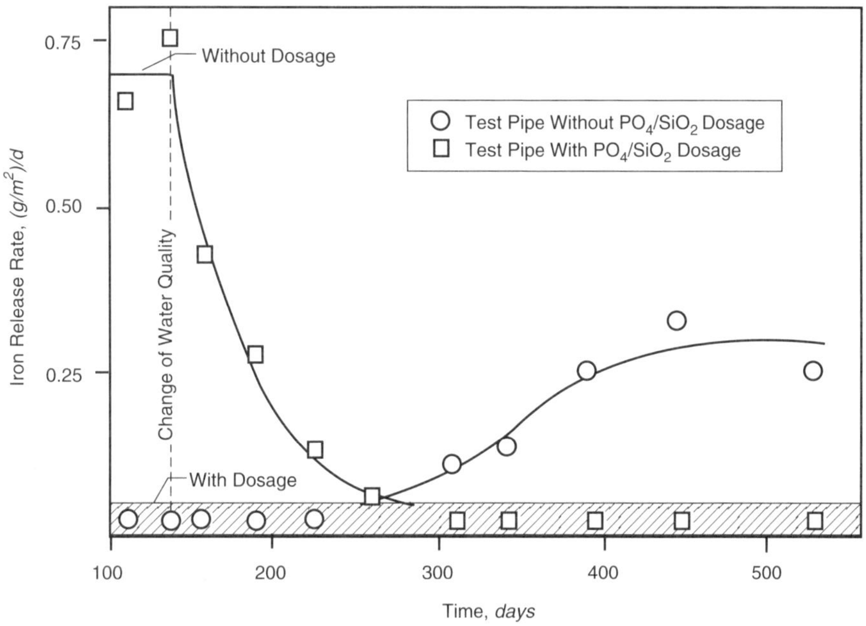

Figure 8-18 Development of iron release rate with exposure times (test loops and test conditions as described in Figure 8-17)

GENERAL CHEMICAL INFORMATION

The chemicals most frequently used to control corrosion are listed in Table 8-24.

For some of the generic inhibitors in water systems, AWWA standards have been established to define the allowable impurities. The European Community has also proposed European Committee for Standardization (CEN) standards for chemicals used in drinking water. These standards include the following corrosion-controlling chemicals:

Chemical Covered	AWWA Standard No.	CEN No.
Soda ash	B201	EN897 (CEN Proposed Standard 1992b)
Quicklime and hydrated lime	B202	
Liquid sodium silicate	B404	
Caustic soda	B501	EN896
Sodium hexametaphosphate	B502	
Sodium tripolyphosphate	B503	
Carbon dioxide	B510	EN936 (CEN Proposed Standard 1992a)
Calcium carbonate	Not covered	EN1018 (CEN Proposed Standard 1993a)
Half-burnt dolomite	Not covered	EN1017 (CEN Proposed Standard 1993b)
Monosodium phosphate, anhydrous	B504	
Disodium phosphate, anhydrous	B505	

Table 8-24 Chemicals used for corrosion control of drinking water

Chemical Name, Formula, and Common Name	Common Forms	Commercial Strength	Bulk Density, lb/ft^3	Solubility, $g/100\ g\ H_2O$
Calcium carbonate $CaCO_3$ High calcium limestone	White or gray; granular or powder	About 95%	100–115	0.0014 at 25°C
Calcium hydroxide $Ca(OH)_2$ High calcium hydrated lime, slaked lime	White powder	72–74% as CaO	25–35	0.19 at 0°C 0.15 at 30°C 0.08 at 100°C
Calcium oxide CaO High-calcium quicklime, burnt lime, unslaked lime	White; lump, pebble, granules, or pulverized	93–98%	55–60	Converts to $Ca(OH)_2$ in solution
Half-burnt dolomite $CaCO_3 \cdot MgO$	White, gray, or tan; granules or powder	About 52% $CaCO_3$ 40% MgO	25–35	0.18
Sodium bicarbonate $NaHCO_3$ Bicarb, baking soda	White powder	99%	135	6.9 at 0°C 16.4 at 100°C
Sodium carbonate $NaCO_3$ Soda ash	Anhydrous; white powder	≥98%	35–65	7.1 at 0°C 18 at 20°C 27.6 at 30°C 45.5 at 100°C
Sodium hydroxide NaOH Caustic soda	Anhydrous; white solid, beads, flakes, or liquid	98.9% NaOH if solid, but 20–50% NaOH in solution form	133	42 at 0°C 347 at 100°C
Liquid sodium silicate $Na_2O \cdot nSiO_2$ Water glass	Opaque, viscous liquid	38–42%		Complete
Zinc-orthophosphates	Liquid	Variable		Complete
Sodium hexametaphosphates	Glassy plates, flakes, lumps, beads, or granular	>80% PO_4	80	Complete
Sodium tripolyphosphates	Powder or granular	>75% PO_4	69	Complete
Monobasic sodiumphosphates	White crystalline solid (granular or powder)	>77.7% PO_4	72	Complete
Dibasic sodiumphosphates	White crystalline solid (granular or powder)	>64.3% PO_4	69	Complete

Adapted from Parsons (1965).

NOTE: To obtain bulk density in kg/m^3, multiply lb/ft^3 value by 16.02.

AWWA standards include a section on chemical additives in water. Trace impurities of these chemicals are limited to values below the contaminant levels for inorganic chemicals listed in the Safe Drinking Water Act. Reference is made to the Water Chemical Codex (*Water Codex* 1982) for guidance regarding maximum allowable levels of impurities.

In addition, a number of chemical manufacturers offer blended corrosion inhibitors that include sodium ortho- and polyphosphates, calcium polyphosphates, potassium ortho- and polyphosphates, zinc, and sodium silicate. The commercial inhibitor blends should be tested and certified by NSF International for potable water use in the United States, with maximum dosage recommendations by the federal government, before they can be legally sold in the United States. Similar approval systems exist in European countries.

Some general comments about chemical handling and properties can provide only modest guidance on use.

Bulk storage and handling of dry chemicals is often chosen for larger water systems, whereas bag packaging is chosen for low-use situations and for most of the polyphosphate inhibitors used in potable water systems. The polyphosphates may be produced and purchased as plates, flakes, lumps, beads, and fine or coarse granular or powdered forms. The material is soluble in 50°F (10°C) water at one part to four parts water, and the pH of a 1 percent solution is in the range of 5.8 to 7.3. Hexametaphosphate is available in crystalline form for dry feeder or solution feeding. It is also available in a slowly soluble vitreous form for use in pot feeders in small water systems.

Several of the corrosion inhibitors have potential problems with regard to handling, storage, or use that should be recognized. Most of the glassy polyphosphates, the crystalline sodium silicates, dry quicklime, sodium hydroxide, and sodium carbonate are very hygroscopic. Unless stored and handled in a low-moisture environment, these chemicals will absorb water and cake, which creates handling and feeding problems.

Blended ortho- and polyphosphates have a pH in the range of 5.8 to 7.5 and can be delivered in bulk or in drums. An advantage is that they need not be stored as hazardous materials, given the more complicated and costly safety precautions now required in the US building codes.

Zinc orthophosphate, which is very soluble, 38 percent by weight, produces an acidic solution with a pH of about 1. The inhibitor is now also provided as a liquid solution that has a zinc content of 7.8 percent and a pH of 0.8. The liquid form can be delivered in drums or in bulk tank trucks.

Aesthetic and Environmental Aspects

While many industrial corrosion inhibitors can cause serious pollution problems, the corrosion inhibitors used for potable water service are of concern mainly for pollution from zinc and, in some instances, phosphate compounds (Ryder 1990). Zinc is also undesirable in the sludge produced at wastewater treatment plants as it limits land disposal application.

Zinc. Maintaining the zinc concentration at less than 0.25 mg/L to prevent possible subsequent pollution problems is ordinarily an objective and can be achieved as zinc concentrations of bimetallic polyphosphate or zinc orthophosphate solutions can be effective in the range of 0.1 to 0.3 mg/L.

However, during the preconditioning period, the inhibitor dosage is often 10 times higher for several days to several weeks. Zinc may exceed desirable pollution levels unless zinc is removed in wastewater treatment.

Zinc in concentrations as low as 0.01 mg/L has been reported to be toxic to trout and certain other fish (McKee and Wolf 1963). For protection of all aquatic biota, the zinc concentration in wastewater effluents should not exceed 0.05 mg/L (Ryder 1990). The prescribed discharge limitation of zinc into a wastewater collection system is in the range of 0.1 to 1 mg/L in many municipal jurisdictions in order to prevent excessive zinc concentration in the effluent or sludge residues and to protect the aerobic biological processes of most secondary wastewater treatment plants. If zinc is added as an inhibitor, discharge limitation for industries may have to be reduced. The toxicity of zinc is related to its chemical speciation, which is sometimes not considered when waste discharge limitations are prescribed by regulatory agencies. Alkalinity, pH, and salt concentration all influence zinc speciation.

Zinc is the only element among the corrosion inhibitors for which there is a secondary drinking water requirement, under provision of the Safe Drinking Water Act. The 5-mg/L secondary maximum contaminant level (SMCL) for zinc is greater than the zinc concentration in formulation or use of the bimetallic corrosion inhibitor compounds, but of course, zinc can also enter water through corrosion of galvanized pipe. A revised limit of 3 mg Zn/L in drinking water is currently being proposed in the United States.

Zinc is currently not allowed in corrosion-inhibitor products in much of Europe and is becoming increasingly restricted in North America in recognition of wastewater discharge and sludge handling limitations.

Phosphates. Phosphate is a nutrient that can stimulate algal growth in water reservoirs and in waters receiving wastewater discharges. The orthophosphate form is most effective in stimulating algal growth in open reservoirs. However, all of the polyphosphates degrade to some degree to orthophosphates with elevated temperatures or a long contact time. In short, there is always a potential for increasing the orthophosphate concentration. Tests conducted by Ryder (1978a) for the Seattle water system found that corrosion inhibitor dosages of both sodium hexametaphosphate and zinc orthophosphate, in the range of 1 to 2 mg P/L, stimulated algal growth significantly. Phosphorus is frequently the limiting nutrient for algal growth in fresh water; in many cases the phosphorus concentration in wastewater effluents discharged to inland waters is not to exceed a range of 0.2 to 1 mg/L. Approximately 25 to 40 percent of the phosphorus can be removed by normal biological secondary treatment processes, but chemical precipitation or biological phosphorus-removal processes are needed to reduce the phosphorus level to 1 mg P/L or below.

A practical limit of dosing phosphate inhibitors at concentrations as low as possible is appropriate. Preconditioning may be most appropriately started in the fall or winter when there is a high effluent dilution flow and when temperature, light, and photosynthesis are reduced, to minimize possible pollution effects. Communication with and approval by the local water quality regulatory agency is recommended during initiation of a corrosion-inhibitor program of any kind in a municipal water supply.

Silicates. Silicates can stimulate certain biologic growths in low-mineralized waters. For example, in the Seattle test, the use of sodium silicate at a dosage of 4 mg/L increased growth of the diatom *Synedra*. Usually, diatoms are not significant water quality problems except in some instances of filter clogging (Ryder 1978a).

Sodium. The possible toxic aspects of corrosion inhibitors have been discussed previously. A recent concern with regard to health is the relationship between sodium intake and the incidence of cardiovascular disease. A 20-mg/L increase for sodium has been suggested as a nonmandatory drinking water requirement. However, many potable waters cannot meet this requirement because of the high background concentrations of sodium in the water sources and the economic impracticability of removing sodium from domestic water sources. This proposed sodium regulation would seem to portend that large additions of sodium should be avoided. For example, caustic soda or soda ash additions should possibly be limited so as not to increase background sodium concentrations by more than 20 mg/L, or lime should be used for increasing pH and adding alkalinity, especially when addition of a sodium compound to water with a high background sodium level would increase that concentration by 10 mg/L or more. In many European countries and in North America, health authorities are recommending use of calcium-based products as opposed to sodium-based products.

CHEMICAL COSTS AND DOSAGES

The costs of inhibitor chemicals are quite varied. Table 8-25 presents relative costs of the different chemicals on a weight basis. A major increase in cost of caustic soda (NaOH) has occurred. It used to be a by-product from chlorine production, but reduced use of chlorine in bleaching of paper has reduced supplies. Costs are expected to increase even further in the years to come.

There are several considerations related to costs of the neutralizing chemicals, as well as safe handling, that are worth mentioning. Some of these are mentioned in Table 8-26. $CaCO_3$ is the cheapest chemical but requires a fairly large investment in an alkaline media filter or a limestone contacter. The dissolution rate is low. Half-burnt dolomite has a much higher dissolution rate and is, therefore, preferred in many cases. This chemical, however, requires a better pH control system.

A comprehensive guide on determining corrosion costs, corrosion-control chemical costs, and benefit–cost analysis in water distribution systems is contained in a report by EES and Kennedy/Jenks/Chilton (1989).

Table 8-25 Comparative costs (1994) of corrosion inhibitors

Type	Dosage Rate, *mg / L*	Inhibitor Cost, *$ / lb*	Treatment cost, *$ / mil gal*
Lime	10–30	0.02	2.00–6.00
Caustic soda	10–30	0.20	10.00–15.00
Soda ash	10–30	0.12	16.70–50.00
Sodium hexametaphosphate	1–4 (PO_4)	0.90 (PO_4)	7.50–30.00
Bimetallic phosphate	0.5–2 (PO_4)	1.50 (PO_4)	6.25–25.00
Zinc orthophosphate	0.1–0.5 (Zn)	2.25 (Zn)	2.00–25.00
Sodium silicate	4–10 (SiO_2)	0.30 (SiO_2)	10.00–25.00
Carbon dioxide	5–10	0.05	2.00–4.00
Phosphoric acid	0.5–3 (P)	0.60 (PO_4)	3.00–18.00
Monosodium phosphate	0.5–3 (P)	1.20 (PO_4)	6.00–36.00
Ortho-polyphosphate blend	0.2–1 (PO_4)	2.50 (PO_4)	4.00–20.00

NOTE: To obtain $/kg, multiply $/lb by 2.20. To obtain $/m^3, multiply $/mil gal by 0.000264.

Table 8-26 Relative cost information for the different chemicals used for corrosion control

Chemical Name, Formula, and Common Name	Installation Costs	Chemical Costs*	Requirements for Operation and Maintenance	Requirement for Process Control
Calcium carbonate materials $CaCO_3$ High-calcium limestone	Very high	Very low	Small	None; overdosing can never occur
Dolomitic materials, half-burnt $CaCO_3 \cdot MgO$	High	3–4 times higher than $CaCO_3$	Relatively small	pH control is necessary; very high pH values can occur
Calcium hydroxide $Ca(OH)_2$ High-calcium hydrated lime	Limemilk; moderate	2.5 times higher than $CaCO_3$; moderate cost	High	pH and turbidity control is necessary
	Limewater; moderately high		Moderately high	pH control is difficult for low-alkalinity water
Calcium oxide CaO High-calcium quicklime	As for $Ca(OH)_2$	1.5 times higher than $CaCO_3$; moderate cost	As for $Ca(OH)_2$	As for $Ca(OH)_2$
Sodium bicarbonate $NaHCO_3$ Bicarb, baking soda	Low	40 times higher than $CaCO_3$; most expensive chemical	Relatively low	pH control is needed
Sodium carbonate Na_2CO_3 soda ash	Low	12 times higher than $CaCO_3$; high	Relatively low	pH control is needed
Sodium hydroxide NaOH Caustic soda	Low	20 times higher than $CaCO_3$; very high	Relatively low	pH control is difficult for low-alkalinity water
Sodium silicate $Na_2O \cdot nSiO_2$ Water glass	Low	5 times higher than $CaCO_3$	Low	

*No transportation information is included.

The inhibitor-chemical costs are relatively minimal. The costs of typical inhibitor chemicals are shown in Table 8-25. Translated into costs per residential consumer, the costs of an inhibitor with a treatment cost of \$20 per million gallons would range between \$1 and \$1.50 per year. The choice of corrosion inhibitor should be determined by relative effectiveness as related to the price. Evaluation of corrosion-inhibitor effectiveness, as tested by pilot-plant studies, is often important because the level of corrosion reduction possible with various inhibitors can vary substantially and should be quantified in terms of percent reduction of corrosion and extension of useful life. A present-worth analysis of water-system corrosion costs with and without inhibitor additions, then, provides a rational basis for choice.

SUMMARY

Corrosion mitigation treatment has received much renewed interest in the United States and Europe. The use of pH, alkalinity, and hardness modifications to control corrosion and metal release is often effective and economical. Conventional methods of adding strong bases and sodium carbonate are widely used in the industry. New technologies for making these adjustments have been developed in recent years, using calcium salts and carbon dioxide. Other inhibitors, including phosphates, zinc, and silicates, are widely used but may be restricted because of water quality impacts or by regulation. Corrosion treatments may have impacts on uses of the water, on other materials in the water system, and on wastewater treatment or other discharges. Possible adverse effects, especially during start-up of corrosion control, should be evaluated, monitored, and mitigated as necessary. The costs of corrosion control are usually moderate or low. If benefits of corrosion reduction can be demonstrated, there usually is good justification for using effective corrosion control treatment.

REFERENCES

AWWA. 1978. *Corrosion Control by Deposition of $CaCO_3$ Films.* Denver, Colo.: American Water Works Association.

AWWA B201-92. 1992. *Standard for Soda Ash.* Denver, Colo.: American Water Works Association.

AWWA B202-83. 1983. *Standard for Quicklime and Hydrated Lime.* Denver, Colo.: American Water Works Association.

AWWA B404-92. 1992. *Standard for Liquid Sodium Silicate.* Denver, Colo.: American Water Works Association.

AWWA B501-93. 1993. *Standard for Sodium Hydroxide (Caustic Soda).* Denver, Colo.: American Water Works Association.

AWWA B502-88. 1988a. *Standard for Sodium Polyphosphate, Glassy (Sodium Hexametaphosphate).* Denver, Colo.: American Water Works Association.

AWWA B503-89. 1989. *Standard for Sodium Tripolyphosphate.* Denver, Colo.: American Water Works Association.

AWWA B504-88. 1988. *Standard for Monosodium Phosphate, Anhydrous.* Denver, Colo.: American Water Works Association.

AWWA B505-88. 1988. *Standard for Disodium Phosphate, Anhydrous.* Denver, Colo.: American Water Works Association.

AWWA B510–89. 1989. *Standard for Carbon Dioxide.* Denver, Colo.: American Water Works Association.

AWWA Membrane Technology Committee Report. 1992. Membrane Processes in Potable Water Treatment. *Jour. AWWA,* 84(1):59–67.

Baldauf, G., and M. Henkel. 1991. Possibilities and Limits in the New Drinking Water Regulations for the Use of Calcium Carbonate Deacidification. *Wasser/Abwasser,* 132(3):132–140.

Ballantyne, D.B. 1986. Design Study for Fort Lewis, Washington, Water Treatment Plant. Kennedy/Jenks/Chilton, job. no. 6715.

Benjamin, M.M., S. Reiber, J.F. Ferguson, E.A. Vanderwerff, and M.W. Miller. 1990. *Chemistry of Corrosion Inhibitors in Potable Water.* Denver, Colo.: American Water Works Association Research Foundation and American Water Works Association.

Boffardi, B.B. 1984. Corrosion Control of Industrial Cooling Water Systems. *Materials Performance,* 23(11):17–24.

Bolze, D., and J. Beyea. 1989. The Citizens' Acid Rain Monitoring Network. *Envir. Sci. and Technol.,* 23(6):645–646.

Boman, E., and E. Vik. 1985. *The Water Supply Situation in Lillesand County. Special emphasis on the rural water supply situation.* NTNFs R&D Committee for Drinking Water Report No. 1518.

Borankovich, J.C. 1973. On-Line Corrosion Monitoring. *Materials Protection and Performance*, 12:20.

Boynton, R.S. 1980. *Chemistry and Technology of Lime and Limestone.* 2nd ed. New York: John Wiley and Sons.

Caldwell, D.H., and W.B. Lawrence. 1953. Water Softening and Conditioning Problem. *Ind. Eng. Chem.*, 45(3):535.

CEN Proposed Standard. 1992a. *Carbon Dioxide Used for Treatment of Water Intended for Human Consumption.* EN936. Brussels: European Committee for Standardization.

———. 1992b. *Sodium Carbonate Used for Treatment of Water Intended for Human Consumption.* EN897. Brussels: European Committee for Standardization.

———. 1992c. *Sodium Hydroxide Used for Treatment of Water Intended for Human Consumption.* EN896. Brussels: European Committee for Standardization.

———. 1993a. *Calcium Carbonate Used for Treatment of Water Intended for Human Consumption.* EN1018. Brussels: European Committee for Standardization.

———. 1993b. *Half-Burnt Dolomite Used for Treatment of Water Intended for Human Consumption.* EN1017. Brussels: European Committee for Standardization.

Cohen, A., and J.R. Myers. 1987. Mitigating Copper Pitting Through Water Treatment. *Jour. AWWA,* 79(2):58–61.

Deffeyes, K.S. 1965. Carbonate Equilibria: A Graphic and Algebraic Approach. *Limnol. Oceanogr.,* 10:412.

Duranceau, S.J., J.S. Taylor, and L.A. Mulford. 1992. SOC Removal in a Membrane Softening Process. *Jour. AWWA,* 84(1):68–78.

DVGW (Deutscher Verein des Gas- und Wasserfachs). 1992. *Water Treatment—Deacidification of Water; pH Value and Calcite Saturation.* DK:628.162.7 Arbeitsblatt W 214, Teil 1, Sept. Bonn: Wirtschafts– und Verlagsges, MBH.

———. 1993. *Water Treatment—Deacidification of Water; Basics for Planning, Design and Operation of Filters.* Arbeitsblatt W 214 Teil 2, Feb. Bonn: Wirtschafts– und Verlagsges, MBH.

EBI (Engler-Bunte Institut der Universität Karlsruhe). 1980. *Untersuchungen zum Einfluss natürlicher organischer Wasserinhaltsstoffe auf die Ausbildung und Korrosionsschutzwirkung von Deckschichten in Trinkwasserinstallationen.* Heft 14 der Veröffentlichungen des Bereichs und Lehrstuhles für Wasserchemie. Bonn: Wirtschafts-und Verlagsges. mbH.

Ebner, B.H. 1975. Red Water Problems: A Thing of the Past. *Public Works,* 106:82.

EC (European Community). 1980. Council Directive of 15th July 1980 Relating to the Quality of Water Intended for Human Consumption (80/778/EEC). *Off. Jour. Eur. Comm.*, 23 (L229/11).

Edwards, M.A., and J.F. Ferguson. 1993. Accelerated Testing of Copper Corrosion. *Jour. AWWA,* 85(10):105–114.

EES (Economic and Engineering Services, Inc.), and Kennedy/Jenks/Chilton. 1989. *Economics of Internal Corrosion Control.* Denver, Colo.: American Water Works Association Research Foundation.

Eikebrokk, B., S.W. Østerhus, and H. Brattebø. 1990. *Corrosion Control Using Micronized Limestone and CO_2.* Royal Norwegian Council for Scientific and Industrial Research's Program on Drinking Water Research, Report No. 36/1989.

Franqué, O. von. 1982. *Korrosionsschäden durch Lochfraß in Kupferrohren.* Final Report, FE KKs Proj. O-F 5.1/2.

Gad, G., and K. Naumann. 1953. Vermeidung von Bleivergiftungen durch neu verlegte Rohre. *Gesundheits-Ingenieur,* 74:85.

Galloway, J.N. 1989. Atmospheric Acidification: Projections for the Future. *Ambio,* 18:161.

Galloway, J.N., and E.B. Cowling. 1978. The Effect of Precipitation on Aquatic and Terrestrial Ecosystem. A Proposed Precipitation Chemistry Network. *Jour. APCA,* 28:229–235.

Gjessing, E., E. Taud, and R.C. Peterson Jr. 1992. HUMOR/HUMEX Experimental Acidification of a Catchment and Its Humic Lake. In *A Special Issue of Environment International.* New York: Pergamon Press.

Gregory, R., and P.J. Jackson. 1983. Unpublished Results. Medmenham, England: Water Research Centre.

Hantge, E. 1992. *Air Pollutants—Mass Balance and Effect on Soil and Water.* Ministry of Environment for Rheinland-Pfalz, Internal Report. Aug. 20, in German.

Hatch, G.B. 1973. Inhibitors for Potable Water. In C.C. Nathan, ed. *Corrosion Inhibitors.* Houston, Texas: National Association of Corrosion Engineers (NACE).

Hedberg, T. 1983. *Testing of Various Types of Alkaline Media for Filter Bed Systems.* Research Report. Publication 6:83. Gothenborg, Sweden: Chalmers Technical University.

Hedberg, T., and E. Johansson. 1981. *Increasing Water Hardness in Alkaline Media Filters.* Industrial Report. Gothenborg, Sweden: Chalmers Technical University.

Hem, L. 1993. Unpublished Results. Oslo, Norway: Aquateam.

Henriksen, A., L. Lien, B.O. Rosseland, T.S. Traaen, and I.S. Sevaldrud. 1989. Lake Acidification in Norway—Present and Predicted Fish Status. *Ambio,* 18:314.

Henriksen, A., L. Lien, T.S. Traaen, I.S. Sevaldrud, and D.S. Brakke. 1988. Lake Acidification in Norway—Present and Predicted Chemical Status. *Ambio,* 17:259.

Illig, G.L. 1957. Glassy Phosphates in Water Treatment. *Jour. AWWA,* 49(6):805–816.

Kelly, T.E., et al. 1978. Zinc/Polyphosphate Combination Control Corrosion in Potable Water Distribution Systems. *Materials Protection,* 17:78.

Kleber, J.R. 1965. Use of Bimetallic Glassy Phosphates for Corrosion Control. *Jour. AWWA,* 57(6):783–790.

Klein, E. 1992. *Operational Possibilities of Deacidifying Aggressive Water to Fulfill the Regulations in Arbeitsblatt W 214.* Stadtwerke Trier, EBI, Karlsruhe, in German.

Lane, R.W., T.E. Larson, and S.W. Schilsky. 1977. The Effect of pH on the Silicate Treatment of Hot Water in Galvanized Piping. *Jour. AWWA,* 69(8):457–460.

Legrand, L., and P. Leroy. 1984. *Influence of Major and Minor Elements in Water on the Corrosivity on Iron Metals.* Paper presented at the Fifteenth Congress of AIDE-IIWSA, Nov.

———. 1990. *Prevention of Corrosion Control and Scaling in Water Supply Systems.* London: Ellis-Horwood.

Leroy, P. 1983. Corrosion Control of Feeders Carrying Potable Water from Desalination of Sea Water. Case of the Yanbu Medina Water Supply System (Saudi Arabia). *Desalination,* 44:317–324.

Letterman, R.D., and S. Kothari. 1992. *Design Tools for Limestone Contactors.* Cincinnati, Ohio: US Environmental Protection Agency Water Engineering Research Laboratory.

Letterman, R.D., M. Haddad, and C.T. Driscoll. 1991. Limestone Contactors Steady State Design Relationships. *Jour. Envir. Eng. ASCE,* 117(3):339–358.

Letterman, R.D., C.T. Driscoll, M. Haddad, and H.A. Hsu. 1987. *Limestone Bed Contactors for Control of Corrosion at Small Water Utilities.* EPA/600/S2-86/099. Cincinnati, Ohio: US Environmental Protection Agency, Water Engineering Research Laboratory.

Loewenthal, R.E., and G.v.R. Marais. 1974. *Carbonate Chemistry of Aquatic Systems.* Ann Arbor, Mich.: Ann Arbor Science Publishers.

McKee, J.E., and H.W. Wolf. 1963. *Water Quality Criteria.* Publication 3A. Sacramento, Calif.: State Water Quality Board.

Mullen, E.D., G.M. Johansen, and A.A. Ubdelraman. 1989. *Effects on Packed Tower Air Shipping on Corrosion Rate.* Paper presented at AWWA Annual Conference, June 18, Los Angeles.

Murray, W.B. 1970. A Corrosion Inhibitor Process for Domestic Water. *Jour. AWWA,* 62(10):659–662.

Neff, C.H., M.R. Schock, and J.I. Marden. 1987. *Relationship Between Water Quality and Corrosion of Plumbing Materials in Buildings.* Illinois State Water Survey Project Report for Grant No. CR808560 10, USEPA, EPA/600/S2-87/036.

Parsons, W.A. 1965. *Chemical Treatment of Sewage and Industrial Wastes.* Washington, D.C.: National Lime Association.

Reiber, S., J.F. Ferguson, and M.M. Benjamin. 1987. Corrosion Monitoring and Control in the Pacific Northwest. *Jour. AWWA,* 79(2):71–74.

Rickard, D., and E.L. Sjöberg. 1983. Mixed Kinetic Control of Calcite Dissolution Rates. *Am. Jour. Sci.,* 283:815–830.

Rodhe, H. 1989. Acidification in a Global Perspective. *Ambio,* 18:155.

Rudek, R. 1975. *Einfluß von Huminsäure auf die Ausfallgeschwindigkeit von $CaCO_3$ aus übersättigtem Leitungswasser.* Diplomarbeit, Universität Karlsruhe, Karlsruhe, Germany.

———. 1979. *Untersuchungen zum Einfluss von natürlichen organischen Wasserinhaltsstoffen auf die Vorgänge bei der Korrosion in Trinkwasserinstallationen.* Ph.D. diss., Universität Karlsruhe, Karlsruhe, Germany.

Ryder, R.A. 1978a. Methods of Evaluating Corrosion. In *Proc. of the Sixth Annual AWWA Water Quality Technology Conference.* Denver, Colo.: American Water Works Association.

———. 1978b. *Seattle Internal Corrosion Study, Phases I, II and III.* Tacoma, Wash.: Kennedy Engineers.

———. 1990. *Effect of Disinfectant Residuals and Phosphate Inhibitors on Corrosion of Water Distribution Facilities.* AWWA Distribution System Symposium Proceedings, 41.

Ryder, R., T. Yeager, and K. Russick. 1994. Pilot Plant and Laboratory Evaluation of Corrosion Inhibitors for Mixed Surface and Groundwater Systems. In *Proc. of the AWWA 1993 Water Quality Technology Conference.* Denver, Colo.: American Water Works Association.

Schindler, D.W., S.E.M. Kaslau, and R.H. Hesslein. 1989. Biological Impoverishment in Lakes of the Midwestern and Northeastern United States from Acid Rain. *Envir. Sci. and Technol.,* 23:573.

Schock, M.R. 1980a. Response of Lead Solubility to Dissolved Carbonate in Drinking Water. *Jour. AWWA,* 72(12):695–704.

———. 1980b. Unpublished Results. Cincinnati, Ohio: US Environmental Protection Agency.

Sheiham, I., and P.J. Jackson. 1981. The Scientific Basis for Control of Lead in Drinking Water by Water Treatment. *Jour. Inst. Water Engrs. and Scientists,* 35(6):491.

Singley, J.E., et al. 1981. Laboratory Studies of Corrosion. In *Proc. of the Eighth Annual AWWA Water Quality Technology Conference.* Denver, Colo.: American Water Works Association.

Skagen, G. 1992. Mandal Water Utility: The Alkalization Project 91–92. Sampling and Monitoring at Selected Test-Sites, City of Mandal, Norway (in Norwegian). Mandal, Norway: City of Mandal.

Stichel, W. 1981. Einfluß von Phosphaten und/oder Silikaten auf die durch Kupferionen ausgelöste Lochkorrosion in verzinkten Warmwasserrohren. *Werkstoffe und Korrosion,* 32:19.

Taylor, J.S., L.A. Mulford, S.J. Duranceau, and W.M. Barrett. 1989. Costs and Performance of a Membrane Pilot Plant. *Jour. AWWA,* 81(11):52–60.

USEPA (US Environmental Protection Agency). 1991a. *Lead and Copper,* Final Rule Correction. *Fed. Reg.* 56(135):32112.

———. 1991b. *Lead and Copper,* Final Rule. *Fed. Reg.* 56(110):26460.

———. 1992. *Lead and Copper,* Final Rule Correction. *Fed. Reg.* 57(125):28785.

Vik, E.A., and T. Hedberg. 1991. Corrosion and Corrosion Control in Drinking Water Systems. In *Proc. of a Corrosion Workshop and Seminar in Oslo, Norway, March 19–21, 1990.* Houston, Texas: National Association of Corrosion Engineers (NACE).

Vik, E.A., R. Storhaug, M. Weideborg, A. Lundar, and A. Lorentsen. 1992. "Aqua-Safe," Establishing a Quality Control Program for Residual Aluminum Reduction. Vannsjø Waterworks, MOVAR, ABB Energy's Aqua-Safe project part 2.1, project report 1, ABB Energy, Oslo, Norway (in Norwegian).

Wagner, I. 1977. Influence of Water Quality and Water Treatment on Corrosion and Coatings in Steel and Galvanized Steel Tubes. In *Proc. EUROCORR '77, Sixth European Congress on Metallic Corrosion.* London: London Soc. of Chem. Ind.

———. 1989. Effect of Inhibitors on Corrosion Rate and Metal Uptake in Drinking Water Systems. In *Proceedings from AWWA's Seminar on Internal Corrosion Control Developments and Research Needs,* pp. 1–17.

Water Codex. 1982. Washington, D.C.: National Academy Press.

Wehle, V. 1979. Einsatz von Silikaten in der Trinkwassernachaufbereitung. *Neue Deliwa Zeitschrift,* 3:94.

———. 1983. Einfluss von Phosphaten und/oder Silikaten auf das Korrosionsverhalten von Installationswerkstoffen gegenüber Trinkwasser. *Brunnenbau, Bau von Wasserwerken, Rohrleitungsbau,* 34(12):427.

Werner, G. 1976. *Untersuchungen zum Korrosionsverhalten feuerverzinkter Installationsrohre in kalten Trinkwässern.* Ph.D. diss., Universität Karlsruhe, Karlsruhe, Germany.

Werner, G., E. Wurster, and H. Sontheimer. 1973. Korrosionsversuche des Zweckverbandes Landeswasserversorgung mit feuerverzinkten Stahrohren. *gwf—Das Gas- und Wasserfach,* 114:105.

———. 1976. Untersuchungen des Zweckverbandes Landeswasserversorgung zum Korrosionsverhalten feuerverzinkter Installationsrohre. *gwf—Das Gas- und Wasserfach,* 117:568.

Wright, R.F. 1988. Acidification of Lakes in the Eastern United States and Southern Norway: A Comparison. *Envir. Sci. and Technol.,* 22(2):178.

9

Corrosion Assessment Technologies

Steven Reiber, Robert A. Ryder, and Ivo Wagner

The objectives of an internal corrosion monitoring program stem from potential health, aesthetic, and economic impacts, which have confronted water utilities in one form or another since the introduction of metal plumbing. Specific objectives of corrosion studies may include

- determining corrosion rates on the variety of distribution materials, including consumers' plumbing materials
- determining the concentration of corrosion by-products in water served to the consumer
- optimizing corrosion control programs to minimize both corrosion and the release of metals from corrosion surfaces
- conducting forensic studies to determine the mechanisms and causes of corrosion-related materials failure
- evaluating corrosion scale, pitting, and tubercle morphologies

Economics and aesthetics aside, the overriding corrosion concern today relates to public health, with an emphasis on the perceived threat generated by the release of metals (particularly lead) from the dissolution of metallic surfaces. The association between chronic low-level lead exposure and adverse health effects is well documented for a variety of public health concerns. Of these, the production of neurologic deficits in the pediatric population has the gravest long-term societal implications;

NOTE: The first edition of this chapter, titled Experimental and Investigation Techniques, was authored by Alfred Kuch, Vernon L. Snoeyink, Robert A. Ryder, R. Rhodes Trussell, and Ivo Wagner.

and because the emerging body of evidence describing the association between these effects and ambient lead levels indicates a stronger causality than previously believed, the US Environmental Protection Agency (USEPA) (as well as the regulatory bodies of other nations) has enacted new regulations substantially reducing permissible drinking water lead levels (USEPA 1989, 1990; Eureau 1991; WHO 1992).

In the United States, the centerpiece of the new regulations has been the 1991 Lead and Copper Rule. It has significantly broadened water utility responsibility and liability, extending beyond the distribution system into the homeowner's residence. Corrosion criteria must now be monitored and regulated throughout the distribution system, up to and including the point of delivery at the consumer's tap. The rule requires that public water systems demonstrate "optimal" corrosion control, a nebulous term, which in practice means that most large water utilities will implement additional treatment strategies to reduce corrosion-induced metal release from plumbing and distribution surfaces (USEPA 1991). The lead concern spotlights corrosion of lead pipe and solders, copper tubing, brasses, and other lead-bearing alloys. This emphasis is moving corrosion monitoring to a higher technological level and giving it a new focus directed at process control and systems optimization.

Corrosion analytic tools vary from simple weight loss procedures to complex and sophisticated impedance spectroscopy. The purpose of this chapter is to introduce the variety of analytic concepts available for corrosion monitoring, discuss their applications and limitations, and provide readers with the appropriate references to guide their analytic efforts.

DIRECT MEASURES OF CORROSION

In its broadest sense, corrosion monitoring falls into one of two categories: direct measurement of the corrosion process (via chemical, electrochemical, or gravimetric means) or some form of indirect measurement relying on water quality information presumably related to the corrosion process. Calcium carbonate saturation indexes, the traditional measures of a water's corrosiveness, fall into this latter category. Although largely discredited as either a corrosion assessment methodology or corrosion control strategy (see chapters 1 and 2), conventional wisdom holds that a properly maintained calcium carbonate deposition layer is protective of the underlying metal pipe. Though not explicitly stated, the assumption is that the calcite scale is impervious, or relatively so, and provides a form of barrier protection that prevents electrolyte contact with the metal surface and restricts diffusion of oxidized species from the corrosion site. A further assumption is the universality of calcium carbonate saturation indexes and that they can be applied regardless of the metal, the nature of corrosion on the metal, or the corrosion products produced by the metal. These suppositions overlook the fact that calcium carbonate scales are both porous and friable and unable to provide meaningful barrier protection unless laid down to a depth that is hydraulically prohibitive. The supposition also disregards the tendency of the corroding metal to develop its own scale, consisting of the oxidized metal (oftentimes in a variety of mineral configurations). Such scales, while not always providing barrier protection, may nonetheless passivate the surface. Calcium carbonate deposition may be incorporated into these scales, but there is little evidence to suggest that its inclusion provides for more effective barrier protection or passivation (Singley 1985; Stumm 1960; Stumm and Lee 1961).

Dissatisfaction with saturation-based indexes has prompted development and refinement of a range of direct corrosion measures with substantially greater analytic utility. The most common direct corrosion measures are grouped into the four categories summarized in Table 9-1, along with applications and limitations. Although the basis of the methodology is different for each category, all techniques relate to a fundamental aspect of the corrosion process and hence give a direct indication of the corrosion rate. A wide variety of technique variations exist within each category, some of which will be discussed in this chapter.

Direct measures of corrosion derive from the rate at which the base metal of the test surface is oxidized. Although the oxidized metal may remain attached to the surface as a corrosion scale, the rate of metal oxidation is commonly expressed in units of a base metal weight loss ($M/L^2 \times t$, where M = standard representation of mass, L = length, and t = time). If the corrosion process is uniformly distributed over the test surface, the weight loss is convertible to a penetration rate (L/t) by dividing the unit area weight loss rate by the metal density (M/L^3). The penetration rate (usually expressed as mils per year [mpy]) describes the rate at which the metal surface is receding due to the corrosion-induced mass loss. It is often used to predict the useful service life of the pipe material. Electrochemical assessment methodologies frequently use current density (A/L^2) as a representation of the electron exchange rate on the corroding surface. Current density can be converted to weight loss by application of Faraday's law and the equivalent weight of the test specimen. Table 9-2 presents corrosion rate conversion factors for a number of different metals.

COUPON TECHNIQUES

Coupon measures represent the cumulative weight loss from a test specimen over an extended exposure period. Of the different analytical methodologies, long-term coupon weight loss is the most useful and readily interpretable corrosion rate measure. Coupon tests give a positive indication of the extent of metal oxidation (the mass of metal lost), as well as visible evidence of corrosion morphology (pitting depth and area). They are useful in the analysis of scale adhesion and mineralogy and are applicable to any type of metal plumbing surface. Coupon exposures are the referee measurement against which most other forms of analysis are calibrated and compared.

Important coupon testing criteria include the following:

- The coupon metal must be representative of the piping material of interest.
- The water chemistry of the coupon exposure must be indicative of the distribution system.
- The hydraulic regime across the coupon must reflect the flow regime across the distribution system pipe wall.
- The duration of the test must allow for development of corrosion scales and/or passivating films that influence the corrosion rate of the underlying metal.

Coupon data are commonly presented as a unit area mass loss per day of exposure (g metal/cm$^2 \cdot$ d). It is common, though not necessarily correct, to assume the metal loss is evenly distributed over the test surface, making it possible to present the data as a surface penetration rate. On metals that generally corrode uniformly (i.e., copper, copper alloys, lead–tin solders) this is acceptable and can be used to calculate an approximation of service life. This type of analysis may, however, be inappropriate for pitting surfaces or other localized forms of corrosion,

Table 9-1 Corrosion assessment options for metal plumbing materials

Methodology	Measurement Form	Applications	Limitations
Coupon exposures	The cumulative measure of metal loss from a well-defined surface over an extended exposure period yields a mass loss rate per unit of surface area.	Metal loss rates generally converted to corrosion penetration rate Provides visible evidence of both corrosion and scale morphology Results regarded as the definitive corrosion measure, frequently used as referee technique for other corrosion measures Can be used to assess pitting propensity and penetration rates	Requires prolonged exposure durations (<90 days) High variability between individual coupon measures Multiple exposures required to establish statistical confidence limits Substantial expense Coupon processing is labor intensive
Electrochemistry	Most electrochemical techniques give a point estimate of the corrosion current density (electron exchange rate per unit area). The current density can be converted to a metal oxidation rate by application of Faraday's law.	Nearly instantaneous measure suitable for on-line process control configuration Useful for screening relative efficacy of large numbers of corrosion control strategies Capable of forensic analysis of corroding surfaces Can be used to develop electrical model of corroding surface	Specimen conditioning critical Generally restricted to uniformly corroding surfaces Absolute correlation with coupon referee techniques can be poor (linear polarization measures)
Oxygen depletion	The reductive consumption rate of dissolved oxygen assumed proportional to the coupled metal oxidation rate.	Yields crude short-term point measure of corrosion rate Analysis performed on substantial pipe section in closed recirculating loop system	Generally limited to rapidly corroding surfaces (i.e., iron and steel) Laboratory-based research tool Competing reductive couples (i.e., chlorine) interfere with measure
Metals release	The corroding surface releases metals to a known volume of contact water. Rate of metals accumulation in solution assumed equivalent to rate of corrosion.	Pipe loops can be designed that represent a laboratory-based facsimile of a plumbing distribution system (tap water metals concentration) Measurements usually made following period of flow stagnation	Test materials require prolonged conditioning in water quality of interest Particulate sloughing from corrosion scales induces high degree of release rate variability Expensive and time consuming Presence of corrosion scales may interfere with test

Table 9-2 Corrosion rate conversion table

Metal	Density, g/cm^3	Equivalent Weight, g	Weight Loss Rates		Penetration Rates			Current Density, $\mu A/cm^2$
			$(g/m^2)/d$	$(mg/dm^2)/d$	$\mu m/year$	mpy*	$mm/year$	
Iron	7,860	27.9	1.0	0.010	46.4	1.81	0.046	3.98
Copper	8,920	31.8	1.0	0.010	40.9	1.60	0.041	3.49
Zinc	7,150	32.7	1.0	0.010	51.0	1.99	0.051	3.39
Lead	11,340	104	1.0	0.010	33.2	1.29	0.33	1.07

*mpy = mils per year

where a representation of uniform metal loss is likely to grossly overestimate remaining service life. In such cases, penetration of the pipe wall will occur in a small fraction of the time predicted by an assumption of uniform corrosion.

Coupon Types and Procedures

There is no single standard regarding coupon geometry, materials, or exposure protocols in drinking water systems. While some coupon techniques have been developed specifically for drinking water distribution systems, others have been borrowed from different industries. The American Society for Testing and Materials (ASTM) has certified two methodological variants for use in the evaluation of metal loss on coupon exposures. The older of these (ASTM D2688-83 method B) relies on flat rectangular coupon specimens mounted on nonmetallic stems and inserted directly into the flowstream of the pipe, usually at an elbow or tee (ASTM 1983a). This technique can be used to make relative assessments of corrosion at different locations in a distribution system and for comparative analyses of corrosion inhibitors. However, since the hydraulic flow lines around a flat coupon positioned midpipe are substantially different from the flow lines at the pipe wall, this coupon technique may be inappropriate when a precise estimate of piping corrosion rates is required. Also, it is frequently difficult to obtain flat coupons that are truly representative of pipe materials.

The second ASTM certified coupon technique (ASTM D2688-83 method C) was developed specifically for water utility applications by the Illinois State Water Survey (ISWS) and was the first to substitute actual pipe sections as the corrosion medium (ASTM 1983b). The advantage of the ISWS approach is that flow conditions within the test section reproduce the effect of pipe hydrodynamics. Thus, the potential effects of impingement corrosion can be studied and the growth and erosion of inhibitor films and oxide scales on the pipe wall can be duplicated. Modifications and adaptations of this technique have been reported, some of which simplify the mounting mechanisms and emphasize the use of multiple coupons.

The basic coupon geometry, conditions of exposure, and duration vary widely in the various protocols. Table 9-3 presents a comparative summary of published coupon protocols, including geometry and exposure conditions. Figure 9-1 provides schematics of some of the coupon exposure rigs and mounting hardware.

Exposure Periods

A rigorous coupon evaluation involves measuring weight loss over an extended period of time and requires coupon sacrifices at multiple points in the exposure cycle. A weight loss against time curve is drawn, and the corrosion rate at any point is the gradient of that curve at that exposure duration. Experience has shown that on most metal surfaces corrosion rates change over the course of the exposure, with the highest corrosion rates occurring at the beginning of the exposure and rapidly decreasing to a lesser and more constant rate. Hence, any comparison of corrosion rates for a particular metal must be standardized to a specific exposure duration. Figure 9-2 presents a weight loss curve typical of cast-iron coupon exposures (Williams et al. 1984). In this example, the weight loss gradient at 120 days was interpreted as the stable long-term corrosion rate. Note that the 120-day gradient is approximately one-half the gradient at 30 days.

The exposure period required to achieve the stable long-term rate will vary as a function of the metal type and water chemistry of the distribution system. In some cases, it may be as low as 3 months, and in others as high as 24 months. Six

Table 9-3 Summary of coupon protocols used for distribution system corrosion measures

Protocol Name	Coupon Geometry	Mounting Arrangement	Exposure Duration	Comments	References
ASTM flat coupon standard Corrosivity of water in the absence of heat in transfer	Flat rectangular coupons stamped from sheet stock (0.5 in. × 4 in. [1.27 cm × 10.2 cm]). Both sides exposed to water flow.	Held on insulating stem and inserted into flow stream of distribution main at elbow or tee.	12 to 24 months	Generally used to monitor corrosion processes on distribution system materials Used primarily with mild steel or cast-iron materials Useful for pitting evaluations	ASTM D2688-83 Method B ASTM G1-81 ASTM G46-76
Illinois State Water Survey Protocol (ISWS) Machined nipple test	Machined pipe nipples (0.75 in. [1.9 cm] ID, 4 in. [10.2 cm] length). Exterior coated with epoxy resin (60 cm^2 [9.2 $in.^2$] exposed surface area).	Fitted into a machined PVC pipe section and plumbed into bypass off distribution main.	Minimum 120 days recommended	Coupon fabrication requires extensive machining Coupon processing is chemically based (acid baths)	ASTM D2688-83 Method C
Modified ISWS Coupon sleeve tester (University of Washington)	Short pipe section (0.75 in. [1.9 cm] ID, 1.5 in. [3.8 cm] length). Exterior phenolic resin coated (20 cm^2 [3 $in.^2$] exposed surface area).	Coupons gasketed and held in clear acrylic sleeve (up to 10 coupons per sleeve). Flow restricted to coupon interior. Easy coupon substitution. Flow controlled to duplicate alternating turbulent and stagnation conditions.	Coupons sacrificed at 30-, 90-, and 180-day intervals.	Coupon processing is mechanically based Measurement precision improved by use of multiple coupon exposures Can be used to monitor corrosion process on residential plumbing materials	Reiber, Ferguson, and Benjamin (1988)
Water Research Center Coupon Rig (WRc, Swindon, United Kingdom)	Flat rectangular coupons (75 mm × 12.5 mm × 4 mm [3 in. × 0.5 in. × 0.15 in.]; 25 cm^2 [3.8 $in.^2$] exposed surface).	Coupons mounted on insulating stems and held in flow control rig.	Coupons sacrificed at periodic intervals for up to 22 weeks.	Developed for use with cast-iron coupons, but suitable for use with other materials Flow rig has high- and low-velocity sections	Williams, Ainsworth, and Elvidge (1984)

ID = internal diameter

Table continued next page

Table 9-3 Summary of coupon protocols used for distribution system corrosion measures (continued)

Protocol Name	Coupon Geometry	Mounting Arrangement	Exposure Duration	Comments	References
Corps of Engineers Research Lab CERL tester	Flat rectangular coupons similar to the ASTM D2688-83 standard	Mounted on insulating stems in special pipe rig, designed to hold a large number of coupons under controlled hydraulic conditions.	Not specified	Coupon surface exposures not typical of pipe wall hydraulics.	Temkar (1988) Temkar et al. (1987)
Ringsäûlen (Engler-Bunte Institut)	Pipe sections (20 mm [0.8 in.] ID, 5–15 mm [0.2–0.6 in.] length).	Coupons (up to 20) with intermediate PVC sections.	Minimum of seven days.	Similar to coupon-sleeve approach described above.	Werner, Wurster, and Sontheimer (1973) Kuch, Sontheimer, and Wagner (1983)
Test pipes TZW Karlsruhe	Pipe originals (≤ 4 in.) excavated 2 m (80 in.) length Test pipes, e.g., steel 40 in., ¾ in. diameter	Direct connection to operational water flow. Short-time closed loop operation for measurement.	3 weeks up to 2 years depending on task	Measured value describes the actual corrosion rate/method release rate. (Not possible for passivating metals.)	Kuch and Wagner (1985) Schumacher, Wagner, and Kuch (1988) Kuch (1988)

ID = internal diameter

months is a common planning period in studies where coupon exposures have not previously been run. In all cases, a statistically meaningful number of coupons must be sampled at representative points in the exposure cycle to determine the variation in corrosion rate from the initial exposure to the more stable long-term condition. Table 9-4 gives estimates of the exposure durations required to develop different types of corrosion data for a number of plumbing material types.

Coupon Processing

The initial processing and polishing steps that a coupon must undergo prior to exposure, coupled with the surface cleaning and restoration steps following exposure, represent a large portion of the total effort necessary to conduct coupon studies. ASTM D2688-83a and ASTM D2688–83b describes a series of procedures that must be carefully performed to obtain reproducible weight losses. The steps to prepare a coupon for exposure include machining and deburring, degreasing, paint stripping, acid wash, application of an exterior epoxy resin coating (nonwetted portion of the coupon), drying, and weighing. Diligence in the preparation steps is critical to ensure the uniformity of all coupon surfaces at the initiation of the exposure cycle. Postexposure processing is conducted to restore the exposed coupon surface to its bare metal condition, while at the same time determining the scale accumulation and metal loss per unit of exposed surface. The ASTM-approved restoration steps include an initial 24-hour desiccator drying period, phenolic stripping of the epoxy resin exterior coating, weighing, scouring of the exposed surface with abrasive powders, exposure to an inhibited acid solution to loosen deposits and etch the surface, ultrasonic bath, water and acetone rinse, 24-hour redry, and final weighing. This combination of physical and chemical processing steps requires several hours of effort for each coupon extended over a two-day period.

An alternative coupon processing approach that eliminates the time-consuming requirements of acid rinses and other chemical procedures has been developed by Reiber, Ferguson, and Benjamin (1988). This approach relies solely on mechanical processing using small high-speed rotating brushes to polish the coupon surfaces prior to exposure, as well as to remove corrosion scales and restore the surfaces postexposure. The processing sequence is performed on

Table 9-4 Estimated exposure duration (months) required for coupon tests on different metals

Metal Type	Development of Long-Term Stable Corrosion Rates	Pitting Rate Evaluation	Comparative Evaluations of Corrosion Inhibitor Relative Effectiveness
Cast iron	6–18	12–24	6–12
Mild steel	12–24	12–24	6–12
Zinc (galvanized iron)	3–6	12–36	1–2
Copper	3–9	12–36	3–6
Lead	6–12	–	6–12
Lead–tin solder	6–8	–	4–6
Red brass	3–9	–	3–6
Asbestos–cement pipe	18–24	–	12–18

— indicates no value.

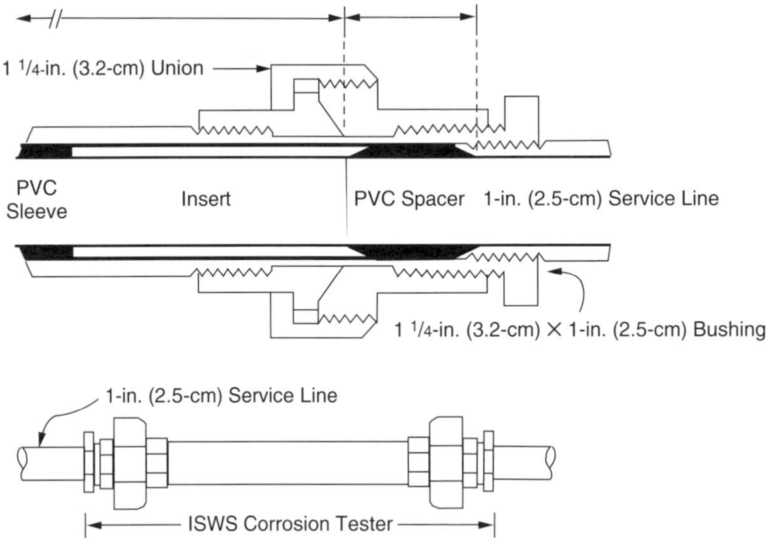

A. Cross section of ISWS machined pipe nipple holders showing coupon inserts, spacers, and assembled corrosion tester.

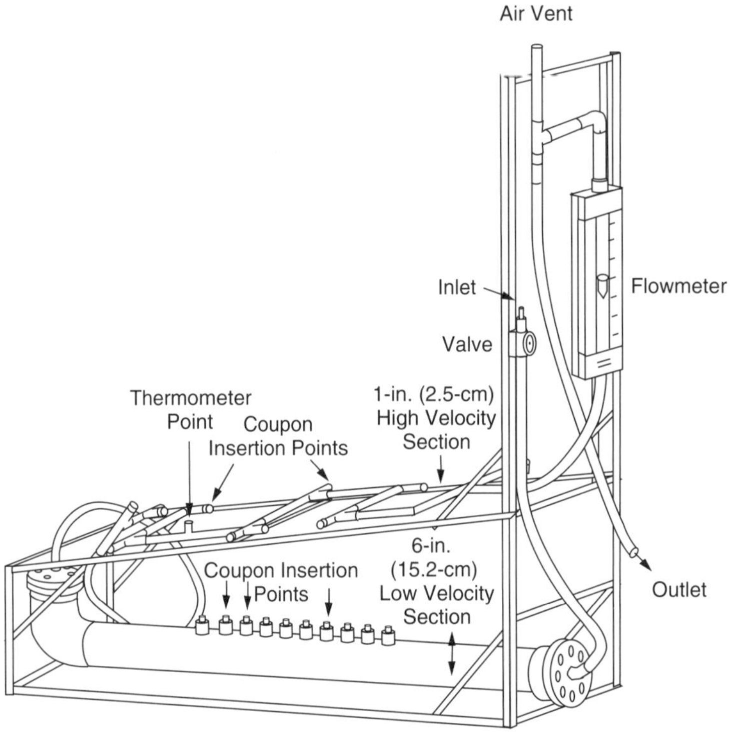

B. WRc corrosion rig with low and high flow sections (uses flat coupon held on an insulating stem).

Source: Figure 9-1A is from ASTM D2688–83 (1983a); Figure 9-1D is from Temkar et al. (1988).

Figure 9-1 Schematics of selected coupon rigs and mounting hardware (*figure continued next page*)

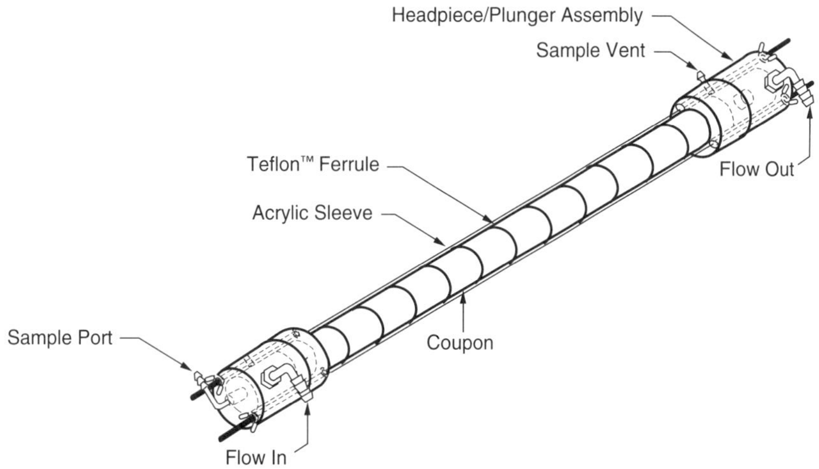

C. Modified ISWS coupon sleeve (University of Washington) mounting up to 10 pipe segment coupons.

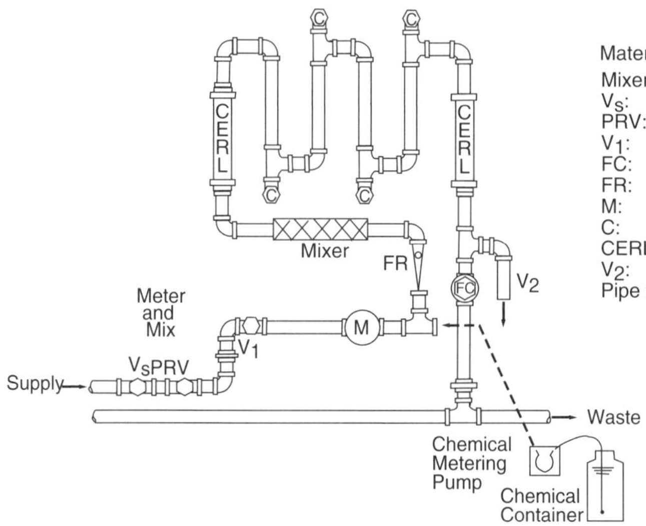

Material listing:

Mixer:	Static
V_S:	Main water supply shut-off valve
PRV:	Pressure regulating valve
V_1:	Test loop isolation valves
FC:	Flow rate controller, PVC, 3 gpm (690 L/h)
FR:	Rotameter, 0–5 gpm (0–1,100 L/h)
M:	Water meter, totalizing
C:	Corrosion coupons
CERL:	Pipe specimens
V_2:	Sample valve, 1/4 in. (0.6 cm)
Pipe fittings:	3/4 in. (1.9 cm), Schedule 40 PVC

D. CERL coupon mounts (flat coupons mounted on insulating stems and inserted at the elbows).

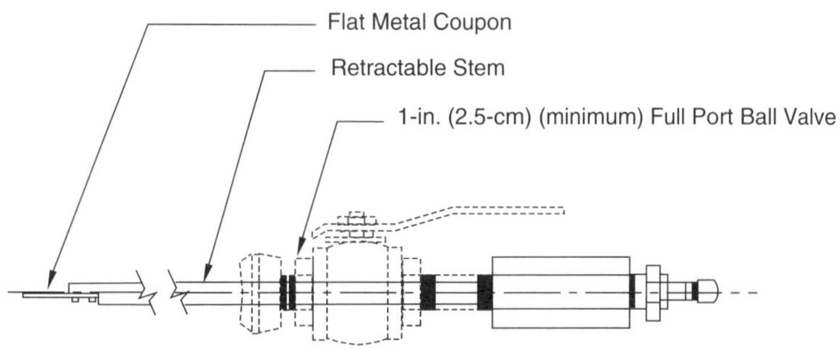

E. Flat coupon mounting hardware with retractable stem (suitable for application in water distribution mains through corporation stop valve).

Source: Figure 9-1A is from ASTM D2688–83 (1983a); Figure 9-1D is from Temkar et al. (1988).

Figure 9-1 Schematics of selected coupon rigs and mounting hardware ***(continued)***

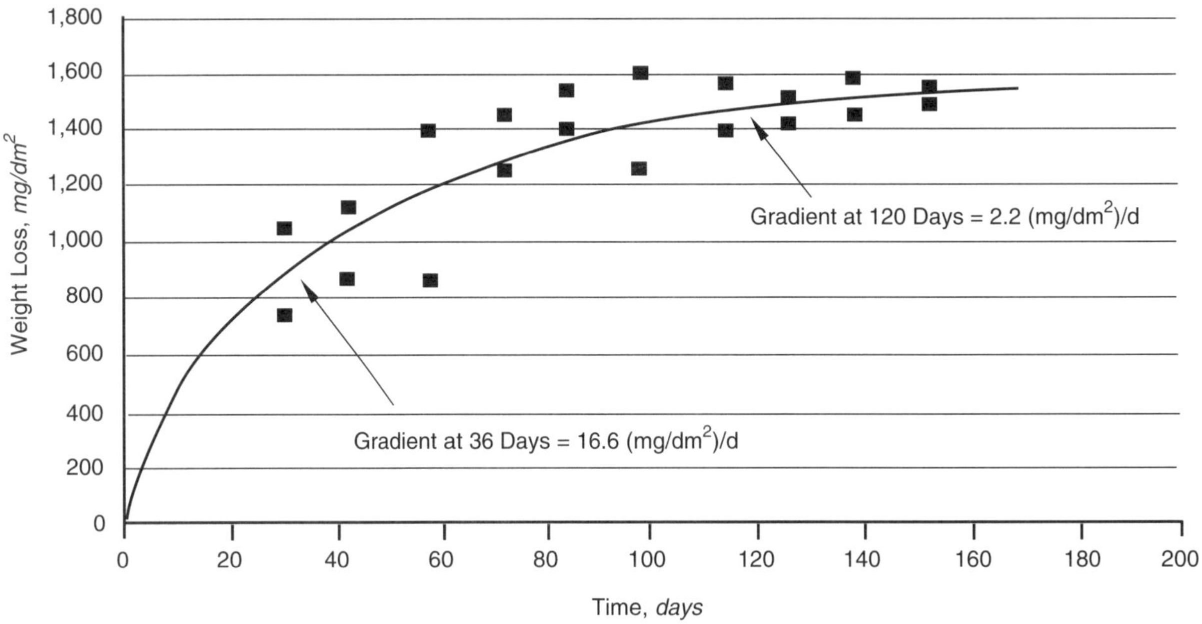

Source: Williams, Ainsworth, and Elvidge (1984). Reprinted with permission of Water Research Centre.

Figure 9-2 Typical weight loss curve for cast-iron coupon exposures; gradient at 120 days approximates the long-term stable corrosion rate

dry surfaces, hence time-consuming chemical baths and redrying are not required. Performed properly, the technique is nondisruptive of the coupon surface and can remove corrosion scales without abrading the underlying metal. The technique gives highly reproducible results, with processing-induced weight loss variations generally less than 0.1 mg/cm^2 of coupon surface area (depending on metal type). The overall processing effort is roughly half that of the chemical.

Precision

The reproducibility of an analytical measure is an indication of its precision. A major drawback to the coupon weight loss analysis has been the high degree of variation between individual coupon measurements, which has limited both the application and acceptance of the technique. This imprecision can result from two sources: first, the complex sequence of handling, preparation, and surface restoration procedures that provide ample opportunity for analyst-induced error; or second, the "natural" variation in corrosion rates produced by differences in metallurgical properties or chemical conditions on the coupon surface during exposure. The first source of error can be addressed by applying the appropriate analytical rigor, but aside from carefully selecting, machining, and exposing the coupon materials, little can be done to alter the effects of the second.

In quantifying the relationship between exposure duration, materials type, and corrosion rate, it was found that a large portion of the measurement variation could be attributed to processing techniques, and that the significance of the variation was related to the duration of the exposure *(shorter exposure durations with lower corrosion-induced weight loss being the most sensitive to processing-induced errors)* (Reiber, Ferguson, and Benjamin 1988). The natural variation in corrosion rate was minor compared to the impact of handling and processing procedures. Depending on material type, the overall dispersion coefficient of the corrosion rate measure (sigma/mean × 100) for a set of identically exposed coupons may be as high as 25 percent, with approximately half of this variation processing induced. This points to an obvious need for analytical rigor in handling and processing the coupons, as well as a need for an overall exposure and sampling plan that considers the number of coupon measures required to achieve the desired level of precision.

Using small-set sampling theory, the number of coupons required to achieve a 90 percent confidence level for a precision of ±10 percent can be readily calculated (Reiber, Ferguson, and Benjamin 1988). Expressed another way, this is the minimum number of coupon exposures required to ensure that the measured mean of the corrosion rate measure comes within 10 percent of the actual mean in 9 out of 10 tests run on the same sample population. Figure 9-3 presents the results of this analysis for coupon sample sets of three different material types. The analysis indicates that for even moderate exposure durations (100 days) multiple coupons (generally more than four) are required to produce even this modest degree of precision. The statistical evaluation also shows that very short-term exposures (<50 days) are impractical because of the excessive numbers of coupons required. In no case could a meaningful level of precision be achieved with a double, or even triple, coupon exposure.

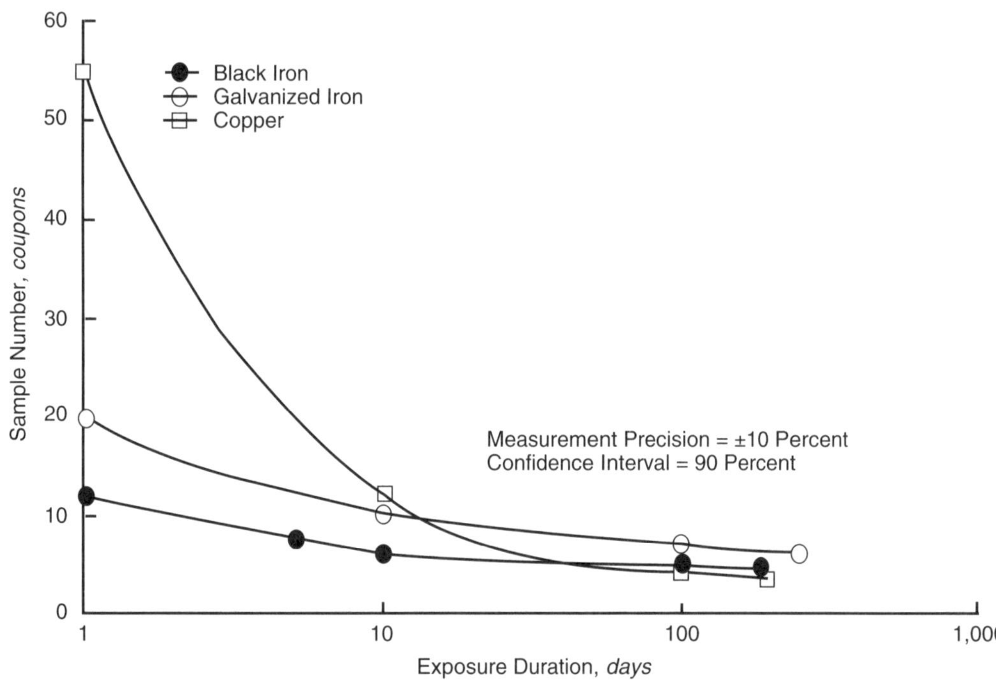

NOTE: The coupon number is that required to give a measurement precision of ±10 percent of the actual mean with a confidence interval of 90 percent. Calculated using a two-tailed *t* test (Student's *t* distribution).

Figure 9-3 Coupon sampling number requirements as a function of exposure duration for copper, black iron, and galvanized iron pipe segment coupons

ELECTROCHEMICAL CORROSION ASSESSMENT

Corrosion is an electrochemical process, and electrochemistry can be a powerful tool in its assessment. Electrochemical techniques can determine the underlying rate of corrosion, as well as characterize the surface reactions that control or limit it. In the past decade, there have been substantial strides in hardware and technique development. The methodology has made the evolutionary adaptation from a purely laboratory-based technology to an automated, operational tool useful for compiling a corrosion history, screening a set of corrosion inhibitors, or optimizing a water quality regime for corrosion control. A decade past, most electrochemical corrosion assessment systems were custom-constructed laboratory units of considerable expense and complexity. It is now possible to purchase an off-the-shelf, computer-driven, research-grade instrument for less than $10,000.

The suitability of an electrochemical methodology is dependent on the corrosion morphology; electrochemical corrosion measures may be inappropriate on surfaces subject to heavy pitting corrosion (i.e., mild steel or cast iron). But, for uniformly corroding surfaces such as copper, solders, zinc, and brass, electrochemical methods can often provide an accurate and nearly instantaneous measure of the corrosion process (Reiber 1989; NACE 1985a). Unlike weight loss methodologies, which are cumulative, the electrochemical techniques give a snapshot of the corrosion process. They define the corrosion rate at a specific point in time and, hence, are of value in determining how short-term changes in water quality and flow conditions influence corrosion processes or how a chemical additives program may be optimized to limit corrosion.

A typical drinking water is a difficult electrolyte in which to make electrochemical measures. Resistivity is high, making the surface potential measurement difficult. The poor buffering capacity and low mineral content of natural waters means that the water chemistry at the corrosion surface may be substantially different than that of the bulk solution, making it difficult to accurately define the chemical conditions of exposure. Corrosion rates in drinking water systems are roughly an order of magnitude less than those encountered in other industries, which means that corrosion current densities are low and difficult to evaluate with a high degree of precision. The measurement problem is compounded by the variety of alloys and surfaces encountered in distribution and residential plumbing systems. While the corrosion character of materials in the petrochemical and other industries is well defined, the electrochemical behavior of plumbing materials in a low-mineral electrolyte is not well understood. Extrapolating corrosion experiences from divergent industries to the water utility's field has been only partially successful (NACE 1985b).

The electrochemical methodologies that have found widest application in distribution system corrosion assessment are those based on polarization measures. Descriptions of three common polarization measures (potentiodynamic scans, linear polarization, and impedance spectroscopy), along with applications, limitations, and references are summarized in Table 9-5. Descriptions of two other nonpolarization electrochemical measures (electrochemical noise and electrical resistance) that have had limited application in distribution systems are also included.

Polarization Fundamentals

On a freely corroding single-metal surface, the electron flow generated via oxidation reactions is consumed by adjacent reduction reactions. This makes it impossible to directly measure the internal rate of electron exchange; if it were possible, a direct measure of the corrosion current could be made. Instead, it is necessary to

Table 9-5 Summary of electrochemical corrosion assessment methodologies

Name	Data Form	Operating Principle	Applications	Precision	Equipment Requirements	References
Electrical resistance	Resistance change in slide wire probe.	Increased electrical resistance on a corroding wire is correlated to a decrease in cross-sectional area due to corrosion-induced metal loss.	On-line continuous corrosion measure.	Poor. Is useful only as a relative indicator of change in corrosion rate.	Slide wire probe Kelvin Bridge Package cost: <$5,000	Rohrback (1956) Cooper (1986)
Potentiodynamic scans	Anodic–cathodic current–potential curve developed from polarization scans on both sides of the freely corroding surface potential.	Electrokinetic interpretation of polarization data—application of the Butler Volmer equation.	Point measures of corrosion current and Tafel constants. Pitting potential.	Yields reliable results on uniformly corroding surfaces. Interpretation of data from pitting surfaces is difficult.	Polarization cell Potentiostat Data logger Data interpretation software Package cost: <$15,000	Dean, France, and Ketcham (1971) Dean (1977)
Linear polarization	Polarization resistance (*Rp*).	Single-point polarization sequence—linear relationship between *Rp* and corrosion rate.	Intermittent on-line corrosion rate measurements.	Variable (dependent upon accuracy of Tafel constant).	Polarization probe (two or three electrode models) Single point potentiostat Package cost: <$5,000	ASTM G59 Mansfeld (1977)
Electrochemical impedance spectroscopy (EIS)	Impedance spectrum (10^2–10^7 Hz).	Polarization frequency response of impressed AC yields impedance spectrum.	Corrosion rate. Electrical model of corrosion surface. Coating effectiveness.	Interpretation of model accuracy is controversial.	Polarization cell Potentiostat Frequency analyzer Data interpretation software Package cost: >$25,000	Silverman (1986) Kendig and Mansfeld (1984) EG&G (1989)
Electrochemical noise	Low-frequency changes in freely corroding surface potential.	Localized corrosion events emit a base signal characteristic of the magnitude and type of corrosion process.	Remains a research tool.	Unproven.	Galvanic coupling cell Frequency analyzer Zero resistance ampmeter Package cost: >$20,000	Hladky and Dawson (1981) Dawson and Uruchurtu (1986)

rely on an artificial perturbation of the corroding surface and, using the associated electrochemical observations, work backward to assess the internal rate of exchange. In essence, the test surface is polarized best by impressing a current on it and the degree to which this shifts the surface potential is measured. By application of fundamental electrochemical principles, information relative to the freely corroding (unperturbed) current density can be extracted.

The fundamentals of polarization were defined 50 years ago in the work of Wagner and Traud (1938), Stern and Geary (1957), and Butler (1924a). The principles are straightforward: a measured current is impressed on a well-defined surface, and the degree to which this alters (polarizes) the surface potential is measured. The reciprocal relationship between potential and impressed current is obtained for a number of potential offsets (in some cases a continuous scan over a limited anodic and cathodic range). The collected data are interpreted via application of the polarization fundamentals developed by Tafel (1904). Figure 9-4 presents a hypothetical polarization curve for a homogeneous metal surface.

The most important recent electrochemical advances have been made in interpretive procedures of polarization data. The utility of these procedures lies in their ability to characterize—with a minimum of analyst effort—the form of the current–potential curve. As the statistical and interpretive power of these programs has

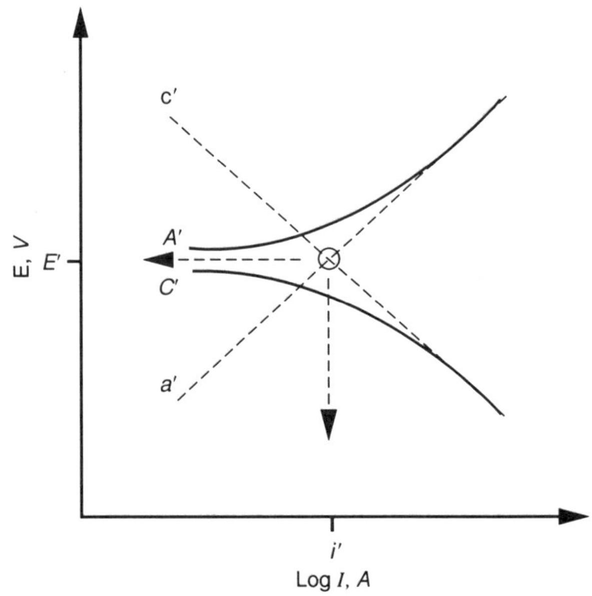

NOTE: The application of graduated current to the corroding surface produces a characteristic polarization curve from which corrosion rate and electrokinetic information can be obtained.

The dashed lines a' and c' give the partial currents of the respective anodic and cathodic processes (the slopes of these polarization curves are termed Tafel slopes). They represent only the current fraction associated with the particular half-reaction. The partial currents are not physically measurable and must be inferred from the linear region of the polarization data.

During a potentiodynamic scan, the total anodic–cathodic exchange current is measured (solid lines A' and C'). These curves represent the sum of the anodic and cathodic partial currents at the indicated potential.

Within any system, the sum of the anodic and cathodic exchanges will equal the sum of cathodic exchanges. For a freely corroding surface, this equivalency is given by the intersection of the anodic and cathodic partial exchange currents (intersection of lines a' and c'). The intersection defines the corrosion current density (i') of the freely corroding surface, as well as the surface potential of the metal (E').

Figure 9-4 Hypothetical polarization curve for a homogeneous metal surface increased, it has become possible to develop accurate representations of the full polarization spectrum using only the polarization data obtained in the non-Tafel region close to the freely corroding surface potential (E_{corr}). This is of significance in a drinking water context, because it allows for minimal perturbation of corroding surfaces that are sensitive to even minor polarization events.

Many of the recently developed software packages provide full graphic functions, sophisticated regression techniques for data fitting, and even experimental simulation options based on modeled parameters. Some popular packages are listed along with applications, limitations, and authors in Table 9-6. Most of these packages are adaptations and refinements on the work of Mansfeld (1973) who developed the original CORFIT software.

Using computers and sophisticated statistical techniques, the analysis packages can be applied to very large data sets, substantially improving the characterization of the current–potential curve. Even data sets with a high noise-to-signal ratio can be accurately analyzed. Figure 9-5 presents a noisy data set and the theoretically defined polarization curve as calculated using Gamry (1992) polarization analysis software.

Electrochemical Limitations

Electrochemical applications in water distribution systems remain hampered by two confounding physical factors: (1) drinking water is a poor electrolyte, and (2) the application of a perturbation current to a slowly corroding system changes the nature of the corrosion surface. The poor conductivity of most drinking waters complicates the polarization analysis by adding an uncompensated internal resistance in the surface potential sensing loop. The problem is most serious at conductivities below 100 µS and for metals with high corrosion rates. The internal resistance (IR) drop distorts the analyst's ability to define the current–potential relationship necessary for an accurate corrosion assessment. Several steps can be taken to limit the IR drop (i.e., polarization cell geometry and placement of reference electrodes), and there are cumbersome procedures that allow estimation of the IR drop and hence quantitative compensation (i.e., current interrupt). Nonetheless, for most conventional approaches, and certainly for continuous on-line measures, IR drop remains a consideration in the application of electrochemical technology to drinking water systems.

The second confounding factor represents a more serious analytic problem that is particularly acute in drinking water systems. In some ways, it is analogous to the "uncertainty principle" of quantum mechanics, wherein, by attempting to measure a physical parameter, that parameter is unwittingly altered in ways that are not readily determinable. The application of a polarizing current to a corroding surface unavoidably alters that surface by changing the rate and extent of the respective electronation–deelectronation reactions. Ideally, this alteration extends for only the period of measurement, after which the surface presumably returns to its prepolarized state. However, if the magnitude of the polarizing current is large relative to the extent of the intrinsic corrosion current (freely corroding current density) and its application is extended, or repeated, it is likely that the applied current may irretrievably change the surface, possibly by promoting or inhibiting passivation processes.

On a rapidly corroding surface this is of minimal concern, since even in the short-term, the cumulative corrosion current will be orders of magnitude larger than any periodic perturbation current. However, for the minimal corrosion current densities of distribution system surfaces (in the range of 0.01 to 1 $\mu A/cm^2$), the perturbation current may be substantial, producing significant surface change, which

Table 9-6 Electrochemical corrosion data analysis software

Software Package	Application	Comments	References
CORFIT	Capable of determining Tafel values and corrosion current densities from data obtained in the non-Tafel region of typical polarization curves.	It is the original and remains the most frequently used routine for polarization data interpretation. Observed data is fitted to the electrokinetic model using a least squares method. The usefulness of the approach is restricted by the limited number of data points the fitting routine can handle. Program written in FORTRAN.	Mansfeld (1973)
POLCURR	Similar application range as CORFIT.	Uses an improved linearized least square method of data fitting. Capable of handling up to 51 data points. Designed to run on VAX systems.	Gerchakov, Udey, and Mansfeld (1981)
POLFIT	Capable of estimating anodic and cathodic Tafel slopes, corrosion current density, and polarization resistance, and of establishing the confidence limits for each of these parameters.	Can provide point-by-point ohmic drop compensation (IR correction) for all observed data points. Provides full graphic representation functions. Capable of simulation modeling of hypothetical polarization parameters. Written in FORTRAN 4.1.	Shih and Mansfeld (1991) Jensen and Britz (1990)
Baukamp EIS software	Designed to aid the interpretation of electrochemical impedance spectra; capable of developing electrical model of corroding surface consisting of distinct electrical building blocks.	Utilizes systems approach. Limited to EIS applications.	EG&G Princeton Applied Research (1991)
Gamry software package	Capable of all standard DC polarization data interpretation; full graphics representation; real-time data collection and interpretation features.	Software provides instrument control, data logging, and analytic functions for Gamry polarization instruments. Capable of handling over 2,000 data points simultaneously. Operates in Windows™ environment.	Gamry Corporation (1992)

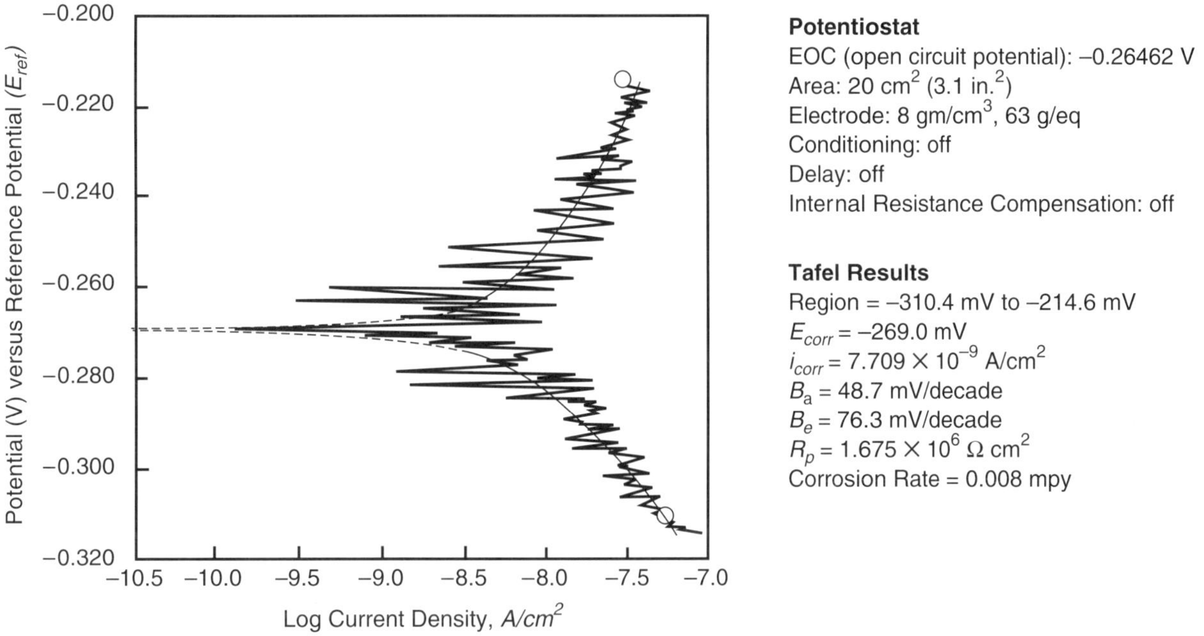

NOTE: The observed data set was generated on a lead–tin solder surface corroding at a minimal rate. Because of the extremely low impressed current densities (nanoamphere range) the scan data have a high noise-to-signal ratio. The excellent fit of the theoretical polarization curve to the observed data set was achieved using Gamry polarization software, which generated the smooth curve by fitting the observed data to the electrokinetic model of Butler (1924b).

Figure 9-5 Experimental polarization data fitted by theoretical electrokinetic model

ultimately invalidates the measure. There are some polarization techniques (i.e., linear polarization) that require only a modest surface potential offset (± 10 mV). However, these are hampered by the fact that as the perturbation signal becomes smaller, the noise-to-signal ratio becomes larger and the accuracy of the measure diminishes. Overall, any electrochemical protocol applied in a drinking water environment should be designed to minimize both the magnitude and the periodicity of the perturbation signal.

Polarization Techniques

The process of forcing the metal surface potential away from the freely corroding state is referred to as polarization. The response of the metal (surface potential versus impressed current) yields a variety of information about the oxidation–reduction reaction occurring on its surface, including the opportunity to model the overall corrosion process. The following polarization techniques describe different methodologies for extracting corrosion rate information.

Potentiodynamic scans. The potentiodynamic scan (PS) is the most fundamental electrochemical corrosion tool. During the PS, the surface potential of the metal test specimen is forced to move in a continuous sweep from a highly cathodic position (relative to the freely corroding potential) to a highly anodic state. Since the corrosion current density i_{corr} cannot be directly measured, it must be estimated from the current-versus-voltage data. This is done by fitting the measured polarization data to a theoretical model of the corrosion process. The electrokinetic model of Butler (1924b) is the most generally accepted model of this process. It assumes that both the anodic and cathodic processes are controlled by the kinetics of the electron transfer reactions on the metal surface. This means that for a freely corroding homogeneous metal surface there are two opposing reactions (the anodic and cathodic exchange reactions) to be considered. The mathematical description of the combination of the cathodic and anodic exchange reactions takes the following form:

$$i = i_{corr} (\exp (2.3 (E - E_{corr})/B_a) - \exp -(2.3 (E - E_{corr})/B_c)) \tag{9-1}$$

Where:

i = measured cell current density, A/cm^2
i_{corr} = corrosion current density, A/cm^2
E_{corr} = corrosion potential, V
B_a = anodic Tafel slope, V/decade
B_c = cathodic Tafel slope, V/decade

The model predicts that at E_{corr} each of the exponential terms equals one, and hence the net overall current is zero and unmeasurable. Slightly removed from E_{corr}, the anodic and cathodic exchange currents are no longer balanced and the net exchange becomes measurable. Each of the exponential terms in this region makes a substantial contribution to the overall current flux and must be considered together. Further removed from E_{corr}, one of the exponential terms predominates and the other term can be ignored. When this occurs, a plot of the log current versus potential (see Figure 9-4) becomes a straight line with a slope equal to the respective Tafel value. The electrokinetic model allows us to extrapolate the linear portions of the log current plot back to their intersection at E_{corr}. The values of the anodic and cathodic currents are equal at that point and represent the freely corroding current density i_{corr}, which also represents the internal rate of electron exchange between anodic and cathodic sites on the unperturbed surface. The rate of electron exchange can be equated with the conversion rate of metal from an elemental to an oxidized state, and hence can be converted to a mass loss rate. The mass loss rate can be presented as a penetration rate (mils per year), which is the conventional representation of a uniform corrosion rate.

The function of the PS is thus to generate sufficient data to apply the electrokinetic model and generate the anodic and cathodic electrokinetic constants (Tafel slopes). When coupled with the polarization scan data, these establish a picture of the electron exchange processes that allows for an accurate estimate of the corrosion current. The important operational parameters for a PS include the scan rate, the maximum potential offset, and the relaxation time between scans. These parameters are specific to the metal specimen and its freely corroding corrosion rate.

Drinking water system corrosion rates are relatively low, and as previously discussed, the possibility of altering the corrosion surface via the measurement process is substantial, hence maximum potential offsets must be limited, and scan rates should be as fast as is consistent with accurate characterization of the polarization response. Relaxation time between scans is necessary for the perturbed surface to regain its prescan corrosion pseudoequilibrium. Table 9-7 presents some critical PS parameters developed on plumbing materials in a low-conductivity drinking water environment.

Linear polarization (on-line measures). Linear polarization (LP) techniques have been used in the water utility field for two decades, principally for on-line corrosion rate monitoring. The technique takes advantage of the nearly linear relationship between the current versus voltage curve when the surface potential offset is in close proximity to the freely corroding surface potential E_{corr}. The slope of this line has units of resistance (Ω) and is referred to as the polarization resistance R_p. The linear relationship usually extends to ±10 mV of the freely corroding surface potential.

An LP measurement involves a single short duration (less than 60 seconds) polarization offset within the linear range. The measured polarization resistance is converted to a corrosion rate by application of a conversion constant derived from the Stern and Geary (1957) equation.

$$i_{corr} = 1/R_p(B_aB_c/2.303\ (B_a + B_c)) \tag{9-2}$$

Where:

i_{corr} = corrosion current density, A/cm^2
R_p = polarization resistance (E_p/i)
E_p = polarization offset (<.01 V)
i = measured current density, A/cm^2
B_a = anodic Tafel constant
B_c = cathodic Tafel constant

The linear polarization measurement provides no information about the values for the Tafel constants, which must be obtained from separate potentiodynamic scans or from experience with the system tested. The largest source of inaccuracy in the LP measurement is derived from the application of a meter constant (B_aB_c /2.303 (B_a + B_c)) that assumes inappropriate Tafel constants. Tafel constants are highly dependent on water quality and test conditions and need to be evaluated for the specific conditions of the test regime.

LP instrumentation is relatively simple and inexpensive. IR drop correction is generally not possible given the simplicity of the instrumentation. Typical commercial LP units suffer from low absolute accuracy but give, nonetheless, a useful indication of relative change in corrosion rate over time. The technique is particularly appropriate for on-line or continuous measurement.

Electrochemical impedance spectroscopy (EIS). An emerging analytic and research tool that may prove useful in drinking water systems is electrochemical impedance spectroscopy (EIS). This technique overcomes some of the electrochemical limitations imposed by drinking water conditions and is well suited for measurement of very low corrosion rates (EG&G 1989). EIS, also referred to as alternating current (AC) impedance, developed rapidly in the past decade, largely due to

Table 9-7 Typical potentiodynamic scan parameters for metals in low-conductivity water

Metal system*	Scan Rate, *mV/s*	Maximum Potential Offset Relative to E_{corr}, *mV*	Minimum Relaxation Time Between Scans, *min*
Copper and copper alloys	0.4	±100	60
Lead	0.2	±100	30
Lead-tin alloys	0.6	±50	240
Zinc (galvanized surfaces)	0.2	±200	30

NOTE: Based on unpublished research of Reiber and Edwards.

*Assumes substantial prepassivation of test surface.

improvements in digital control technology, frequency analyzers, and computer processing of the high frequency waveforms generated by the analysis. Although in its infancy in the water utilities field, it has found a variety of applications in other industries—most notably in the evaluation of protective metal coatings, some of which has carried over into the water utilities field (Kendig and Mansfeld 1984). The basic principle of EIS is similar to that of standard DC polarization techniques: an impressed current perturbs a corroding surface, leading to the interpretation of a current–potential relationship. Instead of a constant direct current (DC) signal, however, an AC is applied, introducing a new analytical variable, the frequency of the excitation signal.

The advantage of EIS is that the AC signal provides more information than a simple DC signal. The EIS perturbation current is applied in a sinusoidal waveform of known amplitude and frequency. The response of the corrosion surface potential in both amplitude and phase (relative to the perturbation current) is analyzed, from which an impedance value can be determined. The system impedance is determined over a broad spectrum of frequencies (hence, electrochemical impedance spectroscopy), and this information is used to construct an electrochemical model of the corrosion surface. In effect, the product of the EIS analysis is an electrical model of the surface constructed of passive circuit elements (resistors, capacitors, and inductors), which, when connected in the appropriate relationship, mimic the polarization behavior of the surface and provide a physical representation of the electrochemical elements that constitute the surface. For example, a simple corrosion surface can be represented as a resistor in parallel with a capacitor, where the magnitude of the resistor represents the polarization resistance of the corrosion surface, and the capacitor represents the double layer capacitance created by the presence of protective scale or passivation layer. Hence, the attraction of EIS technology is that it provides a means of deconstructing a corrosion system into its physical constituencies, in much the same way that we use chromatography to separate complex organic mixtures into individual components. The result is an analytical tool capable of reducing complicated electrochemical systems into a series of more easily studied elements.

EIS offers the potential to advance research on water distribution system corrosion, particularly with regard to passivation phenomena and the application of protective coatings. Because the EIS periodic signal improves the signal-to-noise ratio for very small signals, reliable results can be obtained for surfaces that would otherwise be difficult to measure using a small perturbation current. In effect, EIS allows the polarization resistance to be accurately measured while producing a surface potential shift of only a few millivolts, hence preserving the intrinsic character of the surface and allowing repeated or sustained corrosion measures that do not alter the corrosion rate. This attribute of EIS makes it a promising technique for continuous on-line applications in distribution systems. Additionally, the electrical circuit model that is the end product of the EIS analysis includes as a separate element the internal resistance of the polarization cell electrolyte and, hence, can eliminate this error from corrosion measures in low-conductivity water.

Polarization Cells

Successful electrochemical applications rest on a methodological tripod, two legs of which consist of electronic instrumentation and data interpretation techniques. The third leg is the mechanism and configuration chosen to acquire electrochemical data: the polarization cell. Electrochemical assessment is flexible and can be adapted to a variety of experimental needs, which is underscored by the number and diversity of polarization cells developed for specific applications. In many instances, it has been possible to produce polarization cells that mimic the hydraulics, geometry, and surface conditions of the actual corrosion surface. Although water distribution systems have only recently received attention in this regard, several cell designs are applicable to water research and adaptable for on-line or continuous measurement.

The purpose of a polarization cell is to isolate a test specimen in such a way as to measure its response to an electrical perturbation, while at the same time maintaining a contact environment that closely replicates the freely corroding conditions of its in-service exposure. Common elements of the polarization cell include a test electrode with a well-defined surface area constructed from the material of interest; a stable reference electrode against which the electrical response of the test electrode surface is measured; and a counter electrode that is a part of the forcing circuit to induce electrical change on the test surface (The counter electrode is sometimes referred to as the compliance electrode because its voltage relative to the test electrode is monitored and adjusted to produce the desired current flow to and from the test electrode).

The arrangement of these elements and the geometry of the cell are crucial. To minimize internal resistance, the reference electrode must be in close proximity to the test electrode surface, while the counter electrode should be equidistant from all points on the test electrode surface to ensure even current distribution. Electrolyte circulation in the cell must be maintained to avoid rate-limiting stagnation conditions, and the geometry of the test electrode relative to the electrolyte flow should be representative of in-service hydraulics.

The following cell designs are suitable for laboratory-based research, and some can be adapted to on-line continuous monitoring.

Standard green cell. The standard green cell is typical of polarization cells used to evaluate small metal specimens (Figure 9-6A). The reference, test, and auxiliary electrodes are fixed in place during the testing sequence while the electrolyte is kept in circulation by a mixing apparatus (magnetic stirrer).

Rotating disk electrode. The rotating disk electrode (RDE) is a well-established design that has been used for assessment of numerous materials and electrolytes (Figure 9-6B). It consists of concentric rings (a test and counter electrode) implanted on the end of a nonconducting cylinder. Rotating the cylinder along its axis provides a means of controlling the hydraulics of the electrolyte relative to the corrosion surface. RDE systems are the most widely used cell configurations and can be purchased from a number of vendors. However, they are expensive and the electrode assembly requires special fabrication. RDE systems have been successfully used to investigate corrosion on copper surfaces and some bimetallic lead alloys (Bard 1980).

Rotating cylinder electrode. The rotating cylinder electrode (RCE) is more recent then the RDE (Greef et al. 1985). In this configuration, a conformal test specimen forms a portion of a rotating cylinder wall (Figure 9-6C). Like the RDE, regulation of the cylinder rotational speed provides precise control of the boundary layer hydraulics. Unlike the RDE, the counter electrode in the RCE configuration is removed from the rotating cylinder and exists as a stationary ring encircling the rotating cylinder. Test specimens in the RCE configuration are more easily fabricated than the RDE type. The RCE and RDE systems were both designed to investigate relatively rapid corrosion phenomena, and both systems ensure that mass transport limitations at the corroding test surface can be avoided.

Pipe section flow cell. The pipe section flow cell (PSFC) was specifically designed to make polarization measurements on plumbing materials and uses actual

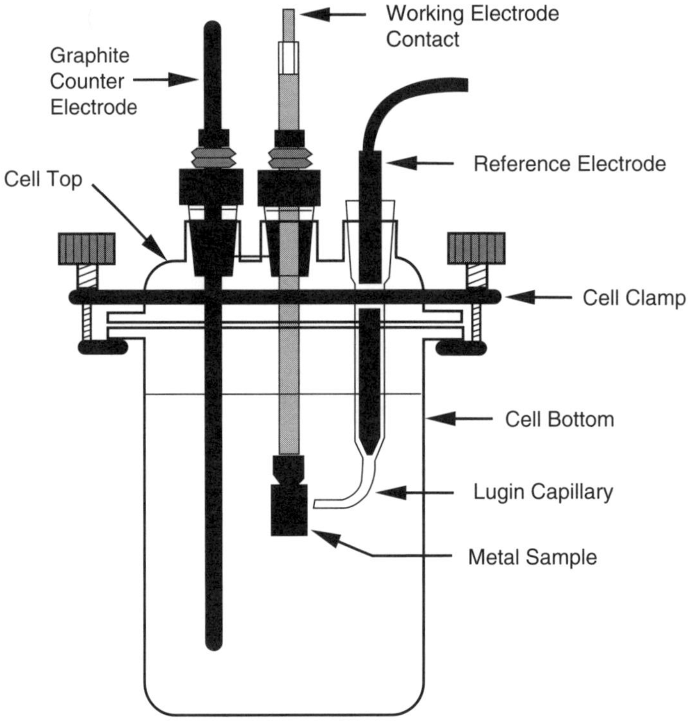

A. Standard Green Cell

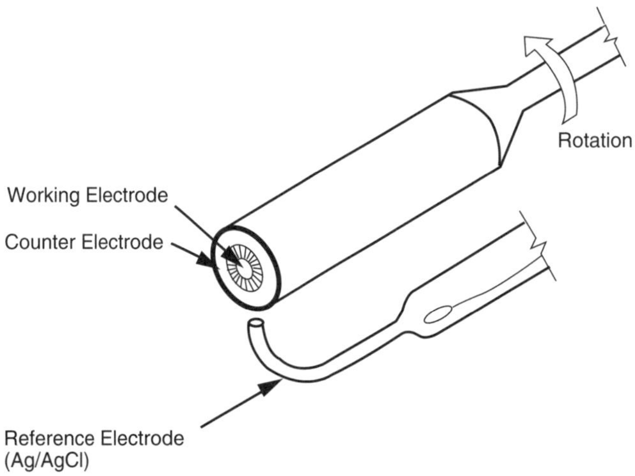

B. Rotating Disk Electrode

Figure 9-6 Polarization cell schematics (*figure continued next page*)

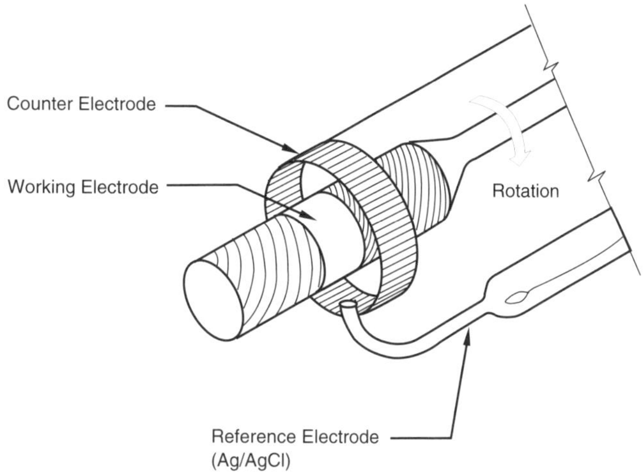

C. Rotating Cylinder Electrode

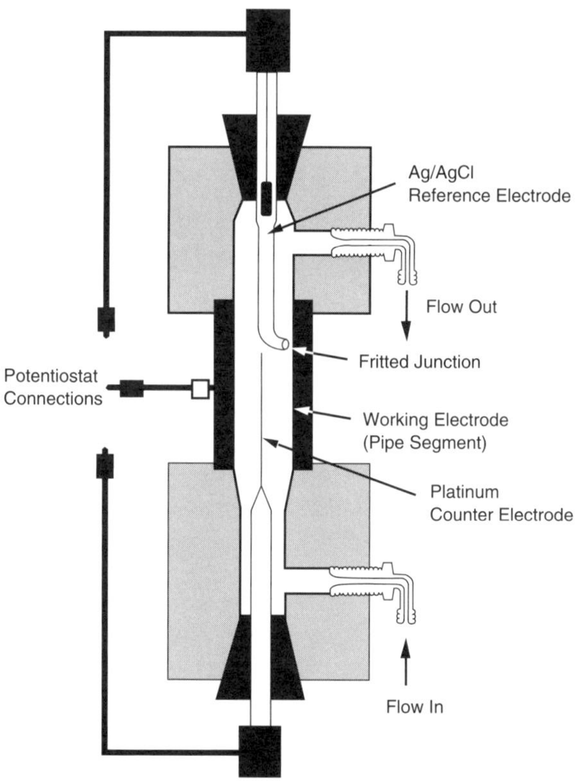

D. Pipe Section Flow Cell (After Reiber)

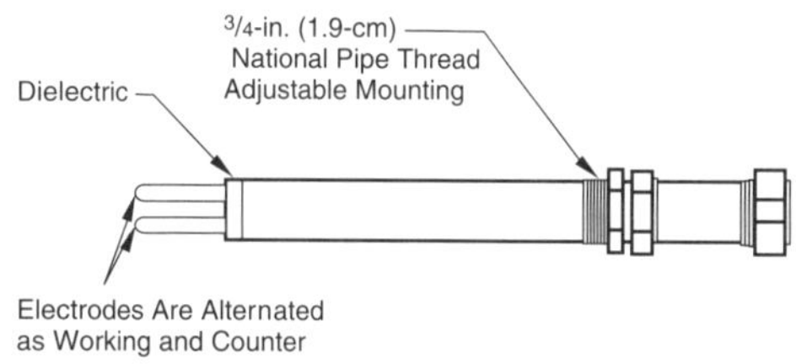

E. Two-Electrode Linear Polarization Probe (Rohrback Cosasco Systems, Santa Fe Springs, California)

Figure 9-6, Polarization cell schematics (***continued***)

pipe sections as the test electrode (Figure 9-6D). The pipe section is mounted between headpieces and water flow is channeled through the pipe section bore; the polarization cell duplicates both the geometry and hydraulics of an actual pipe. A flexible reference electrode extends from the headpiece and can be positioned to the desired point on the test electrode surface; the counter electrode penetrates the cell axially, making it equidistant from all points on the working electrode.

The PSFC has been used to evaluate corrosion on copper, lead–tin solder, and brass surfaces and can be adapted to a variety of pipe diameters (Reiber 1989). Its emphasis on pipe section analysis makes it possible to assess corrosion on residential and distribution plumbing specimens that reflect the true passivation history and surface character of in-service pipe without prolonged conditioning under the water quality in question. The flow cell can be operated as a single pass-through system for on-line assessment of water quality corrosion impacts in the distribution system or can be operated in a closed-loop system with a reservoir, recirculation pump, rotameter, and throttling valve to evaluate specific water qualities (Figure 9-7). In this latter configuration, it is possible to set up test conditions to test the response of in-service pipe materials to anticipated or desired changes in water quality. The precision and validity of laboratory-based screening studies of corrosion control strategies is enhanced by this approach. Ganging polarization cells by mounting different pipe specimens in a single loop facilitates the collection of corrosion data on the variety of plumbing materials that may be affected by a water quality change.

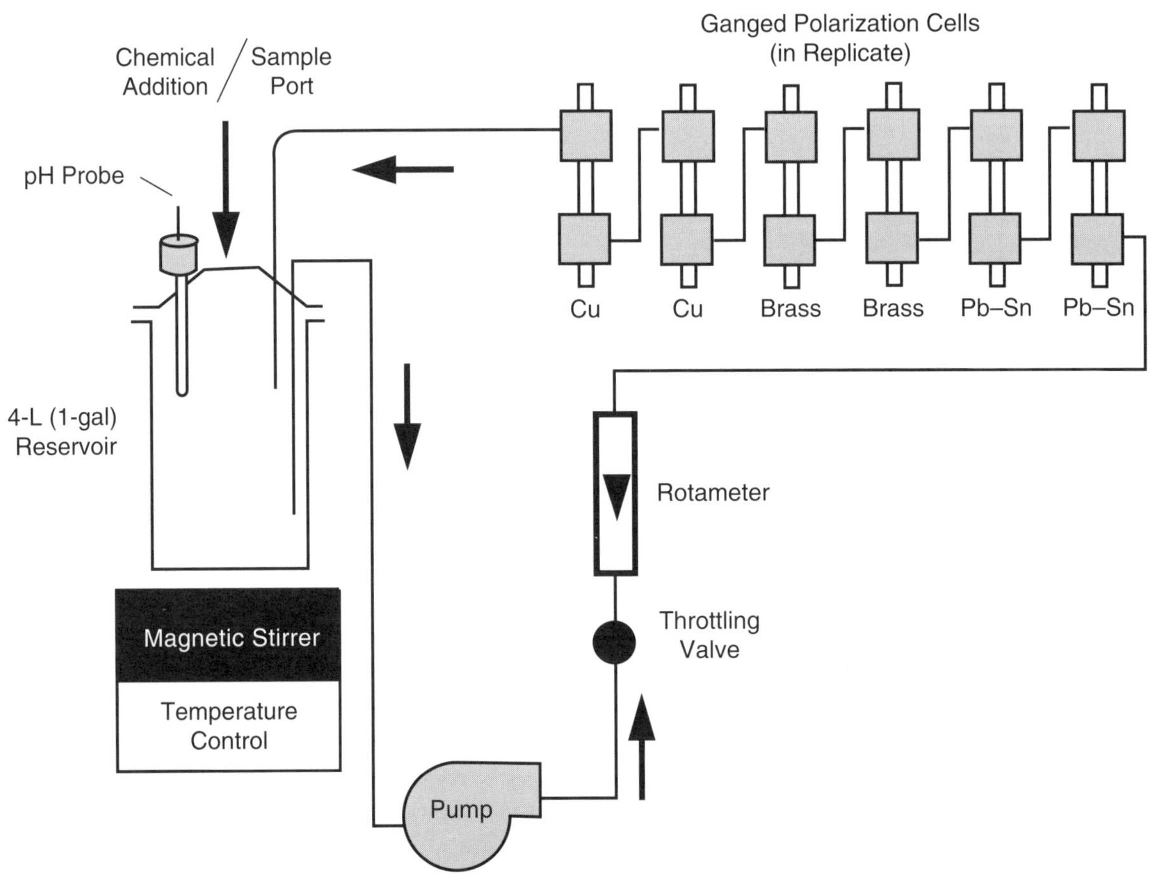

Figure 9-7 Closed recirculation loop with ganged PSFCs

Figure 9-8 shows corrosion rate data collected in an inhibitor screening study on both copper and lead–tin solder specimens using the PSFC. The upper graph presents the water quality regime and inhibitor concentration over the exposure duration. The lower graphs present the electrochemically determined corrosion rates on replicates of copper pipe and lead–tin solder coated pipe segments. The impact of the zinc-orthophosphate inhibitor addition is nearly immediate and produces a corrosion decrease of nearly 40 percent on the copper and over 70 percent on the lead–tin solder surfaces.

Linear polarization probles. Linear polarization probes are available in two- and three-electrode configurations (Figure 9-6E) and can be purchased from several vendors. The probe is intended for on-line monitoring of corrosion processes and can be inserted directly into the flow stream of a plumbing system or a water treatment unit. For simplicity, the reference, test, and auxiliary electrodes are fabricated of the same materials. Absolute accuracy of the corrosion rate measure using these probes is low, but they are useful for measuring relative changes in the corrosion process over time.

Galvanic Polarization Cells

Galvanic cells are used to measure the current flow between dissimilar metals. Cell designs generally consist of two metal specimens held in close proximity (but not touching) and contacting the same fluid stream. The current flow between the dissimilar metals is measured in an external circuit connecting both elements. Galvanic cells can be used to measure both the electromotive differential between the metals as well as current flow. The magnitude and variation in the current-versus-potential relationship are related to the corrosion processes on the test surfaces, which are in turn used to model the galvanic effect of similar couples in the distribution system.

Galvanic cells have been useful in evaluating the corrosion behavior of lead–tin solders joining copper tubing (Oliphant 1983). The electromotive driving force for the solder–copper couple can be as high as 0.45 V. The potential produces a galvanic current that accelerates the solder anode corrosion and induces a higher rate of lead release than would be measured on an independent (uncoupled) solder surface. Because the solder–copper couple is a heterogeneous system, conventional electrochemical analyses are not applicable. Galvanic current assessment is one of the few tools available to explore the coupled surface corrosion behavior.

Figure 9-9 presents a schematic of a galvanic cell consisting of alternate sections of coated solder and uncoated copper pipe connected via an external circuit (Reiber 1991). (The galvanic current flowing in the external circuit is read by an in-series zero resistance ampmeter. The internal reference electrode tracks the changing surface potential on the corroding surfaces as they passivate.) Although not a precise replica of a solder-sweated pipe joint, the cell geometry approximates the configuration of the joint and has similar pipe hydraulics. The configuration allows for multiple pipe sections to be connected in the same flow scheme, thus allowing manipulation of the cathode-to-anode surface-area ratio.

This cell has been used to evaluate passivation phenomena on solder surfaces and has cataloged the impact of important water quality parameters on the solder passivation process. Because it is a direct means of assessing current flow, the galvanic cell offers a simple way to track corrosion currents as a function of exposure duration. Figure 9-10 presents continuously measured galvanic currents on solder–copper facsimile joints as a function of surface aging and exposure pH. Galvanic currents are high on the unpassivated surfaces but diminish as the surfaces

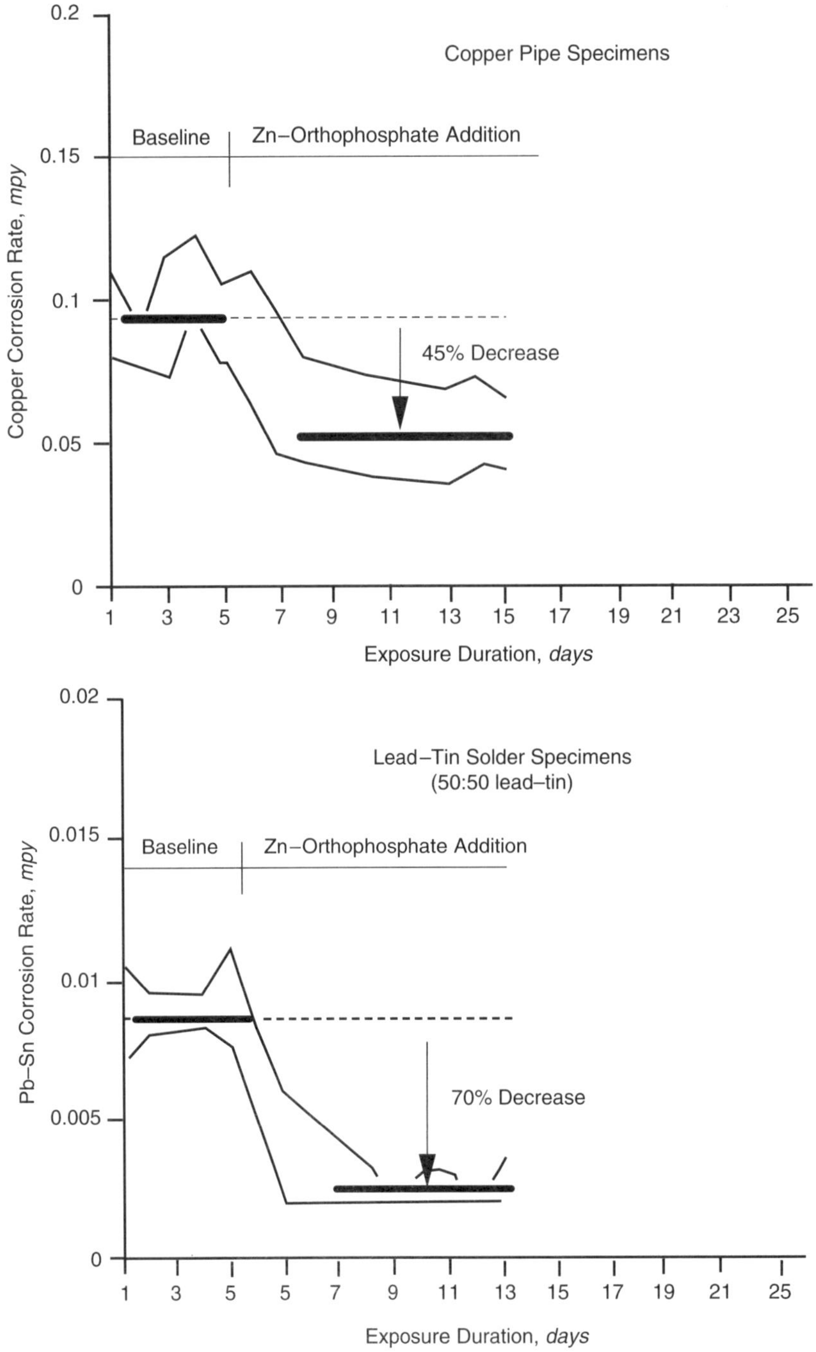

Figure 9-8 Results of a 16-day corrosion inhibitor screening test using the PSFC loop

rapidly passivate. The extent of the passivation is a function of both exposure duration and water quality. The data demonstrate the usefulness of the cell for fundamental investigations of solder corrosion. The cell's simplicity of operation suggests that the cell may also have potential as an on-line process control instrument and may be useful in monitoring the implementation of corrosion control strategies designed to limit lead release.

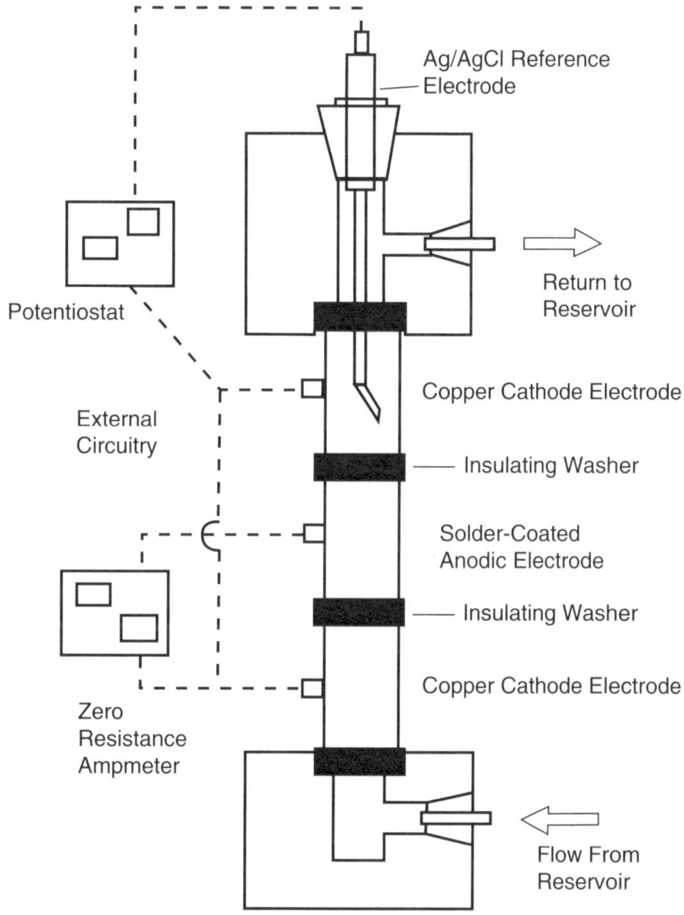

Figure 9-9 Schematic of a galvanic cell designed to stimulate a solder sweated copper pipe joint

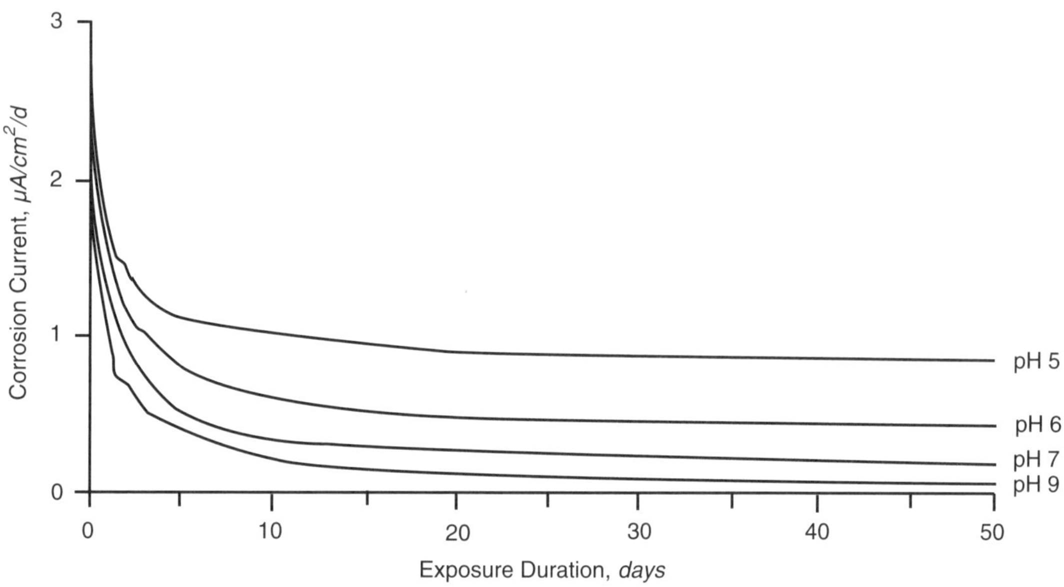

Figure 9-10 Galvanic currents on a lead-tin solder-copper couple as a function of exposure duration and water quality

METAL RELEASE

Metal release tests (sometimes referred to as metal uptake tests) are designed to measure the accumulation of corrosion products in water flowing through a plumbing system or distribution network. When used in a recirculating closed-loop system, the metal release measurement can be interpreted as a point estimate of the corrosion rate. More commonly, metal release measurements support a pipe loop corrosion control demonstration study, where the intent is to simulate a residential plumbing system and evaluate the metals concentrations experienced at the consumer's tap under different corrosion control strategies.

Assessment methodologies that attempt to correlate the underlying metal corrosion rate with the release of corrosion products must recognize the potential for corrosion scale interference. On any metal plumbing surface, the newly oxidized metal must pass through a corrosion scale (or passivation film) before release to the flow stream. The oxide form taken by the metal, the mass of corrosion scale formed on the plumbing surface, and the form in which the metal is released from the scale are functions of water quality, solubility, complexation, and hydrodynamic factors unique to the test system. Residence time within the scale may be substantial, and the scale itself may act as a reservoir of the corrosion products. These factors combined make metal release measurements on scaled surfaces highly variable and sometimes an insensitive indicator of the underlying rate of corrosion.

It should also be recognized that the oxidized metal released from the corrosion scale may be in both soluble and particulate forms (the particulate metal resulting from scale sloughing or hydrodynamic scour). For example, the particulate fraction of the lead measured in standing water samples contacting scaled surfaces often exceeds 50 percent of the total lead release (Hulsmann 1990; Dando and Glasson 1989; Hefny, Badawy, and Egamy 1990). For this reason, the resultant metal concentrations in standing water samples rarely conform to solubility predictions, underscoring the need to take appropriate analytical steps when measuring metal release rates.

Closed-Loop Systems

Closed-loop metal release systems contain a known volume of water in contact with a well-defined corrosion surface, oftentimes a section of pipe removed from a distribution system. The water is recirculated under controlled hydrodynamic conditions and the concentration of metal taken up by the water is tracked over time. The rate of metal concentration increase is used to estimate the loss of metal from the surface of the plumbing material.

Figure 9-11 presents a schematic of a closed-loop recirculation system developed by Kuch, Sontheimer, and Wagner (1983) to measure corrosion rates on mild steel piping. The test pipes that are to be investigated are usually 1 m (3.27 ft) in length and 20 mm (¾ in.) in diameter. These pipes come from pipe rigs operated under controlled conditions, and they are reinstalled after measurements are taken in the closed loop.

For the test, the pipe is inserted into the rig and flushed for approximately 30 minutes with the water quality of interest. The flow-through system is then changed to a recirculation pattern, and the released iron is allowed to accumulate while its concentration is continuously tracked by autoanalyzer. That small portion of the flow that is shunted from the loop for analysis is replaced by an automatic feed system.

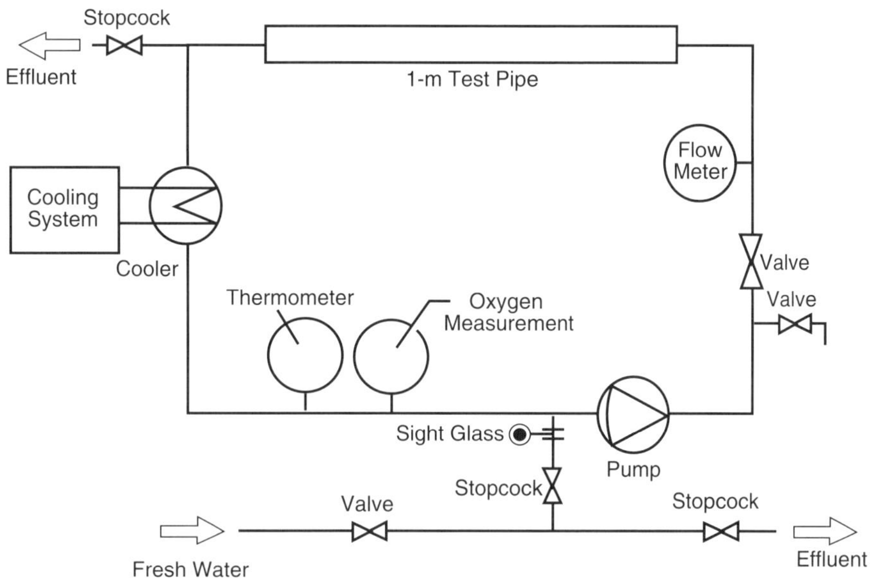

Source: After Kuch, Sontheimer, and Wagner (1983).

Figure 9-11 Schematic diagram of closed-loop device for metal release tests

Figure 9-12 shows a plot of the increasing metal concentration during a 30-minute test cycle. Because the curve is nearly linear, the slope represents a virtually constant metal release rate, which can be interpreted as a steady-state corrosion rate. The iron release rate is calculated as

$$\text{release rate} = dC/dt(V/A)(1.44) \tag{9-3}$$

Where:

release rate = (g Fe/m^2)/d
dC/dt = slope of the concentration versus time curve, (mg Fe/L)/min
V = volume of recirculating water, L
A = exposed interior surface of pipe, m^2

The test for metal release is usually combined with the determination of oxygen depletion and gives additional information concerning the quantity of corroding or corroded metal that is released into the water and not used for the buildup of the corrosion product layer. In the presence of corrosion scales, secondary reactions can lead to metal release without direct connection to the actual metal corrosion (e.g., Kuch mechanism, see chapter 2).

AWWARF Pipe Loop Corrosion Tester

The focus of the 1991 Lead and Copper Rule (LCR) is on limiting the public's exposure to the lead released by corrosion of lead-bearing plumbing surfaces in the home. Under the LCR, most large systems are required to conduct a corrosion control study to determine an "optimal" strategy for reducing lead release rates. In its guidance manual on corrosion control studies, the USEPA emphasizes the use of large pipe loops as both analytic and operational tools to evaluate release rates and select an appropriate control strategy (USEPA1993).

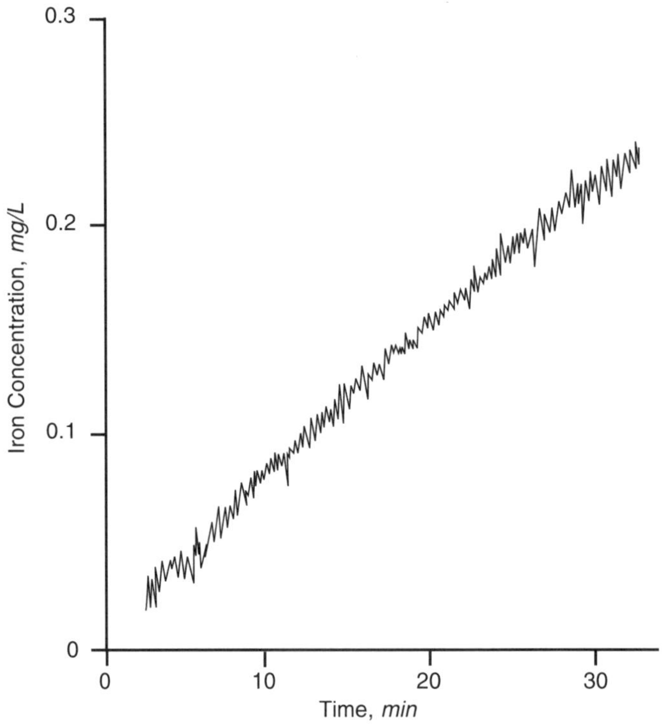

Source: After Kuch, Sontheimer, and Wagner (1983).

Figure 9-12 Original reading of iron accumulation in a metal release test; measurements made on unscaled mild-steel pipe

The American Water Works Association Research Foundation (AWWARF) sponsored a project to develop guidelines for constructing and operating a single-pass pipe loop tester (or pipe rack) representing a facsimile of a residential plumbing network (EES and ISWS 1990). It is designed to provide a surrogate means of evaluating the effectiveness of corrosion control strategies on residential plumbing systems before system-wide implementation of a treatment program. The pipe rack consists of a polyvinyl chloride (PVC) manifold supporting multiple piping material flow paths in parallel (Figure 9-13). Each flow path contains an approximate 60-ft (20-m) length of the piping material of interest. The length and inner diameter of the piping material is selected to ensure that the total water volume held in each path exceeds 1 L (0.26 gal). Flow rate, temperature, pressure, and water quality of the flow streams are controlled. Interrupts are programmed into the flow regime to simulate short and extended stagnation periods (similar to daily household operation). The standard pipe rack contains provisions for coupon inserts so that weight loss measures can be conducted in parallel with the metal release measurement.

The pipe rack may be configured with three replicate flow paths of the same material or it may contain flow paths of different materials, including lead tubing, solderless copper tubing, or lead–tin soldered (jointed) copper tubing. The soldered copper tubing flow path, if present, contains approximately 70 soldered joints, each exposing between 1 and 3 cm^2 (0.15 and 0.45 $in.^2$) of lead–tin solder to the interior flow stream. The solder exposures in contact with the copper tubing constitute a series of galvanic cells. The bimetallic solder alloy is anodic relative to the copper surround; therefore, its corrosion rate is accelerated by galvanic action, which results in a higher release rate for lead. The lead–tin jointed copper tubing is considered

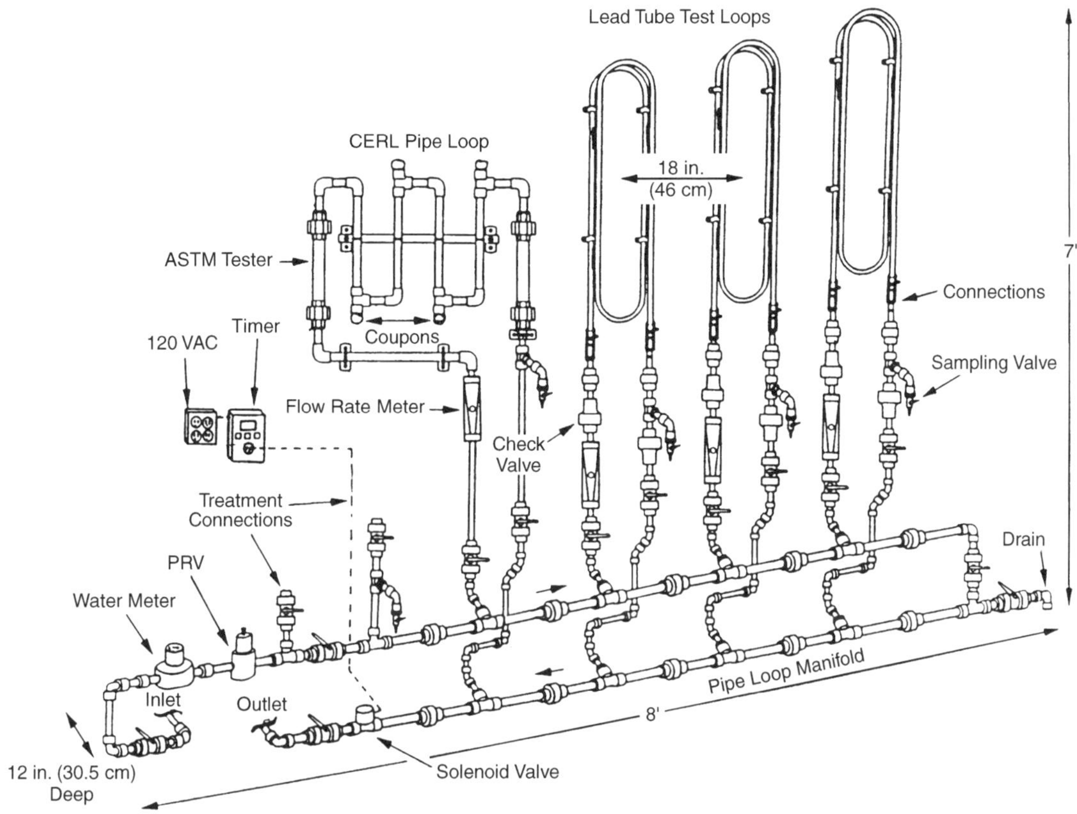

Source: EES and ISWS (1990).

Figure 9-13 AWWARF pipe loop model

to be the flow path most representative of residential plumbing materials and is recommended for residential corrosion impact assessment.

The metal release sampling protocol calls for standing water samples to be drawn from each of the flow paths following a minimum stagnation period of 8 hours. The respective concentrations of metals are used as indicators of likely metal concentrations in standing tap water, which in turn are used as a gauge of water corrosiveness. Metal release measurements are generally made on a weekly basis. The loops require a substantial conditioning period in order to achieve stable metal release rates. Total exposure cycles may run from 3 to 12 months.

A study conducted by Kirmeyer et al. (1993) evaluated the experiences of six utilities using the AWWARF pipe loop protocol to assess corrosion control strategies for controlling tap water lead levels. They found that although the pipe racks require a substantial analytic and fabrication effort on the part of the utilities, most large utilities already have the personnel and expertise on staff to conduct such a study. Moreover, the pipe rack approach can be used to detect differences in the lead and copper levels associated with different types of treatment and can be helpful in determining optimum treatment strategies. They found the pipe rack approach a useful scientific tool for demonstrating the effects of operational changes and treatment modifications on metal levels in the distribution system.

The following important observations relative to the efficacy of the pipe loop approach were reported:

- A high degree of variability exists in the metal release data. Statistical evaluation of pipe rack lead and copper levels shows them to be nonnormally distributed (the closest distribution approximation is lognormal). Nonparametric statistical techniques are necessary to make meaningful predictions of metal release trends resulting from changes in water quality.
- Reproducibility of metal release rates, especially lead, is poor. In studies with three replicate flow paths of lead or lead-soldered copper, one path usually produced statistically different lead levels from the other two loops.
- Lead levels measured from a lead-soldered copper pipe flow path are substantially higher (even after extended conditioning) than the actual residential tap water lead levels measured in the same water. This is also true of lead levels from the lead pipe rack.
- Following an adequate conditioning period, the pipe loop protocol can accurately predict copper levels measured at residential taps.
- Both copper and lead levels in the pipe rack respond quickly to changes in incoming water quality caused by treatment interruptions or modifications.
- Lead levels in the soldered copper loops require three to eight months to stabilize and between six and nine months to stabilize in the lead loops. Copper levels require two to eight months to stabilize in either the soldered copper or pure copper loops.

The principal drawbacks to the pipe loop protocol are the extended conditioning period required of the test system before meaningful data can be obtained and the variable nature of that data. The participating utilities that evaluated treatment alternatives using the AWWARF pipe loop protocol indicate that it can be used to evaluate broad responses in metal release rates to different water quality conditions, but it cannot be used to fine tune treatment processes. Data from the pipe rack should be viewed as a preliminary indicator of treatment effectiveness, rather than as an operational tool.

An alternative to the AWWARF pipe loop system is a pipe loop constructed from modified Illinois State Water Survey coupon sleeve assemblies (Figure 9-14). This system is more compact and allows for concurrent determination of metals release and weight loss on the same plumbing surfaces. Because the water contact surface of each coupon is mechanically polished to a standard finish, there is a high degree of uniformity among sleeves. This improves the reproducibility of metal release measurements among sleeves and decreases both the variability and the stabilization time (Reiber and Lee 1993).

OXYGEN DEPLETION

The depletion of dissolved oxygen within a closed system has been used as a means of measuring corrosion rates on metal surfaces. The test assumes that oxygen reduction is the principal or sole reductive half-cell and is directly coupled to the oxidative half-cell for the metal surface of interest. While generally true for most

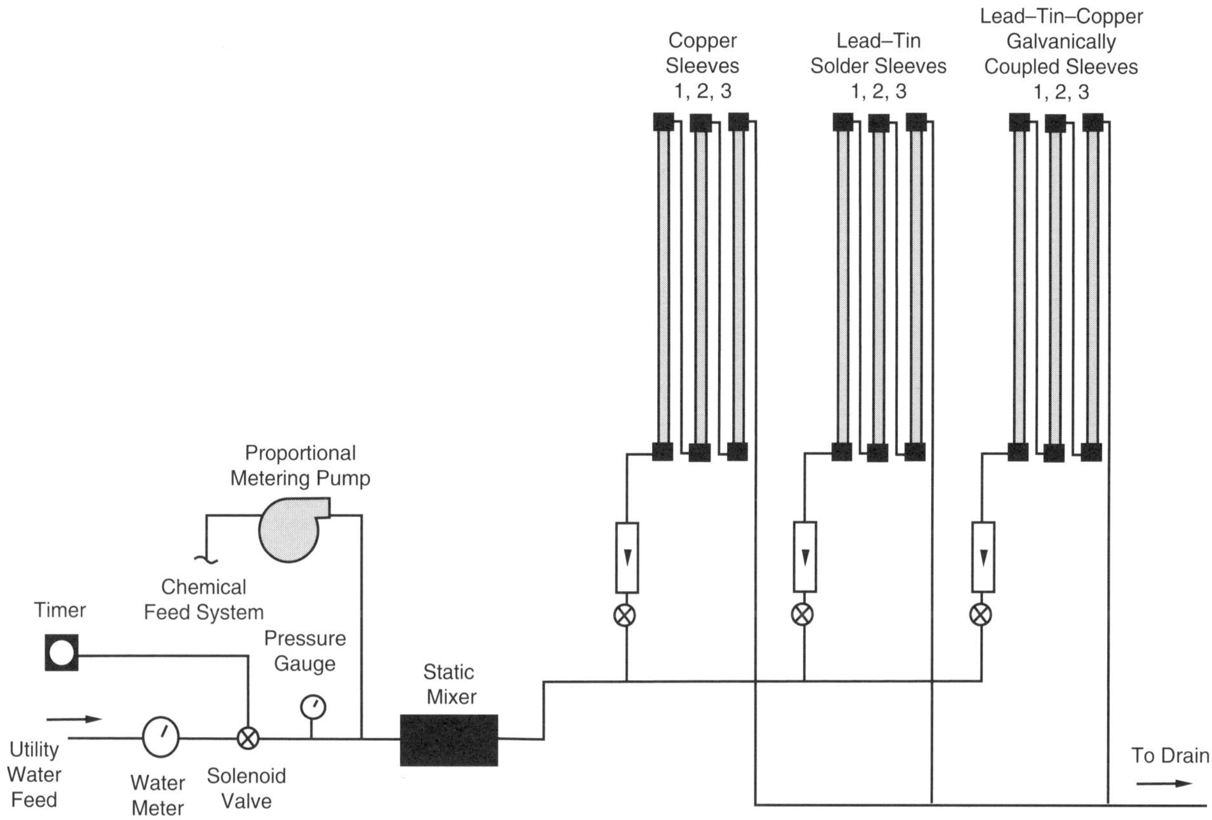

Figure 9-14 Schematic of a coupon sleeve pipe loop; the coupon sleeve pipe loop protocol incorporates both coupon weight loss and metal release analyses

metals in a drinking water environment, this assumption may be inadequate in the case of surfaces undergoing "Kuch-type" corrosion (for exceptions, see chapter 2).

The method was first introduced by Werner (1976) to evaluate corrosion rates for galvanized steel pipes. The technique was modified by Kuch, Sontheimer, and Wagner (1983) who used it to evaluate corrosion rates on mild steel and cast-iron pipes.

The test rig for the oxygen-uptake measurement is similar to the closed-loop recirculation system for metal release measurement (see Figure 9-11). Under controlled conditions of temperature, pressure, and flow, an electrolytic clark-type probe tracks oxygen depletion over time. The rate of oxygen depletion is directly proportional to the rate of the metal oxidation, which can be converted to a metal loss and/or corrosion rate.

$$\mathrm{Me} \rightarrow \mathrm{Me}^{Z+} + Z\mathrm{e}^- \qquad \text{Anodic Site Reaction}$$

$$Z/4\mathrm{O}_2 + Z/2\mathrm{H}_2\mathrm{O} + Z\mathrm{e}^- \rightarrow Z\mathrm{OH}^- \qquad \text{Cathodic Site Reaction}$$

$$\mathrm{Me} + Z/4\mathrm{O}_2 + Z/2\mathrm{H}_2\mathrm{O} \rightarrow \mathrm{Me}^{Z+} + Z\mathrm{OH}^- \qquad \text{Combined Reaction}$$

$$d[\mathrm{O}_2]\,/d\,[\mathrm{Me}^{Z+}] = Z/4 \qquad \text{Oxygen Depletion Rate}$$

The oxygen-uptake technique has the advantage that, unlike the metal release rate measurements, it is not conceptually affected by the presence of an oxide scale. Oxygen consumption proceeds in proportion to the metal oxidation rate irrespective of whether the surface is covered by a scale or whether the scale serves as a reservoir for corrosion products. The technique is, however, limited to corroding systems with relatively high oxygen demands. It has not yet been applied successfully to surfaces with very low corrosion rates, such as the bimetallic solders.

CORROSION MORPHOLOGY ASSESSMENT

All metals corrode. Only morphology and rate are ever in question. Compared to other industries, drinking water utilities experience exceptionally low rates of corrosion in most drinking water, yet the morphologic form of corrosion can be important in determining plumbing material longevity and water quality impacts. A discussion of basic morphologic assessment criteria follows.

Pitting Evaluation

Pitting corrosion is the greatest threat to the service life of plumbing materials. It occurs when an anodic site ceases to randomly migrate over the corroding surface and becomes fixed in place. Pitting is often autocatalytic, in that the rate of corrosion within the pit may increase with the extent of penetration and the development of the pit tubercle. Some metals such as steel or cast iron readily pit because of their metallography; other plumbing materials, such as copper, can be induced to pit under certain water quality conditions. When pitting does occur, it may be important to characterize its morphology so as to distinguish its probable cause and pipe-wall penetration rate. Standard procedures have been developed to evaluate the following pitting characteristics (ASTM G46-76):

Coverage:	Proportion of surface covered by pits
Shape of pits:	Deep, sharp, broad, cratered, or grooved
Pit size:	Maximum and average
Depth:	Maximum, 10 percent deepest, and average
Penetration rates:	Depth/exposure time

Scale Evaluation

Coupon techniques provide information on material lost due to corrosion. They can also be used to determine the types of scales and tubercles likely to form on specific plumbing materials in a particular water quality. The rate of scale–tubercle growth can be determined in multiple-time coupon-exposure tests and used to predict when perceptible clogging and effective loss of flow capacity will occur in small-diameter piping. ASTM defines coupon procedures used to describe the hardness, surface adhesion, color, thickness, surface roughness, and growth rate of corrosion scales (ASTM 1981).

Surface Analyses

Analysis of inhibitor films and corrosion scales is a developing field that utilizes a number of analytic techniques including the following.

X-ray emission spectroscopy. X-ray emission spectroscopy provides a semiquantitative analysis of the elemental concentrations of a corrosion scale or film. An AC arc, or high-voltage spark, is used to excite the chemical elements in the scale, generating an energy spectra display. A characteristic wavelength pattern is produced that can be used to identify both the constituent elements and their approximate proportions.

X-ray diffraction. X-ray diffraction provides a high-resolution quantitative analysis of chemical compounds. It has proven the most reliable method of characterizing the crystalline mineral components of corrosion scales. Amorphous (noncrystalline) compounds cannot be resolved by X-ray diffraction.

Auger electron microscopy. Auger electron microscopy is an analytical technique that bombards the corrosion scale with an argon beam and permits analysis of film composition as it penetrates the film from top to bottom. The corrosion film is decomposed in a progressive manner and provides the spectra of the elements present in each atomic layer. The technique is capable of distinguishing the crystalline structure of the films it penetrates and is useful for characterizing whether the corrosion film is passivating in nature or a barrier-type film. The technique has been used to characterize films as thin as five atoms.

RESEARCH NEEDS

The drinking water community is, arguably, retrograde in the application of corrosion science. Our reticence to adapt new measures is likely attributable to the conservative operations philosophy that public health requirements demand. Whatever the reason, the water supply community has failed to keep pace with advances in corrosion science and has only recently begun to import corrosion technologies developed in other industries. We have developed few methodologies tailored to our own unique analytical requirements and have established even fewer of the methodological standards needed to ensure validity and comparability of corrosion assessment programs. Water purveyors still place far too great an emphasis on outmoded concepts of corrosion assessment, such as calcium carbonate saturation indexes. Moreover, we are faced with regulations that mandate corrosion control efforts based on evaluation of the products of corrosion, rather than the often more sensitive indicator, the corrosion rate itself.

Within the industry, there has been no systematic cataloging of the available information relative to water quality, material type, treatment scenario, metal release, and direct measures of corrosion. Hence, each utility is left, at least partially, to reinvent the wheel when implementing its own corrosion assessment and control program. Methodological and assessment protocols will be served by advances in the following areas.

Standardized Coupon Exposure Procedures

Long-term weight loss measures on coupons composed of actual distribution materials exposed to the drinking water flow provide the most definitive information relative to corrosion rates, surface scales, and pitting propensity. When coupled to water sampling procedures that incorporate flow stagnation periods, coupon exposures can be adapted to yield metal release rates, as well as information predictive of metal concentrations at the consumers' taps.

To serve as a useful regulatory and operational tool and to ensure comparability of results between systems, a standardized coupon methodology is needed. The standard should recognize the statistical variation inherent in weight loss corrosion measures and define the frequency, exposure duration, and sampling locations, as well as specify acceptable coupon handling and processing procedures. A crucial element of this methodology would be the concept of precision, emphasizing the need for multiple coupon exposures coupled to a rigorous and standardized statistical evaluation capable of placing confidence limits about the measured rates.

Development of a Water Quality Based Materials Corrosion Database

As important as any standardized corrosion assessment methodology, and as wanting, is an information database relating empirical corrosion observations on specific material types to water quality conditions. At present, we cannot predict with confidence, relative to specific water quality conditions and material type, the magnitude of the corrosion rate, the degree of scale buildup, the uniformity of the corrosion process, or the degree and type of pitting, if present.

The lack of a coherent database is indicative of the historical overemphasis on indirect measures of corrosion tendencies and shortcomings in current assessment methodologies. It has encouraged the view that every utility's corrosion problems are unique, that results between systems are incomparable, and that each utility requires its own individual and unique approach to corrosion control.

Development of an empirical database derived from direct measures of corrosion rate would allow utilities to compare their water quality conditions and materials considerations to a historical record derived from systems of a similar nature, raw water source, and treatment scheme. It would allow utilities to predict appropriate water quality conditions for minimizing corrosion on particular materials and reduce much of the uncertainty that is now associated with corrosion control implementation. An adequate database would eliminate the confusion brought on by control programs intended to limit corrosion on one metal type, but ultimately enhancing corrosion on a different metal.

The current emphasis on distribution system corrosion and metal release, coupled with the shift toward direct means of corrosion assessment, should provide the incentive to establish and maintain such a database.

Electrochemical Application Standards

Continuing advances in electrochemical measures offer the greatest opportunity for the development of fast, accurate, and inexpensive means of assessing corrosion and its impact on metal surfaces. They present a mechanism for exploring the reactions that drive corrosion, as well as a means of tailoring a corrosion control program to the specific surfaces at risk. Standards for the application of electrochemical measures should be established that reflect the difficulty in making electrochemical measures in a low-conductivity electrolyte and the exquisitely sensitive nature of the corrosion surfaces in question. Standards need to address the following areas:

- geometry of the polarization cell, reference, and counter electrode
- hydraulic and mixing requirements during polarization
- maximum polarization offsets and scan rates
- relaxation times between measures
- IR compensation
- statistical interpretation of polarization data

Linear polarization techniques have the potential to become useful on-line process-control tools for achieving optimal corrosion control treatment. However, the accuracy of the measure relative to other corrosion assessment techniques remains low. Work is required to establish a database of appropriate meter constants for the metal types used in the manufacture of linear polarization probes representing plumbing materials. Moreover, standards are required to define acceptable operating regimes for these instruments.

SUMMARY

Corrosion assessment in water distribution systems is moving away from its historic reliance on indirect measures of corrosion (calcite saturation indices) toward analytic techniques that directly assess the corrosion rate on specific target surfaces, providing a more meaningful and sensitive measure of the corrosion process. The most useful direct measures of corrosion are conducted either by measuring weight loss on test specimens (coupons) following extended exposure periods, by evaluating the electrochemistry of test surfaces during corrosion, or by monitoring the soluble and particulate metals released from a test surface to its contact water. Of these, weight loss techniques are considered the most definitive measure of corrosion and are generally the referee technique against which other assessment protocols are compared. However, weight loss techniques are cumbersome in application, requiring substantial analytic effort, a large number of coupons, and exposure periods often in excess of six months. Metal release methodologies are frequently of more interest to regulators because of the public health considerations related to drinking water lead and copper levels. Several standardized protocols for conducting metal release assessments are now available and are finding widespread application in optimizing corrosion control programs in conformance with the USEPA's Lead and Copper Rule. The analytic protocols with the greatest growth potential are the electrochemical protocols. These techniques have the capability to measure corrosion rates in real time on a variety of materials. The instrumentation and data interpretation procedures have improved substantially in the last decade and will continue to do so. Whereas absolute accuracy has been a problem in the past, new electrochemical methodologies are in development that avoid the common problems of surface perturbation and IR drop. Electrochemical techniques are currently underutilized, but they represent the future of corrosion measurement in water distribution systems.

REFERENCES

ASTM (American Society for Testing and Materials) D2688-83, Method B. 1983a. *Standard Test Methods for the Corrosivity of Water in the Absence of Heat Transfer (Weight Loss Protocol).* Philadelphia, Pa.: American Society for Testing and Materials.

ASTM D2688-83, Method C. 1983b. *Standard Test Methods for the Corrosivity of Water in the Absence of Heat Transfer (Machined Nipple Test).* Philadelphia, Pa.: American Society for Testing and Materials.

ASTM G1-81. 1981. *Recommended Practise for Preparing, Cleaning and Evaluating Corrosion Test Specimens.* Philadelphia, Pa.: American Society for Testing and Materials.

ASTM G46-76. 1976. *Recommended Practise for the Examination and Evaluation of Pitting Type Corrosion.* Philadelphia, Pa.: American Society for Testing and Materials.

ASTM G-59. 1959. *Standard Practise for Conducting Potentiodynamic Polarization Resistance Measurements.* Philadelphia, Pa.: American Society for Testing and Materials.

Bard, A.J., and C.R. Faulkner. 1980. *Electrochemical Methods.* New York: John Wiley and Sons.

Butler, J.A. 1924a. The Kinetic Interpretation of the Nernst Theory of Electromotive Force. *Trans. Faraday Soc.,* 19:729.

———. 1924b. *Kinetic Theory of Reversible Oxidation Potentials at Inert Electrodes. Trans. Faraday Soc.,* 19:(734).

Cooper, G.L. 1986. Proper Electrical Resistance Corrosion Probe Span Selection. In *Electrochemical Techniques for Corrosion Engineering.* Edited by R. Baboian. Houston: National Association of Corrosion Engineers.

Dando, K., and D. Glasson. 1989. Vacuum Microbalance Studies of Lead Deposits From Natural Waters. *Thermochim. Acta,* 152:87.

Dawson, J.L., and J. Uruchurtu. 1986. Noise Analysis of Pure Aluminum Under Different Pitting Conditions. In *Electrochemical Methods in Corrosion Research.* Edited by M. Dupart. Basel, Switzerland: Transtech Publications.

Dean, S.W. 1977. Electrochemical Methods of Corrosion Testing. In *Electrochemical Techniques for Corrosion Testing.* Edited by R. Baboian. Houston, Texas: National Association of Corrosion Engineers.

Dean, S.W., W.D. France, and S.J. Ketcham. 1971. Electrochemical Methods of Testing. In *Handbook on Corrosion Testing and Evaluation.* New York: John Wiley and Sons.

EES and ISWS (Economic and Engineering Services, Inc., and Illinois State Water Survey). 1990. *Lead Control Strategies.* Denver, Colo.: AWWA Research Foundation and American Water Works Association.

EG&G, Princeton Applied Research. 1989. *Basics of AC Impedance Measurements.* Application Note. Princeton, N.J.: Princeton Applied Research Inc.

———. 1991. *Electrochemical Impedance Data Interpretation.* Application Note. Princeton, N.J.: Princeton Applied Research Inc.

Eureau. 1991. European Community Drinking Water Standards, Propositions d'Amenagement de la Directive Eau Potable. 80/778/CEE. Bruxelles.

Gamry Corporation. 1992. *Software Package for Corrosion Analysis.* Langhorne, Pa.: Gamry Corporation.

Gerchakov, S.M., L.R. Udey, and F. Mansfield. 1981. An Improved Method for Analysis of Polarization Resistance Data. *Corrosion,* 37:696.

Grauer, R., P.J. Moreland, and G. Pini. 1982. *A Literature Review of Polarization Resistance Constants.* Houston, Texas: National Assocation of Corrosion Engineers.

Greef, R., R. Peat, L.M. Peter, and D. Pletcher. 1985. *Instrumental Methods in Electrochemistry.* Chichester, England: Ellis Horwood Publishers.

Hefny, M.M., W.A. Badawy, and S.S. Egamy. 1990. Kinetics of Passivation of Lead in Phosphate Solutions. *Electrochim. Acta,* 35(5):799.

Hladky, K., and J.L. Dawson. 1981. The Measurement of Corrosion Using Electrochemical Noise. *Br. Corros. Jour.,* 14:231.

Hulsmann, A.D. 1990. Particulate Lead in Water Supplies. *Jour. IWEM,* 1(1):19–25.

Jensen, M., and D. Britz. 1990. A Linearized Least-Squares Method of Calculation of Corrosion Parameters. *Corrosion,* 46:111.

Kendig, M.W., and F. Mansfeld. 1984. Evaluation of Protective Coatings With Impedance Measurements. In *Proc. of the Ninth International Congress on Metals Corrosion.*

Kirmeyer, G.J., A.M. Sandvig, G.L. Pierson, and C.H. Neff. 1994. Development of a Pipe Loop Protocol for Lead Control. Denver, Colo.: AWWA Research Foundation and American Water Works Association.

Kuch, A. 1984. Untersuchungen zum Mechanismus der Aufeisenung in Trinkwasserverteilungssystemem. Ph.D. diss., University of Karlsruhe, Germany.

———. 1988. Investigations of the Reduction and Re-Oxidation Kinetics of Iron(III) Oxide Scales Formed in Waters. *Corr. Sci.,* 28, pp. 221–231.

Kuch, A., and I. Wagner. 1985. Ursachen der Rostwasserbildung in Trinkwasserverteilungsnetzen und Möglichkeiten zur Sanierung, Veröff. des Bereichs und Lehrstuhls für Wasserchemie, Karlsruhe, Heft 24. Bonn, Germany: Wirtshafts– und Verlagsges. Gas und Wasser mbH.

Kuch, A., H. Sontheimer, and I. Wagner. 1983. Die Messung der Aufeisenung im Trinkwasser—Ein neues Verfahren zur Beurteilung von Deckschichten von Schwarzen Stahlrohren. *Werkstoffe und Korrosion,* 34:107.

Mansfeld, F. 1973. Tafel Slopes and Corrosion Rates From Polarization Resistance Measurements. *Corrosion,* 29:397.

———. 1977. Polarization Resistance Measurements—Experimental Procedure and Evaluation of Test Data. In *Electrochemical Techniques for Corrosion.* Edited by R. Baboian. Houston, Texas: National Association of Corrosion Engineers.

NACE (National Association of Corrosion Engineers). 1985a. *Catalogue of Polarization Constants for Plumbing Materials.* Houston, Texas: National Association of Corrosion Engineers.

———. 1985b. *Distribution Main Corrosion in High-Rise Buildings.* Houston, Texas: National Association of Corrosion Engineers.

Oliphant, R.J. 1983. *Summary Report on the Contamination of Potable Water by Lead from Soldered Joints.* Medmenham, England: Water Research Centre.

Reiber, S.H. 1989. The Electrochemistry of Copper Plumbing Surfaces. *Jour. AWWA,* 81:114.

———. 1991. Galvanic Stimulation of Lead–Tin Solder Corrosion. *Jour. AWWA,* 83(7):97.

Reiber, S.H., and M.M. Benjamin. 1990. A Multiple Set-Point Technique for Electrochemical Corrosion Rate Measurements in Potable Water Distribution Systems. *Corrosion,* 46:11.

Reiber, S., and V. Lee. 1993. Elements of Corrosion Control Demonstration Study. In *Proc. of the AWWA Annual Conference.* Denver, Colo.: American Water Works Association.

Reiber, S.H., J.F. Ferguson, and M.M. Benjamin. 1988. An Improved Method for Corrosion Rate Determination by Weight Loss. *Jour. AWWA,* 79(2):71.

Rohrback, G.H. 1956. Determination of Corrosion Rates in a Variety of Different Processes by Means of the Corrosometer. In *Proc. of the Sixth Regional Conference of NACE.* Houston, Texas: National Association of Corrosion Engineers.

Schumacher, P.G., I. Wagner, and A. Kuch. 1988. Die Trinkwasserversorgung von Göttingen mit Mischwasser. *gwf Wasser/Abwasser,* 129, pp. 146–152.

Shih, H., and F. Mansfield. 1991. Software for Quantitative Analysis of Polarization Curves, Application Note. Los Altos, Calif.: Korros Data Systems Inc.

Silverman, D.C. 1986. Primer on the AC Impedance Technique. In *Electrochemical Techniques for Corrosion.* Edited by R. Baboian. Houston, Texas: National Association of Corrosion Engineers.

Singley, E., R. Pisigian, A. Ahmadi, P. Pisigian, and T. Lee. 1985. *Corrosion and Calcium Carbonate Saturation Index in Water Distribution Systems.* EPA Grant No. Report 805400-01-1. Cincinnati, Ohio: U.S. Environmental Protection Agency.

Stern, M., and A.L. Geary. 1957. *Jour. of Electrochem. Soc.,* 104:56.

Stumm, W. 1960. Investigation of the Corrosive Behavior of Waters. *Jour. San. Engrg. Div.—ASCE,* 86:27.

Stumm, W., and G.F. Lee. 1961. Oxygenation of Ferrous Iron. *Ind. Eng. Chem.,* 53(2):143.

Tafel, J., 1904. *Z. Physik. Chem.,* 50:(641).

Temkar, P. 1988. *Development of a Pipe Loop System for Determining Effectiveness of Corrosion Control Chemicals in Potable Water Systems.* USA-CERL Technical Report N-88/12. Champagne, Ill.: U.S. Army Corps of Engineers Research Laboratory.

Temkar, P., R.J. Schloze, S.W. Maloney, and C.H. Neff. 1987. Pipe Loop System for Evaluating Effects of Water Quality Control Chemicals in Water Distribution Systems. In *Proc. of the Fifteenth Annual AWWA Water Quality Technology Conference.* Denver, Colo.: American Water Works Association.

USEPA (U.S. Environmental Protection Agency). 1989. Drinking Water Regulations; National Primary Drinking Water Coolers That Are Not Lead Free. *Federal Register,* 54(67): 14320–14322.

———. 1990. *Drinking Water Regulations; National Primary Drinking Water Regulations for Lead and Copper.* 40 CFR, Parts 141 and 142. Washington, D.C.: Government Printing Office.

———. 1991. *Lead and Copper Rule Guidance Manual.* Volume I. Washington, D.C.: Government Printing Office.

———. 1993. *Lead and Copper Rule Guidance Manual.* Volume II. Washington, D.C.: Government Printing Office.

Wagner, I. 1984. Corrosion Inhibition in Drinking Water Distribution. Paper prepared for AWWARF/Engler-Bunte Institut corrosion panel.

Wagner, C., and W. Traud. 1938. *Z. Elektrochem.,* 44:391.

Werner, G. 1976. Untersuchungen zum Korrosionverhalten feuerverzinkter Installationsrohre in kalten Trinkwässern. Ph.D. diss., University of Karlsruhe, Germany.

Werner, G., E. Wurster, and H. Sontheimer. 1973. Korrosionversuche des Zweckverbondes Landeswasserversorgung mit Feuerverzinkten Stahlruhren. *gwf—Das Gas- und Wasserfach*, 114:105.

WHO (World Health Organization). 1992. *Health Based Guidelines for Chemicals in Drinking Water.* Geneva, Switzerland: World Health Organization.

Williams, S.M., R.G. Ainsworth, and A.F. Elvidge. 1984. *A Method of Assessing the Corrosivity of Waters Towards Iron.* Swindon, England: Water Research Centre.

10

Organizing Corrosion Control Studies and Implementing Corrosion Control Strategies

Gregory J. Kirmeyer, Ivo Wagner, and Pierre Leroy

This chapter presents a sound technical approach for studying a corrosion problem and implementing a corrosion control strategy. In meeting this objective, the following issues will be addressed:

- framework for corrosion control study
- practical advice and pitfalls of corrosion studies
- presentation and analysis of data
- implementation and monitoring of a corrosion control program

Case studies are presented in appendix C.

IMPACTS OF CORROSION

Corrosion of piping and other water system appurtenances can have a variety of impacts. These impacts can be grouped into these four categories: health and regulations,

NOTE: This chapter did not appear in the first edition.

aesthetics and customer perception, premature piping deterioration and economic impacts, and environmental concerns.

Health and Regulations

From a health standpoint, the most significant metal that can enter the drinking water via corrosion is lead. Lead is known to be a serious cumulative poison and lead intake is known to occur through air, house dust, lead-based paint, newsprint ink, and drinking water. The contribution of drinking water to the total daily intake of lead in the United States is estimated to be 20 percent by the US Environmental Protection Agency (USEPA) (USEPA Lead and Copper Rule 1991). Young children and pregnant women are at highest risk. In drinking water, the primary sources of lead are lead service lines, lead plumbing, brass fixtures, and 50:50 tin–lead solder used to join copper piping. Other potential sources of lead are lead pigtails used to connect the meter to the main, lead counter weights in some meters such as those used for fire service, lead-lined iron piping, lead jointing materials, and sometimes, a lead fraction in galvanizing zinc. Lead enters the water when the water has been standing motionless in contact with the lead source for extended time periods. Federal regulations addressing lead in drinking water came to the forefront in the United States in 1991 with the passage of the Lead and Copper Rule. Other metals of health consequence are cadmium and copper. Asbestos fibers, a documented health concern when inhaled, could also potentially be a health concern when ingested, and debate by the medical community on the role of asbestos fibers in drinking water and cancer is ongoing. Asbestos fibers can enter the drinking water when asbestos–cement (A–C) pipe deteriorates because of solubilization of the cement binder.

Aesthetics and Customer Perception

The metals of most interest from an aesthetic standpoint are copper, iron, and zinc. Copper in drinking water may cause a metallic taste, blue-green particles and discoloration of the water, blue-green staining of porcelain fixtures, staining of swimming pools and hot tubs, and reportedly, green-tinted hair when certain shampoos are used. The primary sources of copper are copper piping used in service lines and consumer plumbing and brass fixtures and faucets.

Iron in drinking water may discolor water, form precipitates, cause rusty water, cause red staining of laundry and fixtures, impart metallic taste to water, and promote the growth of iron bacteria and sulfate-reducing bacteria. Rusty or red water is one of the most common customer complaints received by water utilities. The sources of iron in drinking water related to corrosion are unlined cast-iron piping, unlined steel piping, and galvanized pipe and fittings in plumbing systems that have lost their zinc coating. Approximately 193,566 mi (311,500 km) (22 percent) of all distribution piping in the United States is reported to be unlined cast-iron pipe. Cast-iron piping is also commonplace in Europe. Unrelated to corrosion, iron is also naturally present in many source waters.

Zinc can cause bitter or astringent tastes in drinking water at concentrations of 4 to 5 mg/L. High concentrations may give water a milky appearance at room temperature and a greasy appearance when boiling. Zinc is present in waters largely due to corrosion of galvanized metal in the premise plumbing or small distribution mains.

Premature Piping Deterioration and Economic Impacts

Internal corrosion can have significant adverse economic impacts on the water transmission and distribution systems and consumer plumbing. Typically, consumer plumbing is most seriously affected by internal corrosion because the piping is normally unlined and its diameter is small. Such piping will have a tendency to leak or develop tubercles that reduce both flow and pressure. In recent years, many owners of large-diameter transmission and distribution piping systems typically have protected their investment by installing pipe with internal linings or by cleaning and lining pipe in place. However, there are still many thousands of miles or kilometres of unlined metallic piping in use throughout North America and Europe. Costs to maintain operability over and above the normal accrue when this piping must be replaced prematurely or cleaned and lined in place.

Environmental Concerns

Corrosion of water distribution piping and subsequent leaching of metals raises environmental concerns associated with the presence of lead, cadmium, zinc, and copper in drinking water. The metals enter the wastewater collection system and either pass through the treatment plant and into receiving waters or accumulate in the sludge and end up in a landfill, on croplands, or other locations, depending on the disposal method. Further, the zinc and phosphates present in some corrosion inhibitors are also of concern because they can add to the metal loading to the treatment plant and/or nutrient levels in receiving waters. Metals corroded from water piping are the largest source of these contaminants in the wastewater of many communities. Receiving-water criteria are becoming much more stringent, and the allowable levels of metals in sludge have been lowered depending on its intended use.

DRINKING WATER QUALITY STANDARDS

Corrosion and resulting metals release have caused regulatory agencies to set maximum allowable concentrations for contaminants that include contributions resulting from corrosion or solubilization. Historically, monitoring for metals was performed only in source waters or entry points to the distribution system. In recognition of the source of metals entering the water, monitoring at the consumer's tap is frequently required by many agencies.

North American Regulations

For decades, the United States had a maximum contaminant level of 0.05 mg Pb/L at the entry point to the distribution system. In recognition of the severe health effects of lead, this standard was changed in 1991, and the USEPA mandated that all utilities serving 50,000 persons or more must optimize treatment for control of lead and copper. Further, USEPA established an action level of 0.015 mg Pb/L for lead. The action level is applied by sampling at the consumer's kitchen tap after the water has stood motionless for a minimum period of 6 hours. Because of the variability of lead concentrations that occur by sampling at the consumers' taps, the 90th percentile lead level of all samples collected (100 for the large utilities) should not exceed the action level of 0.015 mg Pb/L. Should the samples exceed the action level at the 90th percentile, the water system, regardless of its size, is required to take several actions, including treatment to control corrosion, additional monitoring, public notification, and public education. In addition, USEPA established an action level of 1.3 mg/L at the 90th percentile for copper.

Concerning asbestos fibers, the USEPA has established a maximum contaminant level of 7.1 million fibers per litre for fibers that exceed 10 μm in length. Corrosive attack on A–C pipe can release asbestos fibers that exceed this limit.

Regarding aesthetic concerns, the USEPA has established secondary maximum contaminant levels of 1.0 mg Cu/L, 0.3 mg Fe/L, and 5 mg Zn/L.

In Canada, the approach is different than in the United States. Health and Welfare Canada has established a maximum acceptable concentration for lead of 0.01 mg Pb/L at the flowing tap of the customer. This is a much less stringent standard than the USEPA regulation, which requires a standing period of at least 6 hours, prior to sample collection.

Drinking Water Guidelines in Other Countries

Different countries and health agencies have proposed finished water qualities to avoid internal corrosion problems in various pipe and house installation materials. These guidelines are summarized in Table 8-8 (chapter 8). One of the main factors in controlling corrosion is pH adjustment; the considerable differences in the pH guidelines are illustrated in Figure 8-3 (chapter 8).

The European Community (EC) has a council directive (EC 1980) relating to the quality of water for human consumption. Relating to corrosion, the EC has a maximum admissible concentration (MAC) for lead of 0.05 mg/L in running water. Where lead pipes are present, the lead content should not exceed 0.05 mg/L in a sample taken after flushing. If the samples are taken either directly or after flushing and the total lead content either frequently or to an appreciable extent exceeds 0.10 mg/L, suitable measures must be taken to reduce the consumer's exposure to lead.

In the German national regulation, calcium carbonate saturation (comparable to the Langelier index [LI] of zero) must be obtained, if necessary by an appropriate treatment procedure. If the pH of equilibrium is above pH 8, it is sufficient to perform pH correction until pH 8 is reached. This regulation is intended to reduce deterioration of drinking water quality in the distribution system as much as possible, especially with regard to heavy-metal release. Exceptions to this rule can only be made if the overall system is composed of stainless steel and plastics.

The national guidelines and consensus views of suitable water quality can be used to guide utilities in corrosion mitigation adjustment. An example developed at a workshop in Oslo, Norway, for a water utility delivering soft water through cement-based and iron materials, with copper and brass plumbing materials in house installations, resulted in the following proposed minimum values for treated water quality (Vik and Hedberg 1991).

pH		7.0
Alkalinity, *meq/L*	>	0.3–0.5
Hardness, *mg Ca/L*	>	10
HCO_3 / (SO_4 + Cl) (*meq/L*) / (*meq/*L)	>	1.5

For Finland, Norway, and Sweden, the recommendation is higher for alkalinity, pH, and calcium concentration (pH 7.5–8.5, hardness 15–25 mg Ca/L, and alkalinity 0.6–1.0 mmol/L).

As seen from Table 8–8 and Figure 8–3, pH, alkalinity, and hardness are considered very important parameters to control. There are several ways to control these parameters and these are described in detail in chapter 8.

STEPS IN ASSESSMENT AND CONTROL OF CORROSION

When conducting a corrosion control study and implementing its findings, there are six logical steps to follow (Figure 10-1):

1. Document extent and magnitude of corrosion.
2. Determine possible causes of corrosion.
3. Develop and assess corrosion control alternatives.
4. Evaluate alternatives and select corrosion control strategy.
5. Document findings in engineering report.
6. Implement corrosion control and monitor effectiveness.

Each of these steps is described in more detail below, drawing on the nine previous chapters along with the experience of the authors for much of its technical content.

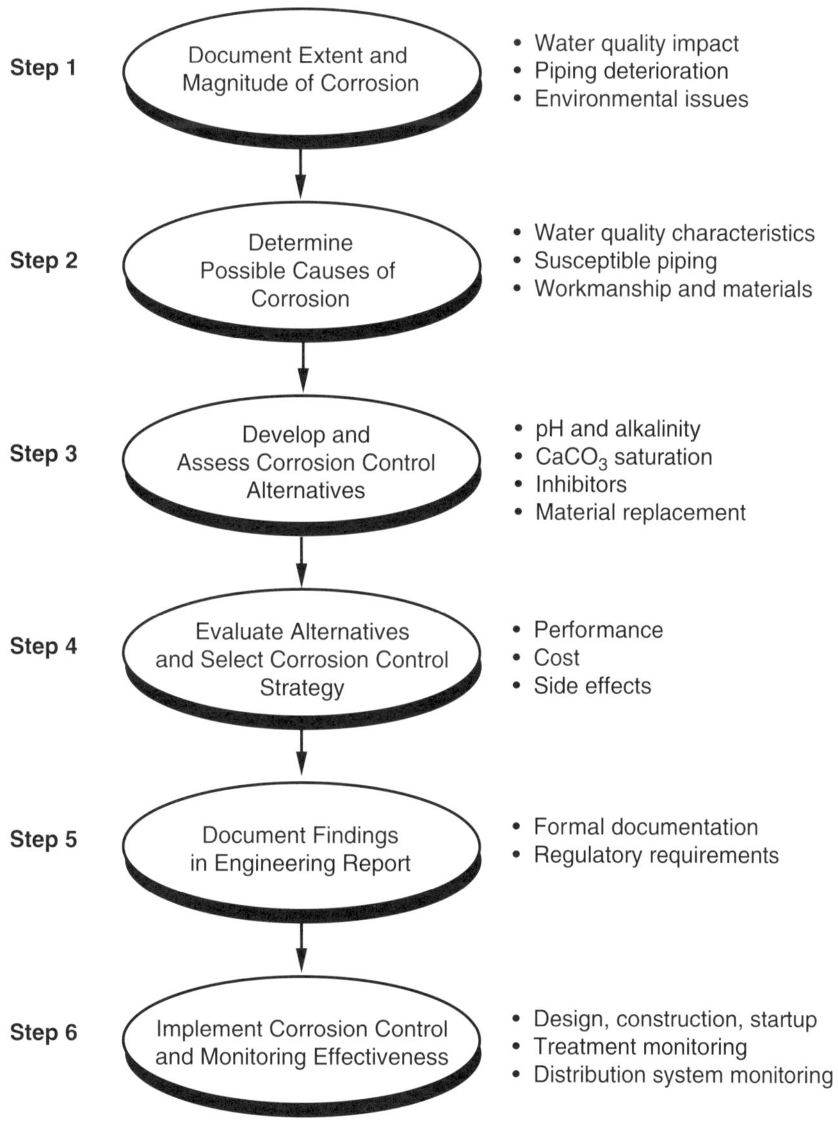

Figure 10-1 Corrosion control program implementation flowchart

STEP 1: DOCUMENT EXTENT AND MAGNITUDE OF CORROSION

This step in the investigation process is aimed at answering the following questions:

- Does the water have a high or low corrosion potential?
- Does the water quality result in significant release of metals such as lead or cadmium, which can have serious health consequences?
- Does the water quality result in release of metals, such as iron, copper, and zinc, which can have aesthetic consequences?
- Does the water result in adverse economic impacts caused by premature deterioration of piping or other appurtenances in the transmission, distribution, or consumer plumbing systems?
- What are the degrees of early failures and lives of distribution system piping, service lines, and consumer plumbing?

This step will likely consist of a combination of office and field investigations, including evaluation of water quality parameters, status with regard to water quality standards, and inspection and analysis of distribution and consumer plumbing system materials.

Corrosion Potential of the Water Supply

The investigator should gather available water quality data from the water sources through the distribution system to the customers' taps and evaluate it for changes in characteristics that would be related to corrosion or deterioration of metals or other materials. For example, changes, particularly increases in the levels of metals such as iron, zinc, cadmium, lead, copper, or calcium, as the water moves through the distribution or consumer plumbing are indicative of corrosion. Collection of first-flush and running water samples from homes and buildings is useful to illustrate metal release from piping as water stands motionless for extended periods in consumer plumbing (refer to Table 10-1). Increases in pH, alkalinity, and calcium content are often caused by the solubilization of calcium, carbonates, and hydroxides present in A–C piping, cement–mortar lining, and concrete structures. This is illustrated in Figure 10-2, which depicts water quality changes as the water stands in a concrete storage reservoir.

It is recognized that almost every water is different in how it responds when exposed to a metal or other material, but there are certain water quality characteristics that are of particular importance. For example, pH has been noted to have a major influence on the potential for leaching of lead, copper, zinc, and cementitious materials. Further, assuming there are some basic water quality data available, corrosion potential can be assessed by comparison of water quality parameters with other supplies that have similar characteristics. Such a comparison is illustrated in Table 10-2, which lists several North American waters, some of which are known to be highly corrosive.

Table 10-1 Percentages of samples that fall within a range of metal values

	Lead, %			Copper, %		
Sample Type	<0.01 mg/L	0.01–0.05 mg/L	>0.05 mg/L	0–1 mg/L	1–2 mg/L	2–4 mg/L
Cold water 1						
First flush, 125 mL	12	64	24	30	6	64
Cold water 2						
Next 1,000 mL	64	36	0	42	9	49
Cold water 3						
Fully flushed	97	3	0	97	0	3

Source: Data from GVWD (1990).

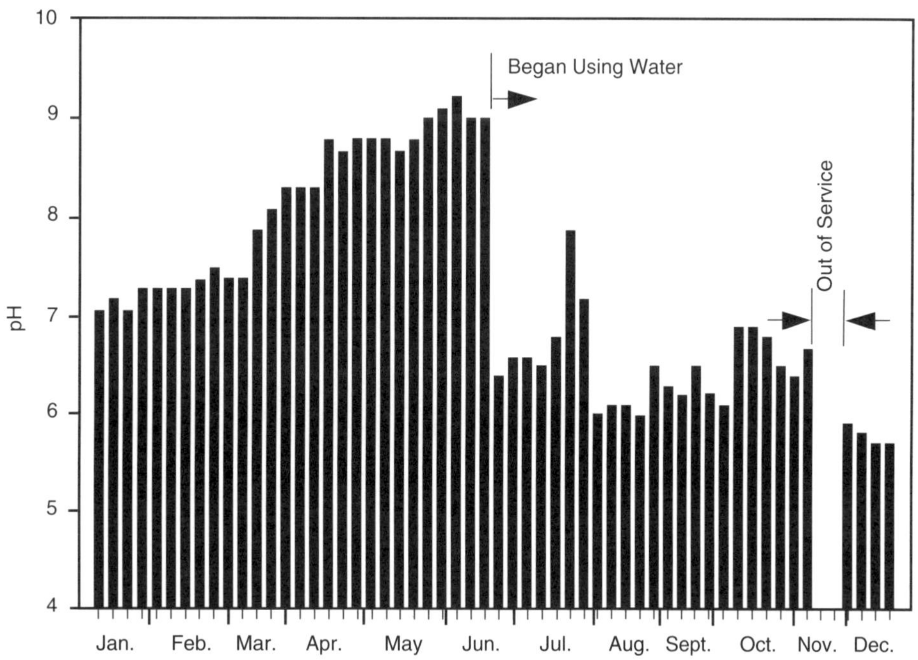

Source: Data from GVWD (1990).

Figure 10-2 The pH rise in GVWD concrete reservoirs as water stands: Central Park 1986

Concerning corrosion indexes, there have been many corrosion indexes developed through the decades, but none has been judged adequate to merit any more than a partial and frequently poor prediction of corrosion. There has generally been too much emphasis on the validity of the indexes to predict and control corrosion. A more appropriate use of these indexes would be to predict whether or not calcium carbonate is likely to precipitate or dissolve under given conditions. Further, some indexes have been used to predict pitting potential. Comparison of the utility's water with an index can be useful if used judiciously and in concert with site-specific testing and other water quality and operational characteristics.

Table 10-2 Characteristics of treated water for several major North American water supplies

Utility	Source	Treatment*	pH	Alkalinity, mg $CaCO_3/L$	Alkalinity, mg Ca/L	TDS, mg/L	LI	Aggressiveness Index	Larson's Ratio	Buffer Intensity, meq/pH unit
Los Angeles (California)	Owens River		8.3	116	25.0	255	+0.3	12.2	0.4	0.12
Metropolitan Water District of Southern California	Colorado River		8.2	125	81.0	697	+0.6	12.6	2.1	0.13
East Bay Municipal Utility District (California)	Mokelumne	A, E, G	9.3	20	8.0	48	−0.4	11.9	0.2	0.08
Metropolitan Water District of Southern California	State Project		8.0	89	32.0	267	0.0	11.9	1.4	0.11
Portland, Oregon	Bull Run	C	7.0	13.4	1.8	23	−3.3	8.8	0.3	0.10
San Francisco, California	Hetch Hetchy	A, E, G	8.0	32	16.0	85	0.2	11.1	0.6	0.03
Seattle Pre 82, Washington	Cedar River	D, F	7.3	13	5.9	40	−2.4	9.6	0.7	0.05
Contra Costa, California	Bollman		8.4	70	17.0	215	0.0	11.9	2.1	0.03
New York City	Catskills		7.0	5–14	3–6.3	40–	−2.7	9.0	1.6	0.08
Seattle Pre 82, Washington	Tolt River	D, F	6.1	3.2	3.4	50	−4.5	7.5	2.8	0.04
Seattle Post 82, Washington		D, E, F	8.2	16	5.0	30	−1.5	10.3	0.5	0.01
Boston, Massachusetts	Wachusett Reservoir	C, E, F, H	8.5	12	3.4	—	—		—	0.01
Tacoma, Washington	Green River	D, G	7.1	19	5.2	—	−2.6	9.5	—	—
Beaumont, Texas	Naches River	A, C, F, H	7.5	25	8.8	40	−1.3	10.2	—	—
Vancouver, British Columbia	Seymour	D	5.8	1.9	1.7	—	−5.3	6.7	2.2	0.08
	Capilano	D	5.7	1.8	1.4	17	−5.5	6.5	2.2	0.08
	Coquitlam	D	5.4	0.5	1.0	16	−6.5	5.5	7.2	0.04
						14				
Recognized Corrosivity			<6.5	—	—	—	<1.0	<10	<0.5	<0.2

Source: Data from J.M. Montgomery (1982).

*Treatment method:
- A = chemical coagulation and filtration.
- B = direction filtration.
- C = chlorination, combined chlorine residual.
- D = chlorination, free chlorine residual.
- E = pH adjustment.
- F = fluoridation.
- G = blending.
- H = corrosion inhibitors.

Health and Regulatory Data

Public health agencies have increased monitoring requirements for health-related corrosion by-products, specifically lead and copper. These data could provide insight into whether the leaching of either one of these metals exceeds the regulatory standards. In the United States, since water standing in consumer plumbing systems must be monitored at the consumer's tap, the data can indicate whether leaching of lead from brass fixtures, 50:50 tin–lead solder, or lead plumbing are problems. Copper can be evaluated in a similar manner.

Presentation of this data in the form of a probability plot as shown in Figure 10-3 can be useful. This enables one to evaluate exceedance levels at various percentiles and whether there are outliers in the data. The histogram of the data presented in Figure 10-4 can be used to provide insight as to whether the data follow a normal distribution or whether it follows some other distribution pattern. The distribution of data is important in the selection of statistical tools to analyze data. Lead data from kitchen taps often follow a lognormal distribution.

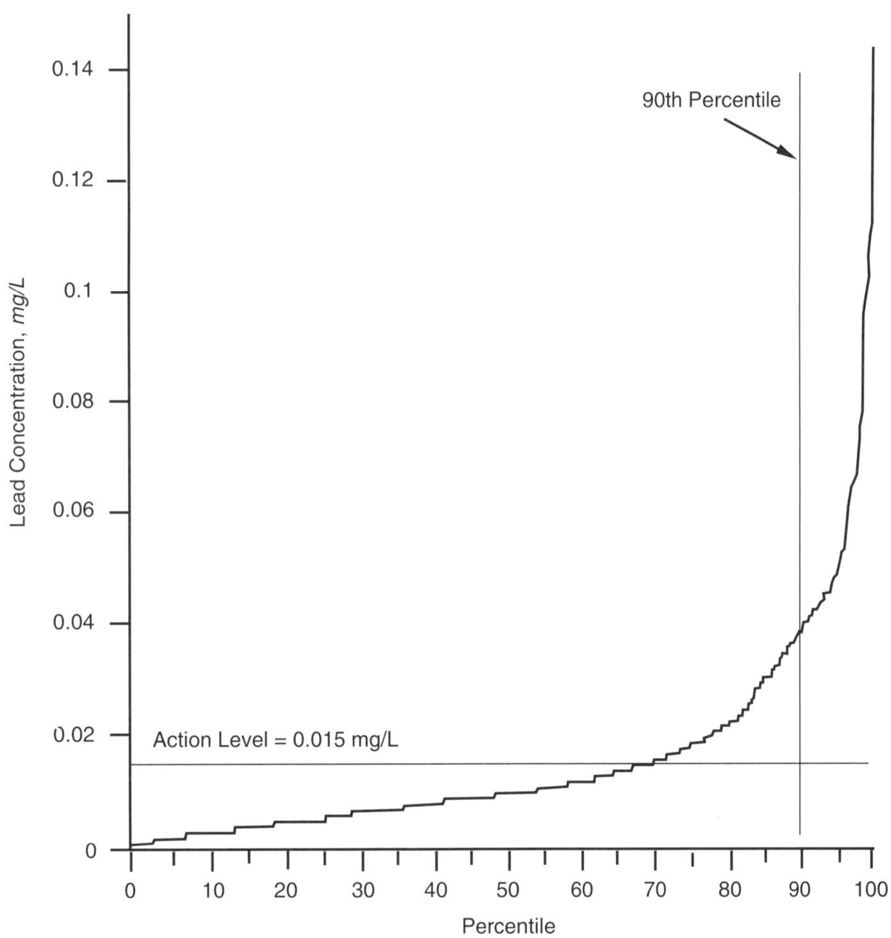

Figure 10-3 Lead probability plot

Aesthetics and Customer Perception

The two metals resulting from corrosion that cause the greatest aesthetic problems are copper and iron. Copper is generally present as a result of dissolution of copper piping in service lines and consumer plumbing and in brass fixtures. Copper and copper alloy corrosion and leaching are thoroughly discussed in chapters 5 and 6, respectively. Iron results from deterioration of unlined cast-iron piping, unlined steel piping, and possibly galvanized iron distribution mains and from galvanized service lines and premise piping. Iron and galvanized pipe corrosion are discussed thoroughly in chapters 2 and 3, respectively.

Copper can be evaluated and data presented in similar fashion to that describing lead above. In addition, review of customer complaint records or a customer survey can be used to determine the extent of aesthetic problems caused by copper. An example of this is a survey (GVWD 1990) conducted by Greater Vancouver Water District (GVWD) in British Columbia, Canada, which indicated that 64 percent of its customers experienced blue-green stains on fixtures and that customers spent an average of 7 hours per year cleaning these stains. Copper levels in the GVWD system ranged from 0.26 mg/L to 3.40 mg/L and averaged 1.62 mg/L in 125-mL first-draw samples. In the second sample, consisting of 1 L, copper ranged from 0.05 mg/L to 2.80 mg/L and averaged 1.51 mg/L.

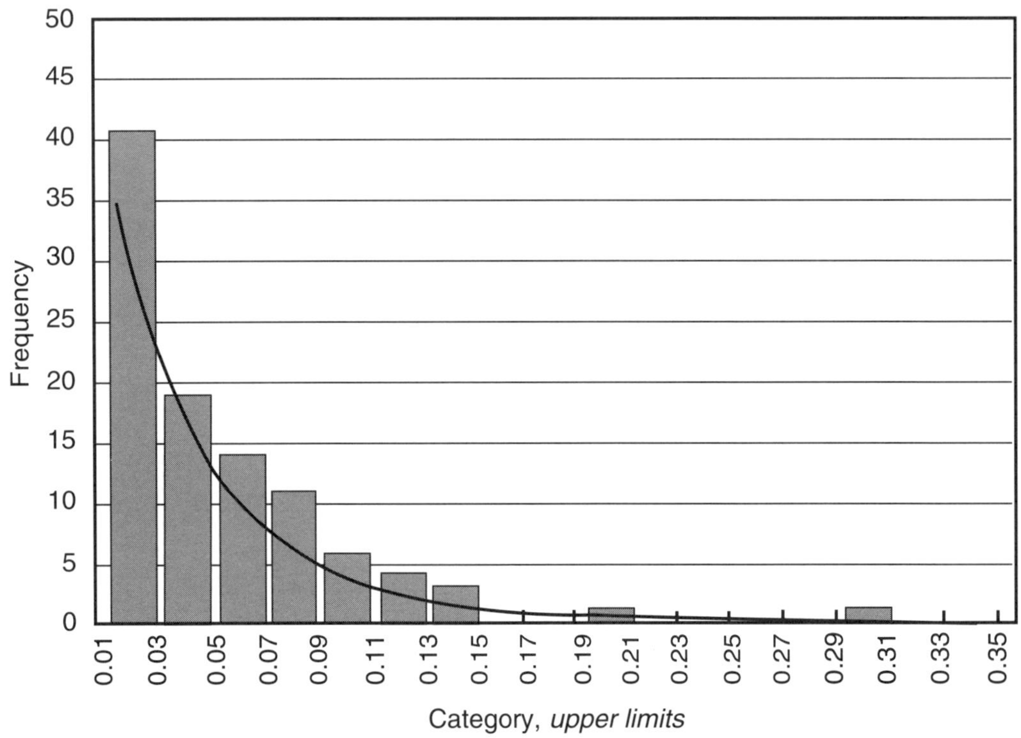

Figure 10-4 Histogram depicting frequency of various lead levels at customers' kitchen taps (one utility)

Concerning iron, the Seattle Water Department (1985), Seattle, Washington, tracked customer complaints related to iron corrosion and developed the data presented in Figure 10-5. In the mid-1960s, a new water supply was brought on-line. As indicated by rusty-water complaints, the corrosion conditions apparently increased significantly with the new water supply.

Premature Pipe Deterioration and Economic Consequences

Deterioration of piping can have serious economic consequences either causing leaks or blockage of flow in piping and reduced pressures. Iron piping will normally tuberculate causing an increase in pipe roughness, increasing head loss and restricting flow as described in chapter 2. Smaller diameter galvanized and iron piping, such as service lines and consumer plumbing, can become so restricted that replacement becomes necessary to maintain water service.

This was illustrated in a study (Seattle Water Department 1976) conducted in Seattle, where the city's low pH, soft surface water source, the Tolt River, resulted in a life for galvanized consumer plumbing of only 20 to 25 years. This investigation is presented as a case study in appendix C. The average "service life" was determined as the age at which three quarters of the original pipe diameter would be obstructed by tubercules in a 6-in.-long (15.2-cm) sample of consumer plumbing. Thus, only one fourth of the original pipe diameter was available for water to flow. Copper piping on the same water source was estimated to have an average service life of 20 to 30 years. Failure occurred due to uniform thinning and pitting in some instances. In a corrosion control study conducted in GVWD, British

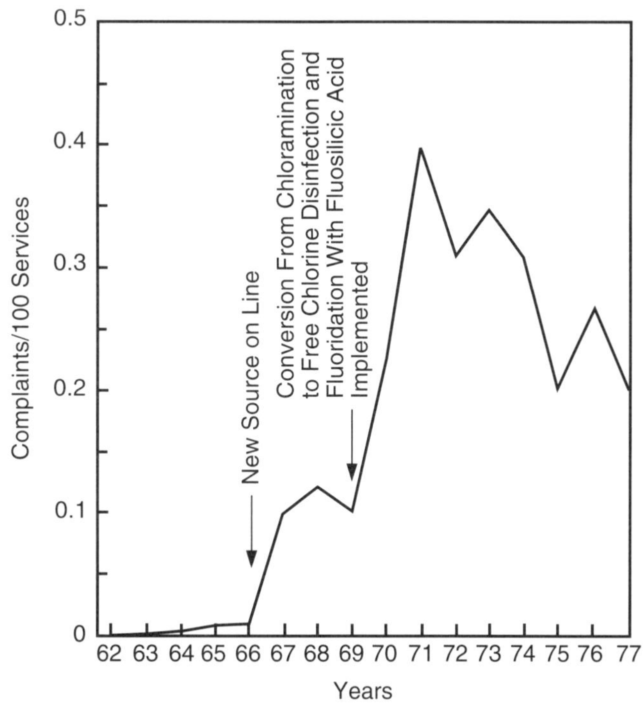

Figure 10-5 History of customer complaints related to iron: Seattle Water Department

Columbia, Canada (GVWD 1990), 42 pipe segments from hot- and cold-water piping were gathered and analyzed. Results from this survey compared favorably with interviews with plumbers and contractors in the area indicating a service life for cold-water copper piping of 20 to 35 years and significantly less for hot-water recirculating systems.

To gather this type of information, removal and analysis of a statistically valid number of pipe segments from distribution and consumer plumbing systems had to be done. Physical measurements to determine the diameter remaining in iron piping or use of micrometers to measure thinning and pitting rates in copper piping were used to gather information on the condition of existing customer piping. The existing service life of piping can be used to determine if pipe life is being significantly reduced by water quality conditions or materials considerations. By comparing existing service life with a target value for service life of 50 to 75 years for new pipe, the cost of premature replacement to maintain serviceability can be estimated. The existing service life can also be used in some cases as a basis for predicting future service life and cost savings after instituting corrosion treatments.

Concerning the economic aspects, the American Water Works Associaton Research Foundation funded a study that was published as *Economics of Internal Corrosion Control* (EES and Kennedy/Jenks/Chilton 1989). This report outlines an 11-step process for developing costs for corrosion and treatments and savings associated with corrosion control. The economic evaluation process is presented in Figure 10-6.

Benefit:cost (B:C) ratios have been developed in a fairly rigorous manner in at least two studies. Seattle developed a B:C ratio for corrosion treatment of over 5:1 (Seattle Water Department 1976), GVWD developed a range of B:Cs of 4:1 up to 13:1 depending on the type of treatment implemented (GVWD 1990). USEPA estimated a B:C ratio of 2.2:1 for full implementation of the Lead and Copper Rule (Levin 1986). Development of the economic aspects of corrosion and corrosion control can be a useful concept in displaying effects of corrosion and determining whether one should implement corrosion treatment.

Environmental Concerns

Metals including copper, zinc, cadmium, and lead can also create environmental problems. These constituents enter the wastewater disposal facilities and end up either in the liquid stream being discharged to a receiving water or in the solids, which end up in a landfill or on croplands. Aquatic life is particularly sensitive to copper, and many regulatory agencies have established very stringent levels for this contaminant. For example, San Francisco Bay (California) Regional Water Quality Board set water quality standards, shown in Table 10-3. A significant portion of the copper, lead, and zinc entering wastewater treatment facilities may originate from excessive corrosion of piping. The usual range of reduction of these metals by aerobic biological wastewater treatment processes is 50 to 90 percent. The metals are absorbed in the sludge and can create further environmental disposal concerns. The metals may then be concentrated by several orders of magnitude as a portion of the dry weight of sludge solids.

Zinc can be present in the potable water supply through corrosion of galvanized piping and from addition of corrosion inhibitors. Zinc can limit the uses of sludges as there are stringent requirements for its levels in sludges applied on crop lands. The zinc concentration that is allowable in sludge is limited to 2,800 mg/kg (2,800 mg/2.2 lb) for disposal to croplands.

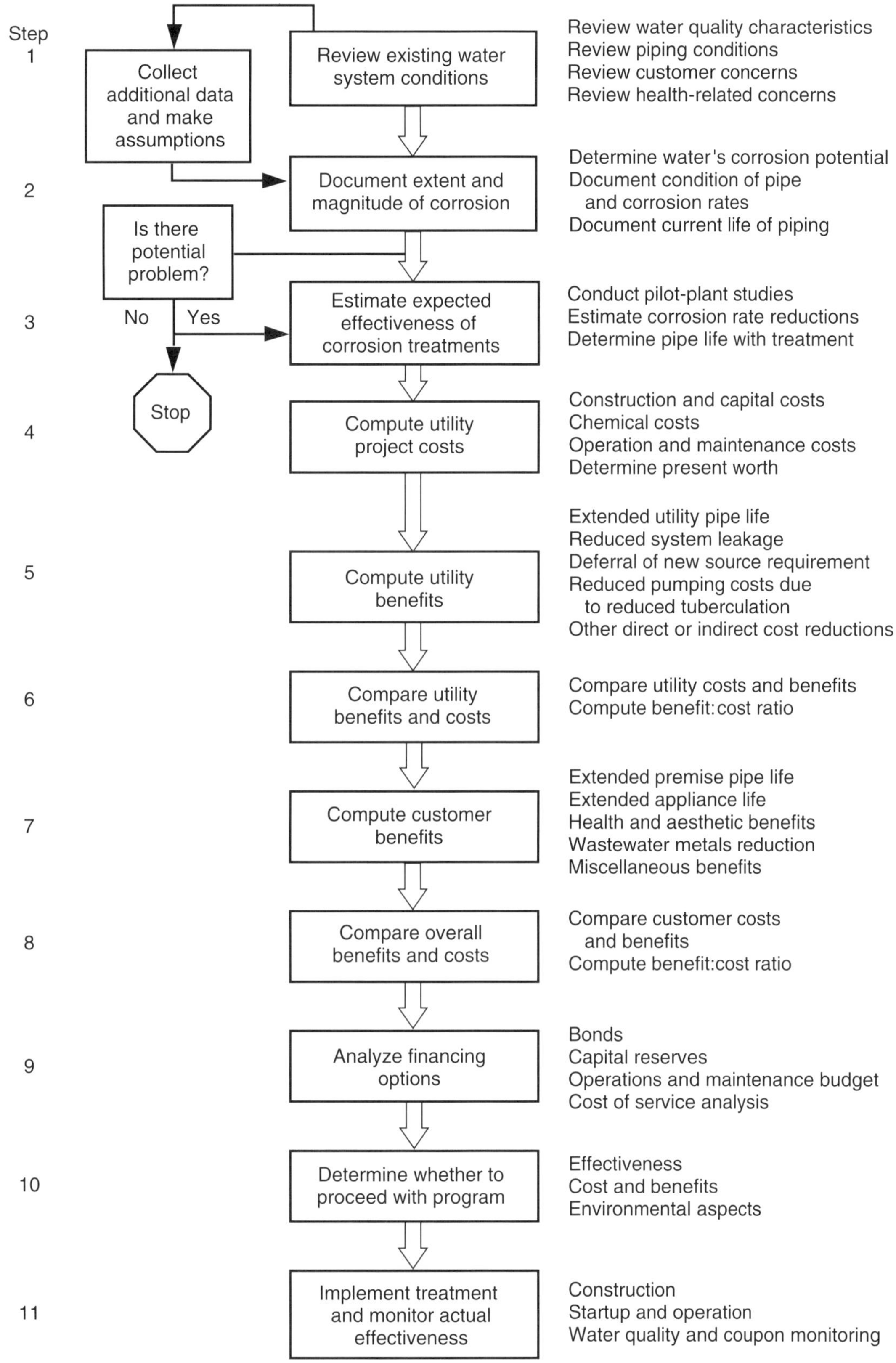

Source: EES and Kennedy/Jenks/Chilton (1989).

Figure 10-6 Flowchart for corrosion economic analysis

Table 10-3 Heavy metal shallow water discharge limits set by San Francisco Bay Regional Water Quality Board

Parameter	Recent Basin Plan, $\mu g/L$	Old Basin Plan, $\mu g/L$
Cadmium	1.1	20
Copper	4.9	200
Lead	3.2	100
Zinc	86	300
Nickel	8.3	100

Source: Santa Clara Valley Water District (1994).

Zinc is present in zinc orthophosphate corrosion inhibitors. Wastewater utilities and regulatory agencies have recently requested that certain water utilities lower or eliminate the use of zinc in their corrosion inhibitors. Santa Clara (San Jose, California) Valley Water District was asked to restrict use of zinc in its corrosion inhibitor. As an interim measure, the utility switched from a formulation containing 0.8 mg Zn/L to one containing 0.3 mg Zn/L. Further, it is conducting pipe loop pilot studies to determine if a nonzinc or low-zinc inhibitor system can be effective (Santa Clara Valley Water District 1994).

Phosphates (PO_4) also present a concern from a treatment standpoint and in receiving waters. Many wastewater treatment plants are required to remove nutrients before discharge to the receiving stream. The amount of PO_4 added in a corrosion inhibitor typically varies from 1 to 5 mg PO_4/L. Although this is fairly minor in relation to typical compositions of domestic sewage, which range from 15 to 60 mg/L, it does represent an additional amount that may need to be removed.

It is advisable to contact the agency responsible for wastewater treatment and disposal and the environmental regulatory agencies dealing with discharges to receiving streams or sludge disposal to ascertain any limitations.

In summary, Step 1: Document Extent and Magnitude of Corrosion should evaluate existing data and may require additional monitoring. Investigations should focus on condition and life of piping materials and on changes that occur in water quality as the water proceeds from source to tap.

STEP 2: DETERMINE POSSIBLE CAUSES OF CORROSION

This step in the investigation process is aimed at answering the following questions:

- What are the primary causative factors of corrosion?
- If more than one metal is involved, are the causes of corrosion the same or different for the different materials?

Each chapter in this book that addresses a material also discusses the primary causative factors of corrosion of that material and the investigator should refer to those chapters for detailed discussion. It is important to note that metals or materials may respond differently to the same treatment and that what is optimum for one material may actually cause an increase in corrosion of another. The following paragraphs attempt to summarize various factors affecting corrosion, including water quality and treatment, and materials and design considerations.

Water Quality and Treatment

Source water quality and subsequent treatments play a significant role in causing corrosion and in the effectiveness of corrosion control strategies in potable water systems. Dissolved oxygen, pH, alkalinity, calcium, calcium carbonate saturation indexes, carbon dioxide content, and temperature are factors that influence the rate of corrosion of iron and copper materials commonly used in water piping systems. Not so well-known factors that play a role in the corrosivity of water include the silica concentration, the presence of natural tannins or other organic materials, ammonia, nitrate, residual chlorine concentrations, chloride, sulfate, and bicarbonate. The relative proportion of chloride, fluoride, sulfate, and bicarbonate can also impact corrosivity (Larson 1975). A listing of several chemical water quality factors affecting corrosion is presented in chapter 8 (Table 8-1), and many of these factors are discussed in the particular chapter where the metal or alloy was described. Also, the reader is referred to Table 8-6 (chapter 8), which describes the materials of concern, the type of deterioration, and the effect on water quality.

Materials Design and Installation

How the materials are manufactured, selected, designed, and installed are equally as important as water quality factors in a corrosion investigation. Some of the key physical and materials factors are discussed below.

Temperature. Water temperature can be an important factor in corrosion. Generally, the oxidation and diffusion rates of metals increase with temperature. Water that is unsaturated with calcium carbonate at temperatures encountered in domestic hot-water heating systems (54.4 to 71.1°C [130 to 160°F]) will usually be more corrosive. Accelerated failures will occur in copper and galvanized steel hot-water piping systems more often than in cold-water systems. However, where carbon dioxide concentration is high, (>10 mg/L), copper will more commonly show pitting failures in cold-water piping (Lyman, Cohen, and Myers 1982). Temperature is also important in affecting the solubilities of various solids that could form passivating films on pipe. Some solids increase in solubility as temperature increases; some decrease. The relationships are usually nonlinear, as well.

Manufacturing process. Copper tubing is produced by heating copper billets and extruding them through a mandrel under pressure. The resulting tubing is then stretched to achieve a cold-worked material of the desired diameter. The product of this process is hard copper tubing. Soft copper tubing is produced by heating the copper in an annealing oven with limited oxygen. Carbon films that have been associated with type I pitting can be formed when the lubricants for the mandrel are "coked" during the annealing process. In the EC, particularly the United Kingdom, specifications call for sandblasting the inside of the finished tube to remove any carbon films. The formation of these undesirable films is also controlled during the manufacturing process by proper choice of lubricants and careful choice of the operating conditions during the annealing process.

Galvanized pipe is manufactured by hot-dipping steel pipe, and features in both the galvanizing and the pipe manufacturing processes can affect the quality of the product. Steel pipe in small diameters is manufactured by continuous butt-welding, electrical resistance welding, or extrusion over a mandrel much like copper pipe. When the pipe is butt-welded, excessive material in the weld can be responsible for crevice corrosion. When the pipe is electrical resistance welded, insufficient material in the weld or improper temperatures can result in a weld that is particularly sensitive to corrosion that forms slot-shaped pits. The surface of the steel pipe must be properly cleaned and free of corrosion products immediately before it is hot-dipped. Such deleterious materials at the zinc–iron interface prevent proper adhesion of the galvanizing layer. Problems also occur with the procedures used to evacuate the excess hot zinc from the pipe surface after the hot dip. Usually steam is used; occasional experiences with the use of air have been unsatisfactory, resulting in a porous, unstable galvanized layer. When the material is allowed to cool too much before it is evacuated, an uneven layer that is too thick can occur. When the material is too hot, the zinc layer may be too thin. Satisfactory performance also depends on the formation of a proper alloy layer. Impurities in both the iron and the zinc can promote the diffusion of the iron–zinc alloy well into the pure zinc layer, reducing corrosion protection. In Germany this process is sometimes referred to by a term that translates to "galvanealing."

Workmanship. Workmanship can definitely affect the susceptibility of materials to corrosion. Bent or deformed piping can cause locally high velocity resulting in erosion corrosion. Mishandling of piping can also cause interior and exterior coatings to be removed or loosened, resulting in a localized area that is subject to high corrosion rates. Corrosive fluxes used to prepare copper piping for soldering can run inside the piping, causing roughness and deposits that can increase corrosion.

Water main cleaning. Many water utilities have water main cleaning and relining programs to increase hydraulic capacity of piping and to extend piping life. Some utilities have cleaned or pigged piping, but did not take the additional step of relining. This is a serious mistake if the piping consists of an unlined, ferrous material. Recently cleaned and exposed ferrous materials will deteriorate rapidly and will likely be subject to increased pitting and tuberculation. If the pipe is to remain unlined, a phosphate or silicate passivation dosage and application of a corrosion inhibitor program should be initiated immediately after pipe cleaning and flushing.

Dissimilar metals. Chapter 1 describes the problems with directly connecting dissimilar metals and provides a table listing the galvanic series. If dissimilar metals must be directly connected, dielectric insulators should be used to separate the metals and reduce the potential for galvanic corrosion. Caution must be exercised inside homes where the water piping may be used for grounding of the electrical system, in which case, interrupting the continuous metallic conductor could cause unsafe conditions.

In summary, a variety of factors can cause corrosion and many have been identified and discussed in previous chapters. Particularly, the descriptions in each chapter and the testing methods described in chapter 9 can be used to ascertain some of the underlying causes of corrosion. Better understanding of the causes of corrosion can enable the investigator to develop better corrosion treatment measures.

STEP 3: DEVELOP AND ASSESS CORROSION CONTROL ALTERNATIVES

This step in the investigation is aimed at answering the following questions:

- What are the feasible corrosion control alternatives?

- What levels and types of testing are needed to compare the effectiveness of alternatives?
- What are the potential secondary side effects of different alternatives?

There are a variety of corrosion control approaches for reducing corrosion and several methods for conducting the testing.

Available corrosion control technologies can be characterized by two general approaches to inhibiting corrosion: (1) forming a precipitate in the potable supply that deposits onto the pipe wall to create a protective coating, or (2) causing the pipe material and the potable supply to interact in such a way that metal compounds are formed on the pipe surface, creating a film of less soluble material. The difference in these two approaches is the mechanism by which the protective film is formed. In the former method, insoluble compounds are formed by adjusting the water chemistry to cause the *precipitation* of the compound, such as calcium carbonate, onto the pipe wall. The success of this method is dependent on (a) the ability to form precipitates in the water column and (b) the characteristics of the deposit on pipe walls, including its permeability, adherence strength, and uniformity.

In the latter approach, the mechanism is the *passivation* of the pipe material through the formation of less soluble metal compounds (carbonates, silicates, or phosphates) that adhere to the pipe wall. In the case of nonmetallic pipe materials, such as A–C pipe, passivation and precipitation mechanisms are also operative. The calcium present in the A–C pipe acts as the metallic component, which may be available to react with the carbonate or phosphate species under passivating conditions. However, when added in small concentrations, zinc, in either zinc chloride or a zinc orthophosphate, can provide a better protective metal film for A–C and concrete pipe subject to aggressive water conditions.

The methods for evaluating and testing for corrosion have been divided into two broad categories: "desk-top" (or paper studies) and "laboratory and field testing," both of which are described in the following sections of this chapter. The benefits and drawbacks of various levels of testing are presented in Table 10-4. The analytical methods and measurement techniques and apparatus used in corrosion testing are described in chapter 9.

Corrosion Evaluations on Paper

The "desk-top" approach relies heavily on use of theory and comparison with similar water systems. Many of the theoretical and practical aspects of corrosion have

Table 10-4 Benefits and drawbacks of various levels of corrosion testing

Consideration	Desktop	Pilot	Full-Scale Field Demonstration
Time	1–3 months	6–18 months	9–24 months
Level of effort	Low	High	High
Cost	Low	High	High
Number of options to be explored	Many/all	Several	One or two
Level of certainty	Low	Modest	High

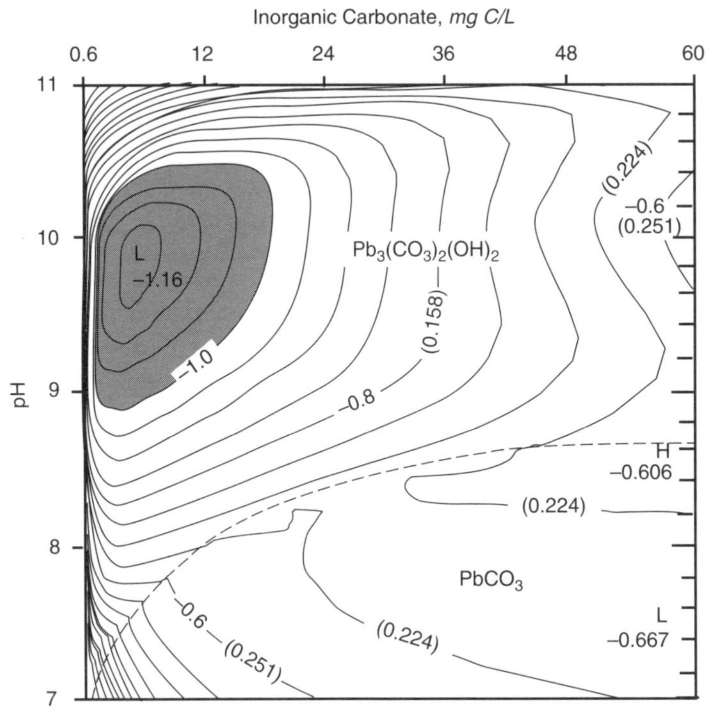

Source: EES and ISWS (1990).

Figure 10-7 Dissolved inorganic carbonate versus pH: No phosphate, I = 0.01, temperature = 25°C

been described in detail in preceding chapters. Solubility models such as LEADSOL or COPPERSOL can be used to predict the theoretical levels of lead and copper under various water quality conditions. An example of the usefulness of the output from such models is presented in Figure 10-7, which indicates that increasing pH levels from 7.5 up to 9.0 would theoretically have a significant effect toward reducing lead levels. There are other solubility models that predict chemical additions needed to result in a positive calcium carbonate precipitation potential or other solubility index. These models can be very useful in paper studies to screen treatments and predict chemical dosages and costs.

A recent survey of 400 US water systems that serve more than 50,000 people focused on lead and copper leaching. It was conducted by the American Water Works Association (AWWA) Water Industry Technical Action Fund, Government Affairs Office (AWWA 1993a) and evaluated first-flush (1-L [1.057-qt]), lead and copper samples collected at customers' taps as part of required regulatory monitoring procedures. Each water system collected at least 100 water samples that had been standing in the consumer plumbing for a minimum of 6 hours. Water quality was also sampled in the distribution system for parameters such as pH, alkalinity, and calcium levels. The results of this survey may be useful to utilities in understanding their lead and copper leaching problem and potential mitigation factors.

Lead survey results. The focus of this discussion will be on water quality and corrosion factors and their apparent influence on 90th percentile lead levels at the tap, as reported by utilities responding to the WITAF survey. The term "percent action level (AL) exceedance" is used and indicates the percentage of utilities in a certain category that exceeded the 90th percentile action level of 0.015 mg Pb/L. Where possible, the data were divided into two categories before further analysis:

waters without an inhibitor (uninhibited) and waters with an inhibitor (inhibited). Corrosion inhibitors were defined as chemicals approved for use in potable water that are designed and used to prevent or reduce corrosion. They included phosphates, polyphosphates, silicates, and blends of these chemicals.

pH. Distribution system pH levels were found to have an influence on the 90th percentile lead levels for both waters with and without corrosion inhibitors. Higher pH levels, i.e., ≥7.0, have lower percent lead AL exceedances when compared with lower pH levels, i.e., <7.0. As indicated in Figure 10-8, the percentage exceedances were 50 percent at pH <7.0, and were typically lower in higher pH ranges. Overall, higher pH levels were associated with lower percent AL exceedances for lead.

Alkalinity. The 90th percentile lead levels were influenced by distribution system alkalinity levels, as depicted in Figure 10-9 for all respondents. Water with alkalinity <50 mg/L had several very high 90th percentile lead levels. As indicated by Figure 10-10, low-alkalinity waters (<25 mg/L $CaCO_3$) had a significantly higher percentage of systems exceeding the lead AL than did water ≥ 25 mg/L ($CaCO_3$). This figure also indicates that as alkalinity increased, the percent of systems exceeding the AL decreased for both inhibited and uninhibited waters. Waters with alkalinities >150 mg/L as $CaCO_3$ had no AL exceedances.

pH and alkalinity. The combination of pH and alkalinity for inhibited and uninhibited waters' influence on lead levels is shown in Table 10-5. Low pH and low alkalinity are most likely to have lead levels that exceed the AL for both inhibited and uninhibited waters. In general, as the pH and alkalinity levels increase, the chance of exceeding action levels decreases, although there are always anomalies evident. For uninhibited waters with pH ≤7.5 and alkalinity <25 mg/L, major benefits could potentially be derived from increasing pH or alkalinity, individually, or in combination, to higher levels. Utilities are reminded, however, that any change

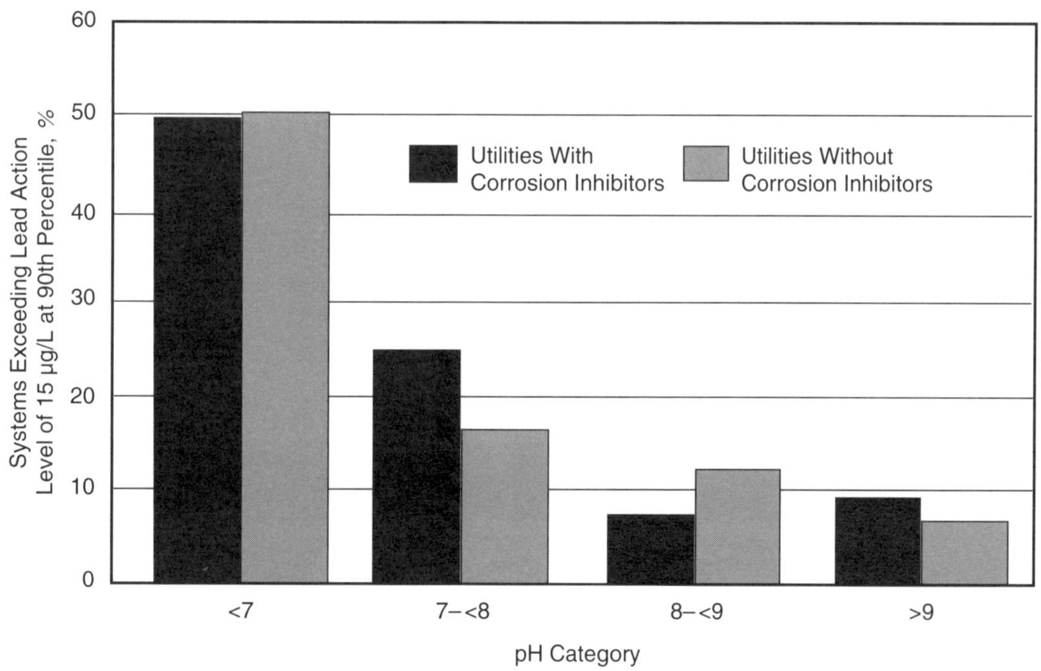

Source: AWWA (1993a).

Figure 10-8 Lead action level exceedance for pH categories

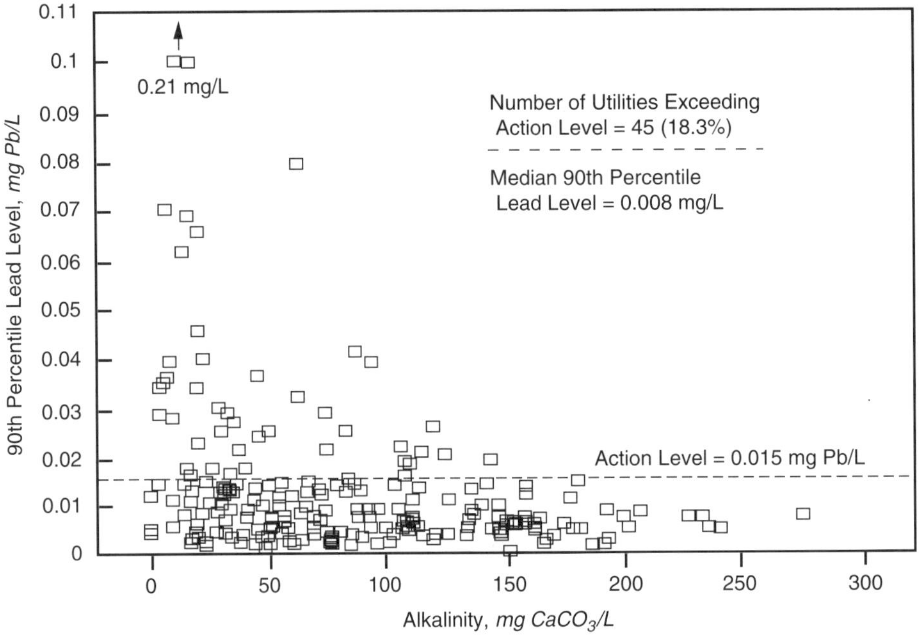

NOTE: Initial monitoring results of systems with more than 50,000 customers. (*n* [number of utilities reporting with sufficient data] = 246.)

Source: AWWA (1993a).

Figure 10-9 Distribution system alkalinity versus 90th percentile lead levels

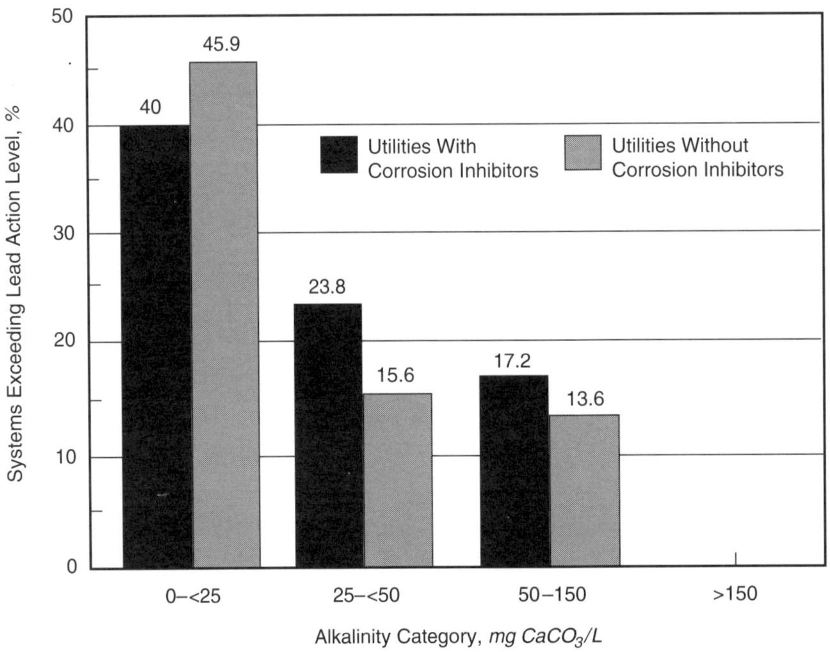

NOTE: Initial monitoring results of systems with more than 50,000 customers.

Source: AWWA (1993a).

Figure 10-10 Lead action level exceedances for alkalinity categories

Table 10-5 Percent and number of lead action level exceedances as a function of pH and alkalinity

	With Inhibitors: Alkalinity, *mg* $CaCO_3$ */L*								Without Inhibitors: Alkalinity, *mg* $CaCO_3$ */L*							
	0–<25		25–<50		50–150		>150		0–<25		25–<50		50–150		>150	
pH	Percent	Number	Percent	Number	Percent	Number	Percent	Number	Percent	Number	Percent	Number	Percent	Number	Percent	Number
≤7.5	46.7	7/15	25	3/12	30	3/10	—	—	83.3	5/6	14.3	1/7	18.2	4/22	0	0/7
>7.5–9	20	1/5	20	1/5	12.5	2/16	0	0/3	46.2	6/13	19	4/21	11.4	7/61	0	0/21
>9.0	—	—	25	1/4	0	0/3	—	—	0	0/5	0	0/4	20	1/5	0	0/1
All pH	40	8/20	23.8	5/21	17.2	5/29	0	0/3	45.9	11/24	15.6	5/32	13.6	12/88	0	0/29
n =	20		21		29		3		24		32		88		29	

Source: Data from AWWA (1993a).

n = number of data points in category.

to the treatment process must be evaluated for possible negative consequences as well, e.g., disinfection efficiency is reduced as pH is increased. Although the database is not large, the combined low-pH and low-alkalinity waters appear to have been aided by inhibitors.

Calcium. For waters without inhibitors and calcium levels below 30 mg Ca/L (75 mg $CaCO_3$/L), the level of calcium seems to have little effect on the percent of systems exceeding the AL. These waters would be considered very soft or soft. Above 30 mg Ca/L, as calcium levels increased, the percentage of AL exceedances decreased (see Figure 10-11).

When evaluating the waters with inhibitors added, the percentage of systems exceeding the AL remains fairly constant in waters with a calcium level below 60 mg Ca/L (150 mg $CaCO_3$/L), which is at the outer boundary of medium hardness. For the levels >120 mg Ca/L, there were only two utilities reporting; thus, the furthest right bar on Figure 10-11 should be interpreted with caution. Between 60 and 120 mg Ca/L, the percent AL exceedances dropped significantly. Figure 10-11 illustrates these findings.

The reason for the difference in AL percent exceedances between waters with and without inhibitors in the 30 to 60 mg Ca/L range could not be determined. The presence of orthophosphate and polyphosphates can undoubtedly affect the solubilization and precipitation potential of calcium, and this may be a factor.

Table 10-6 depicts relationships between alkalinity and calcium levels and lead AL exceedances for both waters with and without inhibitors. For waters without inhibitors, but with alkalinities less than 25 mg $CaCO_3$/L, the percent exceedances are high for all calcium categories. As alkalinity increases beyond 25 mg $CaCO_3$/L, the percent

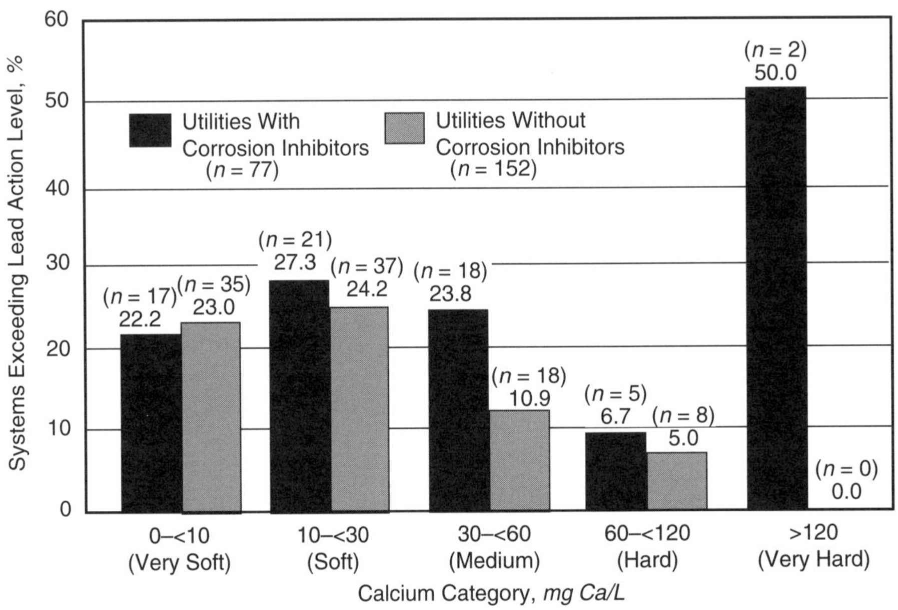

NOTE: Ca (mg/L) = (hardness [mg $CaCO_3$/L]/2.5); *n* = number of data points in category; initial monitoring results of systems serving more than 50,000 customers; inhibitors are of various types.

Source: AWWA (1993a).

Figure 10-11 Lead action level exceedance for calcium categories

Table 10-6 Percent and number of lead action level exceedances by alkalinity and calcium category

	With Inhibitors: Alkalinity, *mg* $CaCO_3$ */L*								Without Inhibitors: Alkalinity, *mg* $CaCO_3$ */L*							
	0–<25		25–<50		50–150		>150		0–<25		25–<50		50–150		>150	
pH	Percent	Number	Percent	Number	Percent	Number	Percent	Number	Percent	Number	Percent	Number	Percent	Number	Percent	Number
0–10	45	5/11	0	0/2	0	0/1	0	0/2	13	1/8	—	—	—	—	—	—
>10–30	55	6/11	17	2/12	10	2/21	—	—	42	5/12	9	1/1	29	2/7	0	0/1
>30–60	67	2/3	0	0/6	13	6/46	0	0/10	50	2/4	33	3/9	36	4/11	0	0/3
>60–120	—	—	0	0/9	9	2/22	0	0/18	—	—	50	1/2	10	1/10	0	0/2
>120	100	1/1	—	—	0	0/8	0	0/9	—	—	—	—	—	—	0	0/3

Source: Data from AWWA (1993a).

AL exceedances drop off greatly for all calcium categories. With an inhibitor present, the pattern is much different. The percent AL exceedances are generally highest when alkalinity is less than 25 mg $CaCO_3$/L, but seem to remain high for alkalinities up to 150 mg $CaCO_3$/L. The cause for the higher levels of exceedance throughout the multitude of alkalinity and calcium categories with inhibitors needs further investigation.

Copper survey results. The focus of this discussion will be on water quality and corrosion factors and their apparent influence on 90th percentile copper levels at the tap, as reported by utilities responding to the WITAF survey. The term "percent action level (AL) exceedance" is used and indicates the percentage of utilities in a certain category that exceeded the 90th percentile action level of 1.3 mg Cu/L. Where possible, the data was divided into two categories before further analysis: waters without an inhibitor (uninhibited) and waters with an inhibitor (inhibited). Corrosion inhibitors were defined as chemicals approved for use in potable water that are designed and used to prevent or reduce corrosion. They included phosphates, polyphosphates, silicates, and blends of these chemicals.

pH. A review of Figure 10-12 indicates the strong influence that pH levels have on the 90th percentile copper levels. None of the 248 reporting utilities exceeded the AL of 1.3 mg Cu/L when pH levels increased above 8. Figure 10-13 further emphasizes this strong relationship between pH and copper exceedances. Elevated pH levels, i.e., above 7.5 to 9.0, are very effective at reducing copper levels to below the AL. For utilities without inhibitors, the percent AL exceedance is 22.5

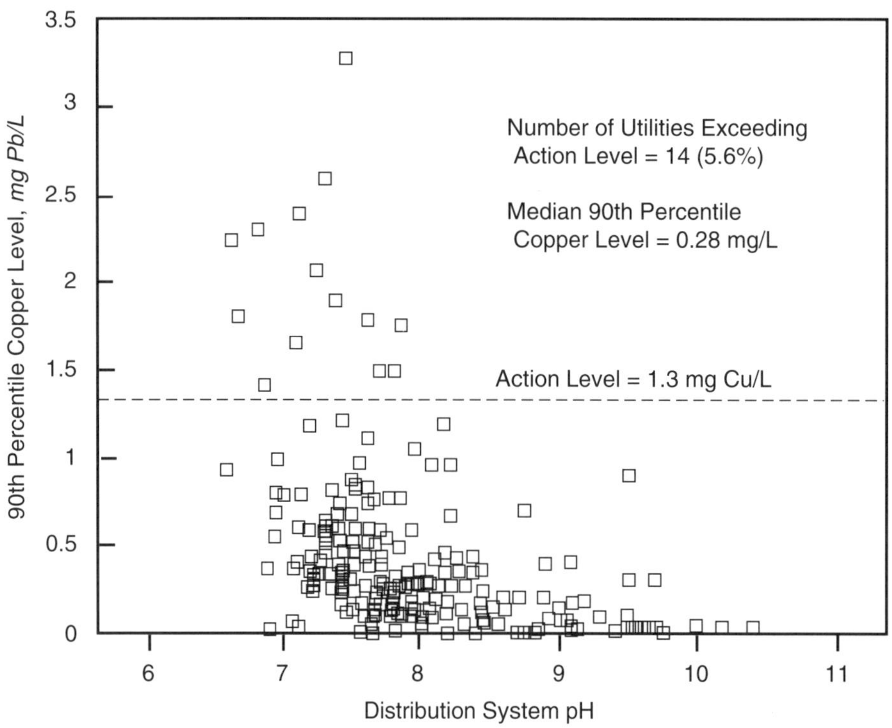

NOTE: Initial monitoring results of systems serving more than 50,000 customers (*n* [number of utilities reporting with sufficient data] = 248).

Source: AWWA (1993a).

Figure 10-12 Distribution system pH versus 90th percentile copper levels

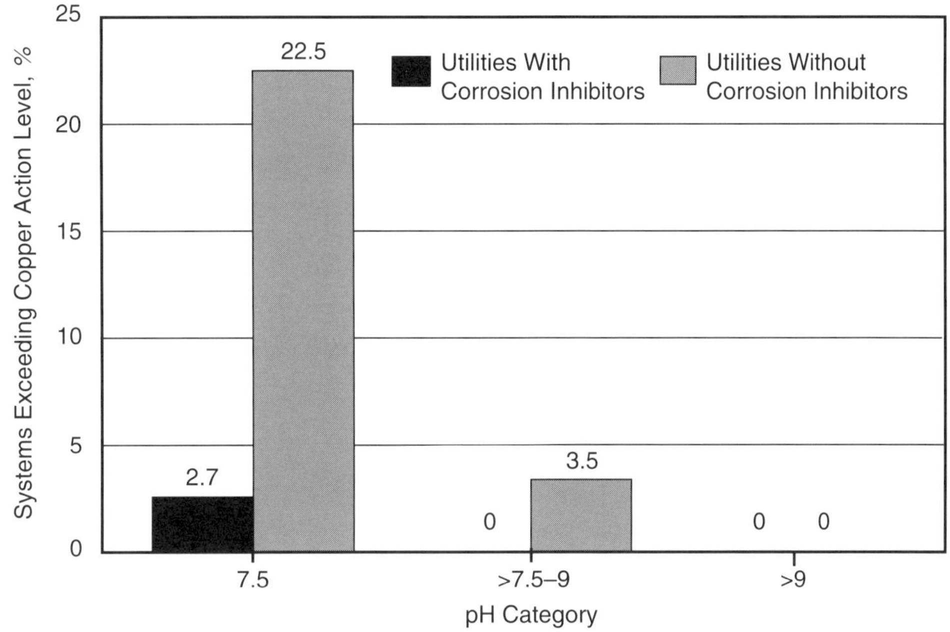

NOTE: Initial monitoring results of systems serving more than 50,000 customers.

Source: AWWA (1993a).

Figure 10-13 Copper action level exceedance for pH categories

percent for the lower pH category (≤7.5) and drops to only 3.5 percent for the pH category (>7.5 to 9.0). It is also interesting to note that of the 155 utilities that adjust pH, none exceeded the copper AL. High pH levels in the distribution system appear to be of great benefit for controlling copper and meeting copper ALs.

Alkalinity. Figures 10-14 and 10-15 indicate some very interesting results relating to copper and alkalinity. It appears that the 90th percentile copper levels and the percent exceeding the AL are higher when alkalinity levels are below 25 mg $CaCO_3$/L and above 75 mg $CaCO_3$/L. The window between 25 and 75 mg $CaCO_3$/L of alkalinity seems to be associated with lower 90th percentile copper levels.

pH and alkalinity. Table 10-7 presents data related to pH, alkalinity, and percent copper AL exceedance levels. For uninhibited waters, this data along with Figure 10-15 illustrates that waters with pH levels of ≤7.5 and alkalinity levels of <25 mg $CaCO_3$/L or >75 mg $CaCO_3$/L have the most trouble meeting the copper AL. Increasing pH levels regardless of the alkalinity level has a major beneficial effect of decreasing percent of copper AL exceedance. However, at the higher alkalinity levels, pH may need to be higher to effect copper reduction when compared with lower-alkalinity waters.

Laboratory and Field Testing

Laboratory and field testing consists of three categories: (1) bench or laboratory scale, (2) pipe loop pilot testing, and (3) full-scale field demonstration (refer to chapter 9). The method and scale of testing selected by a utility will depend on the type of corrosion problem encountered, the time available for testing, expertise of investigators, and resources available. The benefits and drawbacks of bench versus flow-through pipe loops are depicted in Table 10-8. Regardless of the testing method

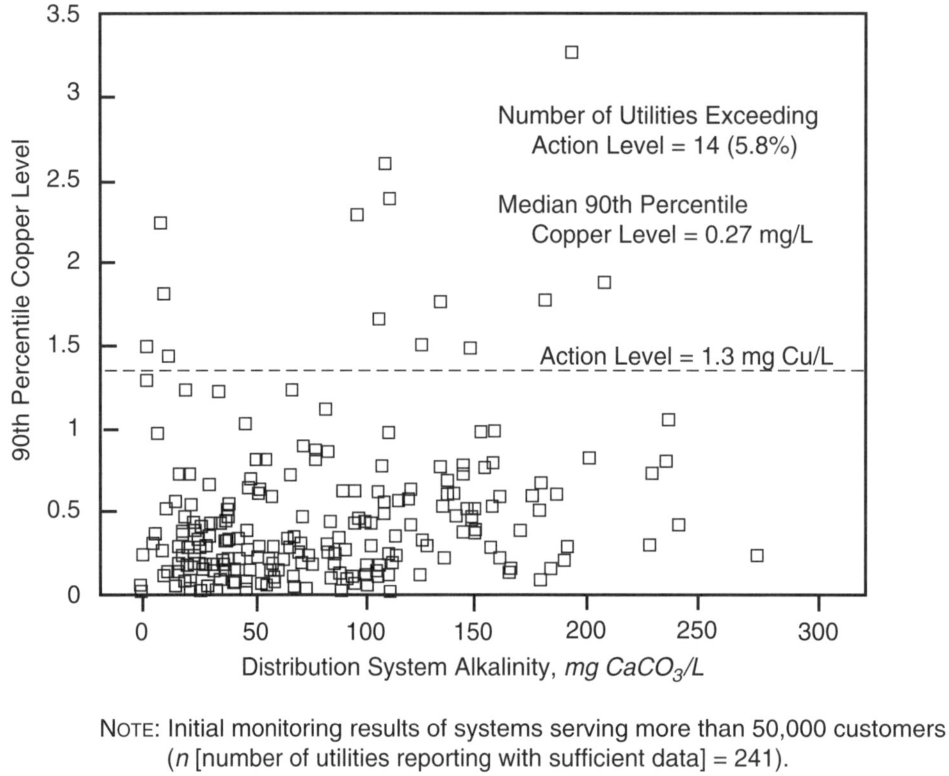

Source: AWWA (1993a).

Figure 10-14 Distribution system alkalinity versus 90th percentile copper level

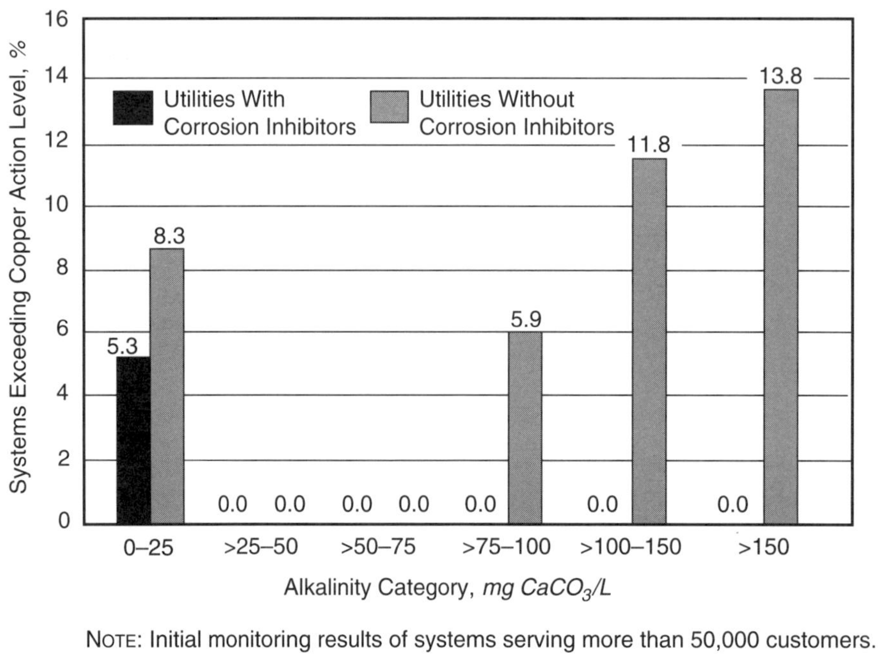

Source: AWWA (1993a).

Figure 10-15 Copper action level exceedance for alkalinity categories

Table 10-7 Percent and number of copper action level exceedances as a function of pH and alkalinity

	With Inhibitors: Alkalinity, *mg* $CaCO_3$ */L*								Without Inhibitors: Alkalinity, *mg* $CaCO_3$ */L*							
	0–<25		25–<50		50–150		>150		0–<25		25–<50		50–150		>150	
pH	Percent	Number	Percent	Number	Percent	Number	Percent	Number	Percent	Number	Percent	Number	Percent	Number	Percent	Number
≤7.5	5.3	1/19	0	0/16	0	0/6	0	0/1	33.3	2/6	0	0/11	25	4/16	42.9	3/7
>7.5–9	0	0/5	0	0/9	0	0/12	0	0/2	0	0/13	0	0/37	7.0	3/43	4.8	1/21
>9.0	—	—	0	0/7	—	—	—	—	0	0/5	0	0/8	0	0/1	0	0/1
All pH	4.2	1/24	0	0/32	0	0/18	0	0/3	8.3	2/24	0	0/56	11.7	7/60	13.8	4
n =	24		32		18		3		24		56		60		29	

Source: Data from AWWA (1993a).

n = number of data points in category.

Table 10-8 Benefits and drawbacks of bench testing versus flow-through pipe loops

Criteria	Static-Bench Scale	Flow-Through Pipe Loops
Length of time	1–3 months	9–15 months
Cost	\$750–\$1,000 per water condition	Five loops: \$60,000–\$75,000 (\$12,000–\$15,000 per condition)
Certainty of data for optimization	Overall low to moderate Good for copper, questionable for lead Poor for iron	Overall moderate to high Very good for copper Reasonable for lead Reasonable for iron
Staff-consultant level of effort	Moderate	High
Expertise required	High	High
Similar condition to home	No, not similar	Yes, more similar than static test

Source: AWWA (1993b).

selected, the laboratory and field testing should be based on a technically sound "desk-top" analysis that clearly identifies the problems and screens alternatives down to the most viable solutions. Based on the regulatory requirements and experience from planning and conducting several bench-scale, pipe loop, and demonstration tests, the following synopsis of issues is offered. The discussion relates specifically to pipe loop testing with the AWWARF pipe rack, but the approach is valid for bench or laboratory testing or other types of corrosion-related testing as well.

Planning for pipe rack studies. The critical first step to performing a corrosion pipe rack study is to allow enough time and commit enough resources to perform proper planning. Successful planning provides for a well-designed and well-constructed pipe rack pilot system, reduces operational problems, and most important, ensures that the data produced are useful and the objectives of the pipe rack study are met.

There are two key elements of the planning process for pipe rack studies: establish the specific pilot study objectives and develop an experimental plan, including pilot plant design requirements.

Development of the pilot plant experimental plan and the design should occur concurrently. A useful experimental plan is specific to the design, while a useful pipe rack layout must be designed to meet the specific testing objectives and requirements of the study.

Establish specific pilot study objectives. Clear, specific objectives are needed so that development of the experimental plan and pilot plant design can proceed. Some objectives may be a higher priority than others. As an example, one primary objective may be to determine the effect of a pH adjustment to 8.5 on leaching of copper. A secondary objective would be to compare trihalomethane (THM) formations at pH 8.5 with the control treatment condition at a lower pH level. Without clear objectives and without prioritization, the pilot plant components quickly get out of hand and the likelihood of excessive costs, schedule delays, conflicting objectives, and a failed pilot program increase dramatically.

Pilot plant siting. Siting of the pipe rack pilot plant (or plants) must be carefully considered prior to the development of the experimental plan and the design requirements. Availability of source waters and housing and support facilities are the two major factors that determine where the pipe racks should be set up.

Influent water to the pipe racks should be easily accessible. Depending on the pilot study objectives, the influent water may be raw, untreated water or it may be finished water. Normally, the influent water (either raw or finished) serves as the control water quality condition in one pipe rack. If there are water sources with significantly different water quality characteristics that enter the distribution system and it is desired to perform corrosion testing for these sources, it may be necessary to locate the pipe racks at separate sites to access the different sources.

Whenever possible, it is usually best to site pipe rack pilot plants within existing facilities. The rack system should be enclosed in a structure that has adequate lighting, power, heating, ventilation, plant water, and drainage. The pipe racks and associated chemical feed equipment should be situated so that they do not interfere with existing water operations. It is desirable to have a dedicated work area for the pilot plant operator(s), including a benchtop space and a sink.

Planning for operations. Planning for pipe rack operations must take into consideration the project timeline, operator personnel, laboratory services, and availability of materials and chemicals.

Project timeline. The length of a pipe rack pilot operation is typically 9 to 15 months. Operation time includes a startup period, which can be up to six weeks in duration. Therefore, the utility should plan on having some degree of operational involvement with the pipe rack study for at least nine months or more.

Operators. The major planning considerations for pilot plant operators are (1) Who will provide the operator or operators? (2) What skills are required of the operator(s)? and (3) How much time will be required of the operator(s)?

Pilot plant operators may be provided by the utility, the utility's consultants, or outside contracted personnel. The level of operator skill required should be approximately that of a water treatment plant/operations manager level II, or a person with some formal training in bench chemistry techniques or water quality engineering operations. The types of tasks required of the operator are the following:

- mix chemical feed solutions and adjust feed pumping rates
- collect samples and perform field analytical tests
- conduct routine maintenance on pumps, mixers, valves, and the like, and troubleshoot as required
- record data and observations

Regardless of the operator's background or level of expertise, training of the operator(s) on specific pipe rack operations should be provided during the startup phase of the study.

The approximate number of operator hours required during routine pipe rack operations is between 10 and 20 hours per week. This does not include laboratory analysis. During the initial startup and training period, it should be assumed that one or more operators will be required nearly full-time.

Laboratory services. The availability of adequate laboratory services is essential, not only for performing the routine analyses required for the pilot study, but also as a major element of the quality assurance/quality control (QA/QC) effort. Most analyses performed with field test kits should be checked periodically against duplicate samples measured by a certified laboratory. It is most desirable to have the services of one laboratory that is certified by the state or approving authority to perform all the types of analyses required for the study.

Materials and chemicals. Availability of materials and treatment chemicals required for the pilot study is also a significant planning consideration. It is best to plan on using equipment components and materials that are known to be reliable and relatively easy to obtain from local vendors. Use of unusual, discontinued, or cannibalized models of components (e.g., pumps, flowmeters, and valves) should be avoided because it may not be possible to find the same or an equivalent model if it becomes necessary to replace them during the study. If lead pipe is a material to be tested, it often takes several weeks to obtain appropriately sized lead pipe.

Treatment chemical availability is generally not a problem in North America or western Europe. It is better to plan to use chemicals available from a reliable local supplier, if possible. If proprietary phosphate and/or silicate inhibitor chemicals are to be tested during the pilot study, it is advisable to select a type of product that is manufactured by more than one chemical company. For instance, zinc orthophosphates with a 3:1 PO_4:Zn weight ratio are available from several manufacturers in the United States. Consideration should also be given to the use of generic inhibitor chemicals for testing, such as phosphoric acid in place of a proprietary sodium orthophosphate product, or blending zinc chloride with phosphoric acid to make zinc orthophosphate.

Development of experimental plan and design requirements. After establishing specific pilot study objectives and resolving pilot plant siting and operational issues, the culmination of the planning phase is to develop a draft experimental plan and the design requirements. Development of the draft experimental plan and the design requirements should be performed concurrently, as they are closely interrelated. Pitfalls in pipe rack operation can be minimized through good planning, design, fabrication oversight, and coordination, and quality control during operation. Some pitfalls and helpful hints for pipe rack design and operation are contained in Figure 10-16.

Experimental plan. The experimental plan should be a "stand-alone" working document primarily for use by the operator and other key project personnel. The experimental plan also documents the specific objectives, methods, materials, and quality control measures of the pipe rack study that can be later included in a corrosion control study report.

The experimental plan should contain all pieces of information the operator needs to successfully operate the pilot plant. Some of the sections may need to be added to and/or modified during the course of startup and operation of the pilot plant. Therefore, it is recommended that the experimental plan be in a three-ring binder format to allow for removal of "old" pages and insertion of revisions, as necessary. Figure 10-17 is an outline of the items that the investigator should consider in an experimental plan.

Design requirements. In developing the specific design requirements for the pipe rack, the following should be considered:

- space constraints of the selected site(s) for the pipe racks
- available and desired water flow rates and pressures
- number of treatment alternatives that will be tested
- chemical feed and equipment and storage requirements
- types of materials to be tested (e.g., lead, copper, steel)

Design

- Ensure that the detailed design meets all requirements of the pilot study objectives and the sampling and testing plan.
- Ensure that any bid design drawings and specifications leave little room for misinterpretation by fabricators.
- Ensure that adequate plant water drainage, power, lighting, heating, and ventilation are available at the pilot plant site.
- Thoroughly mix chemicals in their vats before use and ensure precipitation does not occur in storage vessels. Lime can be a significant problem in this regard.
- Include devices for good mixing of chemical feed solutions with ambient water on all racks (e.g., static mixers, adequate run-of-pipe).
- Include air vents or sample ports near the highest hydraulic points in the pipe rack to allow for venting of air pockets.
- Include an appropriately sized pressure-reducing valve (PRV) on the influent side of the racks; rack pressure should be compatible with chemical pump delivery pressures.
- Include enough sample ports throughout the racks to allow for all required or possible sample collections.

Fabrication

- Provide for good supervision and oversight of the pipe rack fabrication process.
- The same qualified plumber should solder all pipe joints in lead soldered copper pipe loops to ensure consistency in the amount of solder and flux used.
- Have the fabricator document the amount of solder used in each pipe loop.
- Avoid excessive flux deposits on the interiors of soldered pipe loops.
- Leak test all of the pipe loops.
- Ensure that proper shop cleaning of pipe loops is performed.
- If shipping of finished pipe racks is required, ensure careful handling.

Startup and Operation

- Provide for adequate flushing and preconditioning of pipe loops prior to starting operations.
- Ensure that all operators are trained on specific operating and sampling procedures.
- Provide for consistent control and monitoring of pipe rack operations.
- Disinfect sample ports to preclude high bacteria counts in initial water samples.
- Ensure that excessive pressure drops across the PRV do not occur, as air bubbles may result in the pipe rack, causing flow disruption.
- Ensure that the pilot plant site temperature remains above the freezing point of the chemical solutions being used.
- Keep chemical solution tanks covered to reduce evaporation and resulting changes in solution concentrations.
- Ensure that chemical feed systems are operating properly.

Figure 10-16 Steps to avoiding pitfalls in pipe rack design and operation

CONTENTS

Section 1—Overview
- Introduction
- Objectives
- Elements of corrosion control assessment
- Pilot plant physical description
  - Location
  - Treatment alternatives
  - Pipe loop description
  - Chemical facilities

Section 2—Operations
- Pilot plant operations
  - Schedule
  - Startup procedures
    - Valve positions and flow adjustment
    - Chemical systems
    - Flow cycle and control procedures
  - General operating procedures

Section 3—Sampling
- Sample collection procedures
  - General water quality
  - Metal leaching
  - Pipeloop sampling
  - Straight lead pipe sampling
  - Special analyses
    - THMs and haloacetic acids
    - Taste and odor
    - Corrosion rates and pitting
- Sample handling, identification, and custody procedures
  - General water quality
  - Metal leaching samples
  - Special analyses
    - Disinfection by-product samples
    - Weight loss insert specimens
    - Flavor profile
    - Biofilm monitoring
- Analytical procedures
  - General water quality
  - Metal leaching
  - Special analyses
    - Corrosion weight loss
    - Disinfection by-products
    - Flavor profile
    - Biofilm monitoring

Figure 10-17 Outline for an experimental plan (*figure continued next page*)

Section 4—QA/QC
- QA/QC procedures
  - Pipe rack fabrication and startup
  - Pilot plant operations
  - Analytical methods
  - Data management procedures

Appendix 1–A. Chemical solution preparation procedures
Appendix 1–B. Sample identification system
Appendix 1–C. Daily operations data
Appendix 1–D. Metals data
Appendix 1–E. Special analyses data

TABLES

1 Chemical concentration and usage data
2 Overview of sample collection and analysis
3 DBP sample delivery schedule

FIGURES

1 Pilot plant layout
2 Typical pipeloop panel
3 American Society of Testing Materials corrosion tester
4 Mixing panel
5 Metering pump calibration
6 Valve positions for sampling

Source: AWWA (1993a).

Figure 10-17 Outline for an experimental plan (***continued***)

Consideration should also be given to who will construct the pipe racks, where they will be constructed, and how they will be transported from the construction site to the testing site(s). These considerations may dictate specific features of the pipe rack design, including the size and type of pipe rack support system and the size of prefabricated portions of the pipe racks. For example, will the racks be mounted on a wall or is it better to have them freestanding? Once constructed, can the pipe loops or racks be transported without great risk of damage? Will the prefabricated pieces fit through doorways at the pilot plant site?

Water quality sampling and frequency of sampling. Parameters recommended for water quality sampling from the pipe rack are divided into two major categories: standing samples (for corrosion related characteristics) and running samples (for influent water quality characteristics and operational consistency).

Water quality results from standing samples will provide data on metals leaching as well as information on general corrosion processes that can be related to pipe coupon observations of weight loss, scaling, and pitting characteristics. While each utility should develop its study-specific water quality monitoring plan, the following parameters are recommended as a minimum for analysis from first-flush, standing water quality samples: lead, copper, temperature, alkalinity, total and free chlorine, and pH.

The frequency of collecting standing samples should be based on the expected variability of the results and the availability of staff and laboratory resources. The metals levels measured from the pipe racks in some studies exhibit highly variable lead and copper levels, with ratios of the standard deviation to the mean of 1:1 or 1:2. Using these ratios as a guideline, the following number of samples would need to be taken for various confidence and accuracy levels:

	Ratio of Standard Deviation/Mean			
	1:2 Confidence Interval		1:1 Confidence Interval	
Accuracy Level, %	90%	95%	90%	95%
25	11	15	28	39
20	17	24	43	61
15	30	43	77	109
10	68	96	173	246

It is recommended that a minimum of 30 samples be collected for statistical definition.

Running samples should provide the water quality characteristics of the source water both before and after it is treated, as well as information on the accuracy and consistency of the corrosion control chemical feed conditions. Background water quality samples should be taken during the flow period immediately before the standing time, after water has been flowing through the rack for a minimum of 30 minutes, and as close to the shutoff time as is practical. The following parameters are recommended for analysis of running influent samples to provide information on influent water quality to all of the racks and allow for interpretation of the metals levels measured from standing samples: pH, ammonia, alkalinity, dissolved oxygen (DO), orthophosphate, conductivity or total dissolved solids (TDS), heterotrophic plate count (HPC), temperature, free and total chlorine, lead and copper, calcium, iron, color, and total coliforms.

The consistency of the source water will play a significant role in determining how frequently these characteristics should be measured. Treatment interruptions or other events that may radically change incoming water quality may not be documented if these parameters are sampled only on a monthly basis; therefore, sampling weekly or every other week may be prudent.

Running samples should also be collected after chemical treatment to evaluate the consistency of the treated water. Depending on the treatment, the chemical feed system can be monitored by evaluating pH and alkalinity, and concentration of inhibitor, calcium, and disinfectant residual. These characteristics should be monitored every day if possible or at least weekly. On-line, continuous water quality monitoring of pH and alkalinity may be desirable if those parameters are being adjusted.

Thorough records of the mechanical operating conditions should also be maintained, including flow, pressure, and throughput volume. These records will provide day-by-day input on the physical operation of the racks as well as valuable information during the data interpretation phase.

The total length of the pipe rack study must be determined by the time required to attain stable metals levels and the number of samples needed to achieve a desired accuracy and confidence in the data.

Based on results of several pipe rack operations, it can take from three to nine months for lead levels to stabilize in either lead loops or soldered copper loops. Copper levels may stabilize in a shorter time period (three to eight months). Results from several of the pipe rack studies indicated that metals levels were still declining after one year. Iron levels may take over a year to stabilize. In addition, seasonal water quality changes may impact metals leaching.

The approaches and principles described above also apply to full-scale demonstration of corrosion treatment in a distribution system. In addition to these, however, the immediate effects on residential, commercial, and industrial customers must be considered. Further, control systems to feed the chemicals must be carefully considered since corrosion inhibitor chemicals are typically fed just before the water enters the distribution or transmission system, and the problems associated with an overfeed can be serious.

Secondary and Tertiary Effects of Corrosion Treatment

Secondary impacts of treatment for corrosion control are important and need to be an integral part of the approach when developing the treatment strategy. Often times, these secondary impacts either limit or significantly influence the treatment finally selected. The goal should be to develop an effective corrosion control strategy that minimizes undesirable secondary side effects.

Careful consideration must be given to the effects that corrosion control measures will have on treatment process efficiency, on changes in water quality as water is distributed, and on the quality of finished water as it impacts various municipal and industrial uses. The purpose of this section is to document these effects, insofar as they are known, and thus, to provide the information required to guide the selection of a treatment program that has minimal undesirable effects.

Coagulation. The coagulation process is sensitive to pH, and to some extent, to alkalinity and phosphate concentration. If pH must be raised much above 7.5 before coagulation, for example, the coagulant of choice may be a ferric salt or a polyaluminum chloride instead of alum. Use of alum will probably require that the adjustment of pH be made after at least sedimentation, and more likely, after filtration. Any change in pH, alkalinity, or phosphate concentration will require studies to determine the change in coagulation conditions that must be made.

Disinfection. The disinfection efficiency of free chlorine for primary disinfection is affected by pH. Further, the increase in trihalomethane (THM) formation that results when the pH of many waters is increased may cause a utility to change from free chlorine to chloramine. Thus a review of the disinfection efficiency of the different forms of chlorine is appropriate before changes in water quality are made to control corrosion.

Ozone, an alternative disinfectant, is also affected by pH. The ozone concentration in water decreases relatively rapidly with time because it decomposes in water and reacts with dissolved or particulate substances in water. The rate of decomposition is an important function of pH. The reaction half-life in pure water is approximately 1 hour at pH 8, 6 minutes at pH 9, and less than 1 minute at pH 10 (Staehelin and Hoigné 1982). These values may be higher or lower depending on the composition of the water. Because of the decrease in half-life as pH is increased, the required $C \times T$ value (concentration times time) for ozone may be more difficult to achieve (i.e., a higher ozone dose may be required).

In summary, disinfection generally will be better at lower pH. Therefore, pH adjustment after disinfection will usually be preferred.

Chemical oxidation. Common chemical oxidants used in water treatment are oxygen, chlorine, chlorine dioxide, ozone, and potassium permanganate. These are used to oxidize such substances as iron and manganese, taste- and odor-causing compounds, and color and, sometimes, to serve as a coagulation or filtration aid. General statements about the effect of pH on the reactions of chemical oxidants are difficult to make because the reactions are very specific. For example, the reactions of most of the above-listed oxidants with iron and manganese proceed more rapidly at higher pH, thus making a pH increase beneficial. Reactions with other substances, such as ammonia, hydrogen sulfide, and organic matter, occur and are difficult to predict stoichiometrically, thus necessitating at least bench-scale studies for specific waters. At high pH, ozone in water is converted to the hydroxyl free radical, which is very reactive. Although its life is much shorter than the O_3 molecule, it is a more effective oxidant of many organic compounds.

Effect of pH on disinfection by-product formation. The increase in rate of THM formation with an increase in pH is well documented (Symons et al. 1981). Urano, Wada, and Takemosa (1983) analyzed the rate of THM formation and found it to be dependent on free chlorine dose, temperature, total organic carbon (TOC) concentration, and type of TOC, as well as pH. They found an increase of 6 μg THM/L per pH unit increase at 1-hour reaction time, 19 μg THM/L per pH unit at 24 hours, and 34 μg THM/L per pH unit at five days when a commercial humic acid was chlorinated (conditions: 5 mg TOC/L, 5 mg Cl_2/L, at 20°C [68°F]). Other published data for surface water were in general agreement with their findings so these data provide a good estimate of the pH effect. These data indicate that an increase in distribution system pH may have an important effect on the concentration of THMs at the customers' taps. This increase may be sufficient to force a utility to change from free chlorine to combined chlorine for residual maintenance in the distribution system.

For short times, however, such as 1 hour or less, the use of free chlorine might be possible because the expected increase in THM concentration is small. Thus, if a higher pH is to be used, free chlorine might be used for a short contact time for good disinfection, followed by conversion of the free residual to combined chlorine. The free chlorine should be applied as late in the process scheme as possible because TOC is removed by some processes, and THM formation rate typically is directly proportional to TOC concentration.

Because of the importance of THM concentrations and the variability of THM formation rate depending on the type of TOC, any change of pH should be accompanied by an examination of its effect on THM concentration.

Effect of pH and alkalinity on calcium carbonate stability. The precipitation of calcium carbonate is affected by both pH and alkalinity. The calcium concentration and the alkalinity, predominantly, determine the saturation pH, commonly referred to as pH_s, of a water. If the pH is much greater than pH_s, large amounts of calcium carbonate may precipitate and affect the treatment processes and the distribution system. The higher the alkalinity, the lower the pH_s. Thus, care must be taken to avoid precipitating excessive calcium carbonate when pH or alkalinity is increased. Calcium carbonate stability is fully described in chapter 1.

Effect of phosphate on finished water quality. Consideration must be given to the effect of ortho- and polyphosphates on finished water quality before they are used as corrosion inhibitors. Water supplies can be grouped in the following categories based on their potential effects.

- The wastewater from the community is treated and discharged, without treatment for phosphate removal, to receiving waters in which eutrophication is not a problem. In this situation, there apparently is no disadvantage to using phosphate inhibitors.
- The wastewater from the community undergoes treatment for phosphate removal, because of the impact of phosphate containing water on the eutrophication of receiving waters. In this situation, part of the cost of implementing the application of phosphates for corrosion inhibition will be the cost of removing them before treated wastewater discharge. If a community is not removing phosphates, but should be because of their impact on receiving water, application of phosphate inhibitor may make an existing problem worse.
- After phosphate addition, the treated water is put in open, finished water reservoirs. The added phosphate may cause the development of algae, possibly with attendant turbidity and taste-and-odor problems, in these reservoirs. Phosphate inhibitors should not be used in this situation, unless studies can be done to show they do not cause a problem.
- The treated water is mixed with untreated well water and the phosphate dosage then becomes insufficient and aggravated localized pitting corrosion can result.
- The presence of phosphates may be undesirable for selected industrial users. This situation is not likely to be a widespread occurrence, and thus should be evaluated on a case-by-case basis.

Secondary impacts of zinc orthophosphate treatment. An important constraint on the dosages of zinc orthophosphate corrosion inhibitors is the solubility of zinc in the water system. Because the passivating film-forming lead orthophosphate solids tend to be much less soluble than the probable zinc orthophosphate solid that would form (α hopeite, $Zn_3(PO_4)_2 \cdot 4H_2O$), the presence of milligrams-per-litre quantities of orthophosphate can be used effectively to reduce plumbosolvency (Schock 1985). However, just as basic lead carbonates can precipitate given the proper combination of pH and dissolved inorganic carbon (DIC), so can the basic zinc carbonate (hydrozincite, $Zn_5(CO_3)_2(OH)_6$). While the low solubility of this compound has been used successfully to combat A–C pipe corrosion (Schock and Buelow 1981; Schock et al. 1981), it can also act to place an upper limit on pH and phosphate dosages applicable to corrosion control of lead, when the phosphate is added by a formulation containing zinc. The formation of the unwanted precipitate can cause domestic consumer complaints of particulates and turbidity and clog commercial and industrial filters. A pH of 7.8 to 8.0 is the usual upper limit for zinc orthophosphate corrosion inhibitor addition to avoid particulate problems. Whether either the zinc orthophosphate or basic zinc carbonate solid becomes less soluble in hot-water systems has not been studied.

Impact of treatment for corrosion control on industrial and commercial water usage. Industry often uses large quantities of water for cooling, heating, and manufacturing of products. For many of these industrial users, the local public water supply is the only source of water. Water utilities should be aware of the potential affect that a change in water treatment may have on the industrial processes and certain commercial uses, whether the change is made for economic reasons, improved water quality, corrosion control, or to comply with the drinking water regulations. The treatment strategies outlined for lead control or other corrosion control strategies can affect these consumers.

Industrial water usage without preconditioning. The major industrial use of water is for cooling. Although large industrial plants will develop their own water source for cooling and employ cooling towers to conserve water, smaller industrial users generally do not have this option available to them. In many applications, potable water is used to remove heat by once-through cooling of process equipment. Scale (mineral deposits) and corrosion problems may develop when the water temperature is significantly increased by the transfer of heat, although the influent water source has neither corrosive nor scaling tendencies. The heating of potable water for domestic uses in laundries, restaurants, and hospitals may result in similar problems. Phosphates can also increase biofouling and slime accumulations in cooling towers and heat-exchanger piping systems.

The common treatment strategy to reduce the corrosivity of water uses the basic chemical equilibriums associated with the solubility of calcium carbonate and/or corrosion by-products. Alkalinity and pH are important parameters that influence these equilibria. Chemicals are added to adjust pH or alkalinity to provide the most favorable solubility conditions in the distribution system. The effects of pH and alkalinity at higher water temperatures should be evaluated as the corrosivity or scaling tendency may be significantly different and have an adverse effect on industrial processes.

Water may also come in direct contact with the product in many commercial and industrial processes, i.e., production of food, beverages, synthetic rubber, textiles, and other products. Depending on the process, very strict limitations in water quality may apply for one or more chemical parameters. Most industries can employ the public water supply without encountering problems because the drinking water regulations generally exceed the process requirements. However, the most restrictive processes may be affected by minor changes in treatment programs by the water utility. For example, application of a few milligrams polyphosphate per litre to a water supply may complex sufficient iron or manganese to exceed the tolerance for these metals in some processes. Several industrial processes have a low tolerance for alkalinity and hardness, and have a relatively low maximum concentration for usage in the process.

Industrial water usage requiring preconditioning. The quality of the potable water must be modified further for many industrial uses such as in the generation of steam or in the production of high-purity water. Operating and maintenance of the water conditioning equipment used to modify the water quality may be more affected than the product water by changes in the utility treatment programs. Because there is a large diversity in water usage, conditioning equipment, and influent water quality, only a brief summary of the effects of utility treatment programs on the industrial water conditioning methods can be presented. In general, the small changes in water quality for corrosion control will not have a large effect on industrial water conditioning, but in a few special situations it may be significant. For example, the semiconductor industry typically deionizes large quantities of water for wash and rinse water. The addition of silicate inhibitors substantially increases deionization costs.

Softening (ion exchange). The hardness (calcium and magnesium) content of the water source affects the capacity of a water softener to produce soft water. Any increase in hardness concentration will increase the regeneration frequency of softening units. The capacity of a strong-acid resin in softeners can be reduced by resin oxidation and resin fouling. Residual free chlorine, iron, or manganese concentrations exceeding 0.5 mg/L in the influent water will also cause softener difficulties and additional operational costs.

Dealkalinization. Both weak-acid and strong-base ion exchange resins are used to remove alkalinity from the source water. Weak-acid ion exchange methods will remove the calcium and magnesium associated with alkalinity (carbonate hardness) and convert the carbonate alkalinity into free carbon dioxide for removal by degasification equipment. Strong-base resins convert carbonate alkalinity into equivalent amounts of chloride. Both dealkalinization methods are frequently used to condition makeup water for small steam boilers. The alkalinity, hardness, sodium, and free chlorine content of the influent water will affect weak-acid resin capacity. Alkalinity, sulfate, nitrate, phosphate, and organic content will influence the capacity of the strong-base resins. Corrosion treatment strategies that change these components in the distribution water will affect the regeneration of dealkalinization equipment.

Deionization. High- and ultrahigh-purity waters are required by laboratories, semiconductor manufacturers, high-pressure power plants, and other industrial uses. Ion exchange equipment is used to remove the mineral content of potable water for these purposes. Ion exchangers generally employ strong-acid and strong-base resins but various combinations of the resin forms may be used. Since the deionization method removes all cationic and anionic constituents, all chemical treatment programs will have some effect on the capacity of deionizer units. Many dissolved organic substances will be absorbed on anion resin and eventually foul the exchange sites. Free chlorine degrades both anion and cation resins. An activated carbon filter installed ahead of the deionizer will remove both the chlorine and organics but will have no affect on other chemical additives.

Reverse osmosis. Industry uses the reverse osmosis process to reduce the mineralization of potable water for various process uses. Water is forced by pressure through a semipermeable membrane to separate the dissolved minerals from the water. Reverse osmosis units can experience fouling of the membranes by scale, metal oxides, colloids, and biological growths. The membranes vary in their ability to reject various substances depending on numerous factors, including molecular weight, valence, dissociation, and concentration of the substances. Reverse osmosis is, therefore, sensitive to many constituents in potable water whether naturally occurring or additives. Alkalinity, pH, calcium, iron, silica, sulfate, and total dissolved solids are important chemical parameters that will affect membrane life.

Electrodialysis. This treatment process is similar to reverse osmosis in that the mineralization of water is reduced by passage through permeable membranes. However, electrical energy is used to separate the ionized salts instead of pressure as used with reverse osmosis. Membrane fouling can also occur in electrodialysis. Pretreatment is generally required to control fouling. Electrodialysis is sensitive to the same chemical parameters as reverse osmosis.

STEP 4: EVALUATE ALTERNATIVES AND SELECT CORROSION CONTROL STRATEGY

This step is aimed at answering the following questions:

- What statistical methods are available to analyze corrosion-related performance data?
- In addition to performance, what criteria should be used to evaluate alternative treatments?
- What rating and ranking methods should be applied?

Evaluation Criteria

A variety of criteria can be used to evaluate corrosion treatment programs and the most important one is performance. Other criteria include costs, secondary effects, reliability, flexibility, and compatibility with existing and future system.

Performance. The key to selecting an optimum corrosion treatment strategy, performance is normally based on a set of data gathered during desk-top or demonstration studies. The following discussion provides an approach for statistically evaluating the data generated during a pipe rack or other demonstration study and includes the following key components:

- determining data normality
- evaluating data outliers
- determining when lead and copper levels have stabilized during the corrosion study
- determining treatment differences

Determining data normality. When evaluating lead, copper, or other data from a corrosion study, it is important to first assess the distribution of the data that have been collected in order to apply the correct statistical tests. For data that are not normally distributed or for which the data distribution is unknown, nonparametric tests should be applied. Data distributions can be determined with graphical techniques such as histograms, where the y-axis represents the frequency of occurrence of a given variable and the x-axis represents the values of the variable (Figure 10-4). Computer statistical packages can also be used. They offer the added advantage of providing the statistical significance of the underlying distributions in addition to graphical representations.

Evaluating data outliers. After reviewing the distribution of the data generated from a corrosion study, it may seem appropriate to discard several of the very high levels as outliers, or values that are anomalous. By discarding these values, the mean level of the data will be lowered, and there will be less variation in the data. Lead may be particularly variable in both demonstration studies and in home sampling results. It is important to note, however, that lead levels measured at the tap in the distribution system also exhibit occasionally high lead or copper levels. The occurrence of what appear to be unusually "high" lead levels in home tap samples is evidence of the chemical and physical factors that affect the amount of lead that can leach into the water. These factors are also present in pipe loop studies, although an attempt has been made to reduce the variability of lead measurements by controlling these factors to the extent possible. These "high" lead levels may occur at various times during the pipe loop study period and in many cases cannot be attributed to sampling or analytical error. Rather they reflect the highly variable nature of lead corrosion and solubility and particularly the irregular sloughing of particulate lead into the water.

In summary, occasional high lead levels may be measured from corrosion studies as well as from home tap sampling. It is recommended that high lead and copper levels from pipe rack studies not be discarded unless there is documented evidence that the sampling or analytical techniques for those particular samples were in error. Otherwise, the levels should be considered "real" values and incorporated into the entire data set.

Testing stabilization. Pipe loops constructed of new materials and used for corrosion studies of lead and copper leaching will exhibit a decreasing trend in lead and/or copper levels over time as the materials age. These levels may "stabilize" after a certain period of time if the influent water quality is relatively consistent. This stabilization may indicate that the pipe materials have reached equilibrium with respect to corrosion or have reached a state where film formation has become very slow (refer to Figure 10-18). While it may be of interest to evaluate the entire set of data generated from a pipe rack study, it is more appropriate to use only the data that have reached stabilization when determining the effect of various treatments on lead and copper levels. Stabilization can be determined roughly by visual examination of the data, or more specifically through calculating an estimated trend line for the data. Visual examination is likely to be less sensitive to slow changes than numerical techniques and may be less likely to enable determination of the affect of the variability of the data. The most common numerical method for determining trends is application of a best-fit line to the data. A linear trend line is described by the following formula:

$$y = ax + b \tag{10-1}$$

Where:

a = the slope of the line
b = the intercept

If the trend line has a negative slope, the values are decreasing with time. If the line is positive, the values are estimated to be increasing with time. The more negative or positive the slope of the estimated line, the more pronounced is the trend. Estimated trend lines with very flat slopes ($a = 0$) have essentially stabilized.

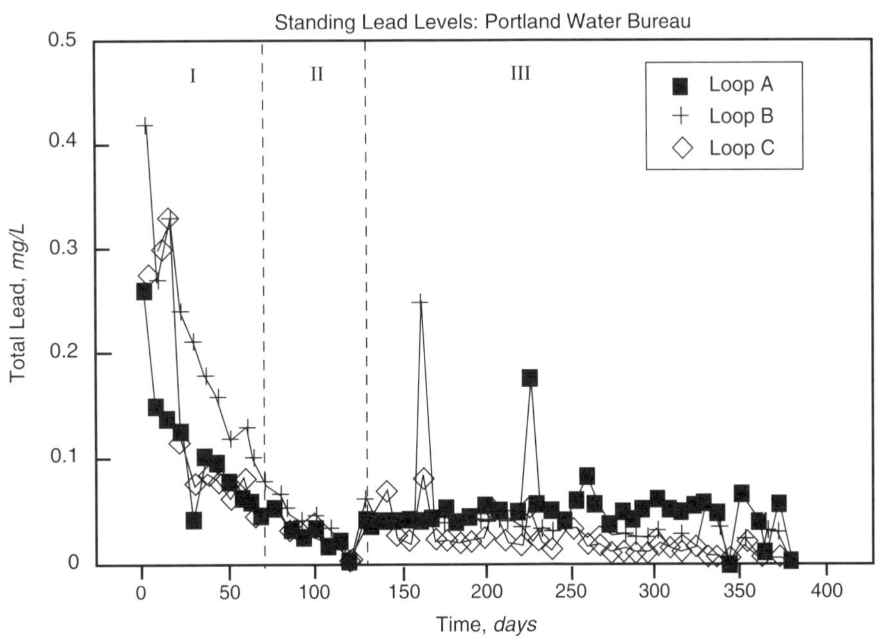

Source: EES and ISWS (1990).

Figure 10-18 Actual pipe rack data exhibiting three periods during a pipe rack study

For parametric data, the Pearson correlation coefficient can be used to determine how close the actual observations (y_i) are to values estimated by the trend line (y^*). Correlation coefficient (r) values of –1 denote a perfect negative correlation and r values of +1 indicate a perfect positive correlation. The correlation coefficient (r) is the positive or negative square root of the coefficient of determination (r^2). Coefficient of determination values of 0 indicate that the equation explains none of the variation in y, whereas values of 1 mean that all of the variation is explained, i.e., all y values lie on the regression line.

For nonparametric data, the Spearman rank correlation coefficient may be used rather than the Pearson correlation coefficient. The Spearman r is the correlation coefficient applied to rank order data and ranges from –1.0 to +1.0. Positive values approaching 1 indicate that increases in one variable are accompanied by increases in the other variable. Negative values near –1 indicate the opposite. With the Spearman r, it is important to note that the r values are calculated on the rank order of the values, rather than the values themselves.

It is important to evaluate the trend in metals levels as the pipe rack study progresses. If levels have begun to stabilize, as determined by evaluating a trend line and correlation coefficient, a decision can be made as to how many more samples will be needed to substantiate the treatment differences.

Determining treatment differences. The ultimate goal of a corrosion control study is to determine whether a particular treated water quality condition results in lead or other corrosion by-product levels that are significantly lower than the control water. Incorporating statistical tests in these determinations provides additional information on the extent and the statistical significance of the difference. Commonly used methods to compare differences between populations are the t-test for parametric data and the Wilcoxon test for nonparametric data. The mean and standard deviation of the data are used in parametric methods and percentiles are used in the nonparametric methods.

Once lead or copper levels have stabilized, these methods can be used to determine differences in metals levels due to various treatments. The significance levels of these differences (r- or p-level) can be calculated to provide information on the level of confidence to expect from the results. A significance level of 95 percent (r = 0.05) is generally considered highly significant, i.e., there is a 95 percent chance the right decision has been made. In addition, if multiple loops within a pipe rack have been used, using data from all three test loops will provide information on the variability observed in lead levels between the control and the various treatments, as well as allow a determination of the degree of confidence that can be assigned to the results.

If after stabilization the data exhibit a normal distribution and there are several data points (i.e., more than 30), a standard t-test can be used to evaluate treatment differences. Metals levels from pipe rack studies have typically exhibited nonnormal distributions. However, particularly when (1) the data have not exhibited stabilization, (2) there are very high metals levels ("spikes") in the data, or (3) very few data points have been collected, nonparametric Wilcoxon test methods are most appropriate for evaluating treatment differences.

An example of determining treatment differences using nonparametric methods is presented below with data from New York City's Catskill–Delaware lead loop results (Kirmeyer et al. 1994).

In the New York City Catskill–Delaware lead loops, the control loop stabilized after approximately 279 days of operation. Loop 5, which was treated with zinc orthophosphate, stabilized after approximately 230 days. Using the Wilcoxon signed rank to compare these two treatments first involves developing a null hypothesis, which will either be accepted or rejected based on the results of the test. In this case, the null hypothesis is that there is no difference between these two water quality conditions, i.e., there is not enough difference in lead levels between the two water quality conditions to say with a certain amount of confidence that the treatment was effective.

The Wilcoxon signed rank test is performed by ranking the difference between matched pairs of data from each of the two loops, summing the positive ranks (T^+), and calculating a Z value, which is then compared to the critical α for 95 percent confidence. If the Z value falls outside the critical region, the null hypothesis is rejected, i.e., the treatment was effective.

In comparing the control loop with loop 5 (zinc orthophosphate), data from day 279 and on is used, and the Wilcoxon signed rank test is accomplished as follows:

Lead Levels					
Control x_i	Loop 5 y_i	$Z_i = x_i - y_i$	R_i	S^*	$R_i(S)$
1.22	0.34	0.88	4	1	4
1.12	1.17	–0.05	1	0	0
1.52	0.43	1.09	5	1	5
0.58	0.22	0.36	2	1	2
1.47	0.35	1.12	6	1	6
0.82	0.45	0.37	3	1	3
					20

Where:

R_i = *rank of* Z_i
S^* = 1 if $Z_i > 0$
S = 0 if $Z_i < 0$
T^+ = 4 + 5 + 2 + 6 + 3 = 20

$$Z = \frac{T^+ - \dfrac{n(n+1)}{4}}{\left[\dfrac{n(n+1)(2n+1)}{24}\right]^{1/2}}$$

in which n = number of samples = 6

$$Z = \frac{20 - 6(7)/4}{\left(\dfrac{6(7)(13)}{24}\right)^{1/2}} = 1.99$$

Since 1.99 is greater than 1.645 (the critical region for 95 percent confidence), the null hypothesis is rejected at the 95 percent confidence level, i.e., the treatment is effective.

This procedure can also be used to determine if one treatment is more effective than another. Comparing loop 6 (zinc orthophosphate) with loop 8 (calcium orthophosphate) lead levels produces the following results:

Lead Levels					
Loop 6	Loop 8	Z_i	R_i	S *	$Ri(S)$
0.11	0.081	0.029	4	1	4
0.083	0.023	0.060	5	1	5
0.071	0.072	–0.001	1	0	0
0.023	0.85	–0.062	6	0	0
0.036	0.024	0.012	2	1	2
0.025	0.044	–0.019	3	0	0
					11

$T^+ = 4 + 5 + 2 = 11$
$Z = 0.105$

The Z value is 0.105, which is less than 1.645; therefore, the null hypothesis is accepted at the 95 percent confidence level, i.e., there is no difference in treatments.

In summary, the appropriate statistical methods should be applied to corrosion data to compare the effectiveness among treatments. The evaluation process begins with collecting an adequate number of samples, ensuring that corrosion or corrosion by-products have stabilized, and applying valid statistical techniques.

Costs. Costs are an important factor in corrosion treatment selection. Life-cycle cost analysis should be performed using both capital and annual operation and maintenance costs. Typically, the cost of chemicals are the predominant portion of the life-cycle cost. Costs are site specific and should be investigated on a case-by-case basis.

Secondary impacts. The potential secondary impacts associated with corrosion control were discussed previously in step 3. The need to consider secondary impacts so that other drinking water quality regulations are met cannot be overemphasized.

Reliability. In selecting the treatment strategy and specifically the corrosion treatment chemicals, the reliability of equipment, availability of chemicals, and consistency of finished water quality is important. Treatment interruptions and highly variable finished water quality are often not conducive to adherent film formation.

Flexibility. It is important that the selected alternative be flexible to be able to handle seasonal variations in water quality and to meet future drinking water quality regulations.

Compatibility with existing and future system. The corrosion treatment strategy needs to be compatible with the water system components, standard operating procedures, and planned improvements. For example, large storage reservoirs or seasonal mixing with groundwaters may affect the preferred choice.

Rating, Ranking, and Selecting the Alternative

A variety of ways can be used to rate and rank corrosion treatment alternatives, including matrix evaluations. These matrixes can be either qualitative or quantitative, and both are somewhat subjective in nature. Since the criteria are not all of equal importance, they can be weighted differently to emphasize those with the greatest importance. An example of the quantitative approach is shown in Table 10-9.

Table 10-9 Quantitative matrix evaluation for evaluation of corrosion control alternative

Treatment Alternative	Corrosion Control Performance	Treatment Reliability	Estimated Costs	Total Points	Ranking
Weighting factors	0.75	0.15	0.1	1	—
Treatment A	7	7	0	6.3	1 (best)
Treatment B	4	0	4	3.4	2
Existing	0	4	7	1.3	3 (worst)

Source: Adapted from USEPA (1992).

A desk-top evaluation of treatments A and B prior to pipe loop testing revealed that these treatment options were equally feasible. In this case, the most important factor for identifying optimal treatment is treatment performance, shown by setting its weighting factor at 0.75. The reliability factor is set as 0.15 and the cost weighting factor is set at 0.10. The reliability of the treatment options is considered more important than the costs since compliance with a lead regulation will eventually be determined by the ability of the water system to consistently produce finished water that meets its treatment objectives. The costs of treatment should be assigned a low weighting factor (here 0.1) to reflect the fact that costs are not directly relevant to selecting the optimal treatment, except in helping to decide between alternative treatments with comparable performance. Based on the results of the final treatment-selection decision matrix, treatment A would be recommended as optimal corrosion control treatment. The use of the matrixes and evaluation criteria should be viewed as tools in the decision-making process; nothing can be entirely substituted for the experience and judgment of the investigator.

STEP 5: DOCUMENT FINDINGS IN ENGINEERING REPORT

This step is aimed at summarizing and answering the following questions:

- What procedures were used to develop the corrosion control program?
- What is the recommended corrosion control program?
- How much will it cost?
- How long will it take to implement?
- What are the design criteria?
- What followup procedures are recommended to monitor effectiveness?
- Does the report need to meet regulatory criteria?

The engineering report may be written for a variety of audiences, which might include plant operators, utility boards and administrators, design engineers, regulatory officials, members of environmental or consumer groups, and the utility's customers. It may be necessary to have certain chapters of a report, such as the executive summary, target the nontechnical audience. Likewise, technical appendixes may be more appropriate for the technical audience, such as state regulators. Information should be presented in an understandable manner that is complete and technically correct.

One example of how such an engineering report might be organized is shown in Figure 10-19. This format contains all of the major elements and justifications previously discussed. When completed, the report should be a "stand alone" document containing sufficient information to document recommendations and enable a final decision to be implemented.

STEP 6: IMPLEMENT CORROSION CONTROL AND MONITOR EFFECTIVENESS

This step is aimed at answering the following questions:

- How should the corrosion monitoring program be implemented full-scale?
- What characteristics and at what frequency should parameters be analyzed in the chemical addition facility and in the distribution system?
- What are the quality control considerations?

In addition to design and construction of facility improvements, implementation of a successful corrosion treatment program requires a significant effort on the part of utility operations staff. The purpose of this section is to identify monitoring requirements and other procedures to facilitate a successful corrosion treatment startup and provide a basis for making treatment adjustments. Monitoring should be completed at both the treatment plant and in the distribution system. Utility staffs need to be trained in the treatment process, emergency response, and other areas. Startup is especially important since problems are more likely to occur during this stage than during routine operations.

Effective monitoring, startup, and treatment adjustment programs are the key to corrosion optimization. Such programs will enable water utilities to adjust treatment in response to seasonal variations in water quality and thus provide water that meets corrosion control objectives, including regulatory and customer needs. In addition, water quality problems can be detected at an early stage and corrective steps taken to prevent more serious conditions from occurring.

Treatment Plant Monitoring

Monitoring programs at the treatment plant fall into two categories: water quality monitoring and chemical and facility monitoring. These two categories are discussed separately below.

Water quality monitoring. Water quality monitoring for a corrosion treatment system should involve sample collection and analysis of water from the process stream immediately before the point(s) of corrosion chemical application as well as of the water leaving the treatment plant and entering the transmission or distribution system. The monitoring plan should identify which parameters will be monitored at what frequency, describe sample collection procedures, and designate who will collect and analyze the samples. It is important that plant operators are aware of the quality of water as it reaches the point of application. This enables them to apply treatment chemicals at dosages necessary to accomplish treatment objectives. This is especially the case where pH and alkalinity adjustments are involved, as these parameters may change seasonally. Utility officials must always be cognizant of the quality of treated water leaving the plant and entering the distribution system.

Section	Title
I	SUMMARY OF CORROSION CONTROL INITIATIVE

- A. Background information
- B. Findings and conclusions
  1. Overview
  2. Impacts of corrosion
  3. Causes of corrosion
  4. Water treatment evaluation
  5. Benefits of corrosion treatment
  6. Other water quality considerations
  7. Compatibility with other system initiatives
- C. Recommended corrosion control program
  1. Three-phased corrosion control approach
  2. Corrosion treatment
  3. Materials selection
  4. Public education

II INTRODUCTION AND BACKGROUND

- A. Purpose
  1. Overall water quality improvement program
  2. Corrosion control initiative
  3. Interrelationship of study initiatives
- B. Description of water system facilities
  1. General
  2. Source
  3. Treatment
  4. Transmission
  5. Service and reservoir storage
  6. Municipal distribution systems
  7. Customer premise piping
- C. Overview of water quality
  1. General
  2. Source reservoirs
  3. Transmission and storage facilities
  4. Distribution system and customers facilities
- D. Initiative approach
- E. Questions to be answered
  1. Causes of internal corrosion
  2. Extent and magnitude of internal corrosion and its impacts
  3. Investigation and development of corrosion control strategies
  4. Other

Figure 10-19 Example Engineering Report Format for Corrosion Control Study (*figure continued next page*)

Section	Title
III	PROBLEM DEFINITION AND STUDY APPROACH

A. Characterization of corrosion problems in the water system
   1. System water quality
   2. Transmission and distribution piping system conditions
   3. Service and premise piping conditions
   4. Tap water quality monitoring information
   5. Review of customer complaint records
   6. Conclusions

B. Corrosion mitigation approaches used by other utilities

C. Background information on internal corrosion
   1. Potential causes of internal corrosion
   2. Impacts of internal corrosion
   3. Comparisons of utility's corrosion-associated water quality conditions to those of other water sources
   4. Regulatory aspects of internal corrosion
   5. Corrosion treatment strategies

D. Project approach
   1. Overview
   2. Water quality investigations
   3. Materials investigation
   4. Pilot plant and field studies

IV LABORATORY AND FIELD METHODS

A. Overview

B. Utility-wide tap sampling programs

C. Utility-wide plumbing pipe sampling programs
   1. Plumbing characterization survey
   2. Plumbing pipe condition sampling program

D. Sampling for lead in school drinking water fountains

E. Corrosion control 12-month pilot-plant study
   1. Purpose of the corrosion control pilot-plant study
   2. Corrosion control pilot-plant study approach
   3. Corrosion pilot-plant study physical description and study schedule
   4. Corrosion pilot-plant study: operation procedures, quality control, and testing methods

V DISCUSSION AND ANALYSIS OF RESULTS

A. Overview

B. Tap sampling programs

Figure 10-19 Example Engineering Report Format for Corrosion Control Study (*figure continued next page*)

C. Plumbing pipe sampling programs
   1. Plumbing characterization survey
   2. Plumbing pipe condition sampling program
D. Lead sampling in regional school drinking water fountains
E. Corrosion control 12-month pilot-plant study
   1. Quality control results
   2. Pipe coupon inserts: results and discussion
   3. Metals concentrations in lead-soldered copper plumbing coil standing samples
   4. Lead concentrations in standing samples from brass faucets
   5. Staining of porcelain plates
F. Corrosion control treatment strategies: a theoretical model approach
   1. Model approach
   2. Application and results of the model
G. Summary of technical results of the initiative
   1. Corrosion conditions of piping materials
   2. Water quality considerations
H. Economic analysis
   1. Overview
   2. Estimation of corrosion control treatment project costs
   3. Estimation of benefits from lower corrosion rates to utility
   4. Comparison of benefits and costs for utility
   5. Estimation of customer benefits from lower corrosion rates
   6. Comparison of overall benefits and costs
   7. Summary of benefit:cost analysis
I. Comparison of corrosion control technical results with those of another municipal water system
   1. Background and purpose
   2. Comparison of corrosion rates found during pilot plant studies
   3. Comparisons in metal levels
   4. Summary

VI CONCLUSIONS AND RECOMMENDATIONS

A. Causes of corrosion in the water system
   1. Source water characteristics
   2. Current water treatment practices
   3. Susceptible materials
B. Effects of corrosion in the water system
   1. Economic effects
   2. Health effects
   3. Aesthetic effects
C. Evaluation criteria and methods
   1. Performance
   2. Costs

Figure 10-19 Example Engineering Report Format for Corrosion Control Study (*figure continued next page*)

3. Secondary effects
4. Reliability
5. Flexibility
6. Compatibility with existing and future system

D. Comparison of treatment alternatives
1. Overview
2. No treatment to reduce corrosion
3. pH and alkalinity adjustments
4. Use of inhibitors
5. Calcium precipitation

E. Recommended corrosion control approach
1. Overview
2. Water quality and treatment objectives
3. Materials selection criteria and practices
4. Public education

VII IMPLEMENTATION MONITORING EFFECTIVENESS

A. Process and design criteria
B. Treatment plant monitoring
C. Distribution system monitoring
D. Startup
E. Personnel and staffing
F. Quality control

Figure 10-19 Example Engineering Report Format for Corrosion Control Study (***continued***)

It is also important that operators are aware of chemical and physical changes that occur as a result of corrosion treatment in order to make adjustments to improve finished water quality. Another objective with monitoring finished water is to establish a baseline for evaluating any subsequent water quality changes that may occur in the distribution system. At a minimum, the following parameters should be monitored and evaluated on a routine basis (quarterly if a surface water supply and semiannually if a groundwater are good starting recommendations):

- flow
- chlorine residual
- pH
- alkalinity
- temperature
- conductivity
- copper
- lead
- zinc
- silicates (if fed subsequently)

- calcium
- hardness
- dissolved oxygen (DO)
- orthophosphate (if fed subsequently)
- total phosphate (if fed subsequently)
- asbestos fibers (if A–C pipe in distribution system)
- iron
- cadmium
- arsenic
- taste and odor
- coliform bacteria
- heterotrophic plate count

Chemical and facility monitoring. Chemical and facility treatment monitoring is necessary to provide assurance that chemicals of acceptable quality are used and applied in correct quantities. This involves accepting chemical deliveries, analyzing feed chemical strengths and purity, noting quantities of chemical stocks, measuring process stream flow, and controlling chemical feed rate and dosage.

Procurement of corrosion treatment chemicals, like other chemicals used as drinking water additives, requires close scrutiny on the part of the utility to ensure that the material meets appropriate acceptance criteria. As required or allowed in the specifications, the utility representative should make the following checks before accepting a given shipment:

- quantity of delivery
- affidavit certifying chemical strength and purity
- visual evidence of contamination
- sample collection and on-site analysis of chemical strength and purity
- general condition and recent hauling history of delivery container
- capacity of off-loading equipment and compatibility with on-site facilities
- suitability and condition of safety equipment required for off-loading

Accurate dosage control requires using feed chemicals of known strength. This typically is not as much of a problem with the corrosion treatment chemicals as it can be with other, less stable chemicals such as sodium hypochlorite. Provided that the utility consistently purchases the same grade, strengths of corrosion treatment chemicals can be expected to remain fairly constant.

Operators should also routinely measure and note the quantity of feed chemicals in stock. This practice can keep personnel aware of when to place new orders, provide a check on delivery quantities, help in quantifying leaks or spills that may occur, and provide data to be correlated with process stream flow records as a check on applied dosage.

Flow measurement in the process stream must be accurate over the full range of plant flows. The flowmeter should display both instantaneous and totalized flow and should be capable of sending an output signal, indicating instantaneous flow, to other components of the treatment plant's monitoring, control, and chemical feed system.

For manual or automatically controlled chemical feed systems, consistent treatment requires that the operator understand and be capable of measuring or calculating three application parameters: dosage, chemical feed rate, and process stream flow.

Distribution System Monitoring

Distribution systems, due to their size and nonhomogeneous nature, present a unique monitoring challenge. Distribution systems typically cover a wide area, which may include business, commercial, industrial, residential, and rural districts, each with different water use patterns and needs. The materials, age, and overall condition of piping in the system generally vary widely from one part of the system to another. Utilities often have more than one source of water supply, each with differing water chemistry and possibly a separate entry point into the distribution system. Some areas are further from the source(s) of supply, in different pressure zones, or may be served from a poorly circulated reservoir or dead-end main. These and many other factors influence the quality of water served to utility customers and must be addressed in establishing an effective monitoring program.

When designing a distribution system monitoring program for the purpose of optimizing corrosion treatment, utility personnel must find a balance between two objectives. One is to evaluate overall water quality throughout the system, while the other is to seek out, monitor, and evaluate problem areas. A suggested approach for accomplishing both objectives is to develop a monitoring program with two sets of sampling sites; one set that provides an overall picture of water quality and another set that targets problem areas.

Monitoring overall water quality. Recommended or required bacteriological monitoring provides a representative sampling of the entire distribution system. This siting plan provides a basis for the utility's monitoring program for such regulated characteristics as coliform bacteria, HPC, and detectable disinfectant residual. In addition, these distribution sites can be used to monitor the same parameters that are monitored immediately after treatment. This will enable the operator to document changes that may have occurred as finished water travels through the distribution and storage system. Monitoring frequency and the number of sampling locations is site specific, but at a minimum, a fingerprint on a monthly or seasonal basis is appropriate.

Monitoring for targeted water quality concerns. Utilities should consider developing one or more specialized siting plans to evaluate specific water quality concerns. For example, the US Lead and Copper Rule requires utilities to select targeted sampling sites most likely to have high lead and copper. This includes sampling small volumes of water from first-flush samples at high risk homes.

Monitoring corrosion in distribution system piping. In addition to monitoring water quality, which measures metal release, it is important to consider direct measurements of corrosion in the system as well. Chapter 9 details methods for measuring and evaluating corrosion rates for various metals. The utility should select and implement one or more of these methods to document the effectiveness of corrosion control. Possible monitoring approaches include collecting and evaluating pipe tap coupons or more formally removing and analyzing pipe segments according to a planned program.

Flat-plate or cylindrical coupons for weight-loss measurements can be inserted into the flow stream before and after treatments and in the distribution system for comparison. Likewise, the Illinois State Water Supply (ISWS) coupon technique, or modified versions, could be used to measure corrosion rates. Electrochemical corrosion assessment techniques, including resistivity or linear polarization monitors, can also be applied to give "snapshots" of the corrosion conditions. The AWWARF pipe racks could be used in a similar manner by locating one rack just before the treatment process, one rack after treatment, and then racks in the distribution system as well. By operating and sampling over a several-year period, a sound picture of the effectiveness of corrosion control can be developed.

Quality Control

Quality control guidelines and procedures should be arranged for all monitoring, startup, and operational activities. Key areas are listed below.

- Field procedures: Identify analyses to be conducted in the field. Establish suitability of sampling sites, sampling equipment and procedures, and analytical methods.
- Treatment plant procedures: Identify analyses to be conducted at the treatment plant. Set schedule and procedures for sample collection. Establish verification checks and analyses for flow, chemical feed, and monitoring equipment.
- Laboratory procedures: Identify laboratory (in-house or outside) responsibilities and analytical methods, detection limits, and quality control for various parameters. Establish sample preservation and handling techniques.
- Data management: Establish how information is going to be logged, organized, and presented. Identify individuals responsible for reviewing data and applying it to make treatment decisions.

SUMMARY

This chapter summarizes the key factors that should be considered in analyzing a corrosion problem and implementing a corrosion-control program. A logical six-step process for conducting such a program is presented and each step is thoroughly discussed. The chapter presents a framework for corrosion-control studies, contains practical advice for conducting corrosion-control studies, gives examples of ways to present data, and presents hints on how to implement the full-scale program. Appendix C presents examples of the successful use of this approach and makes some comparisons between predicted and actual corrosion rate reduction in an operating water system.

REFERENCES

AWWA (American Water Works Association). 1993a. *Initial Monitoring Experiences of Large Water Utilities Under USEPA's Lead and Copper Rule.* Sponsored by the Water Industry Technical Action Fund. Denver, Colo.: American Water Works Association.

———. 1993b. *AWWA Satellite Teleconference. Lead and Copper Rule Compliance: How to Conduct a Corrosion Control Study. Participant Guide.* Denver, Colo.: American Water Works Association.

EC (European Community). 1980. Council Directive of 15th July 1980 Relating to the Quality of Water Intended for Human Consumption (80/778/EEC). *Off. Jour. Eur. Comm.,* 23 (L229/11).

EES and ISWS (Economic and Engineering Services, Inc., and Illinois State Water Survey). 1990. *Lead Control Strategies.* Denver, Colo.: American Water Works Association Research Foundation and American Water Works Association.

EES and Kennedy/Jenks/Chilton. 1989. *Economics of Internal Corrosion Control.* Denver, Colo.: American Water Works Association Research Foundation and American Water Works Association.

GVWD (Greater Vancouver Water District). 1990. Drinking Water Quality Improvement Program—Final Corrosion Control Report. Prepared by Economic and Engineering Services, Inc., for Greater Vancouver Water District.

JM Montgomery Consulting Engineers, Inc. 1982. *Internal Corrosion Mitigation Study Final Report.* Portland, Ore.: Bureau of Water Works.

Kirmeyer, G.J., A.M. Sandvig, G.L. Pierson, and C.H. Neff. 1994. *Development of a Pipe Loop Protocol for Lead Control.* Denver, Colo.: American Water Works Association Research Foundation and American Water Works Association.

Larson, T. 1975. *Corrosion by Domestic Waters.* Illinois State Water Survey Bulletin 59. Urbana: Illinois State Water Survey.

Levin, R. 1986. *Reducing Lead in Drinking Water: A Benefit Analysis.* EPA Report 230-09-86-019. Cincinnati, Ohio: U.S. Environmental Protection Agency.

Lyman, W., A. Cohen, and J.R. Myers. 1982. Causes and Prevention of Pitting Corrosion in Copper Plumbing Systems in the U.S.A. In *Proc. of the International Symposium on Corrosion of Copper and Copper Alloys in Buildings.* Tokyo: Japan Copper Development Association.

Santa Clara Valley Water District. 1994. Internal Corrosion Control Study, Final Report. K/J 92591.03. Prepared by Kennedy/Jenks Consultants for the Santa Clara Valley Water District.

Schock, M.R. 1985. Treatment of Water Quality Adjustments to Attain MCL's in Metallic Potable Water Plumbing Systems. In *Proc. Seminar on Plumbing Materials and Drinking Water Quality.* Cincinnati, Ohio: U.S. Environmental Protection Agency Water Engineering Research Laboratory.

Schock, M.R., and R.W. Buelow. 1981. The Behavior of Asbestos–Cement Pipe Under Various Water Quality Considerations. Part 2: Theoretical Considerations. *Jour. AWWA,* 73(11):609.

Schock, M.R., et al. 1981. Evaluation and Control of Asbestos–Cement Pipe Corrosion. USEPA Report 600/d-d81-06T. Washington, D.C.: U.S. Environmental Protection Agency.

Seattle Water Department. 1976. Seattle Internal Corrosion Study—Phase I, II, and III. Prepared by Kennedy Engineers for Seattle Water Department.

———. 1985. Unpublished Report. Seattle, Wash.

Staehelin, J., and J. Hoigné. 1982. Decomposition of Ozone in Water: Rate of Initiation by Hydroxide Ions and Hydrogen Peroxide. *Environ. Sci. Technol.,* 16:676.

Symons, J.S., A.A. Stevens, R.M. Clark, E.E. Geldreich, O.T. Love, Jr., and J. DeMarco. 1981. *Treatment Techniques for Controlling Trihalomethane in Drinking Water.* Cincinnati, Ohio: U.S. Environmental Protection Agency.

Urano, K., H. Wada, and T. Takemosa. 1983. Empirical Rate Equation for Trihalomethane Formation With Chlorination of Humic Substances in Water. *Wat. Res.,* 17(12):1796.

USEPA (U.S. Environmental Protection Agency). 1991. Drinking Water Regulations; Maximum Contaminant Level Goals and National Primary Drinking Water Regulations for Lead and Copper; Final Rule. 40 CFR, parts 141 and 142. 56 *Federal Register* No. 110, June 7, 1991.

———. 1992. *Lead and Copper Rule Guidance Manual, Volume 2: Corrosion Control Treatment.* Washington, D.C.: U.S. Environmental Protection Agency.

Vik, E.A., and T. Hedberg. 1991. Corrosion and Corrosion Control in Drinking Water Systems. In *Proc. of a Corrosion Workshop and Seminar in Oslo, Norway, March 19–21, 1990.* Houston, Texas: National Association of Corrosion Engineers.

Appendix A

Using Free Energies of Formation to Calculate Equilibrium Constants

Any equilibrium constant for a reaction at 25°C written in the form

$$a\,W + b\,X \rightleftharpoons c\,Y + d\,Z$$

can be computed using the following relationships (Schock 1980; Hunt and Creasey 1980; Snoeyink and Jenkins 1980; Stumm and Morgan 1981):

$$\begin{aligned}\Delta G_R^{\circ} &= \sum \Delta G_f^{\circ}\,(\text{products}) - \sum \Delta G_f^{\circ}\,(\text{reactants}) \\ &= c\Delta G_f^{\circ}(Y) + d\Delta G_f^{\circ}(Z) - a\Delta G_f^{\circ}(W) - b\Delta G_f^{\circ}(X)\end{aligned}$$

and

$$\log K = \Delta G_R^{\circ} / -1.3642$$

Where:

ΔG_R° = the Gibbs free energy change for the whole reaction as written
ΔG_f° = the Gibbs free energy of formation for an individual species

The factor –1.3642 applies to ΔG_f° and ΔG_R° computed in kilocalories per mole.

For example, log K for the complex formation reaction

$$Pb^{2+} + 2H_2O \rightleftharpoons Pb(OH)_2^0 + 2H^+$$

can be calculated as follows:

$$\begin{aligned}\Delta G_R^{\circ} &= \Delta G_f^{\circ} Pb(OH)_2^0 + 2\,\Delta G_f^{\circ} H^+ - \Delta G_f^{\circ} H_2O - \Delta G_f^{\circ} Pb^{2+} \\ &= -96.11 + 0 - 2(-56.690) - (-5.83) \\ &= 23.10 \text{ kcal/mol} \\ \therefore \log K &= 23.10/-1.3642 \\ &= -16.93\end{aligned}$$

For convenience, a large number of equilibrium constants for solubility, complexation, and hydrolysis reactions have been computed from Table 4-11 data and are given in Table 4-12 (chapter 4, this volume).

REFERENCES

Hunt, D.T.E., and J.D. Creasey. 1980. *The Calculation of Equilibrium Trace Metal Speciation and Solubility in Aqueous Systems by a Computer Method, With Particular Reference to Lead.* Tech. Rept. 151. Medmenham, England: Water Research Centre.

Schock, M.R. 1980. Response of Lead Solubility to Dissolved Carbonate in Drinking Water. *Jour. AWWA,* 72(12):695.

Snoeyink, V.L., and D. Jenkins. 1980. *Water Chemistry.* New York: John Wiley and Sons.

Stumm, W., and J.J. Morgan. 1981. *Aquatic Chemistry, an Introduction Emphasizing Chemical Equilibria in Natural Waters.* 2nd ed. New York: Wiley-Interscience.

Appendix B

Determining Carbon in Scales in Copper Tubes

Standard corrosion testing techniques can be used to determine the corrosion of copper tubing in a variety of service applications (see chapter 9). However, two sets of tests specific to the type I pitting phenomena have been developed. Since Campbell (1950) first reported the presence of carbon as a causative factor in the inception of rapid pitting, methods have been developed to detect carbon in tubes. Campbell used classical quantitative analytical techniques. Because these are time-consuming techniques that require a full analytical laboratory, simpler methods that give a qualitative evaluation of the presence and degree of carbon contamination have been developed.

In addition to evaluating the cause of the inception of pitting, an evaluation of the presence or absence of pitting corrosion inhibitor in the water is of great value. Both chemical and electrochemical methods have been developed for this purpose.

BNF EXAMINATION OF COPPER TUBES FOR CARBON FILMS

In the British Non-Ferrous Metals Research Association (BNF) examination method (Campbell 1972), a half-section of copper tube about 1 in. (2.5 cm) long, with the outside cleaned by filing, is placed in a 50-mL beaker and covered with 25 percent nitric acid. The metal surface is viewed with a binocular microscope (25×) until gas evolution occurs uniformly over the entire surface.

With a sample of unused tube containing a carbon film, there is usually an induction period of about 1 minute before gas evolution begins. When the gas bubbles first form, they can be seen to grow beneath a thin, grayish film that breaks away from the metal in fragments and rises to the surface of the acid with the bubbles. When gas evolution is occurring over the entire metal surface, the film fragments may be seen floating on the surface of the nitric acid. The metal specimen should be removed at this stage to avoid visual disturbance due to gas bubbles. It is helpful to have a microscope that will permit both transmitted and reflected natural light to be used for examination of the film fragments; this makes it easy to distinguish between the grayish film fragments and any extraneous oily matter that may be present.

The procedure for analyzing samples from used copper tube is essentially the same. However, it is sometimes complicated by the presence of insoluble deposits of silty or siliceous matter within the tubes. Moderate amounts of siliceous matter can be dissolved after the tube sample has been taken out of the acid by adding ammonium bifluoride and a few drops of hydrochloric acid and warming for about 15 minutes. The loosely adherent silt generally remains in the first beaker; the carbon film, which is less readily detached, floats in fragments on the surface of the acid in the second beaker.

LECO® TEST

A new method for testing new copper tubes, which uses a small furnace capable of accepting small pieces of tubing, has recently been introduced. In this new test, called the Leco® test, the carbon is burned off the tube, and the resulting carbon dioxide is measured by an infrared photometer. This test has been adopted by the German Gas and Waterworks Association (DVGW) to regulate its standards for low levels of carbon. (To test the absence of films, the DVGW uses a simplified BNF examination called the HNO_3 test [see DVGW 1992]).

REFERENCES

Campbell, H.S. 1950. Pitting Corrosion in Copper Water Pipes Caused by Films of Carbonaceous Material. *Jour. Inst. Metals*, 77:345.

———. 1972. *A Review: Pitting Corrosion of Copper and Its Alloys*. British Non-Ferrous Metals Research Association, Miscellaneous Publication 574. London: BNF.

DVGW. 1992. Nahtlosgezogene Rohre aus Kupfer für Gas- und Wasserinstallationen. Anforderungen und Prüfbestimmungen. Bonn: Wirtschafts- und Verlagsges. Gas und Wasser, MBH.

Appendix C

Case Histories—Corrosion Control Planning, Implementation, and Monitoring Effectiveness of pH and Alkalinity Adjustment

CASE STUDY 1—SEATTLE WATER DEPARTMENT, SEATTLE, WASHINGTON

The Seattle Water Department (SWD) provides service to more than 1.35 million people in the Seattle, Washington, metropolitan area. This population is divided approximately equally between direct service customers and customers served indirectly through 30 suburban water districts, municipalities, and nonprofit water associations that purchase water wholesale from Seattle and distribute it to their own customers. Water consumption averages about 175 mgd (661 ML/d). Two primary sources, the Cedar River and South Fork of the Tolt River, originating on the western slopes of the Cascade Mountains, provide Seattle its water. These mountain sources provide extremely high-quality water.

Water Quality Characteristics

Since Seattle's water is derived almost entirely from rainfall and snowmelt in the Cascade Mountain range, it exhibits typical characteristics of such water: soft, low to neutral pH, poor buffering capacity, low mineral content, and dissolved oxygen saturation. These source qualities, so highly valued as indicators of excellent quality water, ironically are the same characteristics that led to corrosion problems. Prior to implementing the corrosion treatment program, source treatment consisted only of intake screening, chlorination, and fluoridation. Since 1982, lime has been added to the Cedar supply, and lime and soda ash to the Tolt. Before exiting 10 uncovered distribution reservoirs, the water is rechlorinated.

Seattle began studying corrosion and corrosion-related problems in 1970, and it took more than a decade of intensive technical studies and review by citizens and public officials before corrosion treatment began in June of 1982. It took this length of time both because of the complexity of the corrosion issues and because at that time there were no universally accepted procedures for measuring corrosion in domestic water systems. Several procedures were tried and modified specifically for this study. In addition, careful study and documentation had to be made of all potential effects—positive and negative—before a change in water quality would be considered by city officials.

An engineering consultant was retained by SWD in December of 1975. The scope of services included defining the causes of corrosion, determining the extent and magnitude of corrosion damage, and developing alternative solutions to mitigate the problems.

Effects of corrosion. The comprehensive corrosion study verified the findings of earlier reports. Corrosion of city mains was not serious, but the corrosion of both galvanized steel and copper piping in consumer plumbing systems resulted in extensive aesthetic and economic problems and created a health concern. The effects were particularly acute in the Tolt service area. General conclusions drawn from the corrosion study regarding effects of corrosion are summarized below:

- Galvanized steel piping comprised 75 percent of the installed plumbing systems within the city of Seattle. Corrosion of galvanized steel was characterized by pitting and rust scale buildup (tubercles) that resulted in reduced flow and blockage of pipe. There was general agreement that most galvanized piping systems more than 30 years old, in all parts of the city, were heavily corroded.
- Copper and plastic piping were routinely used for replacement of galvanized steel for water service connections to the main and in plumbing systems. A uniform etching type of corrosion was occurring in copper piping systems. In general, hot-water pipes were corroding at more rapid rates than cold-water pipes in plumbing systems.
- Discolorations of the water after standing for a period of time in the plumbing system was a prevalent aesthetic nuisance throughout the water service area. "Rusty" or brown-colored water resulted from the corrosion of galvanized piping, and blue-green stains on fixtures resulted from corrosion of copper piping.
- Corrosion caused heavy metals to dissolve from metallic plumbing systems and enter "standing" water; thus, corrosion created a potential health concern.

- Corrosion caused an unnecessary economic burden on the customer. Plumbing life was significantly reduced, causing piping to be prematurely replaced because of rust scale buildup and leaks.
- Cement lining of cast-iron and steel transmission and distribution pipes was providing excellent protection against the corrosive nature of the water.

Specific effects related to plumbing deterioration. Following are specific conclusions related to deterioration and plugging of consumer piping materials.

- The service life of consumer plumbing was being significantly reduced by corrosion. The average estimated life of galvanized pipe was about 35 years in areas serviced by the Cedar and only 20 years for the Tolt. Copper pipe was expected to last about 35 years. In certain instances, the corrosion necessitated repair and replacement in much shorter periods.
- The Tolt River water was more aggressive than the Cedar River water. This was indicated by more extensive deterioration of pipes in the north area of Seattle, which is served by the Tolt. Also evident was the greater release of metals from the pipe into the water resulting from consumer plumbing corrosion.
- The corrosion rate had increased substantially for both the Cedar and Tolt systems since 1970. This increase was attributed to the following: A change to chlorine from chlorine–ammonia treatment to produce a free rather than a combined chlorine residual, an increase in chlorine dosage at the distribution reservoir outlets, and the inception of fluoridation using hydrofluosilicic acid.

Specific effects related to metal leaching. Effects specifically related to metal release are aesthetic problems and health concerns.

Aesthetic problems. The following are the aesthetic problems:

- Red water, a condition associated with "standing" water in galvanized pipe, occurred in the majority of homes in Seattle. The number of complaints directed to the water department represented only a small fraction of the population affected.
- Metal leaching surveys indicated that more than 65 to 75 percent of the "standing" water samples exceeded the secondary standard of 0.3 mg Fe/L.
- The average iron content in "standing" samples was over twice the 0.3-mg/L secondary level on the Cedar and over four times this level on the Tolt.
- Copper was leached from copper plumbing and could result in objectionable taste and blue-green stains on fixtures.
- Data from various metal leaching surveys indicated that about 11 percent of the "standing" water samples exceeded the secondary standard of 1 mg Cu/L.
- Complaint records and corrosion tests indicated that copper leaching was a problem throughout the Seattle service area and that Tolt water was more aggressive toward copper than Cedar water.

Health concerns. Following are the health concerns.

- Results of several metal leaching surveys were analyzed and indicated 13 percent of Tolt and 14 percent of Cedar "standing" water exceeded the maximum contaminant level (MCL) of 0.05 mg Pb/L.

- There was a higher incidence of lead above the MCL in large buildings as compared to smaller ones, probably due to the extended contact time.
- Preliminary metal leaching data indicated that about 2 percent of the "standing" water samples exceeded the MCL of 0.01 mg Cd/L.

Causes of corrosion. The consultant study indicated that the corrosiveness of the water was a result of several related factors, which included the following:

- A dissolved oxygen concentration that was at saturation levels.
- Acidity, which was indicated by the low pH of the water supplies. The raw water pH was approximately 7.4 to 7.7 for the Cedar water and 6.5 to 6.9 for the Tolt water. Chlorination with gaseous chlorine and fluoridation with hydrofluosilicic acid were contributing factors to corrosion, since they lowered the pH to a range of 6.8 to 7.4 for the Cedar and 5.8 to 6.2 for the Tolt water.
- Insufficient natural calcium and bicarbonate alkalinity in the water to form protective calcium carbonate scales on pipe surfaces. Calcium content and alkalinity of the distributed waters were about 17 mg $CaCO_3$/L and 15 mg $CaCO_3$/L, respectively, for the Cedar and 6 mg $CaCO_3$/L and 2 mg $CaCO_3$/L, respectively, for the Tolt.
- A relatively high halogen–alkalinity ratio for the treated water, which was 0.6 to 0.8 for the Cedar and 1.5 to 2.5 for the Tolt. These high ratios resulted in conditions favorable to pitting corrosion, which was prevalent in much of the galvanized pipe.

Recommended Corrosion Control Plan

Based on two years of study, including labor-intensive pilot-plant pipe loop investigations, the engineering consultant proposed a corrosion control program. There were two aspects to the comprehensive plan: (1) a corrosion treatment program and (2) a program relating to selection of corrosion-resistant materials for use in distribution and plumbing systems.

Corrosion treatment program. The corrosion treatment program included increasing the concentrations of minerals that were already present in the existing water supplies. The program proposed to raise the pH and alkalinity of the water and reduce the (halogen + sulfate):alkalinity ratio. Silicate addition to the Tolt supply was also considered for further evaluation. The corrosion treatment program proposed to modify the characteristics of both water supplies to achieve a water quality that balanced corrosivity considerations with other water quality parameters.

Distribution and plumbing system materials selection program. To complement the corrosion treatment program, the recommended corrosion control plan included a program relating to selection of distribution and plumbing system materials. This program encouraged the use of corrosion-resistant materials, including cement lined, copper, brass, and plastic pipe. It also recommended use of thicker walled copper piping and against the use of galvanized steel. Use of low-lead solders (less than 0.2 percent lead) was to be required through plumbing code changes. Asbestos–cement pipe was removed from the plumbing code to prevent further use.

Evaluations Before and After Treatment

As indicated earlier, Seattle's internal corrosion study concluded that the effects of corrosion were serious enough to warrant treatment. How well was corrosion treatment able to mitigate these effects? Various evaluation techniques were used, including monitoring of distribution system pH levels, trihalomethane (THM) formation potential, leached metals from home plumbing, customer complaint records, bacteriological trends, and copper corrosion rate measurements.

Water Quality Changes

Table C-1 shows typical changes in water quality at the treatment plants, before and after corrosion treatment, which consists of addition of 1–2 mg calcium oxide/L at both sources and 9 mg sodium carbonate/L on the Tolt supply.

Although these values approximate the targeted water chemistry adjustments recommended by the study, they do not by themselves give a clear indication of corrosion rate reduction. Better measurable indicators are maintenance of a consistent pH throughout the distribution system and changes in metals release by the water from consumer plumbing systems.

Distribution pH levels. Optimizing distribution system pH levels proved to be an elusive target, particularly because of Seattle's open distribution reservoirs, but due to other factors as well. The pH level may change due to aeration, rechlorination, and the effects of mortar-lined pipe. Adjustment of pH to the 8.0 to 8.3 range was sufficient to create a tendency for the water to absorb atmospheric carbon dioxide as it passed through open reservoirs. This tended to lower pH values by 0.4 to 0.6 units. In addition, rechlorination at these reservoir outlets with gaseous chlorine depressed the pH approximately 0.2 units further, resulting in pH levels of 7.4 to 7.6 in the distribution system.

Further pH changes occurred as a result of dissolving alkaline materials from cement or mortar-lined pipelines. Seattle has been involved in a program of relining large-diameter (up to 78-in. [2-m]) transmission mains since the late 1970s. In addition, although Seattle has virtually no asbestos–cement (A–C) pipe in its system, many of its wholesale customers do. The effect on final pH is not wholly predictable, other than that pH will increase, and relates to the age of the lining, contact time, surface-area-to-volume ratio, and mortar composition and curing process.

Table C-1 Water quality changes since treatment

	Cedar River		Tolt River	
Water Characteristics	Before	After	Before	After
pH, *units*	7.2	8.2	6.0	8.2
Total alkalinity, $mg\ CaCO_3\ /\ L$	16.0	19.0	2.5	13.5
Carbon dioxide, *mg / L*	2.7	0.0	6.0	0.0
Hardness, $mg\ CaCO_3\ /\ L$	21.0	24.0	8.0	11.0
Sodium, *mg / L*	1.8	1.8	1.0	4.8
Total dissolved solids (TDS), *mg / L*	46.0	48.0	25.0	36.0
Specific conductance, *μmhos / cm*	55.0	59.0	27.0	42.0
Larson's ratio	0.65	0.55	3.9	0.8

As a result of the above factors, measured pH values in Seattle's major service zones have ranged from as low as 7.3 to 7.5 in an area served by the Tolt to above 9.0 in wholesale areas with considerable A–C pipe in the distribution system.

Reductions in metals leaching. The metals of most interest in determining the effects of corrosion control are those associated with consumer plumbing materials: copper, iron, zinc, lead, and cadmium. As indicated earlier, Seattle's treatment program was designed primarily as a means of controlling copper, lead, and zinc corrosion. Protection for iron (including galvanized iron) pipes was expected to be minimal.

To measure levels of metals, Seattle established a Residential Water Quality Monitoring (RWQM) program. Test residences were selected from within Seattle's direct service area, with approximately equal numbers from the Cedar and Tolt service areas. About half were random samples; the other half were from "worst-case" situations, defined as those resulting from complaints of aesthetic problems such as rust stains, yellow water, and metallic taste. A relatively large data set was collected both pre- and posttreatment, with the objective to sample as many homes as possible to account for the large variability in pipe type, age, water use patterns, and water sources. This was done in lieu of collecting a large number of replicate samples from a much smaller number of targeted homes. A practical problem occurred in the posttreatment set, when corrosion-related complaints began to dwindle, reducing the pool of available homes to sample.

Sampling was done by supplying the participants with bottles and sampling instructions. Both "standing" (overnight, or morning first-draw) and fully flushed (for 5 minutes) samples were collected. Samples were returned to the laboratory the same day.

The results of this survey (data through March 1987) are shown in Tables C-2 and C-3. Major findings include:

1. All five metals—copper, lead, zinc, iron and cadmium—were found in significantly lower concentrations following corrosion treatment.

2. Standing tap water lead levels have been reduced by approximately 61 percent to 68 percent.

3. Cadmium levels have been reduced by about 38 percent to 58 percent in the overnight standing samples.

4. Copper levels, a cause of numerous complaints because of fixture staining, have been reduced by 50 percent to 74 percent in standing samples. Blue-green staining complaints have disappeared.

5. Levels of iron, Seattle's main aesthetic metal problem, have improved but remain a problem. Average iron reduction is about 17 percent to 26 percent in standing samples.

6. Zinc levels have been reduced by about 41 percent to 54 percent in standing samples.

Accuracy of study projections. Information collected during studies to predict what metal leaching rates might be with corrosion treatment were compared to data from the RWQM program, which gathered samples from homes after treatment began. The accuracy of the predictions varied widely and depended on both the metal being surveyed and the source of the water. Table C-4 presents these results.

Table C-2 Cedar River supply: Metal levels before and after treatment

	Number of Samples		Maximum, μg / L		Average, μg / L		
	Before	After	Before	After	Before	After	Reduction, %
Copper							
Standing	171	214	1,050	750	140	70	50
Running	171	214	220	110	34	15	56
Lead							
Standing	163	214	137	47.5	9.52	3.68	61
Running	163	214	16.5	11	1.54	0.55	64
Zinc							
Standing	171	214	3,100	2,950	537	317	41
Running	171	214	650	900	84	38	55
Cadmium							
Standing	166	214	5.8	10	0.77	0.48	38
Running	163	214	1.1	1	0.18	0.05	72
Iron							
Standing	171	214	6,380	9,000	1,041	860	17
Running	171	214	3,500	4,540	202	153	24

Table C-3 South Fork Tolt River supply: Metal levels before and after treatment

	Number of Samples		Maximum, μg / L		Average, μg / L		
	Before	After	Before	After	Before	After	Reduction, %
Copper							
Standing	143	165	2,550	730	456	118	74
Running	143	165	1,880	210	125	22	82
Lead							
Standing	136	165	133	62	11.84	3.82	68
Running	133	165	16.5	6.5	1.91	0.37	81
Zinc							
Standing	143	165	4,450	1,890	670	310	54
Running	143	165	2,100	810	144	35	76
Cadmium							
Standing	144	165	6.0	2.4	0.69	0.29	58
Running	144	165	5.3	1	0.16	0.03	81
Iron							
Standing	143	165	11,400	6,650	1,417	1,051	26
Running	143	165	2,450	2,100	242	166	31

Table C-4 Metal leaching reductions by treatment

	Cedar River		Tolt River	
	Predicted*	Found†	Predicted*	Found†
Copper, %	46	50	67	74
Zinc, %	–138	41	61	54
Lead (from galvanized pipe), %	49	61	–105	68
Lead (from copper soldered pipe), %	81	61	80	68
Cadmium (from galvanized pipe), %	–32	38	–35	58

*Predicted from pilot studies.
†Found in homes in RWQM program, overnight standing samples.

Copper leaching rates were predicted quite accurately for both water sources. Lead leaching trends from galvanized pipe and from copper soldered joints on the Cedar were reasonable, but were more difficult to explain on the Tolt. Cadmium leaching from galvanized piping would have been predicted to increase based on pilot studies, but in fact decreased in the RWQM program. Leaching rates and sometimes even trends are difficult to predict. More research is certainly needed in this area to advance the state of the art.

Customer complaints. Rusty water and blue-green fixture staining complaints have been a fairly accurate barometer of Seattle's corrosivity problem. Increased numbers of complaints occurred in 1965, after the Tolt supply was started up, and again in 1970, with changes in water chemistry due to fluoridation, higher concentrations of chlorine at distribution reservoir outlets, and a switch from chloramines to free chlorine at the Cedar source. Evaluation of customer complaints over short periods of time, however, unless they can be related to a specific event, is problematical at best. Water supply changes and other operating conditions tend to make trends of relatively small magnitude difficult to evaluate. This was true particularly with rusty-water complaints after startup of corrosion treatment. Overall, there was an approximately 20 percent reduction in rusty-water complaints during a posttreatment evaluation period of 18 months compared to a comparable pretreatment period. However, despite the brief evaluation period, these data, coupled with the iron reduction data from the residential monitoring, demonstrate at least some reduction in rusty water.

More important, copper-staining complaints, very much a problem with Tolt water prior to corrosion treatment, have completely disappeared. And as can be seen from Table C-3, copper levels have decreased to impressively low values, even in "standing" samples, averaging only 118 μg Cu/L.

Environmental Aspects

From an environmental standpoint, high metal concentrations (cadmium, copper, lead, and zinc) can adversely affect receiving waters and can limit the options for disposal of wastewater sludges. In a 1984 study titled "Toxicant Pretreatment Planning Study Summary Report," funded by the regional wastewater agency (the Municipality of Metropolitan Seattle [Metro]), the Seattle water supply was stated to be a major contributor of certain metals to wastewater influents. The water supply was estimated to be the largest source of zinc to two of Metro's treatment plants, 55 percent to West Point and 48 percent to Renton. It was also a major source of copper, 23 percent to

West Point and 29 percent to Renton. The estimated metal loadings and expected reductions with corrosion treatment of the potable water are listed in Table C-5. As indicated in Table C-5, the greatest mass reductions will take place in the West Point system, where most of the water comes from City of Seattle sources of supply.

Cost for Treatment

Table C-6 compares the projected and actual costs for both capital and operating expenses. As indicated by review of Table C-6, the overall costs that were projected versus the actual cost are reasonably close, within 25 percent. If one were to pick an individual cost item, the difference might be significant, but the bottom line figures indicate that reasonable cost projections were made. The average cost is \$8.56/mil gal (\$2.26/ML) for corrosion treatment including amortized capital costs.

Conclusions

Chemical additions, resulting in only minor changes in mineral content and other chemical properties, have resulted in major cost saving to Seattle water customers. Millions of dollars are saved each year in reduced costs for premature replacement of pipes and leak damage. The benefit:cost ratio exceeds 5:1.

Concerning effects of corrosion treatment on water quality, pH levels throughout the service area have ranged from 7.3 up to 9.0. This broad range is caused by open reservoirs, rechlorination, and changes as water travels through cement–mortar-lined and A–C pipes. Even with this range of pH levels, metal leaching rates for copper, lead, zinc, cadmium, and iron in first-flush samples have been significantly reduced by corrosion treatment. Aesthetic improvements have been documented. Blue-green stains on fixtures from copper corrosion have been virtually eliminated and rusty-water complaints have been reduced slightly. Copper levels met the US Environmental Protection Agency (USEPA) action levels of 1.3 mg/L in both the Tolt and Cedar supply areas after treatment. Lead levels met the USEPA action level of 0.015 mg/L at the 90th percentile in the Cedar, but exceeded it in the Tolt service area. Monitoring of the wastewater influents have also documented significant reductions in metals, although precise amounts are difficult to measure because of confounding factors such as pretreatment programs and growth in the service area.

Table C-5 Reductions in metals loads due to Seattle Corrosion Control Program (Metro 1984)

	Renton			West Point		
Metal	Water Supply Load, *lb/d*	Reduction, *lb/d*	Overall Reduction, %	Water Supply Load, *lb/d*	Reduction, *lb/d*	Overall Reduction, %
Cadmium	0.20	0.03	15	0.24	0.14	58
Copper	26.91	11.90	44	45.95	29.68	65
Lead	1.60	0.29	18	2.68	1.00	37
Zinc	58.93	14.44	24	96.98	54.66	56

Table C-6 Comparison of projected and actual costs for Seattle corrosion treatment facilities

	Projected	Actual
Capital Cost	\$1,690,000	\$2,321,600
Annual Operating and Maintenance Costs		
Capital facilities amortized at 7% for 20 years	\$159,500	\$219,100
Corrosion Chemicals*	477,000	128,200
Labor†	47,300	112,400
Maintenance‡	16,900	58,200
Power§	4,230	29,100
Subtotal	\$705,430	\$547,000
Additional chlorine for disinfection	25,400	Not Quantified
Total	\$730,830	\$547,000

*Cedar River, lime; Tolt River, lime and soda ash.

†Operated by existing staff: 4 h/d on Cedar River; 12 h/d on Tolt River.

‡Includes labor and materials.

§Engineering News Record–Construction Cost Index (ENR–CCI) = 4,609.

Most of the predictions and estimates made by the consultants during the planning phase and studies have been reasonably close to actual conditions found in the system since treatment began. Trends were normally easier to predict than actual values, but on the whole, predictions and trends have been borne out in actual practice. When evaluated in light of new USEPA regulations, especially those focusing on lead, the corrosion control program still looks good. In fact, the new regulations may have required such a program, had it not been self-initiated years ago.

Overall, Seattle's corrosion control program has been a success in addressing economic, aesthetic, environmental, and health concerns associated with its corrosive waters.

CASE STUDY 2—CORROSION CONTROL IN DESALINATED WATER USING $CaCO_3$ DEPOSITION IN SAUDI ARABIA

Case of the Water Systems of Yanbu and Medina (Saudi Arabia)

Toward the end of the 1970s, the decision was made to improve water supply conditions for the western Saudi Arabian city of Yanbu and the high Moslem city of

Medina. The technique selected was desalination of seawater. The installations were built between 1979 and 1980. They are composed of

- The production facilities on the Red Sea side, mainly a multistage flash distillation desalination plant followed by potabilization treatment consisting of lime and CO_2 addition, aeration, and disinfection by chlorine
- A transmission system connecting the water desalination plant to each of the cities by two pipelines fed by two pumping stations. The feeders are made of a 24-in. (600-mm) A–C pipe, 25 mi (40 km) long, feeding the city of Yanbu, and a 31.5-in. (800-mm) steel pipe, 109 mi (175 km) long, feeding the city of Medina

The A–C pipe was not lined. The feeder for Medina was made of steel to address the high pressure (90 bars [1,305 lb/in.^{2}]) necessary to cross the 800-m-high (2,620-ft) mountains between the sea and the city. The steel pipe was lined with an epoxy coating during manufacture of straight sections and special parts for the pipeline. In the pumping station, the epoxy lining was applied in situ.

Figure C-1 gives a schematic view of the different treatment steps. Chlorine is prepared by electrolyzation of seawater and is introduced in the water as a 0.5 g Cl_2/L solution in seawater.

The specifications for water composition initially were

- total hardness: 20 mg $CaCO_3$/L
- pH: close to 7.5
- chlorine dosage: 1.0 mg/L

Corrosion problem. After a few weeks of operation, the screens protecting the pumps of the Medina line were full of pieces of lining that had peeled off the pipe walls. A first observation of the inner walls of the pipes in the pumping station indicated that corrosion was occurring very quickly with formation of deep

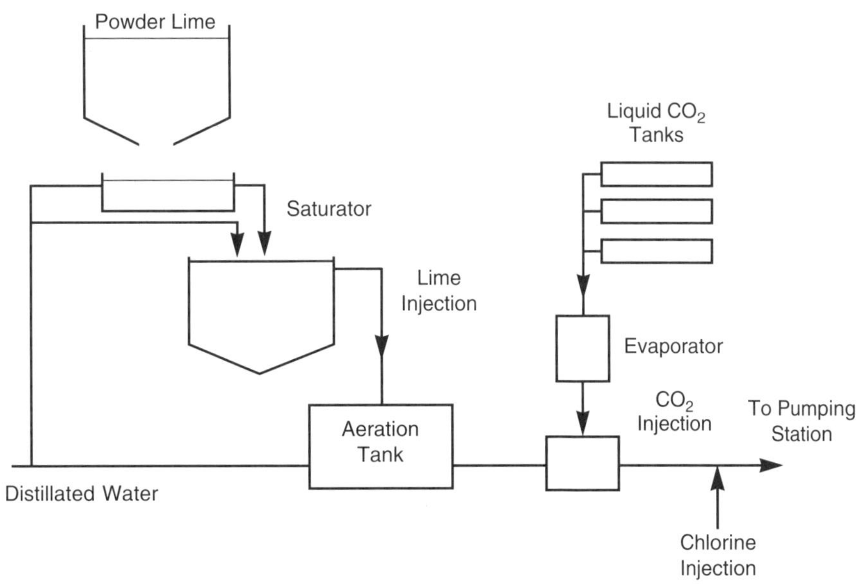

Figure C-1 Potabilization section of the Yanbu and Medina, Saudi Arabia, water treatment plant

craters (0.5 mm [0.0197 in.] in two months) covered by hard corrosion products. The thickness of the deposits covering the craters was about 2 to 5 mm (0.0787 to 0.1968 in.). The same type of corrosion was appearing along the pipeline at the joints between pipe elements where the welding was covered in situ by an epoxy lining.

Although the high temperature of the water (between 95°F [35°C] and 104°F [40°C]) can influence the corrosion rate, it was not sufficient to explain the rapid deterioration caused by corrosion. Another factor influencing the corrosion rate was thought to be the composition of the water. As shown in Table C-7, it appears that high levels of chloride and sodium—certainly too high compared to the bicarbonate and calcium concentrations—were causative factors. Moreover, the water was not saturated with calcium carbonate, and no protective layer could be formed on the inner walls of the pipes.

Remedies adopted. As the system was in operation and the consumption of water very high in Medina, it was not possible to stop the flow of water to apply a new lining in the pipes inside the pumping stations. Furthermore, it would be difficult to obtain a better lining than the first applied, because of the humid conditions and lack of easy access.

A viable alternative appeared to be modification of the water composition leading to reduction of the corrosion rate by decreasing the chloride and sodium levels and to deposition of a protective layer of calcium carbonate on the pipe walls by a calcifying water.

Reducing chlorine and sodium levels was possible by increasing the chlorine concentration of the seawater solution to 2 g Cl_2/L. Such a concentration was the maximum limit for the type of electrolytic cells used.

Deposition of the protective layer required an increase in the calcium hardness to a value near 60 mg $CaCO_3$/L and in the pH to protect the entire pipe wall from near the pumping station to the far reaches of the distribution system in the city of Medina, 175 km (108.7 mi) from the plant. The composition of the water after modifications is shown in Table C-7.

After six months of operation under these conditions, a sufficient protective layer of calcium carbonate was visible on the inner walls of the pipe in the pumping station as well as at the Medina delivery point. Its thickness was about 0.4 mm (0.015748 in.) at the input of the pipe line and 0.2 mm (0.007874 in.) at the output (see Figure C-2).

To limit the growth of the deposit, it was necessary to reduce the calcifying propensity of the water. The lime and CO_2 dosages were progressively reduced to obtain a total hardness about 30 mg $CaCO_3$/L. The pH of the treated water needed

Table C-7 Composition of the water at Yanbu and Medina (Saudi Arabia), before and after treatment

Parameter	Before	After
Ca^{2+}, *meq / L*	0.620	1.340
Mg^{2+}, *meq / L*	0.100	0.050
Na^{+}, *meq / L*	1.585	0.405
K^{+}, *meq / L*	0.1055	0.025
HCO_3^-, *meq / L*	0.636	1.256
SO_4^{2-}, *meq / L*	0.114	0.108
Cl, *meq / L*	1.606	0.422
NO_3^-, *meq / L*	0	0
pH	7.57	8.22
Temperature, *°C*	36	36
Saturation index	–1.0	0.3

Figure C-2 Inner walls of pipes after six months of treatment

to stabilize the deposit was about 8.5 at 36 to 40°C (96.8 to 104°F). Under these conditions, the water was alternatively aggressive and calcifying according to the variations of lime and CO_2 dosages. Since these water quality adjustments, the Medina water system has operated consistently without any corrosion problem.

Case of the Network of Al Khobar (Saudi Arabia)

Toward the end of the 1970s, the decision was made to improve water supply conditions for several towns located in eastern Saudi Arabia. Among these towns situated on the Arabian Gulf were Al Khobar, Damman, and Dhahran. The installations were built between 1979 and 1981. There are two types of installation: production facilities in Al Khobar, mainly composed of a water desalination plant, and a transmission system connecting the water desalination plant to each of these cities.

Seawater is subjected to multistage flash distillation in the water desalination plant. After the salt has been removed, the water is mineralized by adding carbon dioxide and then lime. At this point, the water is disinfected with chlorine and then enters the transmission system (Figure C-3). The chlorine is manufactured by electrolysis of the seawater, specifically NaCl.

The transportation system comprises 149 mi (240 km) of water mains, whose diameters are 47 in. (1,200 mm) at the head of the system and 19.7 in. (500 mm) at the end. The maximum operating pressure does not exceed 8 bar (116 psi).

The mains are made of reinforced concrete pipes. The transmission system includes a pumping station and, for each town involved, a delivery station to regulate the water pressure. In each of these stations, the mains are made of steel with an epoxy lining.

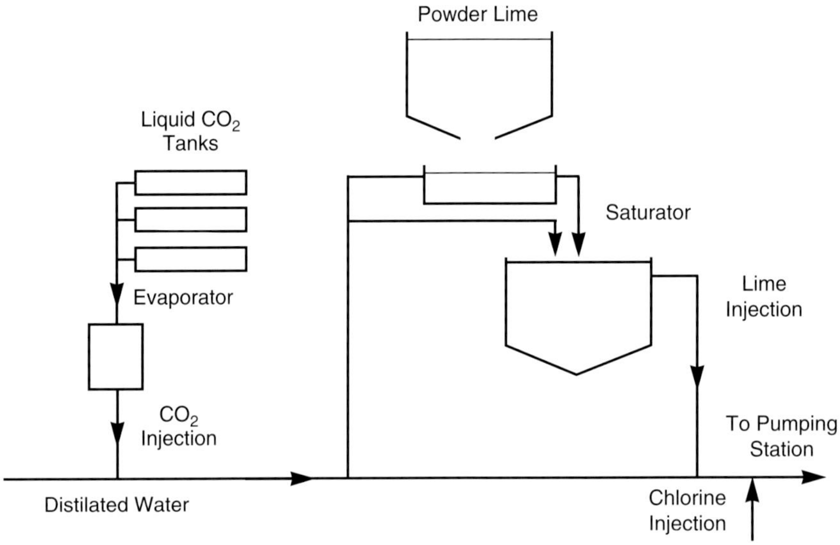

Figure C-3 Potabilization plant at Al Khobar, Saudi Arabia

The composition of the water entering the transportation system during the initial operation is presented in Table C-8. After a few weeks of operation, it appeared that the epoxy lining in the pumping station had peeled off from the underlying steel wall, leading to a rapid corrosion of the metal (Figure C-4). This lining had been installed in situ and not in the manufacturing plant.

Corrosion occurred rapidly for the following reasons: (1) the water temperature was high, and (2) the Cl^- ion concentration was too high compared to the HCO_3^- ion concentration.

In order to reduce the corrosion risks and to protect the pipe walls with a deposit of calcium carbonate, the treatment applied in the potabilization plant was modified to obtain a calcium hardness about 60 mg $CaCO_3$/L and to get a pH high enough to obtain a calcifying water. The increase of calcium hardness was necessitated for the following reasons: (1) the calcium concentration must be high enough to enable formation of $CaCO_3$ at a low pH level, and (2) the pH must be lower than the maximum available value of 9.5 and lower than 8.5 to achieve efficient disinfection with chlorination.

The composition of the water entering the network during this period is given in Table C-8.

During the first year of operation, the modified treatment was maintained in order to produce the calcium carbonate deposit. During this period, it was possible to see the deposit formation, and it appeared white on the cathodic areas and brown on anodic areas, where it is mixed with corrosion products (Figure C-5).

After about one year of operation, the calcium carbonate deposit was thick enough to ensure the protection of the bare pipe walls (about 0.3 mm [0.0118 in.]). After that, it was possible to reduce the calcium hardness, that is to say, the lime and CO_2 dosages to reach a value near 40 mg $CaCO_3$/L, giving an equilibrium pH near 8.6. The dosages were calculated in order to get a water just at the equilibrium.

Figure C-4 Inner wall of pipe where epoxy lining has peeled off

Table C-8 Composition of the water at Al Khobar (Saudi Arabia) before and after treatment

Parameter	Before	After
Ca^{2+}, *meq / L*	0.620	1.270
Mg^{2+}, *meq / L*	0.100	0.280
Na^{+}, *meq / L*	0.350	0.350
K^{+}, *meq / L*	0.050	0.050
HCO_3^-, *meq / L*	0.674	1.290
SO_4^{2-}, *meq / L*	0.080	0.260
Cl, *meq / L*	0.360	0.360
NO_3^-, *meq / L*	0	0
pH	7.75	8.39
Temperature, °C	35	35
Saturation index	–0.8	0.41

As it is impossible to get an absolutely stable pH, the water introduced into the network was sometimes slightly solubilizing and sometimes slightly calcifying. In such conditions the thickness of the deposit did not grow or grew very slowly without scaling risks. The inner wall of the pipes was protected (Figure C-6), even after 10 years of operation.

Figure C-5 Pipe during the formation of deposit

Figure C-6 Inner pipe wall after the deposit has formed; bare surface is protected

Appendix D

Corrosion Units, Other Units, and Abbreviations

CORROSION UNITS

Units Giving Penetration Rates

mils per year (mpy) (= 10^{-3} in./year)
millimetres per year (mm/year) (=10^{-3} m/year)
micrometres per year (μm/year) (=1^{-6} m/year)

Units Giving Mass Loss per Unit Area

milligrams per square decimetre-day (mdd)
grams per square metre per day [(g/m^2)/d]

Units Giving Reaction Rate

microamperes per square metre (μA/m^2) (= 10^{-6} A/cm^2)

Conversions

Multiply:	By:	To Obtain:
mpy	25.4	μm/year
mpy	0.0254	mm/year
μm/year	$(2.74 \times 10^{-6}) \times$ density (kg/m^3)	(g/m^2)/d
mdd	0.1	(g/m^2)/d
mass loss (g)	$\dfrac{1}{\text{area of specimen (m}^2) \times \text{test duration (days)}}$	(g/m^2)/d
μA/m^2	$(8.953 \times 10^{-7}) \times$ gram equivalent weight	(g/m^2)/d

Density (kg/m^3) [20°C]

iron	7,860
zinc	7,150
copper	8,920
lead	11,350

Units and Abbreviations

A	ampere
A/cm^2	ampere per square centimetre
atm	atmosphere
bar	1×10^6 dynes per square centimetre (0.09869 atm)
°C	degree Celsius
cm, cm^2, cm^3	centimetre, square centimetre, cubic centimetre
cm/s	centimetres per second
d	day
dm^2	square decimetre
°F	degree Fahrenheit
ft	foot
ft/s	feet per second
ftu	formazin turbidity unit
g	gram
g metal	gram metal
gal	gallon
gpd	gallons per day
g/cm^2	grams per square centimetre
g/eq	grams per equivalent
g/m^2	grams per square metre
$(g/m^2)/d$	grams per square metre per day
$(g/m^2)/year$	grams per square metre per year
h	hour
°H	Hazen unit
Hz	herz
I	current, ionic strength
in.	inch
kcal/mol	kilocalories per mole
kg	kilogram
kJ/mol	kilojoules permole
km	kilometre
kΩ	kilo-ohm
kPa	kilopascal
L, L^2, L^3	litre, square litre, cubic litre
lb	pound
L/d	litres per day
lin ft	linear foot
L/s	litres per second
m, m^2, m^3	metre, square metre, cubic metre
M	molar (mol/L)
m^3/d	cubic metres per day

meq/L	milliequivalents per litre
mgd	million gallons per day
mg/dm^2	milligrams per square decimetre
mg/L	milligrams per litre
m/h	metres per hour
mi	mile
mil gal	million gallons
mil	million
min	minute
ML	million litre or megalitres
μA	microamp
μA/cm^2	microamps per square centimetre
μg	microgram
μg/L	micrograms per litre
μm	micrometre
mL/h	millilitres per hour
μS	microsiemens
mm	millimetre
m*M*	millimole
m*M*/L	millimoles per litre
mol/L	moles per litre
MPa	megapascal
mg	milligram
m/s	metres per second
mV	millivolt
mpy	mils per year
N/mm^2	newtons per square millimetre
oz/ft^2	ounces per square foot
pE	measure of redox potential
ppm	parts per million
psi	pounds per square inch
s	second
V	volt
yd	yard

Additional Readings

CHAPTER 1

Donahue, F.M. 1972. Corrosion and Corrosion Control. In *Physico-chemical Processes for Water Quality Control.* Edited by W.J. Weber, Jr. New York: Wiley-Interscience.

Hudson, L.D., J.W. Hankins, and M. Battaglia. 1983. Coliforms in a Water Distribution System: A Remedial Approach. *Jour. AWWA,* 75:564.

Stumm, W. 1956. Calcium Carbonate Deposition at Iron Surfaces. *Jour. AWWA,* 48:300.

Trussell, R.R., and J. Thomas. 1971. A Discussion of the Chemical Character of Water Mixtures. *Jour. AWWA,* 63:49.

CHAPTER 2

Asaad, F., and H. Sontheimer. 1991. Untersuchungen zur Stillstandskorrosion bei Stahlrohren. *gwf—Wasser/Abwasser,* 132:537–545.

CHAPTER 4

Friend, J.N., and J.S. Tidmus. 1924. The Relative Corrosion of Zinc and Lead in Solutions of Inorganic Salts. *Jour. Inst. Metals,* 37:77.

Index Publishing Corporation. 1978. *Chapters of the Municipal Code of Chicago Relating to Plumbing, With Amendments to October 19, 1978.* Chicago: Index Publishing Corporation.

NHS, Inc. 1983. *Water Analysis and Evaluation for Communities of the North Slope Borough.* Richland, Wash.: NHS Inc.

Ryder, R.A. 1978. Methods of Evaluating Corrosion. In *Proc. of the Sixth Annual AWWA Water Quality Technology Conference.* Denver, Colo.: American Water Works Association.

USEPA (U.S. Environmental Protection Agency). 1990. National Interim Primary Drinking Water Regulations; Amendments. *Federal Register,* 45(168):57332.

Walker, A.P. 1983. The Microscopy of Consumer Complaints. *Jour. Inst. Water Engrs. and Scientists,* p. 200.

CHAPTER 5

Cruse, H., and O. von Franqué. 1985. Corrosion of Copper in Potable Water Systems. In *Internal Corrosion of Water Distribution Systems.* Denver, Colo.: American Water Works Association Research Foundation/DVGW Forschungsstelle.

Dangle, R.A. 1975. *Study of Corrosion Products in the Seattle Water Department Tolt Distribution System.* EPA 670/2-75-036. National Environmental Research Center. Cincinnati, Ohio: U.S. Environmental Protection Agency.

Ferguson, J.F. 1992. Unpublished Report to the Portland Water Bureau, Portland, Ore.

Lucey, V.F. 1972. *Pitting Corrosion of Copper in Supply Waters: The Effect of Water Composition.* British Non-Ferrous Metals Research Association. Research Report A1838.

Mattson, E. 1980. Corrosion of Copper and Brass: Practical Experience in Relation to Basic Data. *Br. Corros. Jour.,* 15(1):6.

Millero, F.J. 1989. Effect of Ionic Interactions on the Oxidation of Fe(II) and Cu(I) in Natural Waters. *Mar. Chem.,* 28:1.

Schock, M.R., and I. Wagner. 1985. The Corrosion and Solubility of Lead in Drinking Water. *Internal Corrosion of Water Distribution Systems.* Denver, Colo.: American Water Works Association Research Foundation/DVGW Forschungsstelle.

Shoesmith, D.W., T.E. Rummery, D. Owen, and W. Lee. 1976. Anodic Oxidation of Copper in Alkaline Solutions 1. Nucleation and Growth of Cupric Hydroxide Films. *Jour. Electrochem. Soc.,* 123(6):790.

Shoesmith, D.W., T.E. Rummery, et al. 1977. Anodic Oxidation of Copper in Alkaline Solutions 2—The Open-circuit Potential Behavior of Electrochemically Formed Cupric Hydroxide Films. *Electrochimica Acta,* 22:1403.

Snoeyink, V.C., and D. Jenkins. 1980. *Water Chemistry.* New York: John Wiley and Sons.

CHAPTER 6

Lead Industries Association. 1982. *Solders and Soldering, A Primer.* New York: Lead Industries Association.

Oliphant, R.J. 1982. Lead Contamination of Potable Water Arising From Soldered Joints. In *Conference Papers, Special Subject 18.* Zurich: International Water Supply Association.

Sharrett, A.R., et al. 1982. Daily Intake of Lead, Cadmium, Copper, and Zinc From Drinking Water: The Seattle Study of Trace Metal Exposure. *Envir. Res.,* 28:456.

CHAPTER 7

Anderson, C.H., and J.M. Long. 1980. *Interim Method for Determining Asbestos in Water.* EPA 600/4-800-005. Athens, Ga.: U.S. Environmental Protection Agency.

DVGW W 270. 1990. Vermehrung von Mikroorganismen auf Materialien für den Trinkwasserbereich. Wirtschafts– und Verlagsges, MBH.

Hasson, D., and M. Karmon. 1984. Novel for Process Lining Water Mains by Controlled Calcite Deposition. *Corrosion Prev. and Control,* 9:17.

Heinrich, B., R. Künzler, und W. Schwenk. 1983. Untersuchungen über das Korrosionsverhalten von Zementmörteln für Rohrauskleidungen in sauren kalklösenden Wässern. 3R International, 22.

Künzler, R., and W. Schwenk. 1986. Änderung der Wasserparameter bei Kontakt von Trinkwasser mit frischem Zementmörtel. *gwf—Wasser/Abwasser,* 127:11–15.

Legrand, L., and P. Leroy. 1990. *Prevention of Corrosion and Scaling in Water Supply Systems.* London: Ellis-Horwood.

McCabe, L.J., and J.R. Millette. 1979. *Field Studies Division Health Effects.* Cincinnati, Ohio: USEPA Research Laboratory.

McCauley, R.F. 1960. Controlled Deposition of Protective Calcite Coatings in Water Mains. *Jour. AWWA,* 52:1386.

Midwest Research Institute. 1979. Occurrent, Economic Implications, and Health Effects Associated With Aggressive Waters in Public Water Supply Systems. Final Report, MRI Project No. 4552-L.

Parkinson, R.W., I.C. Warren, and K.J. North. 1983. *Recent Experiences With Epoxy Resin Lining of Water Mains.* WRc Engineering Centre, Rept. 106E.

Reed, E.C. 1980. Updating of Pipe Renovation Techniques and Experience. IWSA Conference, Paris.

Schoenen, D. 1980. Makrokolonien auf der Zementmörtelauskleidung eine Trinkwasserleitung. Z. Bakt. Hyg. 1. Abteilung. Ori., B171:466.

Schwenk, W. 1990. Special Problems in Rehabilitation. Water Quality Problems in Distribution Systems With Cement Mortar Linings. *Water Supply,* 8:297–302.

CHAPTER 8

ANSI/AWWA B202-93. 1993. *Standard for Quicklime and Hydrated Lime.* Denver, Colo.: American Water Works Association.

AWWA. 1986. *Corrosion Control for Operators.* Denver, Colo.: American Water Works Association.

———. 1990. *Lead Control Strategies.* Denver, Colo.: American Water Works Association Research Foundation and American Water Works Association.

AWWA B302–90. 1990. *Standard for Ammonium Sulfate.* Denver, Colo.: American Water Works Association.

Benjamin, L., R.W. Green, A. Smith, and S. Summerer. 1992. Pilot Testing of a Limestone Contactor in British Columbia. *Jour. AWWA,* 84(5):70–79.

Bergan, E. 1988. *Corrosion Control of Distribution Systems in Larger Water Works—Countrywise Economic Cost Evaluation.* Norwegian National Institute for Public Health, Report No. 66.

Bricker, O.P., and K.C. Rice. 1989. Acidic Deposition to Streams. *Envir. Sci. and Technol.,* 23:378.

Costello, J.J. 1981. Monitoring for Corrosion Control—The Chicago Experience. In *Proc. of the Ninth Annual Water Quality Technology Conference.* Denver, Colo.: American Water Works Association.

Faust, S.D., and O.H. Aly. 1983. *Chemistry of Water Treatment.* Ann Arbor, Mich.: Ann Arbor Science Publishers.

Ferguson, J.F. 1985. Corrosion Arising from Low Alkalinity, Low Hardness or High Neutral Salts Content Waters. In *Internal Corrosion in Drinking Water.* Denver, Colo.: American Water Works Association Research Foundation.

Gorham, E. 1989. Scientific Understanding of Ecosystem Acidification: A Historical Review. *Ambio,* 18:161.

Hagen, K. 1987. Results of Operation of the CARIX Partial Demineralization Plant. *Vom Wasser,* 69:259–267.

Hanson, H.F., S.S. Hasted, L.M. Mueller, and D.R. Goff. 1987. *Deterioration of Water Quality in Distribution Systems.* Denver, Colo.: American Water Works Association Research Foundation and American Water Works Association.

Hornberger, G.M., B.J. Cosby, and R.F. Wright. 1989. Historical Reconstructions and Future Forecasts of Regional Surface Water Acidification in Southernmost Norway. *Water Resources Res.,* 25:2009.

Kojima, S. 1989. Japanese Municipalities Find Another Way to Add Lime. *Water/Engineering and Management,* 136(August):40–42.

Plummer, N., and E. Busenberg. 1982. The Solubilities of Calcite Aragonites and Vaterite in CO_2–H_2O Solutions Between 0 and 90°C and an Evaluation of the Aqueous Model for the System $CaCO_3$–H_2O. *Geochim. Cosmochim. Acta,* 46:1011.

Sanks, R.L. 1982. *Water Treatment Plant Design. For the Practicing Engineer.* Ann Arbor, Mich.: Ann Arbor Science Publishers.

Schock, M.R. 1990. Internal Corrosion and Deposition Control. 4th ed. In *Water Quality and Treatment. A Handbook for Community Water Supplies.* Edited by F.W. Pontius. New York: McGraw-Hill.

Sjöberg, E.L., and D. Rickard. 1983. The Influence of Experimental Design on the Rate of Calcite Dissolution. *Geochim. Cosmochim. Acta,* 47:2281–2285.

———. 1984. Comparative Dependence of Calcite Dissolution Kinetics Between 1 and 62°C at pH 2.7–8.4 in Aqueous Solutions. *Geochim. Cosmochim. Acta,* 48:485–493.

Skiba, U., and M. Cresser. 1989. Prediction of Long-Term Effects of Rainwater Activity on Peat and Associated Drainage Water Chemistry in Upland Areas. *Wat. Res.,* 23:1477.

Sontheimer, H., W. Kölle, and V.L. Snoeyink. 1981. The Siderite Model of the Formation of Corrosion-Resistant Scales. *Jour. AWWA,* 73:572.

Stericker, W. 1948. Soluble Silicates for Corrosion Inhibition in the Oil Industry. *Corrosion,* 5.

CHAPTER 9

Dawson, J.L. 1991. *Electrochemical Impedance Data Interpretation.* Application Note. Princeton, N.J.: Princeton Applied Research Inc.

Acronyms, Abbreviations, and Variables

ACRONYMS AND ABBREVIATIONS

AC	alternating current
A–C	asbestos–cement
AI	aggressiveness index
AL	action level
ANSI	American National Standards Institute
AOC	assimilable organic carbon
ASTM	American Society for Testing and Materials
AWWA	American Water Works Association
AWWARF	American Water Works Association Research Foundation
B:C	benefit to cost
BDOC	biodegradable organic carbon
BET	Brunaur, Emmett and Teller
BGA	(German Federal Health Office) Bundesgesundlicitsamt
BNC	base neutralizing capacity
BNF	British Non-Ferrous Metals Technology Centre
BS	British Standard
c	material is typically colloidal or amorphous in nature
CCDP	calcium carbonate dissolving potential
CCPP	calcium carbonate precipitation potential
CDA	Copper Development Association
CEN	Comité Européen de Normalisation (European Committee for Standardization)
CEOCOR	Committee of European Organizations on Corrosion
CERL	Corps of Engineers Research Lab
DC	direct current
DIC	dissolved inorganic carbon
DIN	Deutsches Institut für Normung (German Institute of Standardization)
DKI	Deutsches Kupfer-Institut (German Copper Institute)
DN	diameter nominal
DoE	Department of the Environment
DO	dissolved oxygen

DVGW	Deutcher Verein das Gas- und Wasserfachs (German Gas and Waterworks Association)
DWRD	Drinking Water Research Division (USEPA)
EC	European Community
EDXA	energy-dispersive X-ray elemental analysis
EIS	electrochemical impedance spectroscopy
emf	effective electromotive force
EN	Norme Européenne (European Norm)
ESCA	electron spectroscopy for chemical analysis
FTS	fixed time of stagnation
GVWD	Greater Vancouver Water District
HAA	haloacetic acid
HPC	heterotrophic plate count
IAP	ion activity product
ID	internal diameter
ILZRO	International Lead–Zinc Research Organization
IR	internal resistance drop
ISO	International Standards Organization
ISWS	Illinois State Water Survey
IUPAC	International Union of Pure and Applied Chemistry
LCCA	Lead Contamination Control Act
LCR	Lead and Copper Rule
LG	leaded gunmetal
LI	Langelier index
LP	linear polarization
MAC	maximum admissible concentration
MCL	maximum contaminant level
MDC	Metropolitan District Commission
Me	metal
MFL	million fibers per litre
MWD	Metropolitan Water District of Southern California
NACE	National Association of Corrosion Engineers (recently changed to NACE International)
NAS	National Academy of Sciences
NOM	natural organic matter
NSF	National Sanitation Foundation (recently changed to NSF International)
OMS	Organisation Mondiale de la Santé
PB	polybutadiene or polybutylene
pH	negative logarithm of the effective hydrogen ion concentration
PRV	pressure-reducing valve
PS	potentiodynamic scan
PSFC	pipe section flow cell
PVC	polyvinyl chloride
QA/QC	quality assurance/quality control
RCE	rotating cylinder electrode
RDE	rotating disk electrode
redox	oxidation reduction
RO	reverse osmosis

RWQM	Residential Water Quality Monitoring (Program)
S/C	sand-to-cement ratio
SCE	standard calomel electrode
SD	standard deviation
SEM	scanning electron microscope, microscopy
SHE	standard hydrogen electrode
SI	saturation index
SMCL	secondary maximum contaminant level
SWD	Seattle Water Department
T	temperature
TDS	total dissolved solids
TEM	transmission electron microscope, microscopy
THM	trihalomethane
TIC	total inorganic carbon
TOC	total organic carbon
UNC-EQI	University of North Carolina—Asheville Environmental Quality Institute
USEPA	US Environmental Protection Agency
W/C	water-to-cement ratio
WHO	World Health Organization
WITAF	Water Industry Technical Action Fund
WRc	Water Research Centre
XPS	X-ray photoelectron spectroscopy

VARIABLES

a	slope of line
a	internal radius of pipe
a_{Me}^{2+}	activity of the ion Me^{2+}
A	exposed interior surface of pipe
A	total area of the internal pipe wall
b	intercept
B_a	anodic Tafel slope
B_c	cathodic Tafel slope
β_a	mass-transfer coefficient
C	concentration of the ion
C_A	moles of strong acid that are added per litre of solution
C_B	moles of strong base that are added per litre of solution
C_t	total carbonate concentration
C_0	basic inhibition (natural inhibition of water)
C_{04}	calcium carbonate equibrium curve
C_{PO_4}	concentration of orthophosphate
C_T	carbonate carbon
c_{WH}	roughness coefficient
$C \times T$	concentration times time
D	diffusion coefficient
ΔG_f^o	Gibbs free energy of formation (for individual species)
E	effective electromotive force (emf) of solder/copper cell obtained from the Evans diagram
E	replacement costs

E_{corr}	freely corroding surface potential
E_H	redox potential relative to a standard hydrogen electrode
$E_{critical}$	critical potential
f	temperature correction factor
F	Faraday constant
γ	activity coefficient
i	impressed current
i	rate of interest
i_{corr}	freely corroding current density
i_{max}	maximum value of current corrosion
$i_{passive}$	current density required to reach passivation
I	galvanic current with no resistance in the external circuit
I'	galvanic current with a resistance in the external circuit
k	mass-transfer coefficient
K	correction factor, which accounts for the flow velocity
K_{sp}	solubility constant
L	length
M	mass
n	number (of data points in a category)
N	potential offset
n_s	time necessary to operate rehabilitated pipe in order to reach break-even point with the replacement
Pb_{MAX}	plateau "equilibrium" value after long stagnation
Pb_{TAP}	mean lead concentration at the tap in households with lead plumbing
$[Pb]^0$	lead concentration leaving the pipe
$[Pb]^\infty$	the equilibrium lead solubility
$[Pb]^E$	lead concentration of water entering the pipe
$[Pb]^t$	lead concentration after time t
Q	filtration rate
r	correlation coefficient
r^2	coefficient of determination
R	ideal gas constant
R_p	polarization resistance
S	rehabilitation costs
t	time
T	absolute temperature (K)
T	stagnation time
T_{50}	the time at which a concentration of 50 percent of maximum was achieved
T^+	sum of positive ranks
t_F	theoretical contact time
V	volume
$\dot{V}$	volume rate of flow
V_B	volume of alkaline media used
V_F	filter volume
V_i	corrosion rate of the zinc–iron alloy layer
V_M	minimum volume of alkaline media

x	independent variable in line of best fit
y	dependent variable in line of best fit
z	number of electrons transferred in the reaction
Z	test statistic used to define the critical region for evaluating the null hypothesis

Compounds

alpha-hopeite	$Zn_3(PO_4)_2 \cdot 4H_2O$
atacamite, paratacamite	$Cu_2(OH)_3Cl$
bauxite	$Al_2O_3 \cdot 2H_2O$
calcite, aragonite	$CaCO_{3(s)}$
cerussite	$PbCO_3$
chloropyromorphite	$Pb_5(PO_4)_3Cl$
cuprite	Cu_2O
ettringite	a highly hydrated calcium sulfo-aluminate
geothite	α-FeOOH
gypsum	$CaSO_4 \cdot 2H_2O$
half-burnt dolomite	$CaCO_3 \cdot MgO$
hematite	α-Fe_2O_3
hydrocerussite	$Pb_3(CO_3)_2(OH)_2$; also $2PbCO_3 \cdot Pb(OH)_2$
hydroxypyromorphite	$Pb_5(PO_4)_3OH$
hydrozincite	$Zn_5(CO_3)_2(OH)_6$
lepidocrocite	γ-FeOOH
magnetite	Fe_3O_4
malachite	$Cu_2(OH)_2CO_3$
massicot, litharge	PbO
nantokite	CuCl
plattnerite	PbO_2

plumbonacrite	$Pb_{10}(CO_3)_6(OH)_6O$; also $3Pb_3(CO_3)(OH_2) \cdot PbO$
pyrrhotite	Fe_{1-x}
siderite	$FeCO_3$
smithsonite	$ZnCO_3$
tenorite	CuO
vivianite	$Fe_3(PO_4)_2$
willemite	Zn_2SiO_4
zincite	ZnO

Index

NOTE: An *f.* following a page number refers to a figure; a *t.* refers to a table.